Regional Anatomy

# Dissector & Laboratory Companion

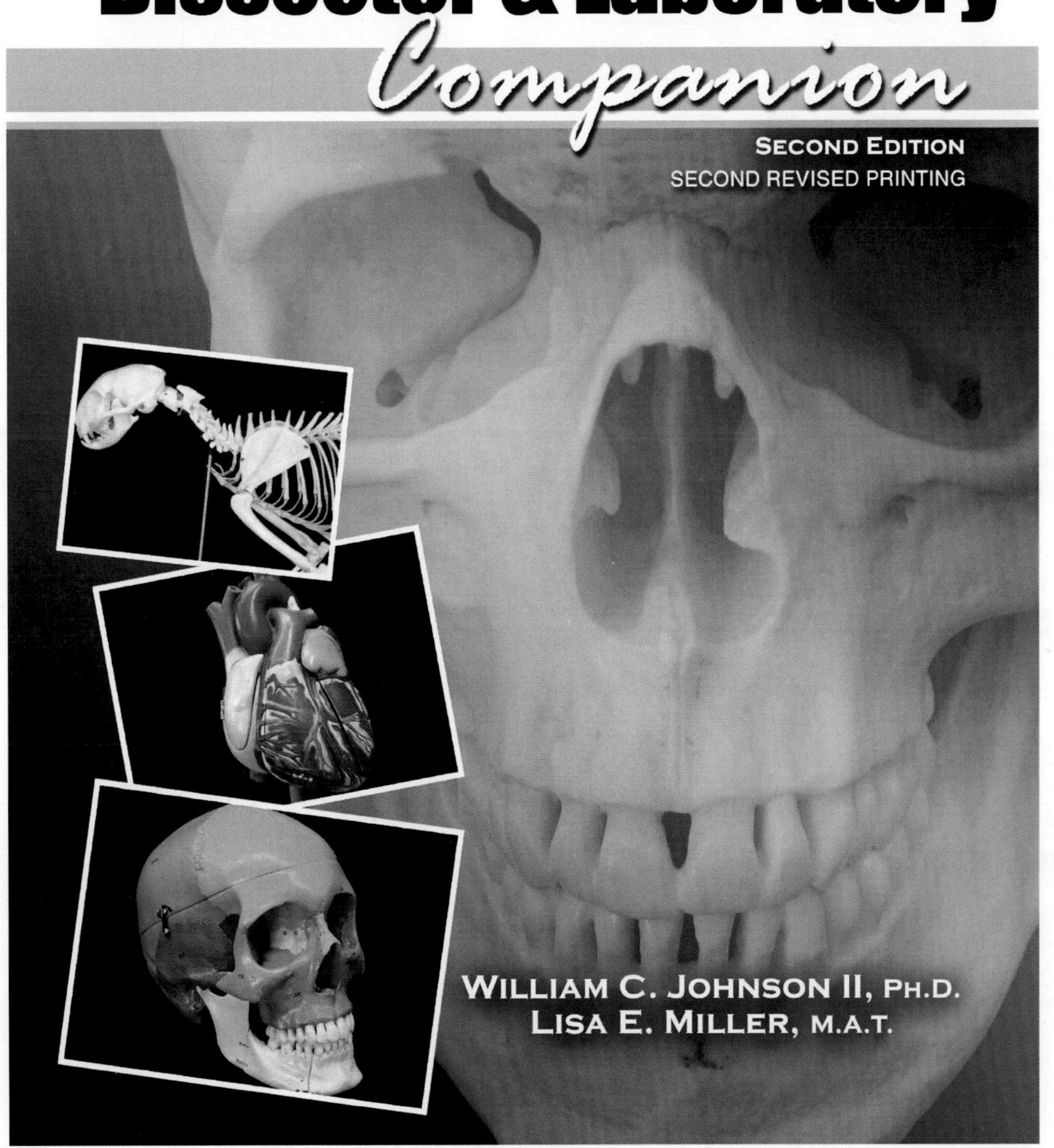

**SECOND EDITION**
SECOND REVISED PRINTING

WILLIAM C. JOHNSON II, PH.D.
LISA E. MILLER, M.A.T.

Illustrations by Jamey Garbett. © 2003 Mark Nielsen.

Lab 2: Pages (top left) 39, (top center) 41, (top center) 43,
(both images) 45, (top left) 47, (top left) 49,

(all three images) 51, (left) 53, (top left) 55, 59.

Lab 5: Page (2 left images) 115.Lab 6: Pages (Anterior and Posterior View-Left) 143.

Lab 7: Pages (Top right and bottom left) 191, (top and bottom right) 193, (right) 195, (right) 197.

Lab 8: Pages (right) 209, (bottom left) 210, (right) 213, (right) 215, (right)
217, (right) 219, (right) 220, (right) 223, (right) 225, (right) 227, (right) 229,
(right) 231, (right) 263, (right) 265, (right) 267, (top right) 269.

Lab 12: Pages (bottom right) 383, (top center, top right, bottom center, and bottom
left) 385, (center and right) 387, (center) 389, (right) 391, (right) 393, (top and bottom
right) 395, (top right) 397, (top right) 399, (bottom right) 401, (top right) 403, (top
right) 405, (top right) 407, (center and right) 409, (right) 411, (center) 413.

Cover images provided by William C. Johnson II and Lisa E. Miller.

First edition title was Regional Anatomy Dissector Companion.

**Kendall Hunt**
publishing company

www.kendallhunt.com
*Send all inquiries to:*
4050 Westmark Drive
Dubuque, IA 52004-1840



ISBN 978-1-4652-7431-1



Printed in the United States of America

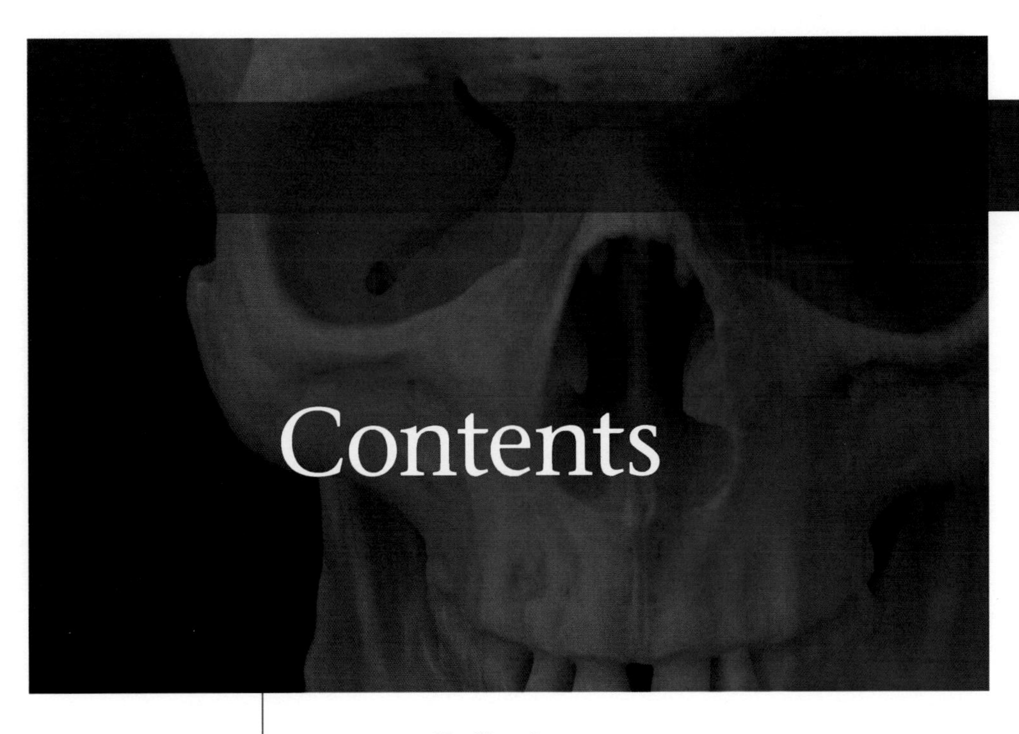



*Dedication* . . . . . . . . . . . . . . . . . . . . . . . . . . . . . . . . . . . . . . . . v

*Acknowledgements* . . . . . . . . . . . . . . . . . . . . . . . . . . . . . . . . . vii

LAB 1 Bones of the Thorax, Shoulder, and Neck . . . . . . . . . 1

LAB 2 Muscles of the Thorax and Neck . . . . . . . . . . . . . . . . . 35

LAB 3 Thoracic Cavity: Viscera, Nerves, and Vessels . . . . . 63

LAB 4 Calf Heart . . . . . . . . . . . . . . . . . . . . . . . . . . . . . . . . . . . 89

LAB 5 Head and Neck . . . . . . . . . . . . . . . . . . . . . . . . . . . . . . 109

LAB 6 Human Skull . . . . . . . . . . . . . . . . . . . . . . . . . . . . . . . 131

LAB 7 Upper Limb . . . . . . . . . . . . . . . . . . . . . . . . . . . . . . . . 171

LAB 8 Upper Limbs: Arm and Forearm . . . . . . . . . . . . . . . . 203

LAB 9 Abdomen . . . . . . . . . . . . . . . . . . . . . . . . . . . . . . . . . . . 257

LAB 10 Abdominal Vessels . . . . . . . . . . . . . . . . . . . . . . . . . . . 293

LAB 11 Pelvis and Reproductive Organs . . . . . . . . . . . . . . . . 331

LAB 12 Lower Limb . . . . . . . . . . . . . . . . . . . . . . . . . . . . . . . . . 367

# DEDICATION

*This text is dedicated to my students: past, present, and future. It is my hope that it will make mastering the subject of human anatomy fun and useful. In other words, do wickett and you will get a stickah.*

Dr. J

# ACKNOWLEDGEMENTS

This lab book is the product of significant effort by many people. It has been made much better because of their generosity, good humor, and enthusiasm for the subject of human anatomy. I would like to acknowledge some of these people here. Dr. Louis J. Zanella provided the context for this work with his lab dissector. He has always been inspirational and very willing to share his time, knowledge, materials as well as being a mentor for me. Many students have donated their work to benefit their peers, too many to compile a complete list. However, this book got its start with encouragement from Al Meehan, who took all of the initial pictures for my anatomy web resource, some of which are included in this text. Jocelyne Van Stolk, Laura Holder, and Tom Marchetti all contributed excellent drawings, which will greatly enhance this book and students' understanding of the material. The professionals at Kendall Hunt Publishing, Michelle Bahr and Samantha Sawyer, have gently guided the development of the final product and that is most appreciated as well! And finally, many thanks to Lisa Miller, who was tireless in photographing, copyediting, and compiling text and imagery. Without that help, this book could not have come to exist.

I would also like to make special mention of my wife Kay (aka Superior Vena Cava) for supporting this effort and tolerating the many weekends that I was working on this text rather than doing things with our family. Thank you, Kay!

Dr. J

■ ■ ■

In addition to those acknowledged by Dr. J, I would like to thank family, friends, and students for their support along the way. In particular, the insights, encouragement, and humor of fellow anatomy tutors Sean Baxter, Judy Colman, and Doug Shehan was and continues to be indespensible. Thank you to Anon (t.a.k.) for keeping errant students informed in a good-humored way, and for Eima Phalen and Justin Thyme for providing fictional entertainment when the stress hits the fan. Thank you to Spot, Fluffy, Snooky, Mittens, and the other named and nameless feline supermodels who have helped us all better understand human antomy. And, of course, a huge thank you to Dr. J for asking me to work on this project and for being such a dedicated, inspirational teacher and friend. It has been a pleasure and honor to work on this book, and I, too, hope it helps many people succeed.

As they say in RI, study LOMG wickett hahd.

Lisa

Lab 1

# Bones of the Thorax, Shoulder, and Neck

## Wish List


**LAB 1 OVERVIEW, pp. 2–3**

**CLAVICLE, pp. 4–5**
- Sternal (Proximal) End, p. 4
- Acromial (Distal) End, p. 4

**SCAPULA, pp. 6–10**
- Acromion Process, p. 8
- Superior Angle, p. 8
- Supraspinous Fossa, p. 8
- Suprascapular Notch, p. 8
- Spine, p. 8
- Metacromion Process (Cat), p. 9
- Infraspinous Fossa, p. 9
- Medial (Vertebral) Border, p. 9
- Inferior Angle, p. 9
- Lateral (Axillary) Border, p. 9
- Subscapular Fossa, p. 10
- Coracoid Process, p. 10
- Supraglenoid Tubercle, p. 10
- Glenoid Cavity, p. 10
- Infraglenoid Tubercle, p. 10

**SKULL, pp. 11–13**
- Nuchal Lines, p. 11
- External Occipital Protuberance, p. 12
- Occipital Condyles, p. 12
- Mastoid Process, p. 13
- Hyoid Bone, p. 13

**VERTEBRAE, pp. 14–27**
- General Landmarks, pp. 14–20
  - Articular Facets, pp. 14–15
  - Articular Processes, pp. 14–15
  - Body, p. 15
  - Spinous Process, p. 16
  - Transverse Processes, p. 17
  - Laminae, p. 18
  - Pedicles, p. 18
  - Vertebral Foramen, p. 19
  - Intervertebral Foramina, p. 20
- Cervical Vertebrae (7), p. 21
  - Transverse Foramina, p. 21
  - Atlas (C1), p. 22
  - Axis (C2), p. 23
  - • Odontoid Process (Dens), p. 23
- Thoracic Vertebrae (12), p. 24
  - Costal Demifacets, p. 25
  - Transverse Costal Facets, p. 26
- Lumbar Vertebrae (5), p. 26
- Sacral Vertebrae/Sacrum, p. 27
  - Sacral Foramina, p. 27
- Coccygeal Vertebrae/Coccyx, p. 27

**RIBS, pp. 28–31**
- True Ribs, p. 29
- False Ribs, p. 29
- Floating Ribs, p. 29
- Landmarks, pp. 30–31
  - Head, p. 30
  - Neck, p. 30
  - Tubercle, p. 30
  - Subcostal Groove (Costal Groove), p. 30
  - Body, p. 31
  - Costal Cartilage, p. 31

**STERNUM, pp. 32–33**
- Manubrium (handle), p. 32
- Sternal Angle, p. 32
- Body (Gladiolus), p. 33
- Xiphoid Process, p. 33

# LAB 1 OVERVIEW

In lab 1 we will study bones of the thorax, shoulder, neck, and a few bones of the head. This is a departure from the organization of the course as a whole. This course is a regional approach, so we normally study **ALL** the structures in an area. But, for the first lab, it is easier if we start with the bones. You should find these bones and landmarks on the articulated human and cat skeletons.

## 1. Bones of the Appendicular Skeleton (Pectoral Appendage)

We will study the pectoral girdle starting anteriorly and working posteriorly:

The **clavicle** is often described as the most easily broken bone in the human body. You will identify the **proximal (sternal) end** and the **distal (acromial) end**, and learn how to differentiate a left from a right clavicle. Although this sounds easy, be advised that the clavicle has caused more than one case of heartburn among Dr. J's students. Read the descriptions for suggestions on how to make this task more manageable and less stressful.

The **scapula** has a longer list of landmarks for you to master. In addition to distinguishing one side from the other, you will observe the **acromion process**, **superior angle**, **supraspinous fossa**, **suprascapular notch**, **spine**, **metacromion process** (quadruped only), **infraspinous fossa**, **inferior angle**, **coracoid process**, **supraglenoid tubercle**, **glenoid cavity**, **infraglenoid tubercle**, **lateral (axillary) border**, **medial (vertebral) border**, and the **subscapular fossa**. The definitions of tubercle, process, fossa, etc., can be found in the glossary. Please note: we are not learning these things just to torture you, although that is a consideration. We are learning these landmarks because generally they relate to muscles that we will soon be studying. There is a context for all of this, and it would be a good idea to include that in the foundation you are building.

## 2. Bones of the Axial Skeleton (Head, Cervical, and Thoracic Regions)

The **skull** has a few features that are relevant to this region. In lab 6 we will study the skull in its entirety, but for now we will observe the **nuchal lines**, **external occipital protuberance**, **occipital condyles** (you guessed it, condyles are defined in the glossary—do you see a trend here?), and **mastoid process**. You should determine the bones on which these landmarks are located. Also observe the **hyoid bone**. You will not be responsible for any specific landmarks on the hyoid bone. Note how similar in appearance it is to the mandible. Coincidence? Probably not. We will talk about that.

Working inferiorly, we will now observe the **vertebrae**. For continuity, we will study the entire vertebral column. We can divide the vertebral column into regions, and you should familiarize yourself with the attributes and number of vertebrae for each of the areas. The landmarks you are responsible for would be: **vertebral bodies**, **articular processes**, **articular facets** (yes, the glossary has those), **laminae** (lamina = singular), **pedicles**, **spinous processes**, **transverse processes**, **vertebral foramina** (foramen = singular and is in the glossary, what a surprise), and **intervertebral foramina**. Also, observe the regional curvatures on the articulated skeleton. They include cervical, thoracic, lumbar, and sacrococcygeal. These curves and their functional importance will be the subject of a lecture, but see if you can determine why there are curves. Does the cat have all the same curves?

The most superior of the vertebrae are the **cervical vertebrae (7)**. You need to know the names and distinguishing characteristics of the two most superior cervical vertebrae: the **atlas (C1)** and the **axis (C2)**. Observe the **odontoid process** of the axis. Also, note the **transverse foramina** of the cervical vertebrae.

The **thoracic vertebrae (12)** are immediately inferior to the cervical vertebrae. In addition to their defining characteristics, note their unique landmarks, including the **transverse costal facets** and **costal demifacets**. Yes, costal is in the glossary.

Observe the **lumbar vertebrae (5)** and their distinguishing characteristics.

The **sacrum** is unusual in that it contains five fused vertebrae. Note that it is part of the sacrococcygeal curve. You should find the following landmarks: the **anterior** and **posterior sacral foramina**. What do they correspond to that you observed in the other regions?

The **coccygeal vertebrae (coccyx, 3-5)** do not have the same remarkable characteristics of the other areas, and this in itself is a distinguishing attribute.

Observe the **ribs**. Find the **head**, **neck**, **tubercle**, **angle**, ***sub*costal groove**, **body**, and **costal cartilage**. Also note that there are three distinct types of ribs. The most superior are the **true ribs (7 pairs)**, followed by the **false ribs (3 pairs)**; and lastly you should study the **floating ribs (2 pairs)**. Be sure you know how these three groups differ from each other.

The **sternum** has distinct areas that are formed by the **sternebrae**. The superior-most portion is the **manubrium (handle)**, and it is usually wide. Observe the **sternal angle (angle of Louis)** between the manubrium and the body. Inferior to the manubrium is the **body (gladiolus)**, which is made up of four fused sternebrae. The most inferior portion of the sternum is the **xiphoid**, or **xiphoid process**.

# CLAVICLE

The clavicle, sometimes referred to as the collarbone, is the only long bone that lies horizontally in the body. It can be palpated and is visible in people who have less fat in this region. The clavicle is S-shaped and sits directly superior and superficial to rib 1. Its sternal (proximal) end is blunt and its distal (acromial) end is curved slightly downward and away from the body. In the cat, the clavicle does not articulate with any bones. In the human, it articulates proximally with the manubrium of the sternum at the sternoclavicular joint and distally with the acromion process of the scapula at the acromioclavicular joint. Functionally, the clavicle acts as a brace to keep the scapula in position so the humerus can hang freely and realize its full range of motion. Its name means "little key" in Latin because of the way it rotates on its axis like a key in a lock when the shoulder is raised (abducted).

## Cavicle (Landmarks)

### *Sternal (Proximal) End*

The sternal (proximal) end of the clavicle can be recognized because it is very blunt. It articulates with the manubrium and lies superficial to rib 1. The superior side of the clavicle is smooth while the acromial (distal) end of the inferior side has a concavity and a tubercle (the conoid tubercle) that face downward. The proximal end of the clavicle forms a saddle joint with the manubrium. The sternal end of the clavicle is of functional importance because its anterior superior half (convex curve) is the origin for:

1. pectoralis major (clavicular head) and
2. sternocleidomastoid muscle.

### *Acromial (Distal) End*

The distal (acromial) end of the clavicle articulates with the acromion process of the scapula, and it points anteriorly (outward from the body). It also looks like a field hockey stick. Remember that the end of the hockey stick that touches the ground is distal to your hands. The superior side of the clavicle is smooth, while the acromial (distal) end of the inferior side has a concavity and a tubercle (the conoid tubercle) that face downward. The acromial end of the clavicle is of functional importance because its superior surface is the origin for:

1. anterior division of deltoid muscle,

and it is the insertion for:

1. superior division of trapezius muscle.

# CLAVICLE

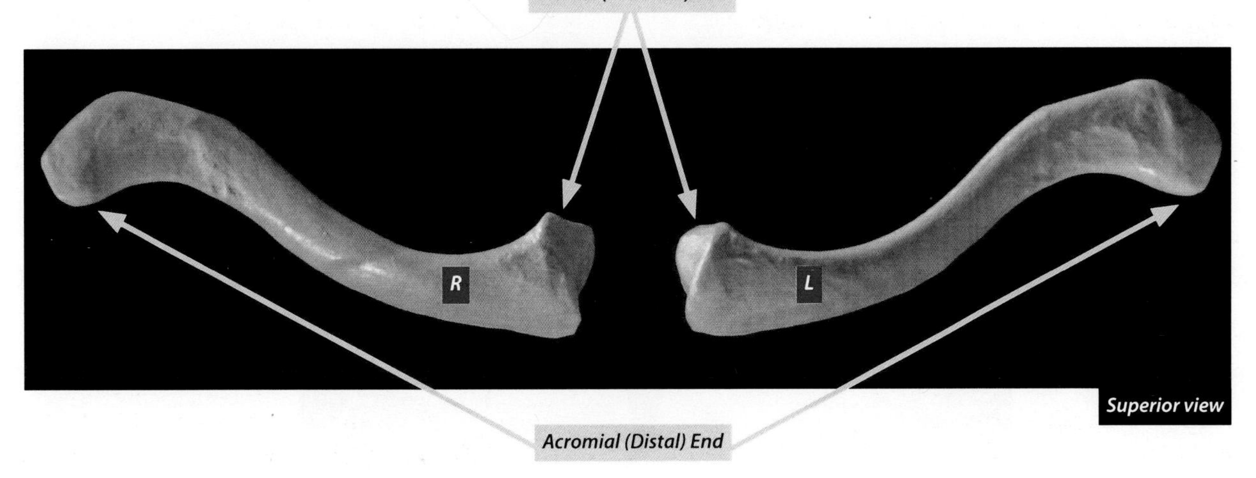


Superior view

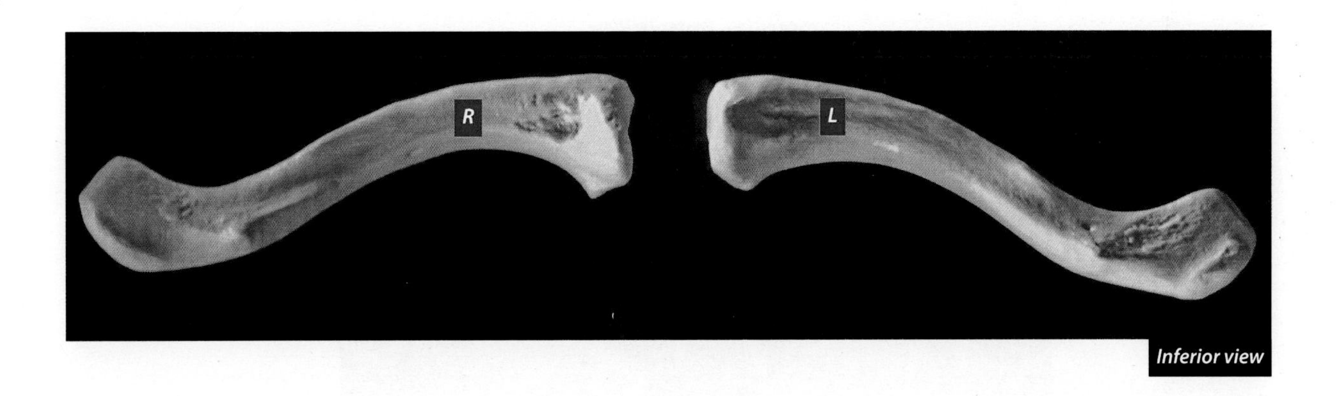


Inferior view

# SCAPULA

The scapula is of functional significance as it forms part of the girdle for the pectoral appendage. The large flat surfaces of the scapula serve as attachment areas for many of the muscles that move the arm or stabilize the girdle. Please see Lab 1, pages 8 through 10 for descriptions of individual features and their muscle attachments.

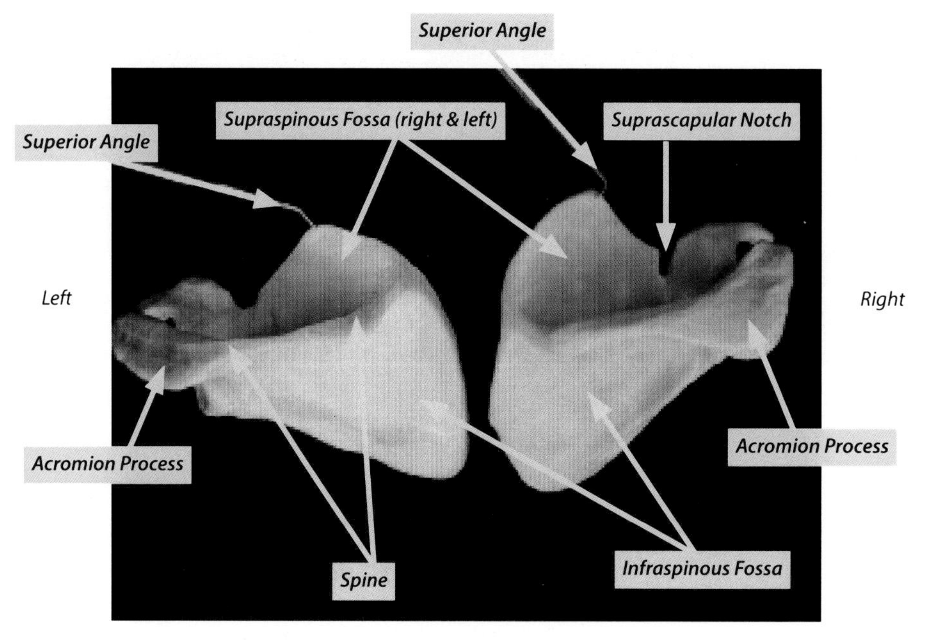


***Scapula—Posterior View***

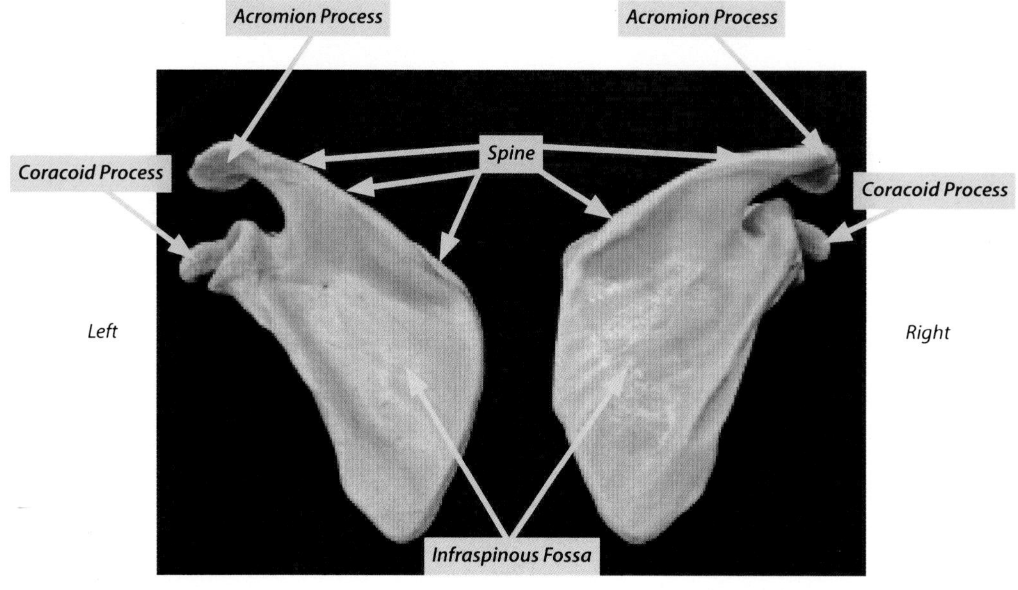


***Scapula—Posterior View***

# SCAPULA

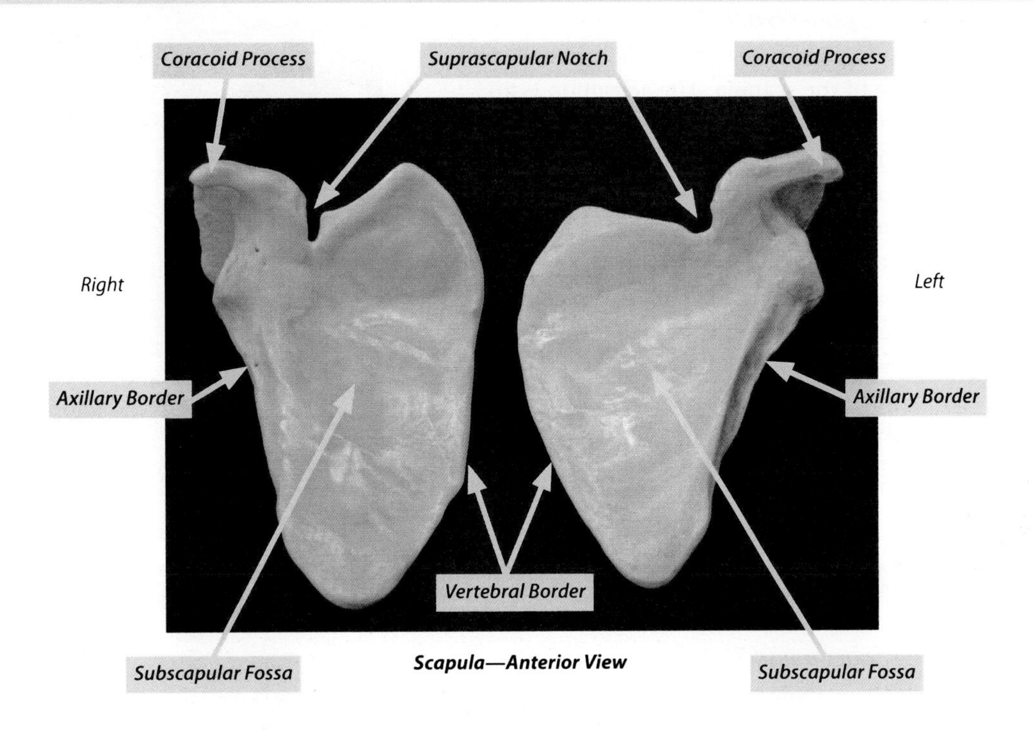


**Scapula—Anterior View**

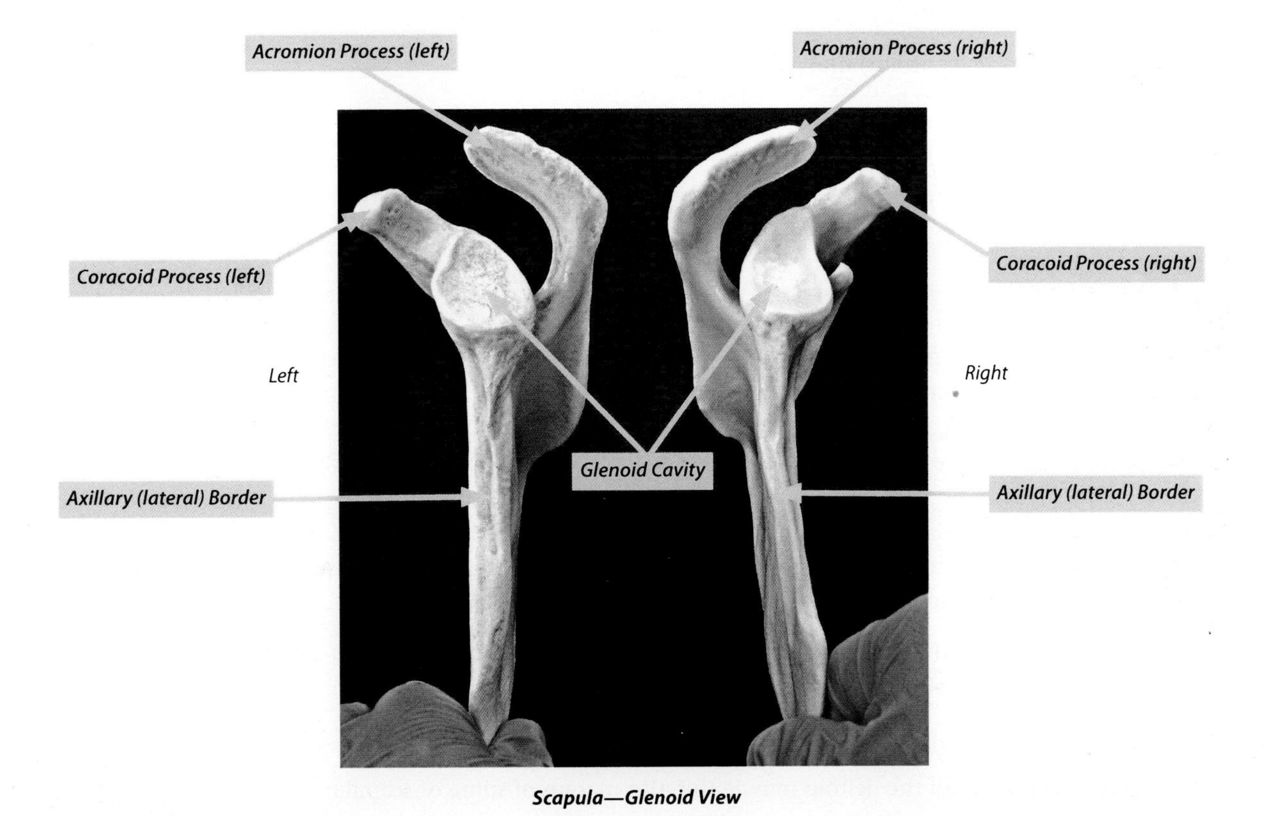


**Scapula—Glenoid View**

# Scapula

### *Acromion Process* (see pages 6 and 7)

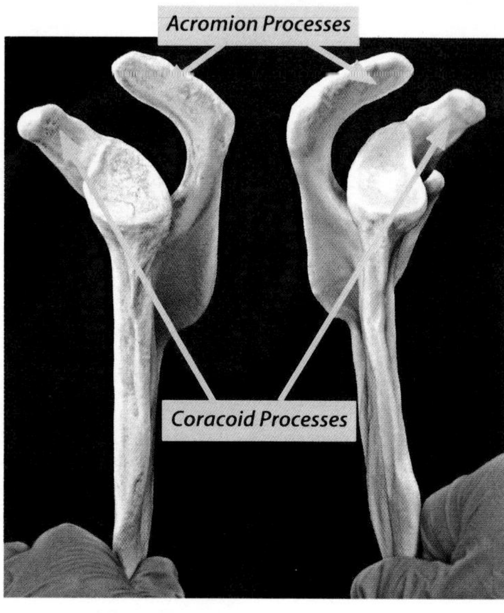


*Left Scapula* *Right Scapula*

Physically, the acromion process (or just the acromion) is a projection of the spine of the scapula. The acromion is an important landmark of the scapula because it articulates with the acromial (distal) clavicle to form the acromioclavicular joint. This is an unusual gliding joint (biaxial) in that it allows for gliding (anteroposterior and vertical) and for some axial rotation of the scapula. The joint capsule of this joint is made stronger by two ligaments, the small acromioclavicular ligament on the vertical surface of the bones and the larger coracoclavicular ligament (two parts) between the coracoid process and the inferior surface of the clavicle. The acromion process of the scapula is the insertion for:

1. superior portion of trapezius;

and origin for:

1. middle portion of deltoid.

### *Superior Angle* (see pages 6 and 7)

The superior angle of the scapula is at the level of the spinous process of thoracic vertebra 2. It is where the medial and superior borders of the scapula meet. The superior angle of the scapula is of functional importance because it is the insertion for:

1. levator scapulae (ventralis) muscle and
2. serratus anterior muscle.

### *Supraspinous Fossa* (see pages 6 and 7)

The supraspinous fossa is a hollow or depression that is superior to the spine of the scapula on its posterior aspect. It is the origin for:

1. supraspinatus.

### *Suprascapular Notch* (see pages 6 and 7)

As its name implies, the suprascapular notch is a small opening located above the spine of the scapula on its superior border. It is of functional importance because the **suprascapular nerve** and the **suprascapular artery** pass through this notch. That would make it a Grant, Grant, Grant thing. It doesn't get better than that!

### *Spine* (see pages 6 and 7)

The spine of the scapula is a long thin projection of bone that runs obliquely across its posterior surface of the scapula. It is the insertion for:

1. middle portion of the trapezius muscle (superior border of spine of scapula) and
2. inferior portion of the trapezius muscle (medial third of spine of scapula);

and the origin for:

1. posterior portion of the deltoid muscle (inferior margin of spine of scapula).

# Scapula

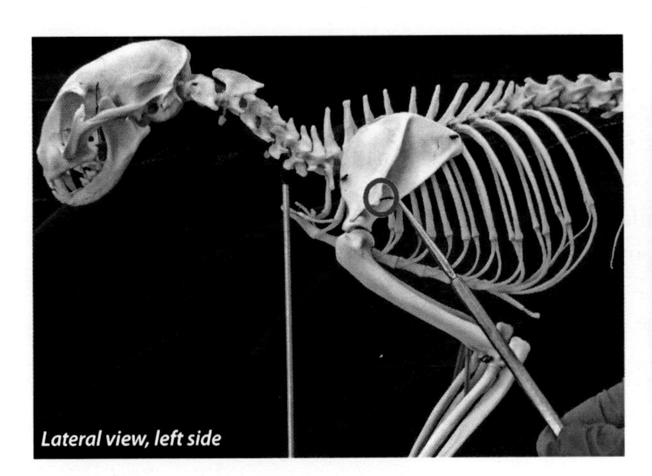
Lateral view, left side

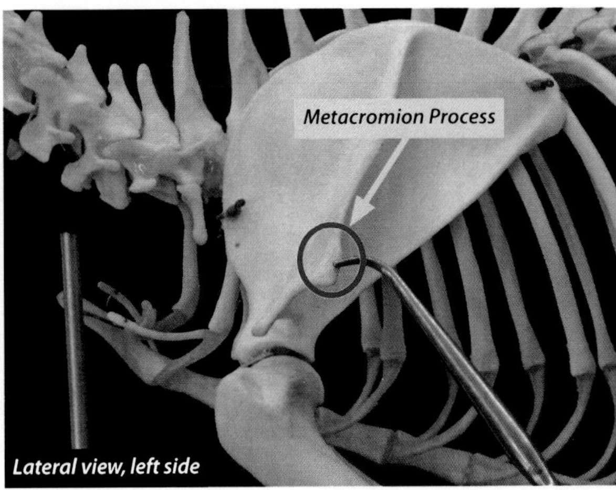

Lateral view, left side

### Metacromion Process (Cat)

The metacromion process of the cat is the insertion point for the acromiotrapezius muscle and the levator scapulae ventralis (omotransversarius) muscle. Humans do not have a metacromion process.

### Infraspinous Fossa *(see pages 6 and 7)*

As its name implies, the infraspinous fossa is located inferior to the spine of the scapula on its posterior aspect. It is the origin for:

1. infraspinatus.

### Medial (Vertebral) Border

The medial border is the edge of the scapula between the superior and inferior angles. In the adult human, the medial (vertebral) border of the scapula is normally about three fingers width lateral to the vertebral column. It is the insertion for:

1. rhomboideus major,
2. rhomboideus minor, and
3. serratus anterior (anterior surface along entire medial border).

### Inferior Angle *(see pages 6 and 7)*

The inferior angle of the scapula is at the level of thoracic vertebra 7, spinous process. The scapular line also passes through it. The inferior angle is functionally important because its posterior surface is the origin for:

1. teres major,
2. teres minor.

and the anterior surface of the inferior angle is functionally important because it is part of the insertion for:

1. serratus anterior.

### Lateral (Axillary) Border

The lateral border is the edge of the scapula between the glenoid cavity and inferior angle. It is the insertion for:

1. teres major (inferior 1/3) and
2. teres minor (superior 2/3).

# Scapula

### ***Subscapular Fossa*** *(see pages 6 and 7)*

The subscapular fossa of the scapula is the origin for:

1. subscapularis.

### ***Coracoid Process*** *(see pages 6 and 7)*

The coracoid process can be palpated in the infraclavicular fossa. The name implies that it looks like a crow's beak, but Dr. J thinks it looks more like Woodstock, Snoopy's friend. The coracoid process of the scapula is the origin for:

1. biceps brachii (short head) and
2. coracobrachialis,

and it is the insertion for:

1. pectoralis minor.

### ***Supraglenoid Tubercle*** *(see pages 6 and 7)*

The supraglenoid tubercle is a small, rounded projection of bone on the superior aspect of the glenoid cavity of the scapula created by a muscle attachment. In looking at the picture of the supraglenoid tubercle, Amy H. (Spring '08) observed that there is another benefit to viewing the scapula from this perspective. If you hold the glenoid cavity in front of you and if the coracoid process looks like a lower-case "r," it's the right scapula (r for right). See the lower picture on page 7. The supraglenoid tubercle is the origin for:

1. biceps brachii (long head).

### ***Glenoid Cavity*** *(see pages 6 and 7)*

The glenoid cavity of the scapula articulates with the head of the humerus to form the glenohumeral joint. This is a ball and socket joint (triaxial) that allows for flexion/extension, abduction/adduction (and therefore circumduction), and rotation of the arm. "Glenoid" is Latin for cavity or socket (to me).

### ***Infraglenoid Tubercle*** *(see pages 6 and 7)*

The infraglenoid tubercle of the scapula is the origin for:

1. the long head of the triceps brachii.

*Supraglenoid Tubercle*

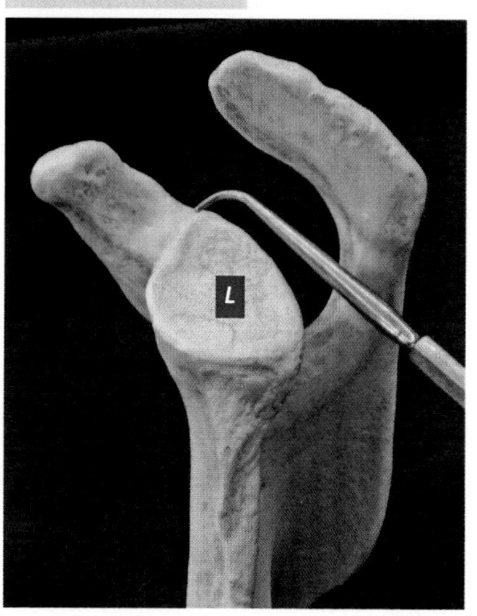


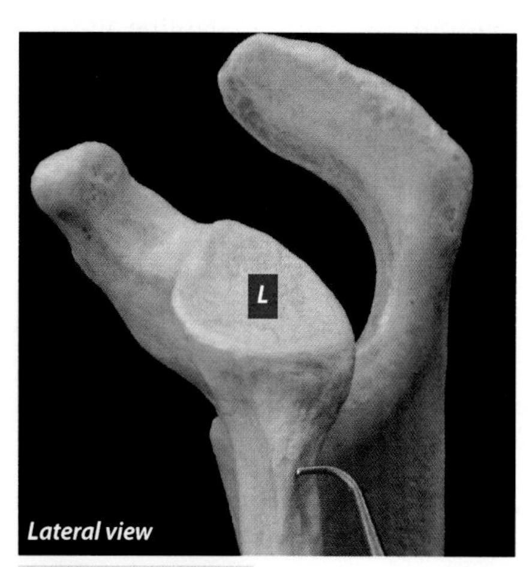


*Infraglenoid Tubercle*

# SKULL

## Nuchal Lines (Occipital Bone)

The nuchal lines of the occipital bone are where many muscles and ligaments of the neck and back attach to the skull. Generally, areas that serve as points of attachment for muscles have raised bone due to the stress on the bone, which stimulates bone growth.

The median nuchal line is also known as the external occipital crest. This is formed because of the attachment of the ligamentum nuchae, which connects the cervical vertebrae to the skull.

The superior and inferior nuchal lines form attachments with the muscles and ligaments that stabilize the articulation of the occipital condyles with the atlas (C1), thereby balancing the mass of the head over the cervical vertebrae. The superior nuchal lines are adjacent to the external occipital protuberance, while the inferior nuchal lines are approximately 2.5 centimeters (1 inch) inferior to the superior nuchal lines. The superior nuchal line of the occipital bone is of special functional importance because it is the origin for:

1. superior portion of trapezius muscle

and the insertion for:

1. sternocleidomastoid (lateral half of superior nuchal line).

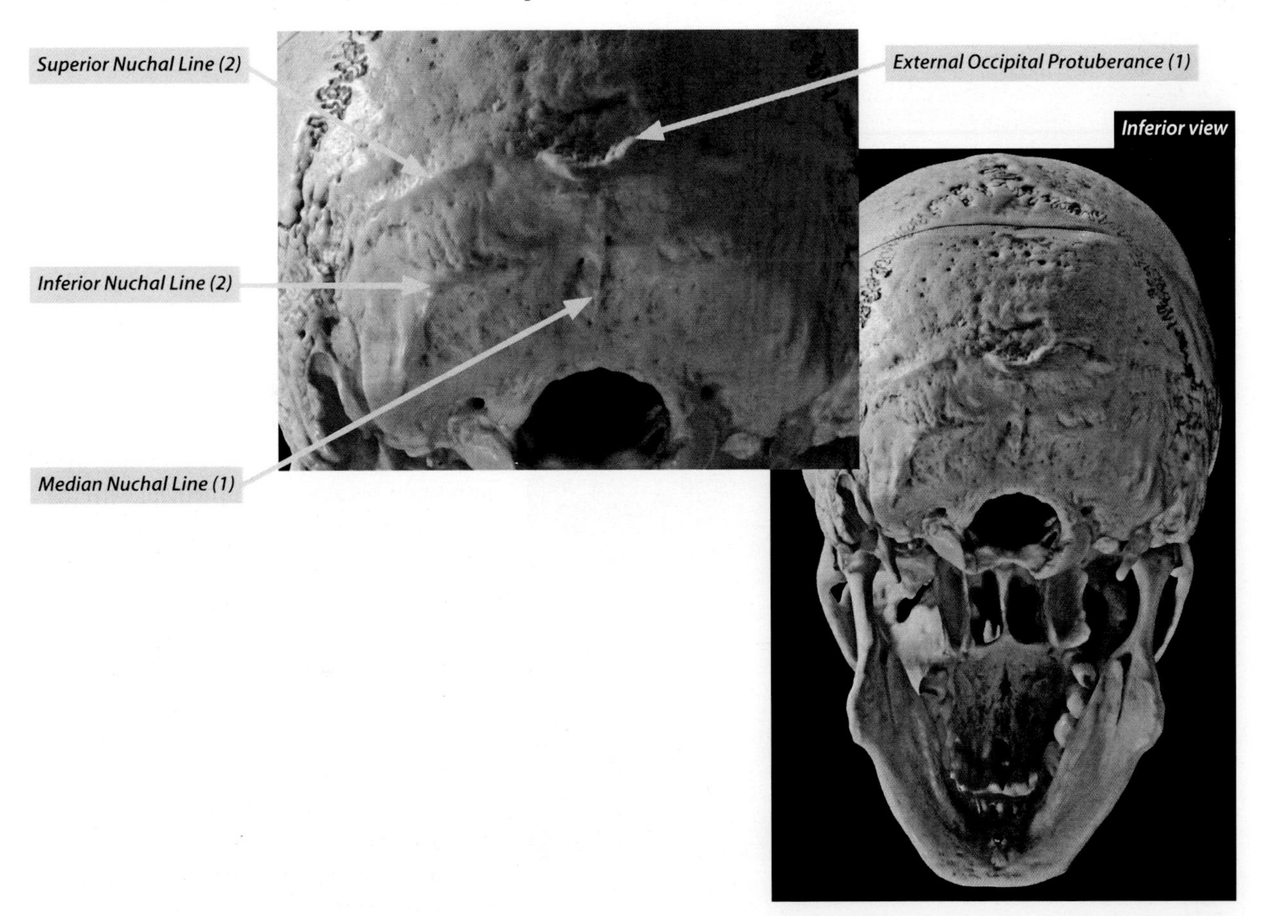

# SKULL

## External Occipital Protuberance (Occiptal Bone)

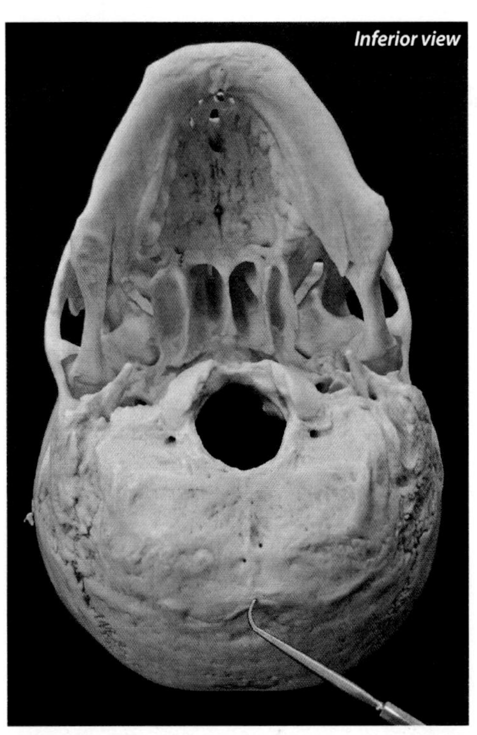
Inferior view

The external occipital protuberance is a raised area on the midline of the occipital bone where the posterior wall meets the base of the skull. It is medial to the two superior nuchal lines. It is at the superior extent of the ligamentum nuchae (a ligament that connects the cervical vertebrae to the skull). The external occipital protuberance of the occipital bone is of functional importance because it is the origin for:

1. superior portion of trapezius muscle.

## Occipital Condyles (Occipital Bone)

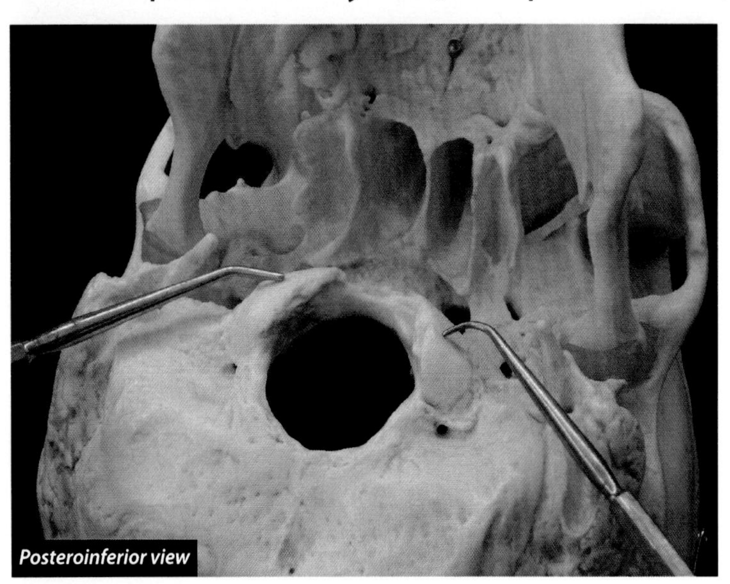
Posteroinferior view

The occipital condyles are two smooth, curved surfaces located at the base of the skull on either side of the foramen magnum (the large opening where the spinal cord exits the cranium). They are functionally important because they articulate with the superior articular facets of the atlas (C1) to form a special type of condyloid joint with a "tongue and grove shape." They allow for flexion and extension (nodding) of the head and a little bit of lateral bending. Dr. J thinks these structures are shaped like the runners of a rocking chair, which facilitate the flexion and extension of the skull relative to the atlas. When one shakes his head "no," the two bones move as one piece. As their name implies, they are landmarks of the occipital bone.

# SKULL

## Mastoid Process (Temporal Bone)

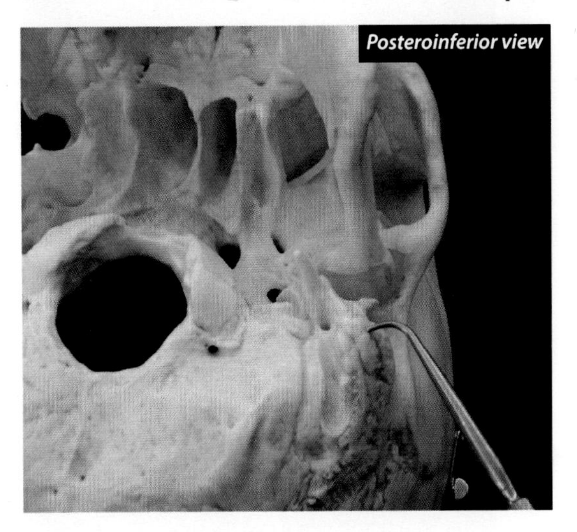


The mastoid process is a feature on the temporal bone. It is the insertion for the sternocleidomastoid muscle. In fact, it is this muscle that causes this landmark to develop. Several other muscles that you will not be responsible for also attach to this landmark. As a group, these muscles are responsible for rotation and/or extension of the head. It gets its name from the similarity it has to the appearance of a breast. Dr. J did not name this landmark—he is just the messenger. The lateral side of the mastoid process is of special functional importance because it is part of the insertion for:

1. sternocleidomastoid.

## Hyoid Bone

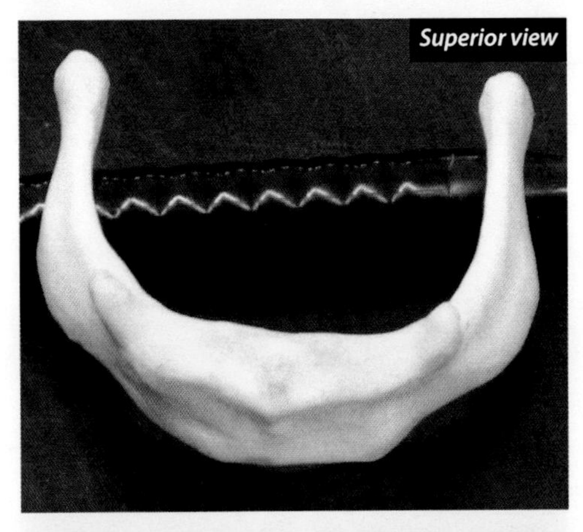


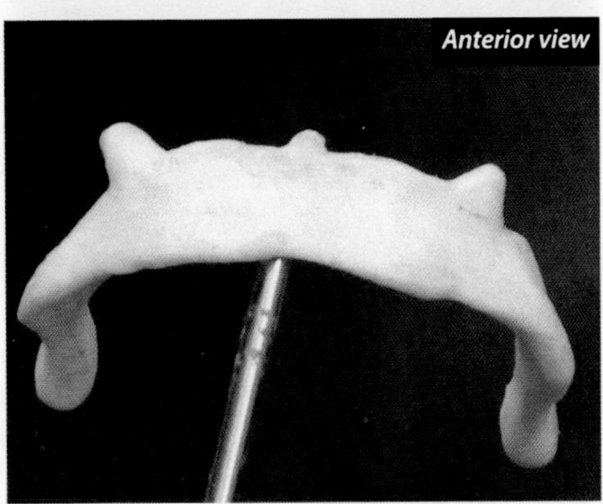


The hyoid bone does not articulate with any other bones. It is held in place by ligaments that connect it to the styloid process of the temporal bone and ligaments that connect it with the thyroid cartilage of the larynx. In spite of the fact that it is not attached to the skull, it is considered part of the skull (and therefore the axial skeleton). The hyoid resembles the mandible, suggesting a common origin. Functionally it is important because it serves as the origin for muscles that move the larynx during the act of swallowing. The hyoid gets its name from early Greek anatomists who thought it resembled the lower case form of the Greek letter upsilon. It is also the favorite bone of the seven dwarves, made famous in their song "Hyoid, hyoid, it's off to work we goid."

υ

*Greek letter upsilon (lowercase)*

# VERTEBRAE

## Vertebrae (General Landmarks)

### Articular Facets and Processes

*Superior articular process and facet, cervical vertebra*

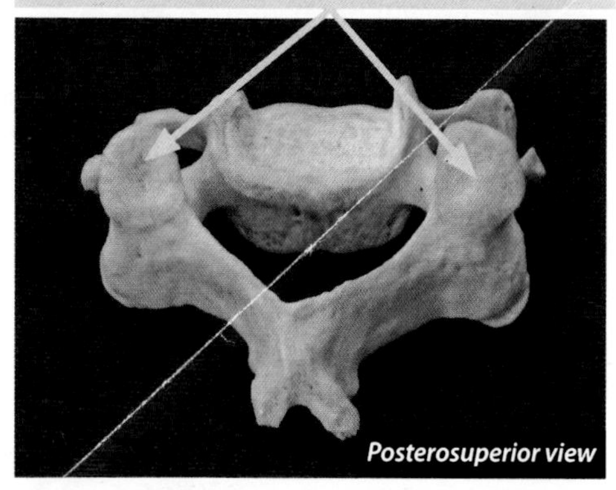


*Superior articular process and facet, thoracic vertebra*

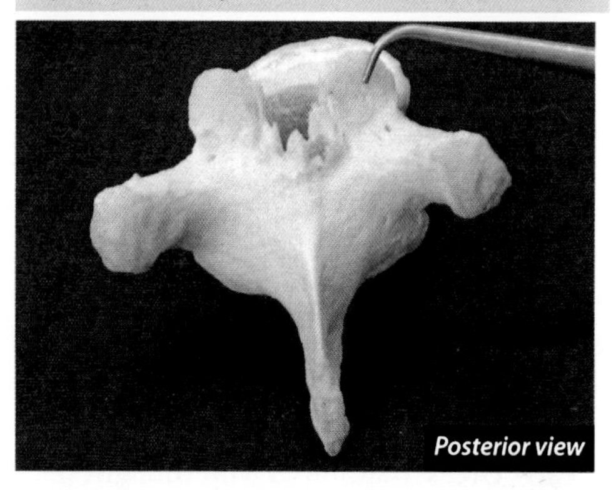


*Superior articular process and facet, lumbar vertebra*

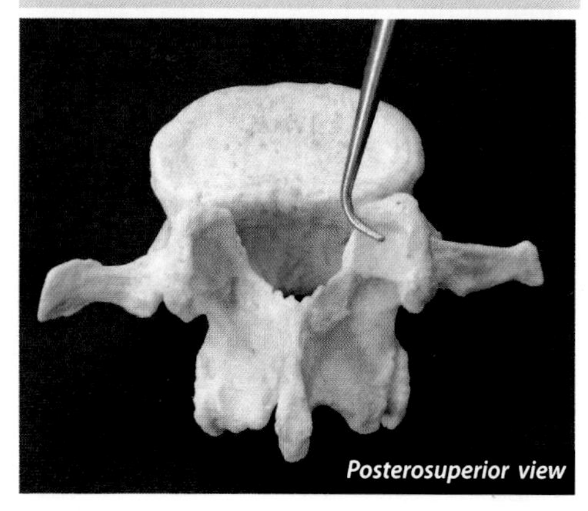


*Superior articular process and facet, sacrum*

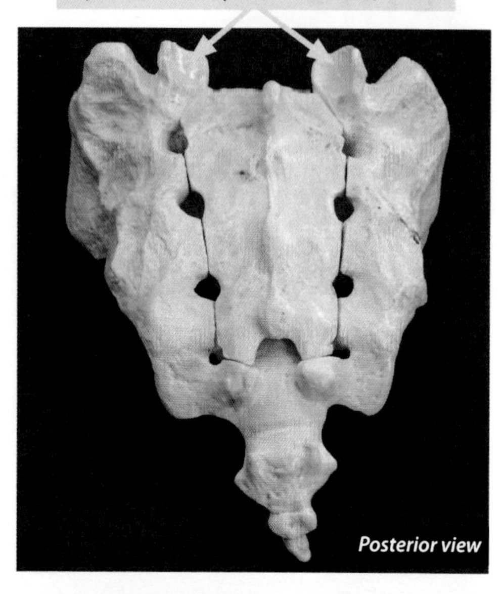


*Inferior articular process and facet*

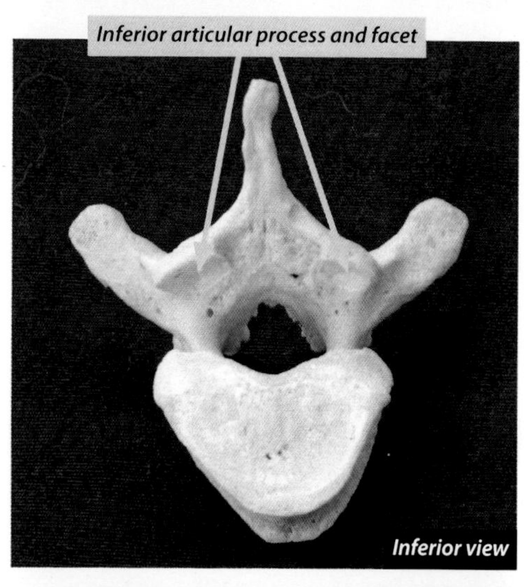


*Inferior articular process and facet, lumbar vertebra*

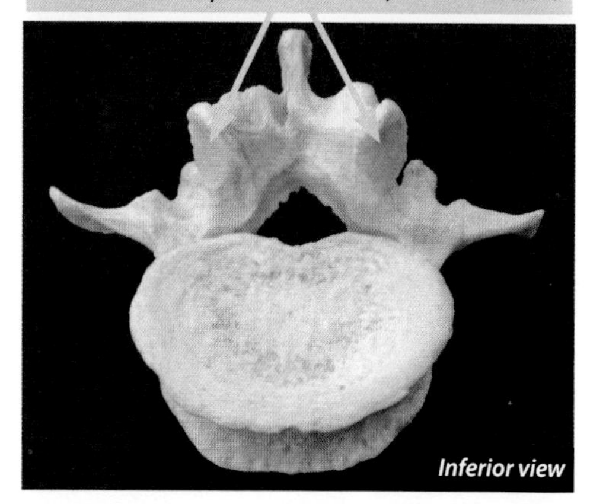

# Vertebrae (General Landmarks)

### *Articular Facet*

A facet is a flat or a nearly flat surface on a bone. The vertebral articular facets are where two vertebrae articulate. There will be one pair of facets on the superior side of the vertebrae and one pair on the inferior side of the vertebrae. These facets are on an oblique plane in the cervical region, on nearly a coronal plane in the thoracic region, and on a sagittal plane in the lumbar region. The oblique plane of the cervical vertebrae increases the range of motion possible for the neck, including rotation, lateral flexion, flexion, and extension. The coronal plane of the thoracic vertebrae effectively prevents flexion and extension, but it does allow for rotation of the vertebral column. Lateral flexion should be possible, but this motion is limited by the presence of the ribs. Functionally, the limited movement of thoracic vertebrae is important to provide a volume in which the lungs and heart can undergo changes in volume without risk of compression. The sagittal plane of the lumbar vertebrae allows for the most flexibility in anterior/posterior direction.

### *Articular Process*

Dr. J tells his students that a process is a part of the bone they can pinch between their fingers. The superior articular process projects in a superior direction from the junction of the lamina and the pedicle while the inferior articular process projects in an inferior direction from that junction. The articular facets are on one side of the process. The angles for the various regions are discussed under the articular facets.

### *Body*

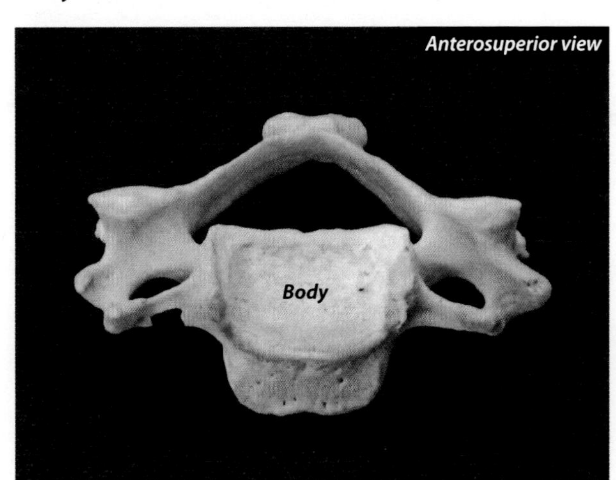


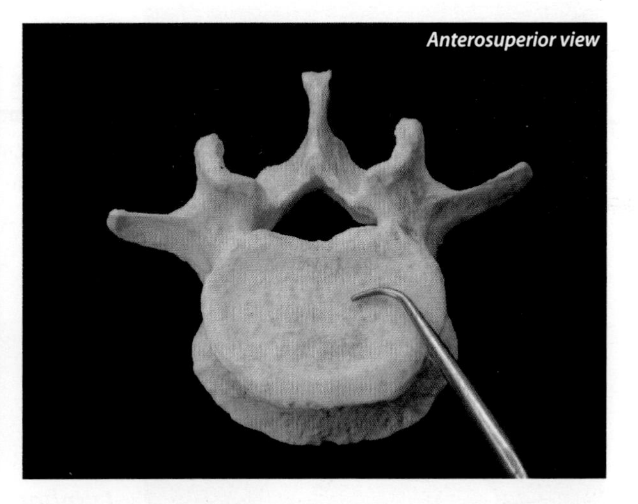


The body (centrum) of the vertebra is the surface that bears most of the weight of the structures superior to it. Since we walk on two appendages, the weight progressively gets greater as one moves from the cervical to the lumbar regions. Therefore, it makes sense that the superior vertebrae have much smaller bodies and that the size of the body progressively gets larger as we move inferiorly. Compare the size of the body of the lumbar vertebra on the right with the size of the body of the cervical vertebra on the left. Not only does the size of the vertebrae change, the shape of the bodies changes as well. The body of a cervical vertebra is wider laterally than it is in the anteroposterior direction, giving it an oval appearance. The thoracic bodies tend to be more heart-shaped and they have depressions where the demifacets are. Lumbar vertebrae have a kidney or oval-shaped body. The depth (thickness) of the lumbar vertebrae is greater than that of the superior areas.

# Vertebrae (General Landmarks)

## Spinous Process

The spinous (spinal) process projects in a posterior direction from the junction of the lamina of a vertebra. Each region has a characteristic shape. When we palpate vertebrae, it is the spinous processes that we detect. One should note that the transition from one type of vertebra to the next is gradual, so the first vertebra from one area looks much like the last vertebra from the preceding area.

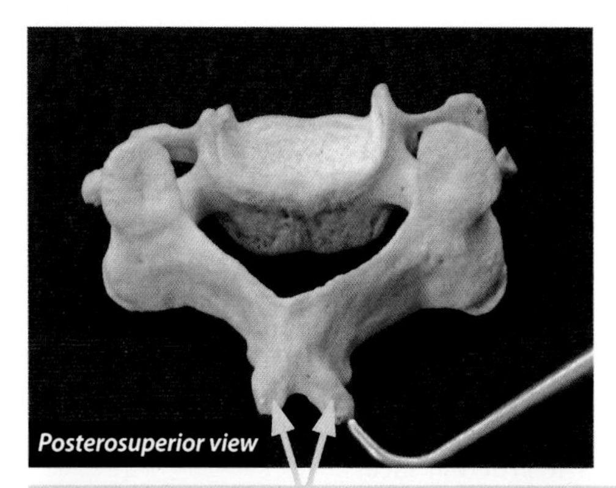


Note that the spinous process of this cervical vertebra is bifid (doubled).

### *Cervical*

We find that most cervical vertebrae have a bifid spinous process. The atlas (C1) lacks a spinous process, and C7 is usually single.

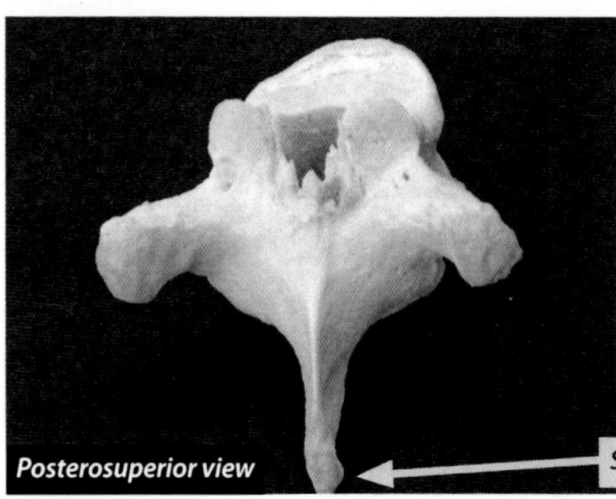


### *Thoracic*

The thoracic vertebrae have long spinous processes that point in an inferior direction, making them look like a giraffe. Remember the movie *Thoracic Park* with all the giraffes!

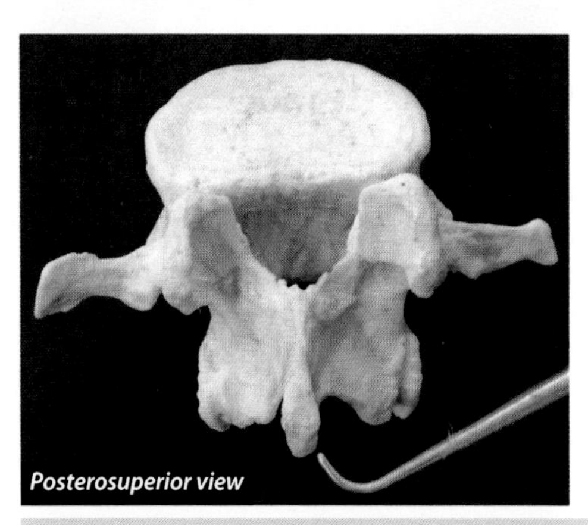


The probe is pointing to the spinous process of this lumbar vertebra

### *Lumbar*

The spinous processes of the lumbar vertebrae are typically short and blunt, and they project in a posterior direction. Muscles that help move the vertebrae attach to the spinous processes. They also serve as points of attachment for ligaments that stabilize the vertebral column.

# Vertebrae (General Landmarks)

## Transverse Processes

The transverse processes project laterally from the junction of the lamina and the pedicle on each side of the vertebrae. They serve as points of attachment for muscles that help move the vertebrae as well as ligaments that stabilize the vertebral column. Additionally, in the superior ten thoracic vertebrae there is a transverse costal facet on each transverse process that articulates with the tubercle of the rib. Each area of the vetebral column has characteristic shapes for this structure.

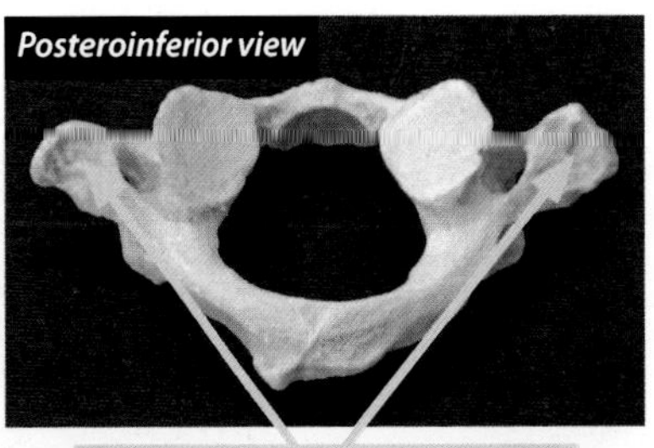


*Transverse Processes of Atlas (C1)*

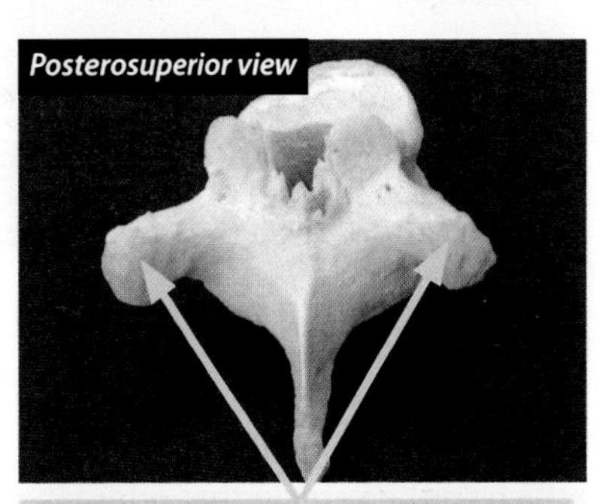


*Transverse Processes of Thoracic Vertebra*

### *Cervical*

In the cervical region, each transverse process has a transverse foramen (see pictures associated with the cervical vertebrae). Additionally, the transverse process of the atlas projects further laterally than other cervical vertebrae, and this extra length serves as a mechanical advantage for the muscles that move the head and atlas relative to the axis. Also, the cervical vertebrae have anterior and posterior tubercles on the transverse processes that serve as points of muscle attachment. The anterior tubercles are formed from the remains of the embryonic cervical ribs.

### *Thoracic*

The thoracic vertebrae have transverse processes that tend to project obliquely, both posteriorly and laterally, making a "V" shape when viewed from the superior or inferior side.

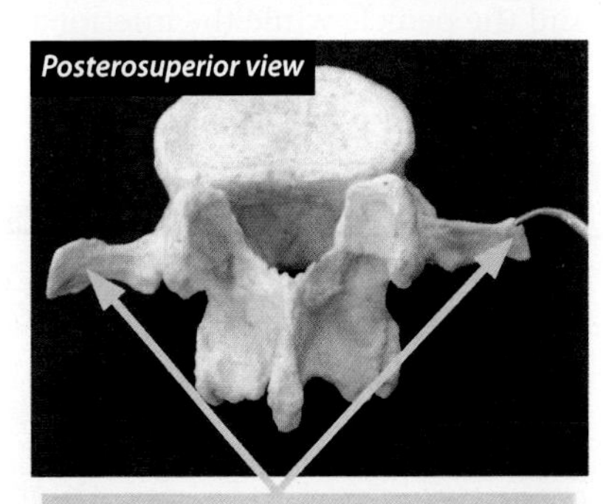


*Transverse Processes of Lumbar Vertebra*

### *Lumbar*

The lumbar vertebrae have long, slender transverse processes that project laterally like the wings of a glider. In fact, the long piece of bone in a T-bone steak is the transverse process of the lumbar vertebra. If that piece were the spinous process, it would have been called an S-bone steak. Hmm. There is one exception in the lumbar region, and that is lumbar five vertebra. It has large cone-shaped transverse processes that serve as points of attachment for the iliolumbar ligaments that attach to the os coxae.

*Fused Transverse Processes of Sacrum—called "ala"*

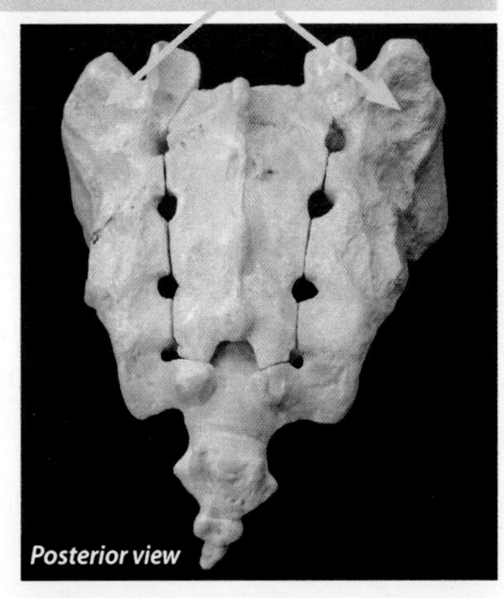


### *Sacral*

The transverse process of the first sacral vertebra is wing-like and named the ala. The other sacral transverse processes fuse laterally, thereby closing off the intervertebral foramina. Lastly, the first coccygeal vertebra has small, poorly developed transverse processes.

# Vertebrae (General Landmarks)

### *Lamina*

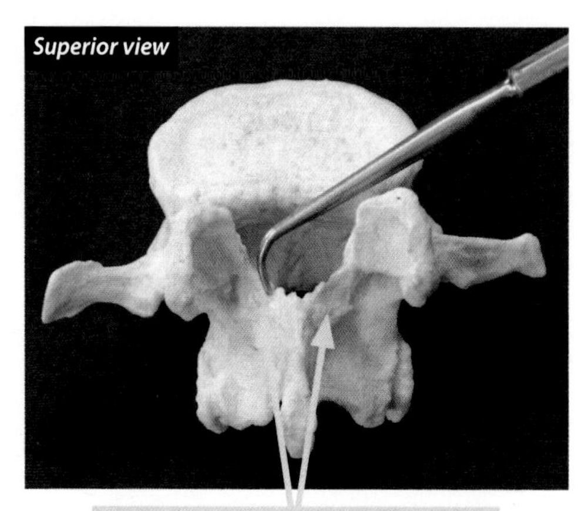


Laminae (plural), posterosuperior view

The lamina forms the posterior portion (roof) of the vertebral arch (neural arch). The left and right laminae join at the midline. The spinous process projects from the midline where the laminae meet. Laterally they attach to the pedicle that originates from the body. The transverse process projects from where the lamina and the pedicle meet. The superior articular process projects in a superior direction from the junction of the lamina and the pedicle while the inferior articular process projects in an inferior direction from that junction. The lamina in the above picture is part of a lumbar vertebra. Dr. J suspects that they were named for the famous movie *Silence of the Lamina*.

### *Pedicle*

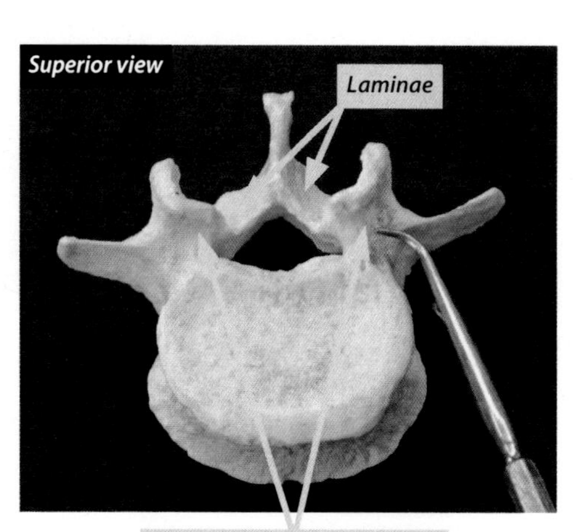


Pedicles, posteroinferior view

The pedicles (meaning, "little feet") are two short, thick processes that project posteriorly and connect the vertebral body to the vertebral arch. It is at the junction of the pedicles and laminae that we find the superior and inferior articular processes. The arches of the pedicles of adjacent vertebrae also form the intervertebral foramen, where spinal nerves exit and enter the spinal column.

# Vertebrae (General Landmarks)

### *Vertebral Foramen*

The vertebral foramen is of functional importance because it provides for the passage of the spinal cord and the meninges. This foramen gets progressively smaller as one moves inferiorly. The large size is not necessary in the inferior regions because the spinal cord becomes smaller in diameter (it tapers like a carrot) as it moves inferiorly. This occurs because the spinal nerves are entering and exiting at each intervertebral foramen. The spinal cord normally terminates in the adult human at the level of the inferior edge of the L1 body, although it may end close to the T12 body or as far inferiorly as the L2 body. In an infant, it usually terminates in the area of the L3 or L4 body. The foramen is triangular in shape for the cervical and lumbar vertebrae. The thoracic vertebrae have a circular vertebral foramen. Together the vertebral foramina form the vertebral canal.

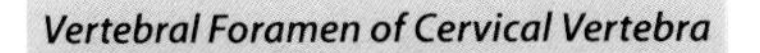


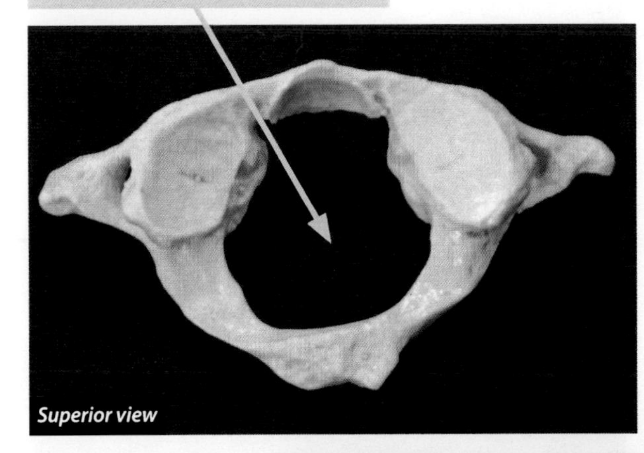


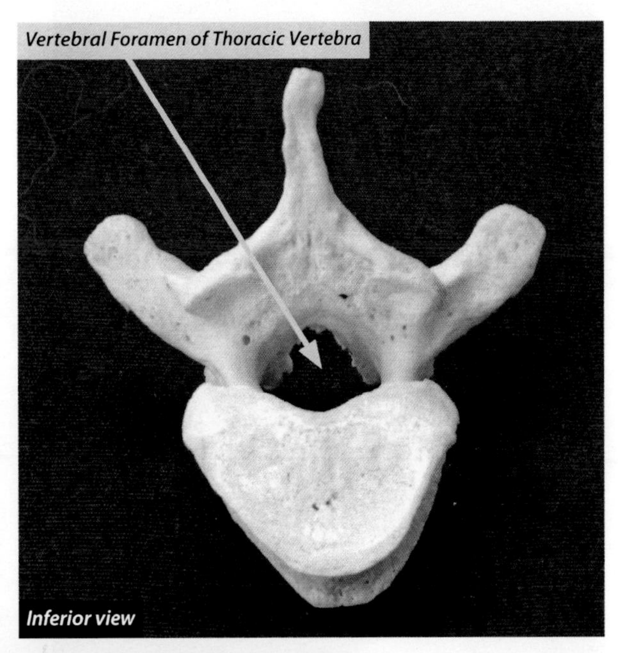


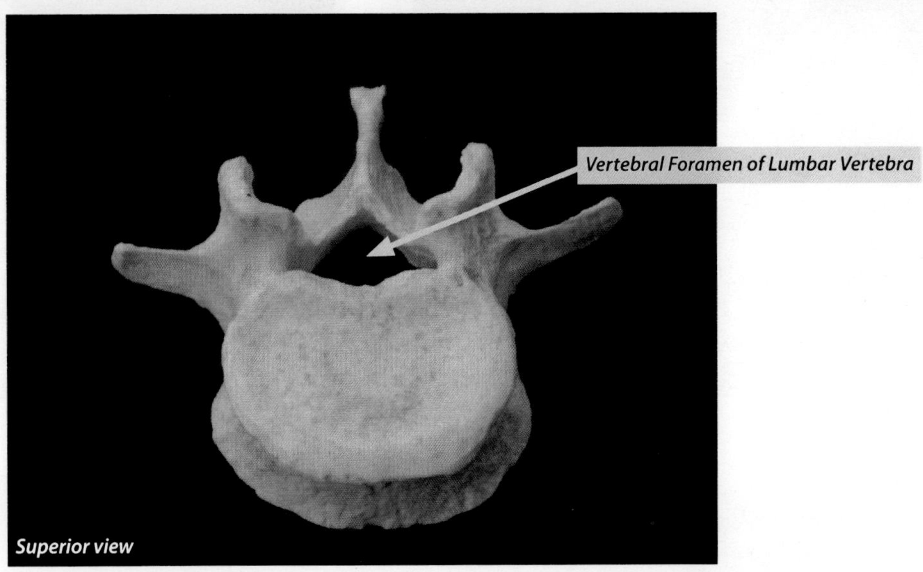

# Vertebrae (General Landmarks)

### *Intervertebral Foramen (Foramina, plural)*

The intervertebral foramina are of functional importance because this is where the spinal nerves, covered with spinal meninges, leave or enter the vertebral foramen. They are formed by the arches of the pedicles of two adjacent vertebrae. Because there can be no support where this opening occurs, it is often a location for injury to the intervertebral disc.

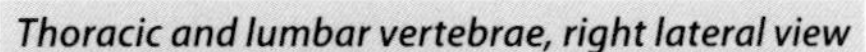
*Thoracic and lumbar vertebrae, right lateral view*

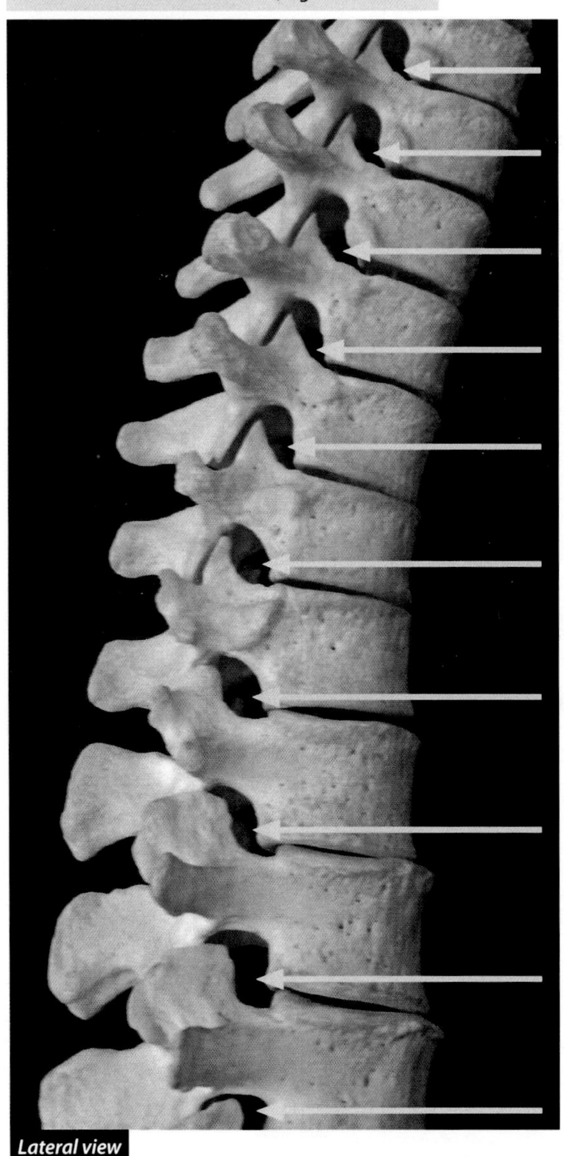
*Lateral view*

*Cervical and thoracic vertebrae, right lateral view*

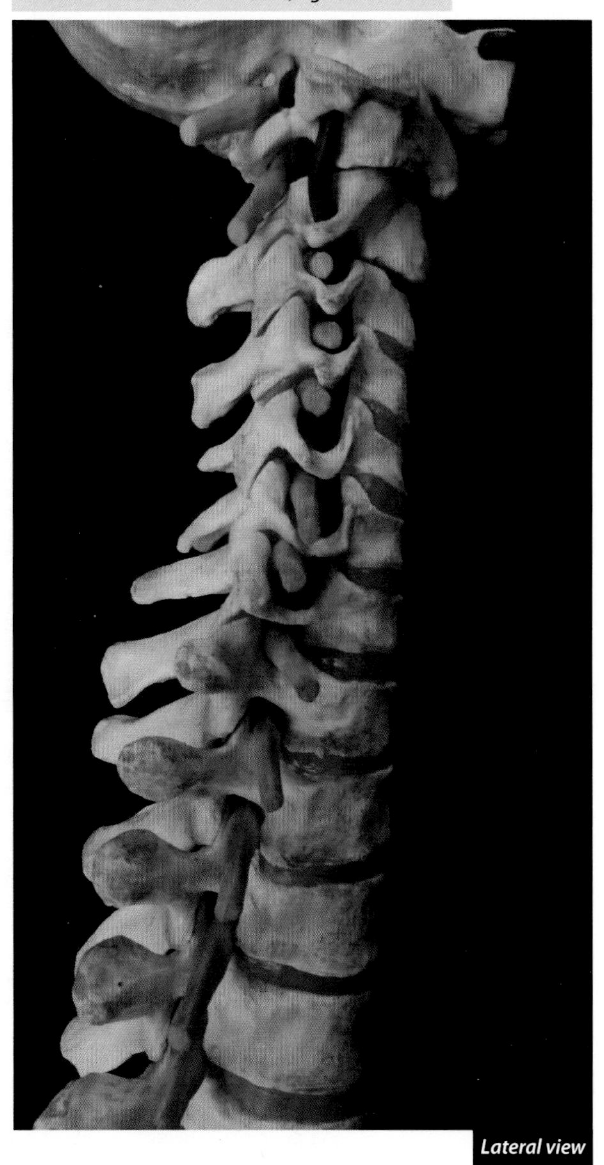
*Lateral view*

*In the picture on the right, we can see spinal nerves entering/exiting the intervertebral foramina and vertebral arteries approaching the skull via the transverse foramina of the cervical vertebrae (7). NOTE: Only the cervical vertebrae have transverse foramina in their transverse processes.*

## Cervical Vertebrae (7)

The superior most seven vertebrae are the cervical vertebrae. They are unique in several ways. Most have a bifid spinous process, although the seventh cervical vertebra has a single spinous process. The body of the cervical vertebrae is wider from side to side than from anterior to posterior. In the adult, it will be smaller than that of the other areas, as it carries less weight. Its vertebral foramen is triangular and large, which is important since the diameter of the spinal cord is greatest at the superior end. They also have transverse foramina (foramen transversarium). The articulations of the cervical vertebrae are in an oblique plane. This is important functionally because it increases the range of motion that we have in the cervical region, which includes rotation, lateral flexion, flexion, and extension.

### *Transverse Foramina*
*(singular = foramen)*

The transverse foramina (foramen transversarium) of the cervical vertebrae are openings that are occupied by the **vertebral arteries** and **veins** in the first six vertebrae and by only the **vertebral veins** in the seventh. The **vertebral arteries** are of particular importance because they serve the brain and spinal cord. The **vertebral veins** do not serve much of the brain, but rather they receive blood from the cervical spinal cord, the cervical vertebrae, and some of the small muscles in the superior portion of the neck. Each transverse foramen is formed by the union of the transverse process posteriorly and the remains of the cervical rib anteriorly. These foramina do not occur in the other regions of the vertebral column.

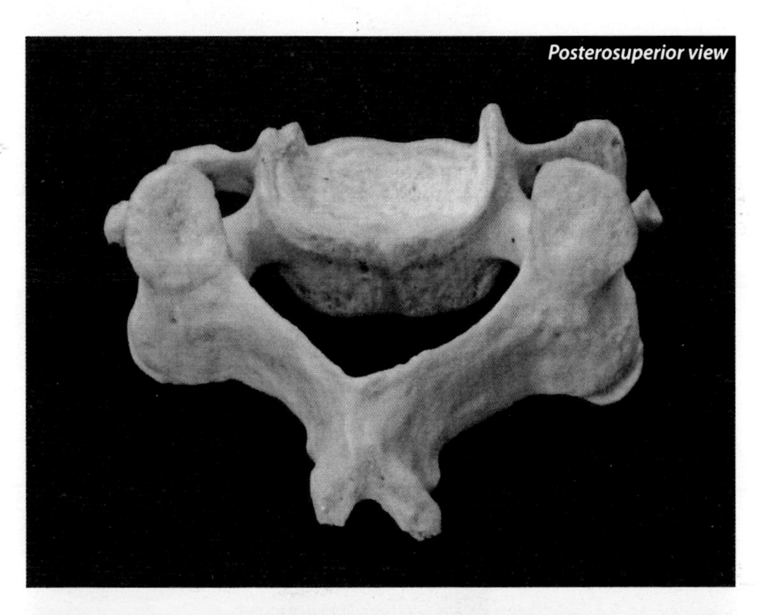


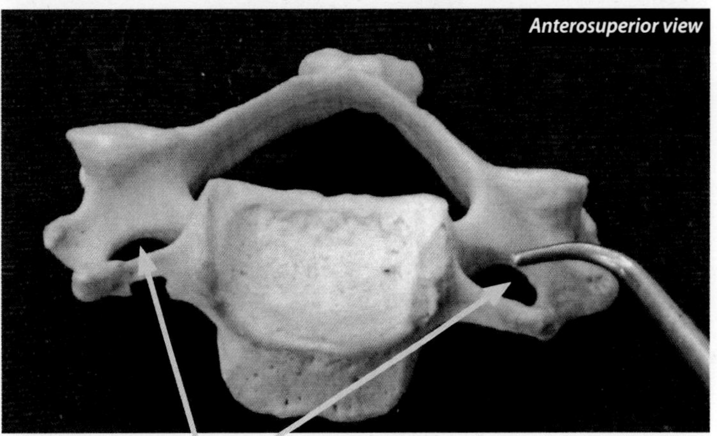


Transverse Foramina of a Cervical Vertebra

# Cervical Vertebrae (7)

### *Atlas (C1)*

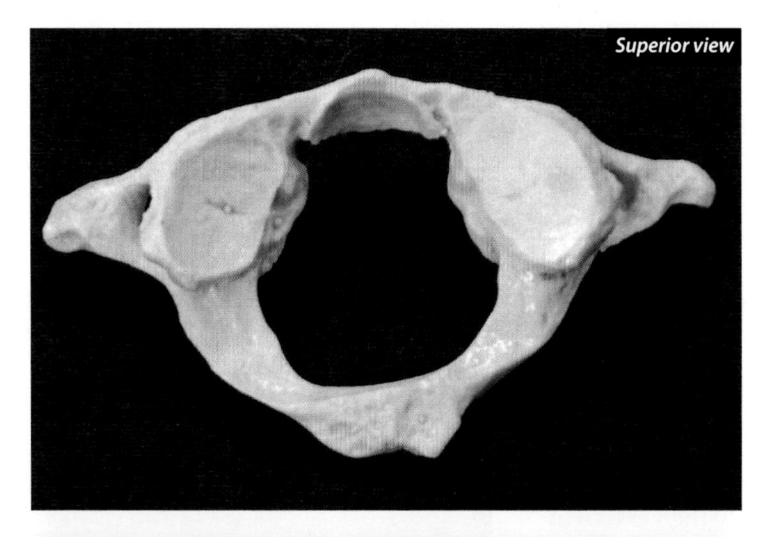


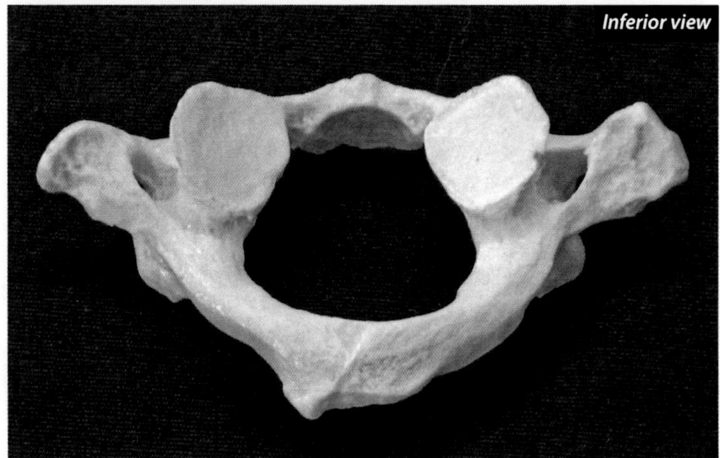


The atlas is the name for the superior most cervical vertebra (C1). It is derived from a reference to the legendary Greek god who held the world on his shoulders, much as the mythological Atlas holds the head (globe shaped) on his shoulders. The occipital condyles of the occipital bone articulate with the superior articular facets of the axis to form a tongue and groove, which is a specialized type of hinge joint that allows for flexion and extension of the skull (nodding the head "yes"). However, when one shakes his head "no," the occipital bone and atlas move as one piece. The atlas is unusual in that it lacks a well-developed spinous process. Another distinguishing feature is that there is no intervertebral disc between the atlas (C1) and the axis (C2). There was a disc, but it becomes part of the dens and body of the axis when they fuse during fetal development or early childhood. The atlas has no body, but it does have an anterior arch and a posterior arch that connect the lateral masses. The lateral masses include the articular surfaces. The fact that it lacks a body led it to write the song "I ain't got no body!" for David Lee Roth.

# Cervical Vertebrae (7)

### *Axis (C2)*

The axis is the name for the second cervical vertebra (C2). It is unusual in that it has the dens (odontoid process) that the atlas rotates around when one shakes her head "no." During this motion, the occipital bone and the atlas move as one piece. This atlantoaxial joint is a pivot joint. There are strong ligaments, the alar ligaments, which connect the dens to the medial surfaces of the occipital condyles. These ligaments are important to prevent excessive rotation of the occipital bone and atlas around the dens. The name of the axis sounds a little like the part of a car (axle) around which the wheel rotates. It was named for the famous rock star, Axis Rose.

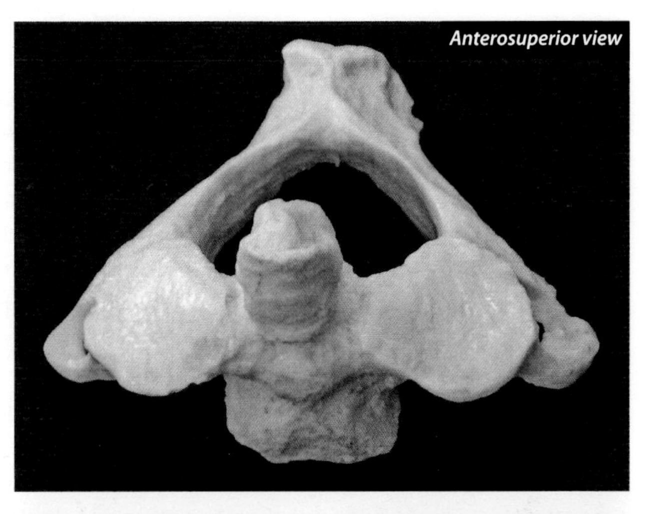


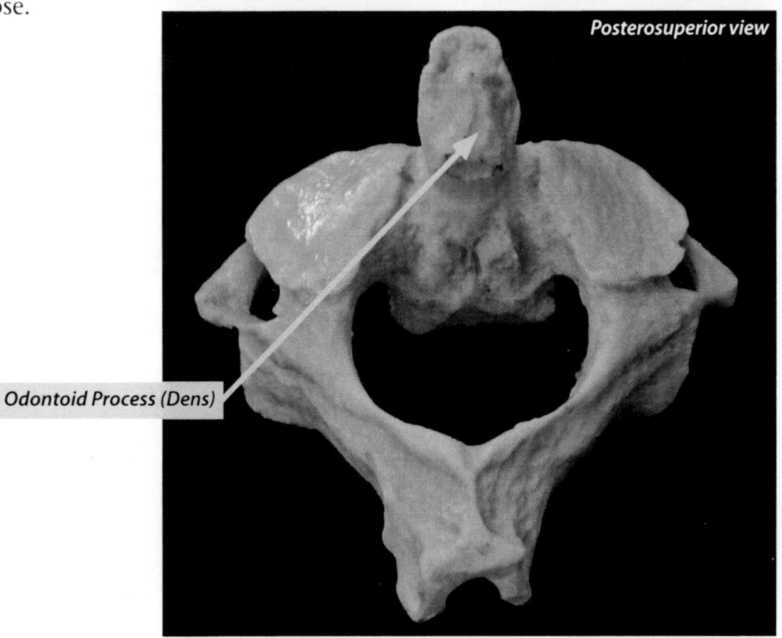


### *Odontoid Process (Dens)*

The odontoid process (dens) gets its name from its resemblance to a tooth. It was the body of the atlas during embryonic development; however, it breaks away. The dens fuses to the body of the axis during fetal development or early childhood. Because it is not yet fused, violent motion of the head can cause the dens to dislocate and damage the spinal cord. Even after it has fused, violent motion of the head can dislocate the atlantoaxial joint and push the dens into the base of the brain or, posteriorly, into the spinal cord. The dens serves as a pivot point for the atlas (C1) at the atlantoaxial joint. It allows for rotation. We find an oval facet on the anterior surface of the odontoid process. This is where articulation with the anterior arch of the atlas occurs. The alar ligaments attach the odontoid process to the medial surfaces of the occipital condyles, and they prevent excessive rotation of the atlas and head relative to the axis. Another ligament, the transverse ligament of the atlas, runs posterior to the odontoid process, and this ligament holds the dens in position.

# Thoracic Vertebrae (12)

The twelve thoracic vertebrae are easily recognized because they look like the head of a giraffe from the posterior side. That is not a good answer for a quiz, but it may help you make a quick identification. Remember the movie Thoracic Park with all the giraffes! The characteristics that you should use to identify a thoracic vertebra on a quiz include its long spinous process that points in an inferior direction and its transverse processes that project posteriorly like a "V."

We also find articulations for the ribs on the thoracic vertebrae, including the transverse costal facets, the superior costal demifacets, and the inferior costal demifacets. Exceptions to this are listed in specific sections that follow.

Lastly, the articulations between adjacent thoracic vertebrae are nearly in a coronal plane. This effectively prevents flexion and extension, but it does allow for rotation of the vertebral column. Lateral flexion should be possible, but the presence of the ribs limits this motion. Functionally, the limited movement of thoracic vertebrae is important to maintain the space in which the lungs and heart can undergo changes in volume without risk of compression.

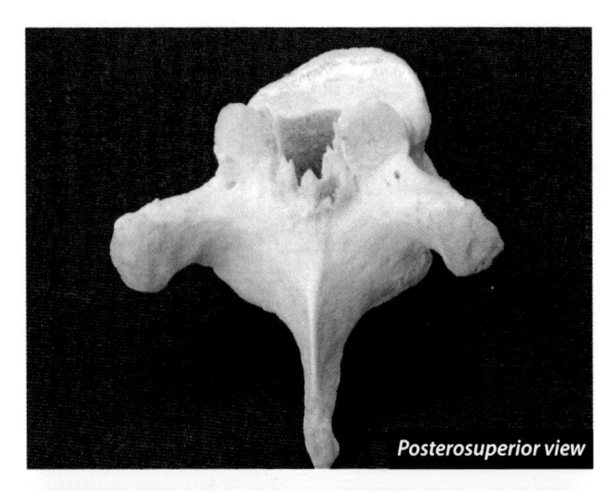

*Posterosuperior view*

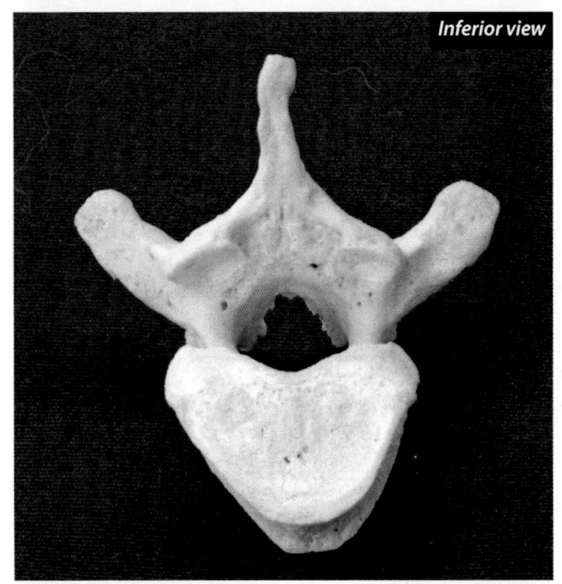

*Inferior view*

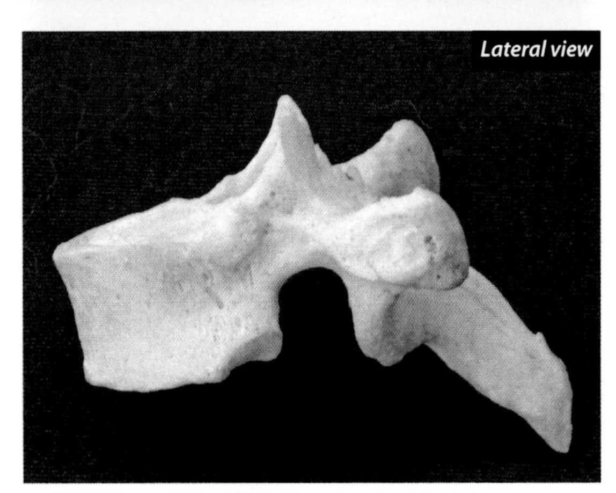

*Lateral view*

# Thoracic Vertebrae (12)

## *Costal Demifacets*

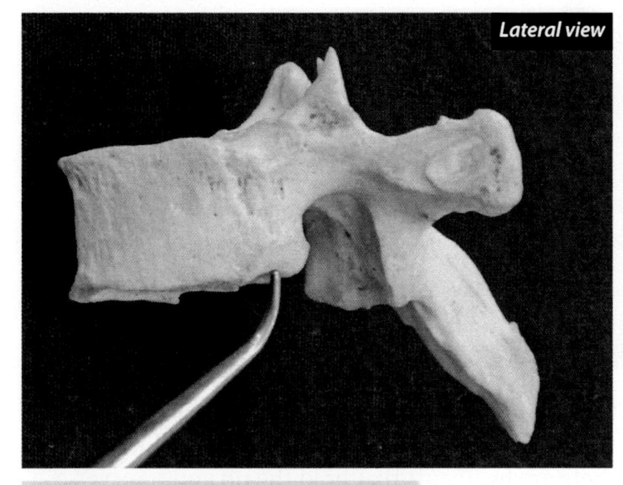


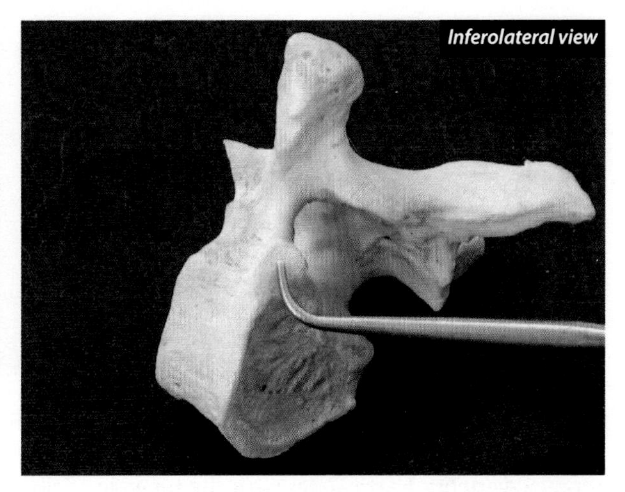


*Inferior costal demifacet of thoracic vertebra*

*Superior costal demifacet of thoracic vertebra*

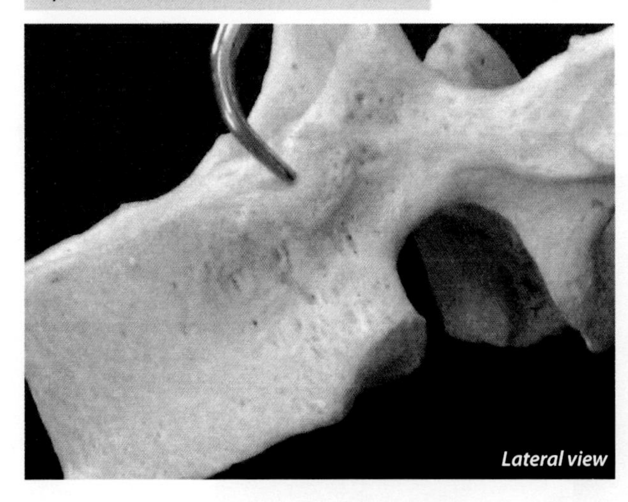


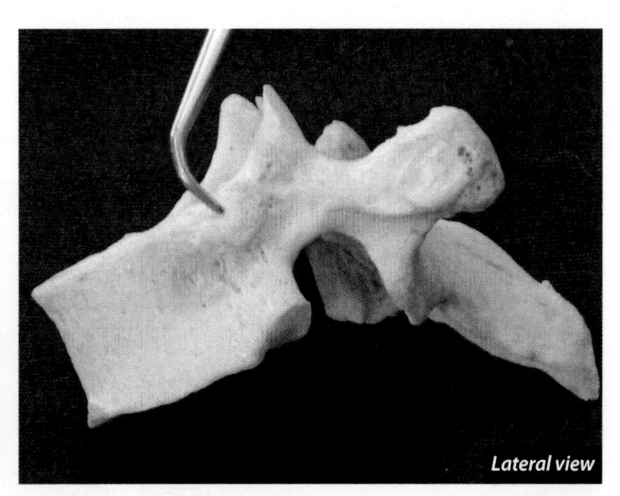


A facet is a flat or nearly flat surface on a bone. A demifacet is actually half of a facet and is where part of the head of the rib or part of the costal cartilage articulates. These demifacets exist for ribs 2 through 9. The remaining ribs articulate directly to the lateral sides of the vertebral bodies in transition areas of the vertebral column. You remember Dr. J explaining that vertebrae transition gradually.

Posteriorly, the superior demifacet of the body of the vertebra receives the head of the rib with the same number as the vertebra of interest. For example, rib 5 articulates with thoracic vertebra 5 at its superior demifacet. The inferior demifacet of the body of the vertebra receives the head of the rib with a number greater than the number of the vertebra. For example, rib 5 articulates with thoracic vertebra 4 at its inferior demifacet. Anteriorly, instead of a single facet for the costal cartilage of rib 7 on the body of the sternum, there may be a demifacet on the inferior end of the body of the sternum, as well as one on the xiphoid process. These demifacets articulate with the costal cartilage of rib 7.

## Thoracic Vertebrae (12)

### *Transverse Costal Facets*

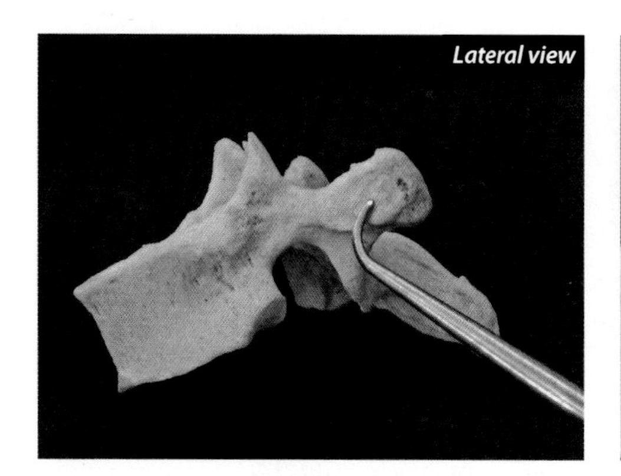
*Lateral view*

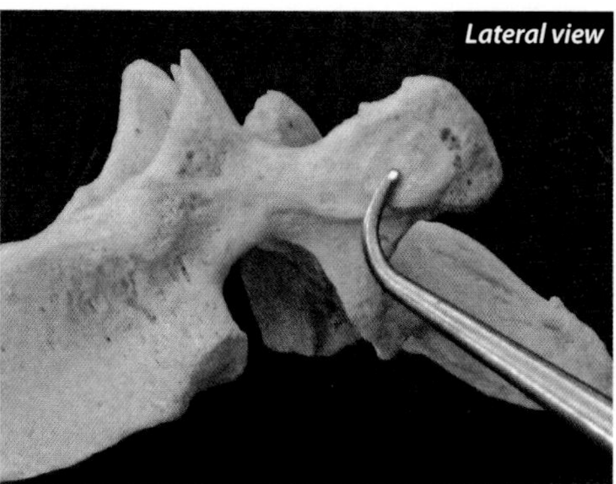
*Lateral view*

The transverse costal facets are cupped surfaces on each transverse process of the first ten thoracic vertebrae. They articulate with the tubercle of the rib of the same number. For example, each transverse costal facet of thoracic vertebra 5 articulates with the tubercle of rib 5. There are ligaments that hold the rib and transverse process together. Functionally, these articulations (in combination with the articulations of the head of the rib and the body of the vertebrae) are important because they limit the mobility of the thoracic vertebrae. Thoracic vertebrae 11 and 12 do not have transverse costal facets because they more closely resemble lumbar vertebrae.

## Lumbar Vertebrae (5)

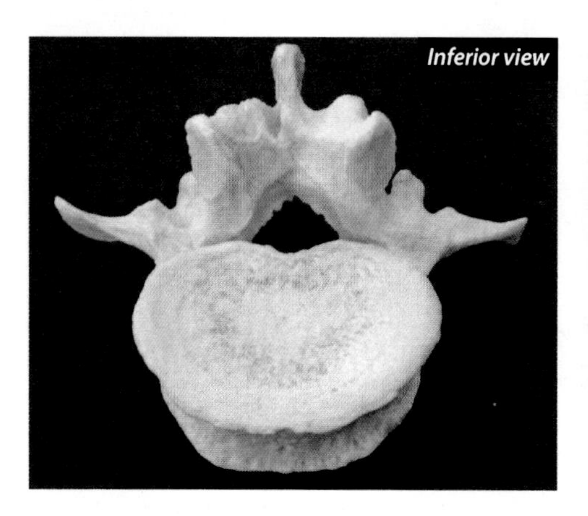
*Inferior view*

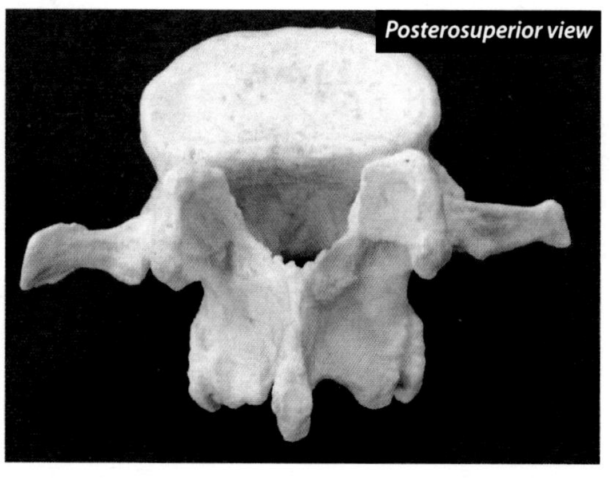
*Posterosuperior view*

The five lumbar vertebrae can be recognized by several features. They have a short, blunt spinous process that projects in a posterior direction. Their transverse processes project laterally and are relatively long. (Dr. J says they look like the wings of a glider.) They have relatively small vertebral foramina since the spinal cord is very thin by the time it gets to the lumbar region. In fact, it normally ends at the inferior edge of the L1 body in the adult human. The sagittal plane of articulation for the lumbar vertebrae allows for the most flexibility in anterior/posterior direction. In the adult, the vertebral body is relatively large when compared to vertebrae from other regions. This size is important because the lumbar vertebrae are supporting more weight than any of the vertebrae superior to them.

## Sacral Vertebrae/Sacrum (5, Fused)

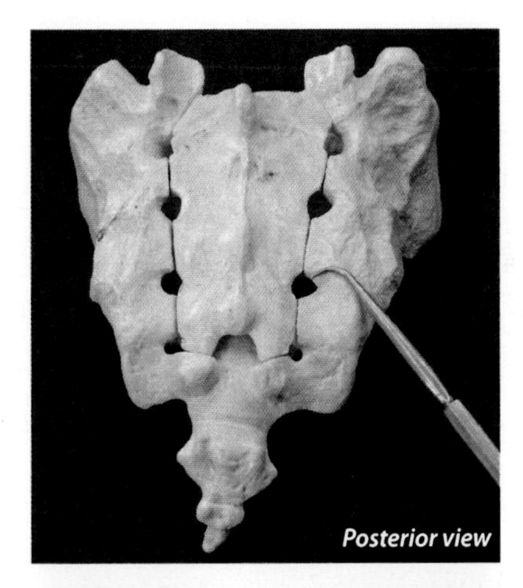

Posterior view

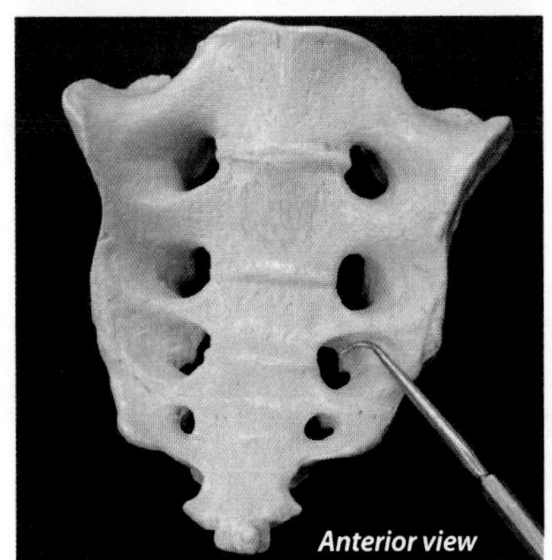

Anterior view

The sacral vertebrae are the most unusual of them all. There are five sacral vertebrae, and they are all fused. This process begins after puberty and is complete by the time a person is thirty years old. Part of the structure is formed when the remains of the sacral ribs fuse to the vertebrae during development. Their vertebral foramina also join to form a canal, terminating inferiorly at the sacral hiatus. There are four pairs of anterior and posterior sacral foramina that allow the sacral rami (nerve roots) to exit from the sacral canal. Superiorly, the sacrum articulates with the fifth lumbar vertebra. Inferiorly, it articulates with the coccyx. Laterally, it articulates with the ilium. The anterior edge of the body at the superior end of the sacrum projects anteriorly into the pelvic cavity. This is called the sacral promontory. The sacrum forms part of the pelvic wall, and it has a concave anterior surface and convex posterior surface (this is called the sacral curve). The flat surfaces of the sacrum are important areas for muscle attachment for the muscles of the thigh and lower back. The sacrum is usually shorter and wider in women than in men. We will spend more time on the sacrum in lab 11.

### *Sacral Foramina*

The sacral foramina occur on the anterior and posterior surfaces of the sacrum. The anterior and posterior rami of the sacral nerves exit from the sacral canal through these openings. They correspond to the intervertebral foramina of the other portions of the vertebral column.

*In these pictures the probes point to the posterior and anterior sacral foramina, respectively.*

## Coccygeal Vertebrae/Coccyx (2-5)

The coccygeal vertebrae are the most inferior of the vertebrae. They form part of the bony pelvis. Their vertebral features are poorly developed or missing. The coccyx usually consists of four vertebrae, although it varies from two to five vertebrae. Normally the inferior vertebrae are fused together, while the first forms an amphiarthrosis with the fifth sacral vertebra. In elderly people (sixty or more years old), it may become fused to the sacrum. Fusion of the coccygeal vertebrae occurs at variable rates, normally beginning by the mid-twenties and reaching completion during later years. It fuses at an earlier age in males than in females. In post pubescent females, it points inferiorly. However, in males the coccyx continues the sacral curve, and it can therefore provide some support for pelvic viscera. The coccyx serves as a point of attachment for some muscles. It can be broken during a fall, or sometimes during the delivery of a baby. In nursery school, we called it the tailbone. It got its name from the fact that some anatomists thought it resembled the beak of a cuckoo! What party were they at?

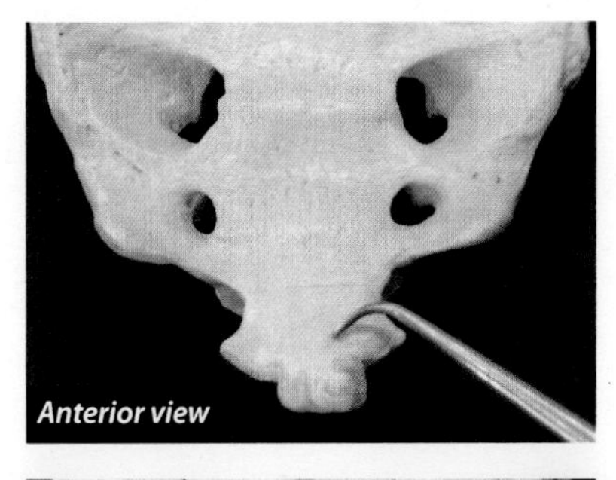

Anterior view

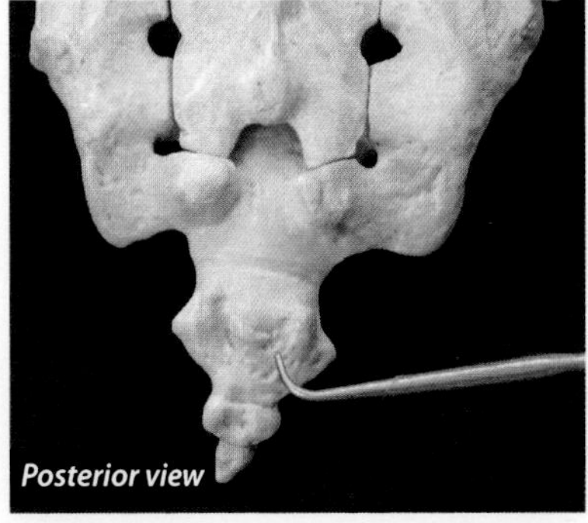

Posterior view

# RIBS

Ribs are the long, curved bones that form the rib cage (boney thorax), which increases in diameter as we move inferiorly. They surround the chest and protect the lungs, heart, and other thoracic organs. Humans have twelve pairs of ribs, and there are three types: true, false, and floating, each of which will be described in turn.

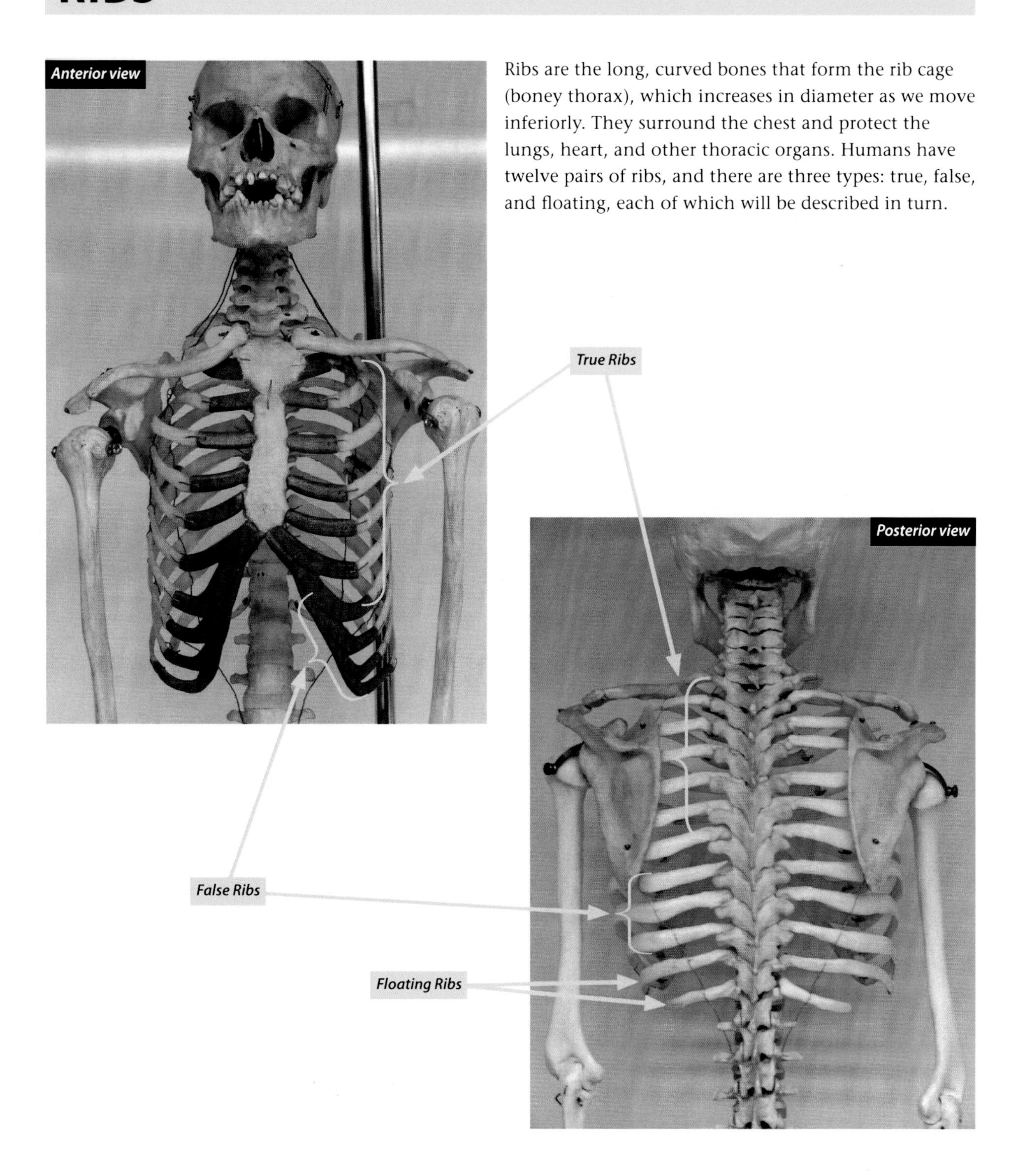

# RIBS

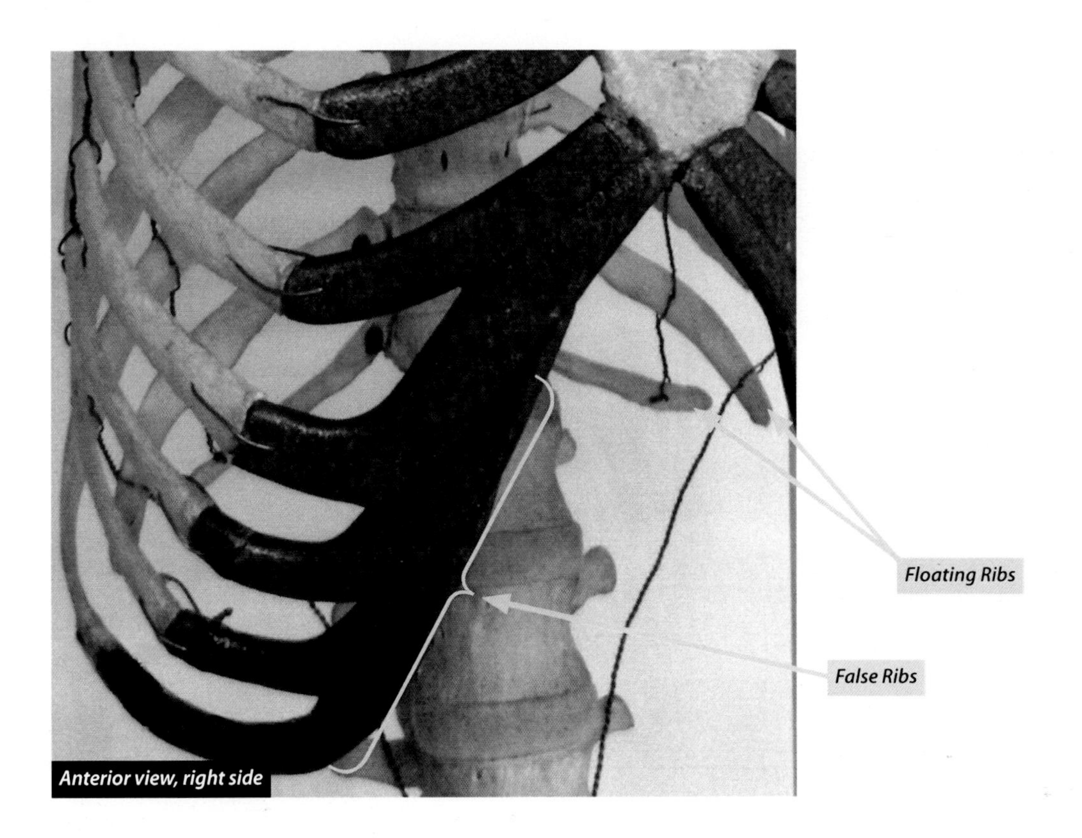


## True Ribs (14)

There are seven pairs of true ribs. They are the most superior of the thoracic ribs and are sometimes called vertebrosternal ribs. They differ from false and floating ribs because they directly articulate with the sternum by means of their costal cartilages. The shortest true rib is rib 1 and all ribs increase in length to rib 7.

## False Ribs (6)

There are three pairs of false ribs. They are located between the true ribs and the floating ribs, and are sometimes called vertebrochondral ribs. False ribs differ from true ribs because they do not directly articulate with the sternum. Their costal cartilages join together, and that fused structure articulates with the costal cartilage of rib 7, which then articulates with the sternum. Many authors include the floating ribs in this group, thereby making five pairs of false ribs, but we will consider the floating ribs a mutually exclusive group in order to simplify things.

## Floating Ribs (4)

There are two pairs of floating ribs, which are the most inferior of the ribs. They are different from the other ribs of the thoracic region because they do not articulate anteriorly with the sternum or with the costal cartilage of other ribs. However, they do articulate with thoracic vertebrae 11 and 12 posteriorly, and they do have costal cartilage that is imbedded within the muscles of the lateral thoracic wall. For this reason, they are sometimes called vertebral ribs.

# RIBS

## Landmarks

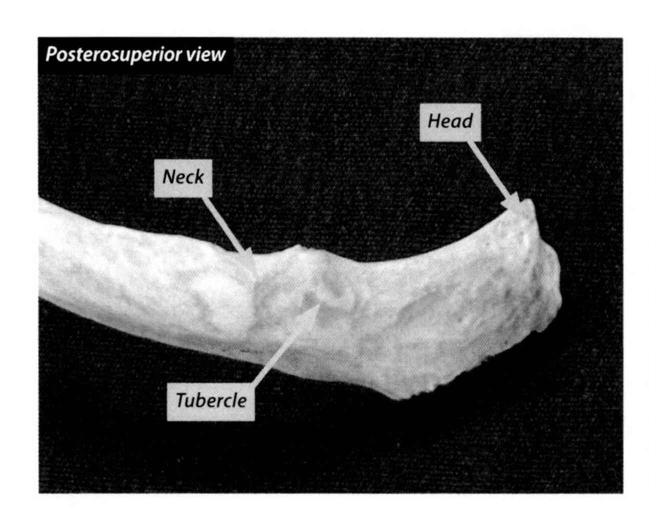


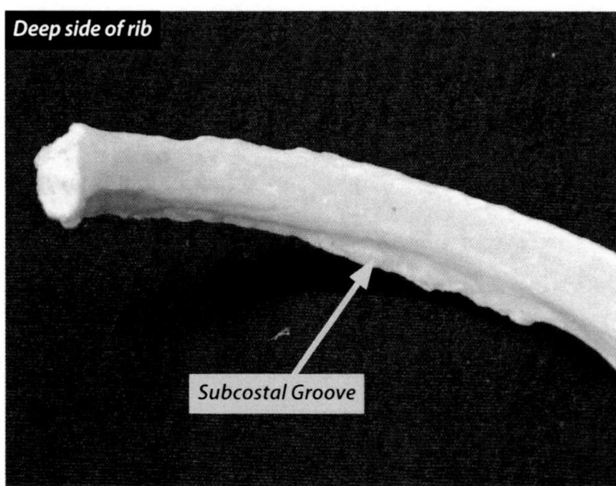


### *Head*

The head of the rib is functionally important because this is where it articulates with the superior demifacet of the vertebra for which it is numbered, and with the inferior demifacet of the vertebra superior to it. For example, rib 5 articulates with thoracic vertebra 5 at its superior demifacet and with thoracic vertebra 4 at its inferior demifacet.

### *Neck*

The neck is a narrow region of the rib between the head and the body.

### *Tubercle*

The tubercle of the rib is functionally important because this is the point of articulation with the transverse costal facet of the vertebra for which the rib is numbered. For example, the tubercle of rib 5 articulates with the transverse costal facet of thoracic vertebra 5. There are ligaments that hold the rib and transverse process together.

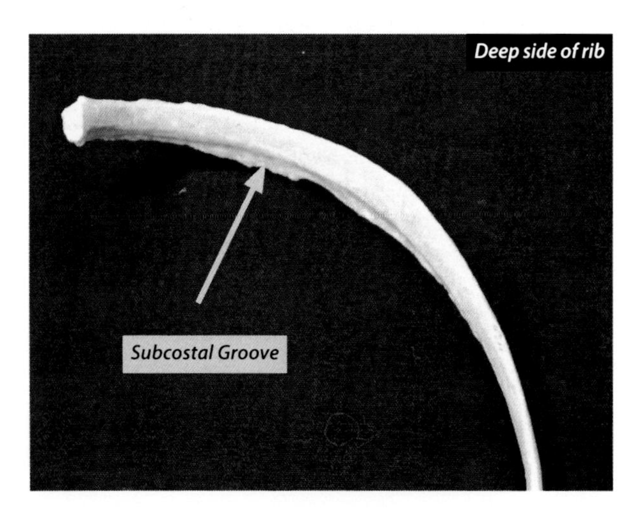


### *Subcostal Groove (Costal Groove)*

The subcostal (costal) groove is functionally important because a neurovascular bundle runs deep to it. This bundle has a **vein** as the superior structure, then an **artery**, and lastly a **nerve** (**VAN** from superior to inferior). If one were to describe its location, it would be on the deep, inferior side of the rib. Dr. J prefers subcostal as the name because it should help remind you of its location.

# RIBS

## Landmarks

### *Body*

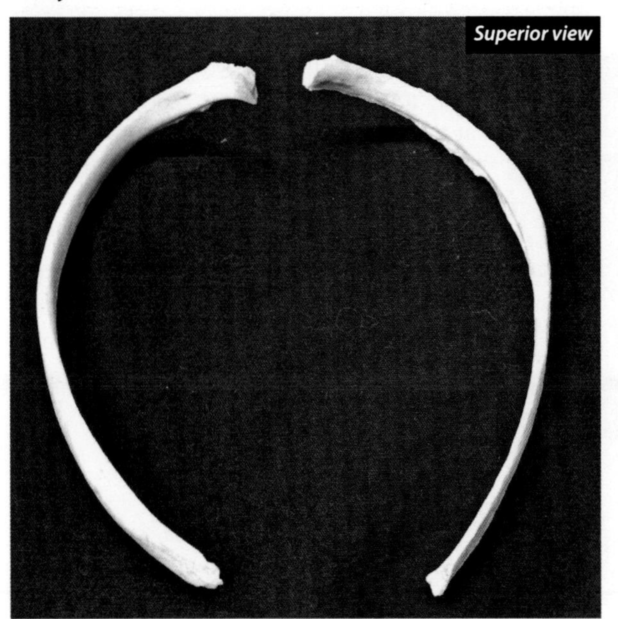


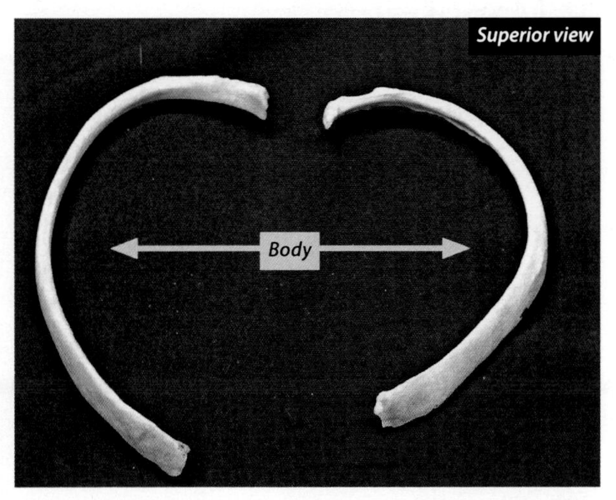


The body of the rib is the attachment point for the intercostal muscles and the origin for:

1. pectoralis minor (ribs 3, 4, and 5 near costal cartilage), and
2. serratus anterior (ribs 1 to 8, superior lateral surface).

### *Costal Cartilage*

The costal cartilage of each rib is functionally important because it forms the attachment of the rib to the sternum, while allowing for some flexibility of the thoracic wall. The joints formed by the articulation of ribs 2 through 7 with their costal cartilages are considered synovial joints (diarthrosis—free moving). The attachment of rib 1, however, is considered a synchondrosis early in life, and it often becomes a synostosis (mostly immoveable) in older people. The costal cartilage is hyaline cartilage.

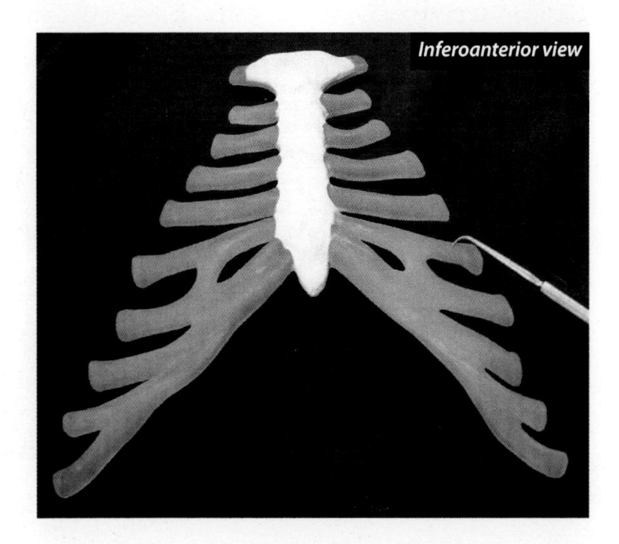

# STERNUM

The sternum, sometimes referred to as the breastbone, is a flat bone located in the anterior midline of the thoracic cavity. It can be palpated and is composed of three parts, whose descriptions follow.

## Manubrium

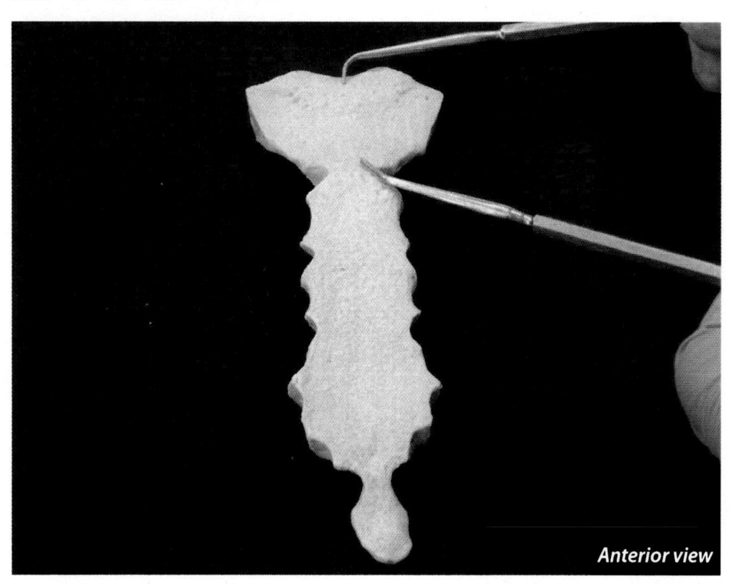
*Anterior view*

The manubrium (handle) is the most superior of the sternebrae and is located at the level of the bodies of thoracic vertebrae 3 and 4. It articulates inferiorly with the body of the sternum (at the sternal angle or angle of Louis). It also articulates with the clavicle, rib 1, and the super or half of rib 2 (demifacet) on each side. Superior to the manubrium, we find the suprasternal (jugular) notch. It forms part of the origin of:

1. pectoralis major, and
2. part of the origin for sternocleidomastoid,
3. sternohyoid, and
4. sternothyroid muscles.

## Sternal Angle

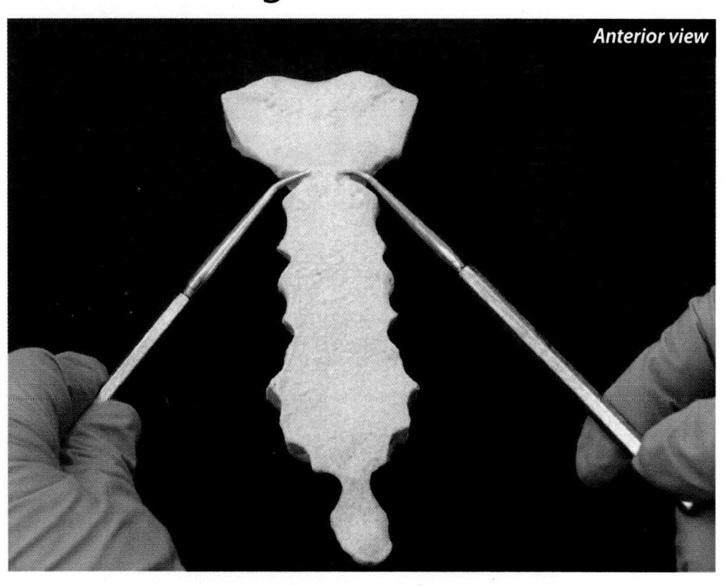
*Anterior view*

The sternal angle (angle of Louis) of the sternum is also called the manubriosternal joint, which is where the manubrium of the sternum articulates with the body of the sternum. It is a fibrocartilaginous joint, allowing for some movement. The sternal angle acts like a hinge so that the body of the sternum can move anteriorly during deep inspiration. It is also a landmark used to identify the boundary between the superior and inferior mediastinal cavities. It is located at the level of the intervertebral disc between thoracic vertebrae 4 and 5. This landmark is also the superior extent of the heart as well as the inferior end of the trachea when a person is supine (lying on the back).

# STERNUM

## Body (Gladiolus)

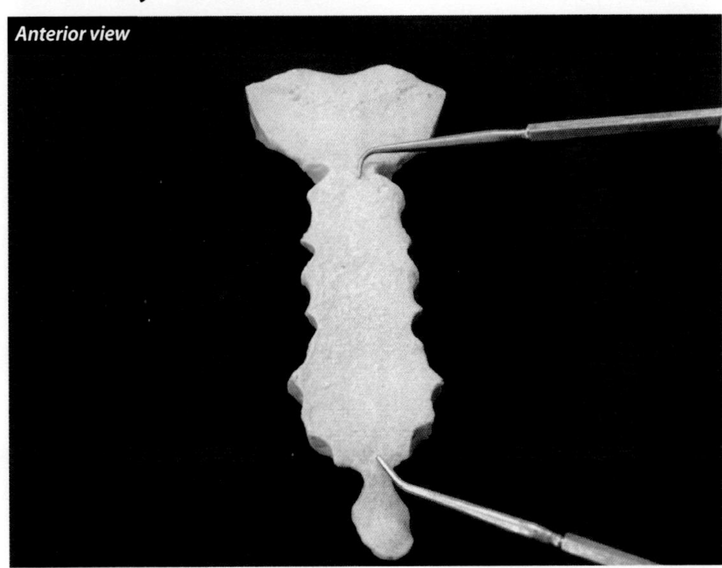


The body of the sternum is formed from four fused bones (sternebrae). The fusion of these bones is complete after puberty. It articulates superiorly with the manubrium (that forms the sternal angle or angle of Louis) and inferiorly with the xiphoid process. It also articulates with the inferior half of rib 2 (this is a demifacet), ribs 3 through 6, and the superior half of rib 7 directly (this too is a demifacet). Ribs 8, 9, and 10 attach to the body indirectly by joining the costal cartilage of rib 7. The body of the sternum forms part of the origin for:

1. pectoralis major, and
2. diaphragm.

## Xiphoid Process

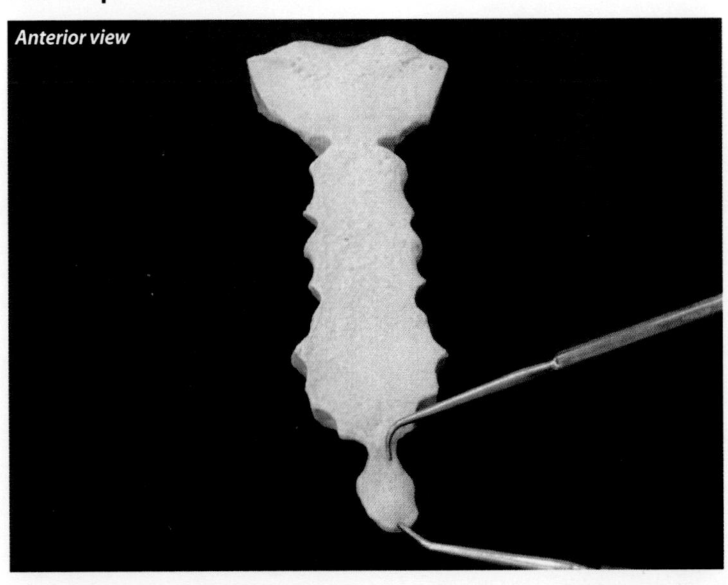


The xiphoid (like a sword) or xiphoid process is the most inferior of the sternebrae. It has no "Zs" or "Ys" in its name. In a young person, it is hyaline cartilage, but it is bone by the time one is forty years old. Although the joint between the xiphoid and the body begins as an amphiarthrosis, it often becomes a synostosis in older people. It is sometimes broken when a person receives CPR. The joint between the xiphoid and the body of the sternum is at the level of the thoracic vertebra 9 body. It also has a demifacet on each side that articulates with the inferior portion of the costal cartilage of rib 7. The xiphoid process is part of the origin for

1. diaphragm;

and the insertion for

1. rectus abdominis.

# Lab 2

# Muscles of the Thorax and Neck

## Wish List


**LAB 2 REVIEW, pp. 36–37**

**MUSCLES, pp. 38–59**

### Dorsal Thoracic Region

- **Trapezius, pp. 38–39**
  - Clavotrapezius (Cat), p. 38
  - Superior Division, Trapezius, p. 39
  - Acromiotrapezius (Cat), p. 38
  - Middle Division, Trapezius, p. 39
  - Spinotrapezius (Cat), p. 38
  - Inferior Division, Trapezius, p. 39
- **Deltoids, pp. 40–43**
  - Clavodeltoid/Clavobrachialis (Cat), p. 40
  - Anterior Division, Deltoid, pp. 40–41
  - Acromiodeltoid (Cat), p. 42
  - Middle Division, Deltoid, pp. 42–43
  - Spinodeltoid (Cat), p. 42
  - Posterior Division, Deltoid, pp. 42–43
- **Levator Scapulae Ventralis (Cat), p. 44**
- **Levator Scapulae, pp. 44–45**
- **Latissimus Dorsi, pp. 46–47**
- **Serratus Ventralis (Cat), p. 48**
- **Serratus Anterior, pp. 48–49**
- **Rhomboids, pp. 50–51**
  - Rhomboideus Capitis (Cat), p. 50
  - Rhomboideus Minor, pp. 50–51
  - Rhomboideus Major, pp. 50–51

### Ventral Thoracic Region

- **Pectoral Muscles, pp. 52–55**
  - Pectoantebrachialis (Cat), p. 52
  - Pectoralis Major, pp. 52–53
  - Pectoralis Minor, pp. 54–55
  - Xiphihumeralis (Cat), p. 54
- **Neck Muscles, pp. 56–59**
  - Sternohyoid, pp. 56–57
  - Sternothyroid, pp. 56–57
  - Sternomastoid (Cat), p. 58
  - Cleidomastoid (Cat), p. 58
  - Sternocleidomastoid, p. 59

**NERVES, pp. 60–61**

- Axillary, p. 60
- Thoracodorsal, p. 60
- Long Thoracic, p. 61
- Spinal Accessory (XI), p. 61

**VESSELS, pp. 60, 62**

**ARTERY, p. 60**

- Thoracodorsal Artery, p. 60

**VEINS, p. 62**

- Cephalic Vein, p. 62
- External Jugular Vein, p. 62

# LAB 2 OVERVIEW

Lab 2 focuses on the muscles, nerves, and vessels of the thoracic wall and cervical regions. This is our first opportunity to view the anatomy similar to the way we would look at a road map. This is a useful perspective to develop as a study tool.

First we must prepare the cat for observation. On the ventral side of the cervical region, you should find a cut through the skin where dye was injected into the blood vessels. Insert the blunt end of the probe into this cut, push it caudally, and lift the skin. Withdraw the probe, carefully cut the skin without going deep with the scissors, and then repeat the process along the ventral midline of the cat. Continue this to the area of the umbilicus. From the end of this incision in the abdominal region, cut the skin on the left side to the vertebral column. Make another cut to the left from the area of the sternum to the point of the elbow. Insert your fingers between the skin and the muscle and reflect the skin toward the vertebral column. If you have a female cat, leave the mammary glands attached to the skin. Make a cut around the elbow. Make another cut from the cervical region up to the middle of the mandible. Cut along the mandible on the left side. When you reach the ear, wrap the cut around the ear toward the midline on the dorsal side. Insert your fingers between the skin and muscles in this area between the head and shoulder. Once all of the skin is reflected on the left side, you can begin your study of the muscles.

## 1. Structures of the Dorsal Neck and Thoracic Region

We begin our observations with the **trapezius muscles** of the cat. The cat has three trapezius muscles that are comparable to the three divisions of the one human trapezius muscle. The most cranial trapezius muscle is the **clavotrapezius**. Moving caudally we find the **acromiotrapezius**, which is shaped like a scallop shell. The most caudal is the **spinotrapezius muscle**. Dr. J's students developed a mnemonic for these muscles from cranial to caudal and it is CAS! Note that is almost how you spell cat. Wow.

The **deltoid muscles** are next, and as with the trapezius muscles, there are three. Again, there is one human muscle with three divisions that correspond to these three cat muscles. At the cranial end we find the **clavodeltoid (clavobrachialis)**. Caudal to that we find the **acromiodeltoid**, and the most caudal of these muscles is the **spinodeltoid**. Dr. J's students developed a mnemonic for these muscles from cranial to caudal and it is CAS! Like oh my God! That is the same mnemonic as the one for the trapezius muscles! Be sure to observe the deltoid muscle on the upper limb model of the human. Running along the margin of the acromiodeltoid, you will find the cephalic vein. It should have been dyed blue, indicating that it contains oxygen-deficient blood. When we get to Lab 8, we will observe it on the surface of the brachioradialis muscle in the forearm also.

Carefully separate the caudal margin of the clavodeltoid and the caudal margin of the clavotrapezius. Reflect that edge of the clavodeltoid cranially to expose the **axillary nerve** (nerves generally look like dental floss) running to the deep side of the clavodeltoid. The **axillary nerve** serves the deltoid muscle and teres minor. You will also see the insertion of the **pectoralis major**. You should observe a groove where the pectoralis major and the craniomedial border of the acromiodeltoid meet. This groove runs distally from the proximal end of the humerus on its cranial side. This is one of three places that the pectoralis major can be tagged on a practical exam.

Observe the **levator scapulae ventralis (levator scapulae)**. It is a relatively narrow muscle that is dorsal to the acromiodeltoid and cranioventral to the acromiotrapezius. It emerges from deep to the caudal margin of the clavotrapezius as it passes to its insertion on the metacromion process of the scapula. Later in this lab you will find this muscle on the lateral side of the neck, deep to the clavotrapezius and the **cleidomastoid**.

Ventral and deep to the spinotrapezius, you will find the relatively large **latissimus dorsi**. Although it is large in surface area, it is thin. When you separate this muscle, be careful as you move toward the cranial end, because it will meet the **xiphihumeralis muscle** that comes from the ventral side. You will recognize a sideways V shape where they meet. Do not try to separate these muscles at the V because it will probably result in cat terrorism to the xiphihumeralis. Also be careful at the caudal end of the latissimus dorsi muscle, where it originates from an aponeurosis (yes, aponeurosis is in the glossary). If you tear this aponeurosis, the muscle will no longer be attached at that end. Once you have observed this muscle, you will transect it where it intersects with the xiphihumoralis. Reflect the humeral end of this muscle to expose the **thoracodorsal nerve** and the **thoracodorsal artery** on its deep side. They both serve the latissimus dorsi.

Deep to the latisimus dorsi you will find a muscle that looks like fingers (or the edge of a serrated knife). That muscle is the **serratus ventralis** (**anterior** in a human). You will be able to see this muscle deep to the scapula from the dorsal side soon. Running nearly horizontally along the surface of the serratus ventralis, you find the **long thoracic nerve**. This nerve serves the serratus ventralis.

Move dorsally and transect the aponeurosis of the acromiotrapezius. Reflect the dorsal edge of the acromiotrapezius to expose the **spinal accessory nerve (XI)**. This nerve serves the sternocleidomastoid and the trapezius muscles. Now gently pull the vertebral border of the scapula away from the thoracic wall. This will expose two of the three rhomboid muscles. Laterally you will see the **rhomboideus capitus** (for this class, it is only found in the cat). It looks like a rubber band. Medial and deep to the rhomboideus capitus you will find the **rhomboideus minor**. The **rhomboideus major** can be found running along the caudal margin of the scapula, deep to the latissimus dorsi and the spinotrapezius muscles. Also, as mentioned previously, you can find the serratus ventralis ventral to the rhomboideus capitus and the rhomboideus minor muscles. This is the second place this muscle can be tagged on a lab practical.

## 2. Structures of the Ventral Neck and Thoracic Region

Place the cat on its dorsal surface. First observe the four pectoral muscles. The most caudal of these pectorals is the **xiphihumeralis** (cat muscle). Be very careful of this muscle as it is extremely thin and is easily cat terrorized. Moving cranially, you will find the **pectoralis minor**. It is superficial to some of the xiphihumeralis. It is the largest of the four pectoral muscles in the cat, while in humans it is smaller than the pectoralis major. Continuing cranially, we find the pectoralis major. We already saw this muscle deep to the clavodeltoid and medial to the acromiodeltoid. Superficial to the pectoralis major you will observe the **pectoantebrachialis** (cat muscle). Because of the position of the pectoantebrachialis, the pectoralis major can be tagged cranial as well as caudal to this narrow muscle.

We will now study the ventral and lateral areas of the neck. We begin immediately lateral to the midline. We will find a very narrow and very thin muscle, the **sternohyoid** muscle. Lisamilla, a former student and author extraordinaire, sometimes refers to this muscle as the "highest" one in the neck to help you remember it better in relation to the other neck muscles. That is not proper anatomical terminology, but it is a play on words. Lateral and deep to the sternohyoid is the **sternothyroid muscle**. Please be gentle with these muscles, as they are not very robust and can be destroyed easily. Moving laterally from these muscles you will observe a muscle that is running obliquely to the temporal bone of the skull. Viewed from the ventral surface this looks like a "V," and it is sometimes referred to as the Peri Como's V neck sweater muscle. This is the **sternomastoid muscle**. Superficial to the sternomastoid muscle you will find a large vein, the **external jugular vein**. Lateral to these structures and deep to the clavotrapzesius muscle you will find the **cleidomastoid muscle**. In humans, these two muscles form one muscle, the **sternocleidomastoid muscle**. Also deep to the clavotrapezius and the cleidomastoid is the **levator scapulae ventralis** (which we have already seen on the dorsal side).

Another reminder: view the structures of the cat as you would view a map. Learn their locations relative to other structures, as well as which muscles are served (innervated) by which nerves. This will put these features in a visual context that will prove useful come exam time.

# MUSCLES

## Trapezius

### Clavotrapezius Muscle (Cat)

Dorsal view, left side

Dorsal view, left side

Ventral view, left side

### Acromiotrapezius Muscle (Cat)

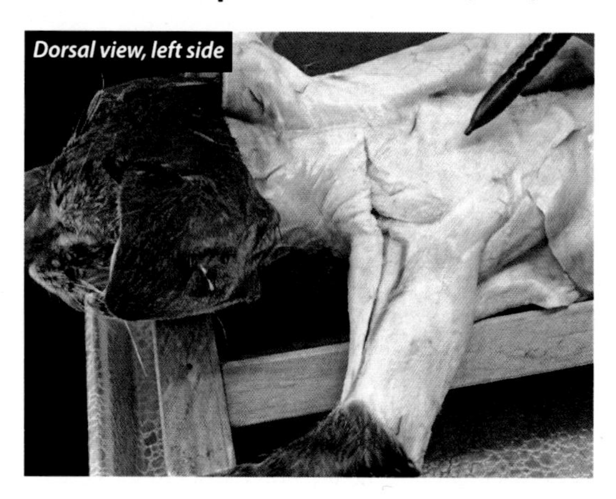

Dorsal view, left side

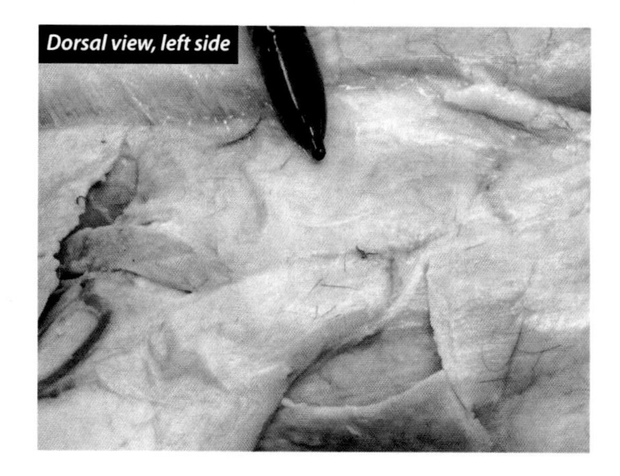

Dorsal view, left side

### Spinotrapezius Muscle (Cat)

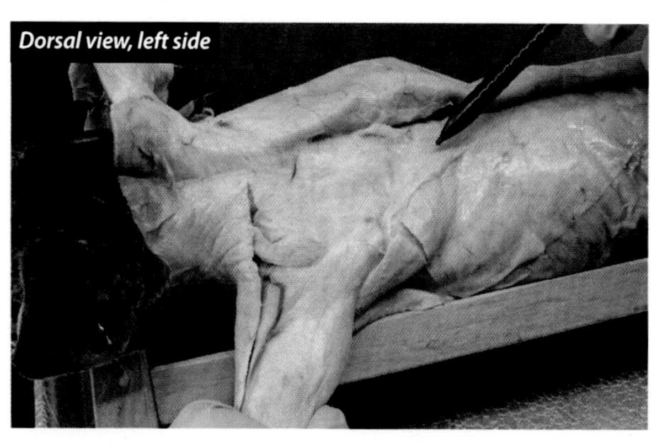

Dorsal view, left side

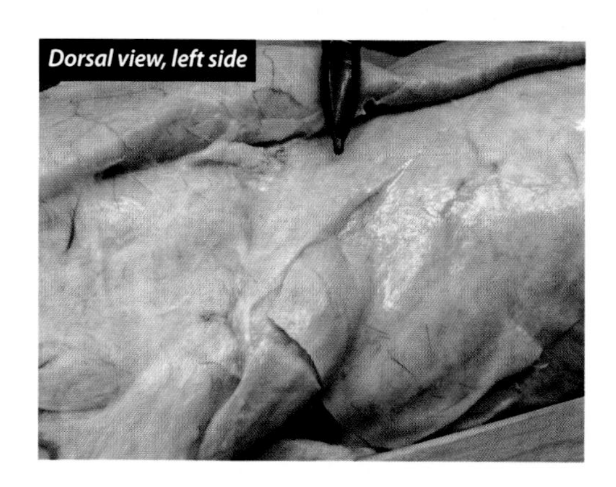

Dorsal view, left side

## Trapezius Muscle (Human)

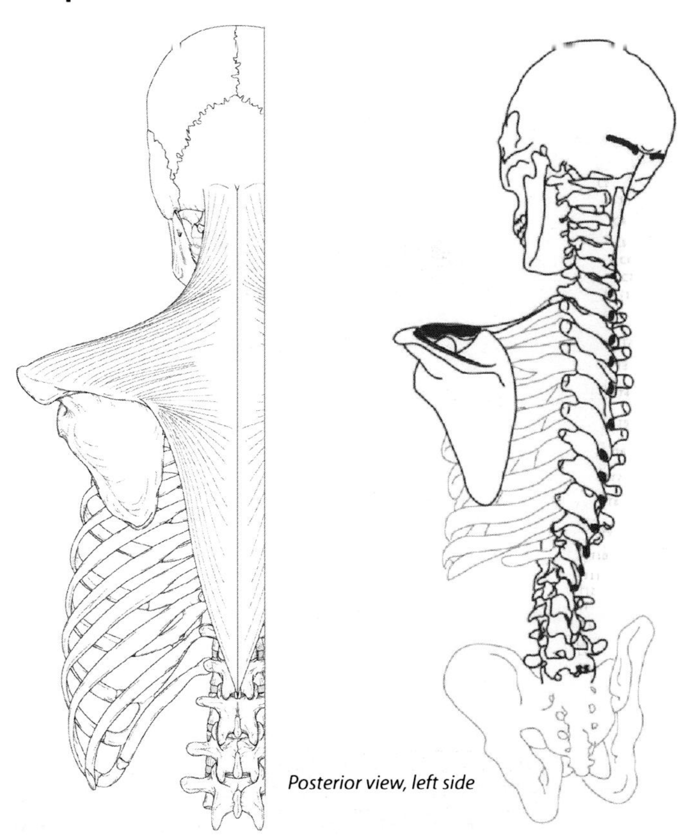

Posterior view, left side

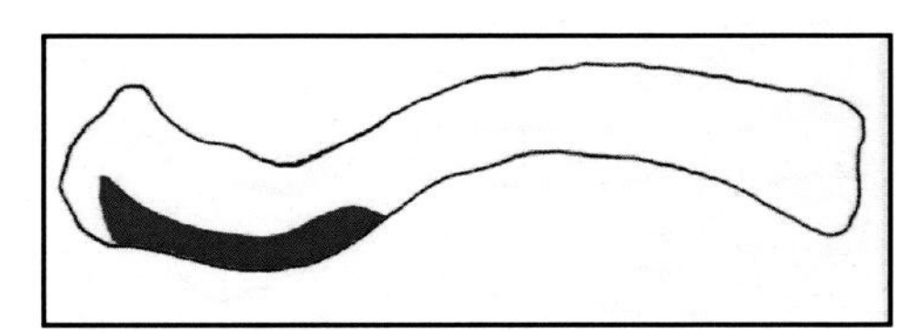

*Superior view of insertion, left side*

*The drawings of the origin and insertion for the trapezius muscle might help you visualize this information (red is the origin and blue is the insertion). Notice that part of the origin for the superior division is a ligament called the ligamentum nuchae, and this portion of the origin is represented by a line between the skull and the spinous process of C7.*

### *Human Information: Trapezius*

***Superior Division***—corresponds to clavotrapezius (cat).

**origin:** external occipital protuberance, medial third of superior nuchal line, ligamentum nuchae, and spinous process of C7

**insertion:** lateral third of clavicle and acromion process

**nerve:** spinal accessory (XI) and anterior rami of C3 and C4 (sensory component)

**action:** elevation and upward rotation of scapula, when contracting with other portions of the trapezius, it retracts (adducts) scapula

***Middle Division***—corresponds to acromiotrapezius (cat).

**origin:** spinous processes of 7th cervical and 1st–3rd thoracic vertebrae

**insertion:** superior border of spine of scapula

**nerve:** spinal accessory (XI) and anterior rami of C3 and C4

**action:** retracts (adducts) and helps elevate scapula

***Inferior Division***—corresponds to spinotrapezius (cat).

**origin:** spinous processes of T6–T12

**insertion:** medial third of spine of scapula

**nerve:** spinal accessory (XI) and ventral rami of cervical spinal nerves 3 and 4

**action:** retracts (adducts), depresses, and rotates scapula

The trapezius is one of five muscles that insert on the scapula and are grouped as muscles of the scapula, or muscles that moor the scapula (shoulder girdle muscles). The other four muscles are: levator scapulae, pectoralis minor, rhomboids, and serratus anterior.

# Deltoids

## Clavodeltoid/Clavobrachialis Muscle (Cat)

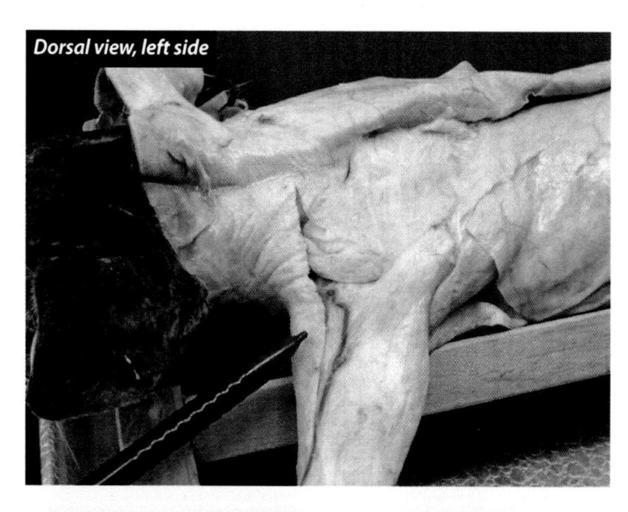

*Dorsal view, left side*

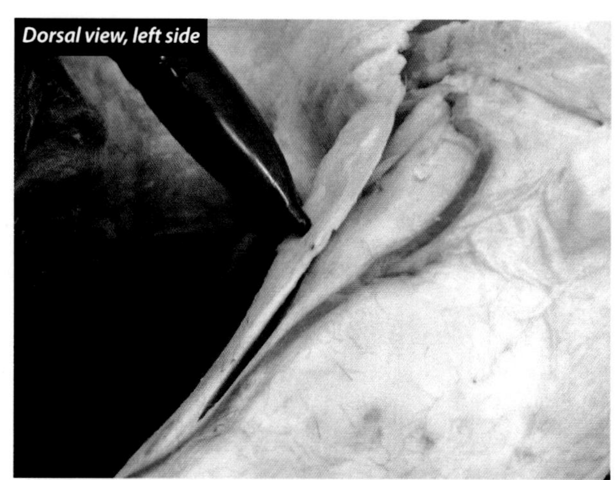

*Dorsal view, left side*

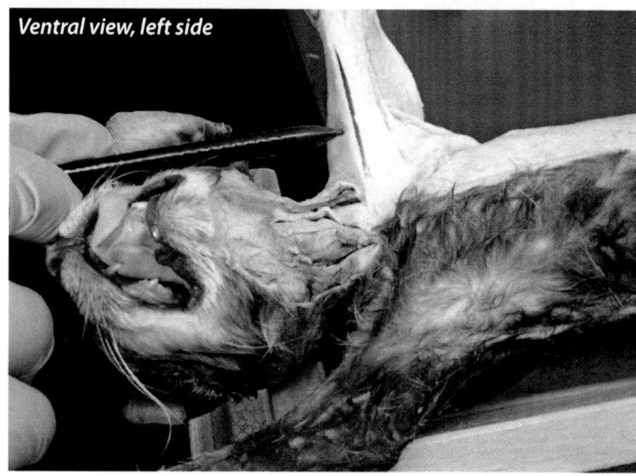

*Ventral view, left side*

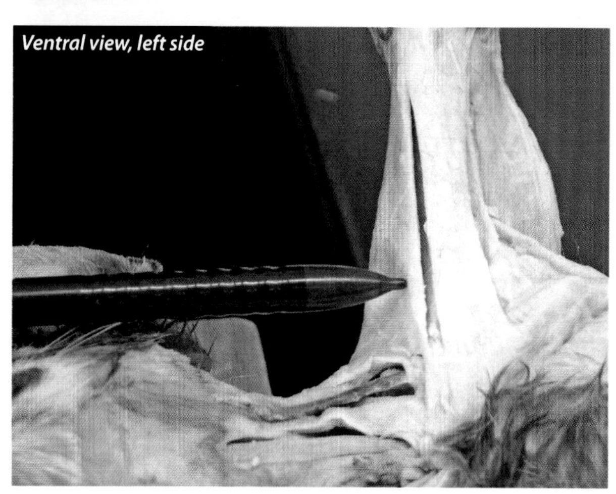

*Ventral view, left side*

### *Human Information: Deltoid, Anterior Division*

The anterior portion of the human deltoid corresponds to the clavodeltoid/clavobrachialis muscle (cat).

**origin:** anterosuperior border of lateral third of clavicle
**insertion:** deltoid tuberosity of humerus on its anterolateral surface
**nerve:** axillary (C5, C6)
**action:** prime flexor of arm; also, abductor and medial (internal) rotator of arm

## Deltoid Muscle (Human)—Anterior Division

**Human deltoid model: anterior view, right side**

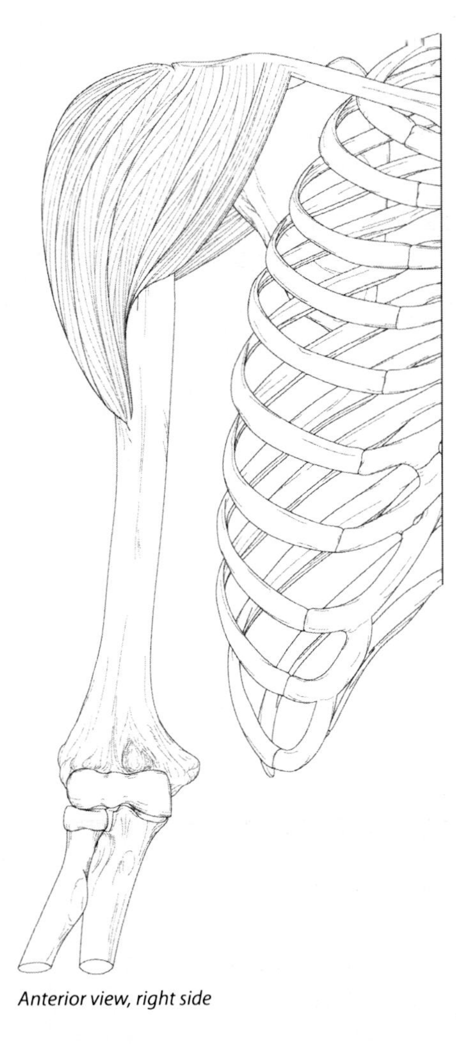

*Anterior view, right side*

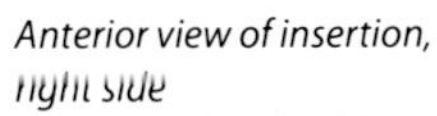

*Anterior view of insertion, right side*

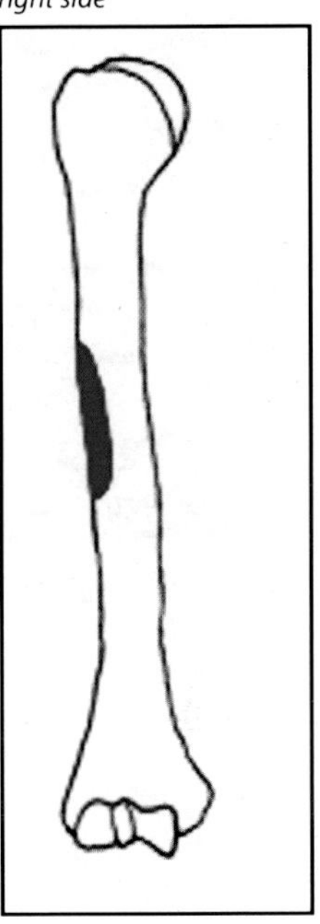

*Posterior view of insertion, right side*

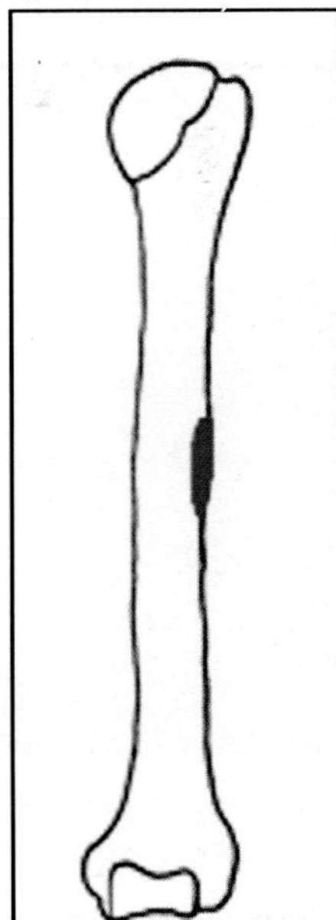

*The above drawings of the origin and insertion might help you visualize this information (red is the origin, blue the insertion). The origin included here is for the anterior division of the deltoid.*

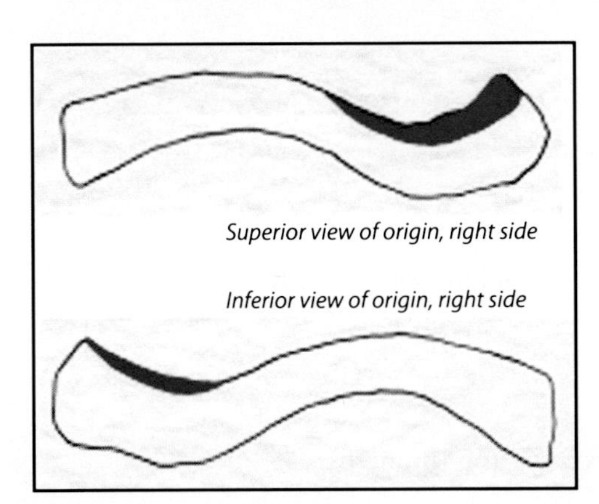

*Superior view of origin, right side*

*Inferior view of origin, right side*

The deltoid muscle is one of the five muscles in the group we refer to as the muscles that insert on the arm and, therefore, move the arm (not including the rotator cuff muscles). They are also called shoulder joint muscles. The remaining four muscles are: coracobrachialis, latissimus dorsi, teres major, and pectoralis major.

## Acromiodeltoid Muscle (Cat)

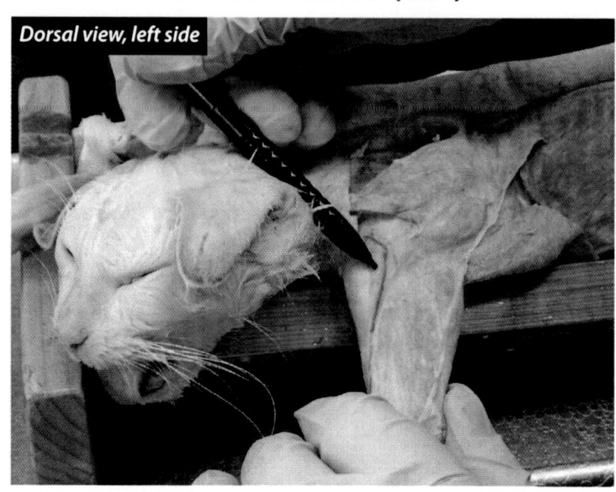


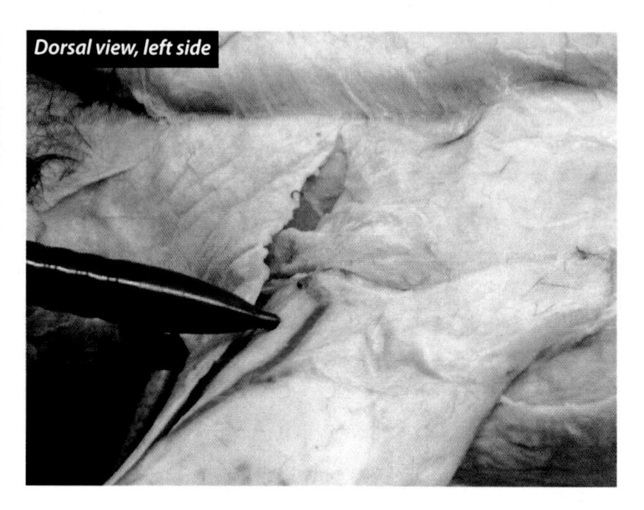


### *Human Information: Deltoid, Middle Division*

The middle portion of the human deltoid corresponds to the acromiodeltoid muscle (cat).

**origin:** acromion process of scapula
**insertion:** deltoid tuberosity of humerus
**nerve:** axillary (C5, C6)
**action:** abducts arm (humerus)

## Spinodeltoid Muscle (Cat)

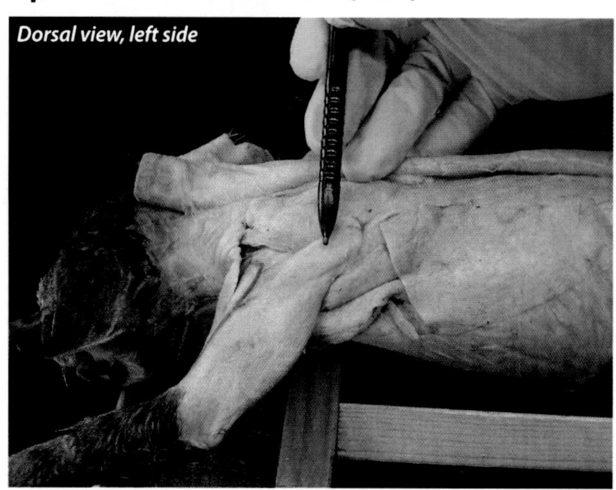


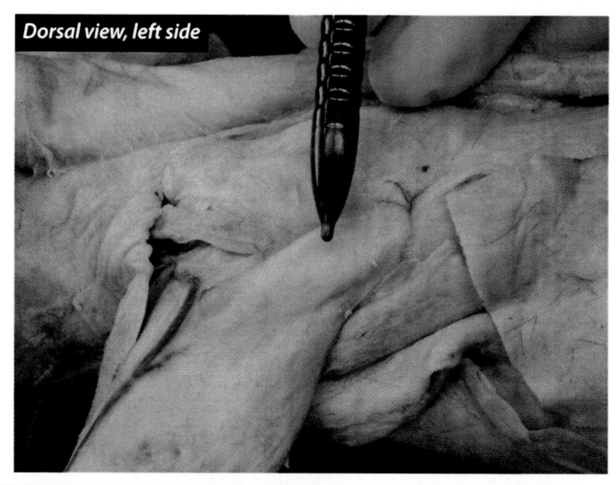


*Dr. J likes to tell his students that the spinodeltoid looks a bit like a football! If you don't like that analogy, you can decide what it looks like to you—a gumdrop, for example.*

### *Human Information: Deltoid, Posterior Division*

The posterior portion of the human deltoid corresponds to the spinodeltoid muscle (cat only).

**origin:** inferior margin of spine of scapula
**insertion:** deltoid tuberosity of humerus
**nerve:** axillary (C5, C6)
**action:** abducts, laterally (externally) rotates, extends, and hyperextends arm (humerus)

## Deltoid Muscle (Human)—Middle Division

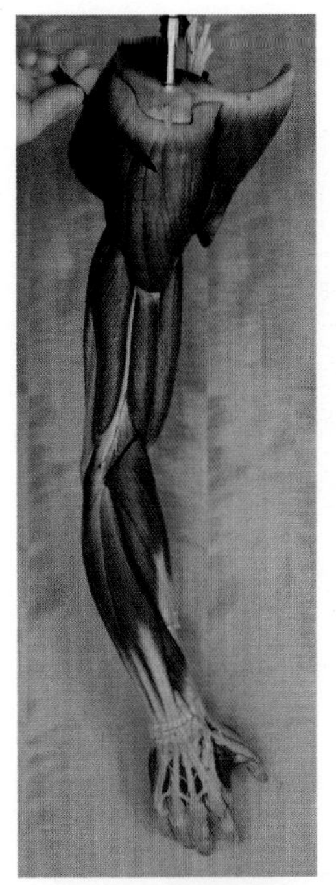

Human deltoid model: lateral view, right side

Middle division

Posterior division

Posterior view, right side

Posterior view of insertion, right side

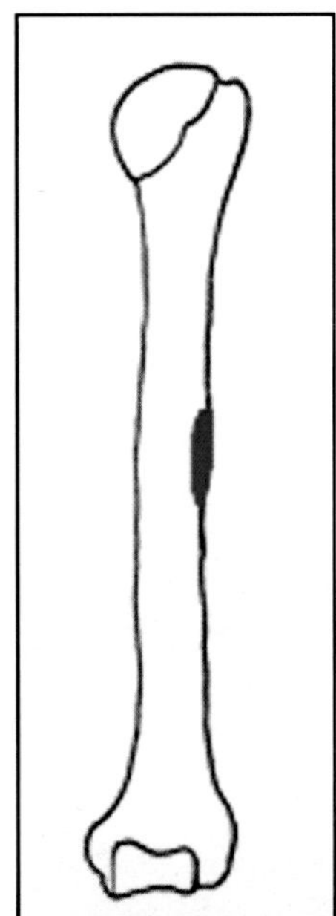

Anterior view of insertion, right side

## Deltoid Muscle (Human)—Posterior Division

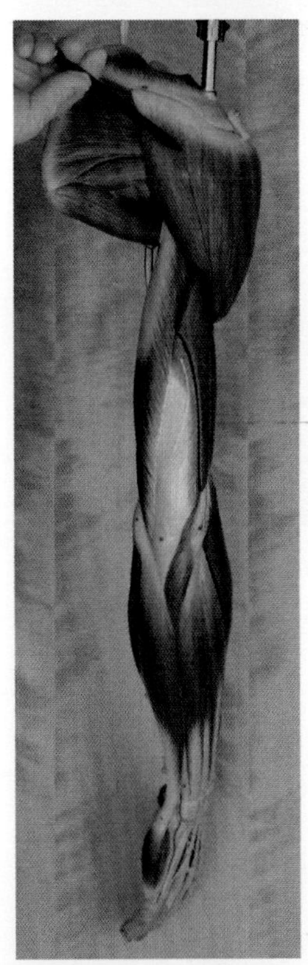

Human deltoid model: posterior view, right side

Posterior view of origin, right side

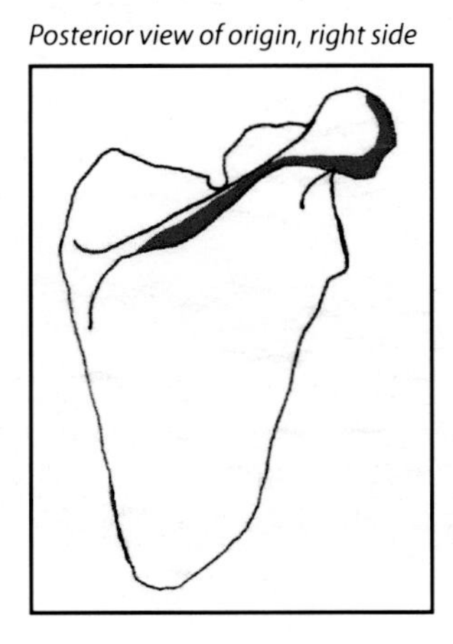

*These drawings of the origin and insertion might help you visualize this information (red is the origin, blue the insertion). The origin included here is for the middle and posterior divisions of the deltoid.*

The deltoid muscle is one of the five muscles in the group we refer to as the muscles that insert on the arm and, therefore, move the arm (not including the rotator cuff muscles). They are also called shoulder joint muscles. The remaining four muscles are: coracobrachialis, latissimus dorsi, teres major, and pectoralis major.

## Levator Scapulae Ventralis Muscle

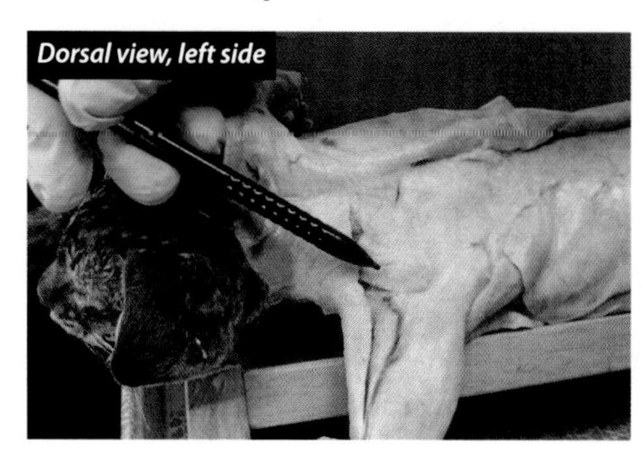


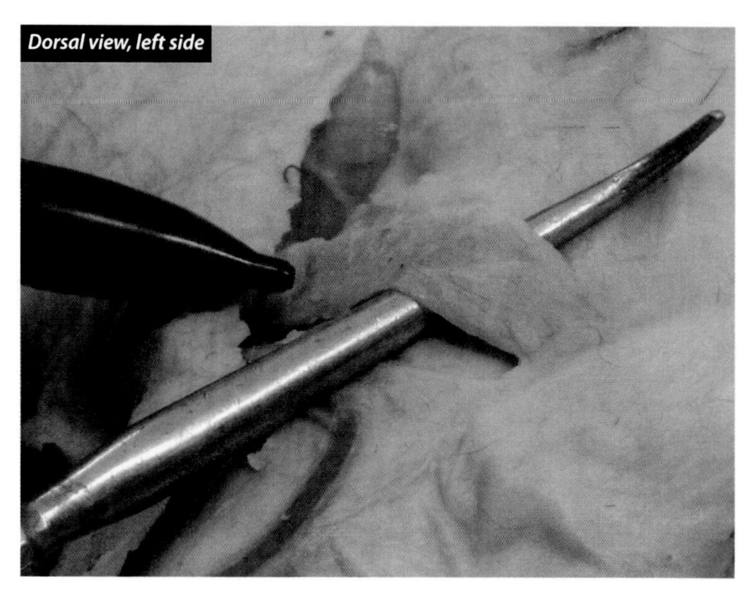


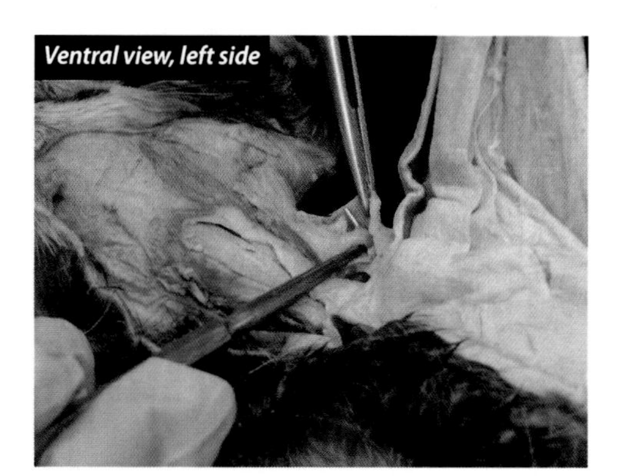


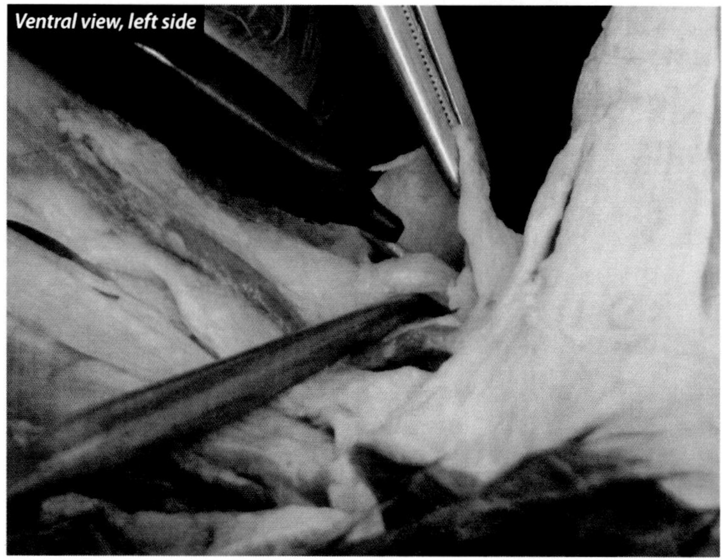


### *Human Information: Levator Scapulae*

The human muscle levator scapulae is referred to as levator scapulae ventralis in the cat.

**origin:** transverse processes of C1–C4
**insertion:** vertebral border of scapula between the superior angle and spine of scapula
**nerve:** dorsal scapular (C5) and anterior rami of C3 and C4
**action:** elevates and rotates scapula; flexes neck laterally

## Levator Scapulae Muscle (Human)

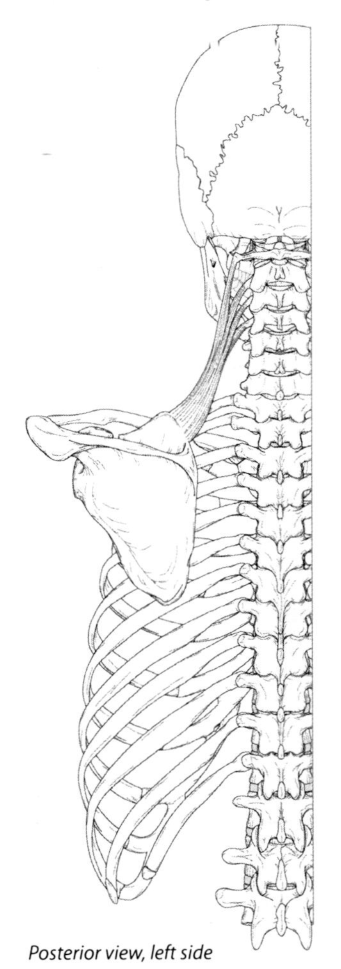

*Posterior view, left side*

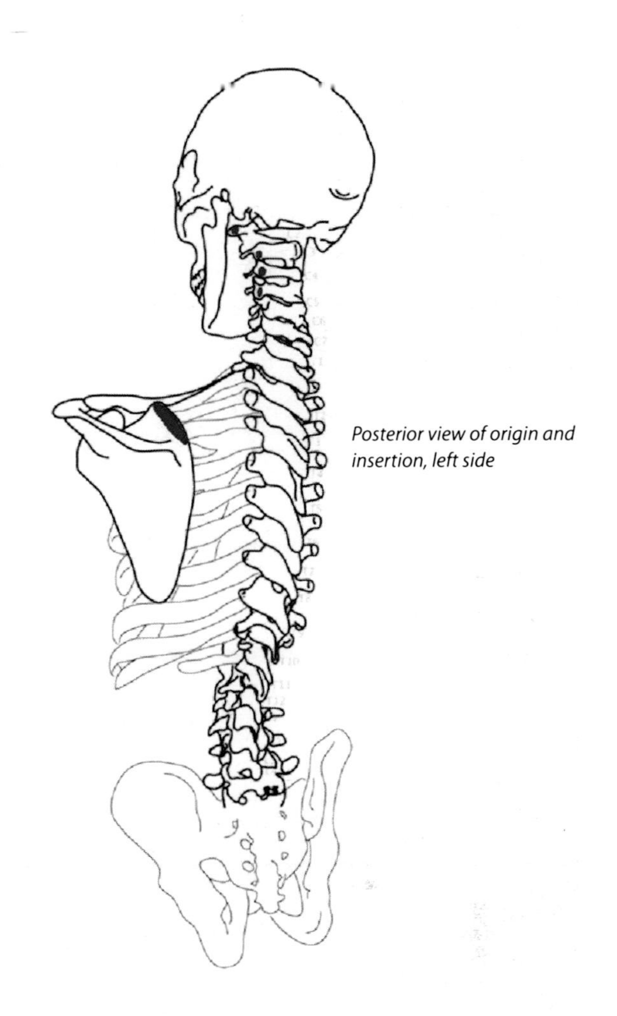

*Posterior view of origin and insertion, left side*

*The above drawings of the insertion might help you visualize this information (red is the origin and blue is the insertion).*

Some authors simply call this the levator scapula, as we did here. It is one of five muscles that insert on the scapula and are grouped as muscles of the scapula, or muscles that moor the scapula (shoulder girdle muscles). The other four muscles are: pectoralis minor, rhomboids, serratus anterior, and trapezius.

## Latissimus Dorsi Muscle

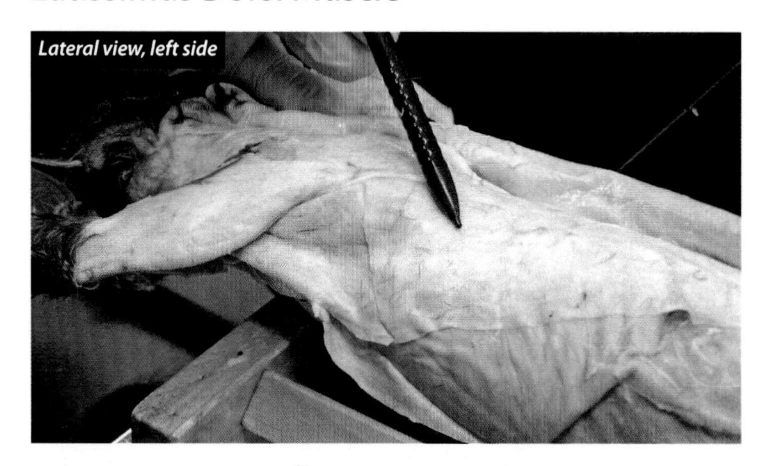
*Lateral view, left side*

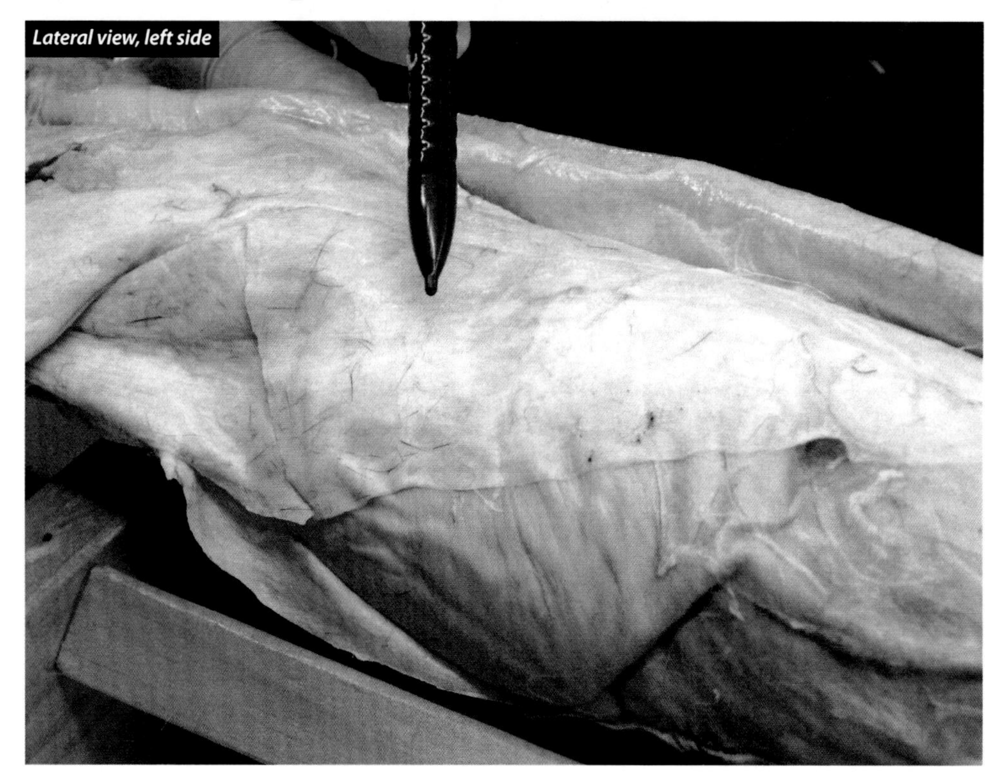
*Lateral view, left side*

### *Human Information: Latissimus Dorsi*

**origin:** aponeurosis from spinous processes of T7–T12 and L1–L5, posterior iliac crest, posterior surface of sacrum, inferior 3 or 4 ribs, and inferior angle of scapula
**insertion:** medial lip of intertubercular groove of humerus
**nerve:** thoracodorsal (C6, C7, C8)
**action:** extends, hyperextends, adducts, and medially (internally) rotates arm (humerus)

## Latissimus Dorsi Muscle (Human)

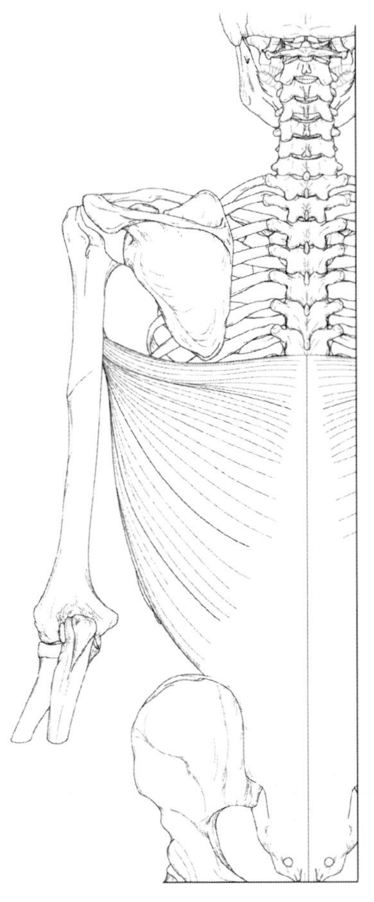

*Posterior view, left side*

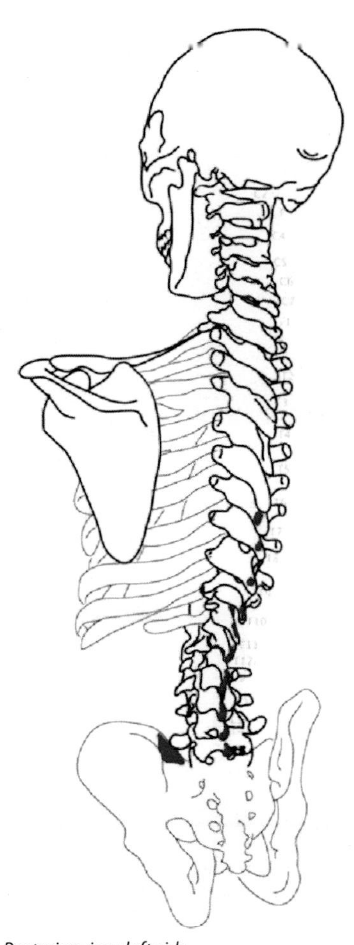

*Posterior view, left side*

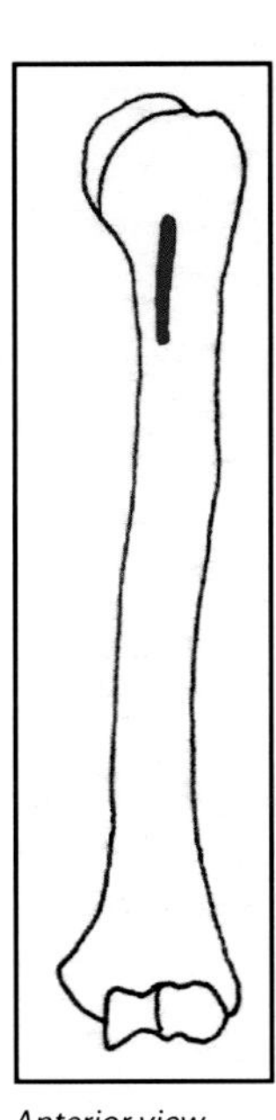

*Anterior view, left side*

*The above drawings of the origin and insertion might help you visualize this information (red is the origin and blue is the insertion).*

Note that the latissimus dorsi is one of the five muscles in the group we refer to as the muscles that insert on the arm and, therefore, move the arm (not including the rotator cuff muscles). They are also called shoulder joint muscles. The remaining four muscles are: coracobrachialis, deltoids, teres major, and pectoralis major. The latissimus dorsi has essentially the same action as the teres major, but the teres major does not hyperextend. Note that the insertion of the latissimus dorsi is between the insertion for the pectoralis major and the teres major.

## Serratus Ventralis Muscle

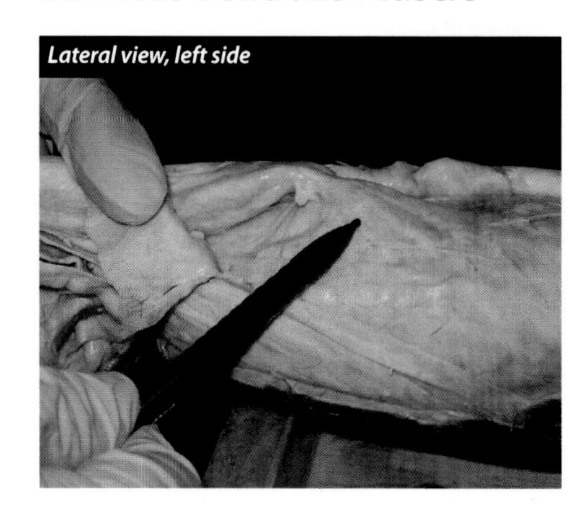
Lateral view, left side

Lateral view, left side

*Note that this muscle can be tagged in two places!*

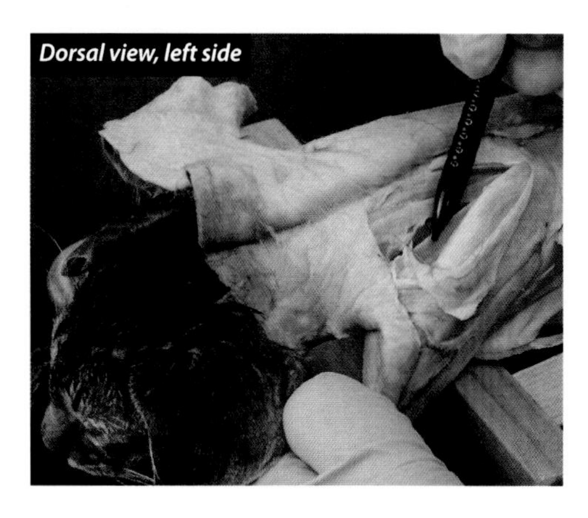
Dorsal view, left side

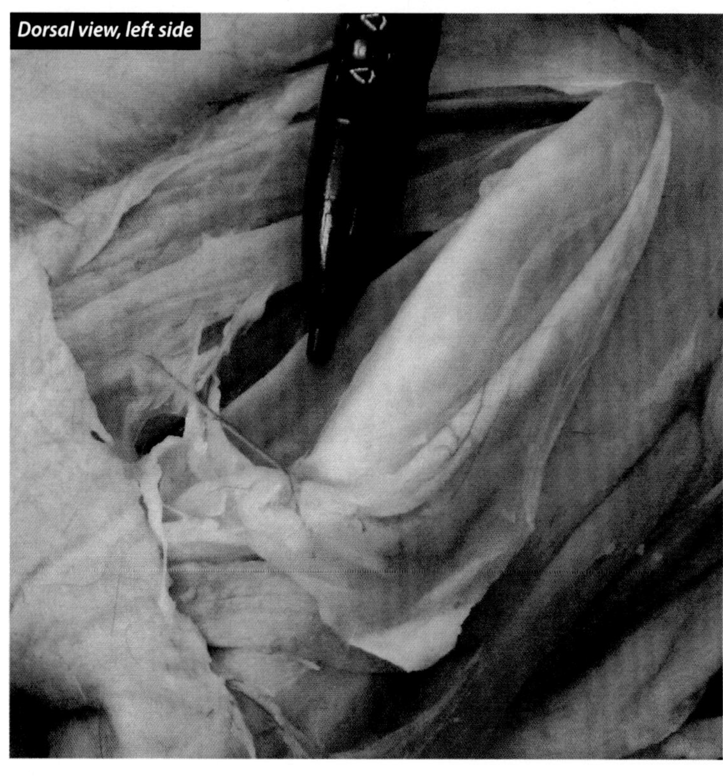
Dorsal view, left side

### *Human Information: Serratus Anterior*

(serratus ventralis in cat)

**origin:** superior lateral surface of ribs 1–8
**insertion:** anterior surface along vertebral border of scapula
**nerve:** long thoracic (C5, C6, C7)
**action:** protracts (abducts) and rotates scapula; prevents scapular winging

## Serratus Anterior Muscle (Human)

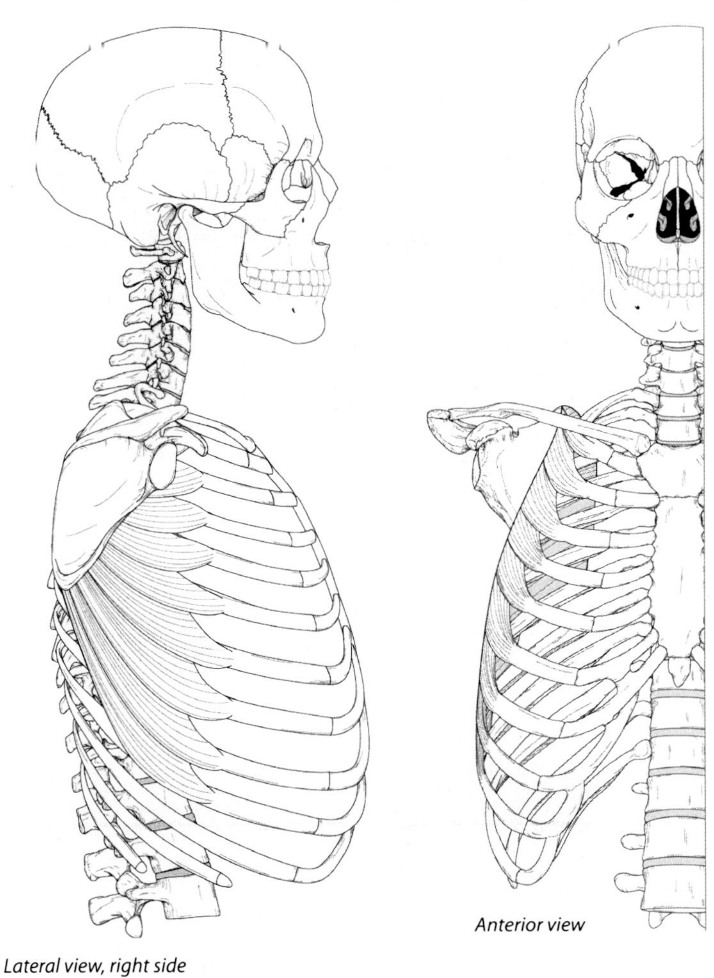

Lateral view, right side

Anterior view

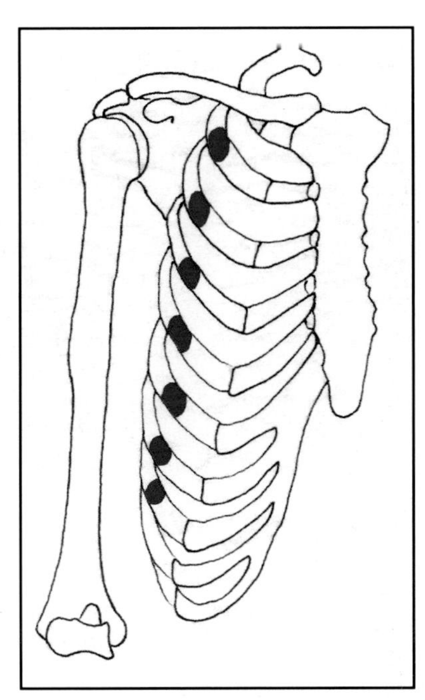

*Anterior view of origin, right side*

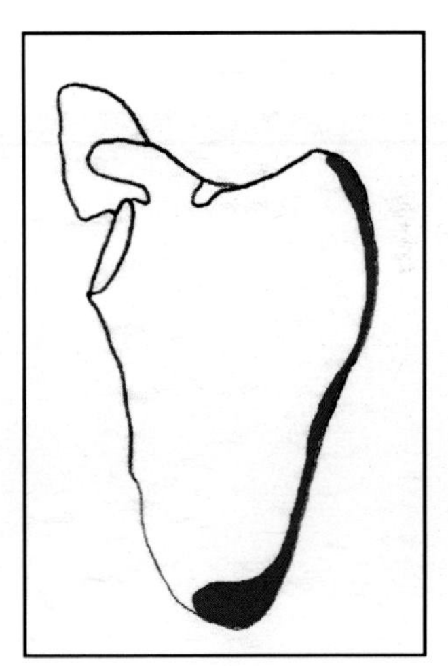

*Anterior view of insertion, right side*

*The above drawings of the origin and insertion might help you visualize this information (red is the origin, blue the insertion).*

The serratus ventralis in the cat (and the serratus anterior in humans) looks like fingers because it attaches to the ribs, but it has nothing to attach to between the ribs. The name implies that it looks like a serrated knife. It is one of five muscles that insert on the scapula and are grouped as muscles of the scapula, or muscles that moor the scapula (shoulder girdle muscles). The other four muscles are: levator scapulae, pectoralis minor, rhomboids, and trapezius.

# Rhomboids

## Rhomboideus Capitis Muscle (Cat)

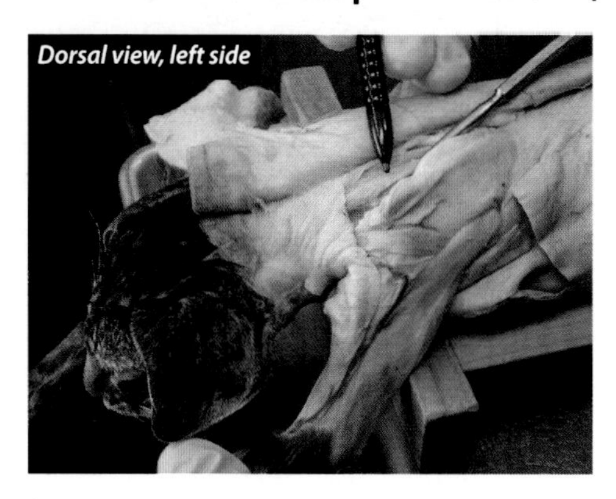


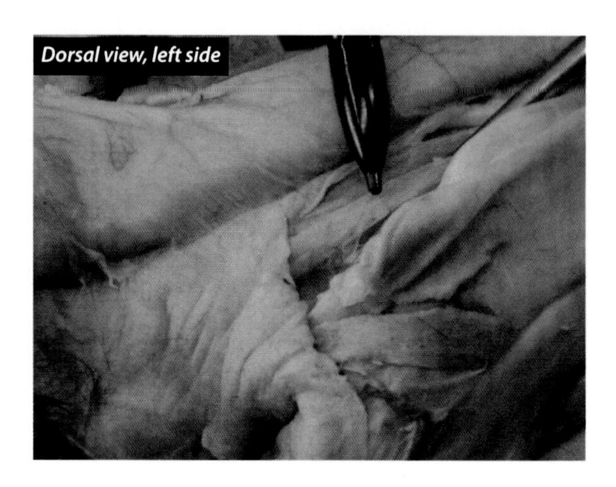


***Rhomboideus Capitis*** (Occipitoscapularis)

**origin:** medial half of cranial nuchal line, not extending quite to midline
**insertion:** scapula near cranial angle
**nerve:** ventral ramus C4
**action:** draws cranially and rotates scapula

*There is no comparable human muscle. Some authors refer to rhomboideus capitis as occipitoscapularis.*

### *Rhomboideus Minor Muscle*

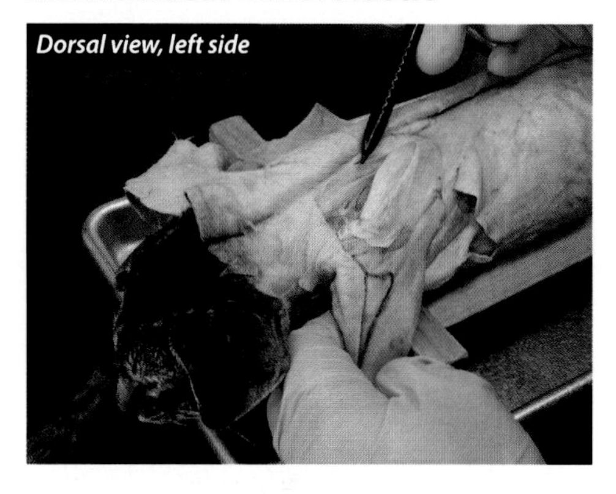


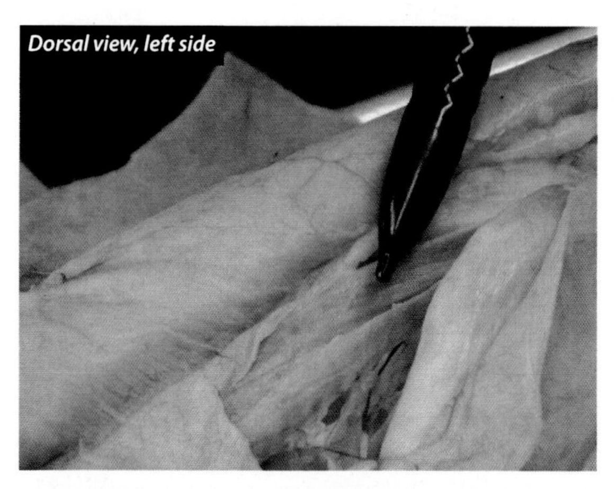


### *Rhomboideus Major Muscle*

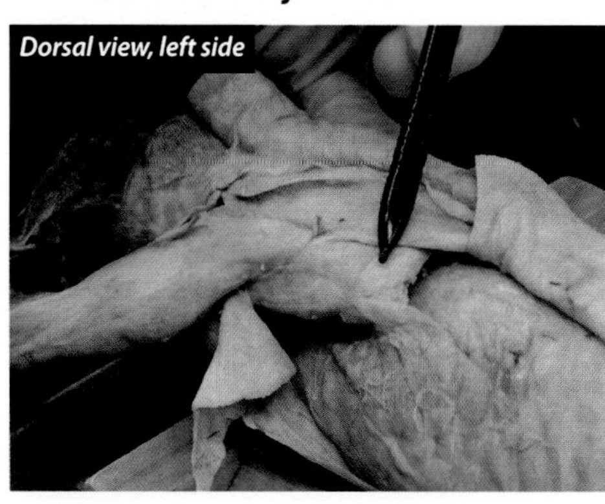


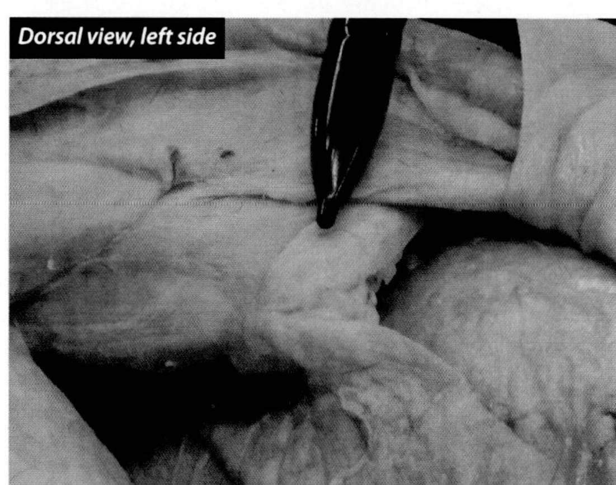

## Rhomboideus Minor Muscle (Human)

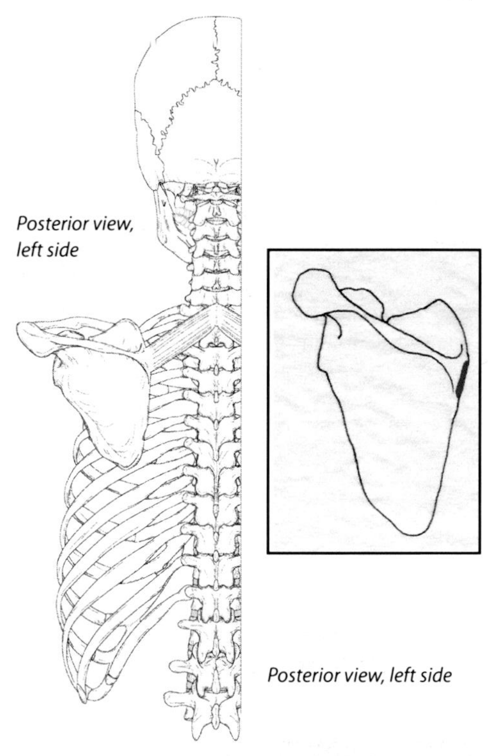

*Posterior view, left side*

*Posterior view, left side*

*Posterior view of origin, left side*

## Rhomboideus Major Muscle (Human)

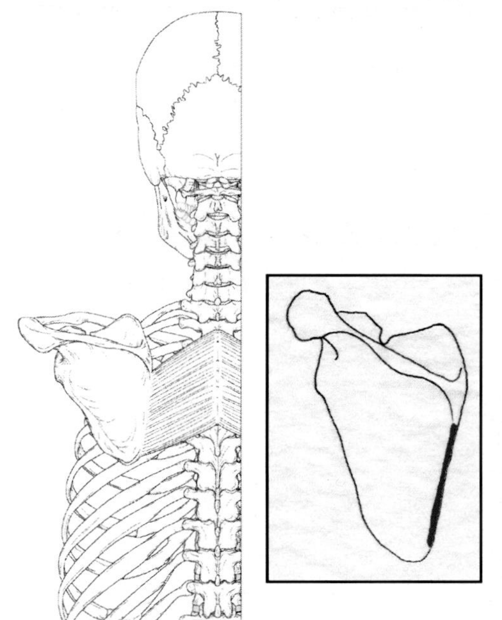

*Posterior view, left side*

*The above drawings of the origin and insertion might help you visualize this information (red is the origin and blue is the insertion). Note that the origin drawing is for both muscles.*

### *Human Information: Rhomboideus Minor*

**origin:** spinous processes of C7 and T1 and lower part of ligamentum nuchae
**insertion:** vertebral (medial) border of scapula at root of spine of scapula
**nerve:** dorsal scapular (C5)
**action:** retracts (adducts), elevates and rotates scapula

### *Human Information: Rhomboideus Major*

**origin:** spinous processes of T2–T5
**insertion:** medial border of scapula from spine to inferior angle
**nerve:** dorsal scapular (C5)
**action:** retracts (adducts), elevates and rotates scapula

The rhomboids, collectively, are one of five muscles that insert on the scapula and are grouped as muscles of the scapula, or muscles that moor the scapula (shoulder girdle muscles). The other four muscles are: levator scapulae, pectoralis minor, serratus anterior, and trapezius.

# Pectorals

## Pectoantebrachialis Muscle (Cat)

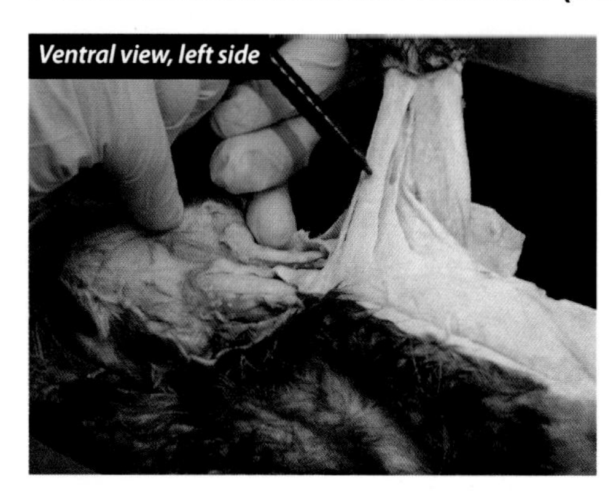


**origin:** lateral surface of manubrium
**insertion:** forearm
**nerve:** ventral thoracic
**action:** assists pectoralis minor—adducts arm

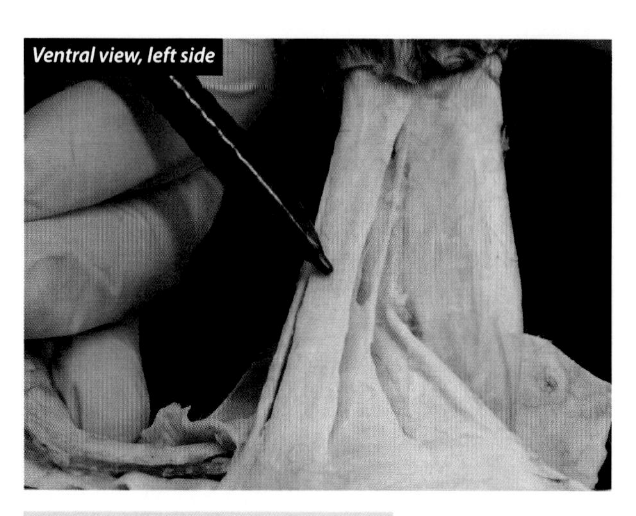


*There is no comparable muscle in humans.*

## Pectoralis Major Muscle

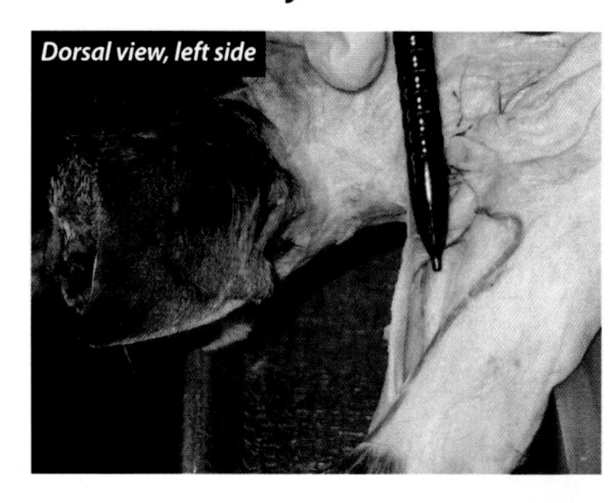


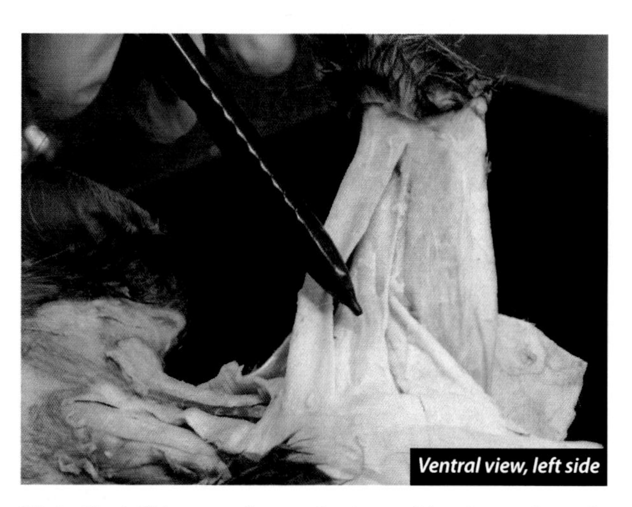


*Note that this muscle can be tagged in three places!*

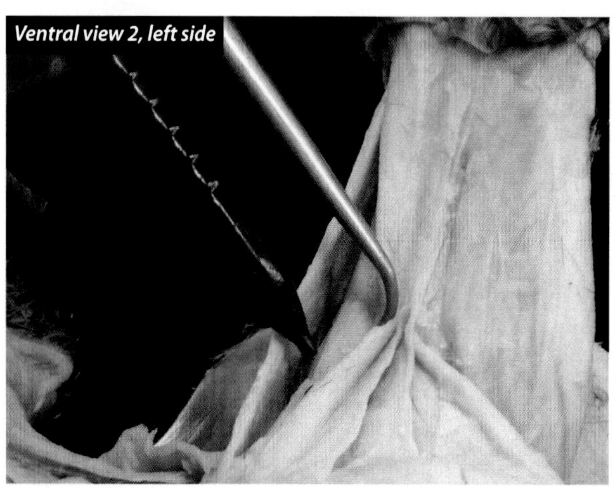

## Pectoralis Major Muscle (Human)

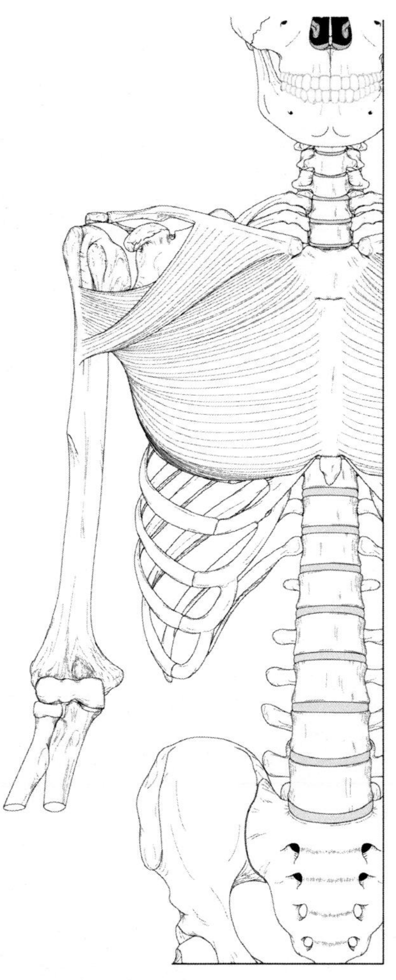

*Anterior view, right side*

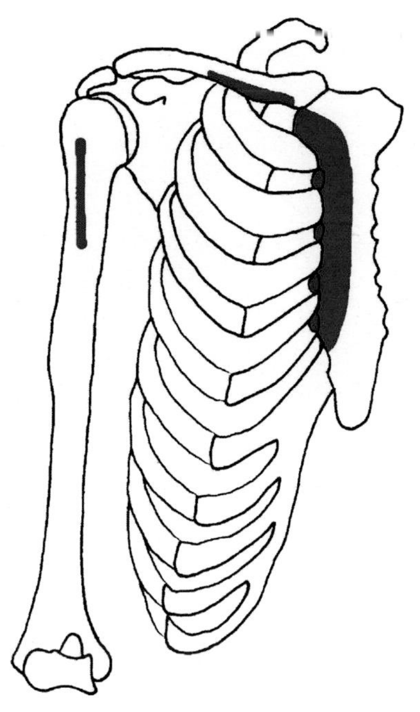

*Anterior view of origin (red) and insertion (blue), right side*

*The drawings of the origin and insertion might help you visualize this information (red is the origin, blue the insertion). Both divisions are on the same origin/insertion drawing.*

### *Human Information: Pectoralis Major (Clavicular Division)*

**origin:** anterior surface of medial third of clavicle
**insertion:** lateral lip of bicipital groove of humerus
**nerve:** lateral pectoral (C5 and C6)
**action:** flexes humerus to approximately 90 degrees
**action of both heads together:** flexion, adduction, and medial (internal) rotation of arm

### *Human Information: Pectoralis Major (Sternal Division)*

**origin:** sternum, costal cartilages of ribs 1–6
**insertion:** lateral lip of bicipital groove of humerus
**nerve:** lateral and medial pectoral (C7, 8, and T1)
**action:** extends humerus to approximately 90 degrees

Note that the pectoralis major is one of the five muscles in the group we refer to as the muscles that insert on the arm and, therefore, move the arm (not including the rotator cuff muscles). They are also called shoulder joint muscles. The remaining four muscles are: coracobrachialis, deltoids, latissimus dorsi, and teres major.

## Pectoralis Minor Muscle

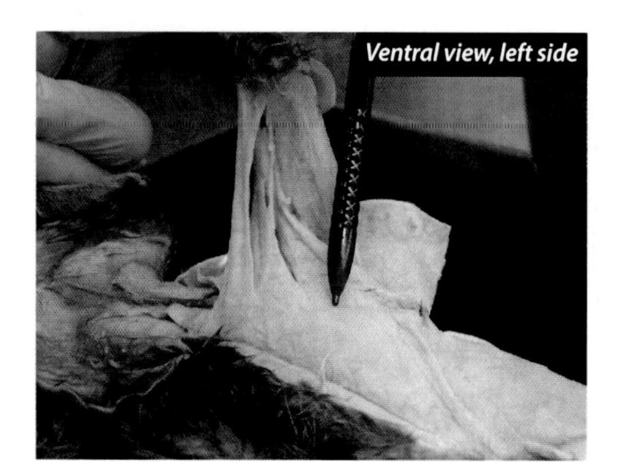
Ventral view, left side

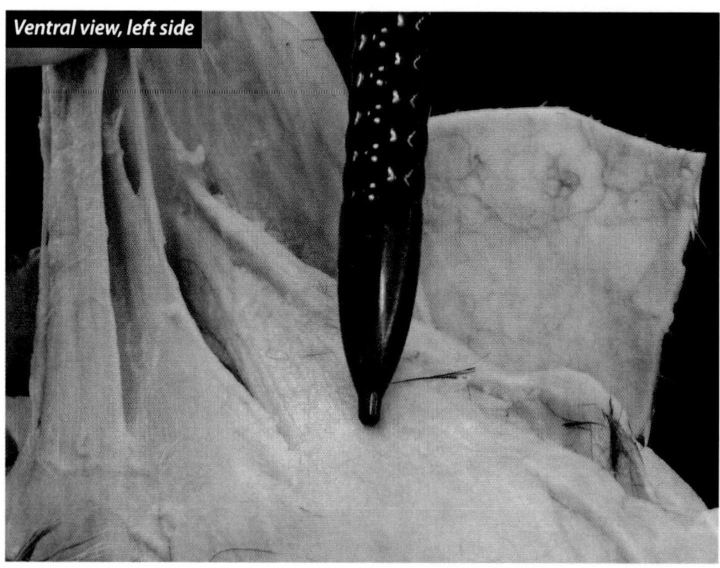
Ventral view, left side

## Xiphihumeralis Muscle (Cat)

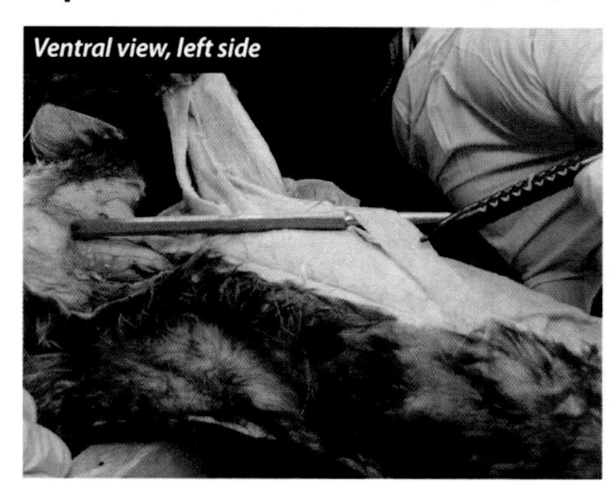
Ventral view, left side

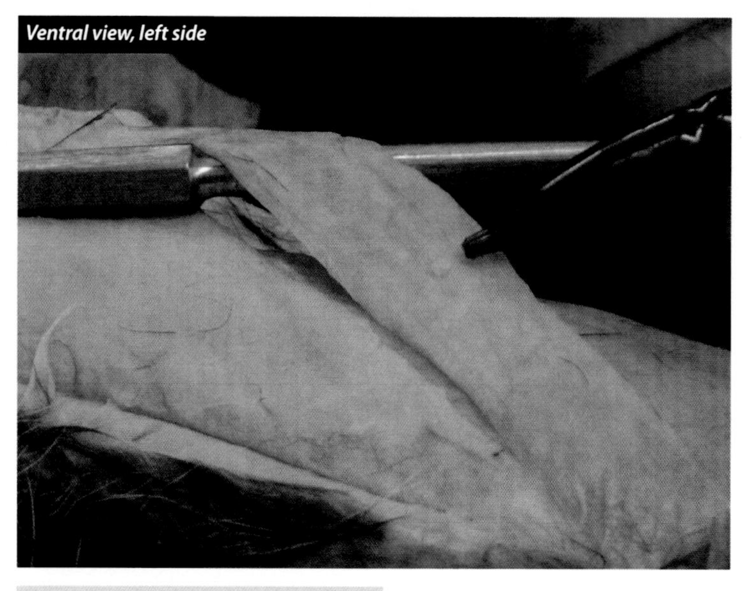
Ventral view, left side

*There is no comparable muscle in humans.*

***Xiphihumeralis (Cat)***

**origin:** a median raphe along xiphoid process or at an angle to the median line on the rectus abdominis muscle

**insertion:** with latissimus dorsi near ventral border of the bicipital groove

**nerve:** cranial pectoral nerves and branches of C7 and C8

**action:** assists pectoralis minor—adducts arm

## Pectoralis Minor Muscle (Human)

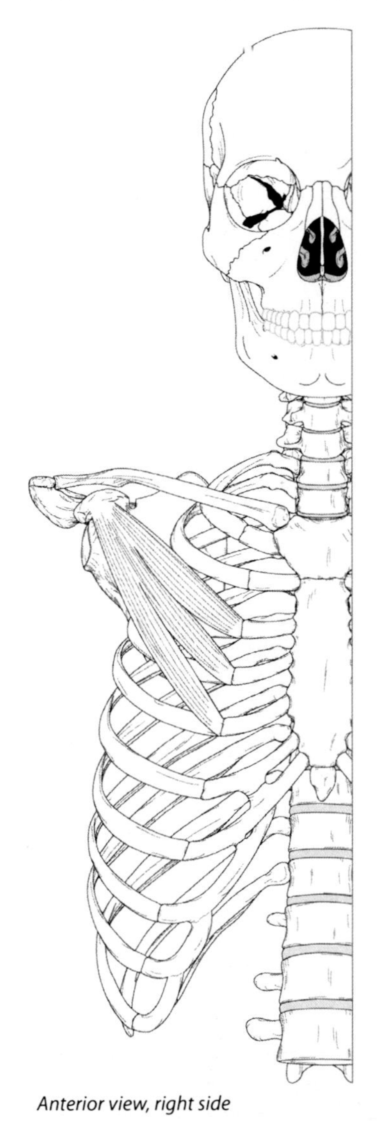

*Anterior view, right side*

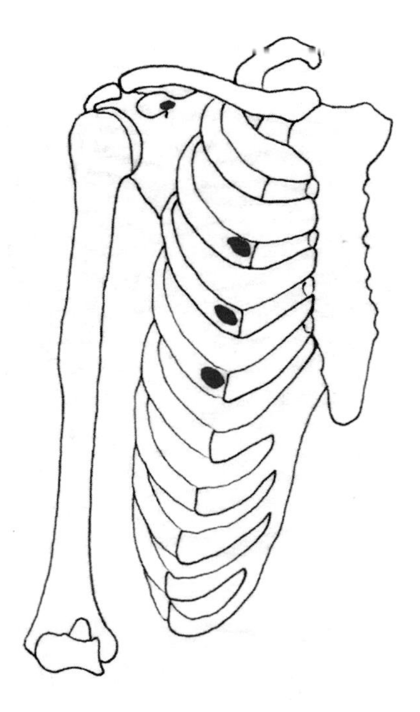

*Anterior view of origin (red) and insertion (blue), right side*

*The drawing of the origin and insertion might help you visualize this information (red is the origin, blue the insertion).*

***Human Information: Pectoralis Minor***

**origin:** anterior surface of ribs 3, 4, and 5 near costal cartilages
**insertion:** coracoid process of scapula
**nerve:** medial pectoral (C8, T1)
**action:** depresses, protracts (abducts) and rotates scapula

The pectoralis minor is one of five muscles that insert on the scapula and are grouped as muscles of the scapula, or muscles that moor the scapula (shoulder girdle muscles). The other four muscles are: levator scapulae, rhomboids, serratus anterior, and trapezius. In humans, it is one of two pectoral muscles. It is smaller than the pectoralis major, but in the cat it is the largest of the four pectoral muscles.

# Neck

## Sternohyoid Muscle

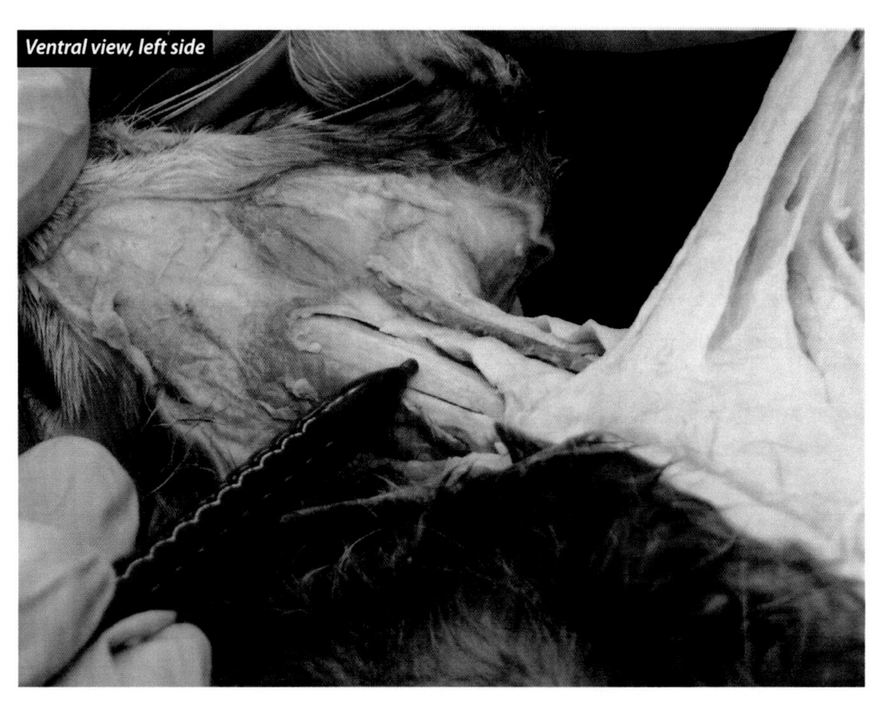


## Sternothyroid Muscle

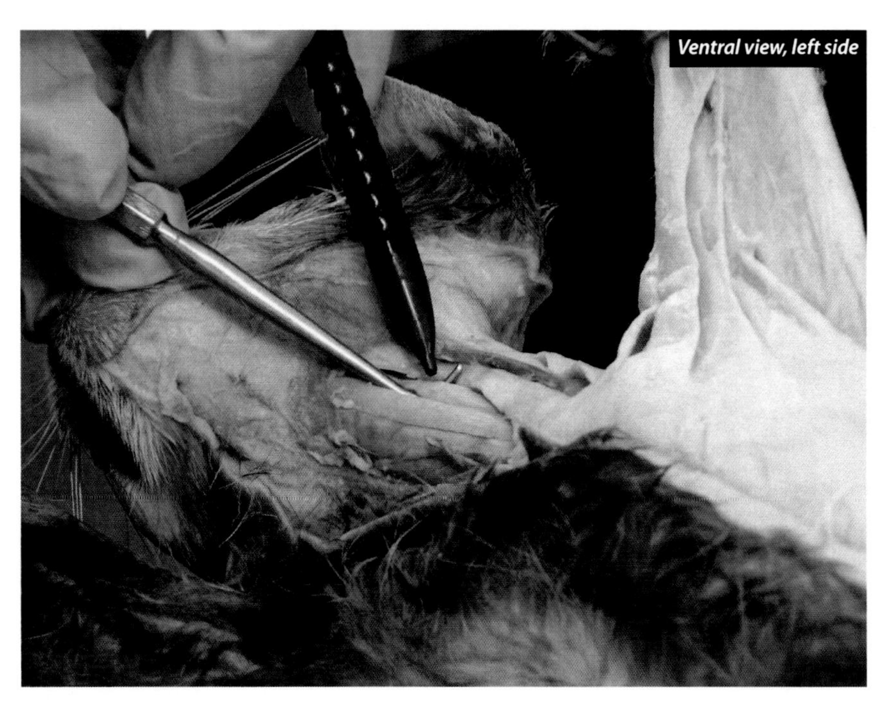

## Sternohyoid Muscle (Human)

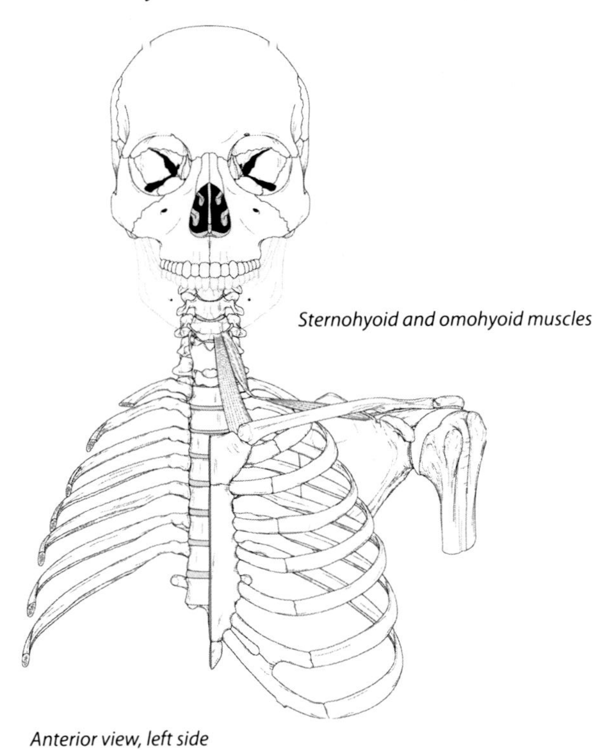


*Anterior view, left side*

### *Human Information: Sternohyoid*

**origin:** manubrium and medial end of clavicle
**insertion:** inferior margin of hyoid
**nerve:** anterior rami of C1–C3
**action:** depress larynx and hyoid if mandible is fixed; may flex skull

## Sternothyroid Muscle (Human)

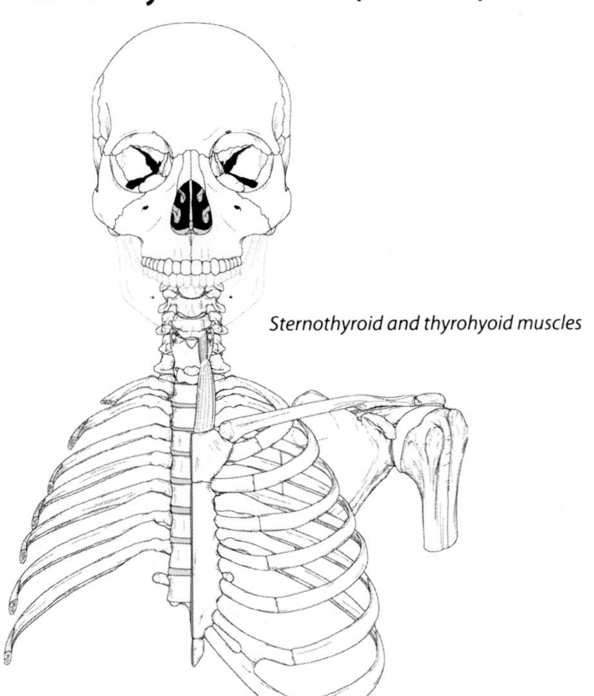


*Anterior view, left side*

*These two muscles affect ventilation activity during sleep by maintaining the opening of the superior respiratory passageways.*

### *Human Information: Sternothyroid*

**origin:** posterior surface of manubrium
**insertion:** thyroid cartilage
**nerve:** anterior rami of C1–C3
**action:** depress larynx and hyoid

## Sternomastoid Muscle (Cat)

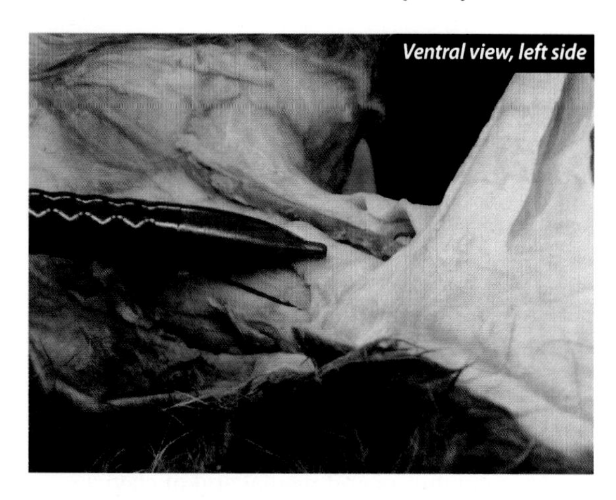


### *Sternomastoid (Cat)*

(Corresponds to human sternocleidomastoid)

**origin:** manubrium and median raphe
**insertion:** lateral half of lambdoidal ridge and mastoid portion of temporal bone as far as mastoid process
**nerve:** spinal accessory (XI) and ventral rami of C1–C3
**action:** bilateral contraction depresses snout, unilateral contraction turns head and depresses snout

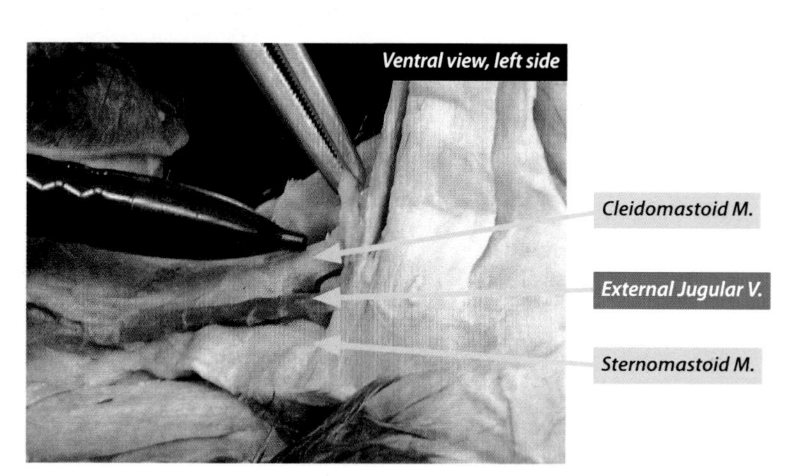


## Cleidomastoid Muscle (Cat)

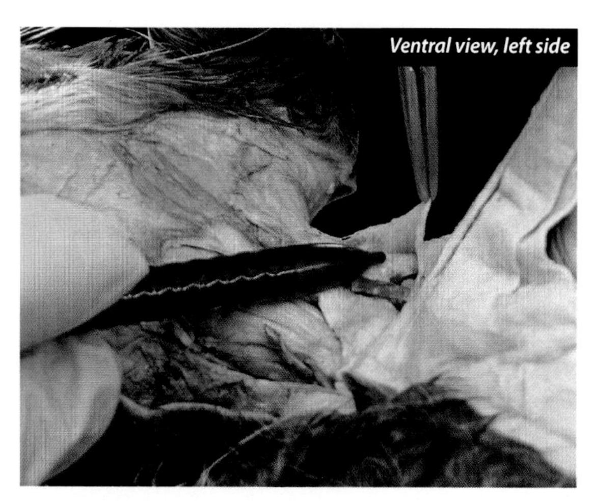


### *Cleidomastoid Muscle (Cat)*

(Corresponds to lateral portion of human sternocleidomastoid)

**origin:** mastoid process
**insertion:** clavicle and raphe lateral to the clavicle
**nerve:** spinal accessory (XI) and ventral rami of C2 and C3
**action:** elevates clavicle when head is fixed; turns head and depresses snout when clavicle is fixed

## Sternocleidomastoid Muscle (Human)

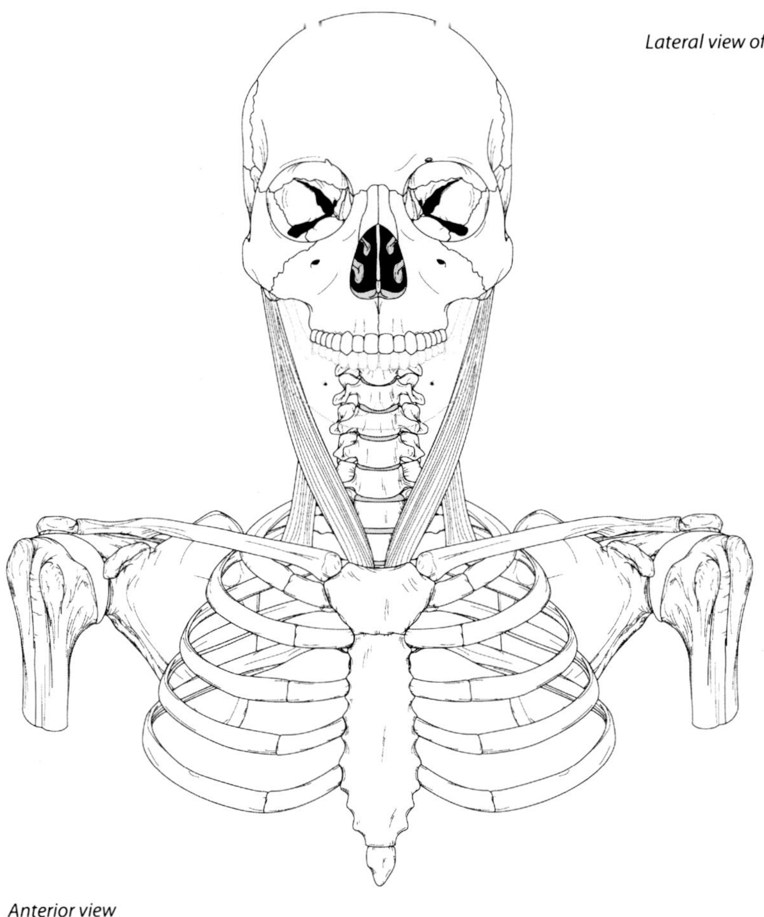
Anterior view

Lateral view of insertion, right side

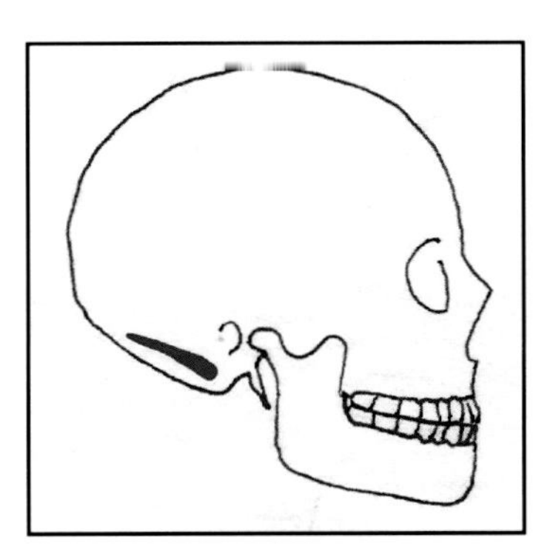

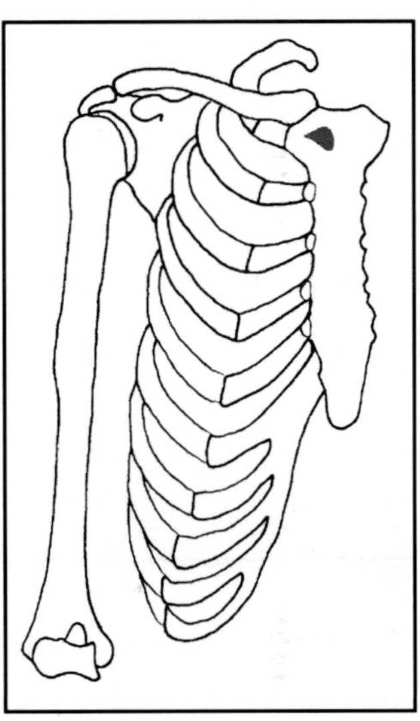
Anterior view

*The above drawings of the origin and insertion might help you visualize this information (red is the origin, blue the insertion).*

### *Human Information: The Medial Portion of the Sternocleidomastoid*

(Corresponds to the sternomastoid in cat)

**origin:** anterior surface of manubrium
**insertion:** lateral surface of mastoid process and lateral half of superior nuchal line
**nerve:** spinal accessory (XI) and anterior rami of C2 and C3
**action:** bilateral contraction leads to flexion of head; unilateral contraction causes rotation of head to opposite side and movement of head toward ipsilateral shoulder

### *Human Information: Lateral Portion of the Human Sternocleidomastoid*

(Corresponds to the cleidomastoid in cat)

**origin:** superior surface of medial third of clavicle
**insertion:** lateral surface of mastoid process and lateral half of superior nuchal line
**nerve:** spinal accessory (XI) and anterior rami of C2 and C3
**action:** bilateral contraction leads to flexion of head; unilateral contraction causes rotation of head to opposite side (contralateral rotation) and movement of head toward ipsilateral shoulder (ipsilateral sidebending)

# NERVES

## Axillary Nerve

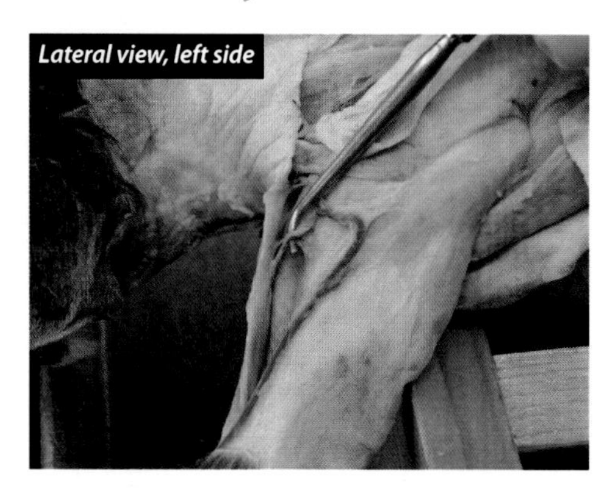
Lateral view, left side

The **axillary nerve** is a branch of the **posterior cord** of the **brachial plexus**. It serves the deltoids and teres minor. In this lab, we will find it on the deep side of the clavobrachialis muscle. In a later lab, we will find it near its origin in the axilla.

## Thoracodorsal Nerve

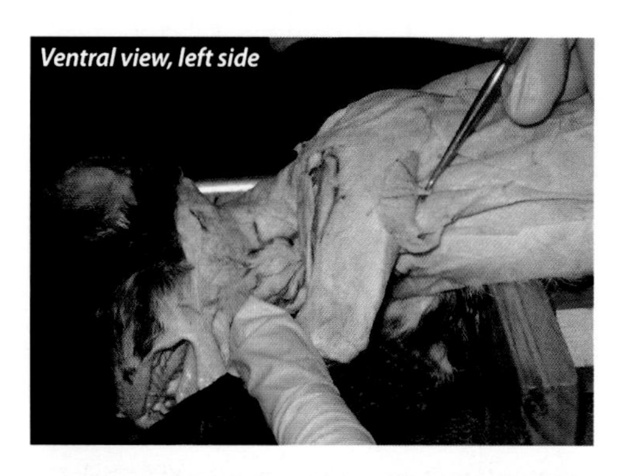
Ventral view, left side

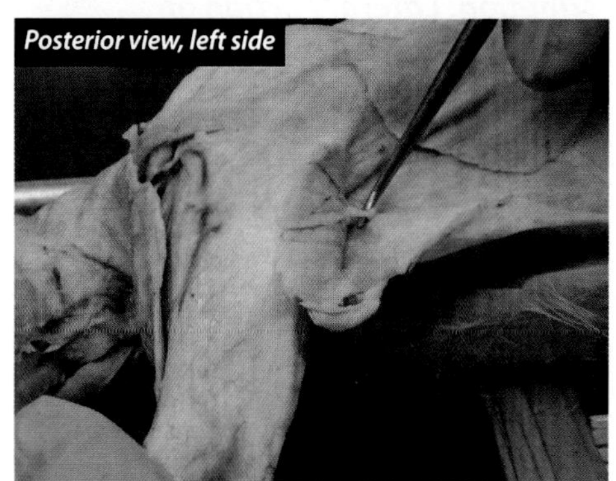
Posterior view, left side

# VESSEL

## Thoracodorsal Artery

The **thoracodorsal nerve** is also called the **third subscapular nerve**. It serves the latissimus dorsi muscle. We will find it on the deep side of the latissimus dorsi at its humeral end running with the **thoracodorsal artery**. There are actually three **subscapular nerves**. The **upper** and **lower subscapular nerves** both innervate the subscapularis muscle. The **lower subscapular nerve** also innervates the teres major muscle. All of these **subscapular nerves** are lateral branches of the **posterior cord** of the **brachial plexus**. The **thoracodorsal artery** is a branch of the **subscapular artery**. It serves the latissimus dorsi muscle and can be observed on the deep side of that muscle at its humeral end where it runs with the **thoracodorsal nerve**.

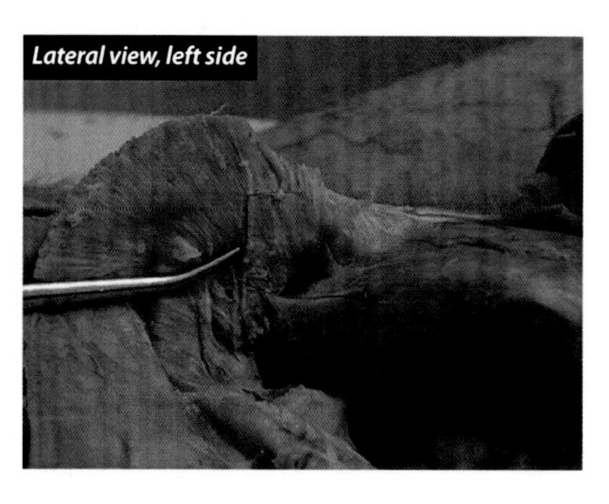
Lateral view, left side

# NERVES

## Long Thoracic Nerve

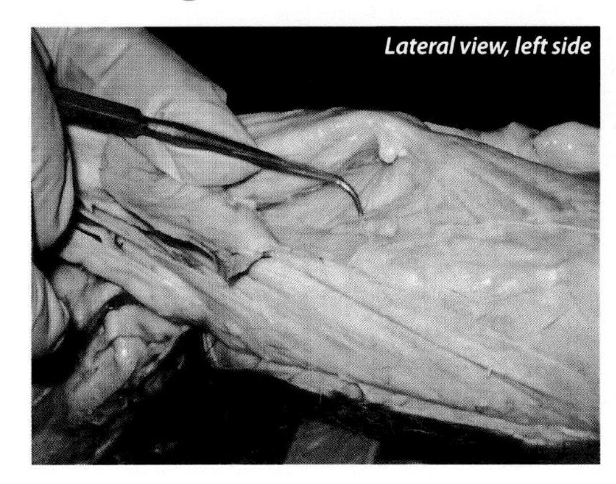
Lateral view, left side

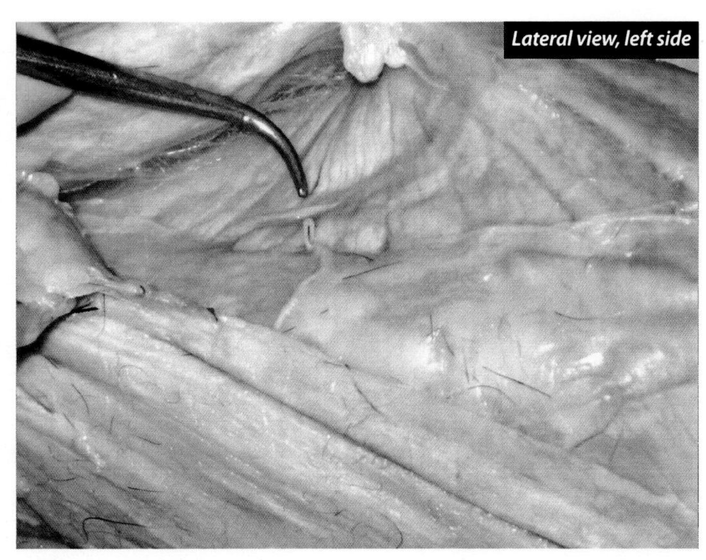
Lateral view, left side

The **long thoracic nerve** serves the serratus ventralis. It is found on the superficial side of this muscle. It is a **lateral branch** of the **anterior rami** of **C5**, **C6**, and **C7**. Students have called this nerve the **Paul McCartney nerve** because he named it for his famous song "The **Long Thoracic Nerve**," which was on an early release of the Beatles album "Let It Be." The name didn't catch on, and I think he may have changed it later.

## Spinal Accessory Nerve (XI)

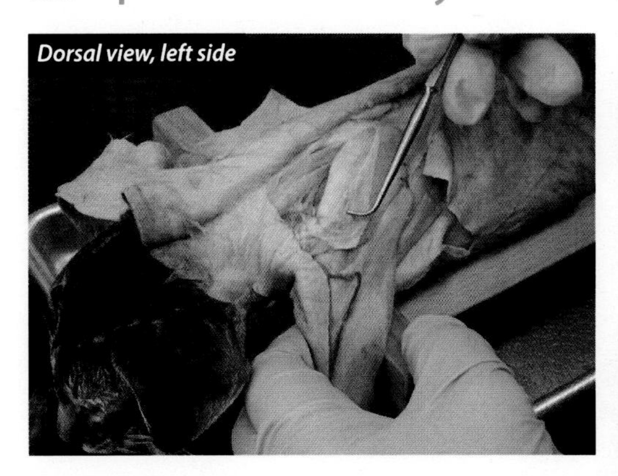
Dorsal view, left side

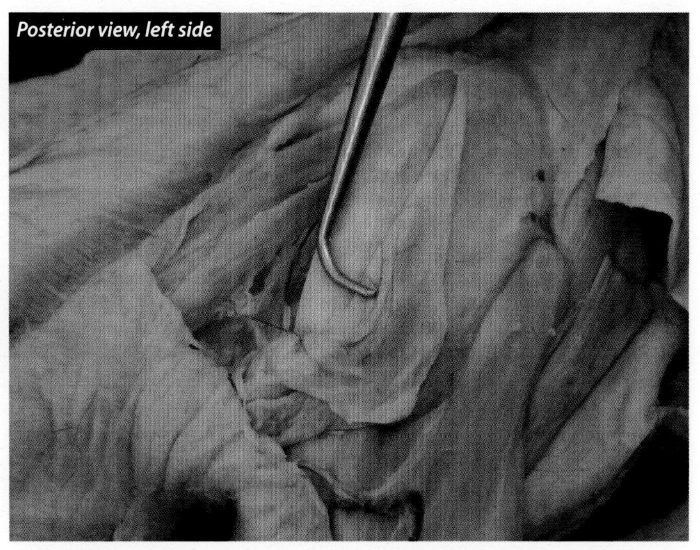
Posterior view, left side

The **spinal accessory nerve** is **cranial nerve XI**. It serves the sternocleidomastoid muscle and the trapezius muscle. We will find it on the deep side of the acromiotrapezius muscle. Sometimes it is called the **accessory nerve**.

# VESSELS

## Cephalic Vein

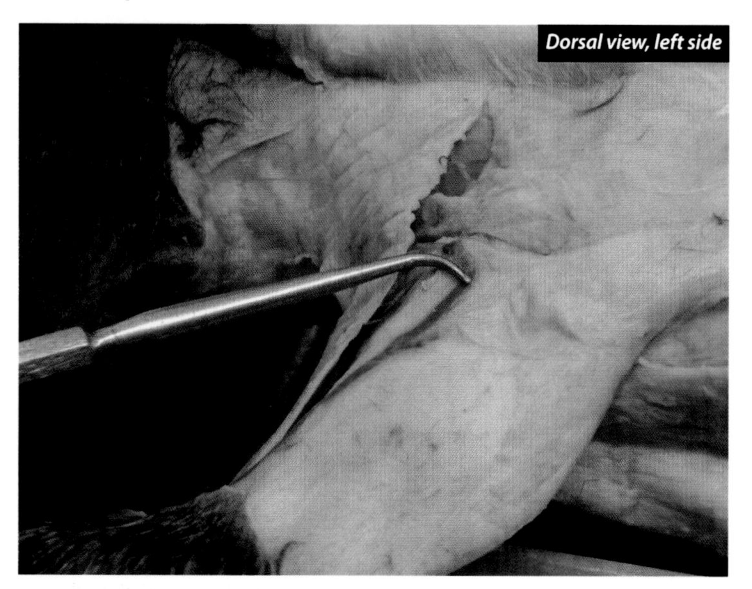


The **cephalic vein** is observed on the superficial, lateral side of the forearm and arm. It is inferior to the clavicle and passes into the axilla where it joins the **axillary vein**. The **axillary vein** passes through the thoracic wall to become the **subclavian vein**.

## External Jugular Vein

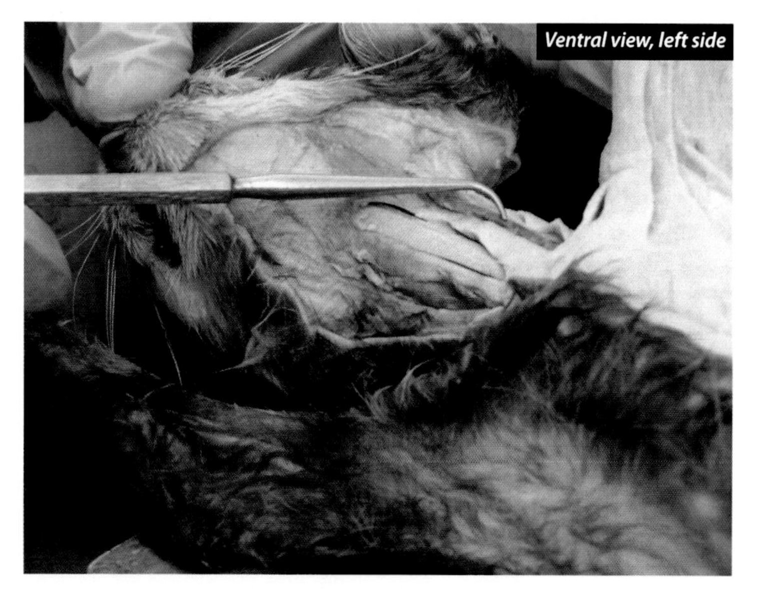


The **external jugular vein** is observed on the superficial, lateral side of the neck. It receives blood from the head region, except for the cranial cavity. It joins the **subclavian vein** to form the **brachiocephalic vein**.

# Lab 3

# Thoracic Cavity: Viscera, Nerves, and Vessels

## Wish List

**LAB 3 OVERVIEW, pp. 64–66**

**ORGANS (VISCERA), pp. 67–73**

- Thymus Gland, p. 67
- Trachea, p. 67
- Bronchi, p. 68
- Lungs (identify all lobes), pp. 68–69
  - Right:
    - Cranial Lobe
    - Middle Lobe
    - Caudal Lobe
    - Accessory (Mediastinal) Lobe
  - Left:
    - Cranial Lobe
    - Middle Lobe
    - Caudal Lobe
- Esophagus, p. 70
- Thoracic Duct, p. 70
- Pleura, p. 71
- Heart, pp. 71–73
  - Pericardium, p. 71
  - Atria (left and right), p. 72
  - Ventricles (left and right), p. 72
  - Interventricular Sulcus, p. 73

**NERVES, pp. 73–75**

- Phrenic (left and right) Nerve, pp. 73, 75
- Vagus (X) Nerve, pp. 74, 75
- Sympathetic Trunk, p. 74

**VESSELS, pp. 76–87**

**DIAGRAMS, pp. 76–78**

- Vessel Diagrams, pp. 76–78

**ARTERIES, pp. 76, 78–83**

- Pulmonary Trunk, p. 79
- Aorta, p. 79
- Brachiocephalic Arteries, p. 80
- Common Carotid Artery (left and right), p. 80
- Bicarotid Trunk, p. 81
- Subclavian Arteries (left and right), p. 81
- Internal Thoracic (Mammary) Artery, p. 82
- Vertebral Artery, p. 82
- Costocervical Artery, p. 83
- Thyrocervical Artery, p. 83

**VEINS, pp. 77, 84–87**

- Internal and External Jugular Veins, p. 84
- Subclavian Vein, p. 84
- Vertebrocostocervical Trunk, p. 85
- Brachiocephalic Veins (left and right), p. 85
- Internal Thoracic (Mammary) Vein, p. 86
- Cranial Vena Cava (precava), p. 86
- Caudal Vena Cava (postcava), p. 87
- Azygos Vein, p. 87

# LAB 3 OVERVIEW

We will study the contents of the thoracic cavity in Lab 3. Collectively, we call this the thoracic viscera.

Preparing the cat for this lab will involve cutting through the costal cartilages on the right side of the thoracic cavity. Begin by inserting the scissors approximately one quarter of an inch to the right of the sternum at the cranial end of the thoracic wall. Cut caudally toward the diaphragm, but **DO NOT CUT THROUGH THE DIAPHRAGM!** You can avoid this act of cat terrorism by inserting your finger into the thoracic cavity and pushing it toward the diaphragm before you cut. When you reach the diaphragm, begin to cut through the thoracic wall that runs parallel to the diaphragm on its cranial side. Cut all the way to the vertebral boarder. The cut should always go diagonally toward the caudal end of the cat. If your cut turns in a cranial direction, you are following a rib and you do not want to do that.

Once you finish the cut to the right side you can cut through the sternum toward the left side. At this point you will need to cut through the mediastinal ligament (it is a pleural ligament) that extends from the cavity to the deep side of the sternum. This ligament is actually two pleural layers that adhere to each other. Now make a cut along the cranial side of the diaphragm on the left side, similar to what you did on the right side. When you reach the vertebral border on the left you are ready to cut the ribs. Using bone-cutters, cut each of the ribs on both sides of the vertebral column. Move the viscera out of the way before you do this so that you will not damage the organs.

The **thymus gland** is in the ventral portion of the thoracic cavity, cranial to the heart. It will be large if the cat is relatively young, but nearly nonexistent if the cat is older. The thymus is an endocrine gland and its most important function is preconditioning T-lymphocytes. Therefore, it is often considered part of the lymphatic system. Loosen the caudal end of the thymus gland from the cranio-ventral end of the pericardium, and peel it cranially. As you do this, try to save the three vessels that are associated with the thymus: the **left internal thoracic (mammary) artery**, the **right internal thoracic (mammary) artery**, and the **internal thoracic (mammary) vein**. We will discuss these vessels soon. Please leave the thymus attached at its cranial end. You probably already suspect that the thymus is named for a famous anatomist who later became a radio talk show host, Don Thymus!

Notice that the **lung** has three lobes on the left side (**cranial**, **middle**, and **caudal lobes**), while it has four lobes on the right side (**cranial**, **middle**, **caudal**, and **accessory** or **mediastinal lobes**). There is one less lobe on each side in humans. The **trachea** can be seen in the neck as it descends caudally toward the lung. It will bifurcate into two **primary bronchi** (one to each side) and those bronchi further divide into **secondary bronchi**, one to each lobe of the lung. You probably won't observe the primary bronchi, but you will see the secondary bronchi when you remove the left lung (next).

Take the left lung in your hand and pull it toward the left side. Observe the root of the lung. It includes the primary bronchus, a **pulmonary artery** (oxygen deficient blood), and two **pulmonary veins** (oxygen rich blood). Carefully cut all three lobes off at the root of the lung. Please be careful not to cut the nerve that is close to the root of the lung, as that is the phrenic nerve which serves the **diaphragm**. There is a left and a right phrenic nerve. The right **phrenic nerve** eventually joins the **caudal vena cava** as it runs to the diaphragm. When you make this cut you will probably cut the secondary bronchi, and that is to be expected. Once you have removed the lung, please insert the scissors into one of the secondary bronchi associated with the lung and carefully cut it open. Reflect the bronchial walls to expose the branches that course out into the lung.

Dorsal to the root of the lung on the left side, you will see a pink tube running toward and then through the diaphragm. This is the (**descending**) **aorta**. **Why would this be pink rather than red?** You will probably be able to observe lateral branches of the aorta that run in the subcostal groove of each rib. You may even see that cranial to the **artery** is a **vein** and caudal to the artery is a **nerve**. Vein, artery, and nerve are arranged cranial to caudal, just as they are in humans. Dorsal to the aorta you may see the **thoracic duct** (quack). This duct directs all lymphatic drainage caudal to the diaphragm and everything on the left side cranial to the diaphragm back to the circulatory system. It will often resemble beads. This appearance is due to the semilunar valves inside the duct. The **esophagus** is located ventral to the aorta and accompanies it through the diaphragm. You should notice two nerves on the surface of the esophagus. Those are the **ventral** and **dorsal vagus (X) nerves**. We will talk about how the name of the two vagus nerves changes during lecture. Lastly, you should observe the **left sympathetic trunk**. This nerve is part of the left vagosympathetic trunk, but it leaves the left vagus near the area of the **left brachiocephalic vein**. Once it is alone, it passes to the border of the vertebral column and continues caudally along the vertebral border to the abdomen. It will be deep to the **parietal pleura**. Some students use the mnemonic "**EATS**" for these structures with the order from ventral to dorsal. E is for esophagus, A is for aorta, T is for thoracic duct, and S is for sympathetic trunk.

Next, make a small incision at the apex of the **heart**; cut along the ventral surface of the **pericardium** in a cranial direction to the cranial end of the heart. Carefully peel the pericardium toward the dorsal side of the heart on both sides and do not remove it from the cat. This will expose the heart. Observe the two **atria** (**auricles**), the two **ventricles**, the **interventricular sulcus**, and the **pulmonary trunk**. We will study many other structures of the heart in laboratory exercise 4.

We can now study the vessels of the thoracic cavity, and we will use the heart as both the origin and the destination for vessels. Please try the road map approach to studying the distribution of these vessels, considering them to be one-way passages for blood to and from the heart. Dr. J sincerely believes that this method will work for you.

We begin with the veins. They will bring blood to the heart and we find vessels coming in from the cranial end of the thoracic cavity. Converging at the left side of the cranial end of the thoracic cavity we have a **left external jugular vein**, a **left internal jugular vein**, a **left subclavian vein**, and a **left vertebrocostocervical trunk** to form the **left brachiocephalic vein**. In most cats, the internal jugular veins, which have relatively small diameters, do not take the dye. **With the knowledge you have, think about why this might be the case.** On the right side, in a little more than half the cats, the comparable right-hand vessels converge to form the **right brachiocephalic vein**. In a little less than half the cats, the **vertebrocostocervical trunk** joins the **cranial vena cava** rather than joining the **right brachiocephalic vein.** The two brachiocephalic veins join to form the **cranial vena cava**. The cranial vena cava receives two other significant veins, the **internal thoracic (mammary) vein** from the sternum and the **azygos vein** from the dorsal portion of the thoracic cavity. The azygos may be observed slightly to the right of the midline, dorsal to the right lung. You should see veins from each of the subcostal grooves draining blood from the thoracic wall to the azygos. It also serves as an alternate path to the caudal vena cava for blood to return to the thoracic cavity from the abdomen.

The **caudal vena cava** brings blood to the right atrium from the caudal end of the thoracic cavity. The vessels that converge to form this major vein do so in the abdomen and pelvis. Remember to observe the **right phrenic nerve** on the lateral side of the caudal vena cava.

There are four **pulmonary veins** that bring blood to the left atrium of the heart. We will study them in more detail in Laboratory 4. Some students use the mnemonic "veins drain" to remind them that veins drain blood back to the atria.

The vessels that carry blood away from the heart are the arteries, and students use the mnemonic "arteries away" to remind them of this. We already mentioned the **pulmonary trunk** when discussing structures you should observe on the heart. That vessel receives blood from the right ventricle and directs that oxygen-deficient blood to the **right** and **left pulmonary arteries**. These, in turn, direct the blood to the lungs.

The **aorta** carries blood to the systemic circulatory system. The first two branches are the **left** and **right coronary arteries**. You will not observe them in this lab, but you will see them in Lab 4. You will observe the third branch of the aorta, the **brachiocephalic artery**. Note, there is only one, so it is neither a left nor a right. Usually, the brachiocephalic artery will have three branches, although occasionally it has only two branches. When there are three, you will observe the **right subclavian artery** passing to the right thoracic outlet, the **right common carotid artery** coursing to the right side of the neck, and the **left common carotid artery** going to the left side of the neck. If there are only two branches, the right subclavian moves to the right thoracic outlet as before, but the other branch is the **bicarotid trunk**. The bicarotid trunk will bifurcate to form the left and right common carotid arteries.

The **left subclavian artery** is the fourth branch of the aorta. It will have four branches in the thoracic cavity that you will be responsible for. The **left internal thoracic artery** is usually the first branch and it is directed toward the sternum. The three other branches have a mnemonic to help you learn the order: "VCT." The V is for the most caudal branch, the **left vertebral artery**. This vessel is usually directed toward the cervical region and it passes into the transverse foramina of the cervical vertebrae 1 through 6, eventually through the foramen magnum of the occipital bone, and into the cranium where it serves the brain. Next, moving cranially, we observe the **left costocervical artery** (C from the mnemonic). It is normally directed medially and its branches serve the muscles of the neck, ribs, and dorsal area. It is quite small and sometimes looks like a comma or an upside down C. The most cranial branch of the "**VCT**" group is the **left thyrocervical artery**. It courses to the neck and will give rise to the **left transverse scapular artery**, which we will observe in Lab 7, as well as branches to the neck.

We will find a nerve that is slightly greater in diameter and runs medial to the phrenic nerve as it enters the thoracic cavity from the neck. This is the **left vagus (X) nerve**. Because it is in such close proximity to the phrenic nerve, a method for verification of its identity is desirable. Go into the lateral area of the neck and find the left common carotid artery. The **left vagosympathetic trunk** accompanies this vessel. If you pull the left vagosympathetic trunk, you will see the left vagus nerve move, while the left phrenic nerve will not move. The dorsal and ventral vagus nerves were mentioned earlier. They were found along the esophagus as it approaches the diaphragm.

# ORGANS (VISCERA)

## Thymus Gland

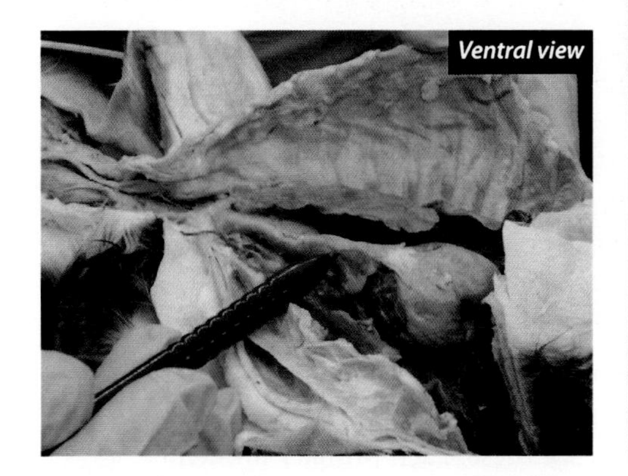


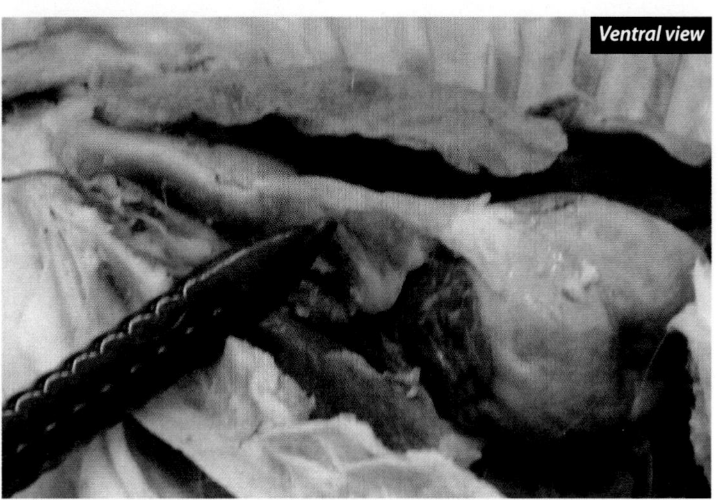


The thymus gland is part of the lymphatic system, and it is located cranial and ventral to the heart. It also functions as an endocrine gland. It reaches maximum development at puberty and then becomes replaced by connective tissue. Functionally it is important because it preconditions T-lymphocytes. It was named for Don Thymus, who is famous for his radio program "Thymus in the Morning."

## Trachea

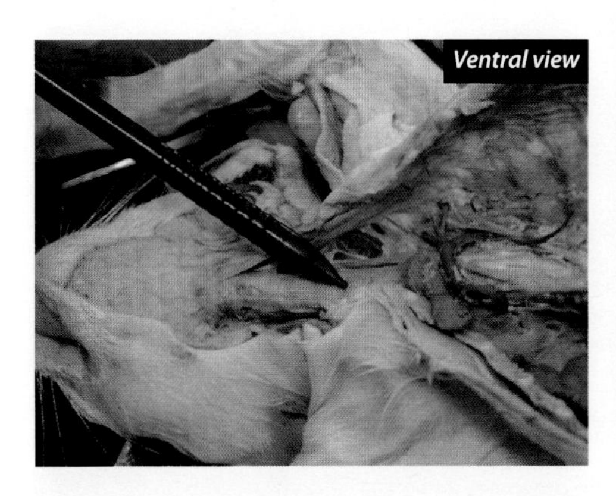


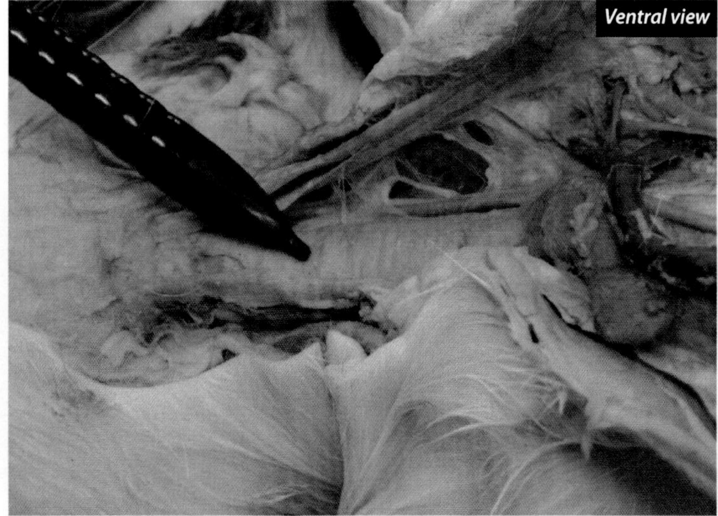


The trachea begins at the inferior end of the larynx (level of C6 body) and extends inferiorly to where it bifurcates into the left and right primary bronchi in humans. The inferior end is at the level of the sternal angle of humans in a supine position and to the body of T7 in the anatomical position. It is recognizable because of its 16 to 20 cartilaginous rings that normally prevent it from collapsing. It is about 1.5 inches in diameter in young males (smaller in females) and is lined with ciliated epithelial cells that sweep mucous out of the trachea and into the pharynx.

## Bronchi

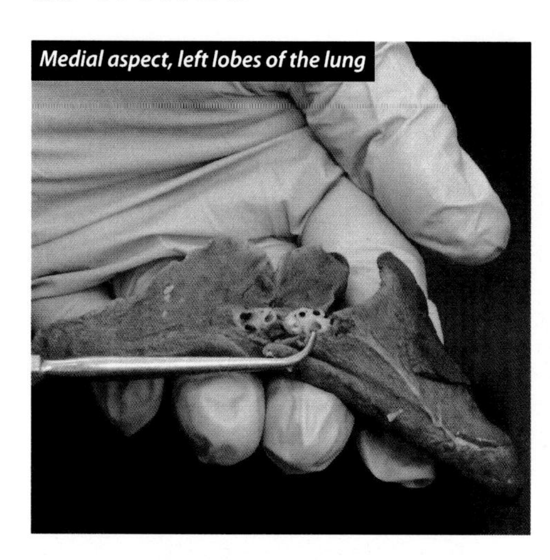
Medial aspect, left lobes of the lung

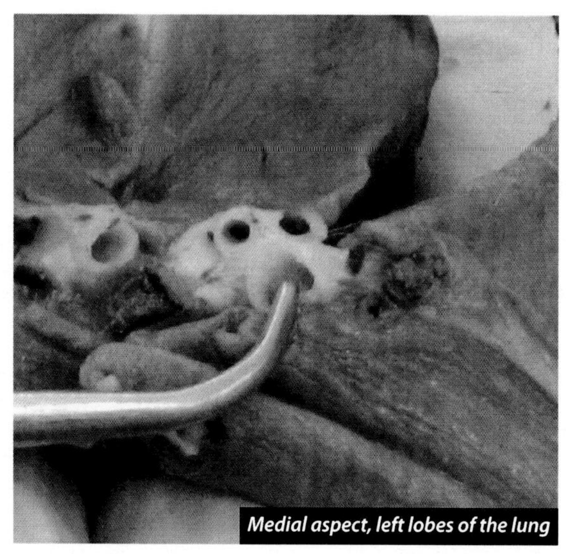
Medial aspect, left lobes of the lung

The bronchi are the large branches on both sides of the respiratory tree and are recognizable because they have cartilaginous rings in their walls that help keep them from collapsing. There are two primary bronchi that branch from the trachea. You will probably not see these. Each primary bronchus branches to a number of secondary bronchi. There is a secondary bronchus for each lobe of the lung; thus, there are seven in a cat, but only five in a human. The secondary bronchi also subdivide as they move through the lung. The above dissection is at the level of the secondary bronchi of the left lung and the probe is covering the opening of the secondary bronchus that serves the caudal lobe.

## Lungs

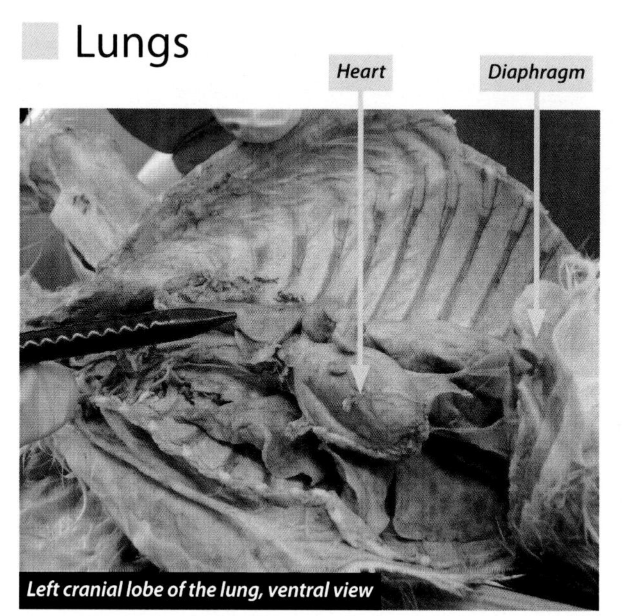

Left cranial lobe of the lung, ventral view

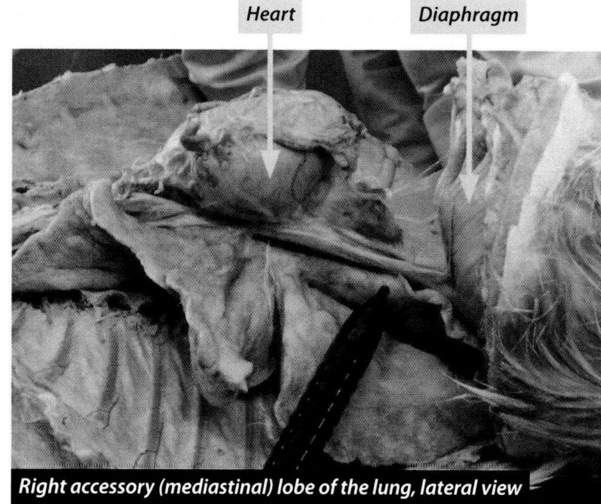

Right accessory (mediastinal) lobe of the lung, lateral view

The lungs are the major respiratory organs of the body. They are responsible for the exchange of the respiratory gases, oxygen, and carbon dioxide. They function as a major excretory organ in getting rid of carbon dioxide. To facilitate the gas exchange, they have undergone miniaturization that effectively increases the surface area to volume ratio. The functional unit of the lung is the microscopic alveolus.

# Lungs

## Lobes of the Lung

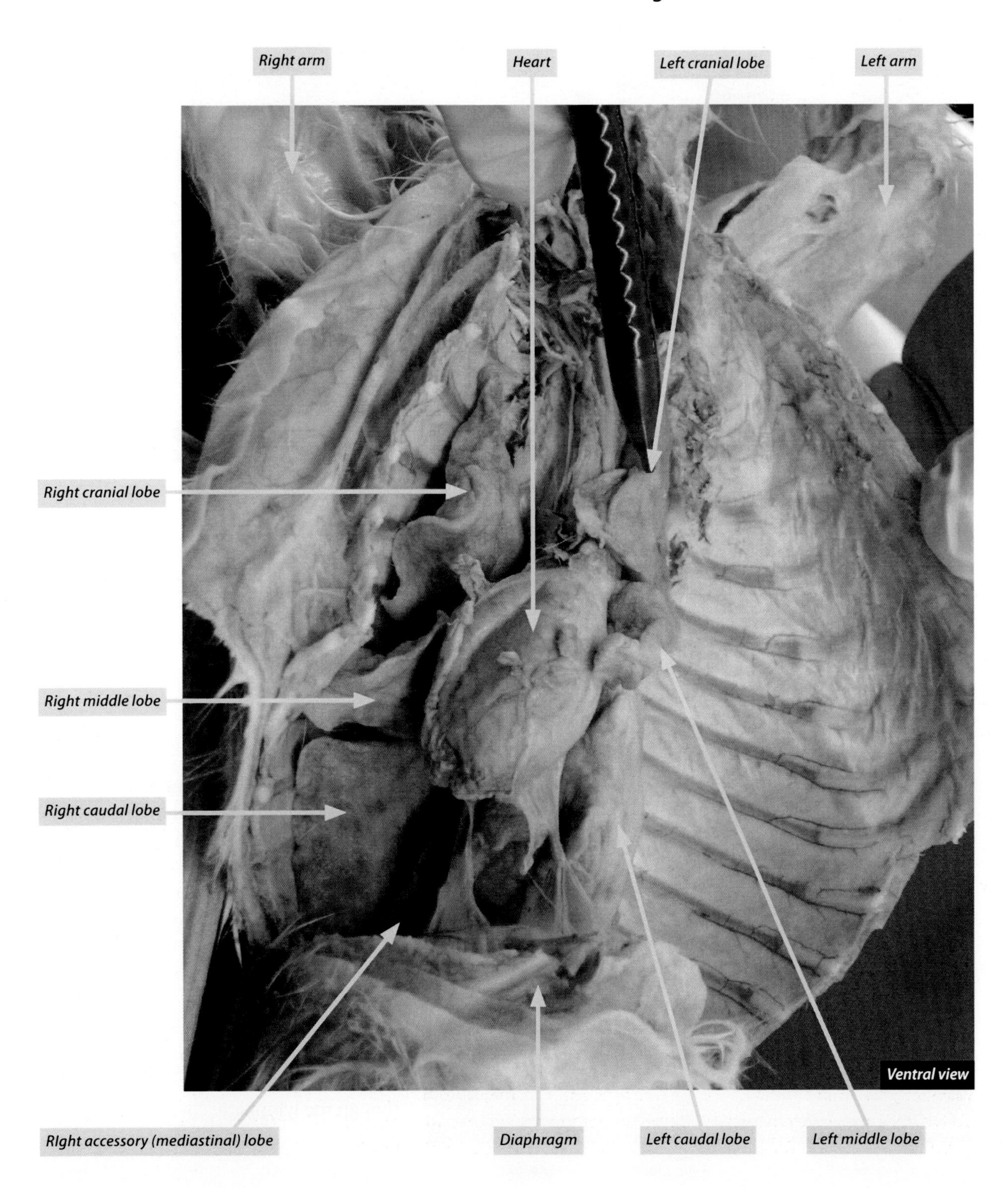

## Esophagus

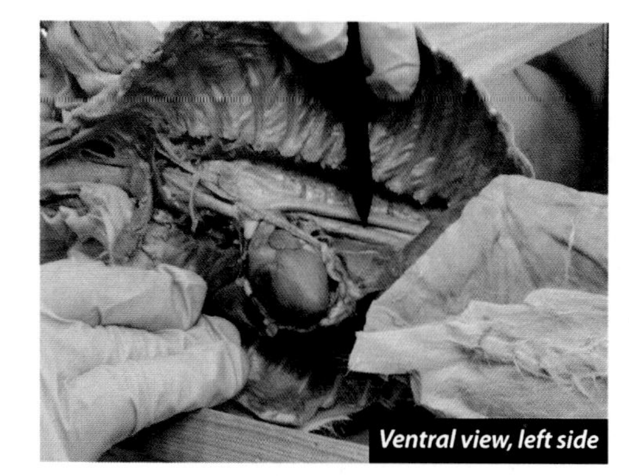
Ventral view, left side

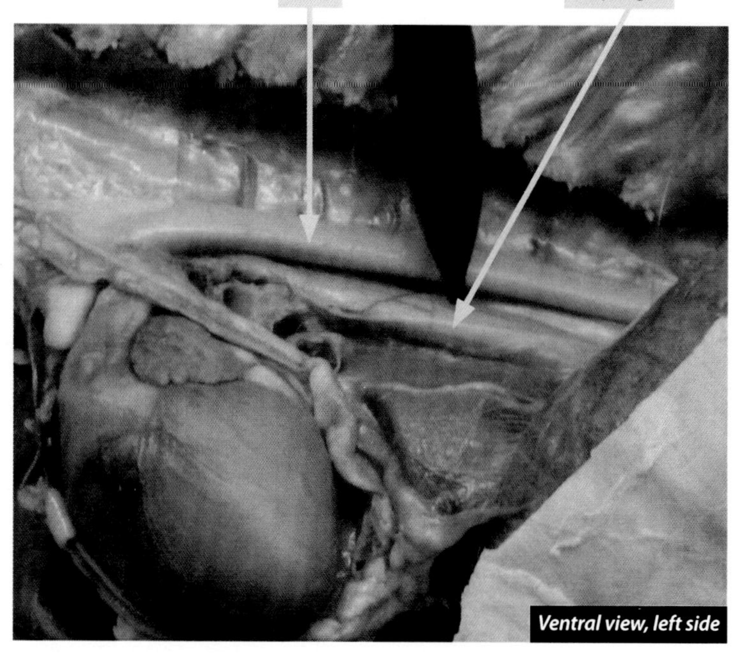

Ventral view, left side

The esophagus is an organ that is primarily smooth muscle tissue. It is essentially a tube that extends from the pharynx to the stomach. Functionally it is important because it directs food from the pharynx to the stomach during swallowing. It also provides for the opposite flow during regurgitation. Note that if you carefully examine the picture on the right you can see the **dorsal vagus nerve**. Both the **ventral vagus nerve** and the **dorsal vagus nerve** join the esophagus caudal to the heart and run with it to the diaphragm.

## Thoracic Duct

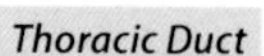

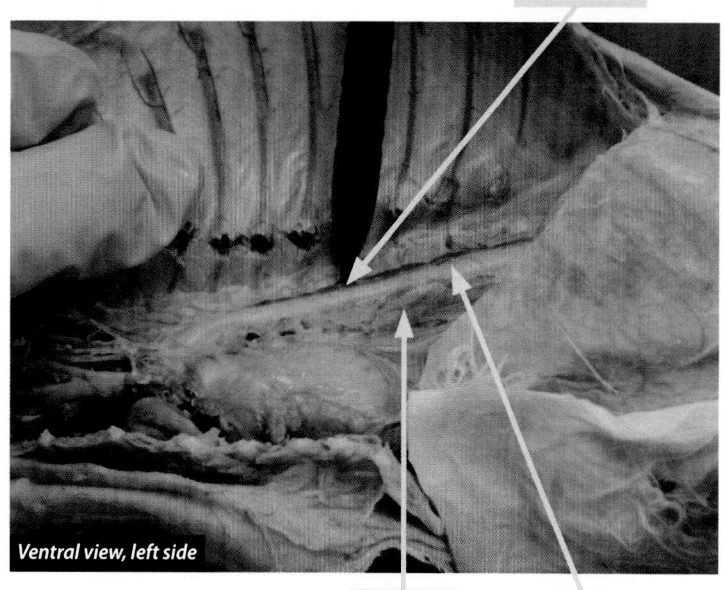

Ventral view, left side

The thoracic duct (quack) is part of the lymphatic system. It resembles a string of beads. The constrictions in its walls are where the one-way (semilunar) valves are. We find the thoracic duct dorsal to the **descending aorta**. It empties its contents into the **brachiocephalic** or **subclavian vein** on the left side of a human. Since there is no dedicated pump to create pressure gradients to move the lymphatic fluid, we rely on the skeletal muscle pump and the one-way valves to direct the flow though the lymphatic channels.

## Pleura

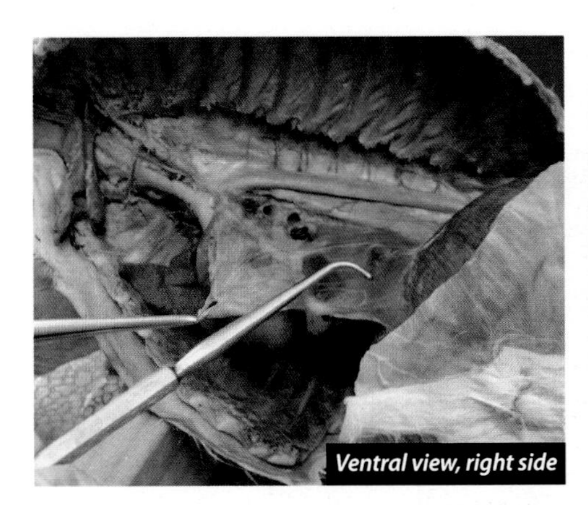


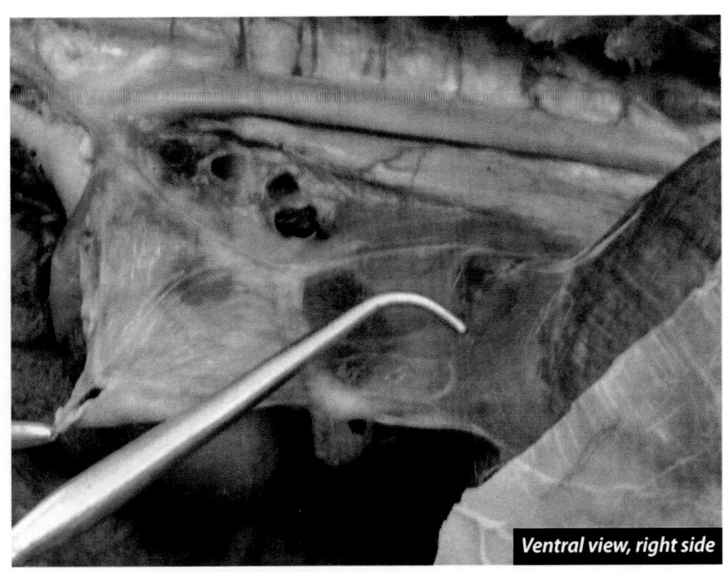


The pleura is an organ that is primarily epithelial tissue. It covers the thoracic organs and lines the thoracic cavity. Functionally it is important because it secretes fluid that lubricates the surfaces with which it is associated, helps reduce heat buildup, and is responsible for surface tension that holds the lungs against the thoracic wall.

## Heart: Pericardium

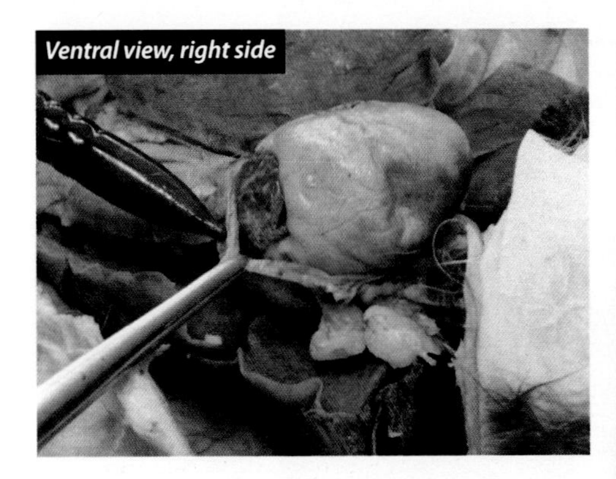


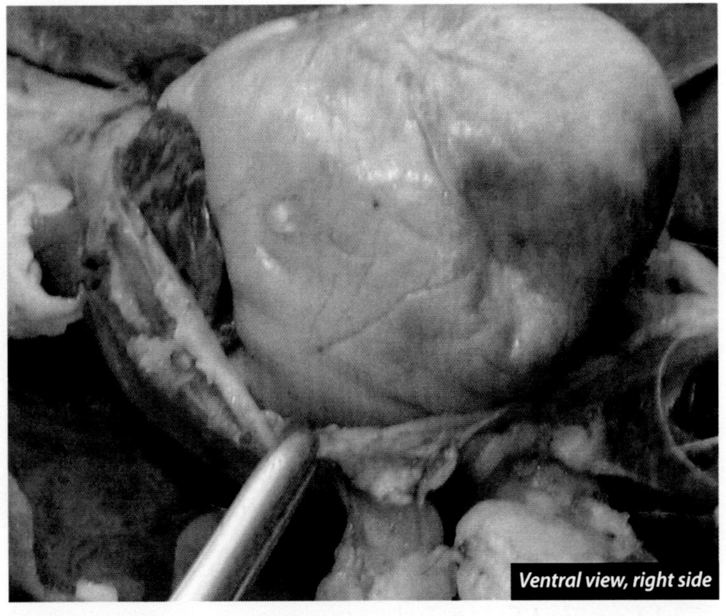


The pericardium surrounds the heart. It consists of modified pleurae (the plural of plura—wow!) and incorporates fibrous tissue as well as the normal pleural epithelial cells. It includes two layers rather than the usual single layer. It was named for Peri Como, a famous singer who in later years was lead singer for Pearl Jam, the Bangles and INXS.

## Heart: Atria

***(singular = atrium)*** The atria (left and right) are often described as the receiving chambers of the heart. Blood enters the right atrium from the **cranial** and **caudal vena cava** and the **coronary sinus** and the left atrium from the four **pulmonary veins**. The atria are found at the superior (cranial) end of the heart. They are sometimes referred to as auricles. They have thin walls compared to the ventricles.

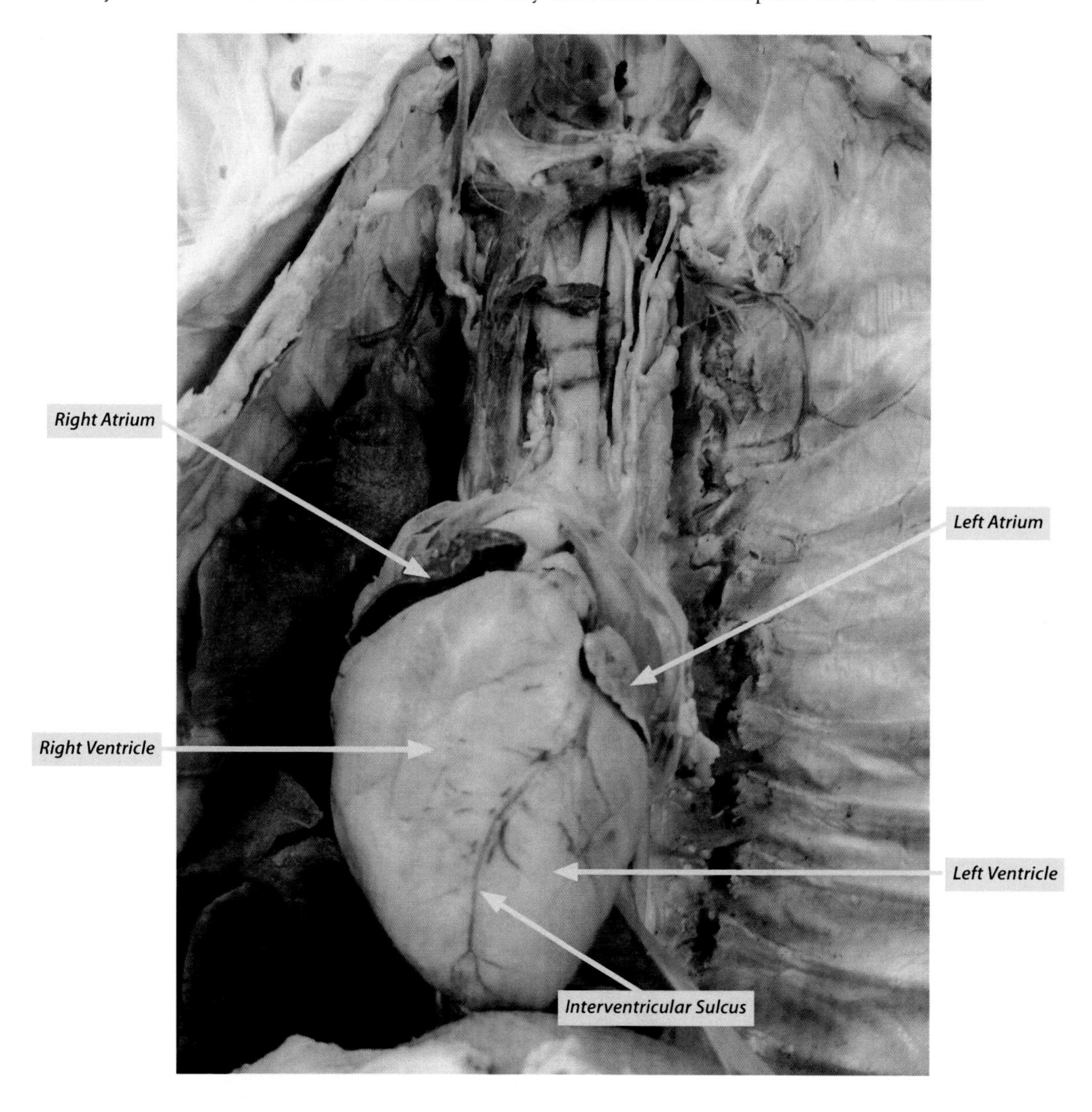


## Heart: Ventricles

The ventricles (left and right) are often described as the pumping chambers of the heart. Blood enters the ventricles from the atria. They provide most of the pressure necessary to move blood through the vessels of the body. They are found at the inferior (caudal) end of the heart. They have thick walls compared to the atria.

## Heart: Interventricular Sulcus

The interventricular sulcus is a superficial groove or depression between the ventricles. It is of importance because this is where some of the large **coronary arteries** and **veins** run. Those vessels carry blood to and from the walls of the heart.

# NERVES

## Phrenic Nerve

The **phrenic nerve** is formed from the union of branches of the **anterior rami** of **cervical spinal nerves 3**, **4**, and **5**. It serves the diaphragm. Note that on the right side it runs along the **caudal vena cava** as it passes to the diaphragm. When you remove the left lung, please be careful not to cut the **phrenic nerve** (no cat terrorism!). Remember the mnemonic, "**C3**, **4**, **5**—stay alive—**phrenic**, **phrenic**, **phrenic**!"

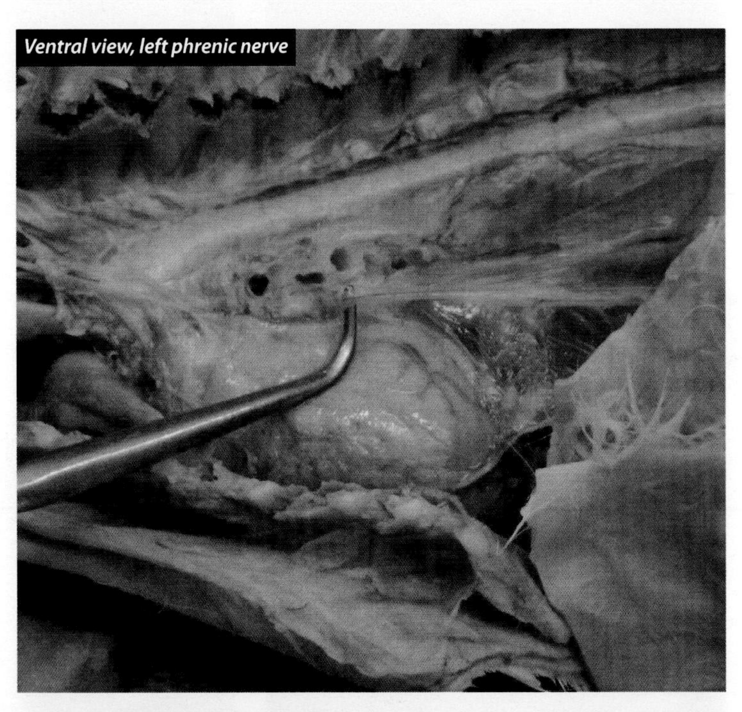
*Ventral view, left phrenic nerve*

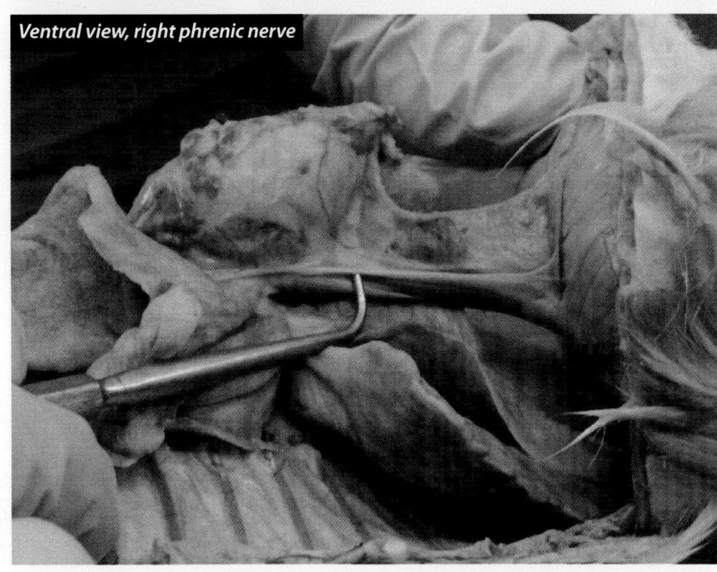
*Ventral view, right phrenic nerve*

## Vagus Nerve

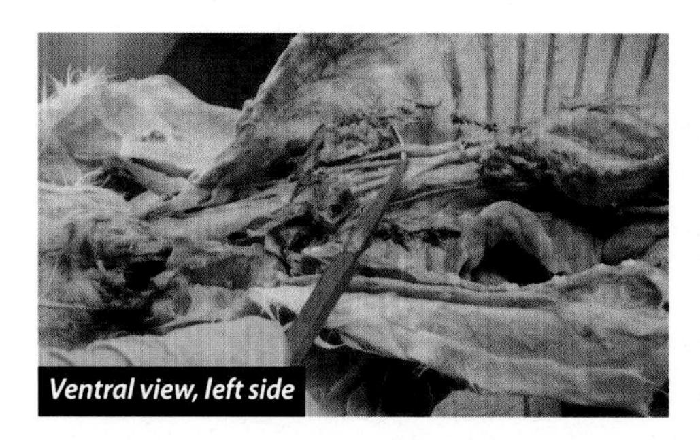
Ventral view, left side

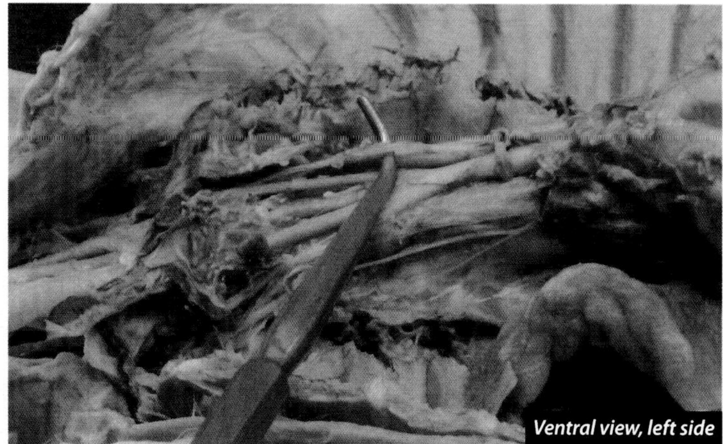
Ventral view, left side

The **vagus nerve** is **cranial nerve X**. It is the only **cranial nerve** to pass into the body cavities below the neck. It is primarily parasympathetic in nature and is the major **nerve** affecting the heart and most of the gastrointestinal tract. It slows the heart down and speeds up the activity of the gastrointestinal tract. As the **left** and **right vagus nerves** approach the heart, they run parallel to the respective **phrenic nerves**. After leaving the heart, they reorganize as **ventral** and **dorsal vagus nerves** that run along the esophagus to and through the diaphragm. Spelling counts on this nerve so be sure to spell **vagus** correctly.

## Sympathetic Trunk

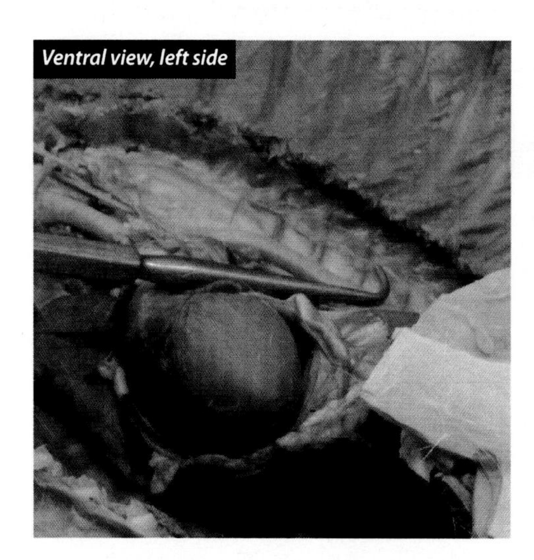
Ventral view, left side

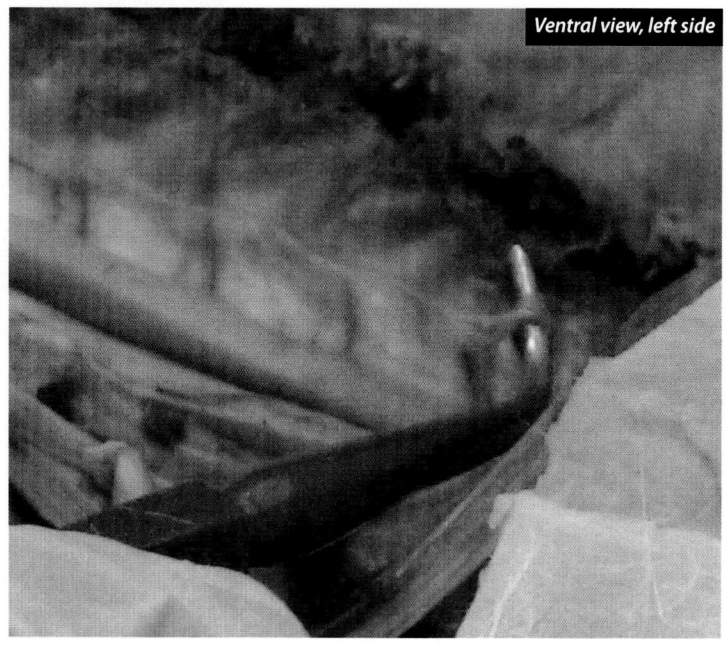
Ventral view, left side

The **sympathetic trunk** runs along each side of the vertebral column. It receives neurons from **anterior rami** of the spinal nerves via the **sympathetic trunk ganglia**. It serves many organs from the head to the pelvis. Because it receives nerves from the intervertebral foramina, it is difficult to lift away from the body wall. Please be gentle with this nerve or it will break (no cat terrorism!). In a later lab we will see it in the neck where it runs with the **vagus nerve** (**X**). That bundle is called the **vagosympathetic trunk**.

## Phrenic Nerve and Vagus Nerve

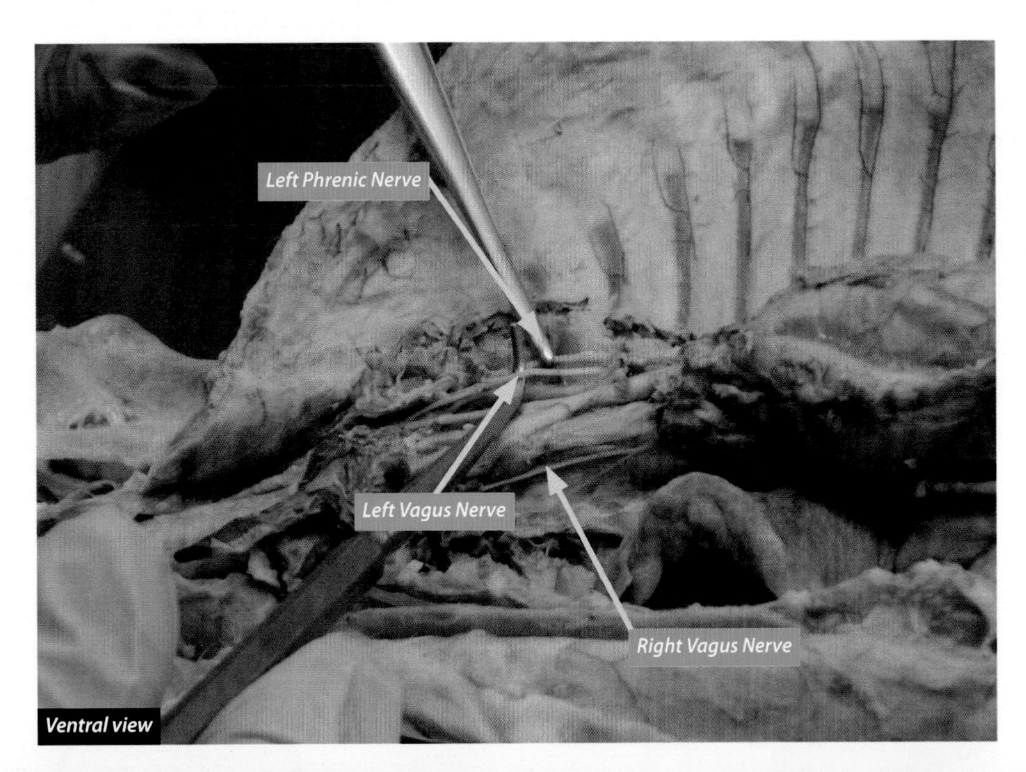


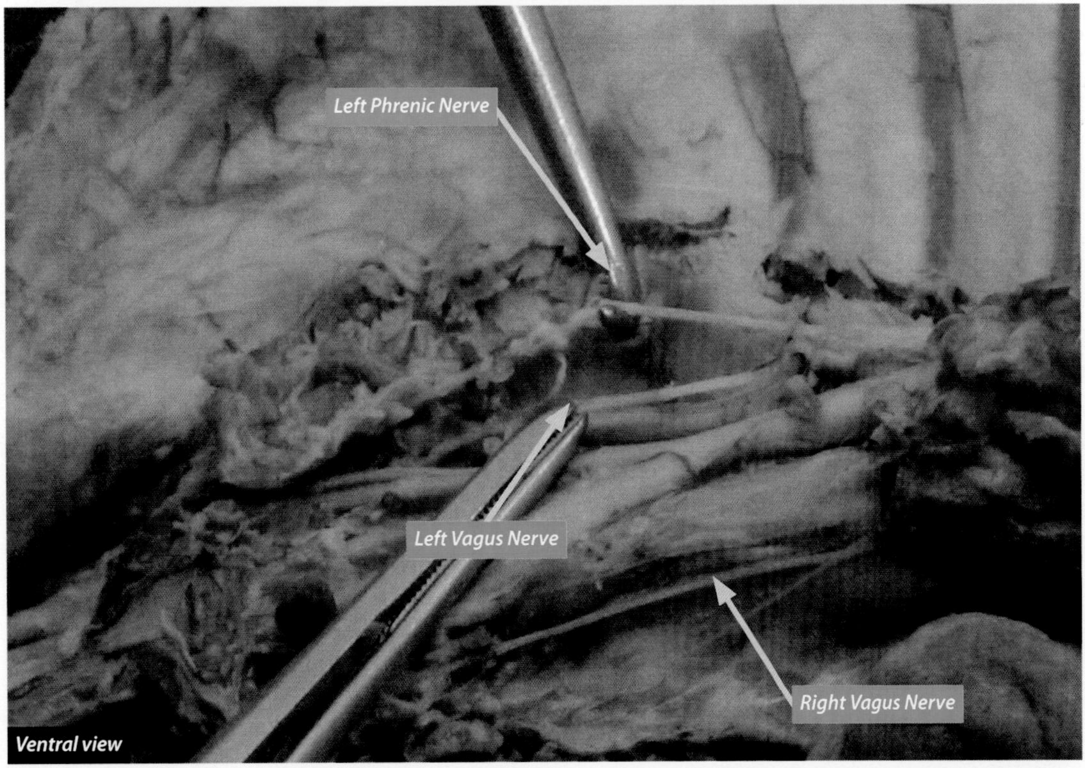

# VESSELS OF THE THORACIC CAVITY

The drawing on the left (below) shows the arrangement of the major thoracic **arteries** in the cat. Note the three possible arrangements of the **brachiocephalic artery** and its branches. Only one of those variations has a **bicarotid trunk**. You will be responsible for all three since they are found commonly in cats. There are no pictures of the **left** and **right pulmonary arteries** since they are very small. We will study those vessels in the calf heart lab (Lab 4). Note that the cat heart is happy! This drawing on the right (below) shows the normal human configuration of the arteries we study in Lab 3. Note that the left **common carotid artery** comes directly from the **aorta** rather than from the **brachiocephalic** or **bicarotid trunk** as it did in the cat. Otherwise the arrangement is very similar to that of the cat.

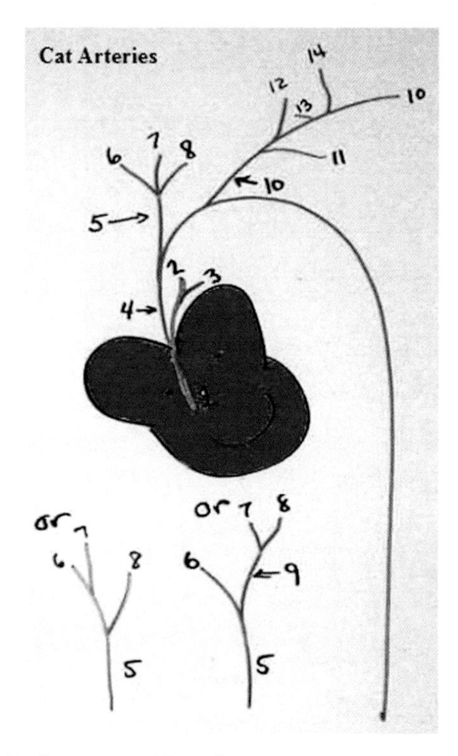


1. **Pulmonary Trunk**
2. **Right Pulmonary Artery**
3. **Left Pulmonary Artery**
4. **Aorta**
5. **Brachiocephalic Artery**
6. **Right Subclavian Artery**
7. **Right Common Carotid Artery**
8. **Left Common Carotid Artery**
9. **Bicarotid Trunk**
10. **Left Subclavian Artery**
11. **Left Internal Thoracic (Mammary) Artery**
12. **Left Vertebral Artery**
13. **Left Costocervical Artery**
14. **Left Thyrocervical Artery**

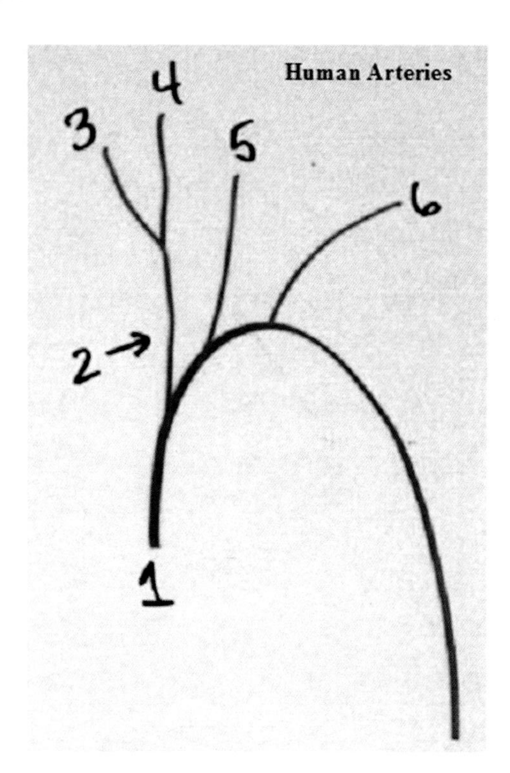


1. **Aorta**
2. **Brachiocephalic Artery**
3. **Right Subclavian Artery**
4. **Right Common Carotid Artery**
5. **Left Common Carotid Artery**
6. **Left Subclavian Artery**

# VESSELS OF THE THORACIC CAVITY

The drawing below shows the arrangement of the major veins in the thoracic cavity of the cat. Of particular interest is the **right vertebrocostocervical trunk**, which can have two configurations. The most common path of this vessel is directly to the **right brachiocephalic vein**, although it may go directly to the **cranial vena cava** instead. Also, you should note that the heart is still happy!

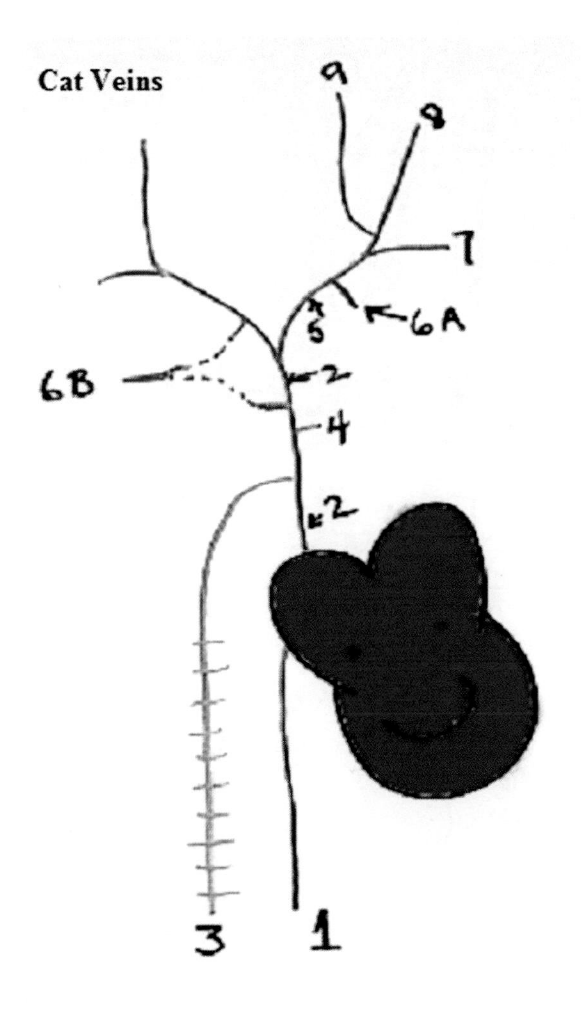


1. **Caudal Vena Cava**
2. **Cranial Vena Cava**
3. **Azygos Vein**
4. **Internal Thoracic (Mammary) Vein**
5. **Left Brachiocephalic Vein**
6a. **Left Vertebrocostocervical Trunk**
6b. **Right Vertebrocostocervical Trunk**
7. **Left Subclavian Vein**
8. **Left External Jugular Vein**
9. **Left Internal Jugular Vein**

# VESSELS OF THE THORACIC CAVITY

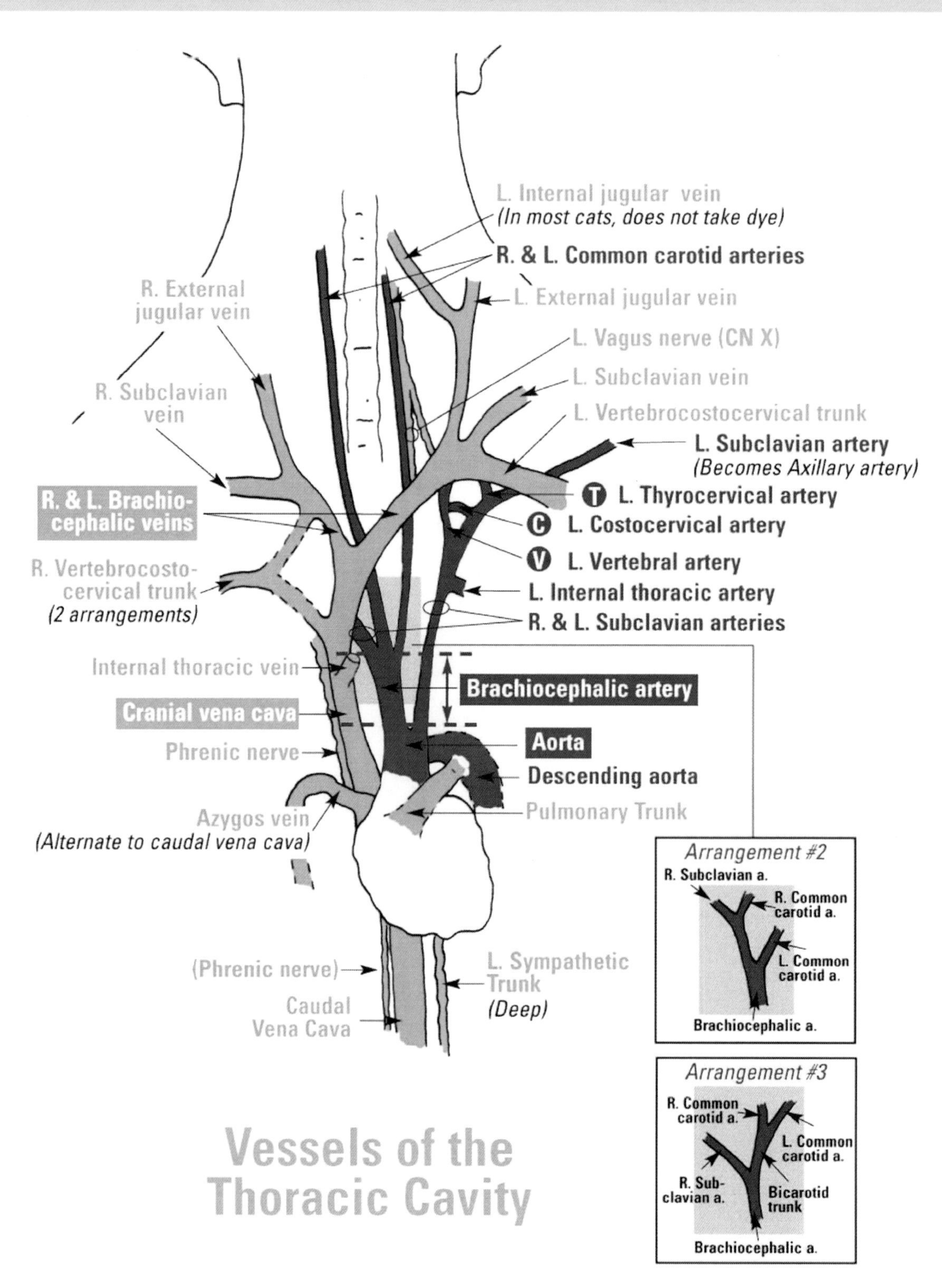


Vessels of the Thoracic Cavity

# VESSELS

## Pulmonary Trunk

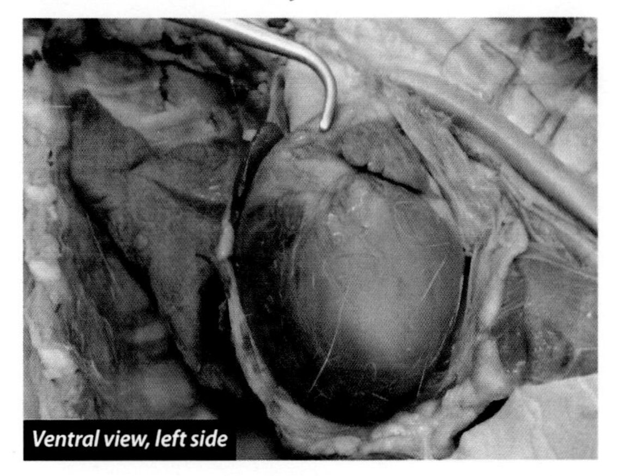


The **pulmonary trunk** is the great artery that carries oxygen deficient and carbon dioxide rich blood toward the lungs from the right ventricle. The **pulmonary trunk** bifurcates to form the **left** and **right pulmonary arteries**. It is typically blue in cats. Remember the blue color indicates low oxygen and high carbon dioxide levels. It is sometimes called the **pulmonary artery** or **pulmonary aorta**. Dr. J recommends the unique name pulmonary trunk.

## Aortic Arch

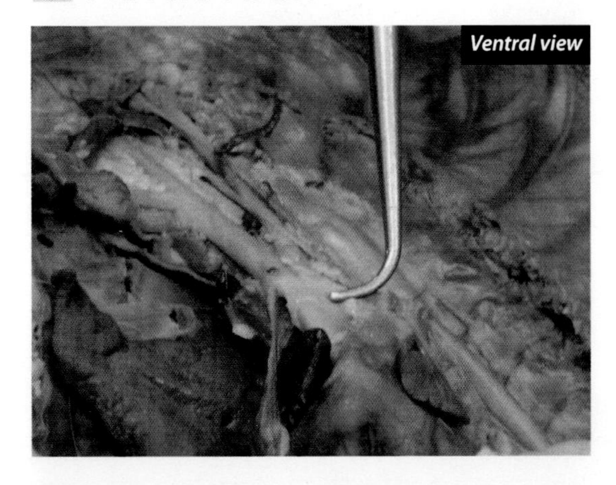


The **aorta** is the **great artery** that carries blood away from the left ventricle to all the systemic arteries of the circulatory system. The blood in this **artery** is normally enriched with oxygen and deficient in carbon dioxide. You can see the thoracic duct dorsal to the aorta in this picture.

## Aorta (Descending)

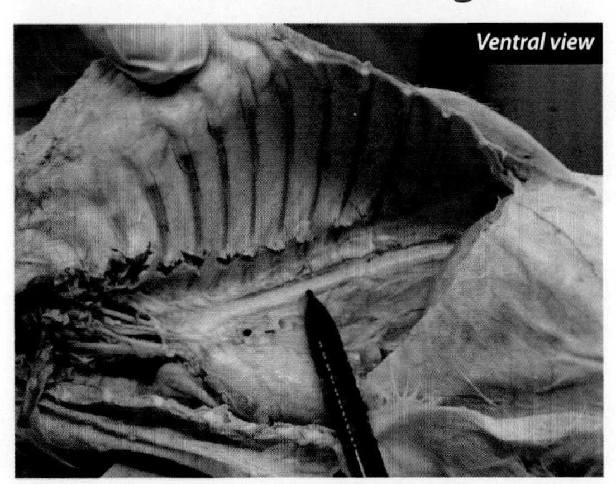

## Brachiocephalic Artery

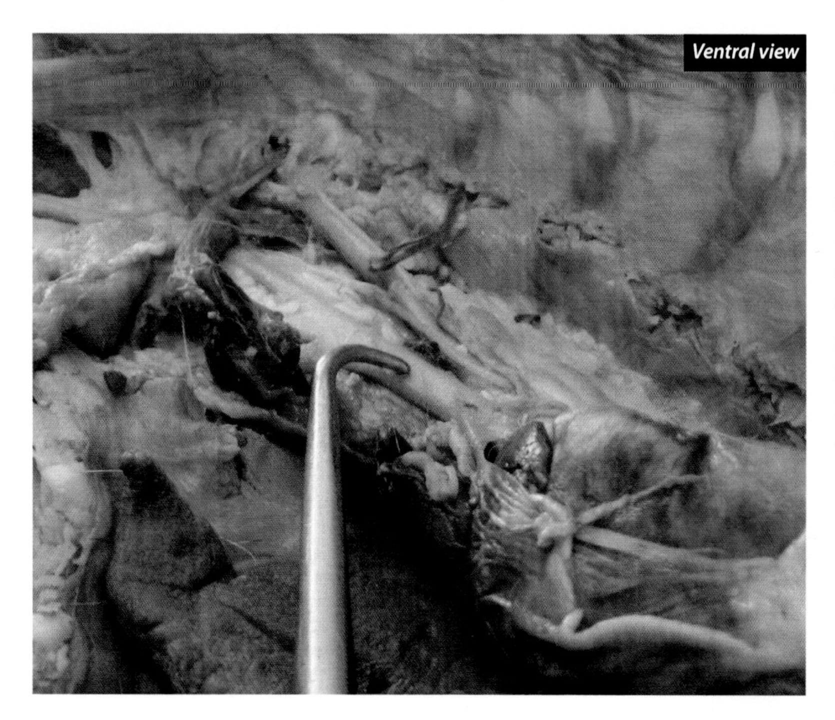
Ventral view

There is only one **brachiocephalic artery**. It is the third branch off the aorta. It gets its name because it serves the arm and head. It gives rise to the **right subclavian artery** and to the **right** and **left common carotid arteries** in most cats. In some cats it gives rise to the **bicarotid trunk** instead of to the **left** and **right common carotid arteries**.

## Common Carotid Arteries

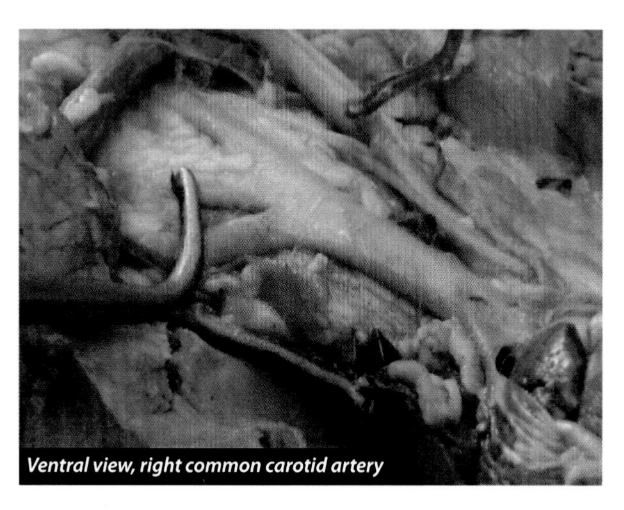
Ventral view, right common carotid artery

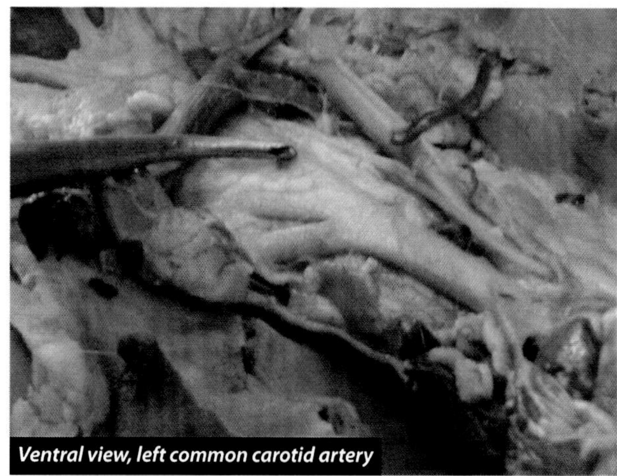
Ventral view, left common carotid artery

The **common carotid arteries** (**left** and **right**) are normally branches of the **brachiocephalic artery**, or less frequently branches of the **bicarotid trunk**. They have a number of branches, which will be the object of study in the laboratory.

## Bicarotid Trunk

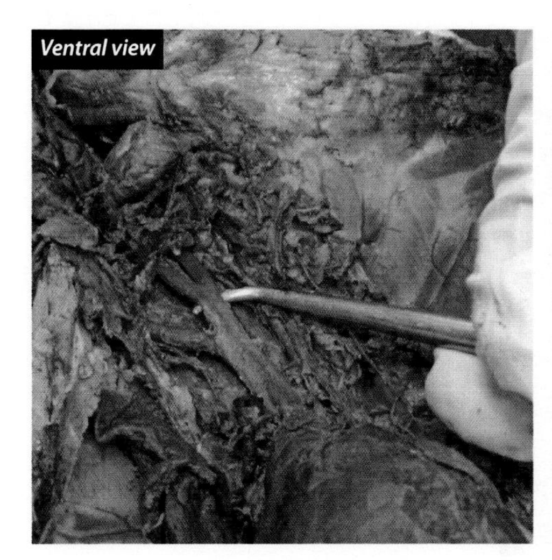


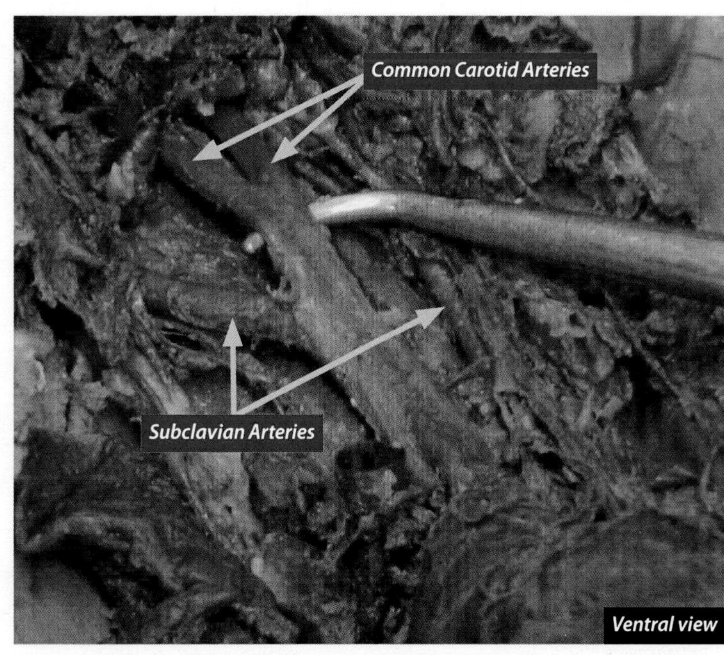


The **bicarotid trunk** is not found in all cats. It occurs when the **brachiocephalic artery** bifurcates into a **right subclavian artery** and the **bicarotid trunk**. The **bicarotid trunk** later bifurcates to form the **left** and **right common carotid arteries**.

## Subclavian Arteries

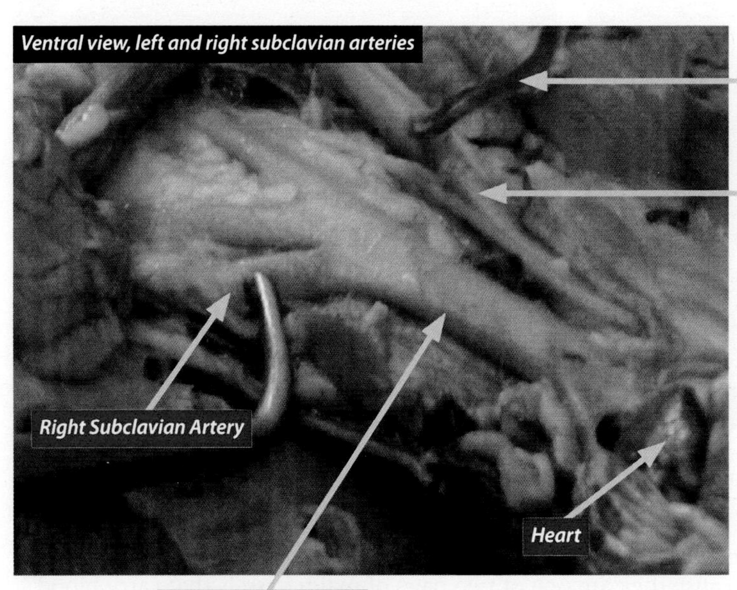


The **right subclavian artery** is a branch of the **brachiocephalic artery**. The **left subclavian artery** is the fourth branch of the **aorta**. They run through the thoracic outlet to become the **axillary arteries**. They have a number of branches that this lab will cover.

## Internal Thoracic (Mammary) Artery

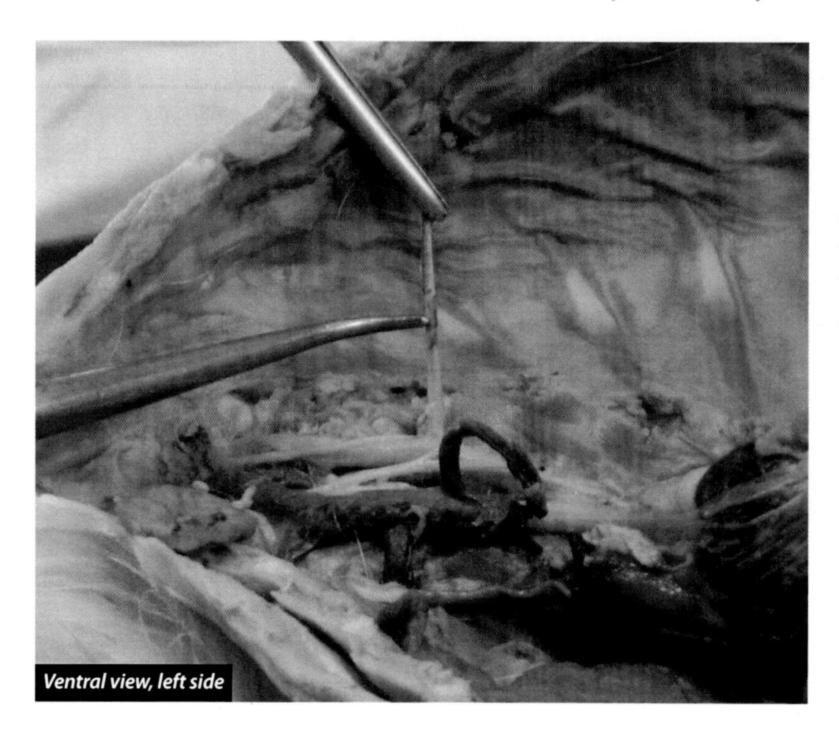


The **internal thoracic (mammary) artery** is usually the first branch of the **subclavian artery** on both sides. It serves the ventral (anterior) thoracic wall. There is one on the left and one on the right. They normally run along both sides of the sternum.

## Vertebral Artery

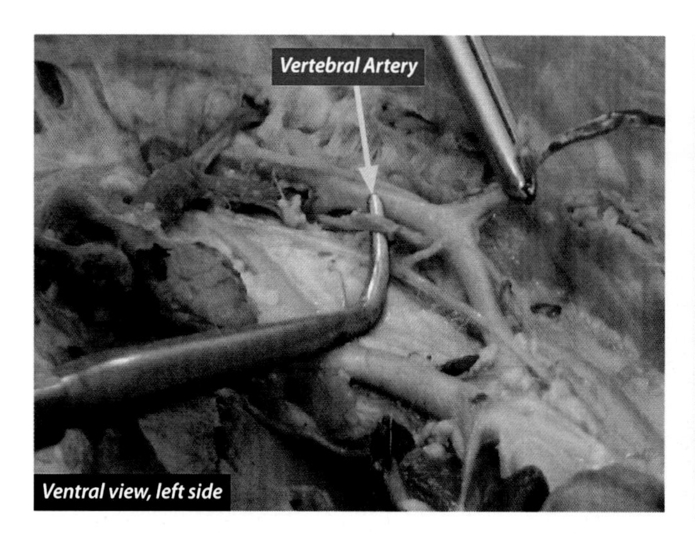


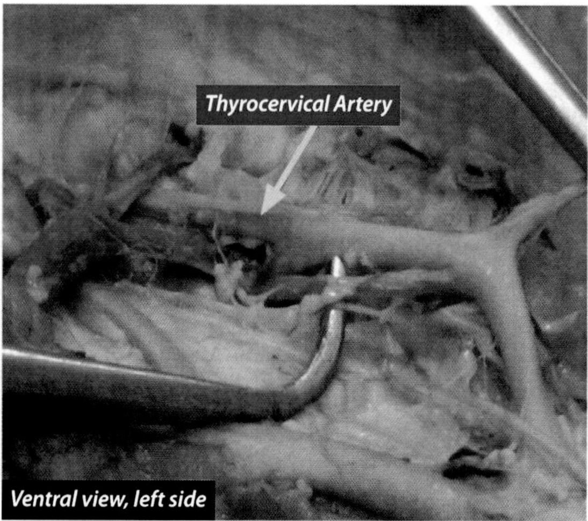


The **vertebral artery** is usually the second branch of the **subclavian artery** on both sides. It usually passes toward the head while the **costocervical artery** passes more medially. The **vertebral arteries** run through the transverse foramina of the cervical vertebrae. They enter the cranium by passing through the foramen magnum. Functionally they are important because they are one of two pairs of major vessels that carry blood to the brain on each side. The **vertebral artery** is the "V" of the mnemonic VCT.

## Costocervical Artery

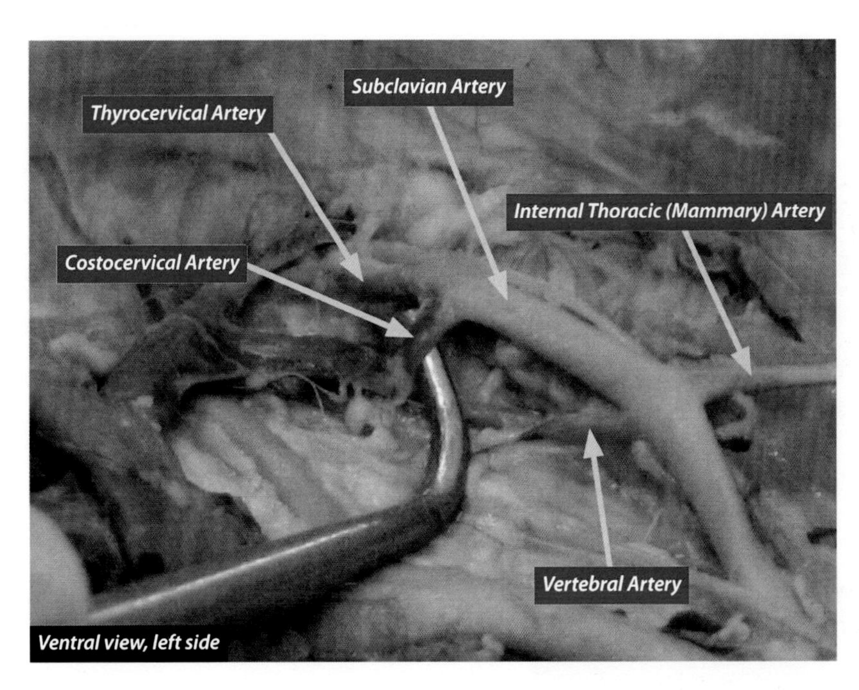


*Ventral view, left side*

The **costocervical artery** is usually the third branch off of the **subclavian artery** on both sides. It usually passes medially while the **vertebral artery** passes toward the head. As the name implies, it serves the ribs and cervical region, as well as the back. It is the "C" of the "VCT" mnemonic.

## Thyrocervical Artery

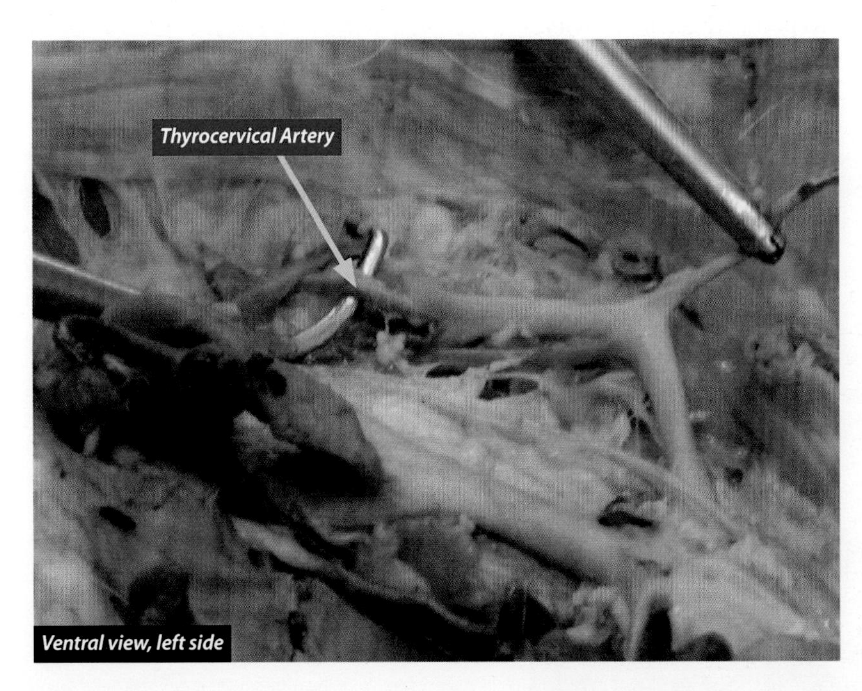


*Ventral view, left side*

The **left** and **right thyrocervical arteries** are usually the fourth branches of the left and right **subclavian arteries**, respectively. One of the branches of the subclavian artery serves the thyroid gland. A second branch enters the shoulder region and becomes the **transverse scapular artery**. We will find the **transverse scapular artery** deep to the scapula, and again when it passes through the suprascapular notch and gives rise to the **suprascapular artery**. This is the "T" of the VCT mnemonic.

# VEINS

## Internal and External Jugular Veins

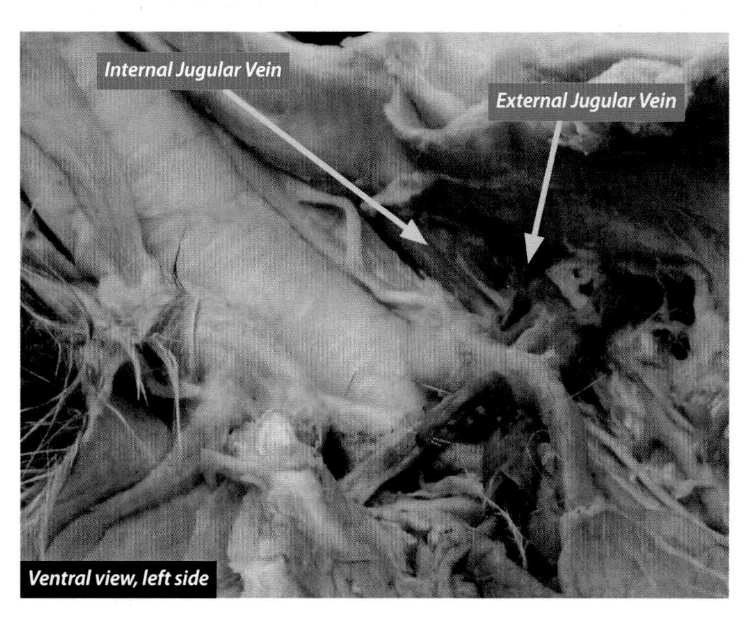


### *Internal Jugular Vein*

Although the **internal jugular vein** is found in virtually all cats, it rarely takes the dye and is relatively small. Because of this it is difficult to find. Typically, it is found in the **carotid sheath**, which includes the **common carotid artery** and the **vagosympathetic trunk**. It usually joins the **external jugular vein** before the junction with the **subclavian vein**.

### *External Jugular Vein*

This lab observes the **external jugular vein** as it joins the **subclavian** near the cranial (superior) end of the thoracic cavity. This union forms the **brachiocephalic vein** on each side. The **external jugular vein** receives blood from the head region except for the cranial cavity.

## Subclavian Vein

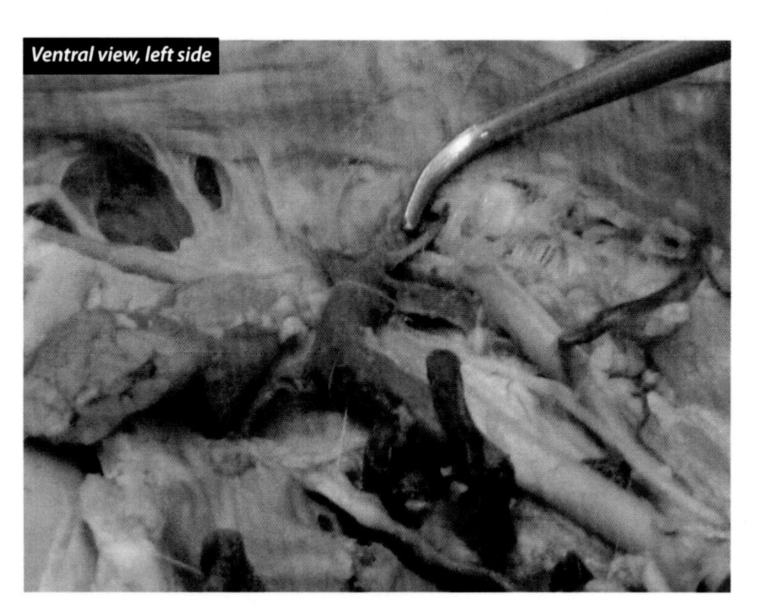


The two **subclavian veins** join to help form the two **brachiocephalic veins** when they join the **external jugular veins**. They drain blood from the upper limb back toward the heart. The **subclavian veins** begin when the **axillary vein** enters the thoracic cavity through the thoracic outlet.

## Vertebrocostocervical Trunk

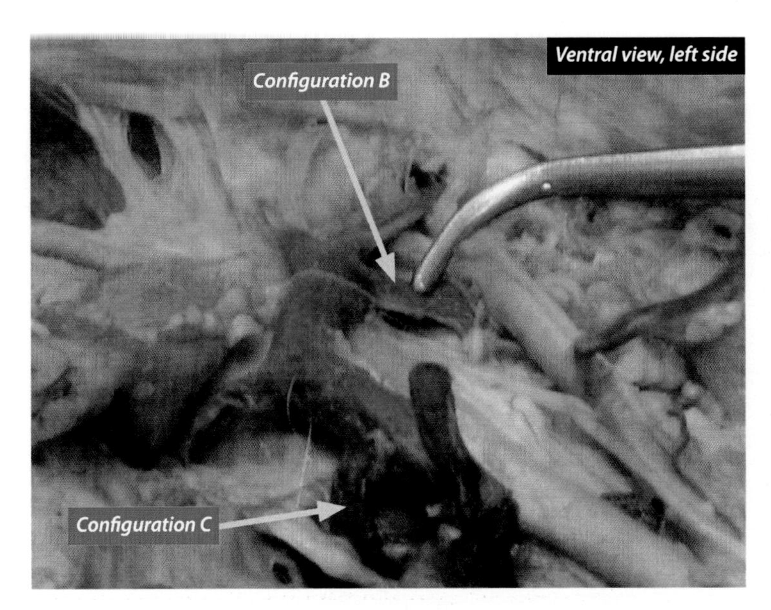


This **vein** is formed when the **vertebral** and **costocervical veins** join. In nearly all cats, it joins the **brachiocephalic vein** on the left side. In the majority of cats, it joins the **brachiocephalic vein** on the right, but in many cats it goes directly into the **cranial vena cava** between the **brachiocephalic** and **azygos veins**. Both "B" and "C" configurations are shown to the left.

A: This configuration occurs a little more than half the time—the **vertebrocostocervical trunk** enters the **right brachiocephalic vein**. This configuration is not in any of the pictures.
B: It is on the left side of the **vertebrocostocervical trunk** and nearly always enters the **left brachiocephalic vein**.
C: This configuration occurs a little less than half the time—the **vertebrocostocervical trunk** enters the **cranial vena cava** directly.

## Brachiocephalic Veins

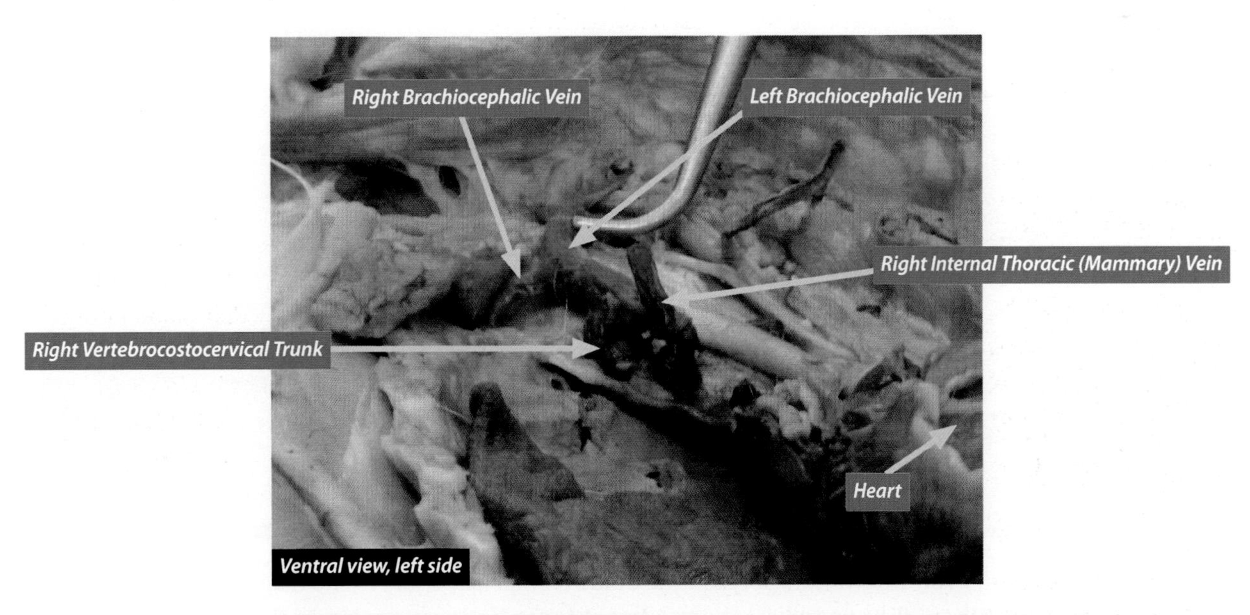


There are two **brachiocephalic veins**. They get their name because they serve the arm and head. They are formed by the union of the **external jugular vein** and the **subclavian vein** on each side. This vessel also receives blood from the **vertebrocostocervical trunk** in more than 50 percent of the cats in lab.

## Internal Thoracic (Mammary) Vein

Ventral view

Normally, there is only one **internal thoracic (mammary) vein**. It drains the blood from the ventral (anterior) thoracic wall into the **cranial (superior) vena cava**.

## Cranial Vena Cava

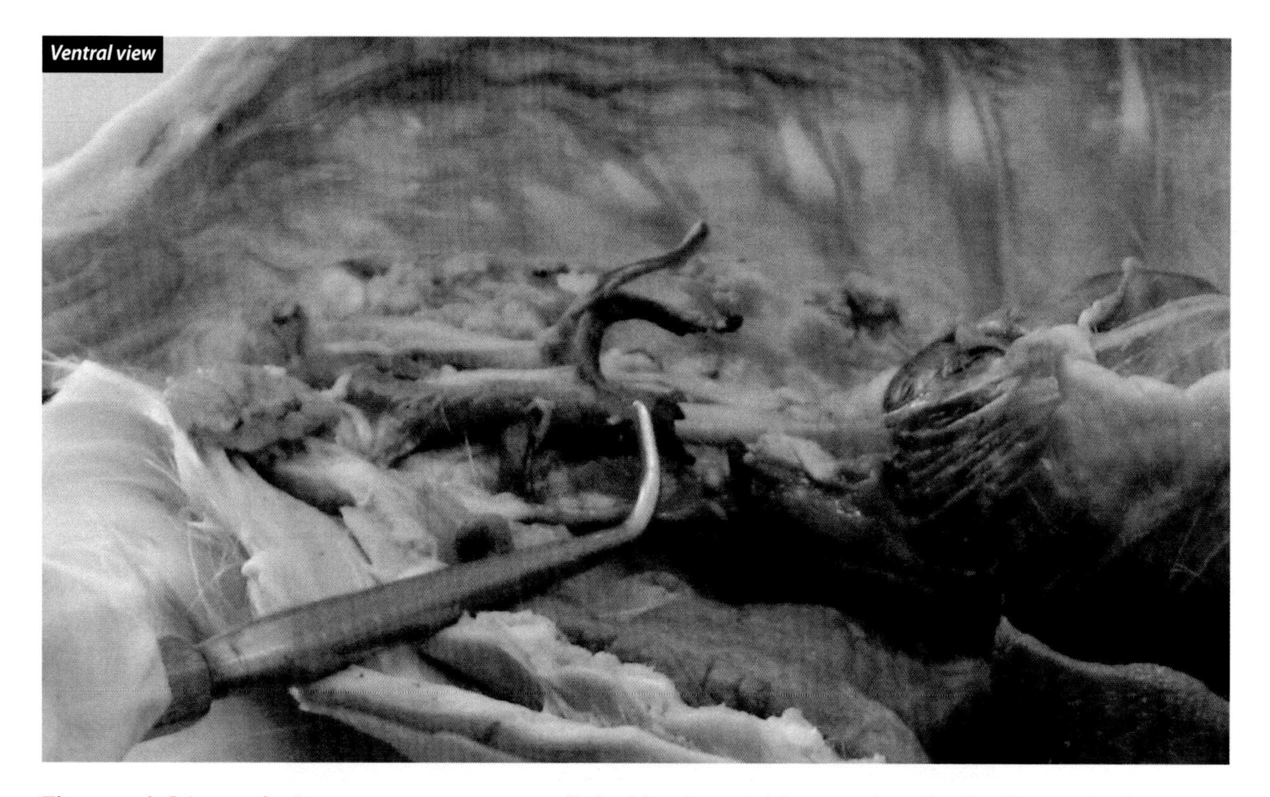
Ventral view

The **cranial (superior) vena cava** transports all the blood cranial (superior) to the diaphragm back to the right atrium of the heart. It begins where the two **brachiocephalic veins** join in the cranial (superior) thoracic region.

## Caudal Vena Cava

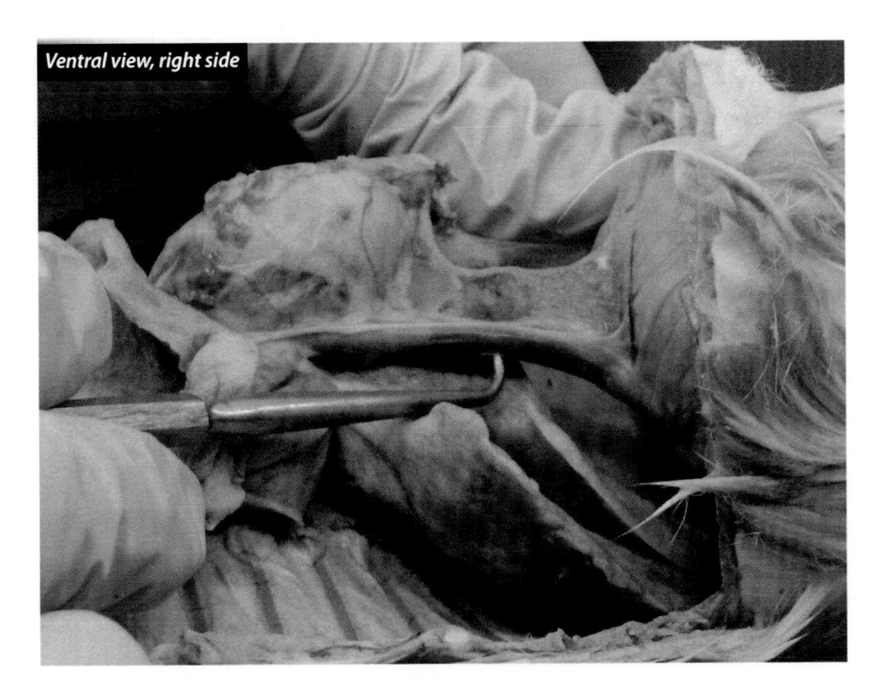
Ventral view, right side

Note that the **caudal vena cava** is medial to the right lung. Therefore, it should not be confused with the **azygos vein** that is found dorsal to the right lung. The **caudal vena cava** transports all the blood from the caudal (inferior) to the diaphragm back to the right atrium of the heart. It begins where the two **common iliac veins** join in the caudal (inferior) abdominal region. In the thoracic cavity, the **right phrenic nerve** runs to the diaphragm along the **caudal vena cava**.

## Azygos Vein

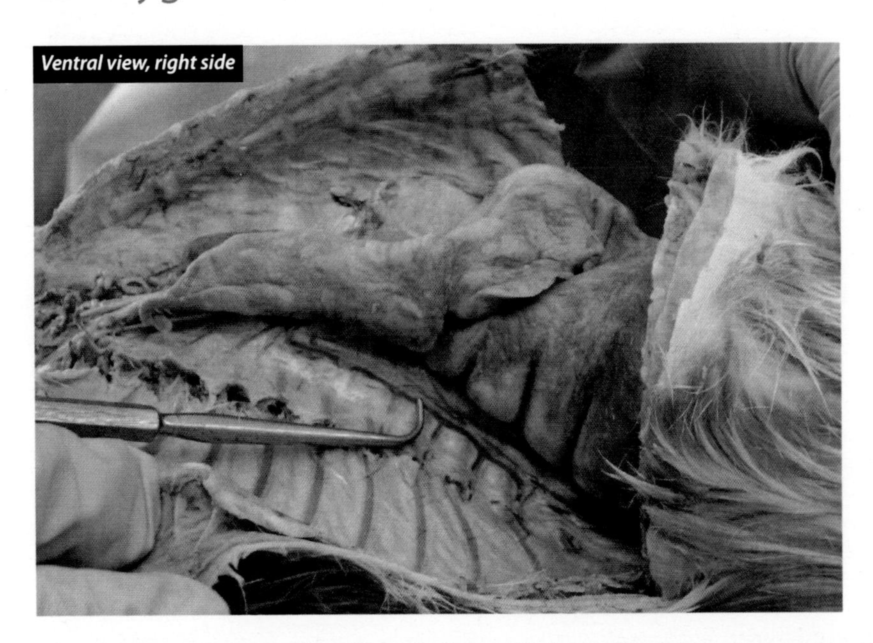
Ventral view, right side

Note that the **azygos vein** is dorsal to the right lung. Therefore, it should not be confused with the **caudal vena cava** that is found medial to the right lung. The **azygos vein** drains the blood from the dorsal (posterior) thoracic wall into the **cranial (superior) vena cava**. It also can serve as an alternate path for the return of blood from the caudal (inferior) to the diaphragm if the **caudal (inferior) vena cava** were to be blocked.

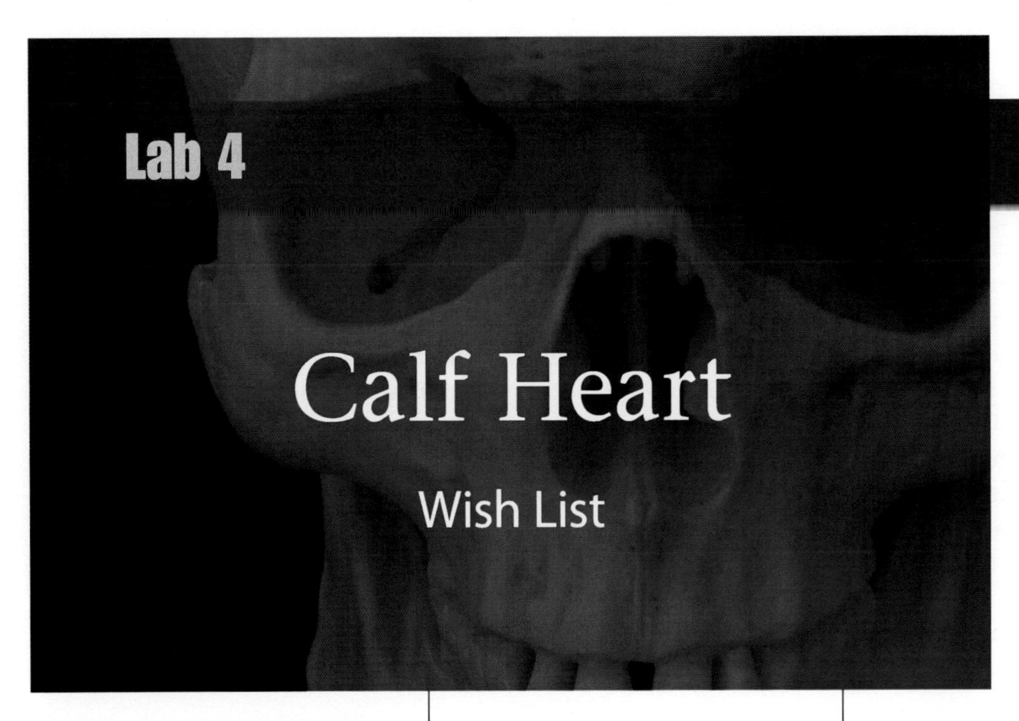

# Lab 4

# Calf Heart

## Wish List


**LAB 4 OVERVIEW, pp. 90–91**

**DIAGRAM, p. 92**
- Blood Circulation Diagram

**SUPERFICIAL STRUCTURES, pp. 93–102**
- Atria (left and right), p. 93
- Ventricles (left and right), p. 94
- Coronary Sulcus, p. 95
- Interventricular Sulcus, p. 95
- Cranial Vena Cava, p. 96
- Caudal Vena Cava, p. 96
- Pulmonary Trunk, p. 97
- Pulmonary Arteries (left and right), pp. 97–98
- Pulmonary Veins (2 left and 2 right), p. 98
- Bronchus, p. 99
- Aorta, p. 100
- Brachiocephalic Artery, p. 101
- Ligamentum Arteriosum, p. 101
- Coronary Arteries and Veins, p. 102
- Coronary Sinus, p. 102

**DEEP STRUCTURES pp. 103–107**

**Atria, p. 103**
- Pectinate Muscles, p. 103
- Fossa Ovalis, p. 103

**Ventricles, pp. 104–107**
- Tricuspid Valve, p. 104
- Chordae Tendineae, p. 104
- Papillary Muscle, p. 105
- Trabeculae Carneae, p. 105
- Moderator Band, p. 105
- Interventricular Septum, p. 106
- Semilunar Valves, pp. 106–107
  - Aortic, p. 106
  - Pulmonary, p. 107
- Bicuspid (Mitral) Valve, p. 107

# LAB 4 OVERVIEW

We will study the heart and the associated structures in Lab 4.

The hearts will most likely be pre-dissected for you. But if they are not, you can prepare the heart by following these directions. If the **pericardium** is still surrounding the heart, cut a slit up the ventral side. You should remember from Lab 3 that the pericardium is a special type of **pleura**. You can determine where ventral is by looking at the great vessels associated with the heart. On the dorsal side, you should be able to locate the **caudal vena cava** entering the **right atrium**. It will have very thin walls. The **aorta** will be leaving the heart at the cranial end. It is easily recognized because of its thick walls. Those two vessels will help with a cranial/caudal and ventral/dorsal orientation.

When the pericardium is pulled back, use a large scalpel to make a coronal cut. Begin at the apex of the heart and slowly work cranially, cutting through the **left** and **right ventricles**. Be sure to make the cut so that both the **right** and **left atria** are cut open near the middle of each chamber. This will allow you to observe structures inside the atria.

## 1. Superficial Structures

You should use superficial structures to orient the heart. On the ventral side, you will find the **interventricular sulcus** running obliquely across the surface, while on the dorsal surface it will run in more of a cranial caudal orientation. The interventricular sulcus is a groove between the left and right ventricles. You should see some of the coronary vessels running in this groove. Also on the ventral surface, you will find the **pulmonary trunk** to be cranial to the ventricles. It passes obliquely toward the left side from the right ventricle. Dorsal to the pulmonary trunk you will find the **aorta**. It is recognizable because of the very thick walls it has. At the base of the aorta, where it leaves the heart, you may find an opening with thick walls. This will be the **brachiocephalic artery** (there is only one). Occasionally there will be some of this vessel remaining. If you carefully remove the connective tissue between the pulmonary trunk and the aorta, you will discover the **ligamentum arteriosum**. We will discuss the developmental history of this structure in lecture.

Rotate the heart so that you are looking at the dorsal surface. You may see tubular structures with a gray colored, hard material in the walls. These will be part of the **bronchial tree** and the hard material is hyaline cartilage that makes up the cartilaginous rings. The **caudal vena cava** was mentioned earlier. You can also see the remains of the **cranial vena cava**. Place the blunt end of the probe in this structure and it will lead into the right atrium. You will observe the **coronary sulcus**. It is a groove, similar to the indentation where an adult male wears his belt. Carefully and gently insert the blunt end of the probe caudal to the caudal vena cava where it enters the right atrium. You should find an opening that leads to the coronary sulcus. Covering the coronary sulcus, much like a Quonset hut, is the **coronary sinus**. It directs blood from the wall of the heart back into the right atrium.

When you find an opening on the dorsal side, gently insert the blunt end of the probe and find out where the opening leads to. This will allow you to identify the structure. It will be a **pulmonary vein** (it leads to the left atrium), a **pulmonary artery** (it leads to the **pulmonary trunk**), a **bronchus** (it will have cartilaginous rings), one of the **vena cavae** (they lead to the **right atrium**), or the **aorta** (it comes from the **left ventricle**).

## 2. Deep Structures—Right Side

Begin your investigation of the deep structures of the heart in the right atrium. Identify the three openings that direct blood into the right atrium. Observe the **pectinate muscles** on the walls of the atrium. They are also found in the left atrium. There is an oval depression in the **interatrial septum** and that is the **fossa ovalis**. We will discuss the origin of this feature during lecture.

Blood normally leaves the right atrium by way of the **right atrioventricular orifice**, and as it enters the right ventricle it passes over the **tricuspid valve** (three cusps). The cusps remind Dr. J of a trampoline. Attached to the edges of the valve are the **chordae tendineae**. The other end of the chordae tendineae attaches to the **papillary muscles** (they look like pimples). When they are stretched during the closing of the tricuspid valve, they contract and pull on the valve, thereby preventing prolapse of the valve. Both the chordae tendineae and the papillary muscles exist in the left ventricle as well.

Within the right ventricle, in several areas, you will see the **trabeculae carneae**. They too look like tree roots, but their name translates into meaty beams. They have a similar benefit to the papillary muscles in that they pull on the chordae tendineae when the valve is closing. These same structures exist in the left ventricle too. Also in the right ventricle, on the ventral side, you will observe the **moderator band**. It extends from a papillary muscle on the ventricular wall to the **interventricular septum**. If you cut through the ventricular wall at the beginning of the pulmonary trunk, you will be able to observe the **pulmonary semilunar valves**. Please do not make this cut unless directed to do so by your instructor. Normally several examples of hearts with this cut are already prepared for you in the container that holds the hearts.

Observe the **pulmonary trunk**. If the heart has been prepared for you, there will be a cut that runs through its wall on the ventral side. If it has not been prepared, you can make a cut that is about 2 inches in length along its length. Carefully insert the blunt end of a probe into that cut and carefully push the probe away from the ventricle. The end of the probe should appear as it comes out of one of the **pulmonary arteries**. There is a right and a left pulmonary artery. The right pulmonary artery makes a U turn and passes to the right lung. The left pulmonary artery courses to the left lung.

## 3. Deep Structures—Left Side

Oxygen rich blood returns to the left atrium by way of four **pulmonary veins** (two from each lung). You will probably not find all four of these veins. The best way to find one, however, is to gently insert the blunt end of a probe into the left atrium and press against the dorsal wall of the left atrium. The left atrium has all the structures that the right atrium has. You should observe the pectinate muscles and palpate the fossa ovalis. Blood will pass from the left atrium to the left ventricle by flowing through the **left atrioventricular orifice**. In doing so, it moves over the **bicuspid (mitral) valve**. The bicuspid valve has chordae tendineae, which are attached to papillary muscles as they were in the right ventricle. There are also trabeculae carneae, as there were in the right ventricle, and the functionality of these structures is the same. You should remember that the left ventricular wall is significantly thicker than that of the right ventricle. The two ventricles pump the same amount of blood. We will discuss this difference and the functional significance in class.

Blood passes from the left ventricle over the **aortic semilunar valve** as it enters the aorta. You can observe the two **coronary arteries** immediately cranial to the aortic semilunar valve. If you investigate the coronary sulcus, you should be able to find two openings that lead to a coronary artery and a **coronary vein**. You will be able to distinguish one from the other based on the thickness of their walls. The coronary artery will have thicker walls than the coronary vein.

# BLOOD CIRCULATION DIAGRAM

**Label the human heart for practice!**

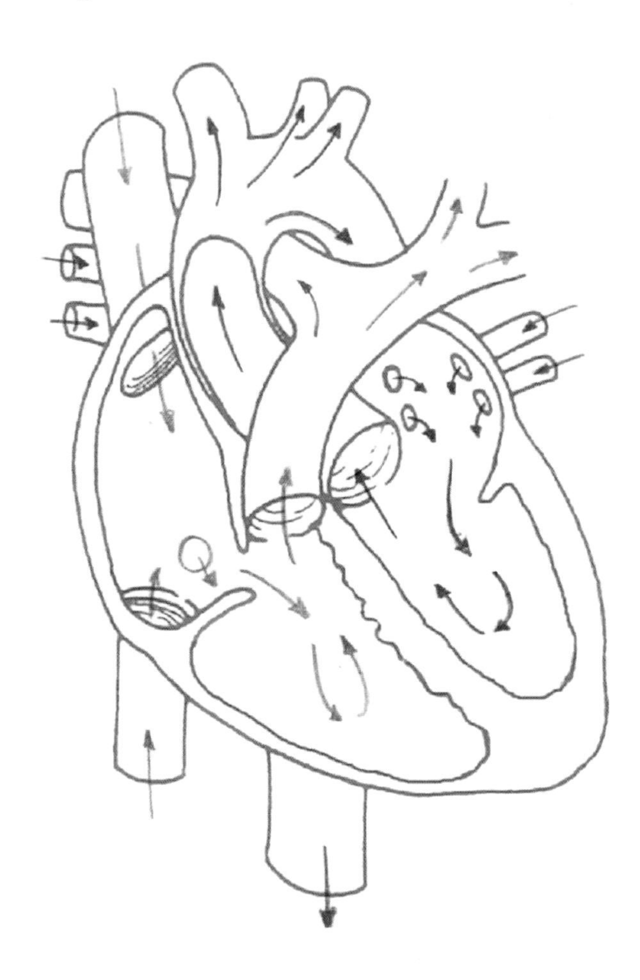

***PULMONARY CIRCULATION***

***Deoxygenated blood pumped to the lungs.***

**Cranial Vena Cava**
**Caudal Vena Cava**
**Coronary Sinus** (receives blood from 2 coronary veins)
**Right Atrium**
**Tricuspid Valve**
**Right Ventricle**
**Pulmonary Semilunar Valve**
**Pulmonary Trunk**
**2 Pulmonary Arteries** (1 to each lung)

***SYSTEMIC CIRCULATION***

***Oxygenated blood pumped to the entire body.***

**4 Pulmonary Veins** (2 from each lung)
**Left Atrium**
**Bicuspid (mitral) Valve**
**Left Ventricle**
**Aortic Semilunar Valve**
**Aorta**
**Brachiocephalic Artery**
**Left Common Carotid Artery**
**Left Subclavian Artery**

# SUPERFICIAL STRUCTURES

## Atria

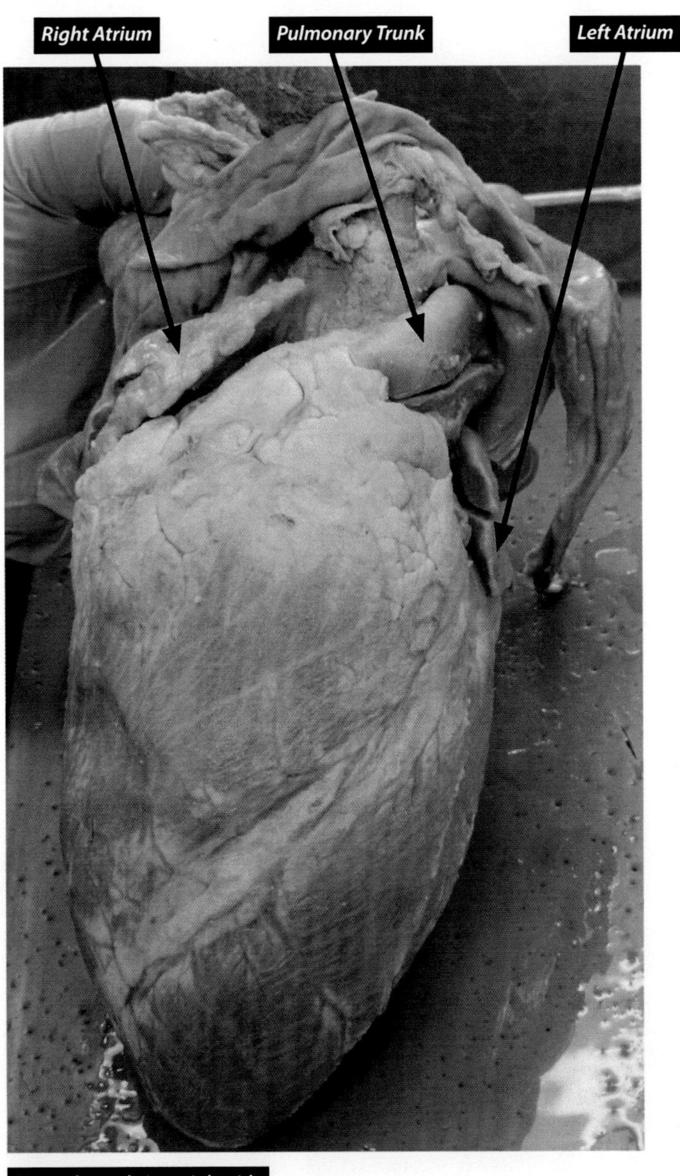


Ventrolateral view, right side

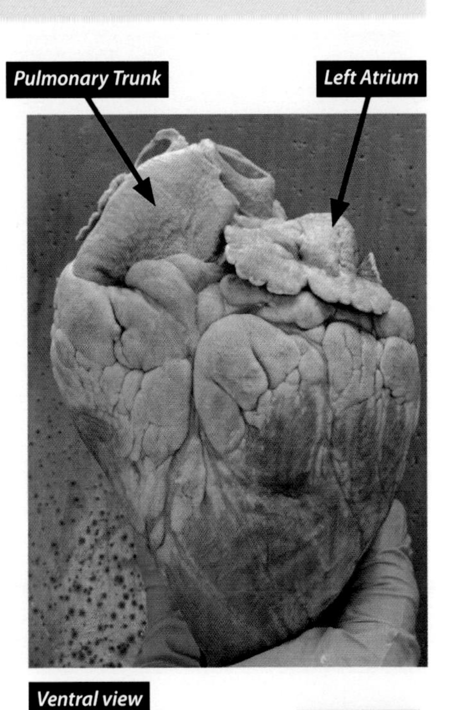


Ventral view

Dorsal view

### *Atria (right and left) (atrium = auricle)*

The atria (left and right) are often described as the receiving chambers of the heart. Blood enters the atria from the veins and the **coronary sinus**. These chambers are found at the superior (cranial) end of the heart. They are sometimes referred to as auricles. They have thin walls compared to the ventricles.

## Ventricles

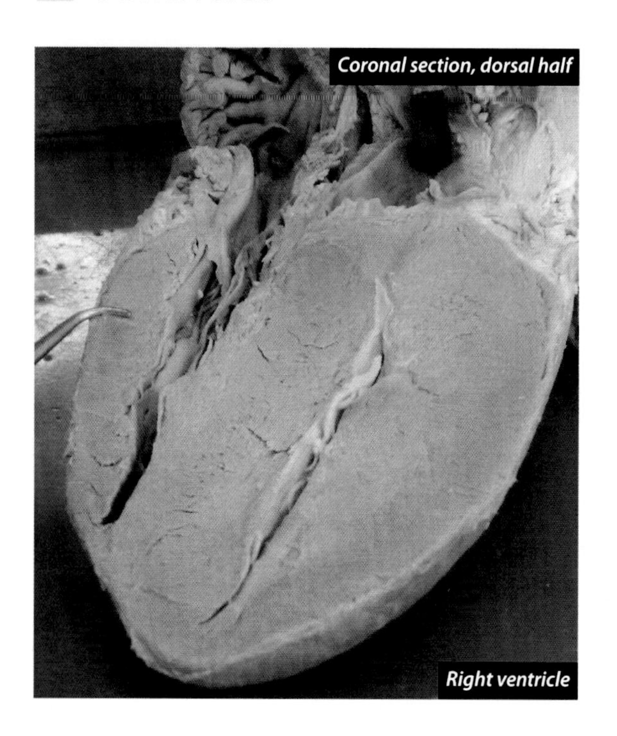
Coronal section, dorsal half
Right ventricle

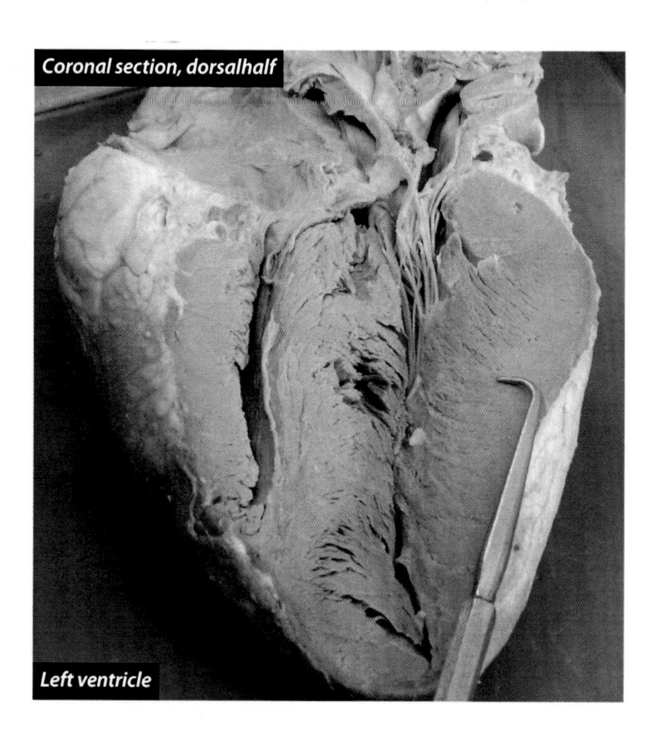
Coronal section, dorsalhalf
Left ventricle

### *Ventricles (left and right)*

The ventricles (left and right) are often described as the pumping chambers of the heart. Blood enters the ventricles from the atria. They provide most of the pressure necessary to move blood through the vessels of the body. They are found at the inferior (caudal) end of the heart. They have thick walls compared to the atria. Note that the left ventricular wall is thicker than the right ventricular wall.

Ventral view, right ventricle
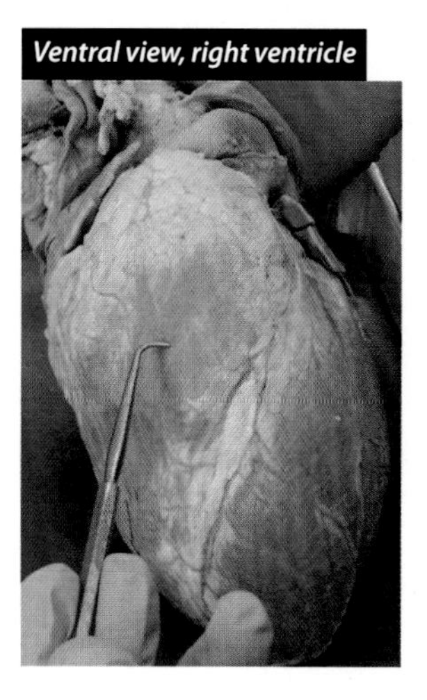

Ventral view, left ventricle
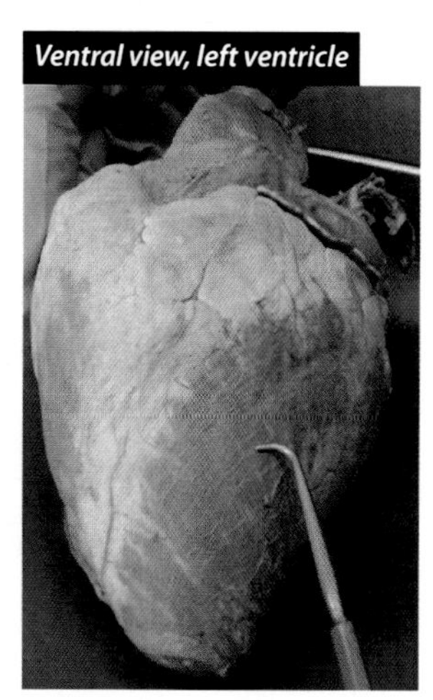

Dorsal view, right ventricle
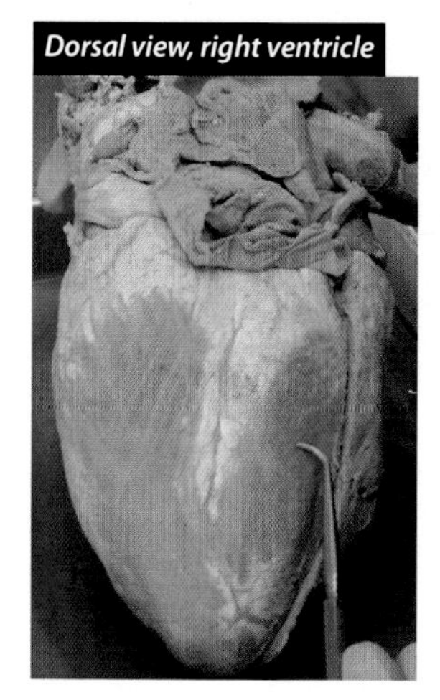

Dorsal view, left ventricle

## Coronary Sulcus

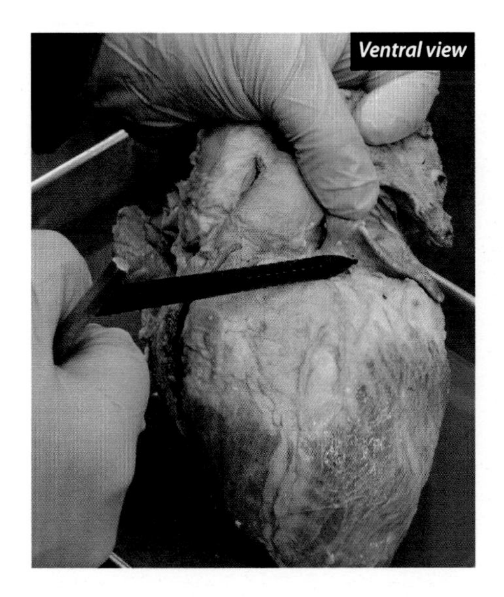


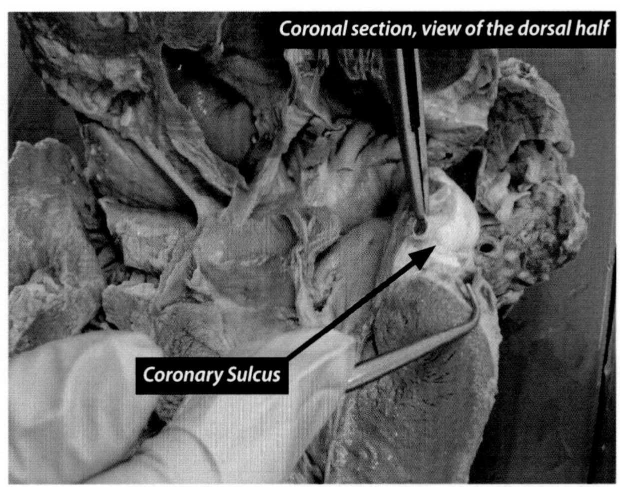


*Coronary vein and coronary artery are in the coronary sulcus.*

The coronary sulcus is a large groove that is formed between the atria and the ventricles. Anatomically it is important because this is where we find some of the **coronary arteries** and **veins** that serve the heart.

## Interventricular Sulcus

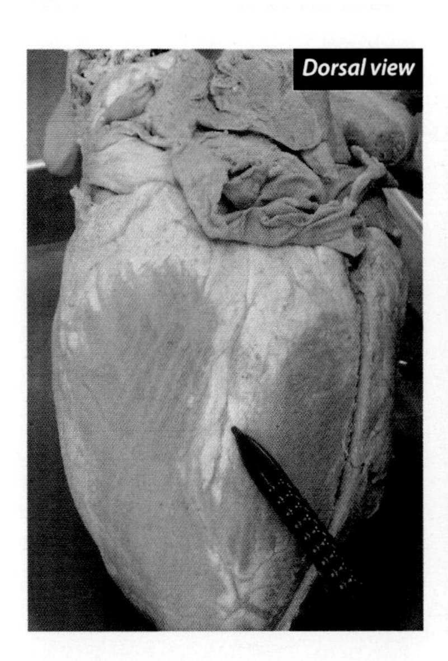


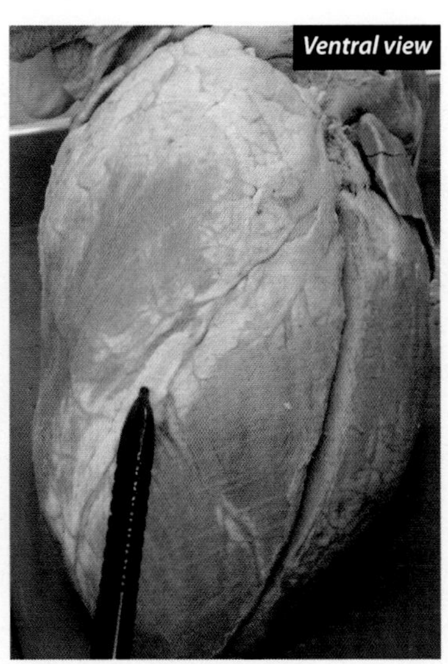


The interventricular sulcus is a large groove that is formed between the two ventricles. Anatomically it is important because this is where we find some of the **coronary arteries** and **veins** that serve the heart.

## Cranial Vena Cava

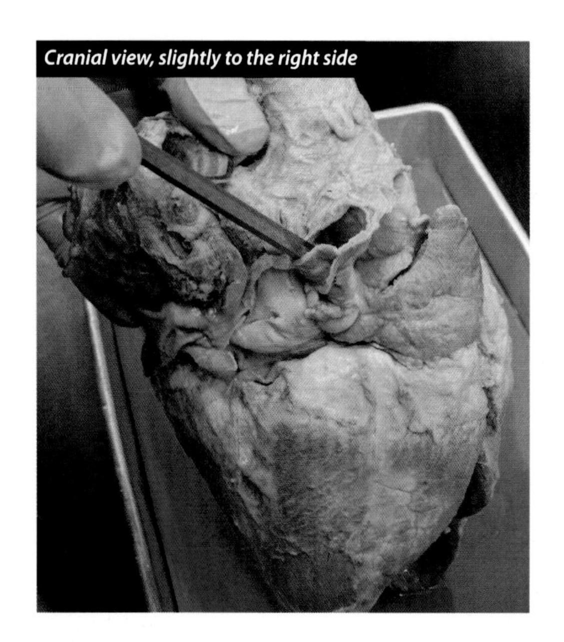
Cranial view, slightly to the right side

Cranial view, slightly to the right side

The **cranial vena cava** transports all the blood from cranial (superior) to the diaphragm back to the right atrium. It begins where the two **brachiocephalic veins** join in the cranial (superior) thoracic region.

## Caudal Vena Cava

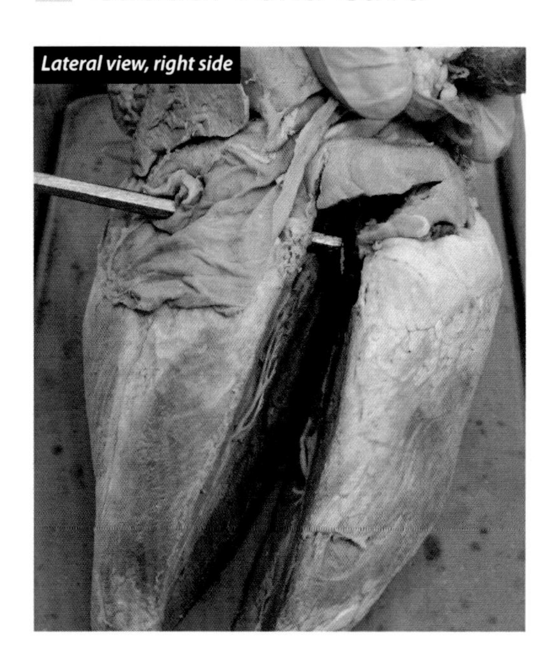
Lateral view, right side

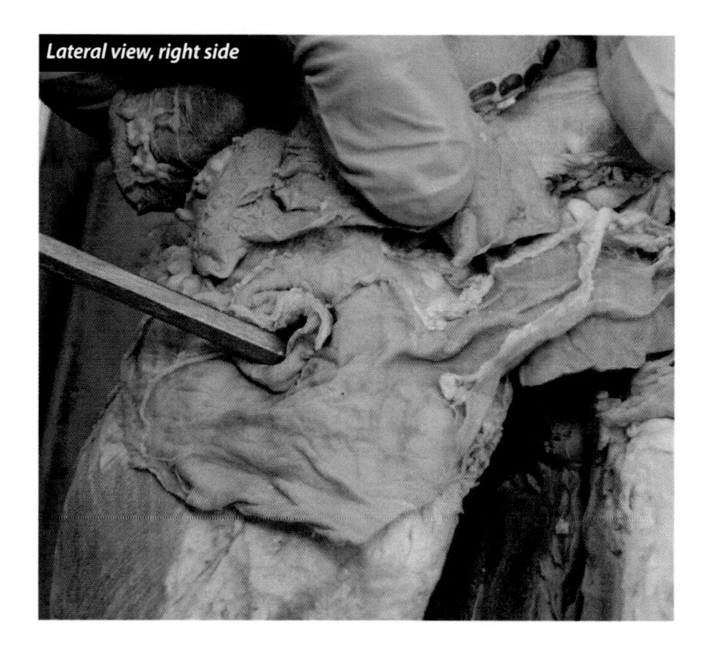
Lateral view, right side

The **caudal vena cava** transports all the blood caudal (inferior) to the diaphragm back to the heart. It begins where the two **common iliac veins** join in the caudal (inferior) abdominal region. It enters the right atrium. Note the very thin walls that the probe is holding open. These often collapse, which makes the **caudal vena cava** challenging to find.

## Pulmonary Trunk

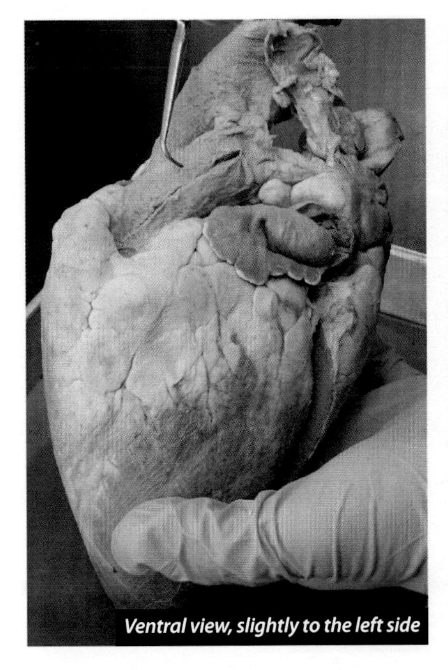
*Ventral view, slightly to the left side*

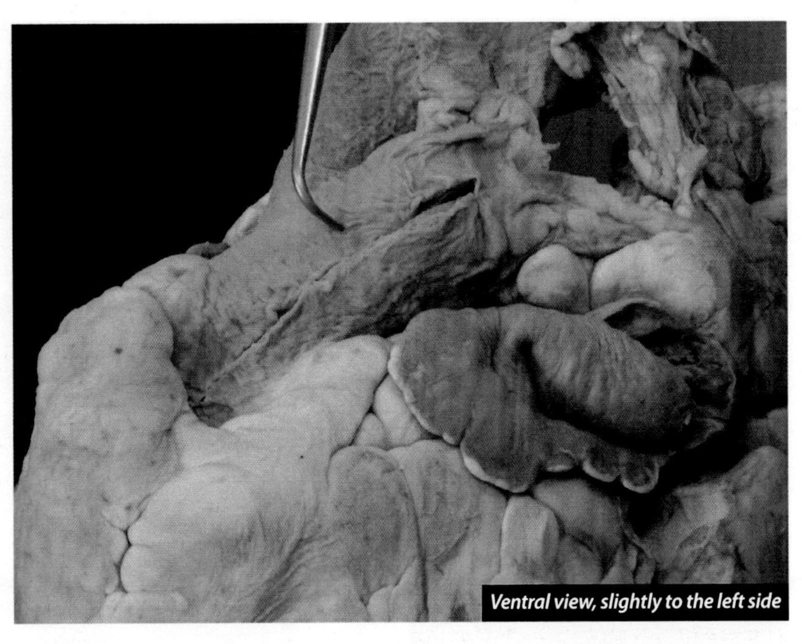
*Ventral view, slightly to the left side*

The **pulmonary trunk** is the great artery that carries blood that is oxygen deficient and carbon dioxide rich toward the lungs from the right ventricle. The **pulmonary trunk** bifurcates to form the **left** and **right pulmonary arteries**. It is typically blue in cats.

## Pulmonary Arteries

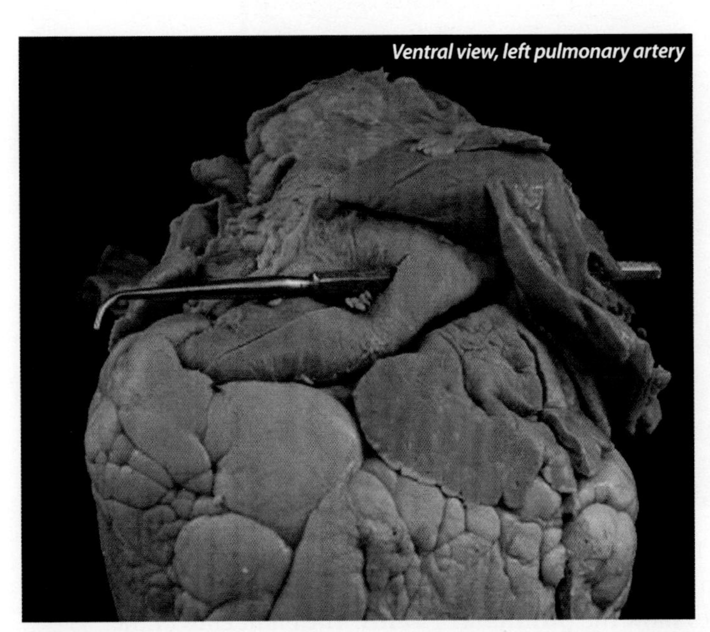
*Ventral view, left pulmonary artery*

The **pulmonary trunk** bifurcates to form the **left** and **right pulmonary arteries**. The blunt end of the probe is extending through the left **pulmonary artery**. These **arteries** are unusual in that they carry oxygen-deficient blood and deliver it to the lungs. They are typically blue in the cat. There are two **pulmonary veins** on each side. They are difficult to find in the cat because of their small size. They carry blood that is oxygen rich and carbon dioxide deficient back to the left atrium. We will identify these in the calf heart.

## Pulmonary Arteries

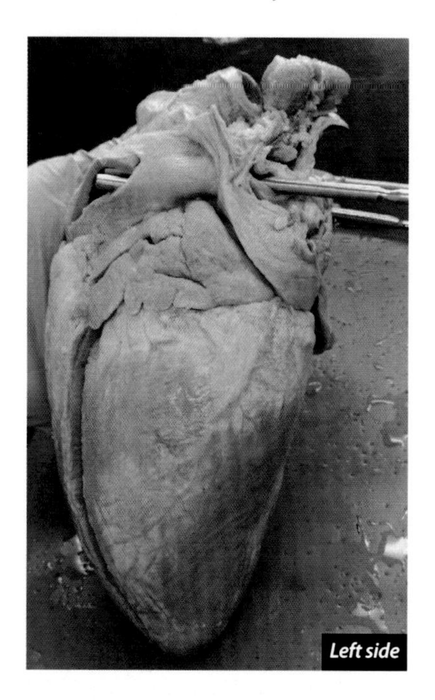


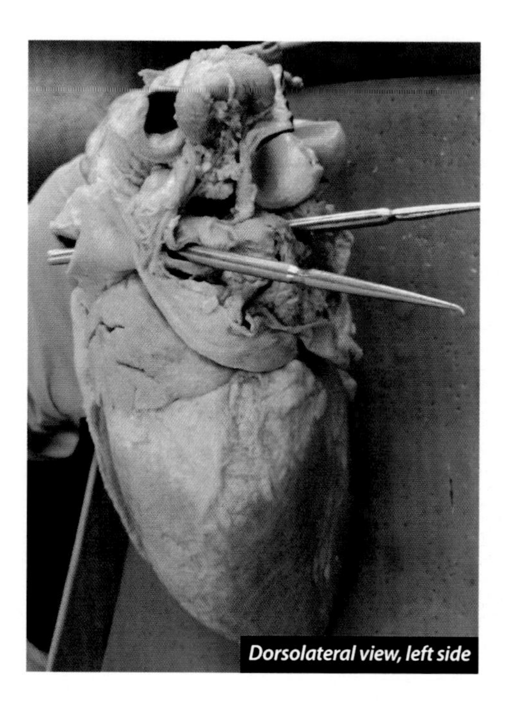


## Pulmonary Veins

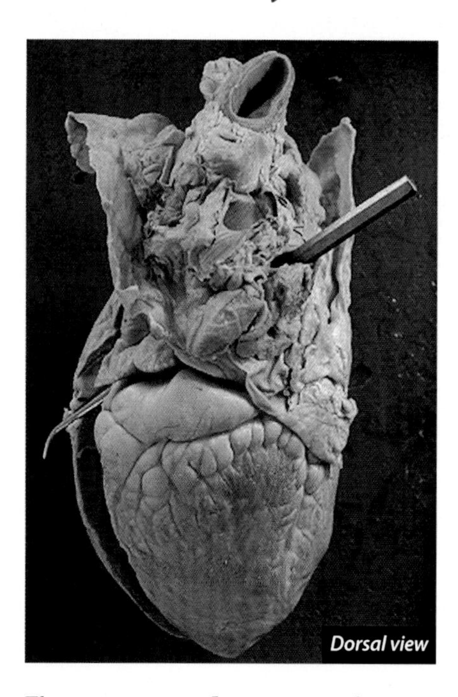


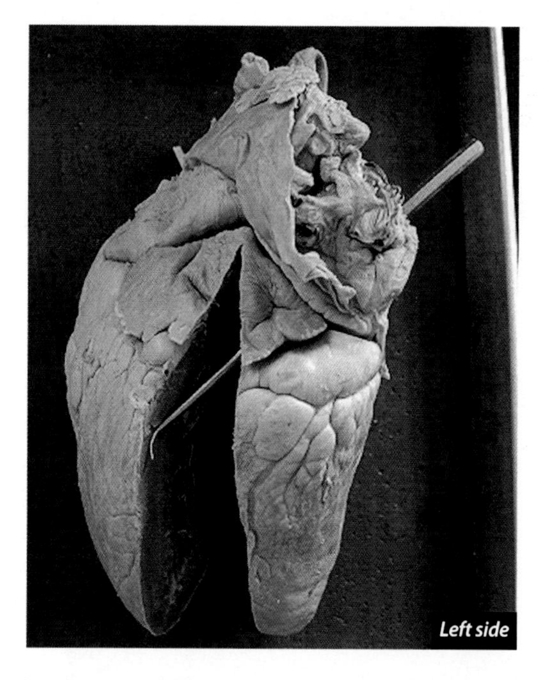


There are two **pulmonary veins** on each side. They are difficult to find in the cat because of their small size. They carry blood that is oxygen rich and carbon dioxide deficient back to the left atrium. We will identify these in the calf heart. These two pictures show the **pulmonary veins**. The blunt ends of the probes are coming out of the veins, while the sharp end of the probe is coming out of the left atrium.

## Bronchus

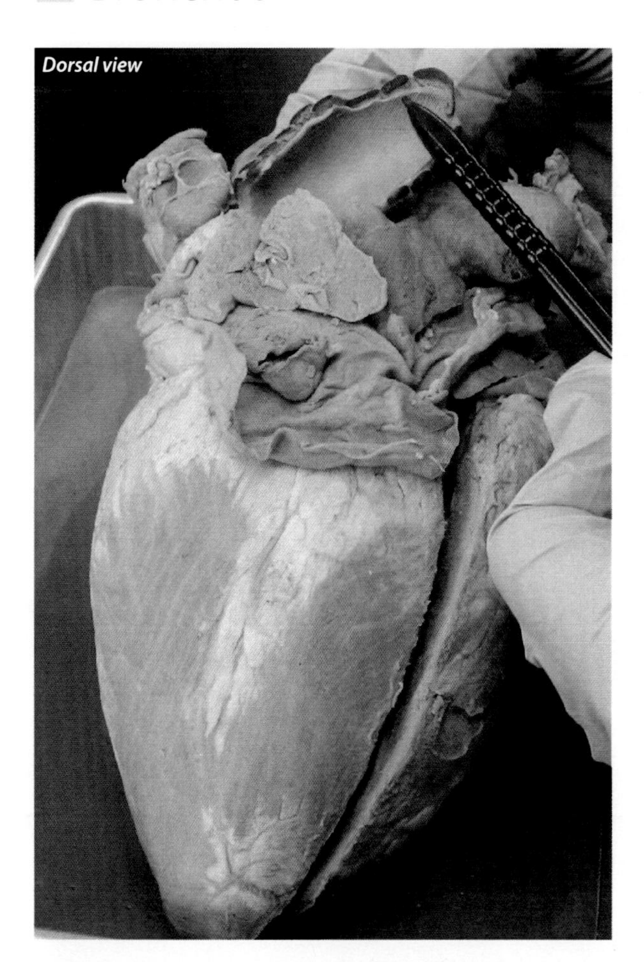

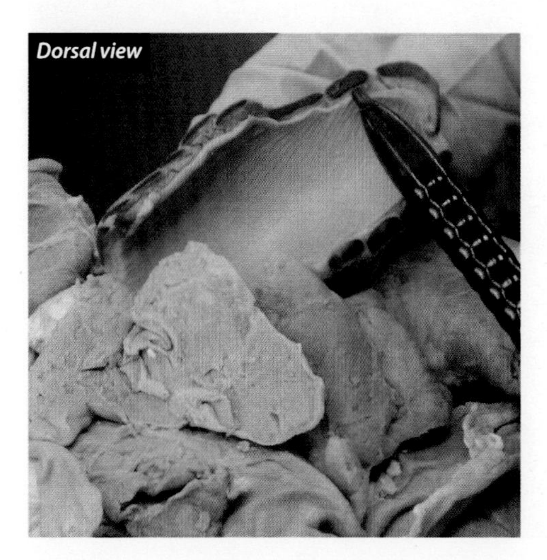


The bronchi are the large branches on both sides of the respiratory system. There are two primary bronchi that branch from the trachea. You will probably not see these. Each primary bronchus branches to secondary bronchi, one to each of the lung lobes. Thus, there are seven secondary bronchi in a cat. They are recognizable because of the cartilaginous rings in their walls that help keep them from collapsing. They, too, subdivide as they move through the lung. In both of the pictures above you can see the cartilaginous rings that were cut when the heart was removed from the calf.

## Aorta

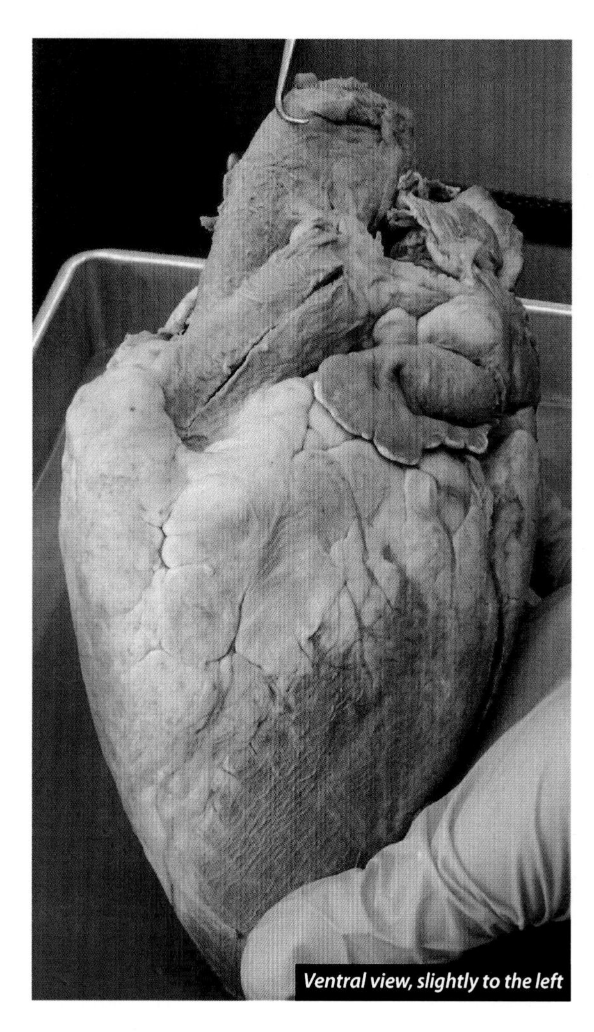

*Ventral view, slightly to the left*

The **aorta** is the great artery that carries blood away from the left ventricle to all the systemic arteries of the circulatory system. The blood in this artery is normally enriched with oxygen and is deficient in carbon dioxide. The cranial view shows that the aorta is used by restaurants as a source of calamari.

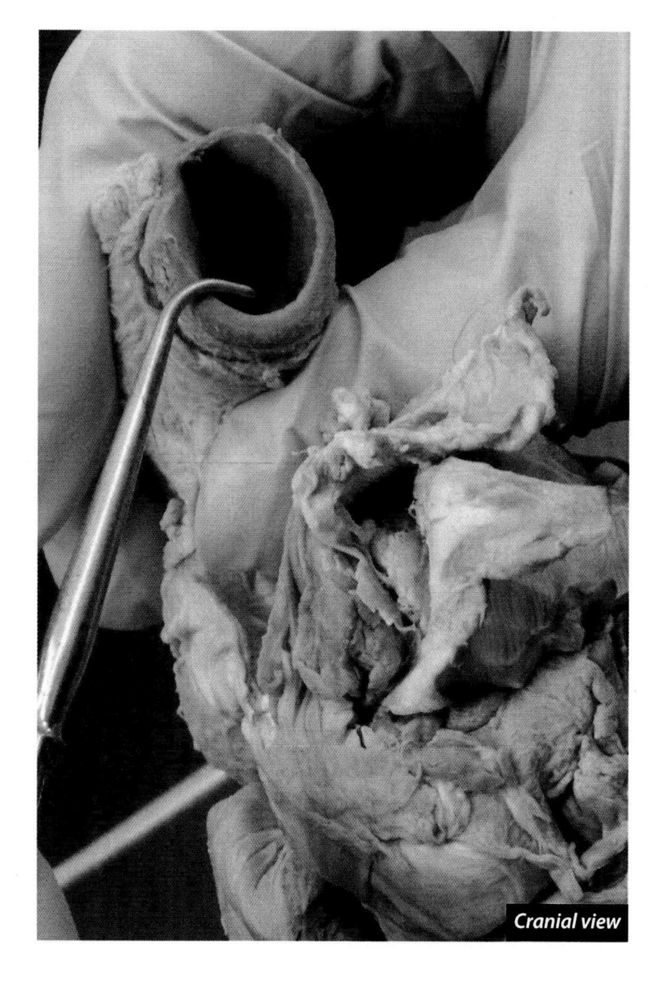

*Cranial view*

## Brachiocephalic Artery

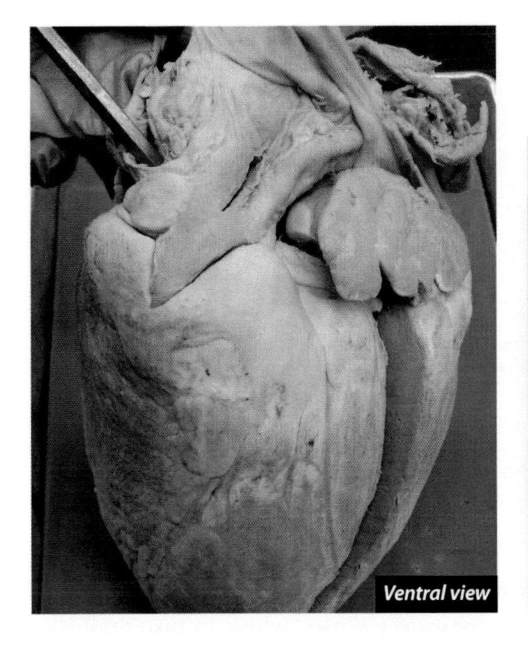


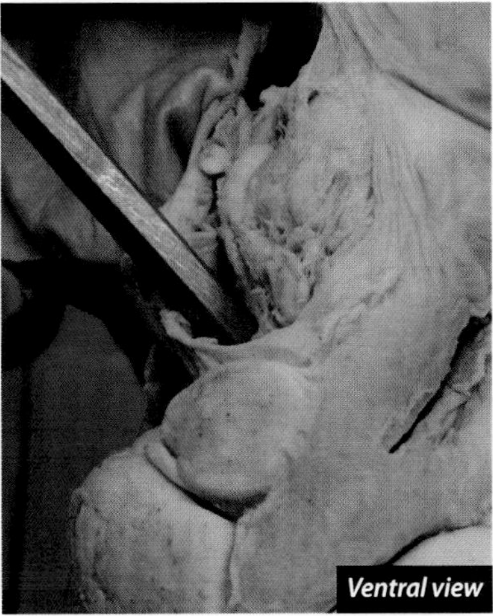


There is only one **brachiocephalic artery**. It is the third branch of the **aorta**. It gets its name because it serves the arm and head. In the calf heart, it often is found to be a hole in the wall of the **aorta** because of the way the heart is trimmed.

## Ligamentum Arteriosum

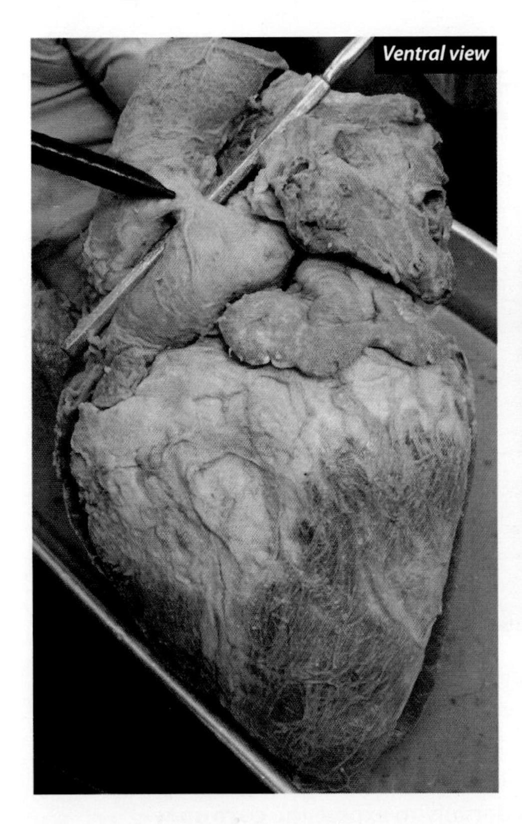


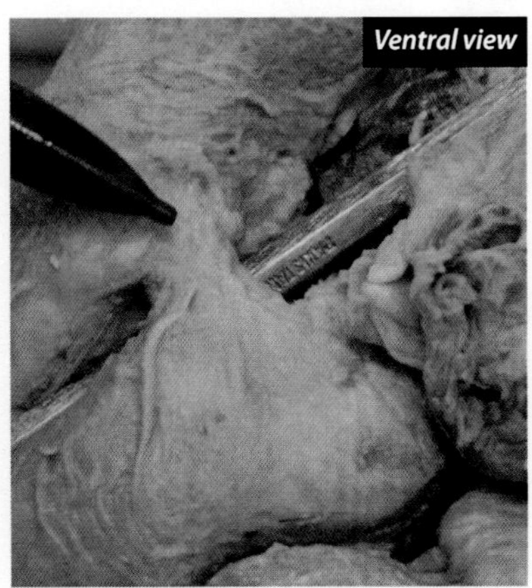


The ligamentum arteriosum is a fibrous bridge between the **pulmonary trunk** and the **aorta**. In the fetus, it was formerly the **ductus arteriosus**, which allowed **fetal blood** to flow directly from the **pulmonary trunk** into the **aorta**, thereby allowing for a bypass of **pulmonary circulation**. This is one of two pulmonary bypasses in **fetal circulation** that we study.

## Coronary Arteries and Veins

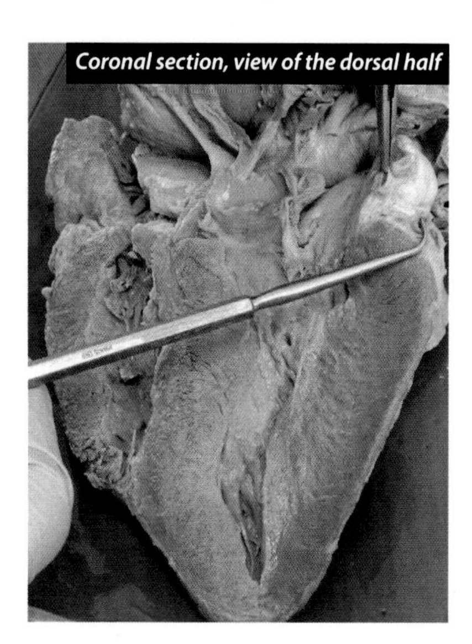


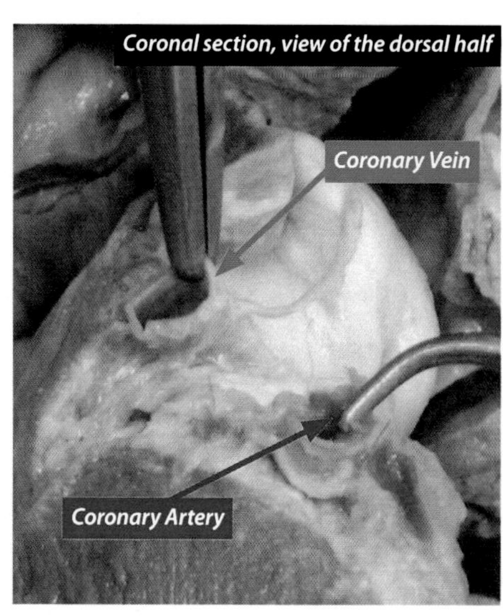


Coronary vein *and* coronary artery *in the coronary sulcus.*

The **coronary arteries** and **veins** are part of the systemic circulatory system. The **coronary arteries** are the first two branches of the **aorta**. The **coronary veins** drain into the **coronary sinus**, which in turn drains into the right atrium. Functionally these vessels are very important because they serve the walls of the heart.

## Coronary Sinus

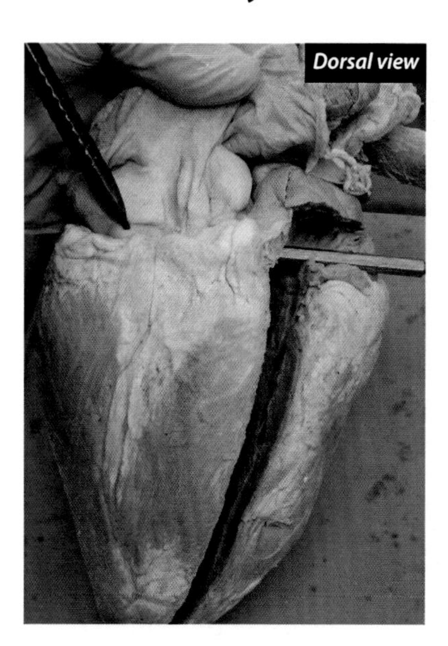


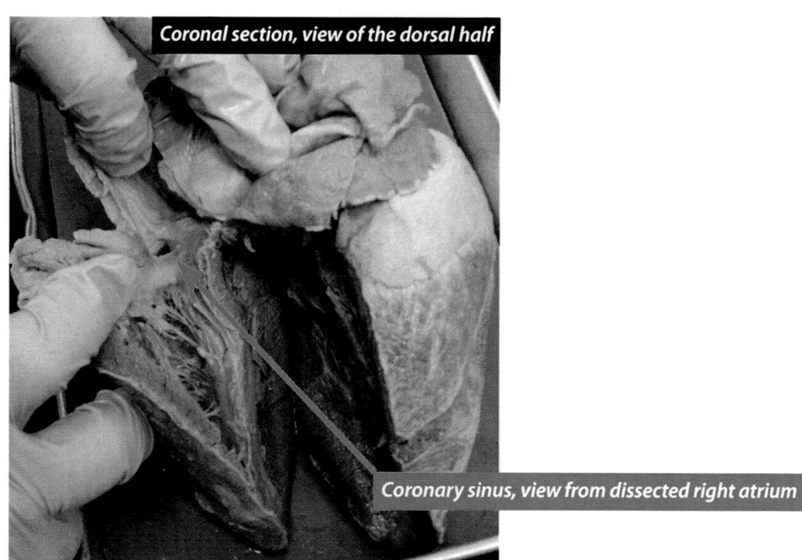


The **coronary sinus** is not a vessel because it does not have the anatomy of a vessel. The walls do not make a complete enclosure, and its floor is actually formed by the wall of the heart in the coronary sulcus. It carries deoxygenated blood from the muscle in the walls of the heart back to the right atrium. In the picture on the left, the blunt end of the probe has been placed into the **coronary sinus**. In the picture on the right, the right atrial wall has been cut and reflected dorsally to expose the **coronary sinus**. You can also clearly see the cusps of the tricuspid valve caudal to the right atrium.

# DEEP STRUCTURES

## Atria

### Pectinate Muscles

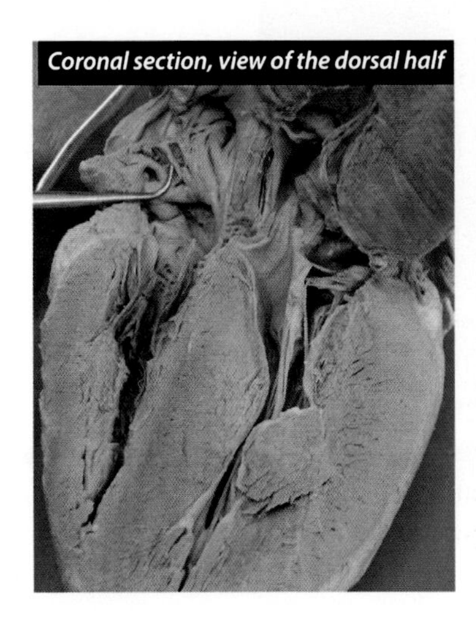

Coronal section, view of the dorsal half

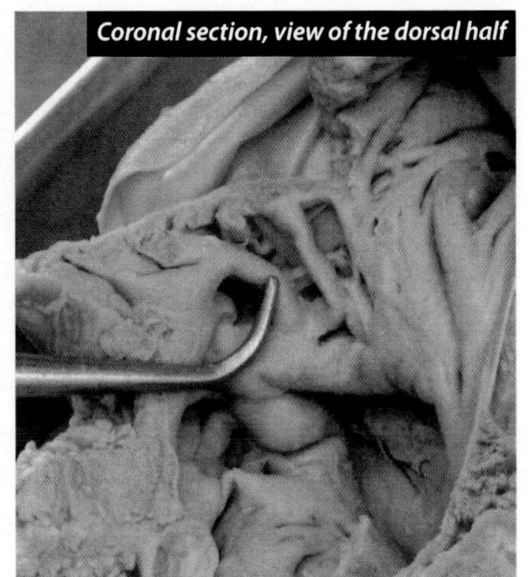

Coronal section, view of the dorsal half

The pectinate muscles are ridges on the walls of the atria. They are most easily observed in the right atrium. Dr. J likes to draw the analogy between the pectinate muscles and tree roots. Isn't he silly!

### Fossa Ovalis

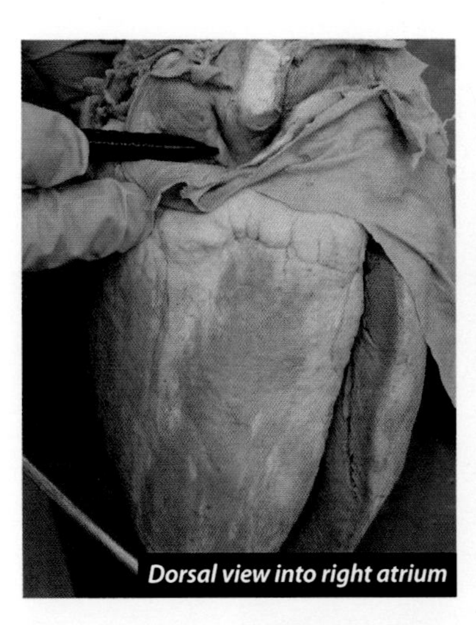

Dorsal view into right atrium

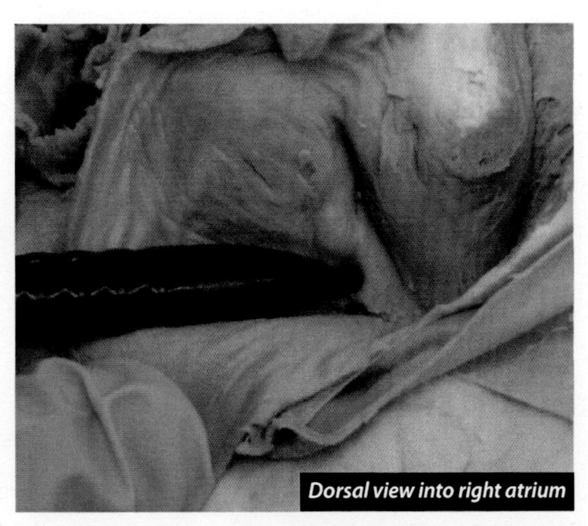

Dorsal view into right atrium

*The fossa ovalis is deep inside the atria of the heart. In these pictures the superficial wall of the right atrium has been cut away to expose the fossa ovalis.*

The fossa ovalis is part of the interatrial septum and is located close to the opening of the **coronary sinus**. It begins as a structure in the fetus called the foramen ovale, which allows fetal blood to bypass pulmonary circulation. This is one of two pulmonary bypasses in fetal circulation that we study. At birth, changes in pressure in the heart cause two small flaps to close off the foramen. Scar tissue fixes the flaps in position, and they become the fossa ovalis, which prevents the bypass from continuing. A little known fact is that it was named for Frankie O'Vale, a famous Irish rock and roll singer from the Four Seasons ("Walk Like a Man") and later a member of White Snake.

# *Ventricles*

## Tricuspid Valve

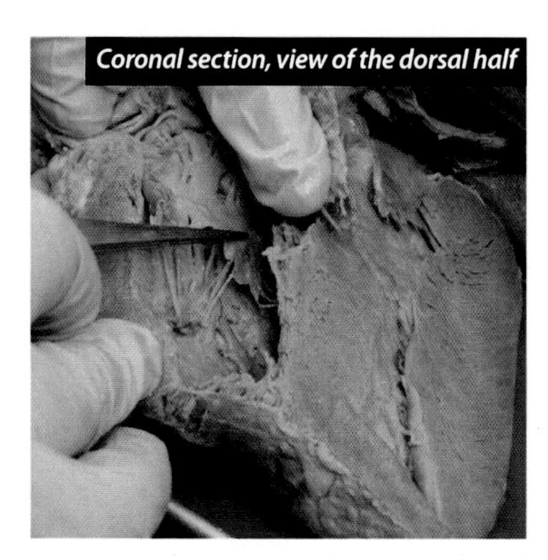


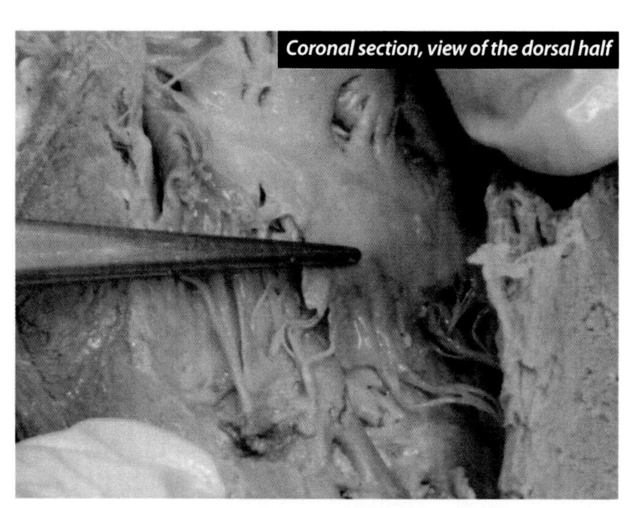


The tricuspid valve (three cusps) is found between the right atrium and the right ventricle. Functionally it is very important because it prevents the backflow of blood from the right ventricle to the right atrium during ventricular contraction. Such backflow is very dangerous. Remember the mnemonic "**Tri right** before you **Bi**."

## Chordae Tendineae

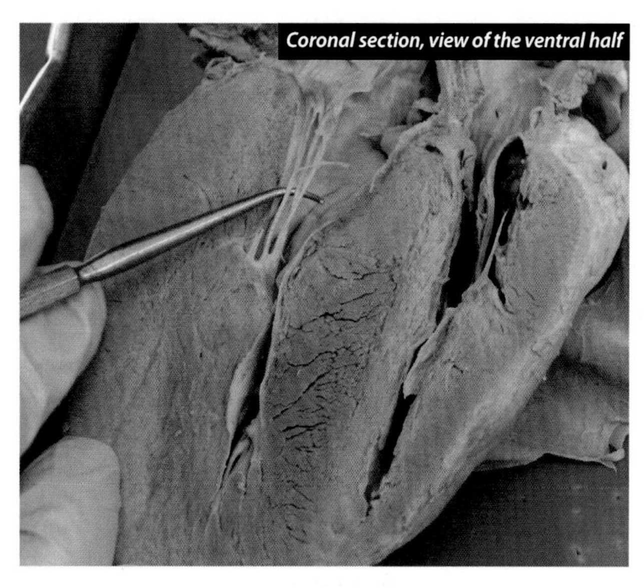


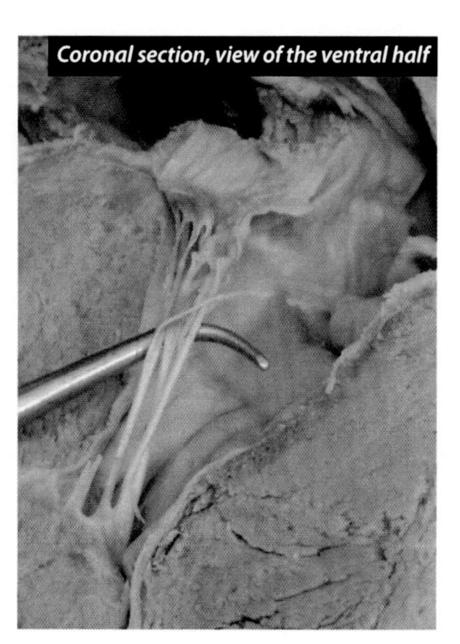


The chordae tendineae are attached to the atrioventricular valves and the walls of the ventricles. They help anchor the cusps of those valves so that they do not prolapse into the atria and allow the regurgitation of blood back into those chambers.

## Papillary Muscle

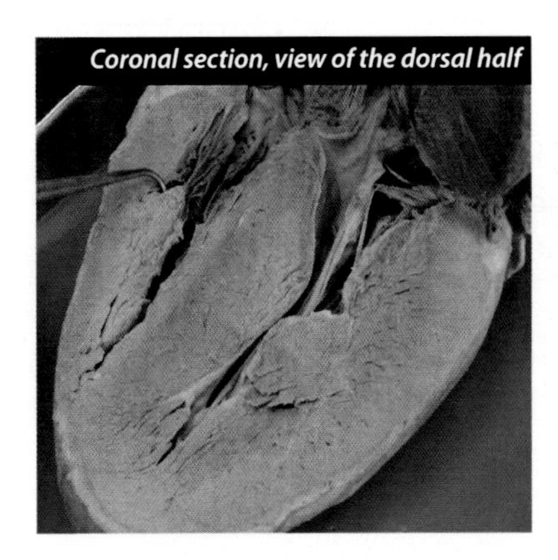
Coronal section, view of the dorsal half

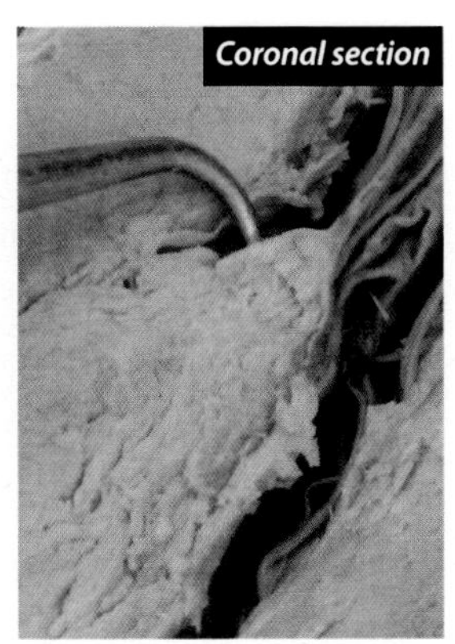
Coronal section

The papillary muscles are raised areas on the walls of the ventricles. The chordae tendineae attach to them. Functionally they are important because when the chordae tendineae pull on them and cause them to stretch, they contract against the stretch. This helps prevent prolapse of the atrioventricular valves. The papillary muscles are a special type of trabeculae carneae.

## Trabeculae Carneae

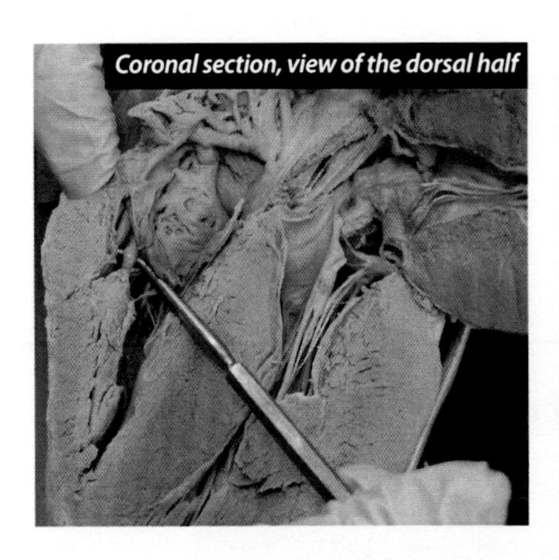
Coronal section, view of the dorsal half

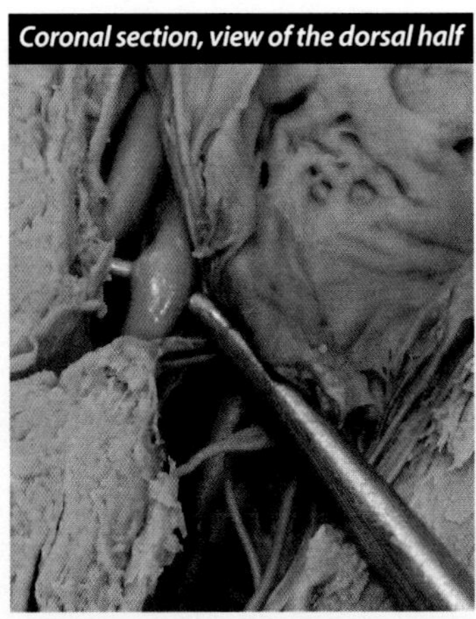
Coronal section, view of the dorsal half

The trabeculae carneae are similar to the pectinate muscles of the atria. They are muscular ridges on the walls of the two ventricles and are most easily seen in the right ventricle caudal (inferior) to the attachment of the tricuspid valve at the atrioventricular orifice. If one were to translate the roots of this name, one would come up with "meaty beams" as the meaning. Like the pectinate muscles, they also resemble tree roots. Functionally they appear to prevent prolapse of the tricuspid and bicuspid valves into their respective atria. They may also improve blood flow in the ventricles by reducing the suction that smooth walls would cause.

## Moderator Band

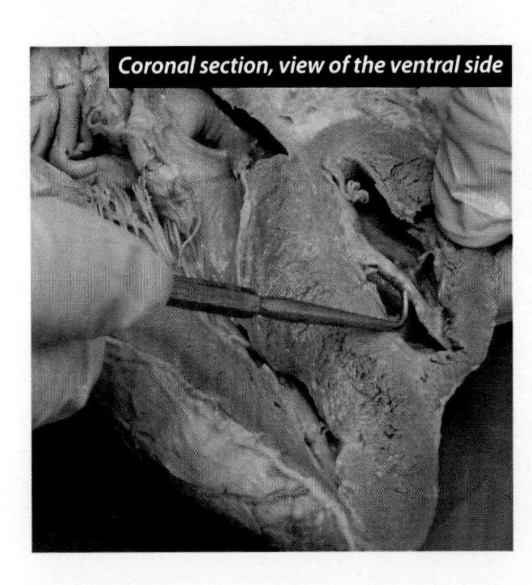
Coronal section, view of the ventral side

Coronal section

The moderator band stretches from a papillary muscle on the interventricular septum to the wall of the right ventricle. Functionally it is important because it conducts impulses between these two regions on the heart, thereby coordinating the contraction of the cells. You may remember that the moderator band toured with Eric Clapton last summer.

## Interventricular Septum

Coronal section, dorsal half

The interventricular septum is formed primarily from cardiac muscle tissue. Functionally it is important because it separates the two ventricles and prevents their blood from mixing.

## Semilunar Valves (Aortic and Pulmonary)

### Aortic

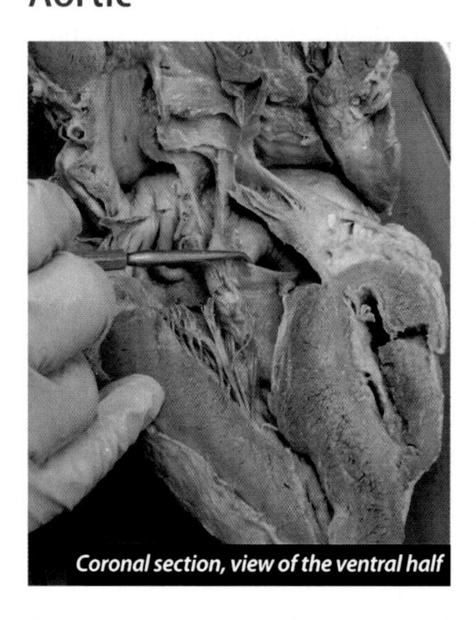
Coronal section, view of the ventral half

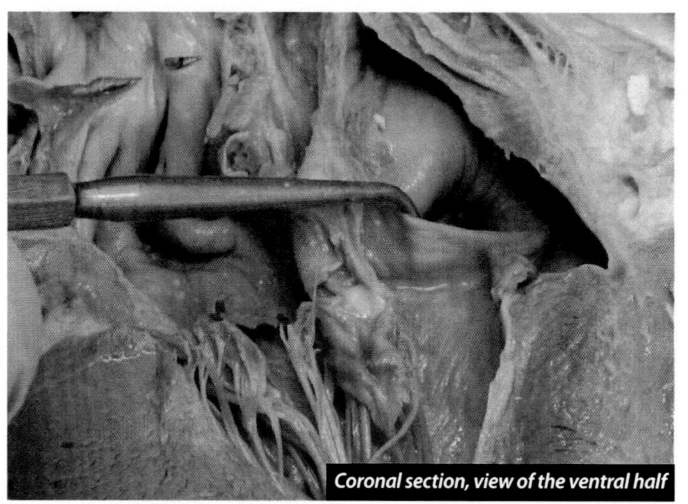
Coronal section, view of the ventral half

The semilunar valves are typically found in **veins** and lymphatic ducts. They are also found at the beginning of the **aorta** and the **pulmonary trunk**. They have three cusps. They are functionally important in the heart because they prevent the backflow of blood from those vessels into the ventricles. Dr. J likes to draw the analogy that each cusp is shaped like a shirt pocket, and when the blood tries to move the wrong way it opens the pocket away from the wall. Together, the three cusps normally block the backflow of blood in an effective manner.

## Semilunar Valves (Aortic and Pulmonary)

### Pulmonary

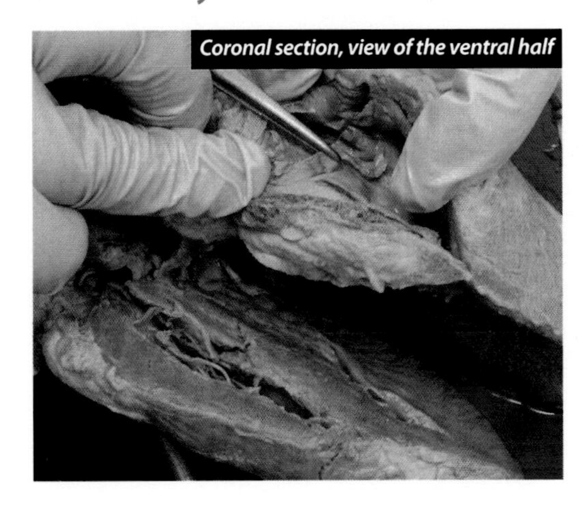
Coronal section, view of the ventral half

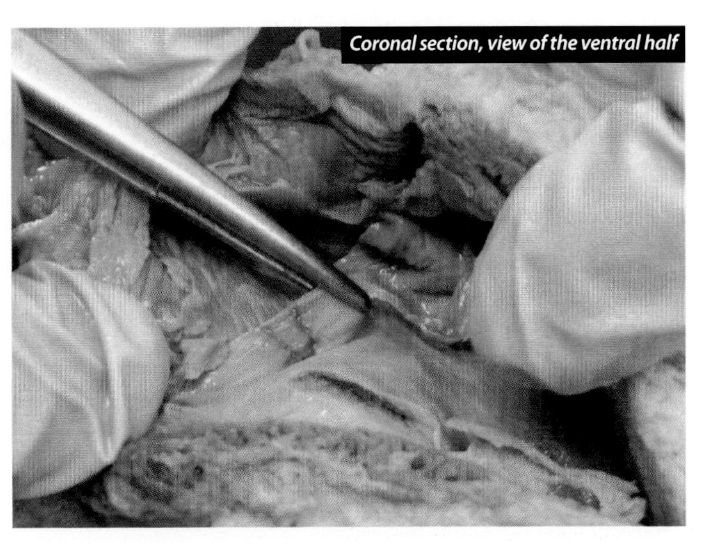
Coronal section, view of the ventral half

The description for these structures can be found under the aortic semilunar valve entry.

## Bicuspid (Mitral) Valve

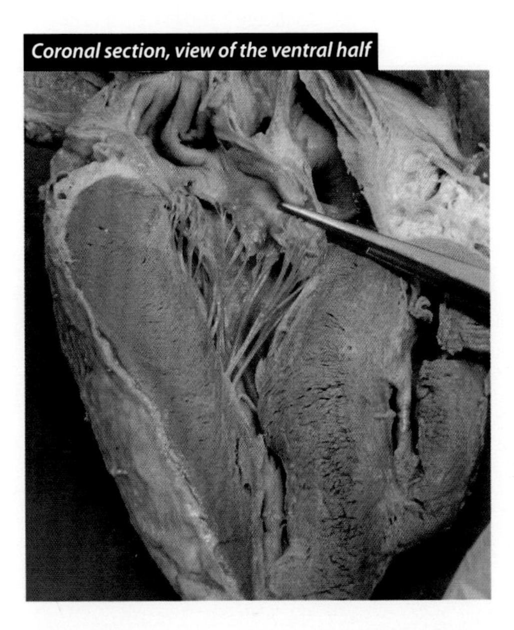
Coronal section, view of the ventral half

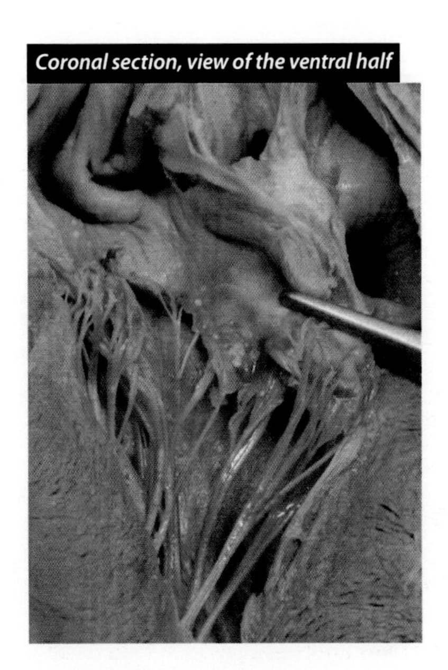
Coronal section, view of the ventral half

The bicuspid valve (two cusps) is found between the left atrium and the left ventricle. It is sometimes referred to as the mitral valve because of its angular shape. Functionally it is important because it prevents the backflow of blood from the left ventricle to the left atrium during ventricular contraction. Such backflow is very dangerous because it increases the blood pressure in the lungs. This high pressure may lead to pulmonary edema.

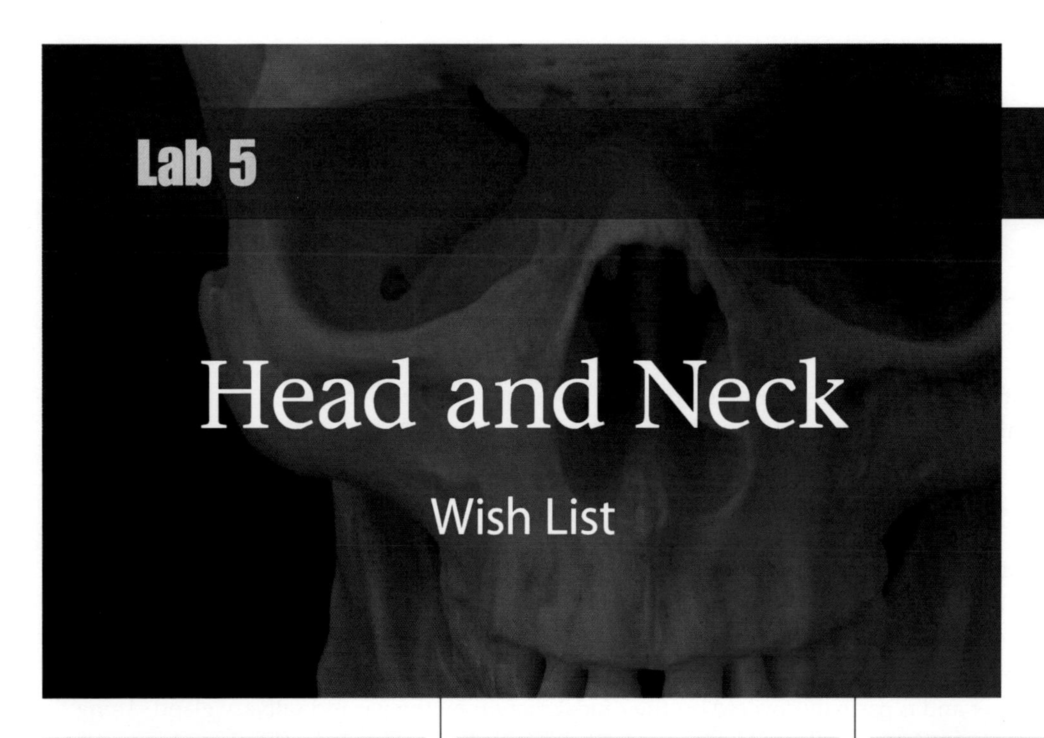

Lab 5

# Head and Neck

## Wish List


**LAB 5 OVERVIEW, pp. 110–112**

**BONES, p. 113**
- **Tympanic Bulla** (Temporal Bone), p. **113**
- **Hyoid,** p. **113**

**MUSCLES, pp. 114–115**
- **Masseter,** pp. **114–115**
- **Digastric,** pp. **114–115**

**ORGANS, pp. 116–120**
- **Parotid Salivary Gland and Duct,** p. **116**
- **Mandibular (Submaxillary) Salivary Gland and Duct,** pp. **116–117**
- **Sublingual Salivary Gland,** p. **117**
- **Thyroid Gland,** p. **118**
- **Trachea,** p. **118**
- **Sheep Larynx,** pp. **119–120**
  - Epiglottic Cartilage, p. 119
  - Arytenoid Cartilage, p. 119
  - Cricoid Cartilage, p. 120
  - Thyroid Cartilage, p. 120

**AREA DIAGRAM, p. 121**
- **Anterior Structures of the Neck, 121**

**NERVES, pp. 122–124**
- Facial Nerve (VII)—(Dorsal and Ventral), p. 122
- Hypoglossal Nerve (XII), p. 122
- Spinal Accessory Nerve (XI), p. 123
- Vagosympathetic Trunk, p. 123
- Vagus Nerve (X), p. 124
- Sympathetic Trunk, p. 124

**VESSELS, pp. 125–129**

**ARTERY, pp. 125–128**
- Sublingual (Lingual) Artery, p. 125
- External Maxillary Artery, p. 125
- External Carotid Artery, p. 126
- Common Carotid Artery, p. 126
- Cranial Laryngeal Artery, p. 127
- Cranial Thyroid Artery, p. 127
- Muscular Artery (Branch), p. 128

**VEINS, pp. 128–129**
- Internal Jugular Vein, p. 128
- External Jugular Vein, p. 129

# LAB 5 OVERVIEW

Lab 5 focuses on the muscles, nerves, vessels, and other structures of the facial and cervical regions. This is another opportunity to view the cat anatomy similar to the way we would look at a road map. This continues to be a useful perspective to develop as a study tool.

We prepare the cat for observation by removing the skin from the left side of the face. This will expose the side of the face, ventral to the eye.

## 1. Structures of the Facial Region

With the skin removed, you should be able to see the structures on the side of the face. You should notice the object just ventral to the ear. This is the **parotid salivary gland** and it is the largest of the three salivary glands. It looks like cauliflower. In fact, that may be where cauliflower comes from. It is served by the **glossopharyngeal nerve** (IX). Cranial to that you will see a large muscle, the **masseter muscle**. It is about the size of a quarter and is partially covered on the caudal margin by the parotid salivary gland. Along the dorsal margin of this muscle you will see the **dorsal facial nerve** (VII), and along the ventral margin you should find the **ventral facial nerve** (VII). Between these two nerves you will find the **parotid duct** running obliquely across the surface of the masseter muscle. Occasionally you will see an artery accompanying the duct, but more often it will resemble a nerve (dental floss). You can tell it is not a nerve because there is no nerve in that area (back to road maps).

You will find two mandibular lymph nodes on either side of the anterior facial vein. You will not be responsible for these structures, but you should remove the two lymph nodes so that other structures in the area will be more easily observed. Ventral to the parotid salivary gland you will find the **mandibular** (**submaxillary**) **salivary gland**. It strongly resembles a gumdrop without sugar on it or a candy dot. It has its own duct that runs deep to the digastric muscle and superficial to the **sublingual salivary gland**. Both of these glands are controlled by the **facial nerve** (VII). You will not observe the ducts for the sublingual salivary gland. They empty into the oral cavity, on the inferior side of the tongue.

## 2. Structures of the Cervical Region

We begin the study of the cervical region by transecting the left **digastric muscle** that runs along the medial side of the mandible. Be careful not to cut too deeply as there are several structures immediately deep to the digastric muscle. Close to the mandible you will find the **mandibular duct**. It lies superficial to the sublingual salivary gland, which is usually triangular and looks like the end of a Bic pen. There is a mnemonic for the three salivary glands—PMS. Wow. Now you know what that stands for!

While we are in this region, you should observe the **left sublingual** (**lingual**) **artery** and the **left hypoglossal nerve** (XII) running together toward the tongue. Notice that these two names translate to the same meaning, that being "under the tongue." That may help you link that information. Follow the **left sublingual artery** back toward the cervical region to observe where it branches from the **left external carotid artery**. Then follow the **left external carotid artery** cranially to see the **left external maxillary artery** branch from it and pass deep to the sublingual salivary gland. The **left external maxillary artery** serves the masseter muscle.

If you reflect the digastric muscle away from the external carotid you will be able to observe the **tympanic bulla**. It is a landmark of the **temporal** bone.

Next we move caudally in the cervical region and transect the left **sternohyoid** and **sternothyroid muscles**. Reflect them toward their origin and insertions. This will expose the left **thyroid gland**. You will note the remarkable resemblance of the thyroid gland to rice peel-offs. Unlike the human thyroid that has an isthmus connecting the two lobes, the cat thyroid exists as two separate lobes.

Slightly lateral and dorsal to the thyroid gland you will see a relatively large artery, the **left common carotid artery**. We will study the branches of the **common carotid artery**, starting at the caudal end of the neck. The first medial branch of the **left common carotid artery** is the **left cranial thyroid artery**, which serves the cranial end of the thyroid gland. As is the case with nearly all of these branches, the destination is part of the name. Close to where the cranial thyroid artery branches you will observe a lateral branch of the left common carotid artery; it is the **left muscular artery**. This vessel serves the muscles of the neck. Moving cranially, you will find the **left cranial laryngeal artery** that serves the cranial end of the larynx. The **left common carotid artery** ends where it bifurcates to form the **left internal carotid artery** and the **left external carotid artery**. You will not be responsible for observing the **internal carotid artery** because it is very small. The **external carotid artery** and two of its branches, the **sublingual** and the **external maxillary arteries**, we have discussed previously.

While you are in the area of the cranial end of the **left common carotid artery**, remove a relatively large lymph node that is lateral to the larynx. Emerging from the same area that the **hypoglossal nerve** comes from, you will see the **left spinal accessory nerve (XI)** moving laterally to the **cleidomastoid muscle**. You should remember that you observed this nerve in Lab 2 on the deep side of the **acromiotrapezius muscle**. The spinal accessory nerve serves the trapezius muscles as well as the cleidomastoid muscle.

In some of the cats, you will see the **left internal jugular vein** running with the **left common carotid artery** in a caudal direction, and eventually it will be directed laterally to where it joins the **left external jugular vein**. This union forms the **left jugular vein**, which is short and joins the **left subclavian vein** to form the **left brachiocephalic vein**. Unfortunately, the **left internal jugular vein** often does not have dye in it. **Using what you know about vessels, see if you can think of a reason this would be the case**. The **left** and **right internal jugular veins** are of functional importance because they are the only vessels of significance that drain blood from the cranium. They leave the cranium by way of the jugular foramen.

You will observe the **left vagosympathetic trunk** running with the **left common carotid artery** and the **left internal jugular vein**. Collectively, this neurovascular bundle is called the **carotid sheath**. This is actually two nerves, the **left vagus** and the **left sympathetic trunk**, that run together with the same connective tissue wrapping. These two nerves are **NOT** physically joined. The **vagosympathetic trunk** will split into a **left vagus** and a **left sympathetic trunk** at about the level where they run dorsal to the **left brachiocephalic vein**. Please be careful not to separate them cranial to the point where they divide.

## 3. Sheep Larynx

You should observe the sheep larynx that has been prepared for you. You will see the **epiglottic cartilage** (**epiglottis**). Functionally it is important because during the act of swallowing, it folds over and covers the glottis, which helps reduce the risk of food entering the trachea. A cut should have been made through the dorsal wall of the larynx. Reflect the sides laterally to expose the glottis. The glottis is the space between the vocal folds (cords). At the cranial end of the larynx you will observe the **left** and **right arytenoid cartilages**. Functionally they are important because the vocal chords are anchored to the arytenoid cartilages, and when they are moved the vocal cords move as well. Notice in this dorsal wall that a relatively large piece of cartilage was cut, the **cricoid cartilage** (**signet ring cartilage**). This is significant because the cricoid cartilage is the only cartilage that makes a complete ring, and the arytenoid cartilage is the only paired cartilage that we study. This will be discussed further in lecture.

Close the larynx and examine the superficial surface. Cranially, you should be able to palpate the **hyoid bone** that surrounds the epiglottic cartilage. Moving caudally, you will see the large **thyroid cartilage** that projects ventrally. We called this the Adam's apple when we were in nursery school. Very close to the caudal margin of the thyroid cartilage on the ventral side is the relatively thin part of the cricoid cartilage. These two are connected by the **cricothyroid ligament**. Caudal to the cricoid cartilage, you will find a portion of the **trachea**. Notice that on the dorsal side, the rings are not complete, as we discussed in lecture. You may find a portion of the **esophagus** on the dorsal side of the trachea. If it is there, it has been cut with a sagittal cut. Reflect the walls of the esophagus and notice the longitudinal ridges inside the esophagus that have the appearance of corduroy. These are **rugae**, and their functional importance will be reviewed by your instructor. Along the lateral sides of the thyroid cartilage you may find the **thyroid glands** and you may also see remains of the **sternothyroid muscle**.

# BONES

## Tympanic Bulla

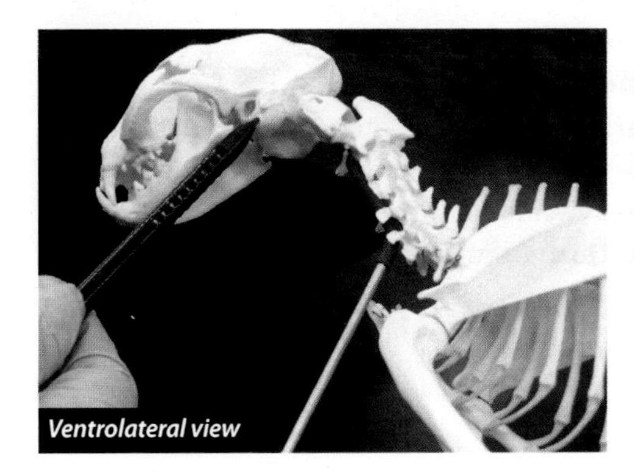
Ventrolateral view

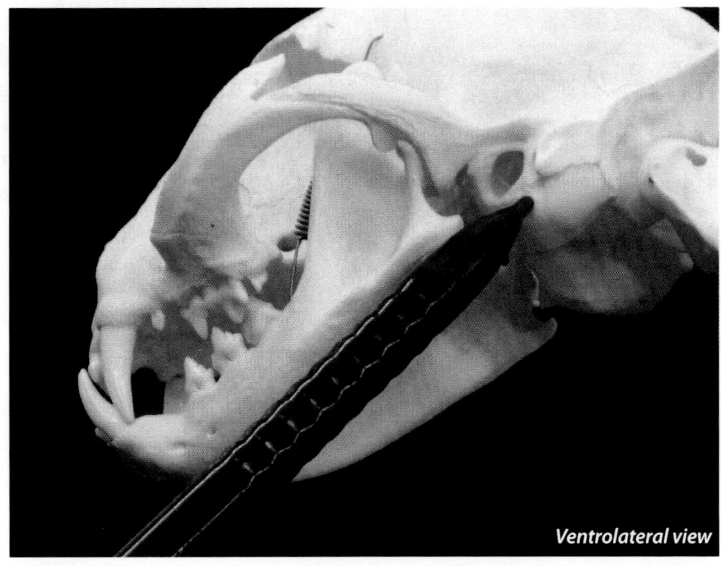
Ventrolateral view

The tympanic bulla is hollow and houses the middle ear. The *external auditory meatus* opens on its lateral surface. It is part of the temporal bone.

## Hyoid Bone

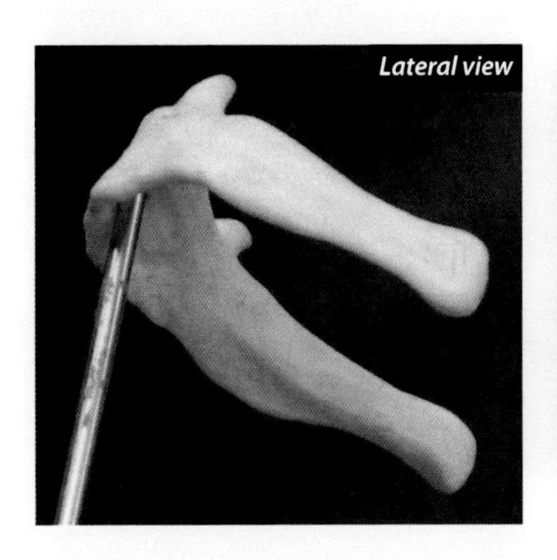
Lateral view

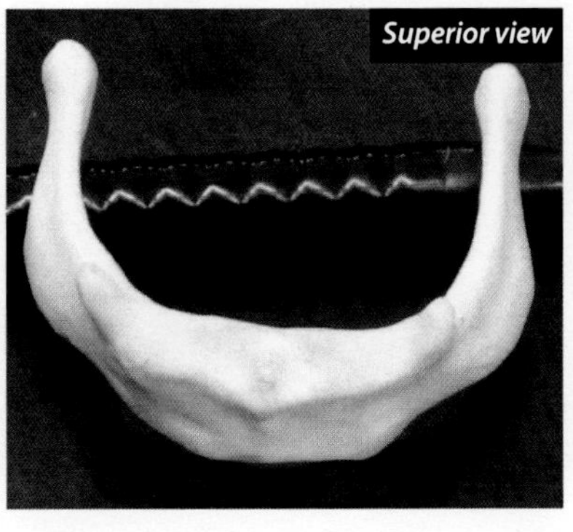
Superior view

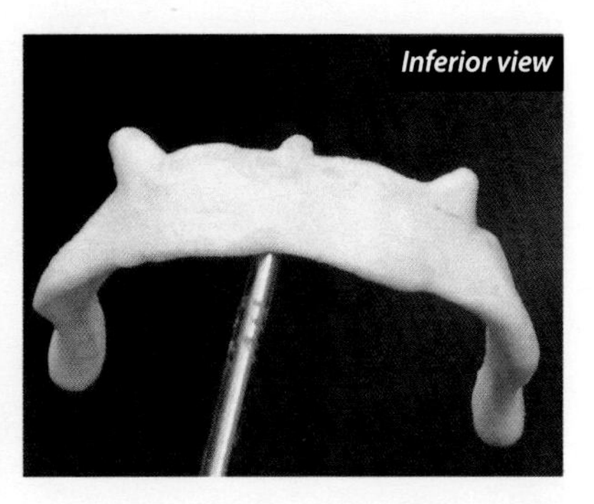
Inferior view

The hyoid bone does not articulate with any other bones. It is held in place by ligaments to the styloid process of the temporal bone and the thyroid cartilage of the larynx. It also has muscle attachments. In spite of the fact that it is not attached to the skull, it is considered part of the axial skeleton, specifically part of the skull. It has a shape similar to the mandible, suggesting a common origin. Functionally, it is important because it is the origin of muscles that move the larynx during the act of swallowing.

# MUSCLES

## Masseter Muscle

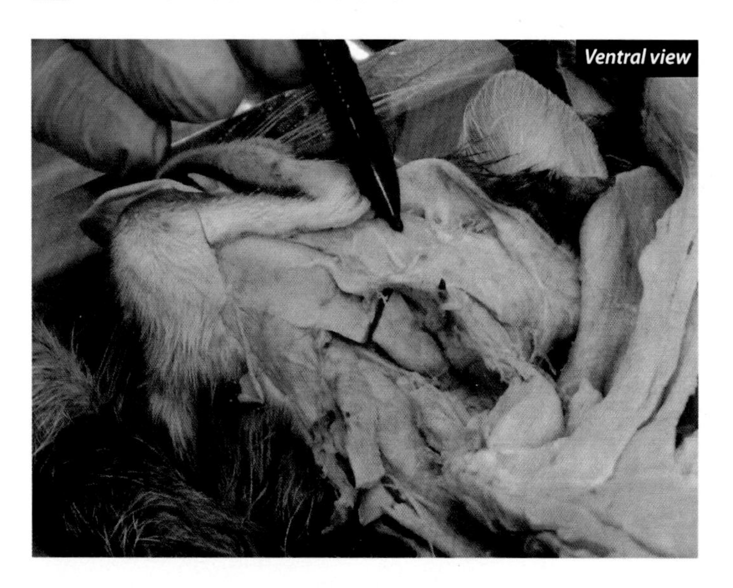


***Human Information: Masseter Muscle***

**origin:** zygomatic arch and zygomatic bone
**insertion:** angle and ramus of mandible
**nerve:** mandibular ($V_3$)
**action:** prime elevator of mandible

## Digastric Muscle

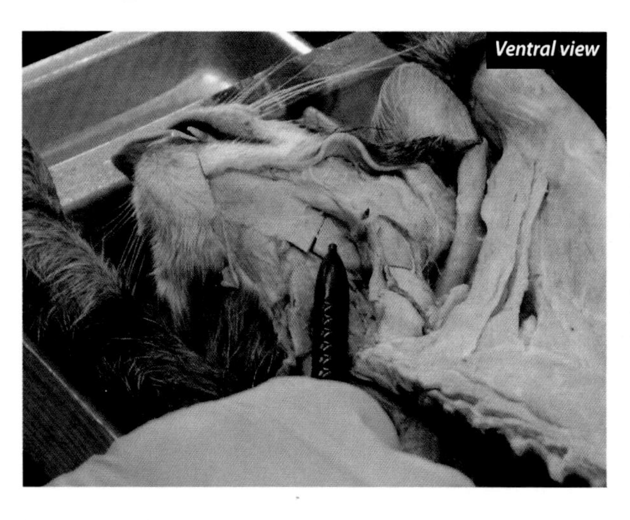


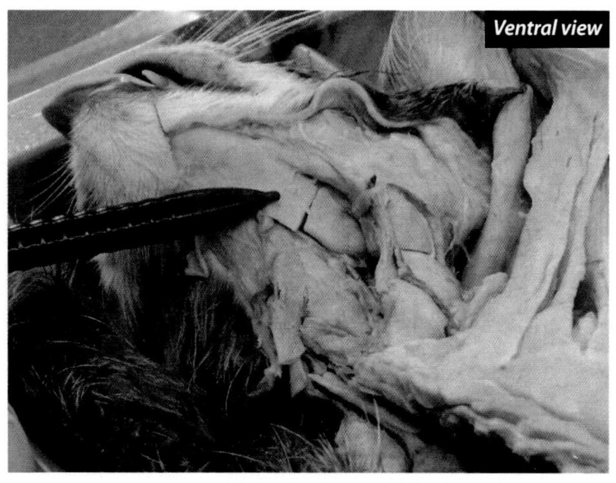


***Human Information: Digastric Muscle***

**origin:** mandible and mastoid process of temporal bone
**insertion:** by connective tissue loop to hyoid bone
**nerve:** mandibular ($V_3$) and facial (VII)
**action:** elevate hyoid bone, depress mandible (open mouth)

## Masseter Muscle

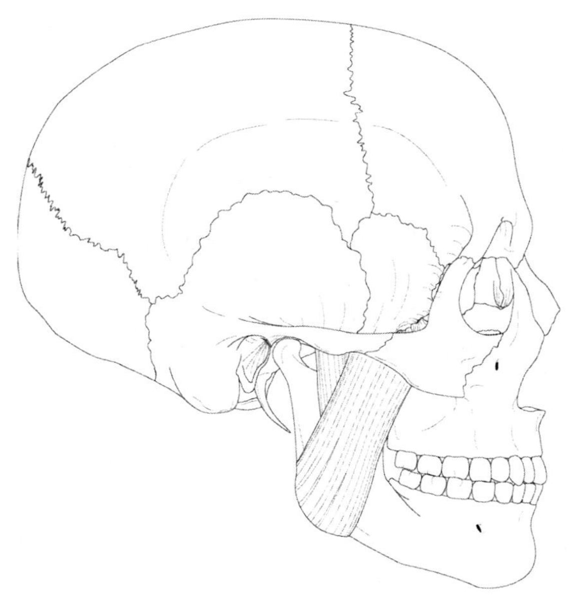

*Lateral view, right side*

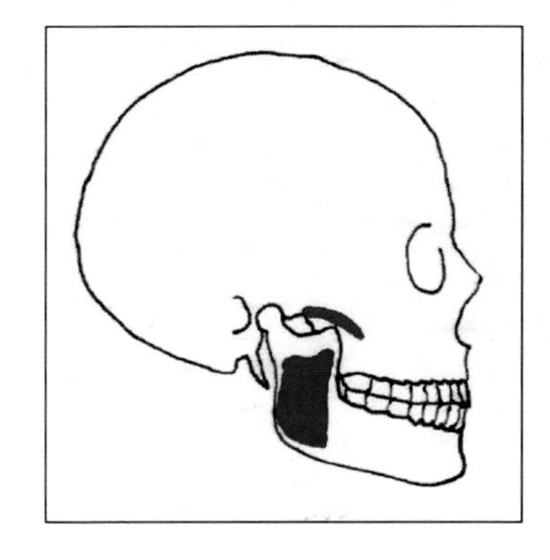

*This drawing of the origin and insertion might help you visualize this information (red is the origin, blue the insertion).*

## Digastric Muscle

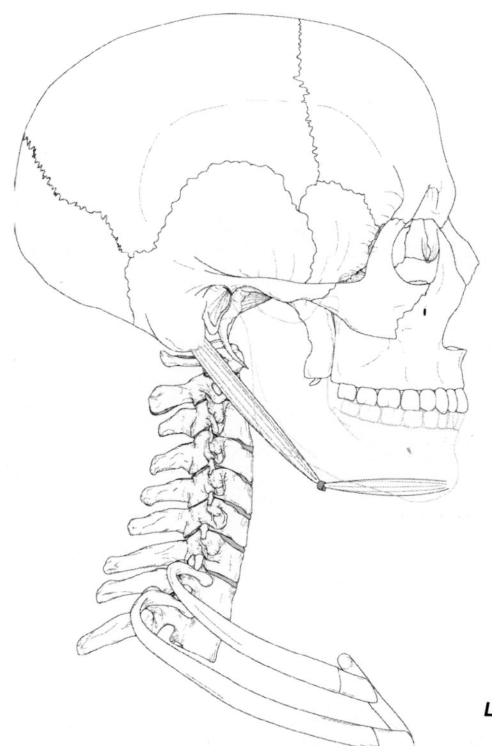

*Lateral view, right side*

# ORGANS

## Parotid Salivary Gland and Duct

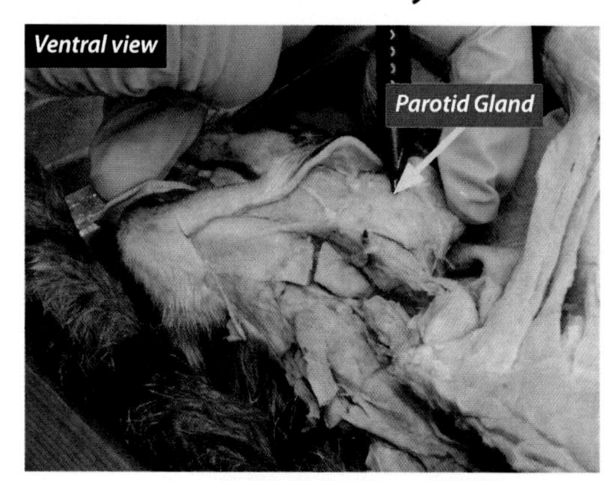


*Ventral and dorsal facial nerves (VII) are found on either side of the parotid*



The parotid salivary gland is the largest of the salivary glands. It lies superficial to the masseter muscle, and its duct passes across the masseter to deliver the saliva lateral to the second upper molar. As with most glands, it has a rough texture. Some students think it looks like the surface of cauliflower. It is an exocrine gland. Its secretions are controlled by the **glossopharyngeal nerve (IX)**.

## Mandibular (Submaxillary) Salivary Gland

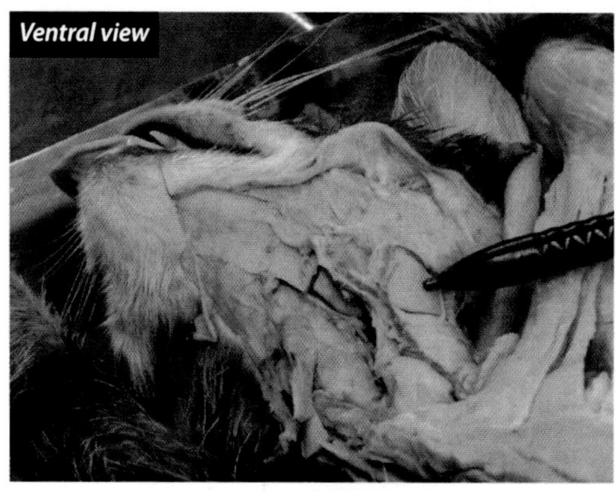


*Mandibular Salivary Gland*



The mandibular salivary gland lies deep to the mandibular body anterior to the angle of the mandible in humans. Its duct opens lateral to the frenulum of the tongue. Its secretions are controlled by the **facial nerve (VII)**. In the cat, the duct runs on the deep medial side of the digastric muscle, superficial to the sublingual salivary gland. The gland looks like a gumdrop without the sugar, or a dot (a dart in some places in Rhode Island!).

## Mandibular (Submaxillary) Salivary Duct

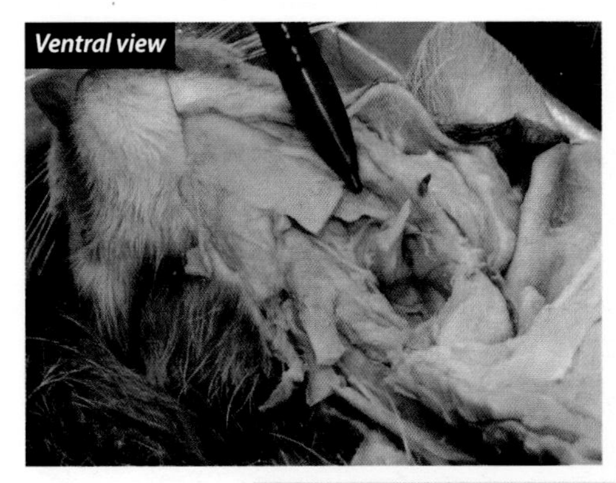


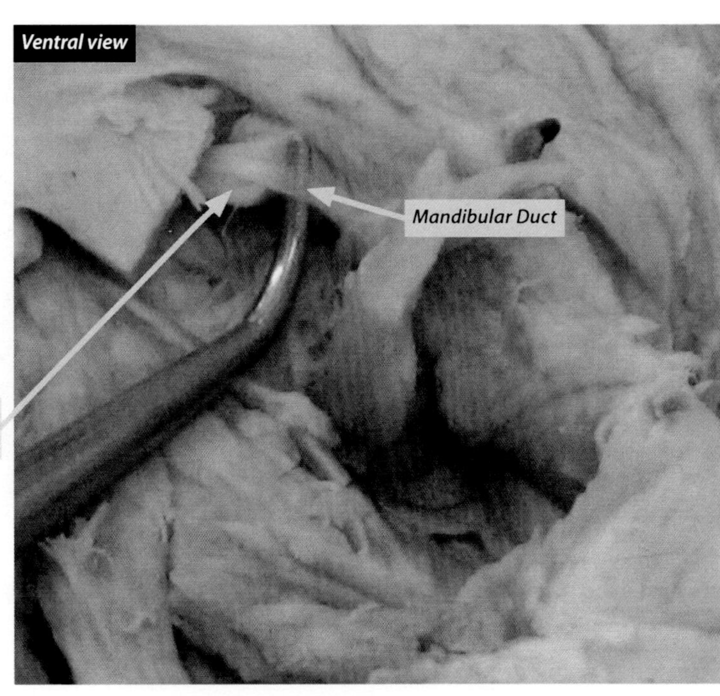


Note that the sublingual gland lies deep to the mandibular duct.

## Sublingual Salivary Gland

We can observe the **Mandibular Duct** running superficially over the **Sublingual Salivary Gland**.

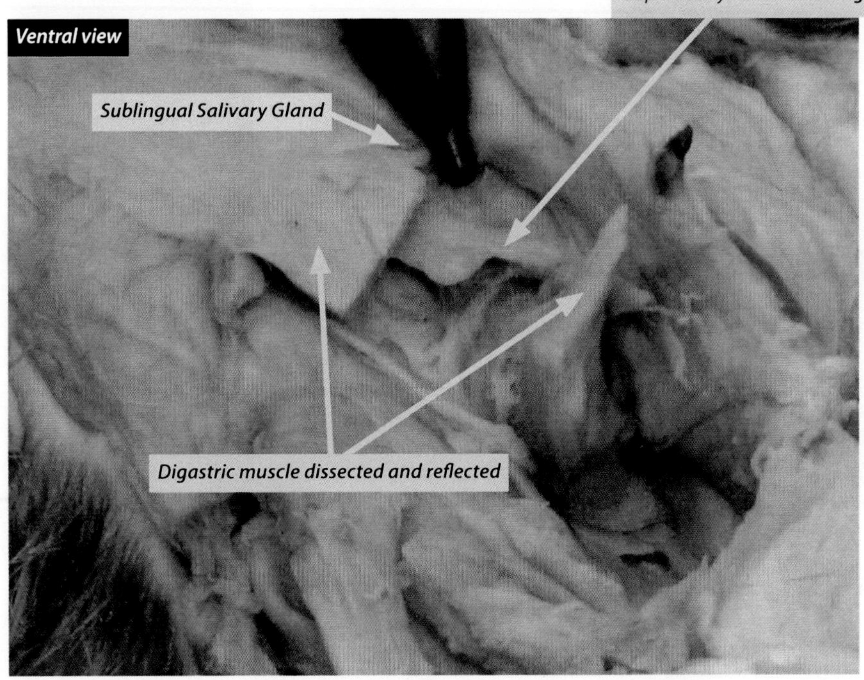


Have you ever been gleeked? If so, you were the recipient of exocrine secretions from the sublingual salivary gland. In the cat, it is situated deep to the submaxillary duct and the digastric muscle. It has its own ducts that open ventral to the tongue. It is the smallest of the salivary glands. In the human, it is inferior to the tongue with 10 to 12 ducts opening into the oral cavity. Its secretions are controlled by the **facial nerve** (VII). It is about the size and shape of the tip of a Bic pen—in fact this may be where Bic pens come from!

## Thyroid Gland

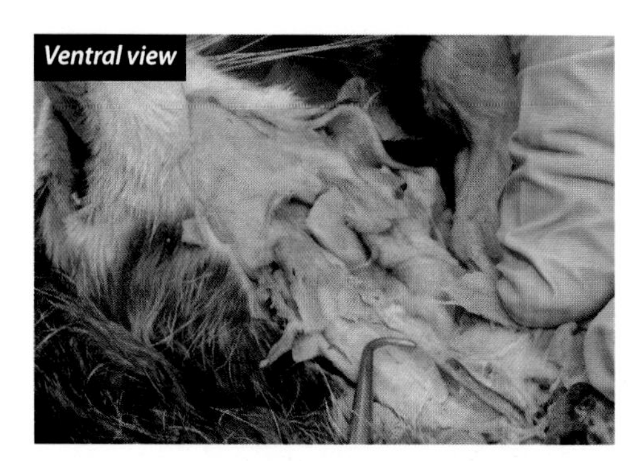


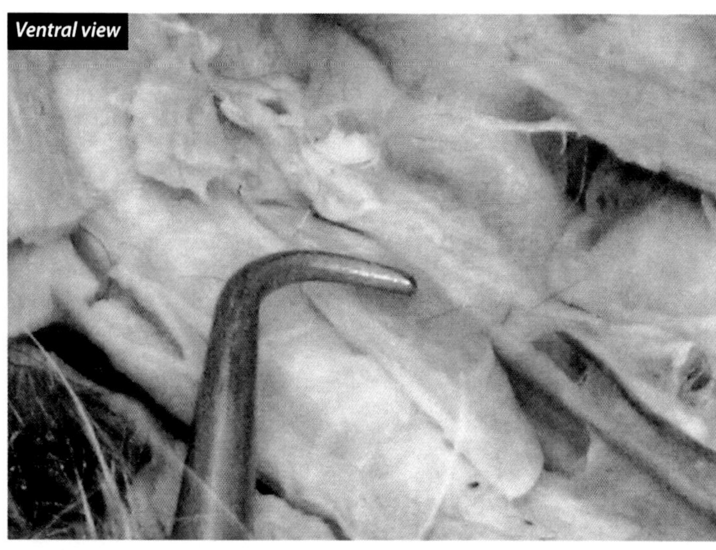


The thyroid gland is a bilobate endocrine gland. It is located inferior to the thyroid cartilage of the larynx and lateral to the trachea. It is the largest purely endocrine gland in the body. It produces two major hormones, thyroid hormone and calcitonin. They will be a topic for physiology. Notice: In the cat they look like rice peel-offs. In fact, this may be where rice peel-offs come from!

## Trachea

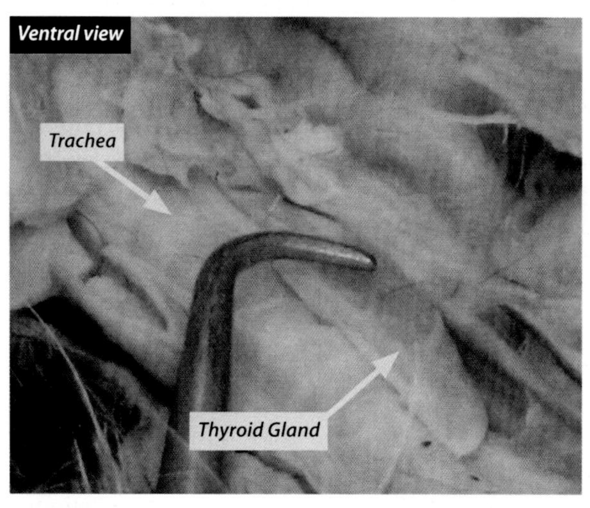


Cat trachea

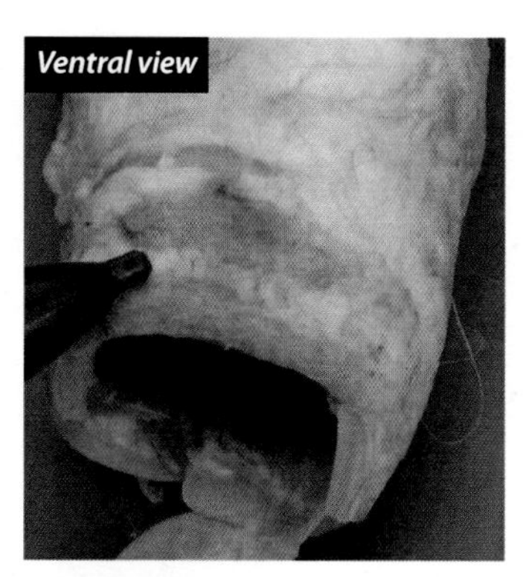


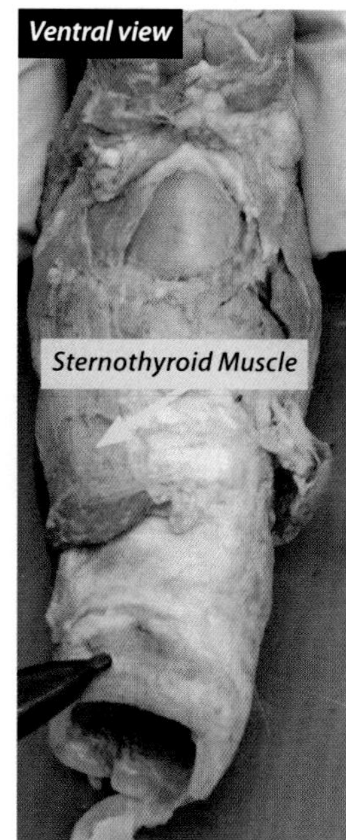


Sheep trachea

The trachea begins at the inferior end of the larynx (level of C6 body) and extends inferiorly to where it bifurcates into the left and right primary bronchi in humans. The inferior end is at the level of the sternal angle of humans in a supine position and to the body of T7 in the anatomical position. It is recognizable because of its 16 to 20 cartilaginous rings that normally prevent it from collapsing. The rings are not complete but are held together posteriorly by dense connective tissue. The advantage of this will be discussed in lecture. It is about 1.5″ in diameter and is lined with ciliated epithelial cells that sweep mucous out of the trachea and into the pharynx.

# Sheep Larynx: Specific Structures

When we were in nursery school, we called the larynx the voice box. It is made up of four major cartilages and houses the vocal cords. Functionally, it is important because it connects the pharynx with the trachea, as well as being the structure where sound is produced. It extends between C4 and C6. Superiorly, it is attached to the hyoid by the thyrohyoid ligament and inferiorly it is continuous with the trachea.

## Larynx: Epiglottic Cartilage

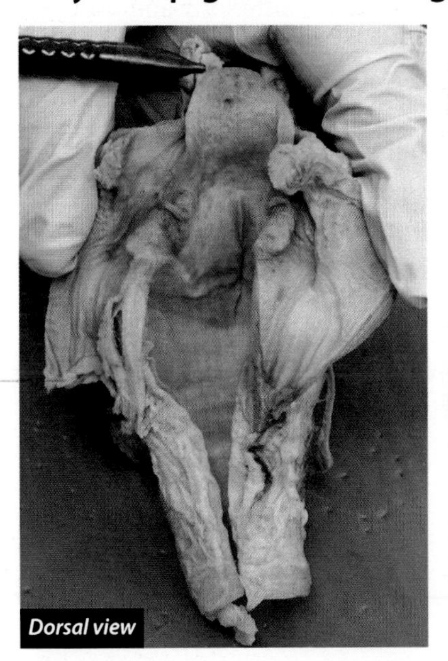
*Dorsal view*

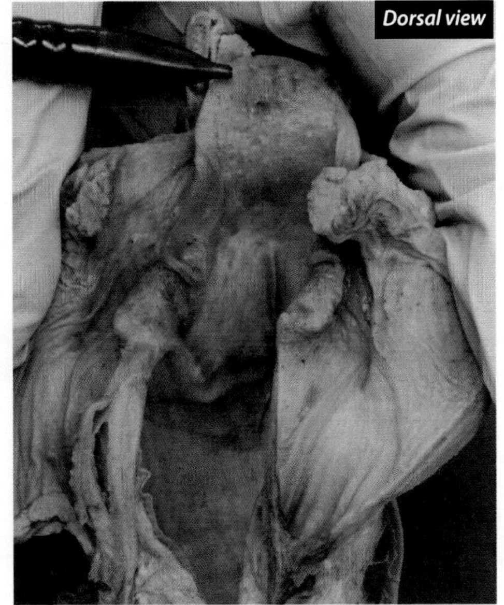
*Dorsal view*

The epiglottis or epiglottic cartilage is superior to most of the laryngeal cartilages. It projects superiorly and attaches to the tongue. Functionally, it is important in humans because during swallowing it covers the opening of the larynx when the larynx moves superiorly. It has an obvious resemblance to Mick Jagger's tongue, so many students refer to it as the Mick Jagger's tongue cartilage.

## Larynx: Arytenoid Cartilage

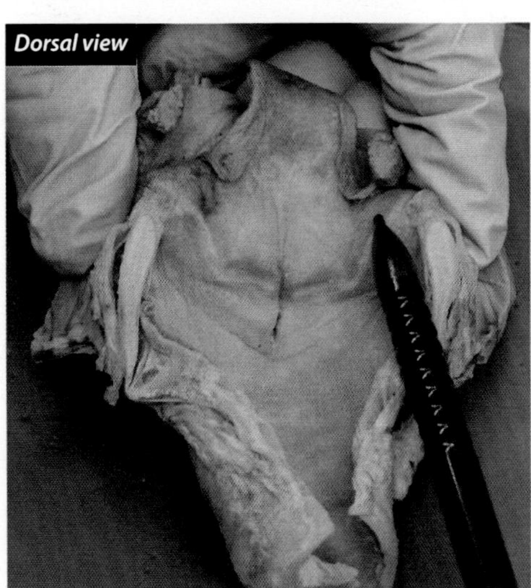
*Dorsal view*

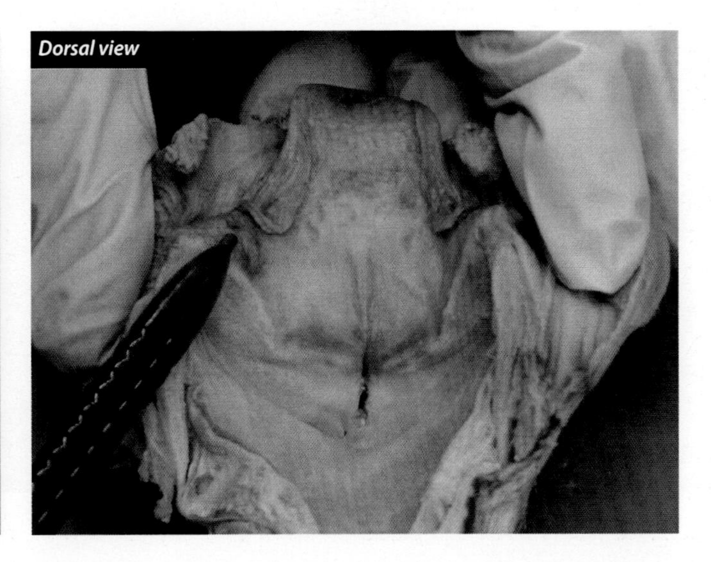
*Dorsal view*

The arytenoid cartilage is the only cartilage of the larynx that we will study which is split into two pieces. It forms the walls of the glottis and anchors the vocal cords. They cannot be seen from the outside of the larynx.

# Sheep Larynx: Specific Structures

## Larynx: Cricoid Cartilage

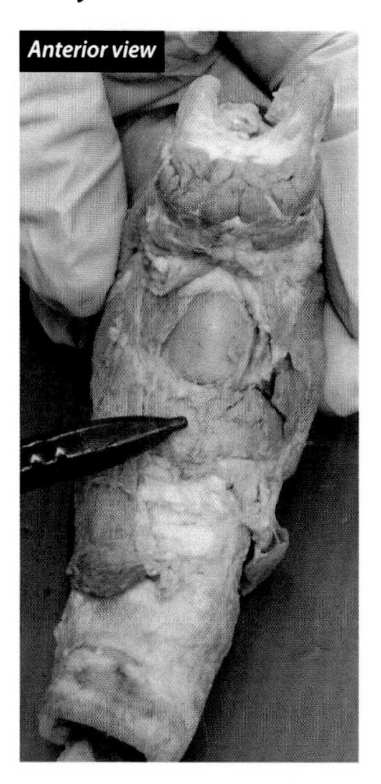


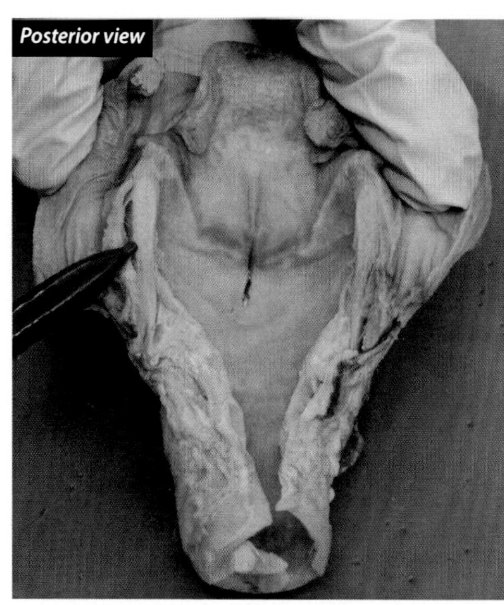


The cricoid cartilage is also known as the signet ring cartilage. It is the only cartilage that forms a complete ring. It is narrow anteriorly and wide posteriorly. It is connected superiorly to the thyroid cartilage by the cricothyroid ligament and inferiorly to the trachea. It forms most of the posterior wall of the larynx. The most probable advantage to the complete ring is that it provides stability for the muscles that move the vocal cords.

## Larynx: Thyroid Cartilage

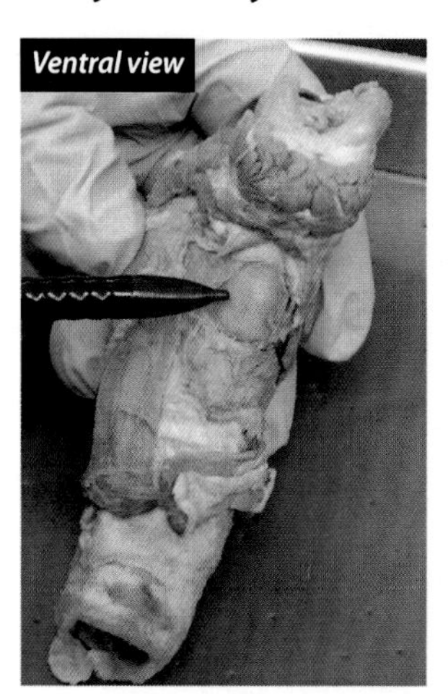


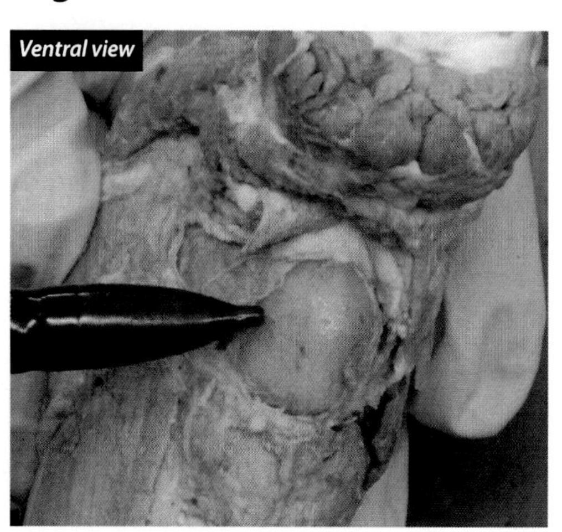


When we were in nursery school, we called the thyroid cartilage the Adam's apple. It is the largest of the laryngeal cartilages and larger in sexually mature males due to the effect of sex hormones. It is sometimes described as shaped like a plow or a shield.

# AREA DIAGRAM

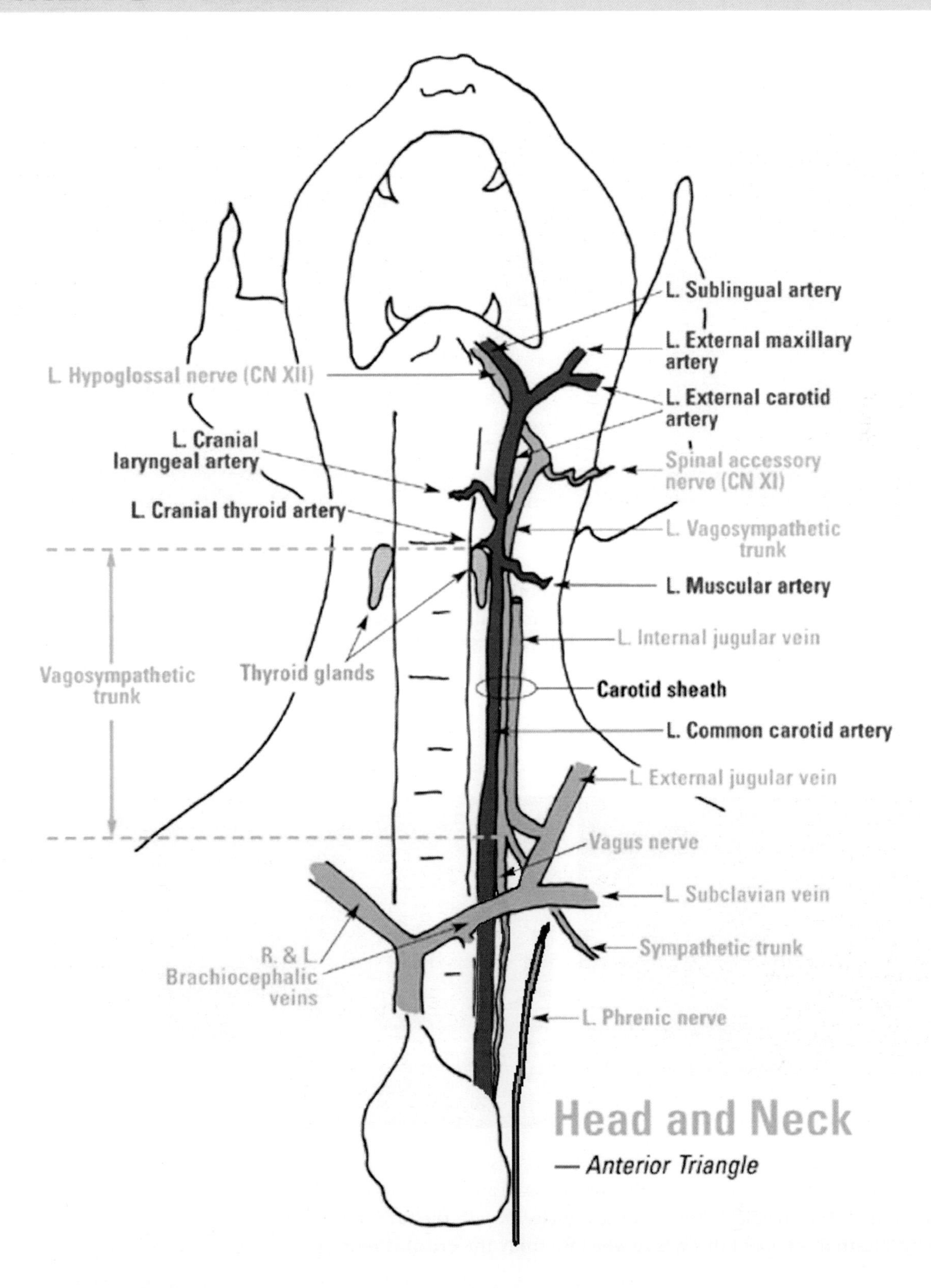

# NERVES

## Facial Nerve (VII)—Dorsal and Ventral

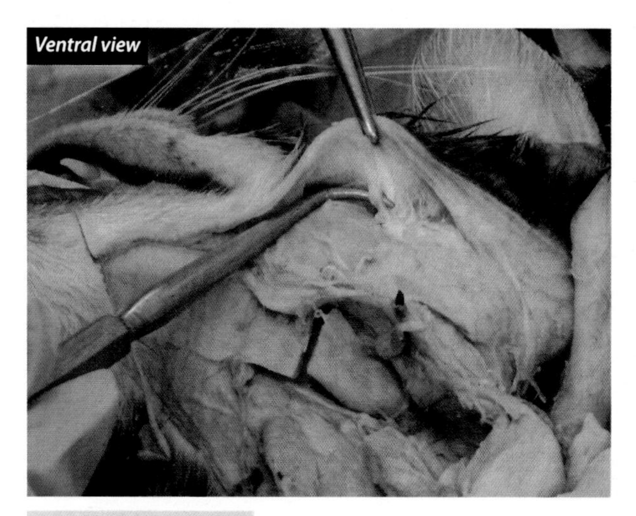


Dorsal Facial Nerve (VII)

Ventral Facial Nerve (VII)

The **facial nerve** is **cranial nerve VII**. It is both motor and sensory. It also has parasympathetic functions controlling the lacrimal gland, as well as the mandibular and sublingual salivary glands. It exits the skull via the stylomastoid foramen. We will observe this nerve on the lateral portion of the face where the two branches bracket the masseter muscle. We will learn more about this **nerve** when we study the **cranial nerves**.

## Hypoglossal Nerve (XII)

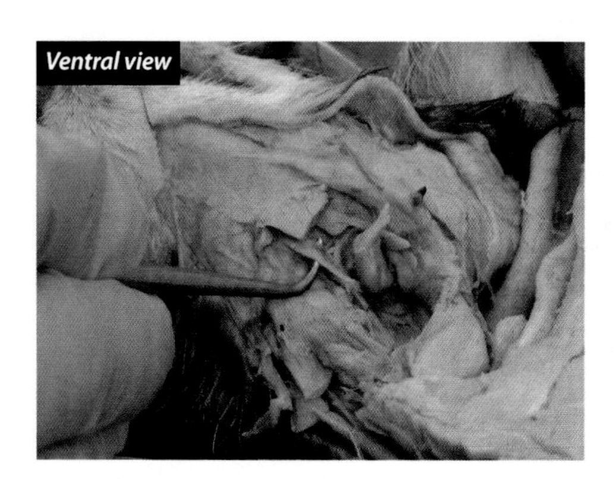


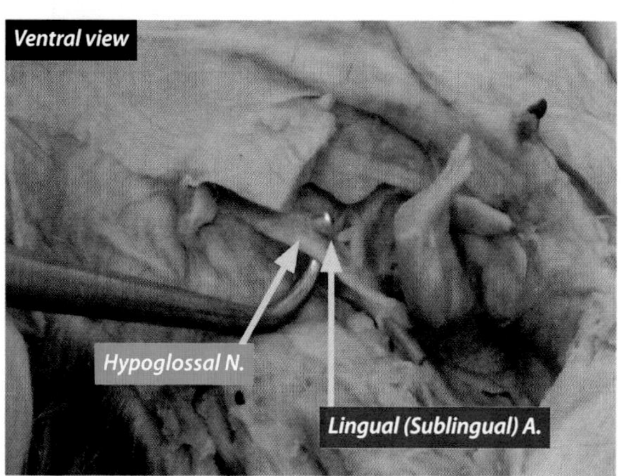


The **hypoglossal nerve** is **cranial nerve XII**. It is a motor nerve serving the muscles of the tongue. It exits the skull via the hypoglossal canal and runs with the **sublingual (lingual) artery** medial to the digastric muscle. As with the **facial nerve (VII)**, we will learn more about this **nerve** when we study the **cranial nerves**.

## Spiral Accessory Nerve (XI)

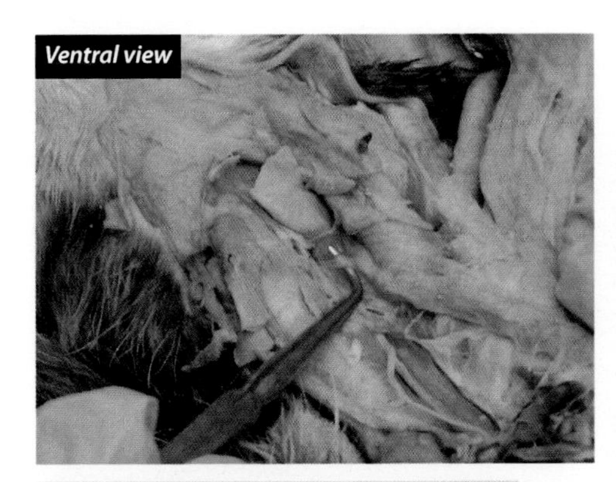


*This nerve is caudal to the digastric muscle and sublingual salivary gland.*

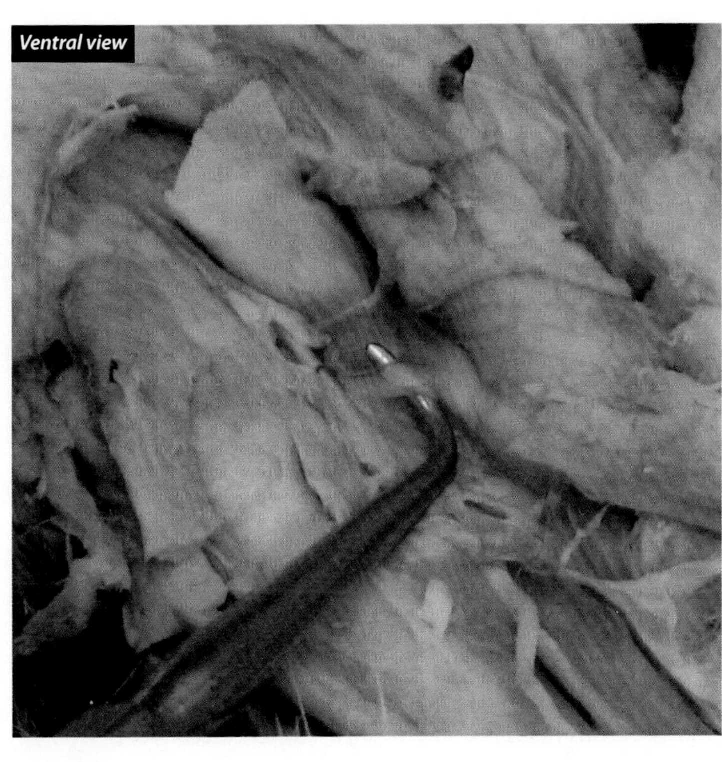


The **spinal accessory nerve** is **cranial nerve XI**. It is a motor nerve that serves the sternocleidomastoid and the trapezius muscles in humans. In the cat, it appears along with the **hypoglossal (XII)** and **vagus (X) nerves** lateral to the larynx and deep to the mandibular lymph node. From here it passes to the deep side of the cleidomastoid muscle.

## Vagosympathetic Trunk

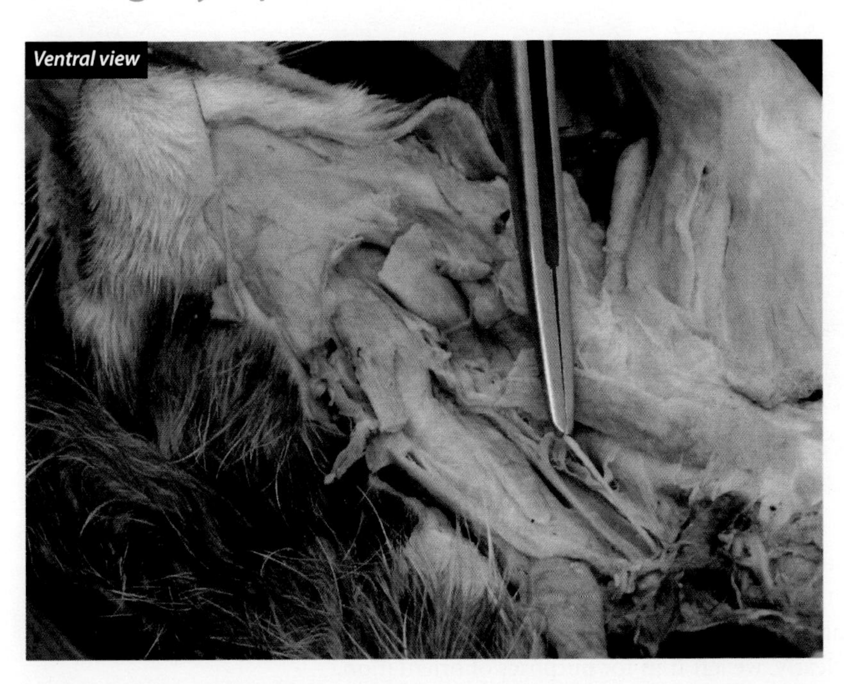


The **vagosympathetic trunk** runs along each side of the trachea in the carotid sheath, which also includes the **common carotid artery** and the **internal jugular vein**. It is formed where the **vagus nerve** and the **sympathetic trunk** run adjacent to each other. They do not actually become one structure, and a probe can separate them. PLEASE DO NOT SEPARATE THEM!

## Vagus (X) Nerve

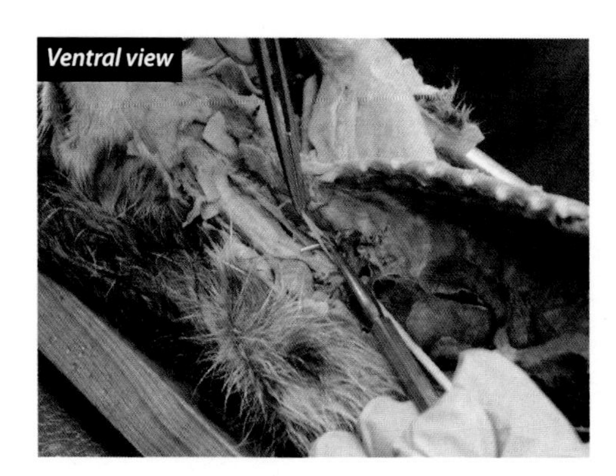


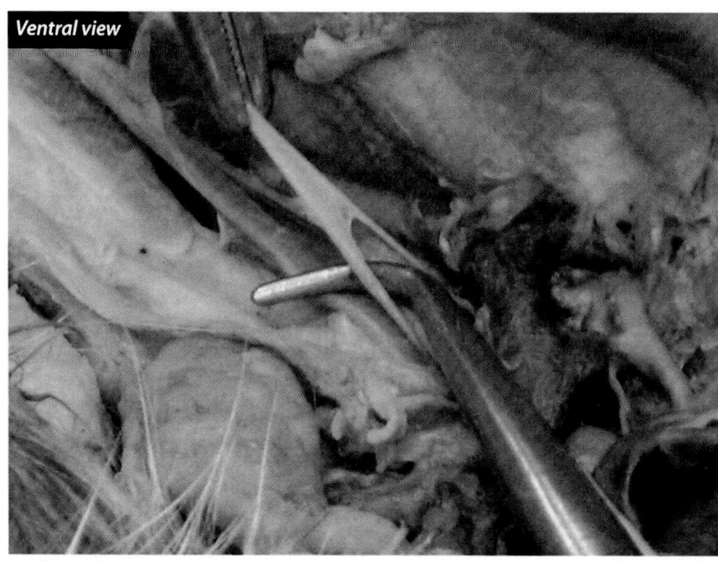


The **vagus nerve** is **cranial nerve X**. It is the only cranial nerve to pass into the body cavities below the neck. It is primarily parasympathetic in nature and is the major nerve affecting the heart and most of the gastrointestinal tract. It appears in the cat along with the **hypoglossal** and **spinal accessory nerves** lateral to the larynx and deep to the mandibular lymph node. From here it runs into the carotid sheath where it is joined by the **sympathetic trunk** to become the **vagosympathetic trunk**. The above pictures show this **nerve** where it has just split away from the **sympathetic trunk** as it enters the thoracic cavity. Spelling counts on this nerve; be sure to spell **vagus** correctly!

## Sympathetic Trunk

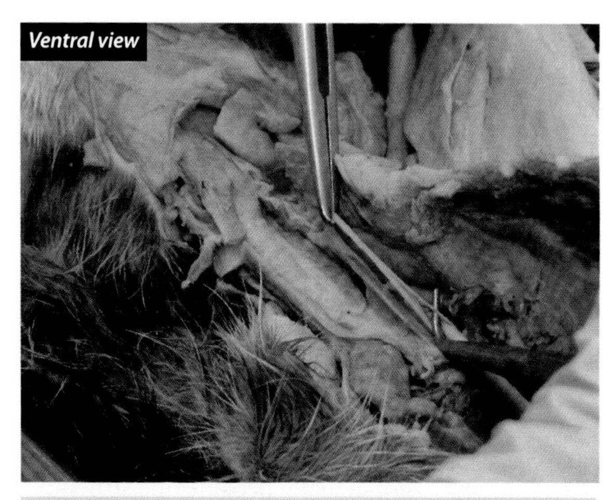


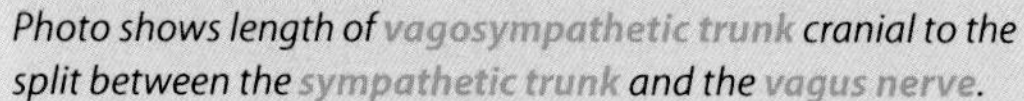
*Photo shows length of vagosympathetic trunk cranial to the split between the sympathetic trunk and the vagus nerve.*

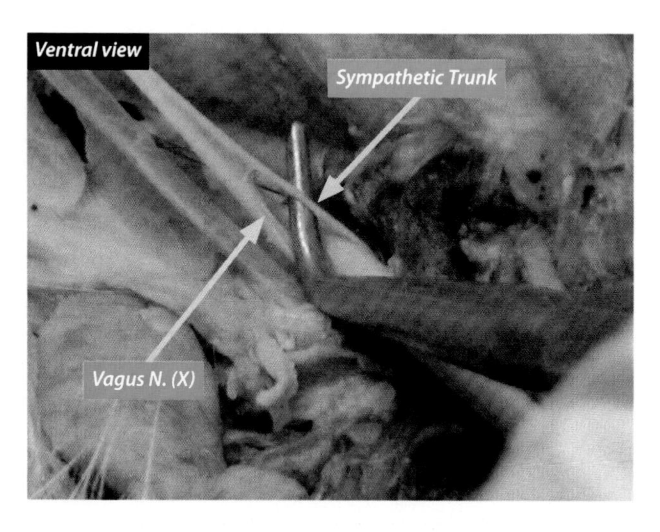


The **sympathetic trunk** runs along each side of the vertebral column. It receives neurons from the **anterior rami** of the **spinal nerves** via the **sympathetic trunk ganglia**. It serves many organs from the head to the pelvis. It first appears in the cranial end of the thoracic cavity. These pictures show where it first separates from the **vagus nerve**. Although the picture on the left does not show it clearly, we left it in for purposes of orientation.

# ARTERIES

## Sublingual (Lingual) Artery

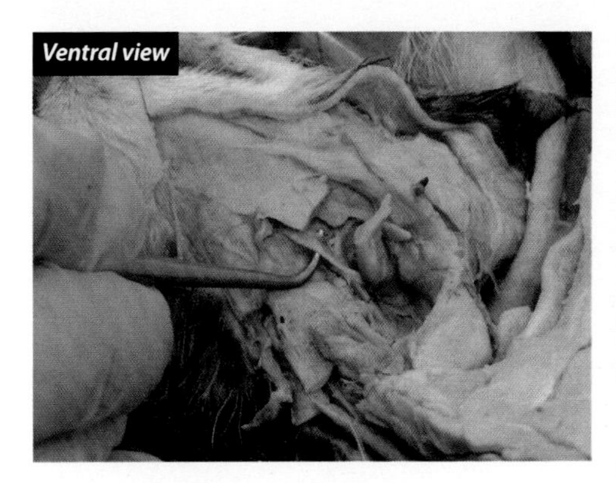


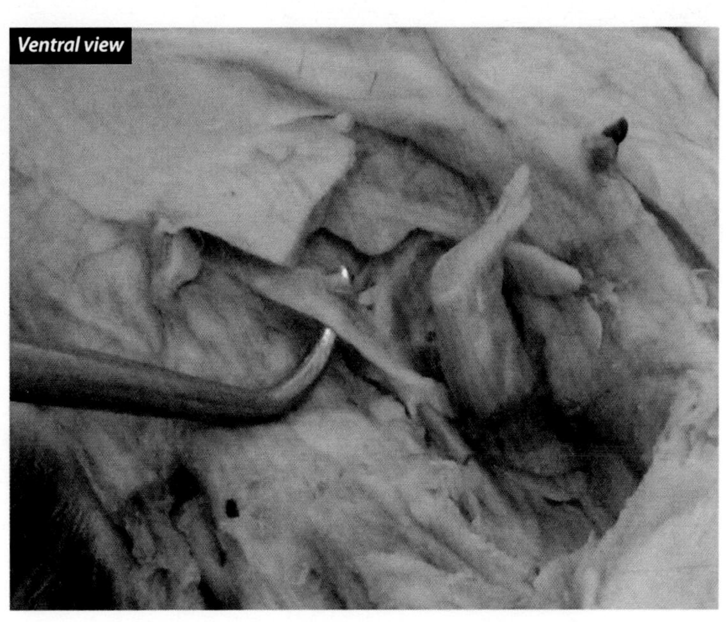


The **sublingual (lingual) artery** is a medial branch of the **external carotid artery**. It runs with the **hypoglossal nerve (XII)** along the medial margin of the digastric muscle. Functionally it is important because it supplies the tongue with blood.

## External Maxillary Artery

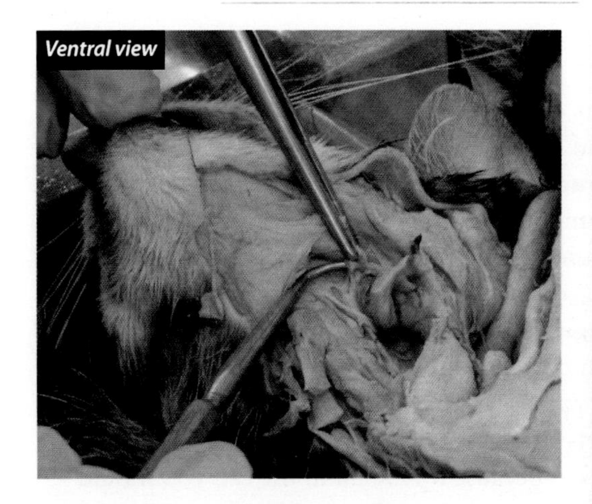


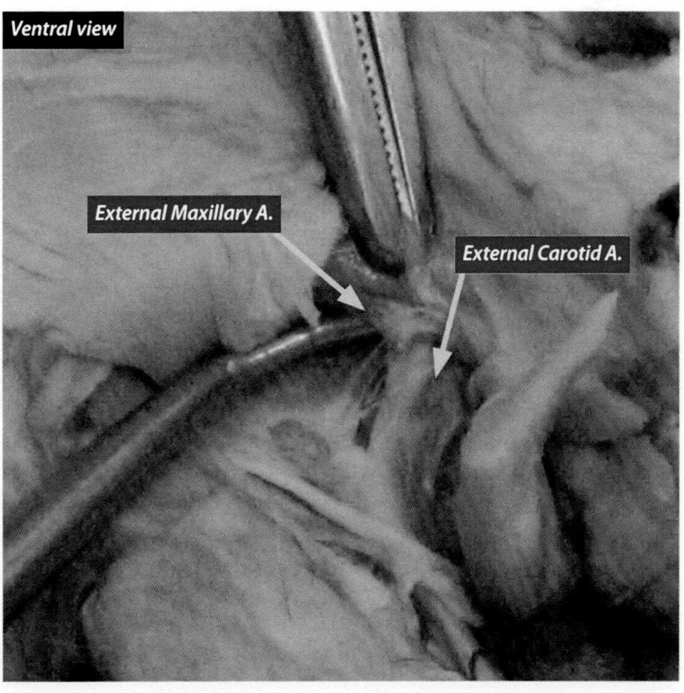


The **external maxillary artery** is a medial branch of the **external carotid artery** that arises cranial to the **lingual (sublingual) artery**. This **artery** runs deep to the digastric muscle and supplies blood to the salivary glands and the masseter muscle.

## External Carotid Artery

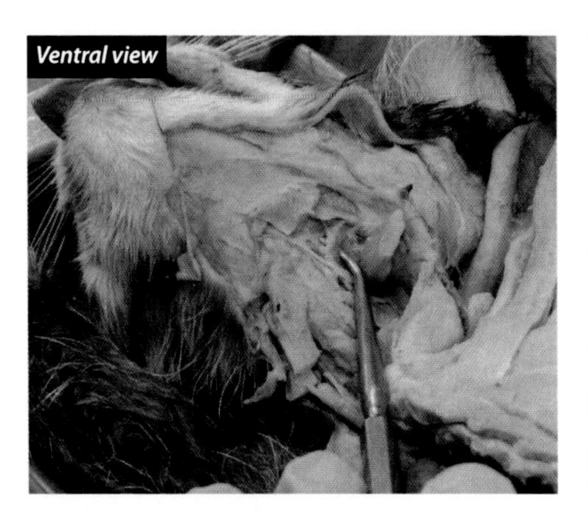


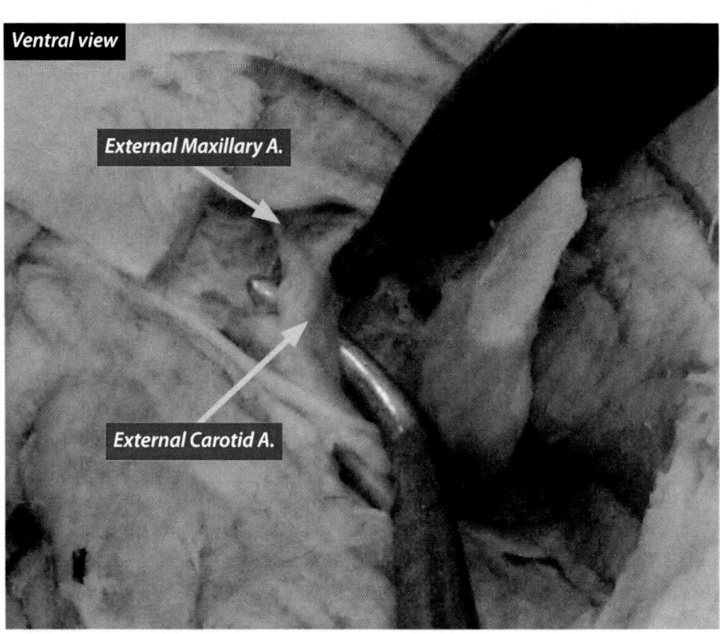


The **external carotid artery** and the **internal carotid artery** are the last two branches of the **common carotid artery**. It supplies blood to the structures of the head external to the cranial cavity. There are several branches of the **external carotid artery** that will be discussed in turn. The **internal carotid artery** will not be observed because it is very small in the cat; however, it is of significance for two reasons. First, the **common carotid artery** bifurcates, becoming the **internal carotid** and **external carotid arteries**; therefore, it marks the end of the **common carotid artery**. The **internal carotid artery** is one of two arteries from each side that serves the brain. The other is the **vertebral artery** that we discussed in the first and third labs.

## Common Carotid Artery

The **common carotid arteries** (**left** and **right**) are normally branches of the **brachiocephalic artery**, or, less frequently, branches of the **bicarotid trunk**. They are found in the carotid sheath on each side with the **vagosympathetic trunk** and the **internal jugular vein**. They have a number of branches, which will be discussed in turn. One branch of particular interest is the **internal carotid artery**. Although we will not observe this vessel because it is small in the cat, you should be aware of its significance. Eventually the **common carotid artery** bifurcates, becoming the **internal carotid** and **external carotid arteries**. The **internal carotid artery** is one of two arteries from each side that serves the brain. The other is the **vertebral artery** that we discussed in the first and third labs.

## Cranial Laryngeal Artery

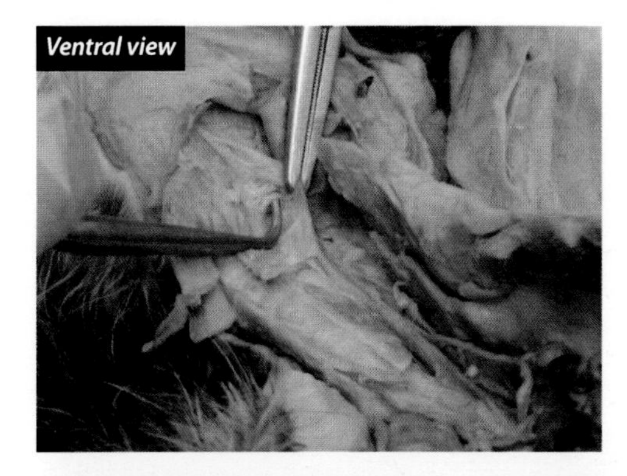


The **cranial laryngeal artery** is a medial branch of the **common carotid artery**. It serves the cranial end of the larynx.

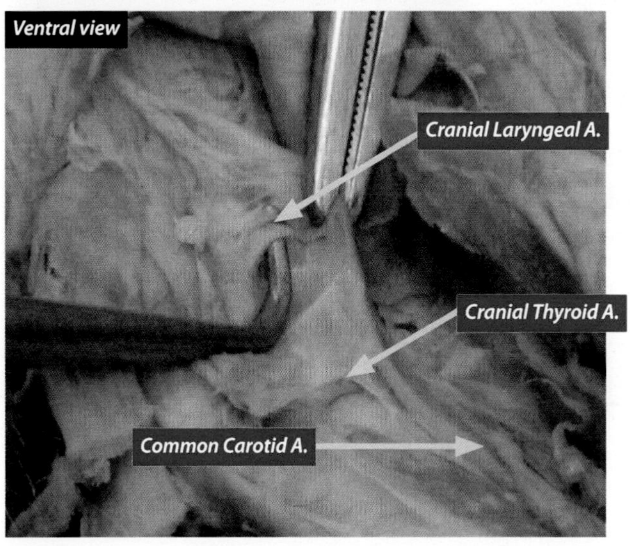


## Cranial Thyroid Artery

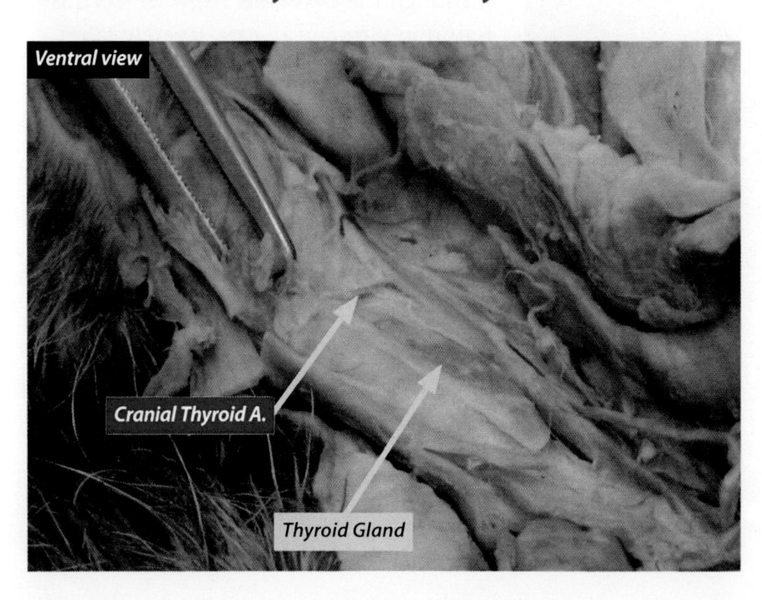


The **cranial thyroid artery** is the first medial branch of the **common carotid artery** that we will study. It serves the thyroid gland.

## Muscular Artery

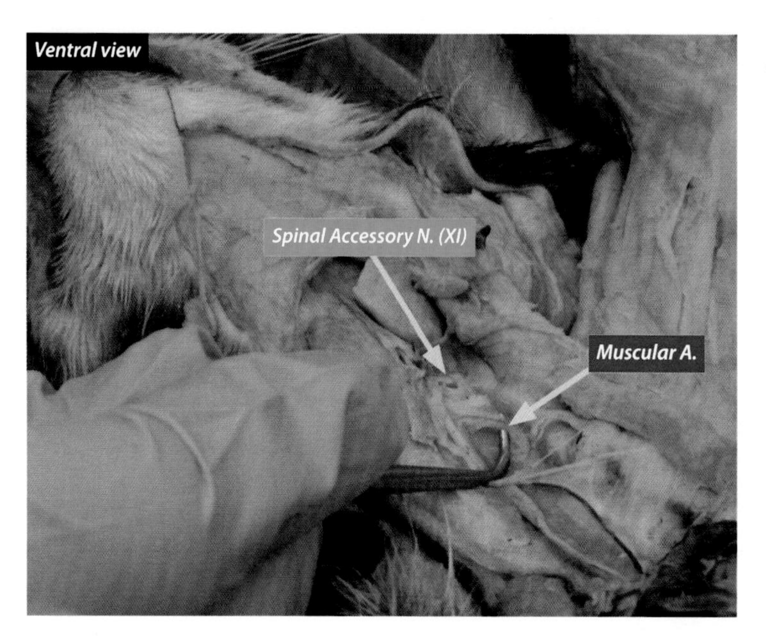


The **muscular artery** (**branch**) is a lateral branch of the **common carotid artery**. It branches off the **common carotid artery** close to the **cranial thyroid artery** and supplies blood to the muscles of the cervical region.

# VEINS

## Internal Jugular Vein

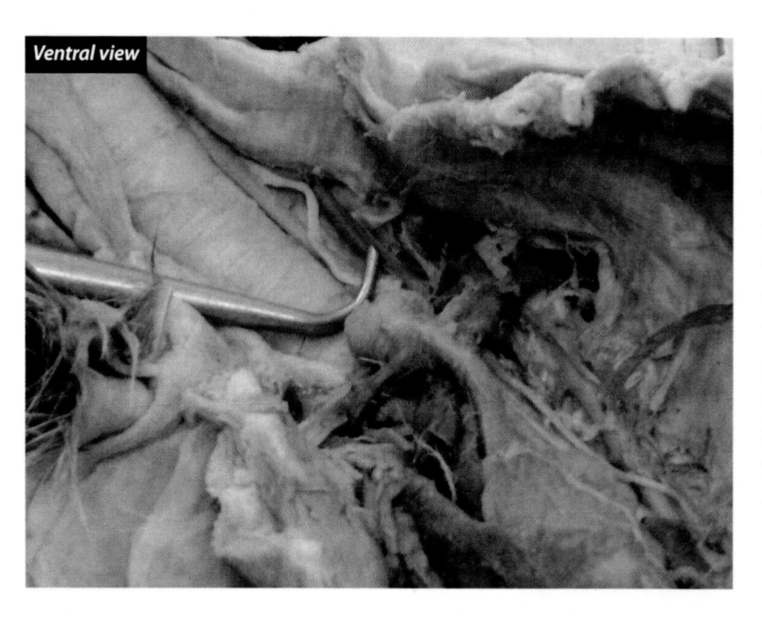


Although the **internal jugular vein** is found in virtually all cats, it rarely takes the dye and is relatively small. Because of this, it is difficult to find. Typically, it is found in the carotid sheath, which also includes the **common carotid artery** and the **vagosympathetic trunk**. It usually joins the **external jugular vein** before the junction with the **subclavian vein**. Functionally it is important because the **left** and **right internal jugular veins** are the major vessels draining blood from the brain.

# External Jugular Vein

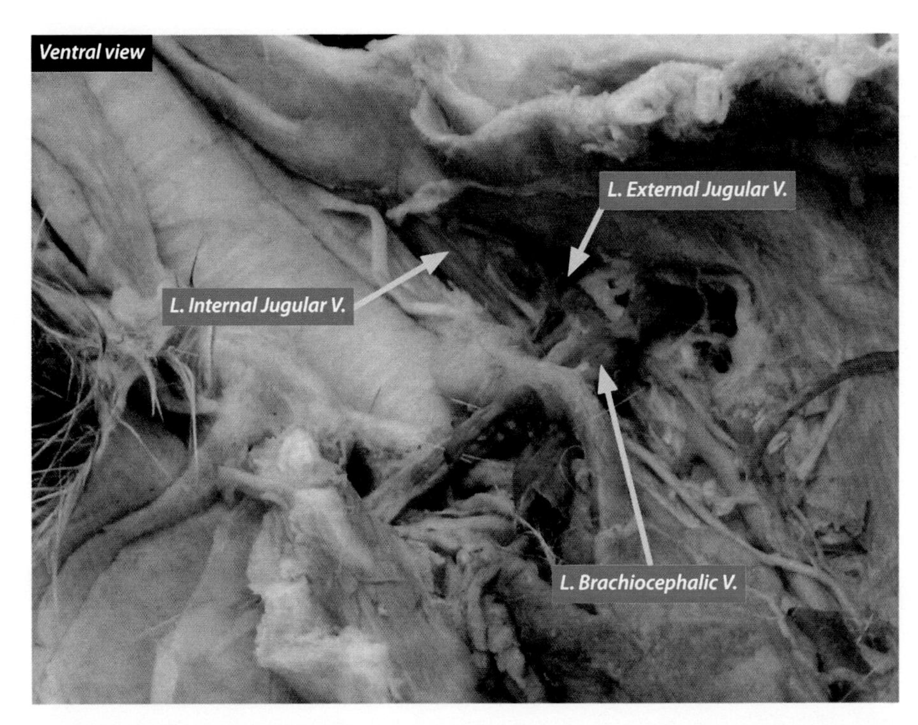


The **external jugular vein** in this lab is observed as it emerges from the deep side of the muscles in the neck and runs caudally until it meets the **subclavian vein** on the same side. This union forms the **brachiocephalic vein** on each side. Before that union occurs, the **internal jugular vein**, which often does not take the dye, joins the **external jugular vein**. The **external jugular vein** receives blood from the head region, except from the cranial cavity.

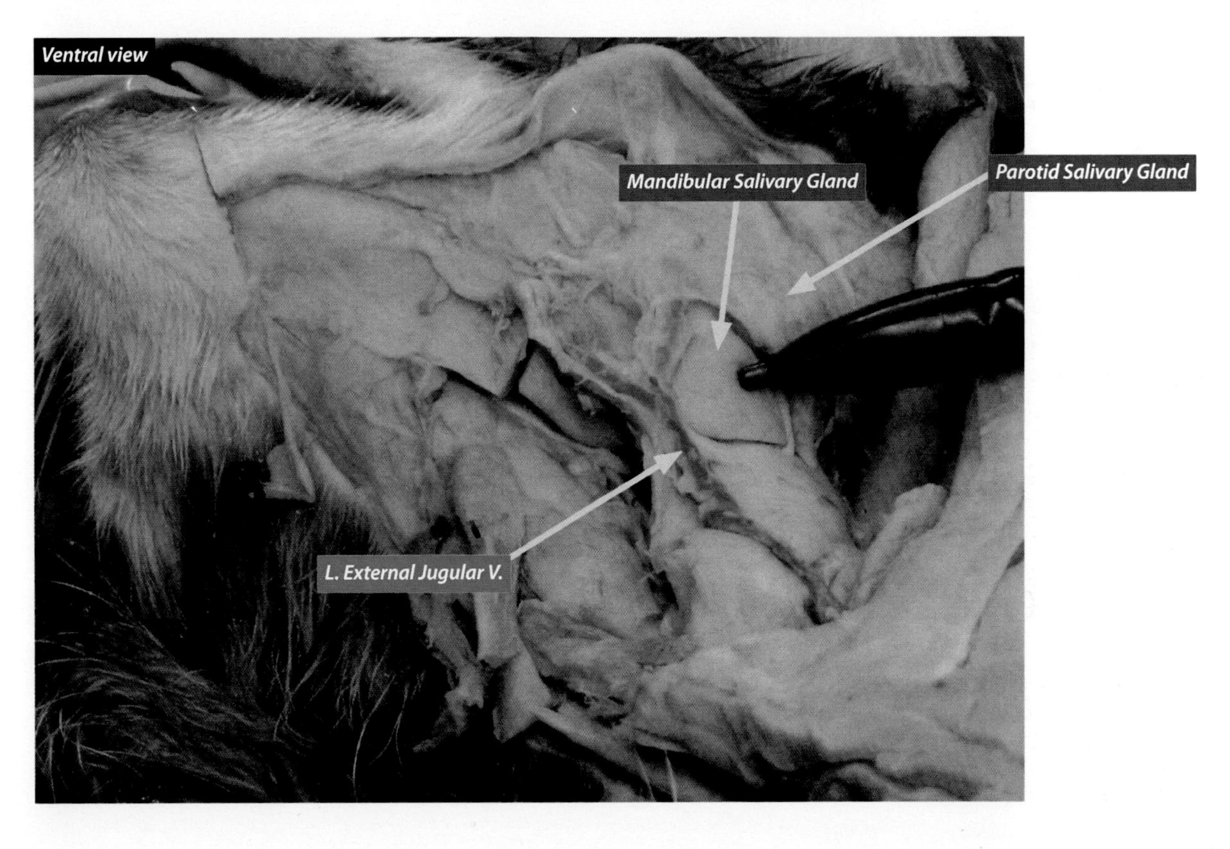

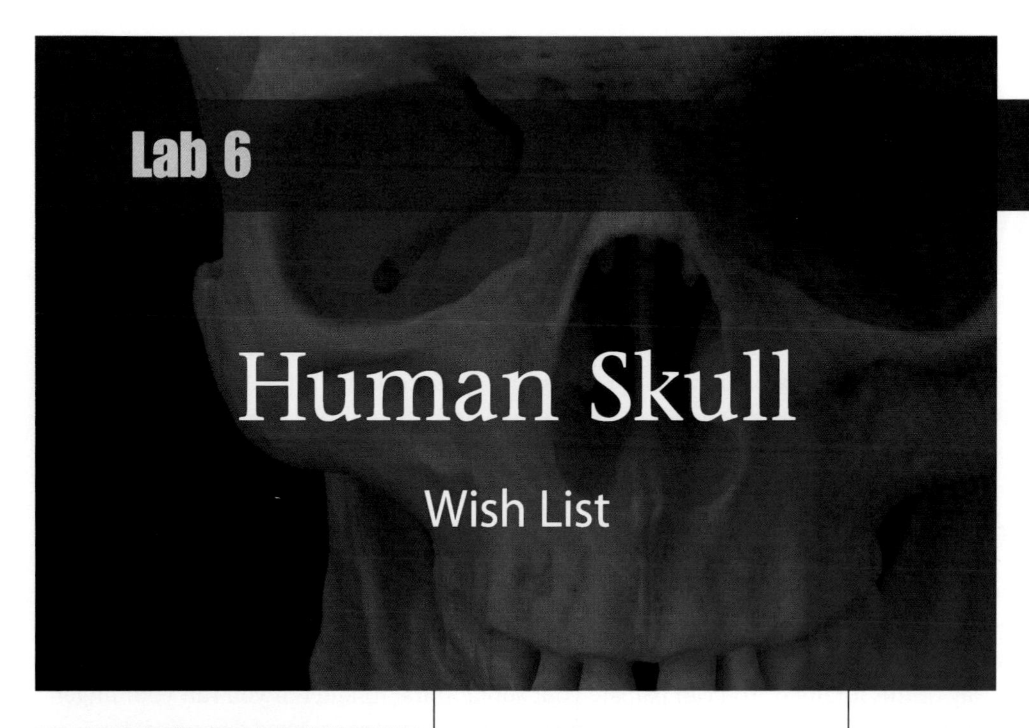

# Lab 6

# Human Skull

## Wish List


**LAB 6 OVERVIEW, pp. 132–134**

**BONES, pp. 135–143**

Bones of the Cranium, pp. 135–137
- Frontal, p. 135
- Parietal, p. 135
- Occipital , p. 136
- Temporal, p. 136
- Sphenoid, p. 137
- Ethmoid, p. 137

Bones of the Face, pp. 138–141
- Nasal, p. 138
- Lacrimal, p. 138
- Inferior Nasal Conchae, p. 139
- Maxilla, p. 139
- Mandible, p. 140
- Zygomatic, p. 140
- Palatine, p. 141
- Vomer, p. 141

Other Bones of the Skull, pp. 142–143
- Auditory Ossicles, p. 142
  - Incus, p. 142
  - Malleus, p. 142
  - Stapes, p. 142
- Hyoid, p. 143

**LANDMARKS, pp. 144–159**
- Fontanels (Fetal Skull), p. 144
- Sutures, pp. 144–145
  - Coronal Suture, p. 144
  - Lambdoidal Suture, p. 145
  - Sagittal Suture, p. 145
  - Squamosal Suture, p. 145
- Occipital Bone Landmarks, p. 146
  - External Occipital Protuberance, p. 146
  - Nuchal Lines, p. 146
  - Occipital Condyles, p. 146
- Temporal Bone Landmarks, pp. 147–148
  - Mandibular Fossa, p. 147
  - Mastoid Process, p. 147
  - Petrous Ridge, p. 147
  - Styloid Process, p. 148
  - Zygomatic Arch (Temporal and Zygomatic), p. 148
  - Zygomatic Process, p. 148
- Sphenoid Bone Landmarks, pp. 149–150
  - Lesser Wing, p. 149
  - Greater Wing, p. 149
  - Pterygoid Process, p. 150
  - Sella Turcica, p. 150
- Ethmoid Bone Landmarks, pp. 151–153
  - Cribriform Plate, p. 151
  - Crista Galli, p. 151
  - Superior Concha, p. 152
  - Middle Conchae, p. 152
  - Perpendicular Plate, p. 153
  - Nasal Septum (Ethmoid and Vomer), p. 153
- Maxilla Landmarks, p. 154
  - Palatine Process, p. 154
- Mandible Landmarks, pp. 154–156
  - Alveolus (Mandible and Maxilla), p. 154
  - Body, p. 155
  - Condyloid Process, p. 155
  - Coronoid Process, p. 155
  - Mandibular (Sigmoid) Notch, p. 156
  - Ramus, p. 156
- Zygomatic Bone Landmarks, p. 156
  - Temporal Process, p. 156
- Palatine Bone Landmarks, p. 157
  - Horizontal Plate, p. 157
- Vomer Landmarks, p. 157
  - Nasal Septum (Vomer and Ethmoid), p. 157
- Sinuses, pp. 158–159
  - Frontal Sinuses, p. 158
  - Sphenoid Sinuses, p. 158
  - Ethmoid Sinuses, p. 159
  - Maxillary Sinuses, p. 159

**FORAMINA, pp. 160–167**
- Optic Canal, p. 160
- Superior Orbital Fissure, p. 160
- Inferior Orbital Fissure, p. 161
- Lacrimal Canal, p. 161
- Foramen Rotundum, p. 162
- Foramen Ovale, p. 162
- Internal Auditory Meatus, p. 163
- Jugular Foramen, p. 163
- Foramen Magnum, p. 164
- Hypoglossal Canal, p. 164
- Stylomastoid Foramen, p. 165
- Mandibular Foramen, p. 165
- External Auditory Meatus, p. 166
- Supraorbital Foramen (Notch), p. 166
- Infraorbital Foramen, p. 166
- Mental Foramen, p. 167

**OVERVIEW—FORAMINA, p. 167**

**CRANIAL NERVE OVERVIEW, p. 168**

**CRANIAL NERVE CHART, p. 169**

# LAB 6 OVERVIEW

In Lab 6 we will study bones of the skull and its foramina and landmarks. The focus is to make bridges with information from Lecture Quiz 9 on the cranial nerves, as well as with the attachments for muscles that move the skull. We will use two physical divisions: the bones of the cranium and the bones of the face. With that as a framework, we will move from anterior to posterior and from superior to inferior for each division.

## 1. Bones of the Cranium

We begin with the **frontal bone**. The **frontal sinus** can be seen in the skull that is cut in a sagittal plane. It can also be seen in the x-ray of the skull that we have in lab. Be sure and ask to see the x-ray on a light table. The **supraorbital foramen** (or **supraorbital notch**) is superior to the orbit. The **ophthalmic nerve** ($V_1$) passes onto the face by way of this foramen. You can palpate your supraorbital foramen if you run your finger across where your eyebrow is or used to be. The frontal bone is singular and it is considered a bone of the cranium, as well as a bone of the face. There are two frontal bones in the fetal skull and you will see that they are separated by a **fontanel**. That fontanel extends between the frontal bones and the parietal bones and later becomes the **coronal suture** in an adult.

The two **parietal bones** are separated by the **sagittal suture**. Also, the majority of the **squamosal suture** is located between the parietal bone and the temporal bone. There are no other landmarks that you are responsible for on the parietal bones.

The **occipital bone** is a single bone found at the posterior portion of the skull. It is separated from the parietal bone by the **lambdoidal suture**, which was a fontanel in the fetal skull. The next several landmarks were already studied in Lab 1. The **external occipital protuberance** is at the midline of the occipital bone. Adjacent to it are the **superior nuchal lines**. The **inferior nuchal lines** are inferior to the superior nuchal lines. The **occipital condyles** are located lateral to the **foramen magnum**. In the anterolateral wall of the condyles you will see the **hypoglossal canal**, which transmits the **hypoglossal nerve** (XII). As a point of information, a canal is a foramen with appreciable depth, like a tube.

The **temporal bones** are inferior to the parietal bones, along the lateral side of the cranium. The **zygomatic process** of the temporal bone articulates at its anterior end with the **temporal process** of the **zygomatic bone**. These two processes make up the **zygomatic arch**. Note, often a process is named for the bone with which it articulates rather than the bone of which it is a part. The **mandibular fossa** of the temporal bone is on the inferior side of the cranium and it is where the **condyloid process** of the **mandible** articulates with the temporal bone. The **styloid process** of the temporal bone extends inferiorly and is posterior to the mandibular fossa. Lateral and posterior to the styloid process you will find the **mastoid process** of the temporal bone. Between the styloid process and the mastoid process is the **stylomastoid foramen**, through which the **facial nerve** (VII) passes to the outside of the skull.

Inside the cranium you will see the **petrous ridge** of the temporal bone, which is an irregular bony ridge between the middle fossa of the cranium and the posterior fossa of the cranium. There is a relatively vertical face of the temporal bone between the occipital bone and the petrous ridge. Near the center of that vertical surface you will find the **internal auditory meatus (canal)**. The **facial nerve** (VII) and the **vestibulocochlear nerve** (VIII) pass out of the cranium together through this canal.

Externally, you will observe the **external auditory meatus (canal)**, which is a dead-end opening into the skull. Along the suture between the temporal bone and the occipital bone you will find the **jugular**

**foramen**. The **glossopharyngeal nerve (IX)**, the **vagus nerve (X)**, the **spinal accessory nerve (XI)**, and the **internal jugular vein** pass through this foramen.

The **sphenoid bone** is singular and can be seen both inside and outside of the cranium. The **lesser wing** of the sphenoid bone is at the posterior margin of the anterior fossa. The **optic canal** is part of the lesser wing and the **optic nerve (II)** passes through this foramen. Immediately posterior to the lesser wing you will observe the **sella turcica**. Lateral to the sella turcica is the **greater wing** of the sphenoid bone. This can also be seen on the outside of the skull, anterior to the temporal bone and inferior to the parietal and frontal bones. There are two foramina on the greater wing for which you are responsible, the **foramen rotundum** and the **foramen ovale**. The **maxillary nerve ($V_2$)** passes through the foramen rotundum, and the **mandibular nerve ($V_3$)** passes through the foramen ovale. Between the lesser wing and the greater wing you will find the **superior orbital fissure**. It transmits the **oculomotor nerve (III)**, the **trochlear nerve (IV)**, the **ophthalmic nerve ($V_1$)**, and the **abducens nerve (VI)**. You can observe the **sphenoid sinus** in the skull that is cut with a sagittal cut. It can also be seen in the x-ray of the skull. Turn the skull so that you can see the inferior side and observe the **medial** and **lateral pterygoid processes**.

There is only one **ethmoid bone**, and it is about the size and shape of a walnut. Superiorly, you will find the **crista galli**, which is a sail-like shape that extends between the frontal lobes of the brain. The **cribriform plate** of the ethmoid bone is adjacent to the crista galli and is in a relatively horizontal plane. The cribriform plate has the **olfactory foramina** that the **olfactory nerve (I)** passes through. This is an unusual situation, because rather than one foramen on each side, there are many foramina on each side. The **perpendicular plate** of the ethmoid bone extends into the nasal cavities and forms the superior portion of the **nasal septum**. The inferior portion of the nasal septum is formed by the **vomer**, and the anterior portion of the nasal septum is formed by septal cartilage. The septal cartilage is hyaline cartilage and will be missing from the skulls in lab. The **ethmoid sinuses** are lateral and superior to the nasal cavities, and they will be seen in an x-ray of the skull or in an individual model of the bone. The ethmoid bone has three conchae: the supreme conchae, the superior conchae, and the **middle conchae**. The only pair you will observe in the plastic skulls is the pair of middle conchae.

## 2. Bones of the Face

The **nasal bones** are relatively small and are found at the bridge of the nose. They do not have landmarks that you are responsible for. The **lacrimal bones** are also relatively small and they are found in the anteromedial corner of the orbit of the eye. The **lacrimal canal** passes through and provides for communication between the orbit and the nasal cavity, and it contains a duct that transmits tears. The lacrimal canal and the external auditory meatus are the only foramen we study that do not have a nerve passing through them.

The **maxillary bone** is larger than it appears to be at first glance, because not only is it on the anterior portion of the face, it also forms the inferior portion of the orbit and the anterior portion of the hard palate. Inferior to the orbit you will see the **infraorbital foramen**, which transmits the **maxillary nerve ($V_2$)** onto the face. The **palatine process** of the maxillary bone forms the anterior portion of the hard palate and is named for the bone with which it articulates. The **alveoli** (alveolus is the singular form) are depressions in the maxillary bone (and mandible) into which the teeth are anchored. Examine the individual model of the maxillary bone and you can see the **maxillary sinus**. This sinus can also be seen in the x-rays of the skull.

The **inferior nasal concha** is part of the lateral wall of the nasal cavity. It is the only one of the conchae that is a separate bone.

The **mandible** is inferior to the maxillary bone. As was the case for the maxillary bone, it contains **alveoli** for the teeth. The **body** is the portion of the mandible with the alveoli in it. At the anterolateral surface of the body you will find the **mental foramen**. This is the third foramen through which the

**mandibular nerve** ($V_3$) passes. Posteriorly, the **ramus** projects superiorly from the body. At the superior end of the ramus there is a **coronoid process** anteriorly and a **condyloid process** posteriorly. They are separated by the **mandibular (sigmoid) notch**. As a point of interest, this notch is named for a famous anatomist, Sigmoid Freud! Wow! The condyloid process articulates with the mandibular fossa of the temporal bone. On the deep side of the ramus you will observe the **mandibular foramen**, which transmits the **mandibular nerve** ($V_3$).

The **zygomatic bone** is posterior to the maxillary bone and makes up what is commonly referred to as the cheek bone. The **temporal process** of the zygomatic bone articulates with the temporal bone and this process forms the anterior end of the **zygomatic arch**.

The **palatine bone** is found posterior to the palatine process of the maxillary bone. The **horizontal process** of the palatine bone forms the posterior portion of the hard palate.

The **vomer** forms the inferior portion of the **nasal septum**, particularly the posterior part of the nasal septum. In fact, all of what is seen at the posterior portion of the septum where it meets the pharynx is the vomer.

The **inferior orbital fissure** is not a feature of one bone, but rather is formed by the union of the maxillary bone, the palatine bone, the sphenoid bone, and the zygomatic bone. The **maxillary nerve** ($V_2$) is the only nerve passing through this feature.

## 3. Other Bones of the Skull

The **auditory ossicles** are within the middle ear. They include the **incus**, the **malleus**, and the **stapes**. They will be found in a separate container.

The **hyoid bone** was covered in Lab 1, but will be covered again here because it is considered a bone of the skull, although it is set apart from it and located inferior to it.

# BONES

## *Bones of the Cranium*

### Frontal Bone

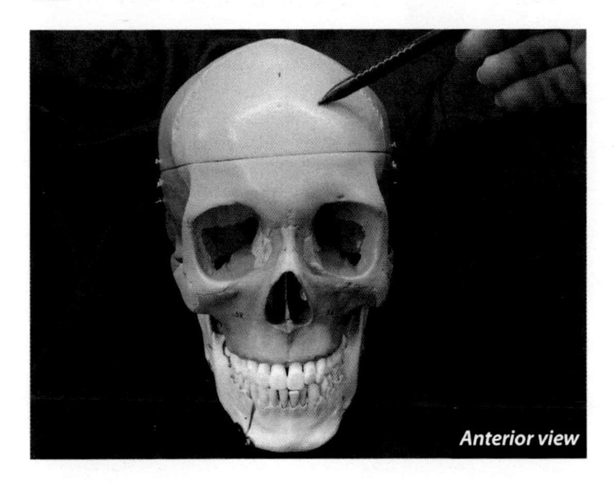
*Anterior view*

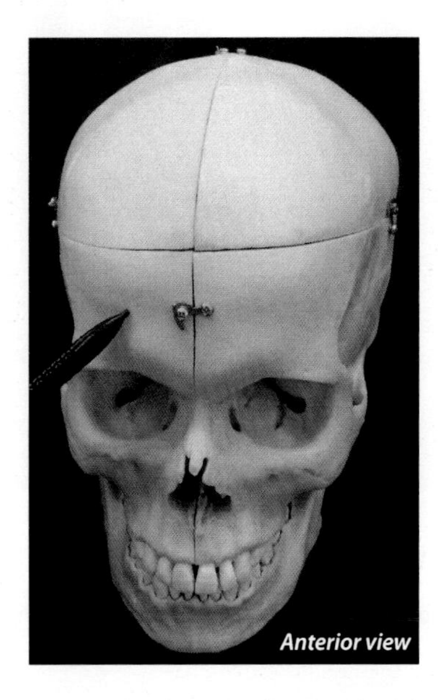
*Anterior view*

The frontal bone is a single bone of the cranium and face, although when we look at the fontanels we will see that it begins as two bones in the fetus. The frontal bone forms the forehead, the roof of each orbit, and the majority of the anterior cranial fossa. Superiorly, it forms the coronal suture with the parietal bones; laterally, it forms a suture with the zygomatic bones, and anteriorly it meets the nasal bones. Landmarks found on the frontal bone include the frontal sinus and the supraorbital foramen (notch). These will be discussed in turn.

### Parietal Bones (2)

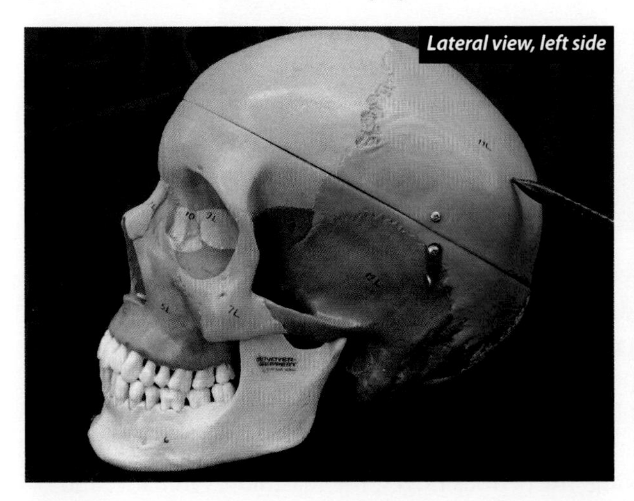
*Lateral view, left side*

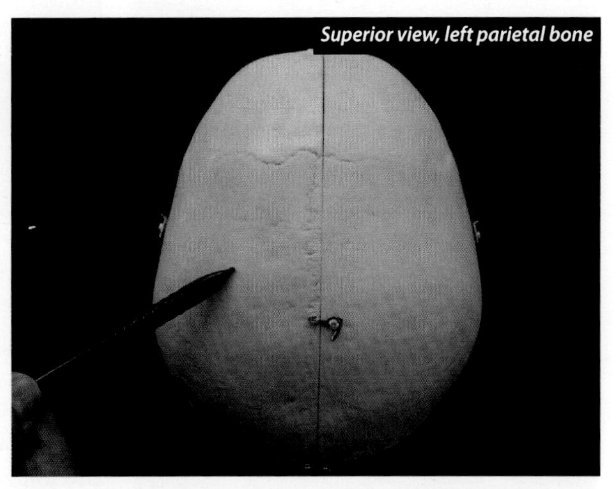
*Superior view, left parietal bone*

The two parietal bones are cranial bones that articulate anteriorly with the frontal bone at the coronal suture, posteriorly with the occipital bone at the lambdoidal suture, and inferiorly with the temporal bone at the squamosal suture. They also articulate with each other along the sagittal suture on the superior surface of the skull.

## Occipital Bone

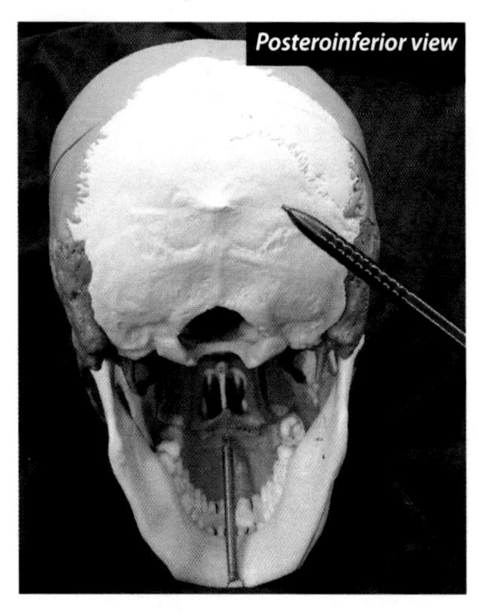


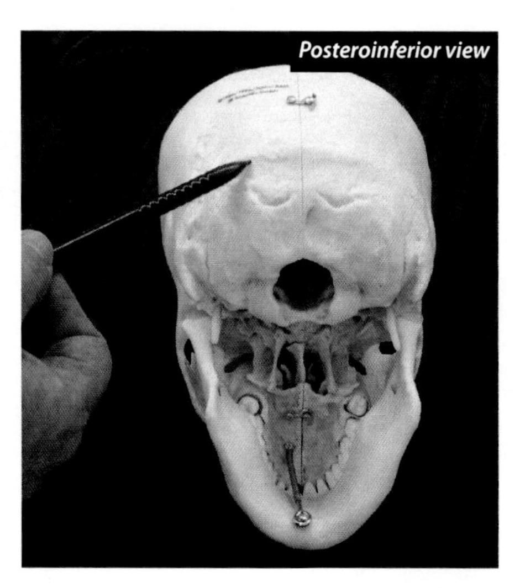


The single occipital bone is part of the posterior cranium. It articulates with the sphenoid bone, temporal bones, atlas, and parietal bones (at the lambdoidal suture). The occipital bone has a number of landmarks, including the following: the external occipital protuberance, nuchal lines, condyles, foramen magnum, hypoglossal canals, and part of the jugular foramina. Each of these will be discussed in turn.

## Temporal Bones (2)

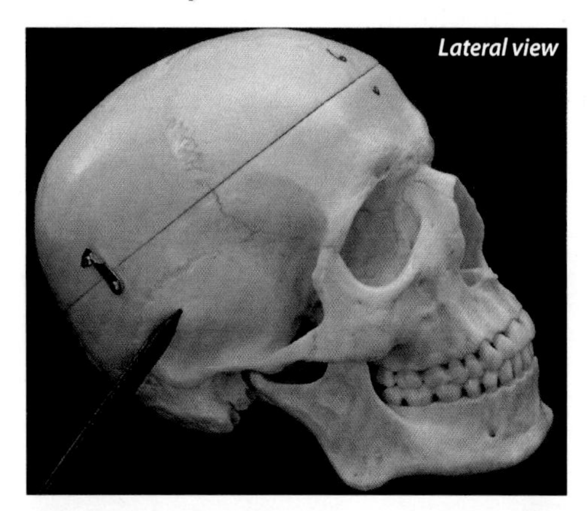


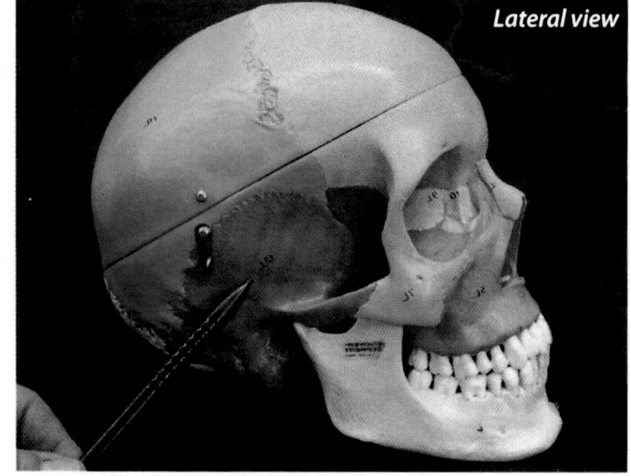


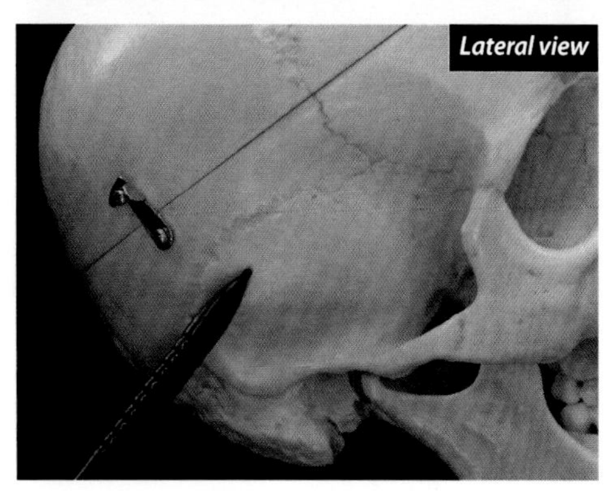


The two temporal bones are bones of the cranium. They are positioned inferior to the parietal bone and articulate with it at the squamosal suture. They articulate anteriorly with the zygoma, posteriorly with the occipital bone, and inferiorly with the mandible. The temporal bones form most of the middle fossa of the cranium. They contain a number of landmarks, including the following: the petrous ridges; the mastoid, zygomatic, and styloid processes; the mandibular fossae, stylomastoid foramina, internal auditory meatus and external auditory meatus; and part of the jugular foramen. Each of these will be discussed in turn.

## Sphenoid Bone

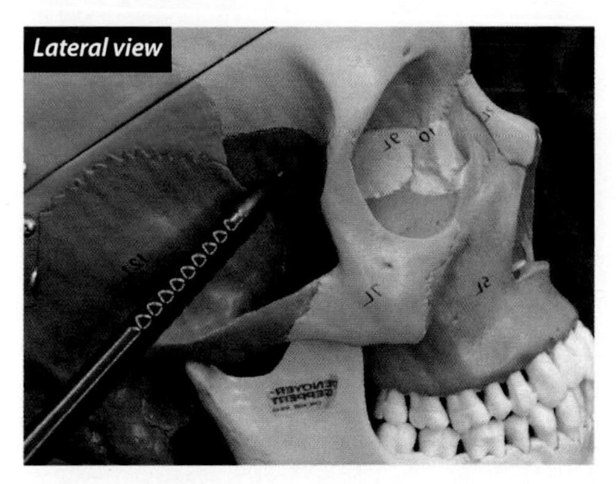
*Lateral view*

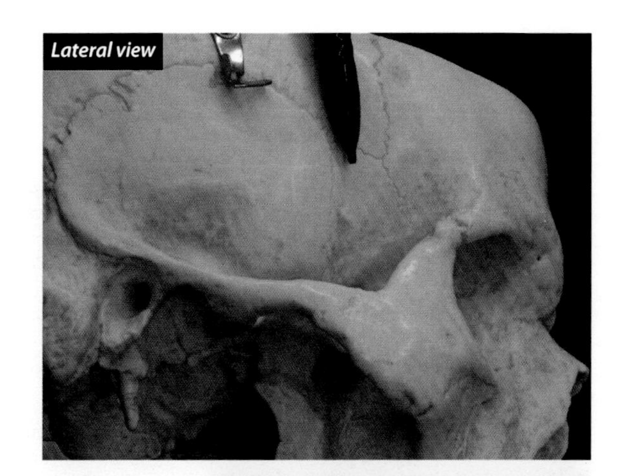
*Lateral view*

The single sphenoid bone is considered a cranial bone. It articulates with every other cranial bone, as well as with the zygomatic, vomer, maxillary, and palatine bones (facial bones). It has many important landmarks, including the following: the greater and lesser wings, the sinuses, the sella turcica, and the pterygoid processes. It also has a number of significant foramina, including the rotundum, the ovale, the optic canals, and the superior orbital fissures, as well as others. These will be discussed in turn.

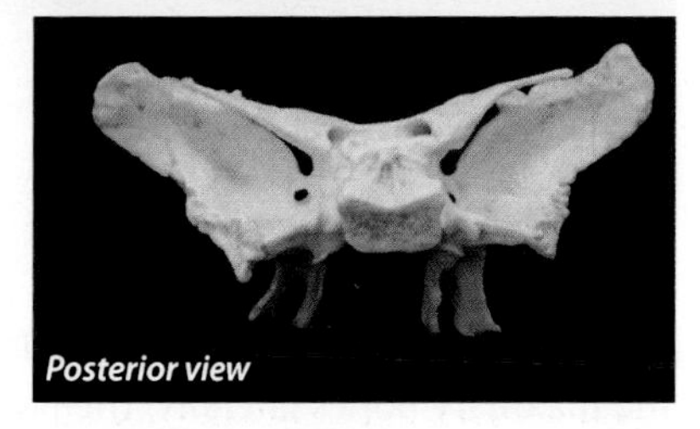
*Posterior view*

## Ethmoid Bone

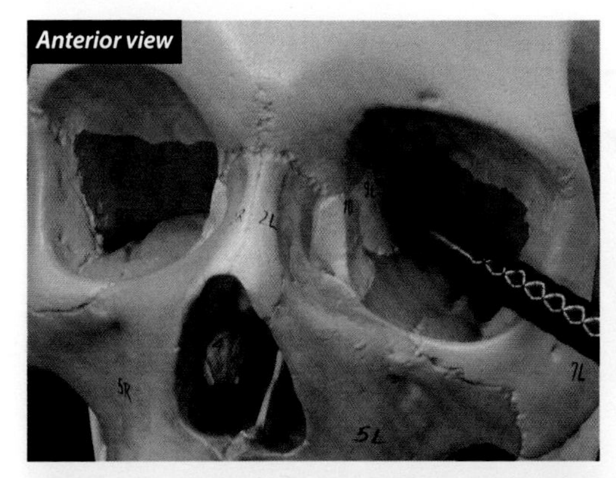
*Anterior view*

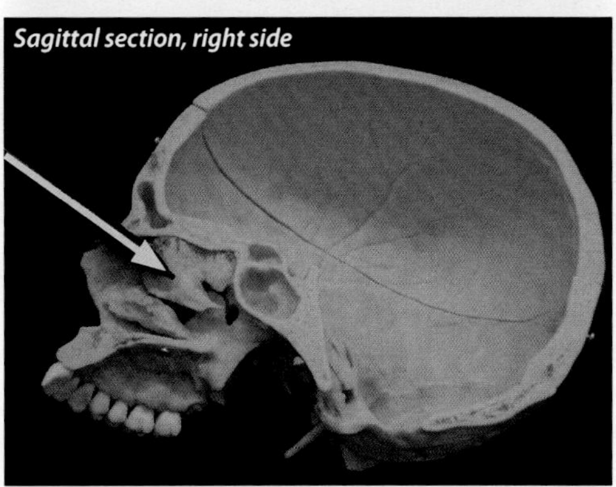
*Sagittal section, right side*

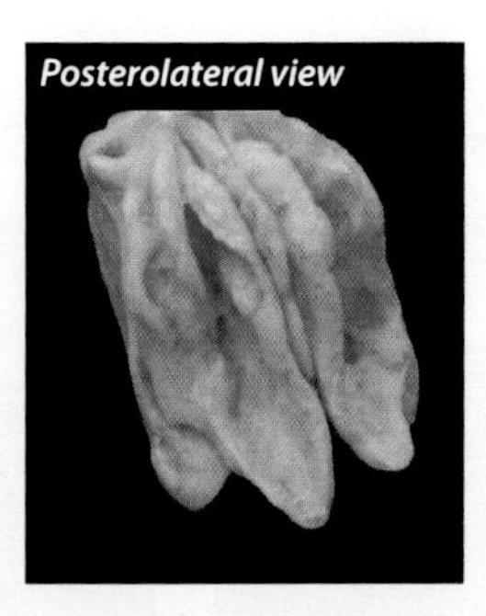
*Posterolateral view*

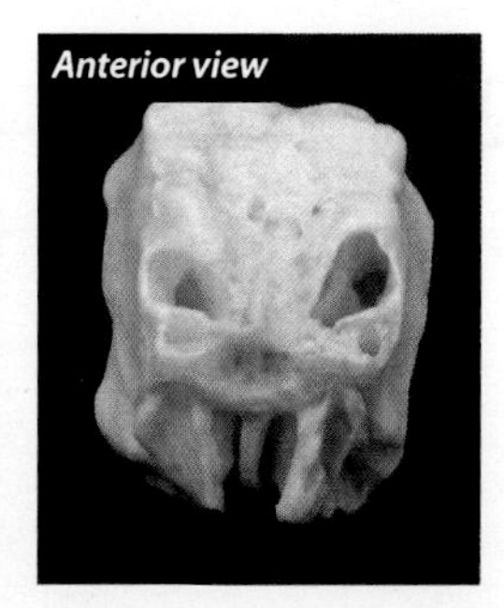
*Anterior view*

The ethmoid bone is a single bone of the cranium. It is anterior to the sphenoid bone and posterior to the nasal bones. It forms most of the area between the nasal cavity and the orbit of the eye. It has a number of landmarks associated with it, including the following: the crista galli, the cribriform and perpendicular plates, the ethmoid sinuses, and the superior and middle conchae. Each of these will be discussed in turn. The word ethmoid comes from the Greek term *ēthmoeidés*, which means sieve-like.

# *Bones of the Face*

## Nasal Bones (2)

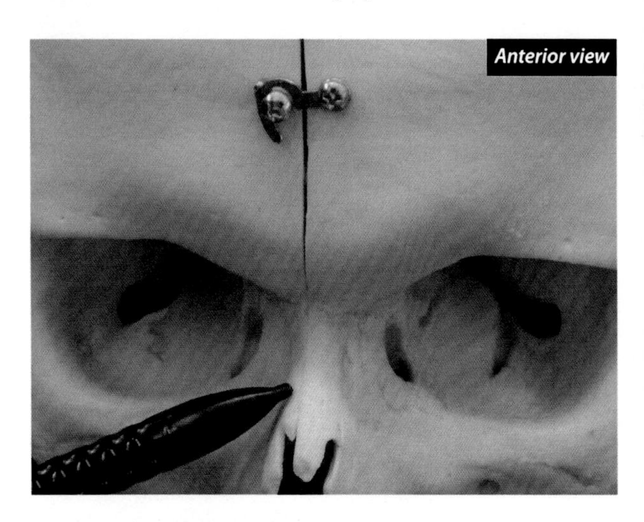


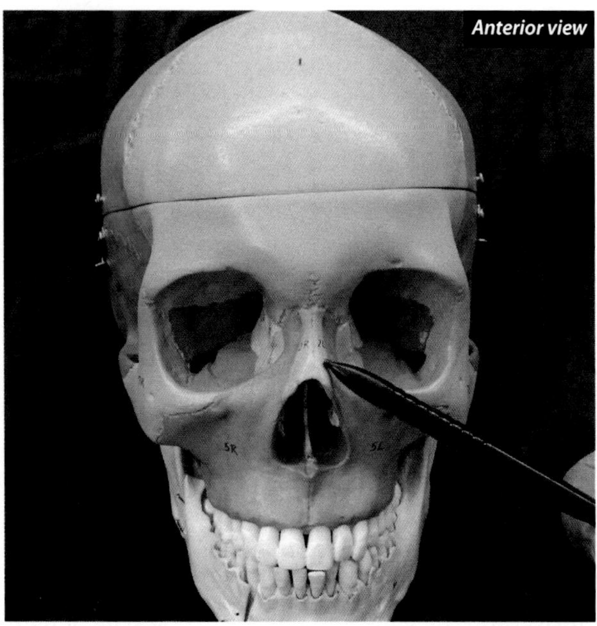


The two small nasal bones are considered facial bones. They form the bridge of the nose. They articulate with the maxillary bones laterally, the perpendicular plate of the ethmoid bone posteriorly, and the frontal bone superiorly. Inferiorly, cartilage that forms most of the external nasal skeleton attaches to the nasal bones.

## Lacrimal Bones (2)

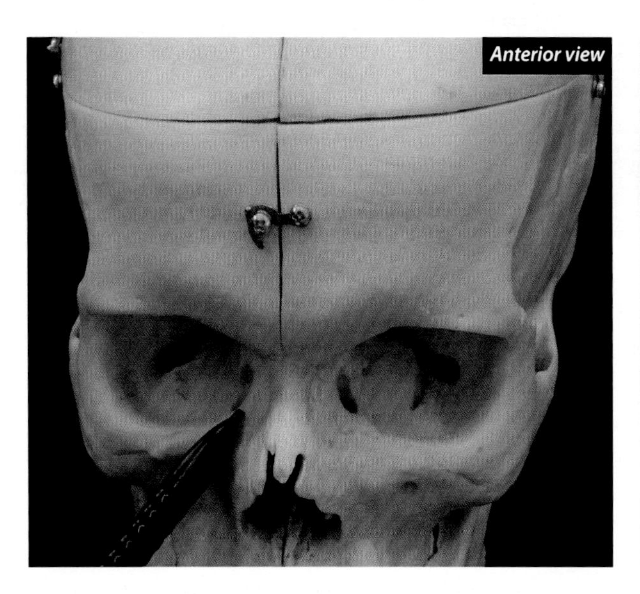


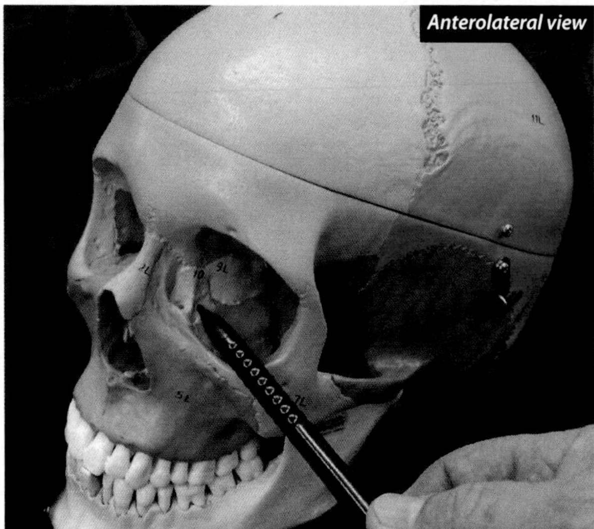


The two lacrimal bones are bones of the face. They make up part of the anterior medial wall of the orbit, and they contact the maxillary bone anteriorly, the ethmoid bone posteriorly, and the frontal bone superiorly. Each lacrimal bone contains a lacrimal canal, which is functionally important as the passage for the tear duct. The canal terminates in the nasal cavity.

## Inferior Nasal Conchae (2)

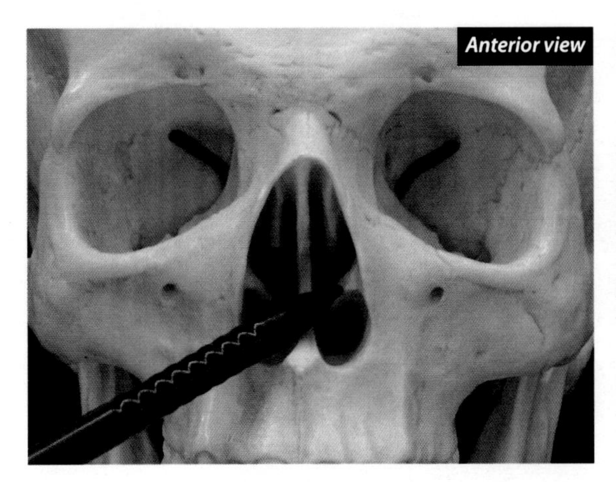


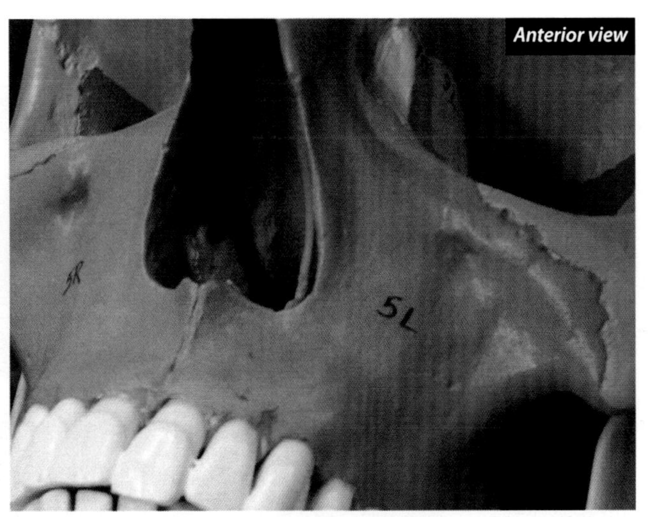


There are two inferior nasal conchae (inferior turbinate), one on the lateral wall of each nasal cavity. Similar structures are features of the ethmoid bone. Functionally, conchae are important because their curved surface causes turbulence as air moves into the nasal cavity, helping to warm, moisten, and filter the air before it enters the lower respiratory tract. They also help recover moisture and heat during expiration. This is especially important in cold and dry conditions and may be one of the reasons why at certain times the mucosa covering these structures becomes swollen on one side, and then after a few hours the swelling reduces on that side, while the other side becomes swollen. Another possible advantage of this variation in air flow between the two nasal cavities would be to allow more intense stimulation of the olfactory nerves on the swollen side, by slowing the flow of air.

## Maxilla (2)

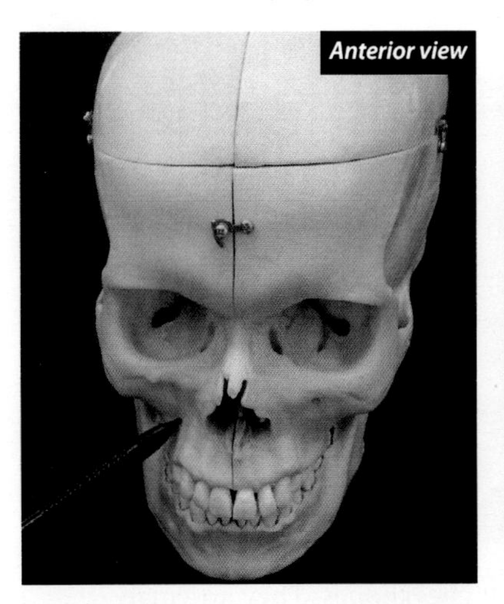


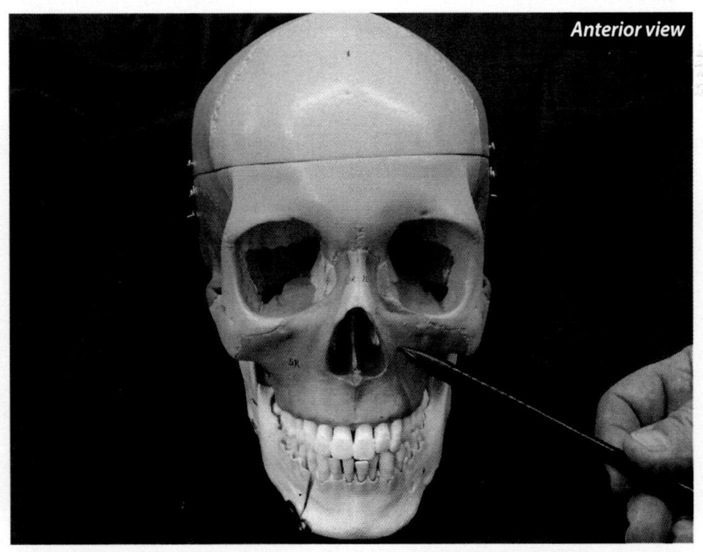


The two maxillary bones are considered part of the face and they form the upper jaw, as well as the anterior two thirds of the hard palate. They also form the lateral walls of the nasal cavities, and a portion of the bony orbit of each eye. They have a number of landmarks, including the following: the sinuses, alveoli and palatine processes, the infraorbital foramina, and part of the inferior orbital fissures. These will be discussed in turn. All the facial bones except the mandible articulate with the maxillary bones. These bones were named in honor of the famous spy, Maxilla Smart.

## Mandible

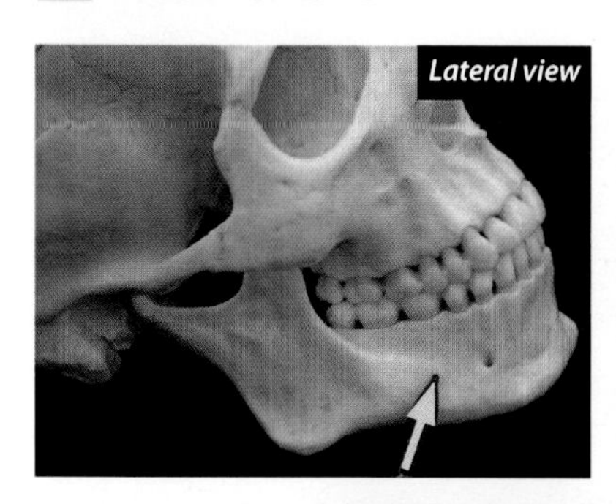


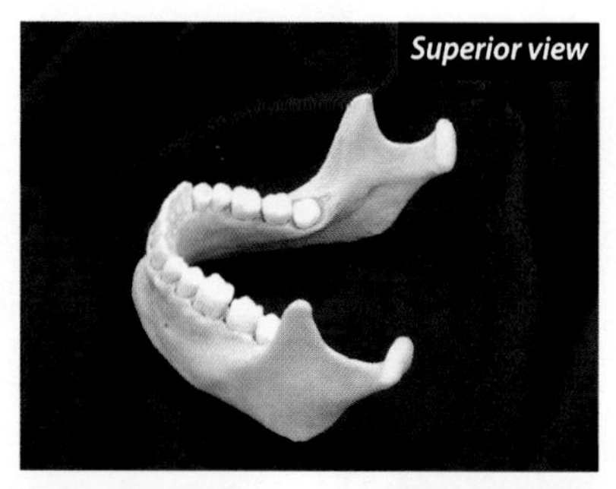


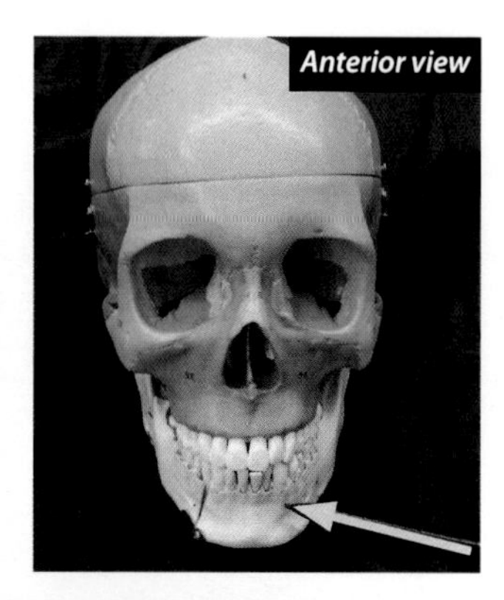


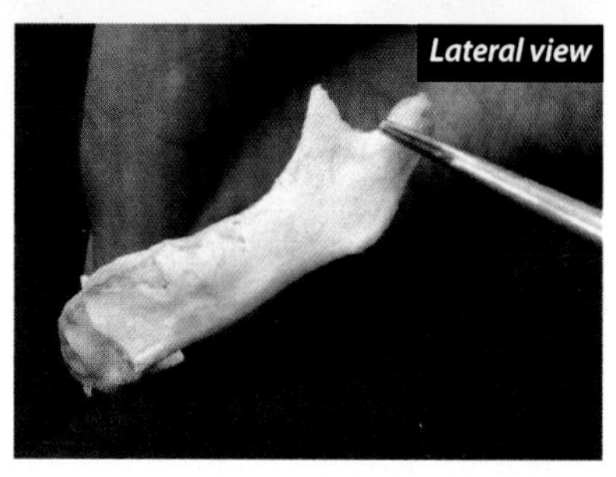


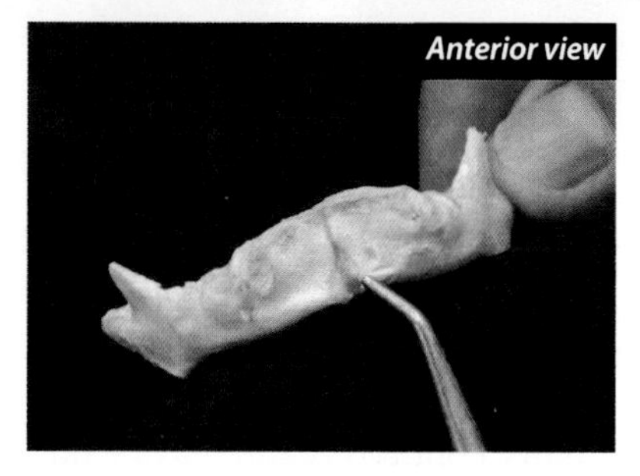


The mandible is a single bone of the face. It articulates with the mandibular fossa of the temporal bone. Irritation of this joint leads to the condition known as TMJ syndrome. When we were in nursery school we called the mandible the lower jaw. It has a number of landmarks associated with it, including the following: the body, rami, alveoli, coronoid and condyloid processes, and mandibular (sigmoid) notches. It also houses two important foramina, the mandibular and mental foramina. These landmarks will be discussed in turn. We find that, like the frontal bone, the mandible is actually two bones in the fetus. The two pictures above of a fetal mandible show where the two portions have not yet fused.

## Zygomatic Bones (2)

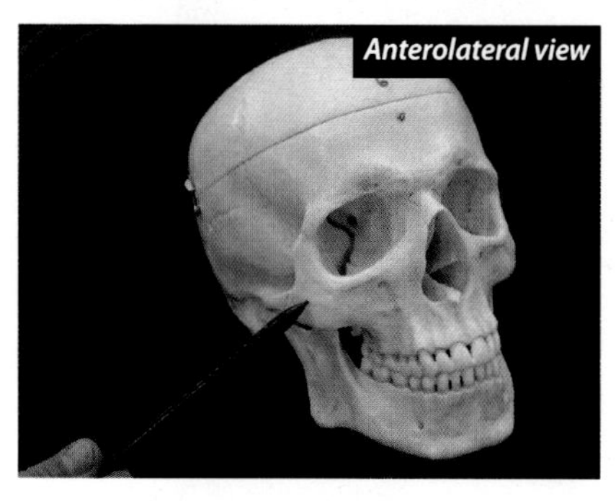


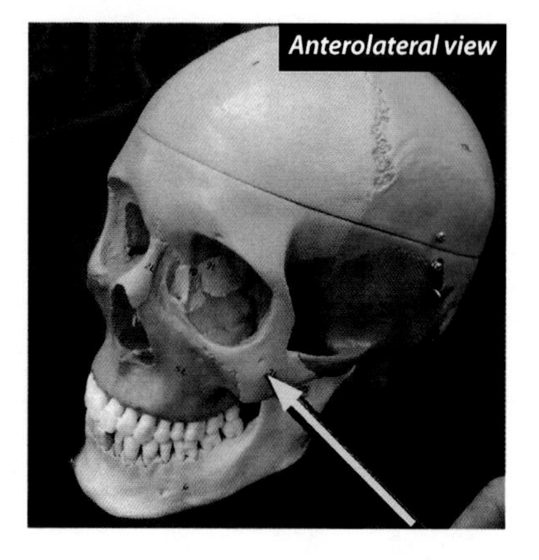


The two zygomatic bones are facial bones. In nursery school we called them cheekbones. They articulate posteromedially with the sphenoid bone, posterolaterally with the temporal bone, superiorly with the frontal bone, and anteriorly with the maxillary bones. The zygomatic bones make up a portion of the bony orbit of the eye. The temporal processes are the only landmarks on these bones that we will study, and they will be discussed in turn. The maxillary process of the zygomatic bone (you are not responsible for this landmark) is of special functional importance because it is part of the origin for:

1. the masseter muscle.

## Palatine Bones (2)

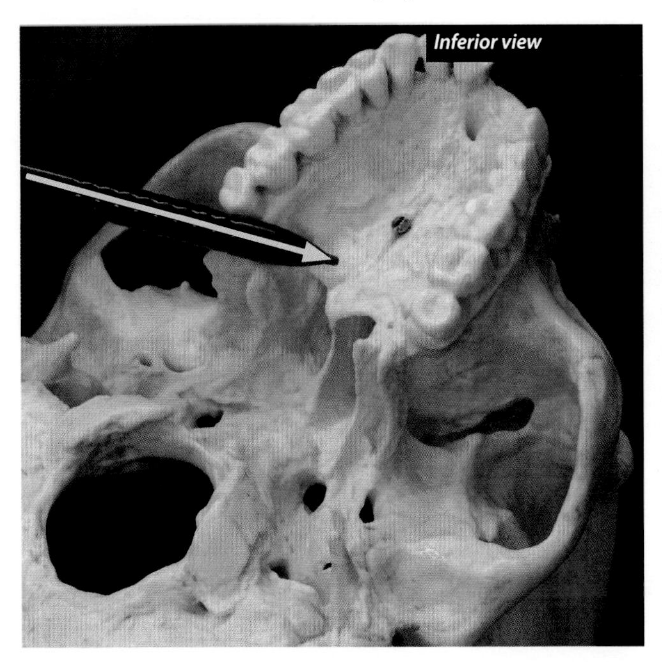


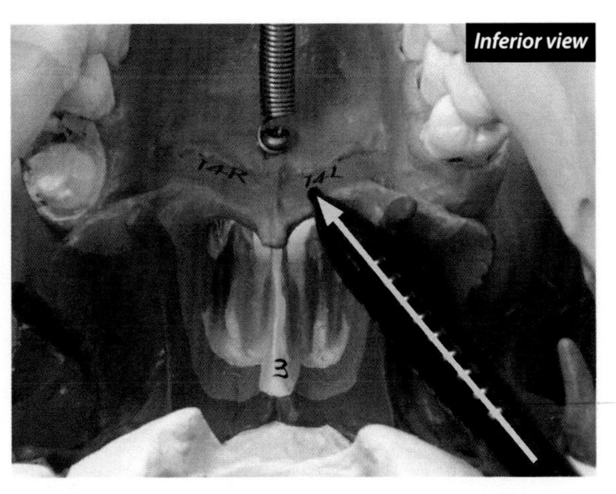


The two palatine bones are facial bones that form the posterior third of the hard palate, where they articulate with the palatine process of the maxillary bones. The portion of those bones that forms the hard palate is called the horizontal plate. Each bone also has a perpendicular (vertical) plate, which you will not be responsible for on the lab practical. You will, however, be responsible for the perpendicular plate of the ethmoid bone, which will be discussed in turn.

## Vomer

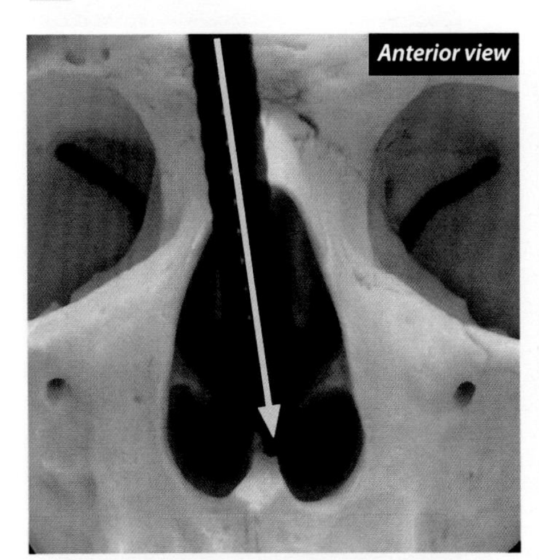


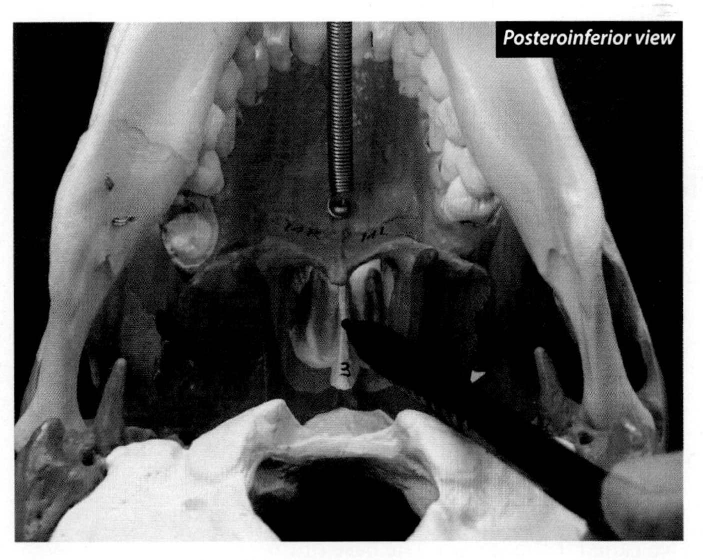


The vomer is a single bone of the face forming the inferior portion of the nasal septum. It starts as a large portion of that septum posteriorly and then narrows to a point at its anterior end. For this reason, it is often referred to as being "plow-shaped." Given what plows look like on trucks that drive on Rhode Island roads, Dr. J thinks wedge-shaped is a better description. Anteriorly, the nasal septum is composed of cartilage.

# *Other Bones of the Skull*

## Auditory Ossicles (3)

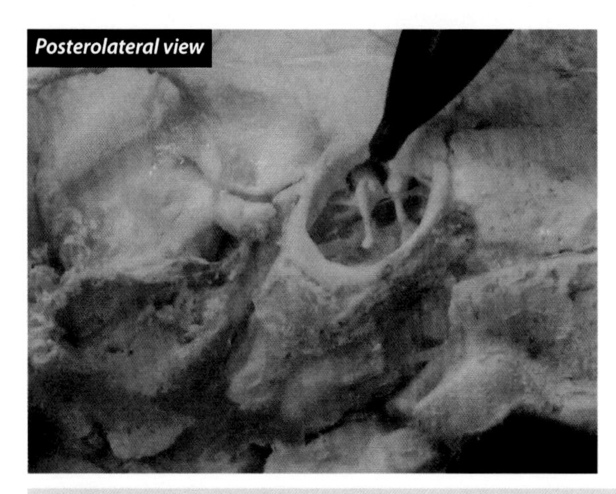


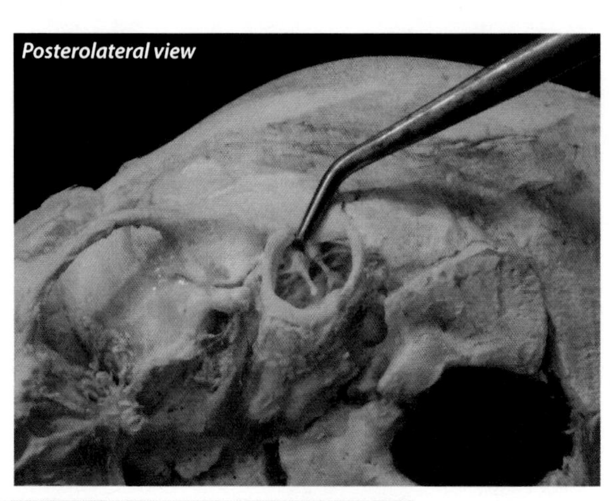


*Note that these are ossicles that can be seen in the fetal skull—we are looking into the external auditory meatus.*

### Incus

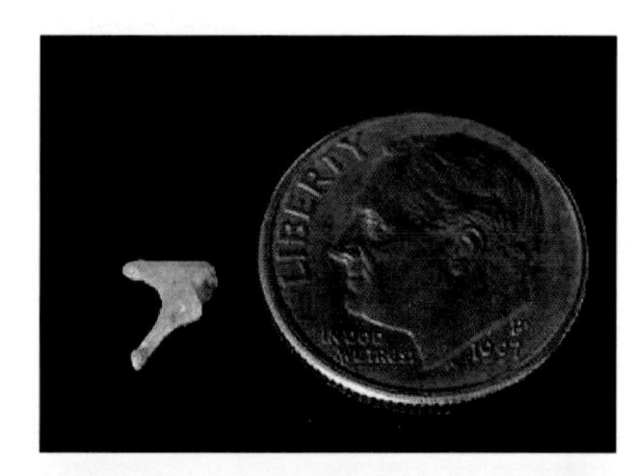

The incus is one of the three auditory ossicles. In nursery school we called it the anvil. Functionally it is important because in association with the other ossicles it helps amplify the pressure of vibrations associated with sound by about twenty times. These three bones are located in the middle ear, which is within the temporal bone.

### Malleus

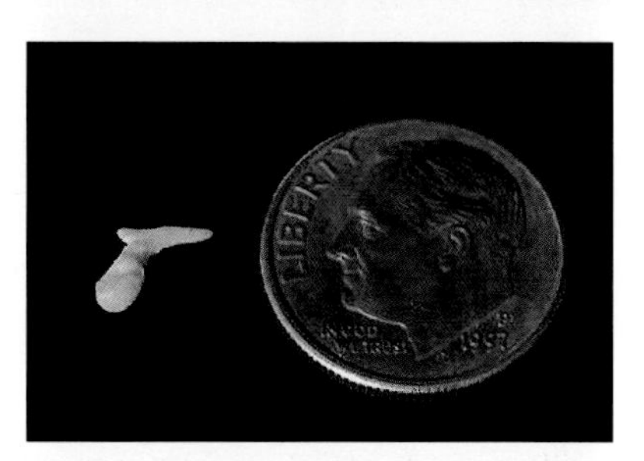

The malleus is one of the three auditory ossicles. When we were in nursery school, we called the malleus the hammer. The "handle" of the malleus attaches to the eardrum.

### Stapes

The stapes is one of the three auditory ossicles. In nursery school we called it the stirrup. The base of the stapes contacts the oval window, a hole in the medial wall of the middle ear.

# Hyoid Bone

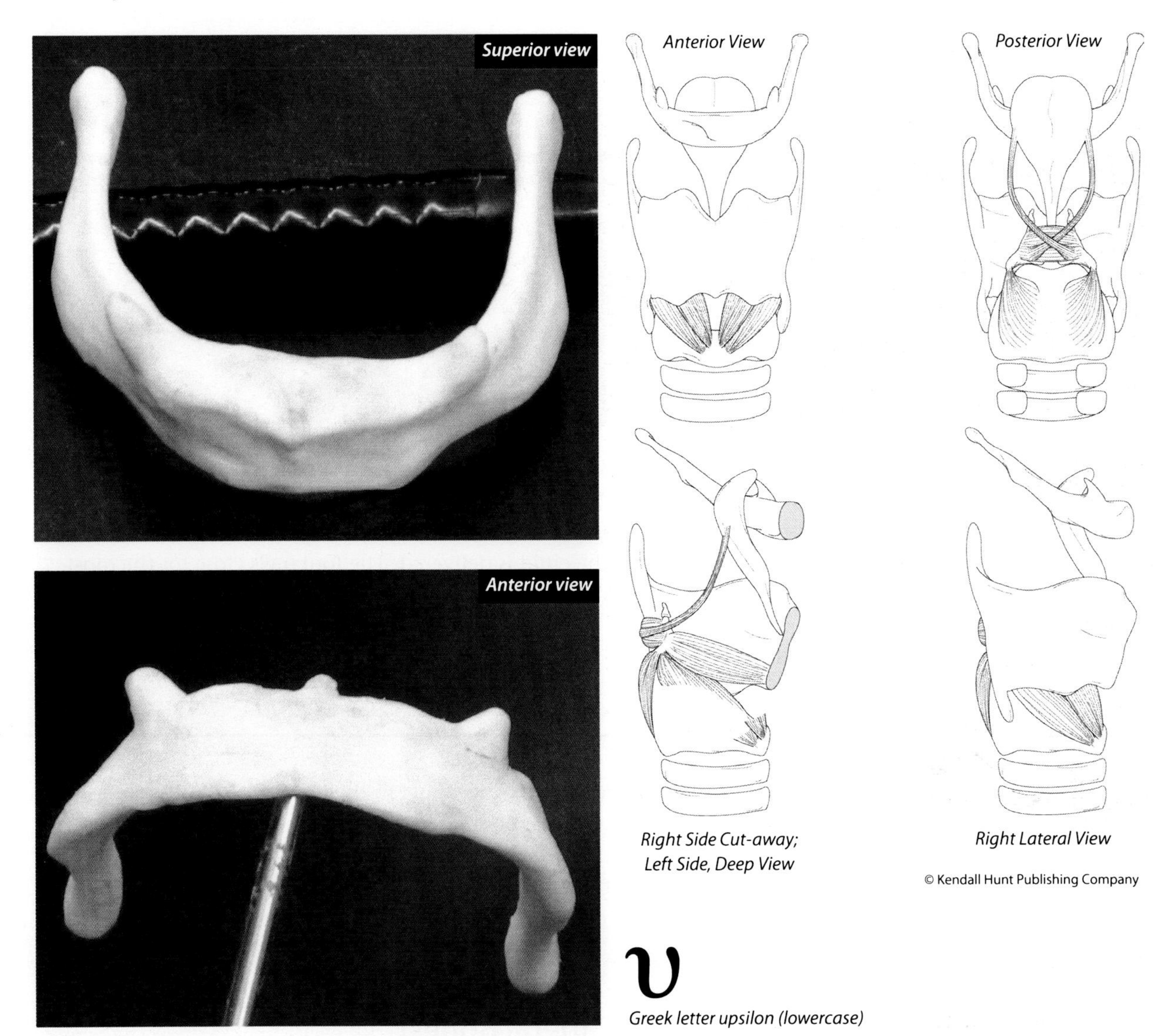


Right Side Cut-away; Left Side, Deep View

Right Lateral View



Greek letter upsilon (lowercase)

The hyoid bone does not articulate with any other bones. It is held in place by ligaments that connect it to the styloid process of the temporal bone and ligaments that connect it with the thyroid cartilage of the larynx. In spite of the fact that it is not attached to the skull, it is considered part of the skull (and therefore part of the axial skeleton). The hyoid resembles the mandible, suggesting a common origin. Functionally it is important because it serves as the origin for muscles that move the larynx during the act of swallowing. The hyoid gets its name from early Greek anatomists who thought it resembled the lower case form of the letter "upsilon." It is also the favorite bone of the seven dwarves, made famous in their song "Hyoid, hyoid, it's off to work we goid."

# LANDMARKS

## Fontanels (Fetal Skull)

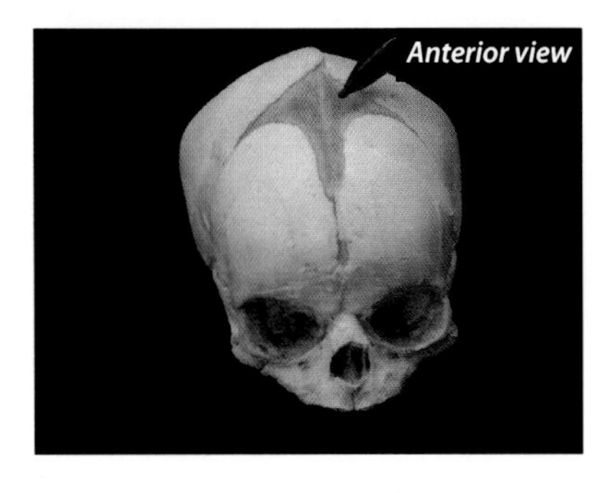


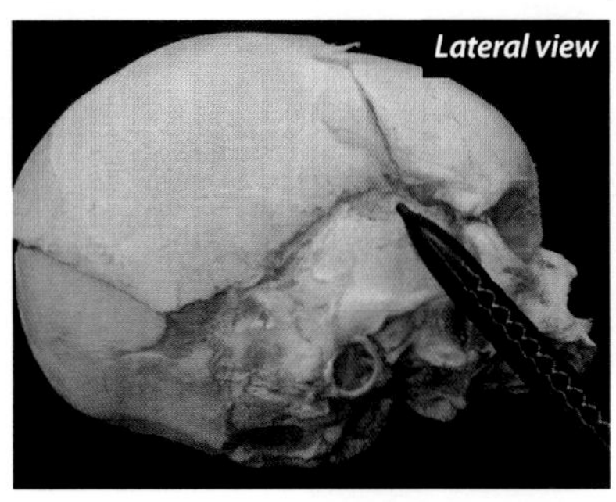


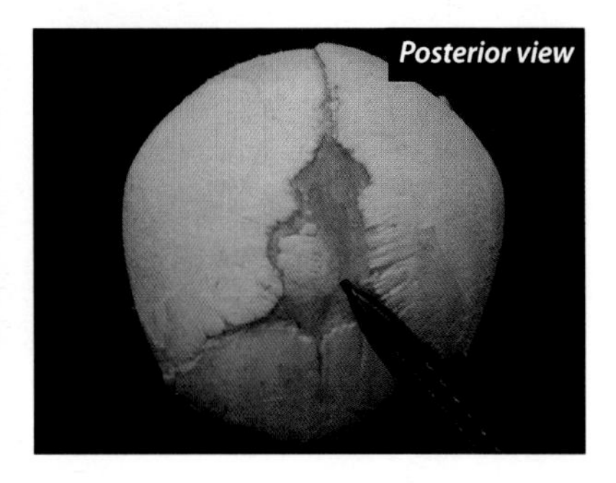


The word fontanel means "little fountain." In nursery school we referred to these as the soft spots of the skull. Eventually they become sutures. Technically, these are fibrous joints or ligamentous unions. The flexibility that comes about because of these structures is important during birth so that passage through the birth canal will be easier and so that growth can occur after the child is delivered. The fontanels close for the most part by the time the child is a year old. However, complete ossification of the ligaments separating the bones doesn't begin until the individual's late twenties and is not complete until the person's fifties. This is a liability because if a newborn always sleeps with its head in the same position, the shape of the head may be affected.

## Sutures (4)

Each suture is an example of a synostosis. This structure starts as a fontanel in the fetus. Technically these are fibrous joints or ligamentous unions. The flexibility that comes about because of these structures is important during birth so that passage through the birth canal will be easier and so that growth can occur after the child is delivered. The fontanels close for the most part by the time the child is a year old. However, complete ossification of the ligaments separating the bones doesn't begin until the individual is in her late twenties and is not complete until she is in her fifties!

### Coronal Suture (between Frontal and Parietal Bones)

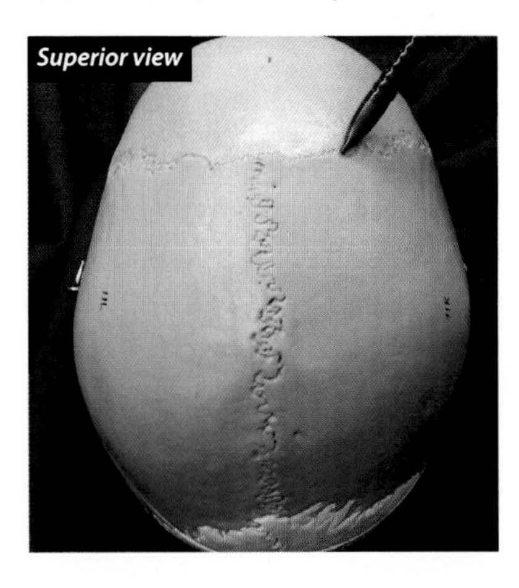


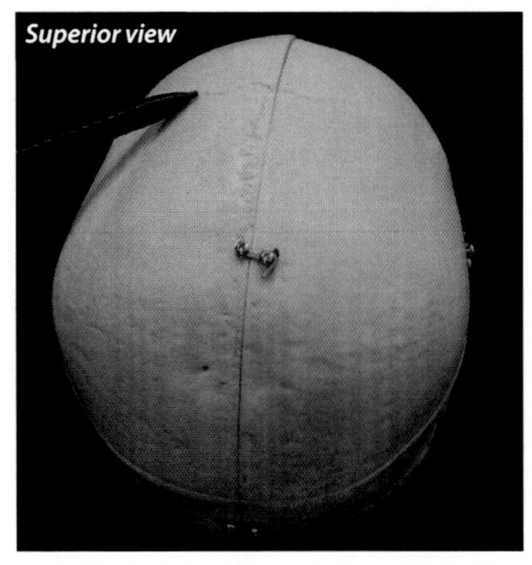


The coronal suture is found between the frontal bone and the two parietal bones.

## Lambdoidal Suture (between the Occipital and Parietal Bones)

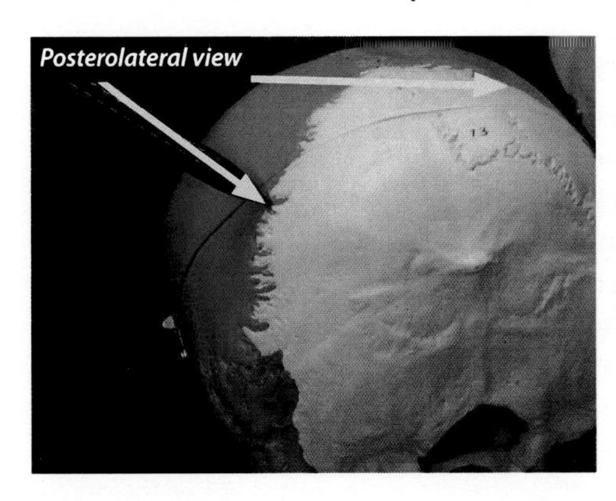


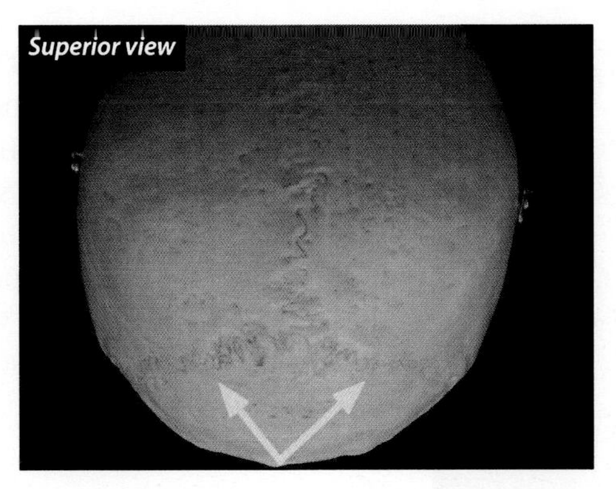


The lambdoidal suture is found between the occipital bone and the two parietal bones. Some people think that the name of this suture comes from the fact that it looks like a capital Greek letter lambda. Dr. J suspects it was named for the famous movie *Silence of the Lambdoidals!*

## Sagittal Suture (between Parietal Bones)

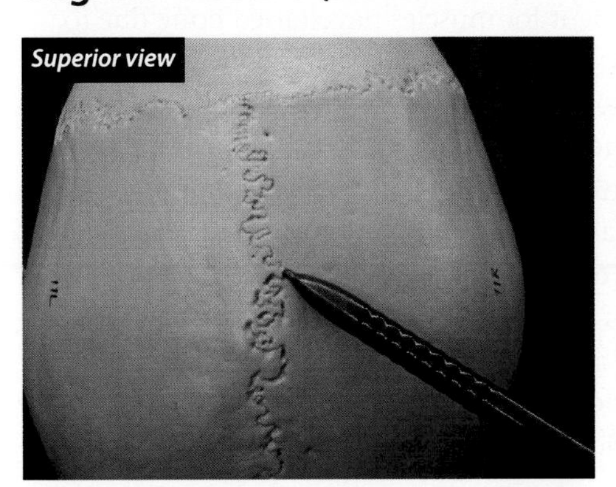


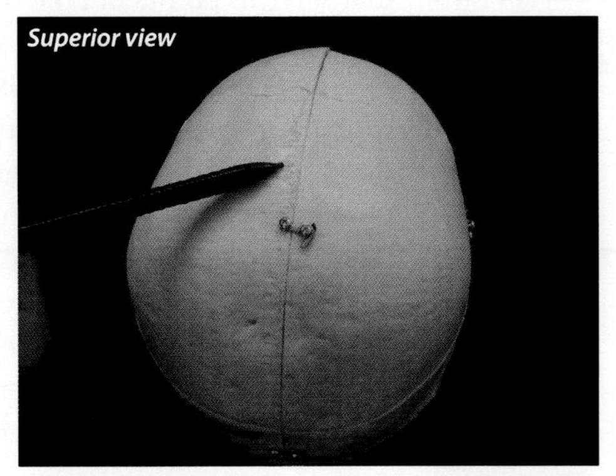


The sagittal suture is found between the two parietal bones.

## Squamosal Suture (between Temporal and Parietal Bones)

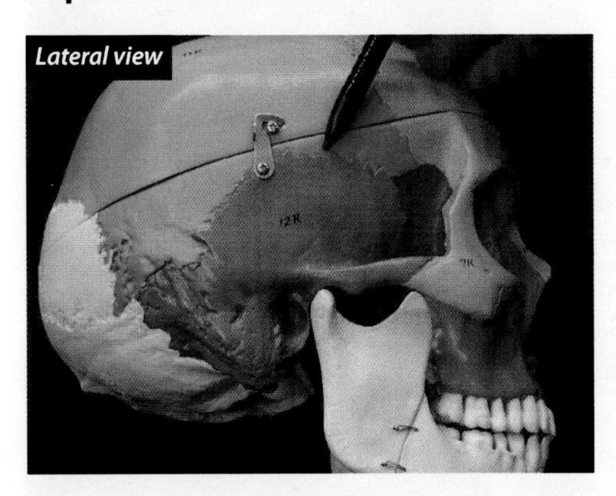


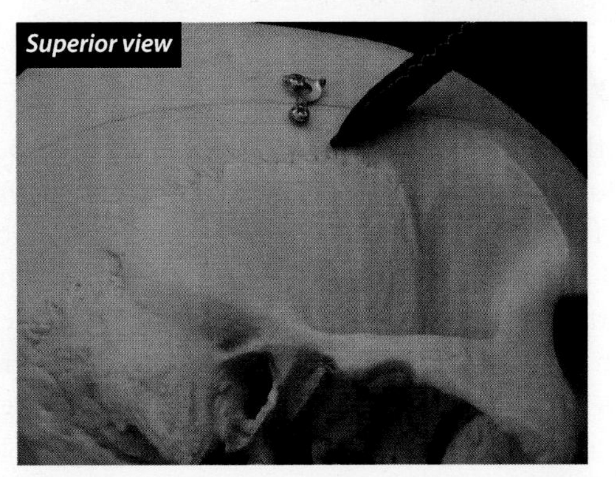


Most of the squamosal suture is found between the temporal bone and the parietal bone on each side of the skull.

# Occipital Bone Landmarks

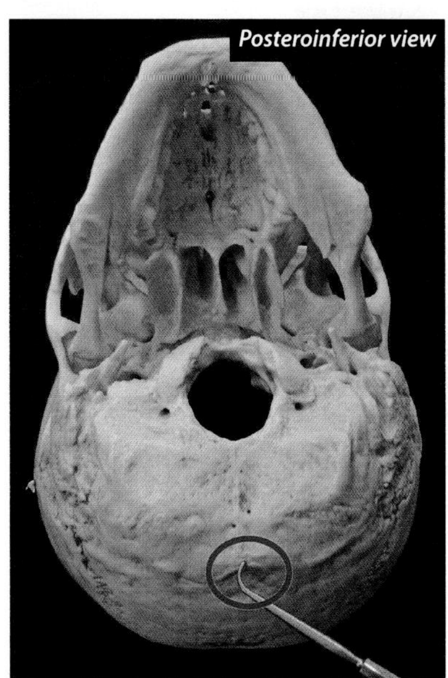


## External Occipital Protuberance (Occipital)

The external occipital protuberance is a raised area on the midline of the occipital bone where the posterior wall meets the base of the skull. It is medial to the two superior nuchal lines. It serves as the origin for the superior division of the trapezius muscle and is an attachment point for the superior extent of the ligamentum nuchae, which connects the cervical vertebrae to the skull. Its highest point is sometimes called the inion or the "bump of knowledge."

## Nuchal Lines (Occipital)

The nuchal lines of the occipital bone are where many muscles and ligaments of the neck and back attach to the skull. Generally, areas that serve as points of attachment for muscles have raised bone due to the stress on the bone and the stimulation that causes bone growth. The median nuchal line is also known as the external occipital crest, which is formed because of the attachment of the ligamentum nuchae that connects the cervical vertebrae to the skull. The superior and inferior nuchal lines form attachments with the muscles and ligaments that stabilize the articulation of the occipital condyles with the atlas, thereby balancing the mass of the head over the cervical vertebrae. The superior nuchal lines are adjacent to the external occipital protuberance, while the inferior nuchal lines are approximately 2.5 centimeters (1 inch) inferior to the superior nuchal lines.

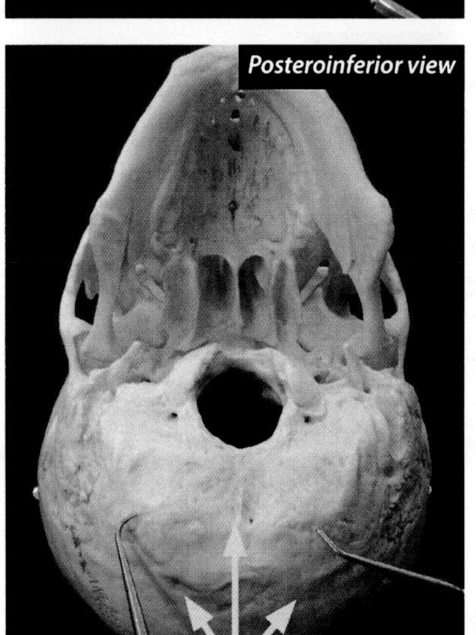


*Inferior (probes), median, and superior nuchal lines (arrows).*

## Occipital Condyles (Occipital)

As the name implies, the occipital condyles are landmarks of the occipital bone. They are functionally important because they articulate with the superior articular facets of the atlas (C1). This joint allows for flexion and extension of the head (nodding the head "yes"), as well as a little bit of lateral bending. Dr. J thinks these structures are shaped like the rockers of a rocking chair, which facilitate the flexion and extension of the cranium relative to the atlas. When one shakes his head "no," the two bones lock together and move as one piece. We find the hypoglossal canal in the lateral base of the occipital condyles, immediately superior to them. This foramen is important as it transmits the **hypoglossal nerve (XII)**.

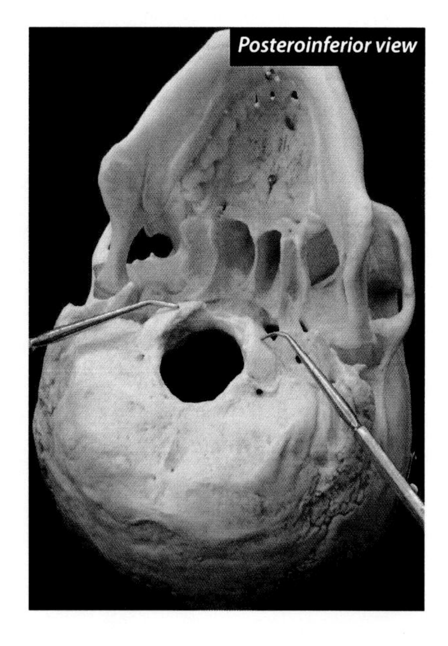

# Temporal Bone Landmarks

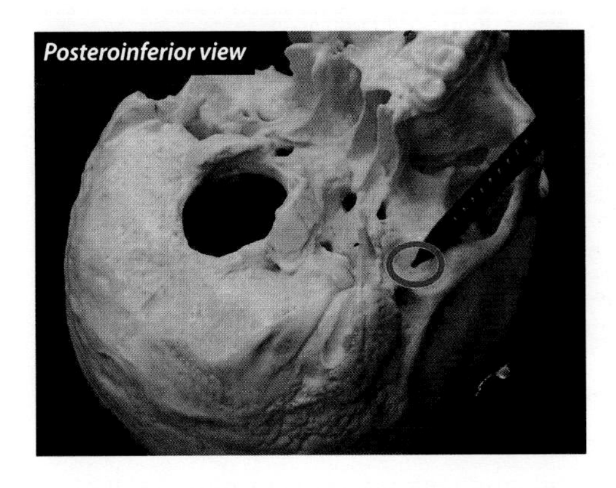
*Posteroinferior view*

## Mandibular Fossa (Temporal)

The mandibular fossa of the temporal bone is an important landmark because this is one articular surface of the temporomandibular joint. This is the surface with which the head of the condyloid process of the mandible articulates. This is an atypical hinge joint in that in addition to elevation and depression of the mandible, it also allows for protraction, retraction, and lateral deviation. These additional motions are helpful when positioning food on the occlusal surfaces of the teeth. There is a meniscal cartilage associated with this joint. This joint may become problematic for some people, resulting in TMJ syndrome.

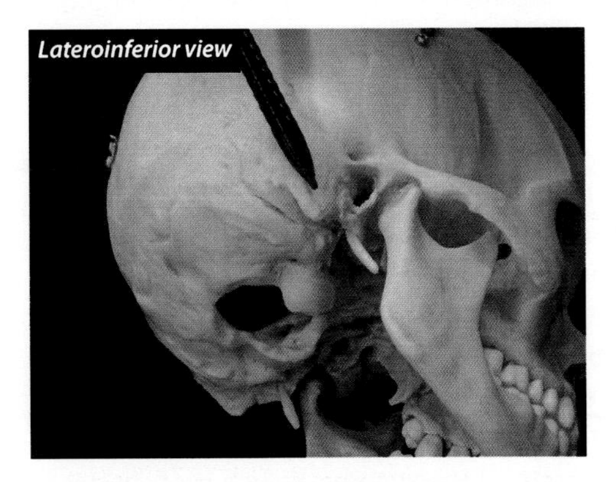
*Lateroinferior view*

## Mastoid Process (Temporal)

The mastoid process is part of the temporal bone. It is the insertion for the sternocleidomastoid muscle. In fact, it is this muscle that causes this landmark to develop. Several other muscles for which you will not be responsible also attach to this landmark. As a group, they produce rotation or extension of the head. The mastoid process gets its name from the similarity it has to the appearance of a breast. Dr. J did not name this landmark—he is just the messenger.

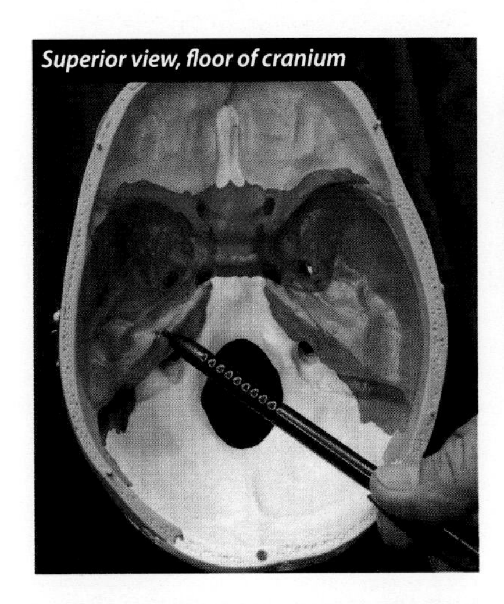
*Superior view, floor of cranium*

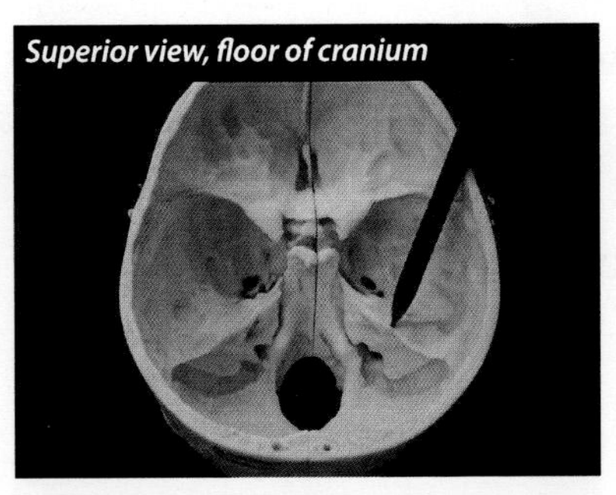
*Superior view, floor of cranium*

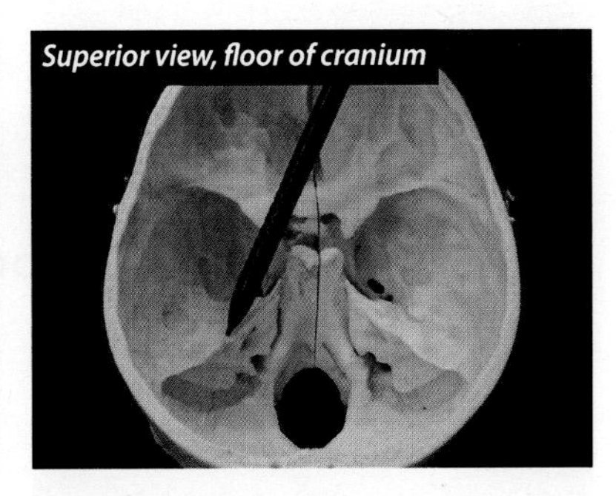
*Superior view, floor of cranium*

## Petrous Ridge (Temporal)

The petrous ridge of the temporal bone is an important landmark, as it is the border between the middle fossa and the posterior fossa of the cranium. Within this portion of the temporal bone are cavities that form the inner and middle ear. The sharp portion of the ridge is a point of attachment for a membrane that stabilizes the brain. It gets its name from the fact that it is irregular, like a rocky wall. Actually, it was also named for Petrous the flying squirrel, a famous friend of Bullwinkle the moose.

## Styloid Process (Temporal)

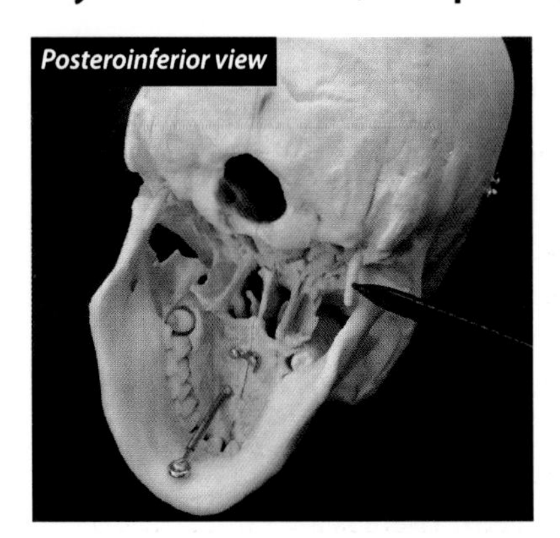


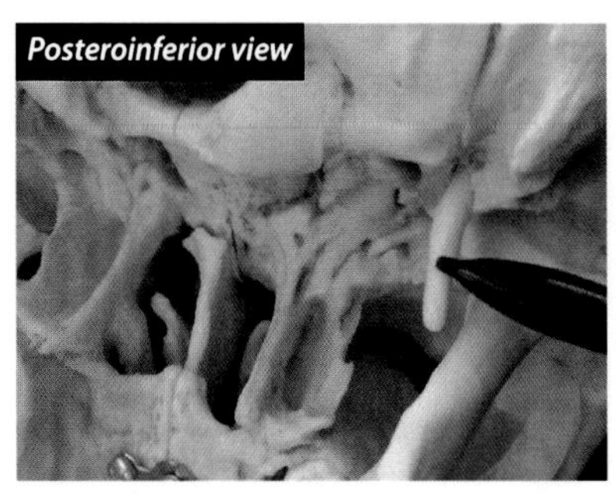


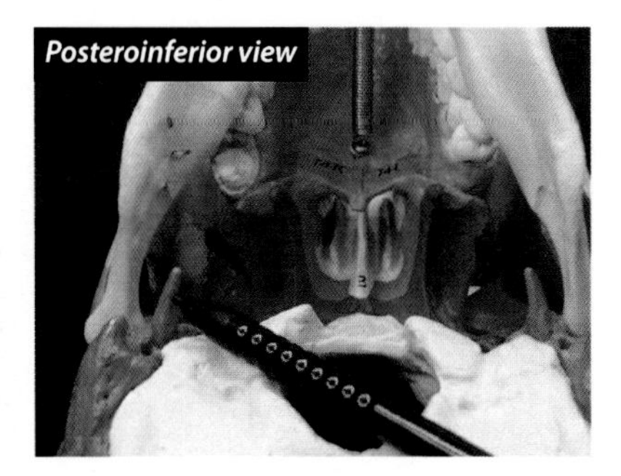


The styloid process is part of the temporal bone and is anteromedial to the mastoid process. As is the case with many processes, it forms where muscles attach to the bone. It is the origin of the styloglossus, stylohyoid, and stylopharyngeus muscles. It also serves as a point of attachment for the stylohyoid and stylomandibular ligaments. You will not be responsible for these muscles or these ligaments. Whew, right?

## Zygomatic Arch (Temporal and Zygomatic)

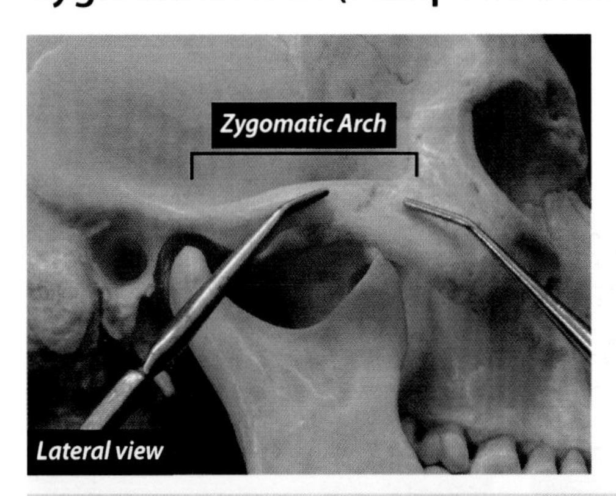


The zygomatic arch is made up of two processes from two bones, the zygomatic process of the temporal bone posteriorly and the temporal process of the zygomatic bone anteriorly. Dr. J likes to draw the analogy to a drawbridge with a part coming from each end of the bridge. The tendon of the temporal muscle passes deep to the zygomatic arch. It is sometimes fractured in automobile crashes or from a blow to the side of the face. The inferior margin of the zygomatic arch is of special functional importance because it is part of the origin for:

1. the masseter muscle.

*The two probes in this picture are pointing to the zygomatic process of temporal bones and the temporal process of zygomatic bone—both structures make up the zygomatic arch. They are named for the bone that they touch.*

## Zygomatic Process (Temporal)

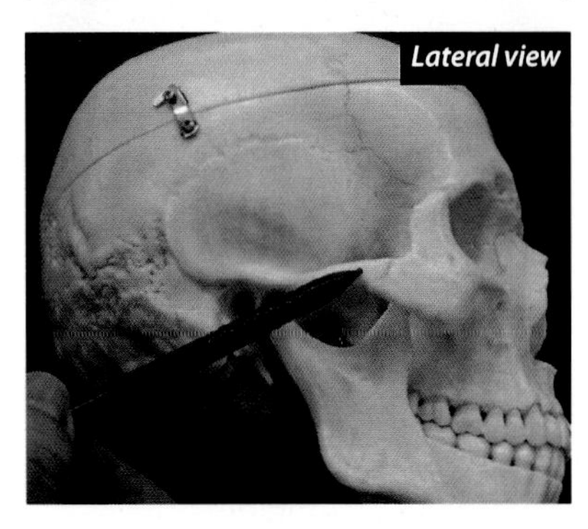


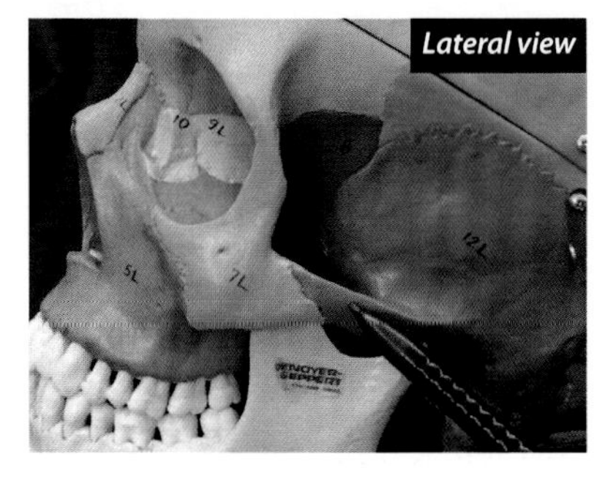


As is often the case, the zygomatic process of the temporal bone is named for the bone with which it articulates rather than the bone of which it is a part. The zygomatic process forms the posterior portion of the zygomatic arch and projects anteriorly to where it articulates with the temporal process of the zygomatic bone. This projection originates from the squamous portion of the temporal bone. The masseter muscle, which is the prime elevator (flexor) of the mandible (closes the mouth), attaches to the inferior margin of the temporal process.

# Sphenoid Bone Landmarks

## Lesser Wing (Sphenoid)

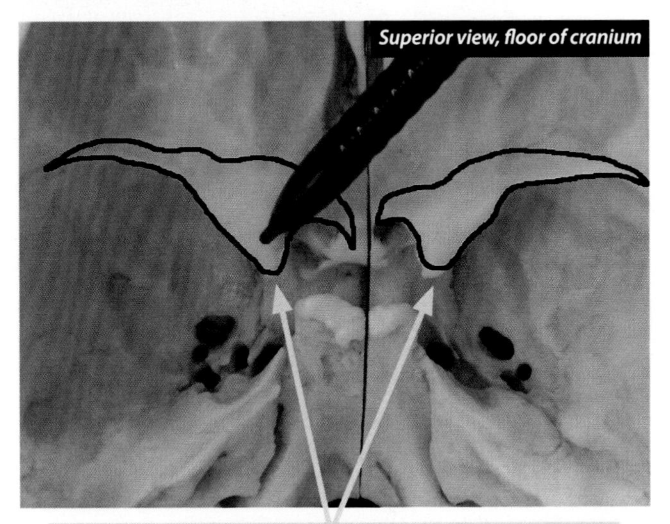


Arrows are pointing to left and right lesser wings of the sphenoid.

The lesser wing of the sphenoid bone forms part of the anterior fossa of the cranium, as well as part of each orbit. Although some authors describe it as horn-shaped, Dr. J thinks it resembles the cranial end of a manta ray. The projections that form a border with the middle fossa are known as the anterior clinoid process. This process serves as the point of attachment for a sheet of dura mater that separates the cerebellum from the posterior portion of the cerebral hemispheres.

## Greater Wing (Sphenoid)

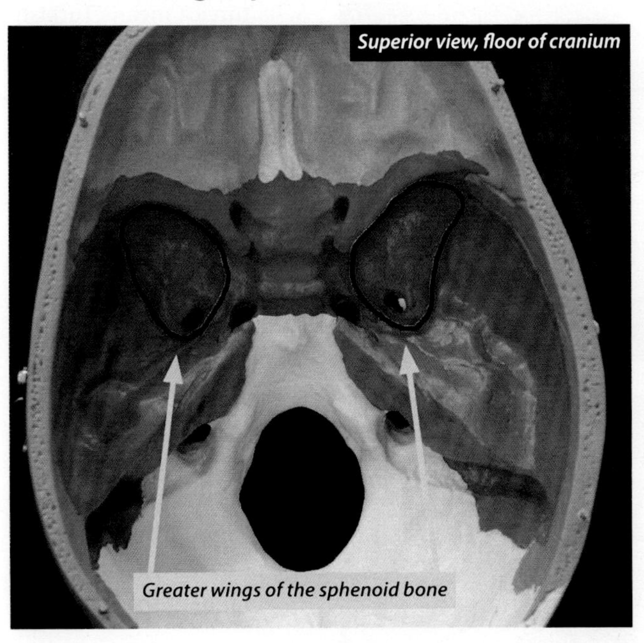


Greater wings of the sphenoid bone

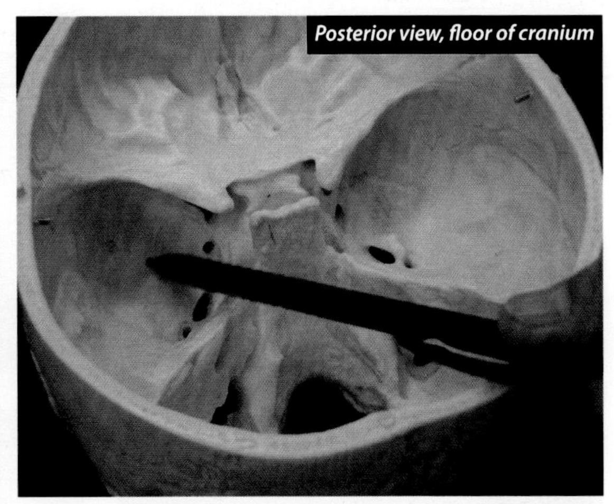


The greater wing of the sphenoid bone can be seen on the outer surface of the skull, as well as in the middle fossa of the cranium. It also makes up a large portion of the medial wall of the orbit. Several of the foramina that we study are associated with the greater wing of the sphenoid bone. They include the foramen rotundum (**maxillary nerve**), the foramen ovale (**mandibular nerve**), and one edge of each superior orbital fissure (**oculomotor nerve**, **trochlear nerve**, **ophthalmic nerve**, and **abducens nerve**).

# Sphenoid Bone Landmarks

## Pterygoid Process (Sphenoid)

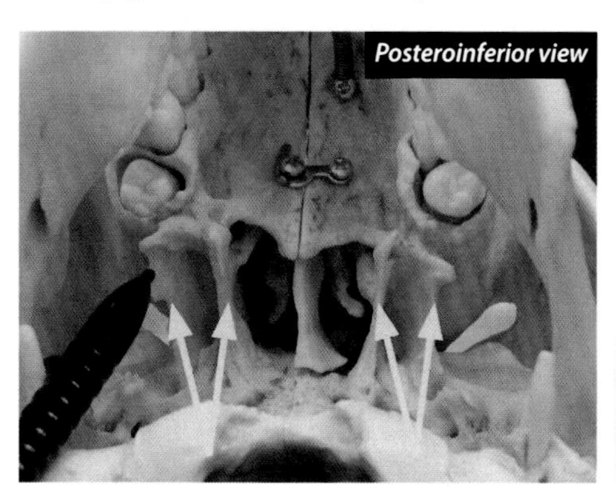
*Posteroinferior view*

*Arrows are pointing to left and right medial and lateral pterygoid processes.*

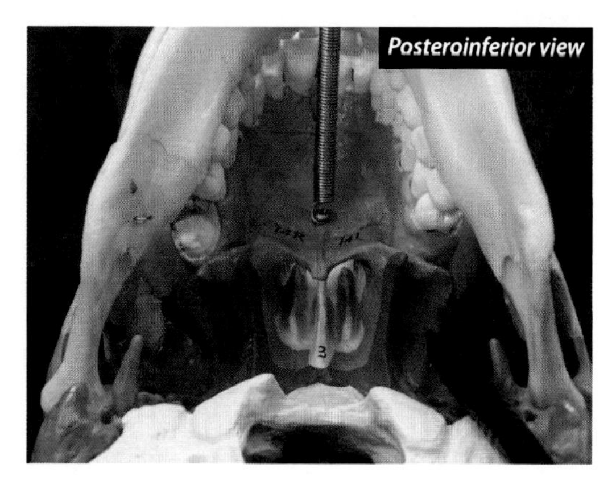
*Posteroinferior view*

There is a medial and a lateral plate for each of the pterygoid processes. They project inferiorly from the greater wing. The medial pterygoid processes of the sphenoid bone articulate with the perpendicular plates of the palatine bones anteriorly. They are the origins for the pterygoid muscles that elevate the mandible. They are also associated with the soft palate. You will not be responsible for the pterygoid muscles.

## Sella Turcica (Sphenoid)

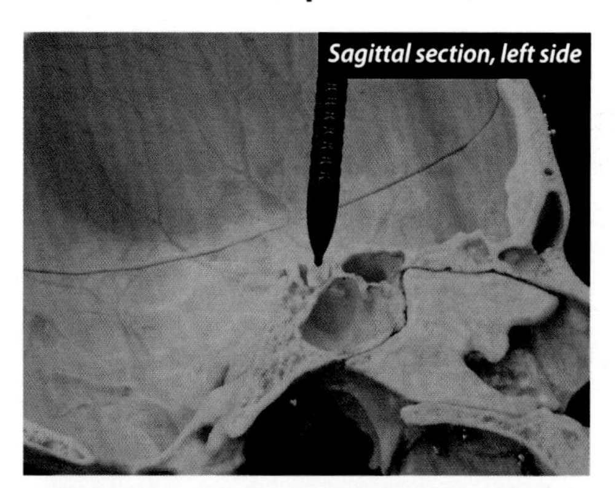
*Sagittal section, left side*

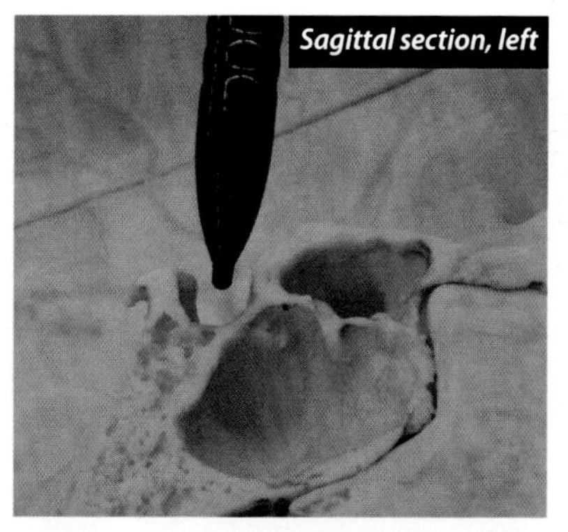
*Sagittal section, left*

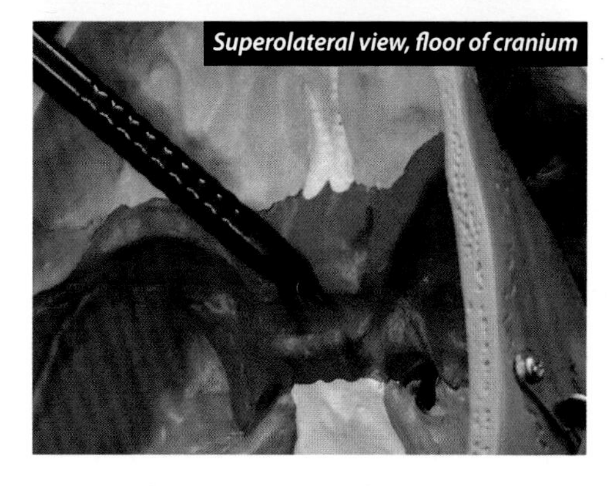
*Superolateral view, floor of cranium*

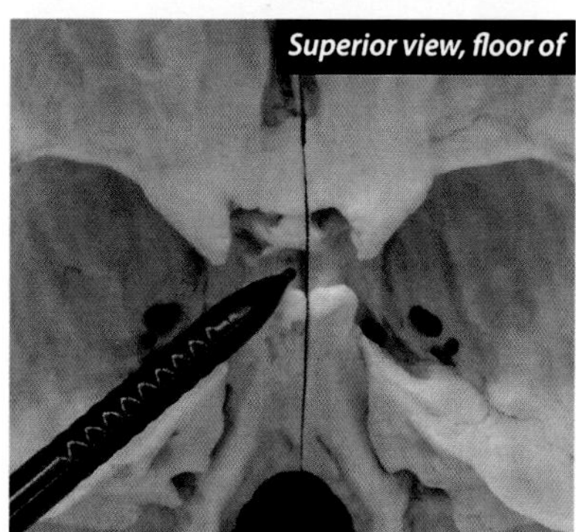
*Superior view, floor of*

Sella turcica means Turkish saddle. It is called that because it looks like a saddle. It is part of the sphenoid bone and is of importance because this is where the pituitary gland (hypophysis) is found. It is part of the middle fossa of the cranium. A sheet of dura mater attaches to the posterior margin of the sella turcica. This sheet is called the tentorium cerebelli, which separates the cerebellum from the posterior part of the cerebral hemispheres.

# Ethmoid Bone Landmarks

## Cribriform Plate (Ethmoid)

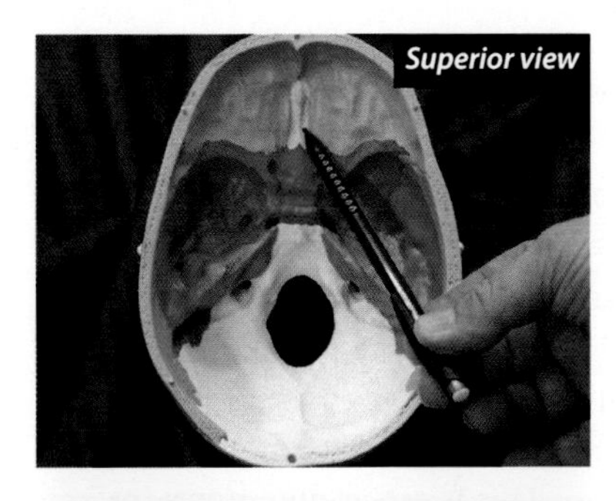


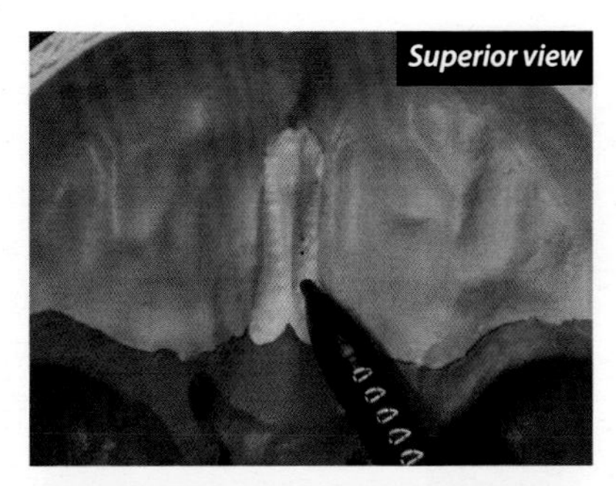


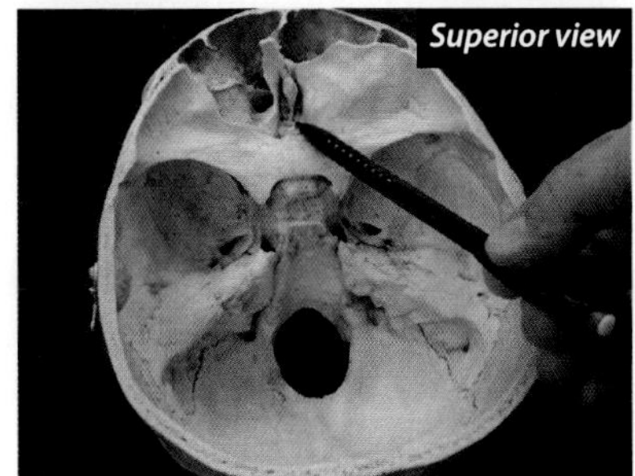


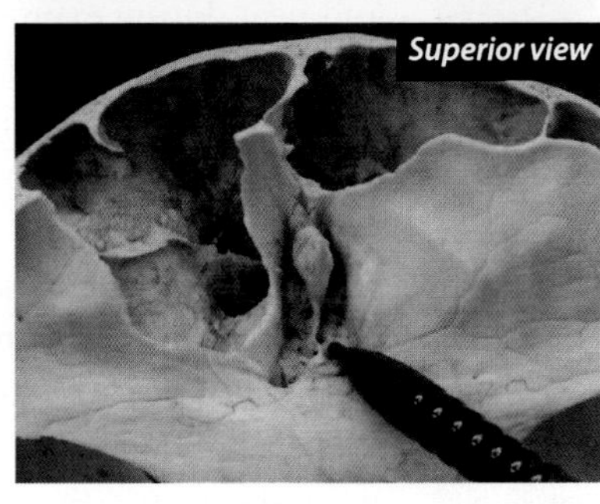


The cribriform plate of the ethmoid bone has the foramina through which fibers of the **olfactory nerve (I)** pass into the nasal cavities on either side. This is somewhat unusual, as instead of there being one foramen on each side, there are many foramina, giving the cribriform plate the appearance of a sieve (strainer). We find olfactory receptors in the epithelium covering the inferior surfaces of the cribriform plate. The cribriform plate is found on either side of the crista galli, thereby forming part of the anterior fossa of the cranium. It resembles the deck of a sailboat. Remember that Dr. J likes to draw an analogy between parts of the ethmoid bone and a sailboat.

## Crista Galli (Ethmoid)

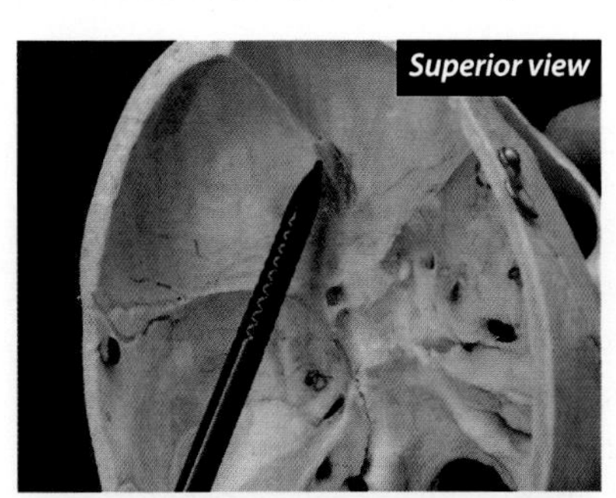


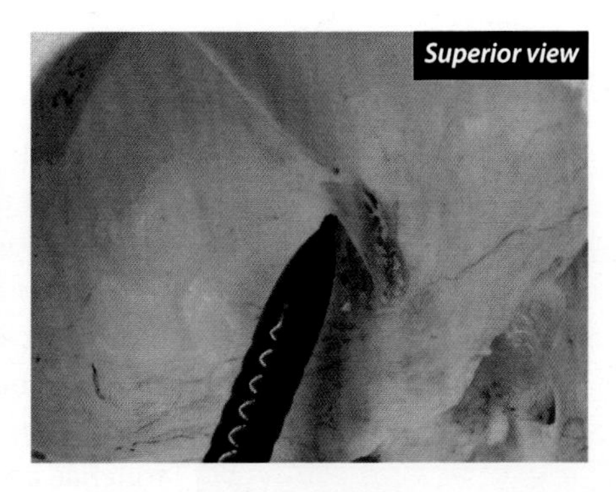


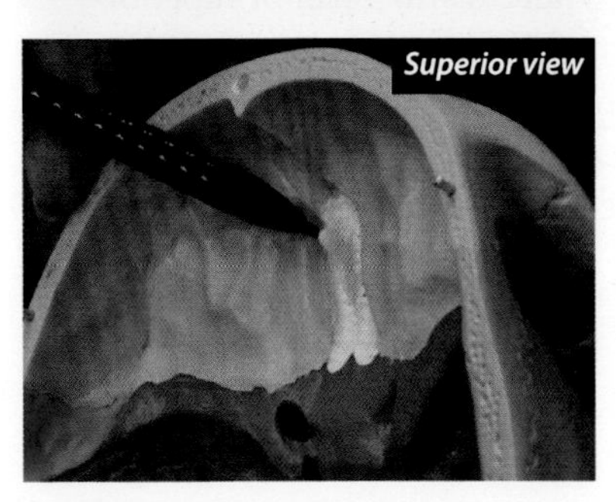


The crista galli translates to chicken's comb, the fleshy ridge on the top of a chicken's head. It rises superiorly from the cribriform plate, which can be seen to its left and right sides. It extends superiorly between the frontal lobes of the brain. Functionally, it is important because an extension or fold of the dura mater, the falx cerebri, is anchored to it, which helps stabilize the position of the brain. The crista galli is like the sail of the sailboat, the cribriform plate is like the deck of the sailboat, and the perpendicular plate of the ethmoid is like the keel of a sailboat.

# Ethmoid Bone Landmarks

## Superior Concha (Ethmoid)

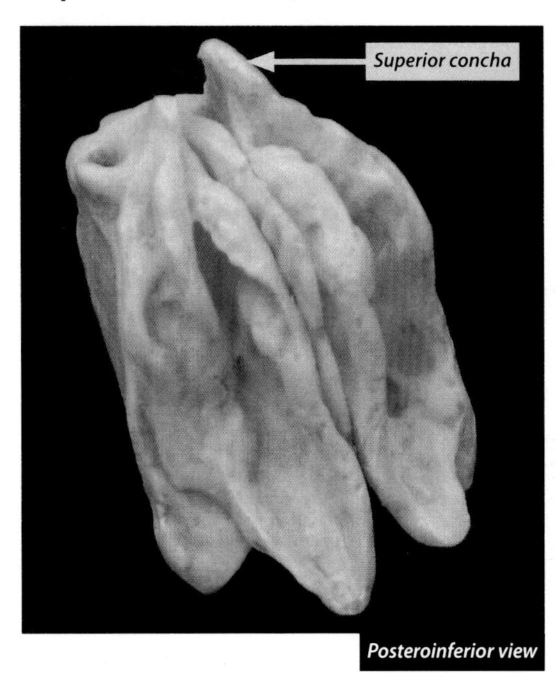


The superior nasal conchae (superior turbinate) are part of the ethmoid bone, and there is one on the lateral wall of each nasal cavity. You will not see this on the white plastic skulls, but it can be seen in a good example of a real skull. All conchae are functionally important because their curved surface causes turbulence as air moves into the nasal cavity. The turbulence helps to warm, moisten, and filter the air before it enters the lower respiratory tract. In addition, the conchae help recover moisture and heat during expiration. This is especially important in cold and dry conditions. An additional benefit has been proposed based on a recent finding that the airflow is different for the two nasal cavities. It has been shown that the mucosa that covers the conchae becomes swollen on one side, and then after a few hours the swelling reduces on that side and the other side becomes swollen. A possible advantage of this would be that the sense of smell is better on the swollen side because the air moves more slowly and there is more intense stimulation by scents when airflow is slow. We find olfactory receptors in the epithelium covering the medial surfaces of the superior nasal conchae. There is also a pair of middle conchae and supreme conchae on the ethmoid bone.

## Middle Concha (Ethmoid)

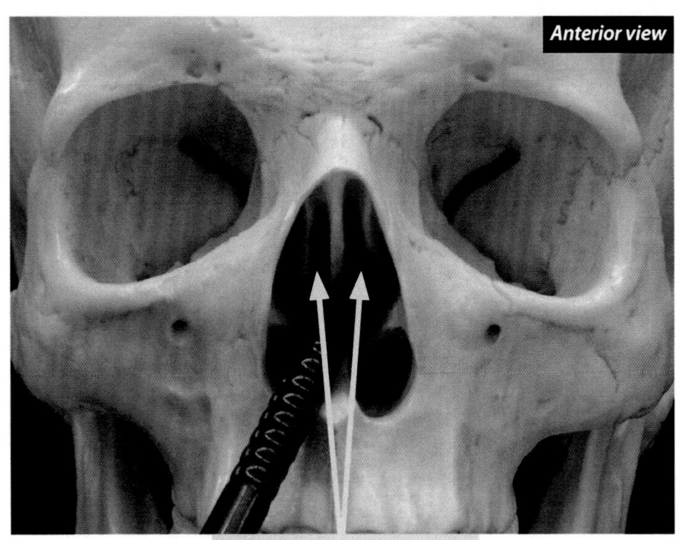


The middle nasal conchae (middle turbinate) are part of the ethmoid bone and are found on the lateral wall of each nasal cavity. As with the inferior conchae, they are functionally important because their curved surface causes turbulence as air moves into the nasal cavity. There is also a pair of superior conchae and supreme conchae on the ethmoid bone. However, the middle conchae are the only ones that can be tagged on the white plastic skulls for a lab practical.

# Ethmoid Bone Landmarks

## Perpendicular Plate (Ethmoid)

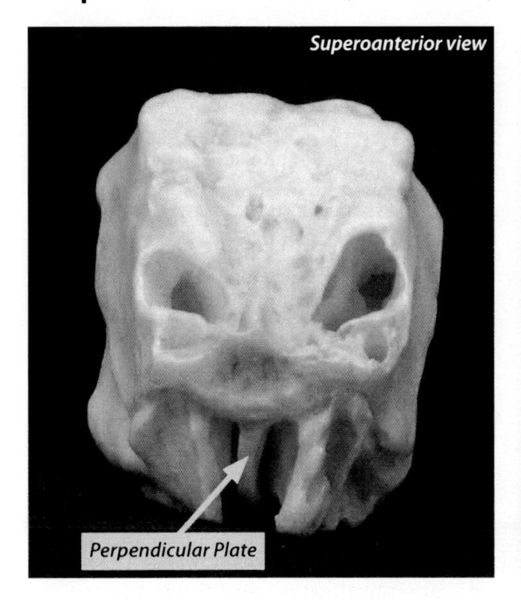


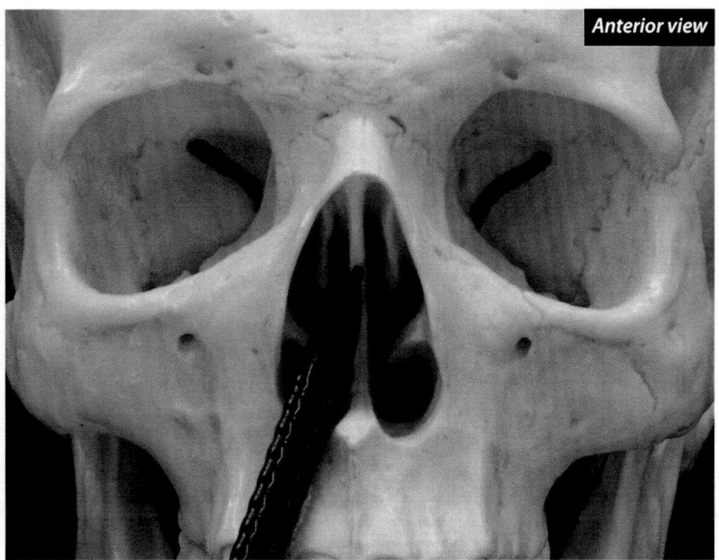


The perpendicular plate of the ethmoid is the superior portion of the bony nasal septum, which separates the left and right nasal cavities. It extends inferiorly in the median sagittal plane from the rest of the ethmoid bone. It is also of significance because olfactory receptors are found in the epithelium that covers the superior surfaces of this structure. Remember that Dr. J says the crista galli is like the sail of the sailboat and the cribriform plate is like the deck of the sailboat. Well, the perpendicular plate of the ethmoid is like the keel of the sailboat. There is also a perpendicular plate on each palatine bone, a vertical extension that blends in with the medial pterygoid processes of the sphenoid. You will not be responsible for the perpendicular plate of the palatine bone.

## Nasal Septum (Ethmoid and Vomer)

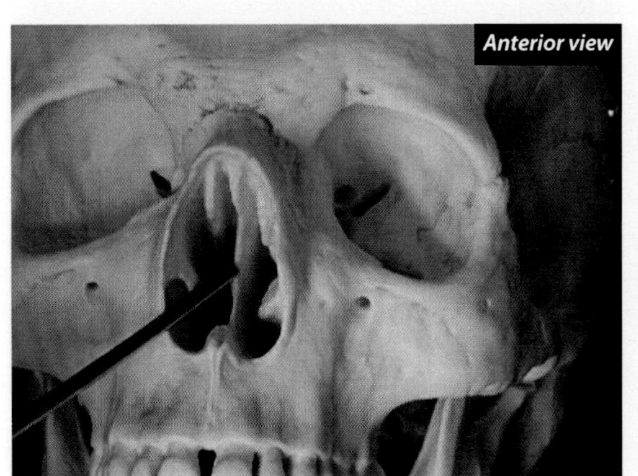


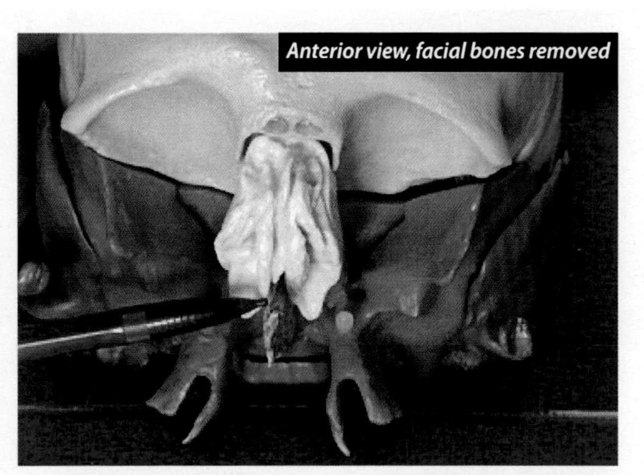


The nasal septum is primarily made up of two bones and a septal cartilage. The superior bone is the perpendicular plate of the ethmoid. The inferior bone is the vomer. The septal cartilage is hyaline cartilage, and it projects anteriorly from its connection to the vomer and the perpendicular plate of the ethmoid. This structure separates the two nasal cavities. The superior portion of the perpendicular plate of the ethmoid bone is covered with epithelial tissue that includes olfactory receptors.

# Maxilla Landmarks

## Palatine Process (Maxilla)

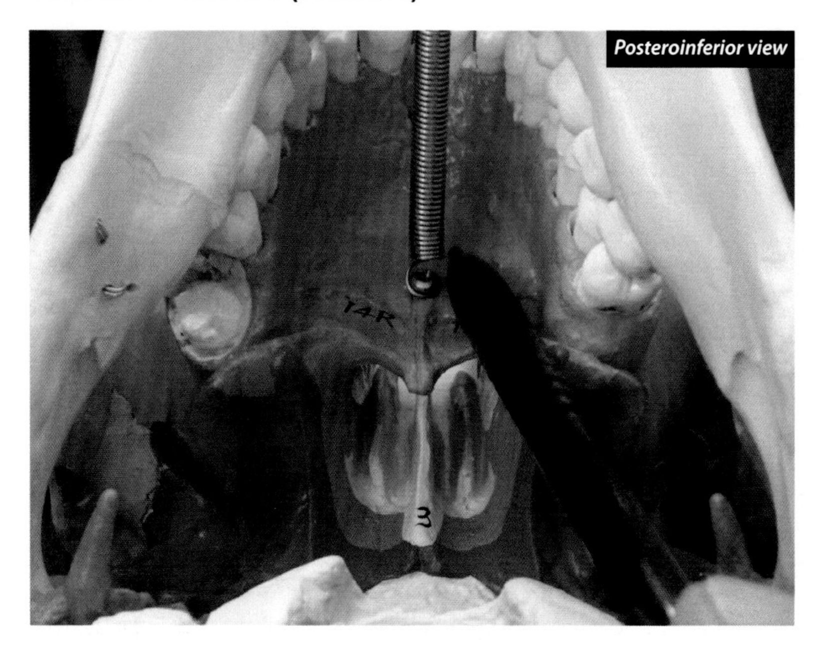
*Posteroinferior view*

The palatine processes of the maxillary bones make up the majority (about 67 to 75 percent) of the hard palate. Along the midline, the two palatine processes of the left and right maxillary bones articulate with each other. They articulate posteriorly with the horizontal plate of the palatine bone on each side. The shelf that they form is part of the roof of the oral cavity as well as part of the floor of the nasal cavities. There is a canal on each side anteriorly, and they accommodate small vessels and nerves. It is a passageway between the nasal cavity and the hard palate. This is the incisive canal, but you will not be responsible for it on any quizzes or exams.

# Mandible Landmarks

## Alveolus (Mandible and Maxilla)

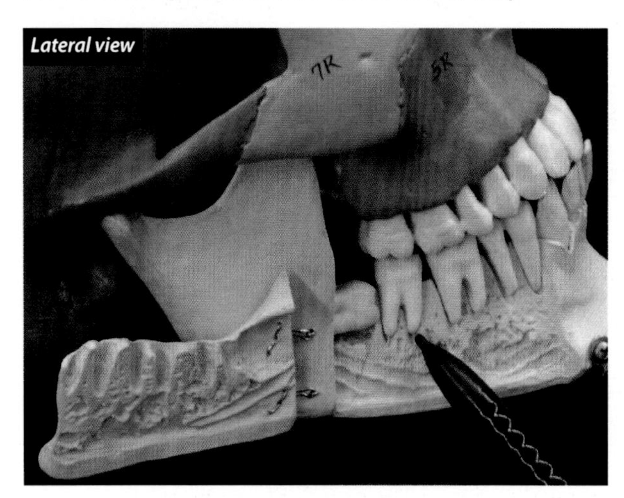
*Lateral view*

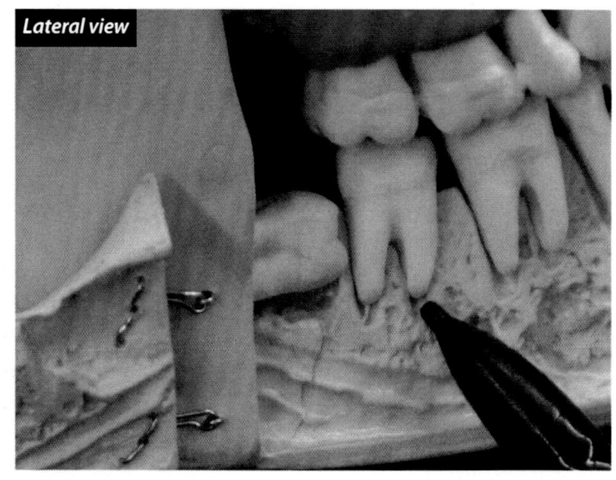
*Lateral view*

The alveoli are found in the mandible and the two maxillary bones. They are what we called sockets when we were in nursery school. These are depressions in the bone into which the teeth are anchored.

# Mandible Landmarks

## Body (Mandible)

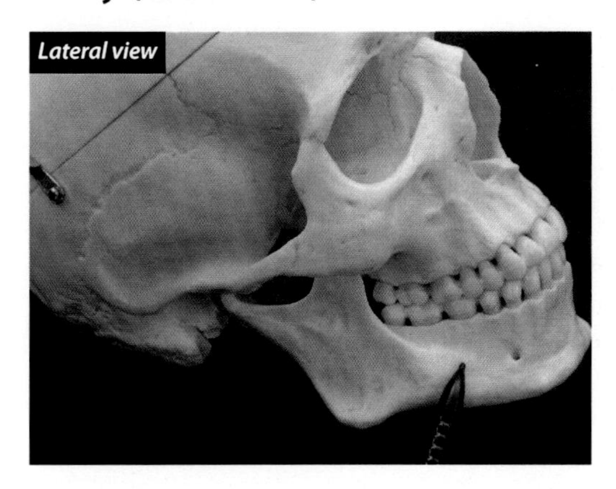


The body of the mandible has alveoli that secure the teeth. It is made up of two portions: inferiorly, the base of the mandible; superiorly, the alveolar part of the mandible. Remember that although we consider this one piece extending to the left and right sides, it was two bones in the fetus. We find the mental foramen in the anterior portion of the body of the mandible. After passing through the foramen ovale and the mandibular foramen, the **mandibular nerve** ($V_3$) emerges onto the face by passing through the mental foramen.

## Condyloid Process (Mandible)

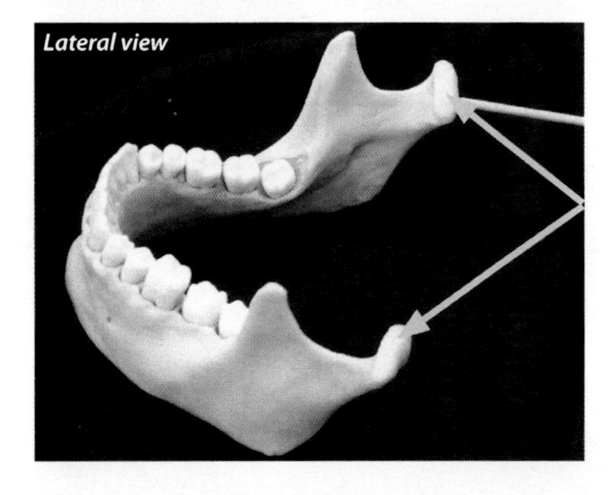


The condyloid (condylar) process of the mandible extends superiorly from the ramus of the mandible. The condyloid process of the mandible includes the head of the mandible, which is functionally important, as this is where the mandible articulates with the mandibular fossa of the temporal bone. This is known as the temporomandibular joint. This is an atypical hinge joint in that in addition to elevation and depression of the mandible, it also allows for protraction, retraction, and lateral deviation. These additional motions are helpful when positioning food on the occlusal surfaces of the teeth. The neck of the mandible is also part of the condyloid process, and it serves as the insertion for the lateral pterygoid muscle. To distinguish this structure from the coronoid process of the mandible, you might want to remember that it is a rounded, knuckle-like projection of bone, because the word "condyle" comes from the Greek word for "knuckle."

## Coronoid Process (Mandible)

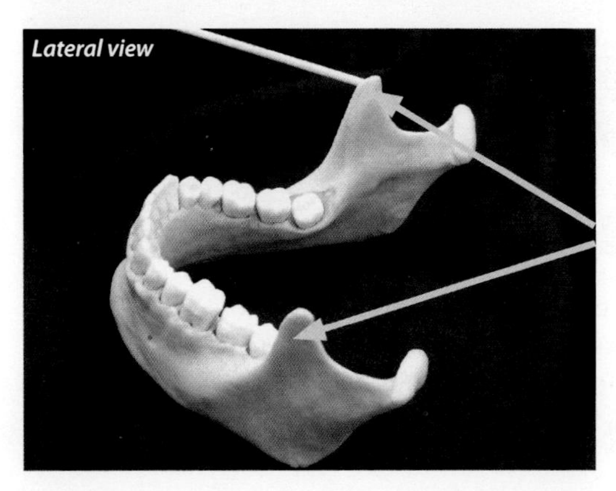


The coronoid process of the mandible extends superiorly from the anterior portion of the ramus. It is functionally important as the point of attachment for the temporalis muscle. Although we will not study this muscle, it is important in the elevation and retraction of the mandible. Technically, coronoid translates to mean resembling a crown. Dr. J thinks it looks more like a shark's tooth, so he would have called it the shark's toothoid process. But they didn't ask him! What's up with that?

### Mandibular (Sigmoid) Notch (Mandible)

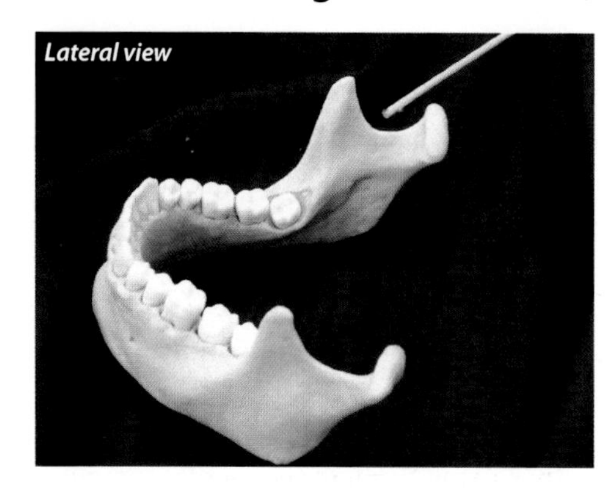


The mandibular (sigmoid) notch is a concavity on the mandibular ramus between the condyloid and coronoid processes. It gets its name from the famous anatomist who later became a psychiatrist, Sigmoid Freud!

### Ramus (Mandible)

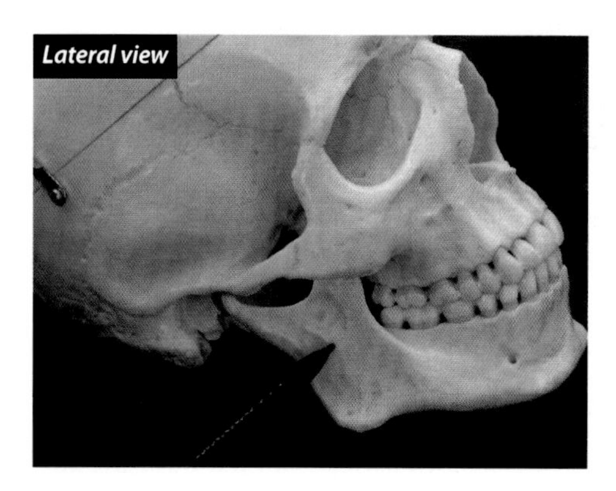


We have seen the word ramus used in a number of places in the body. It means a branch-like structure of some larger structure. The ramus of the mandible is a posterior process that meets the body of the mandible at the mandibular angle. It is nearly in a sagittal plane. The coronoid and condyloid processes project superiorly from the superior edge of the ramus. The mandibular (sigmoid) notch is located between these two processes. The mandibular foramen is located on the medial (deep) surface of the ramus and transmits the **mandibular nerve ($V_3$)** into the mandibular canal. The medial pterygoid muscle attaches to the medial surface posteroinferior to the mandibular foramen. The lateral surface of the ramus is of special functional importance because it is the insertion for:

1. the masseter muscle.

*You will be responsible for the masseter muscle, but not the medial pterygoid muscle. Dr. J thinks the ramus actually may have been named for Uncle Ramus.*

## Zygomatic Bone Landmarks

### Temporal Process (Zygomatic)

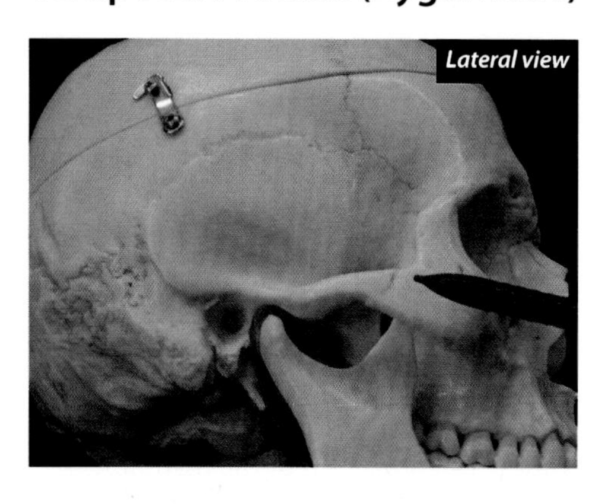


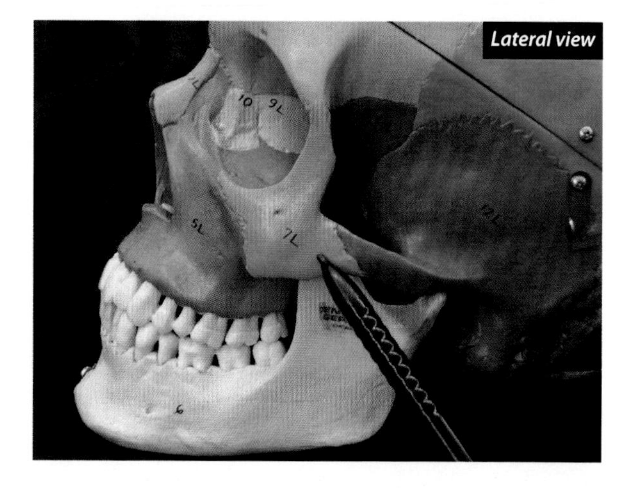


The temporal process of the zygomatic bone is another example of a structure that is named for the bone with which it articulates rather than the bone of which it is a part. The temporal process forms the anterior portion of the zygomatic arch and projects posteriorly to where it articulates with the zygomatic process of the temporal bone. The masseter muscle, which is the prime elevator (flexor) of the mandible (closes the mouth), attaches to the inferior margin of the temporal process.

# Palatine Bone Landmarks

## Horizontal Plate (Palatine)

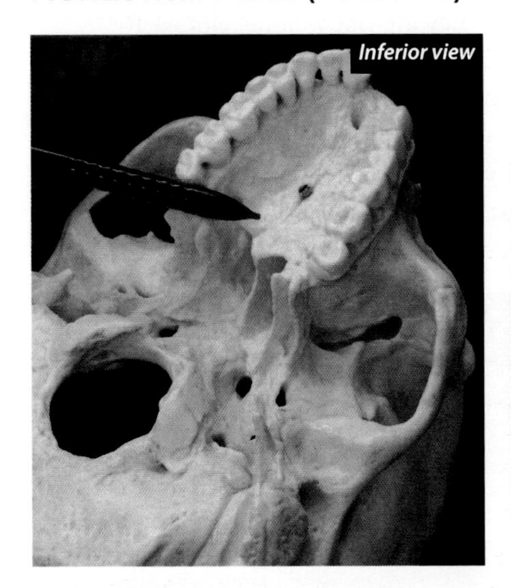
*Inferior view*

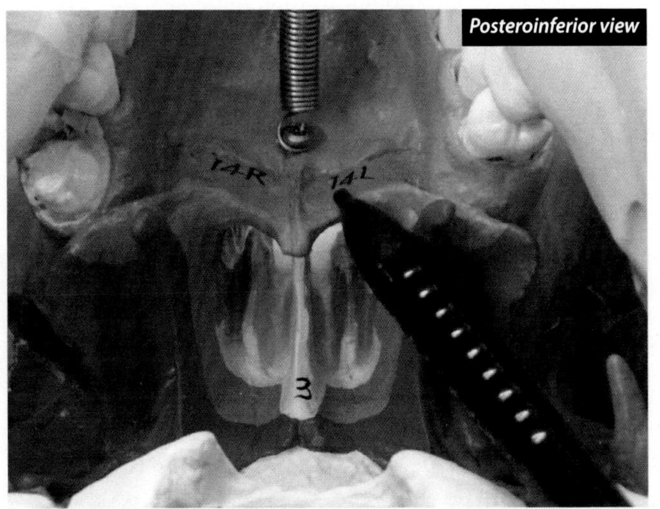

*Posteroinferior view*

The horizontal plate of the palatine bone makes up the posterior portion of the hard palate, where it articulates anteriorly with the palatine process of the maxillary bone on each side. Along the midline, the two horizontal plates of the left and right palatine bones articulate with each other. The shelf that they form is part of the roof of the oral cavity as well as part of the floor of the nasal cavities. Their contribution to the hard palate is between 25 and 33 percent of its surface.

# Vomer Landmarks

## Nasal Septum (Vomer and Ethmoid)

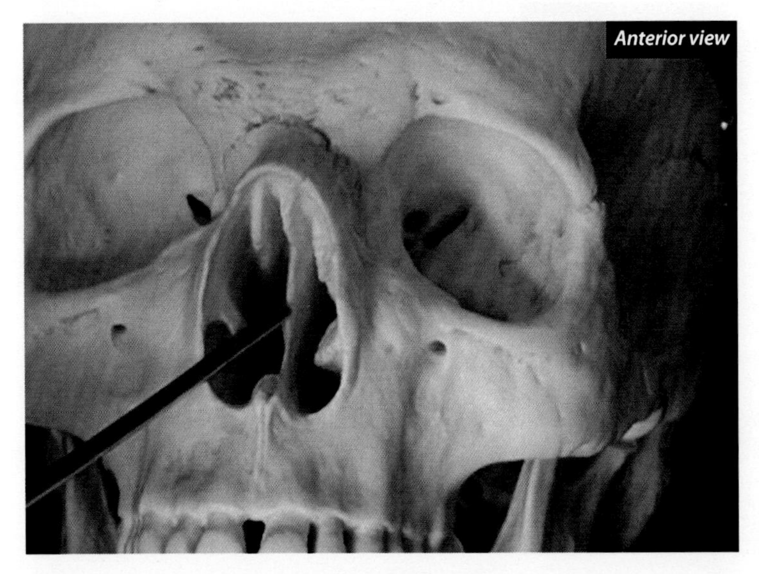
*Anterior view*

The nasal septum is primarily made up of two bones and a septal cartilage. The superior bone is the perpendicular plate of the ethmoid. The inferior bone is the vomer. The septal cartilage is hyaline cartilage, and it projects anteriorly. This structure separates the two nasal cavities. The superior portion of the perpendicular plate of the ethmoid bone is covered with epithelial tissue that includes olfactory receptors.

# Sinuses

The sinuses associated with the bones of the skull are air-filled cavities lined with mucous membranes. There are four major sinuses: ethmoid, frontal, maxillary, and sphenoid. They are not well-developed at birth, but they enlarge as the child grows and reach full size in adolescence. There are small openings from these sinuses into the nasal cavities. The mucus is propelled into the nasal cavities by cilia. Functionally, they are important in the quality of the voice and they also reduce skull weight. The sinuses may also be called paranasal sinuses. All of the sinuses are named for the bone in which they reside.

## Frontal Sinuses

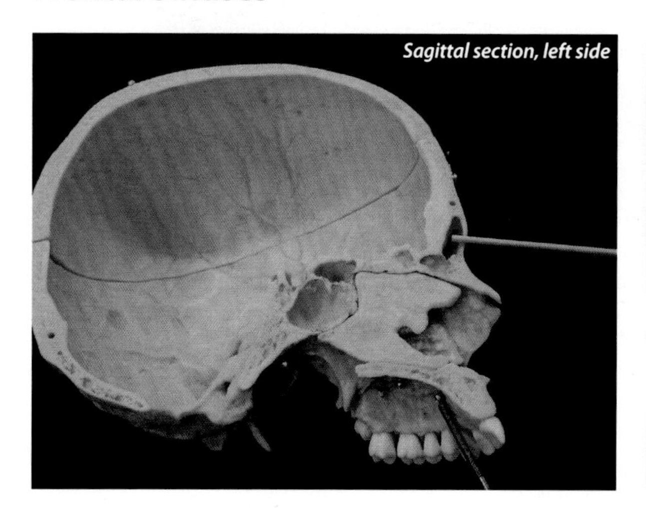


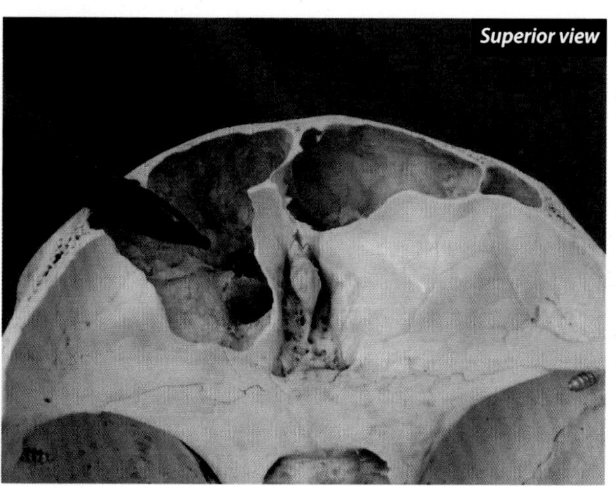


The frontal sinus is immediately superior and medial to the orbits and, as the name implies, it is in the frontal bone. Another Grant thing!

## Sphenoid Sinuses

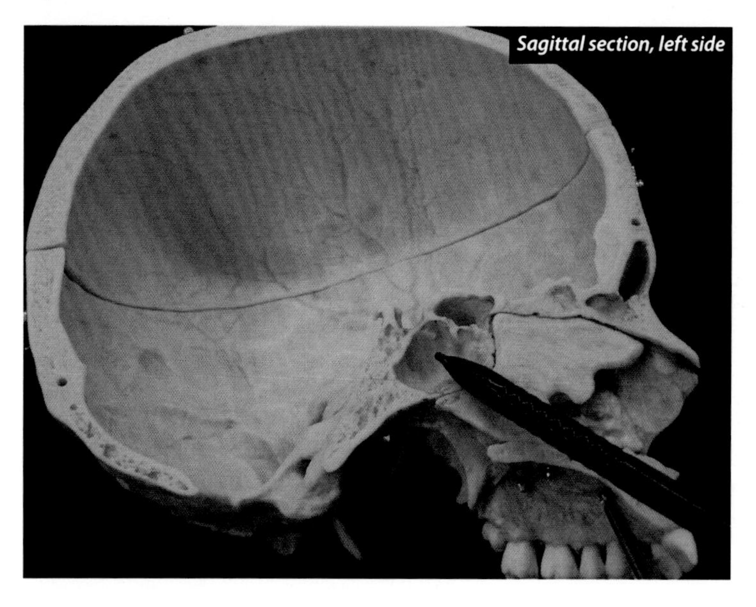


The sphenoid sinus is inferior to the sella turcica of the sphenoid bone. It is very close to the geographic center of the skull. It can be seen on the skull that has been cut in a sagittal plane, as well as in the x-rays of the skull.

## Ethmoid Sinuses

The two ethmoid sinuses are lateral to the nasal cavities and they straddle the nasal cavities in a manner similar to saddlebags on a horse. You can see this on the model of the ethmoid bone (below), as well as in the x-rays of the skull.

Foreground: ethmoid bone, superior view

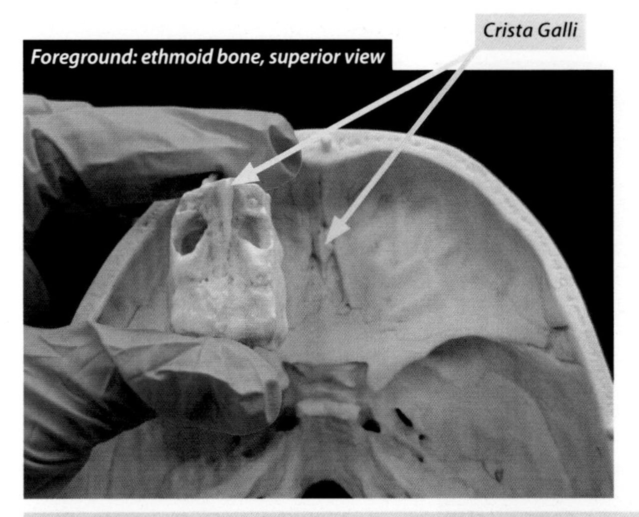


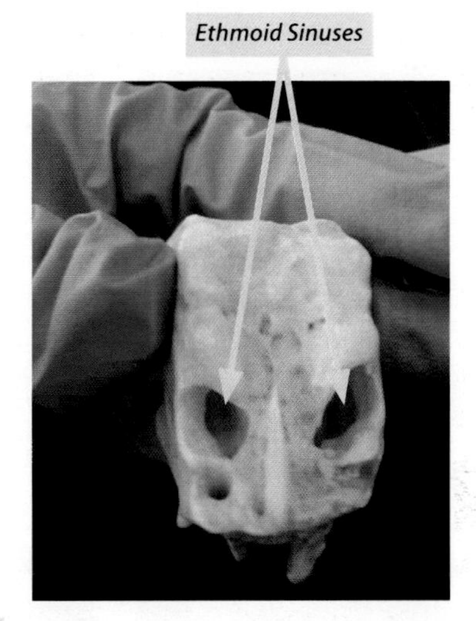


This picture is intended to show you how the ethmoid bone is oriented in the cranium.

## Maxillary Sinuses

The two maxillary sinuses are lateral to the nasal cavities and they are inferior to the orbit of the eye. You can see this on the model of the maxillary bone, as well as in the x-rays of the skull.

Foreground: left maxillary bone, anterior

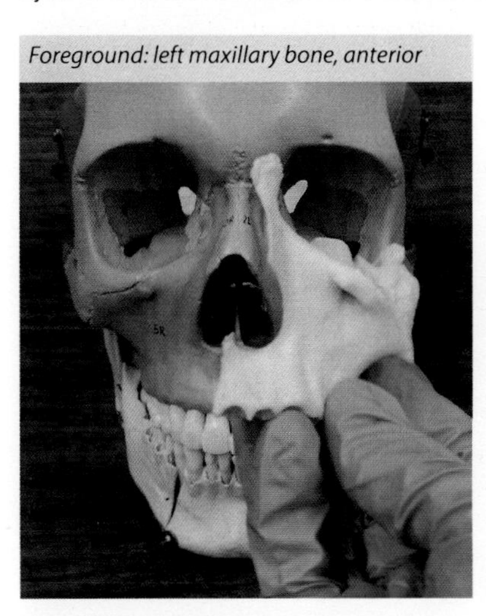

Foreground: left maxillary bone, 45 degree rotation to the left

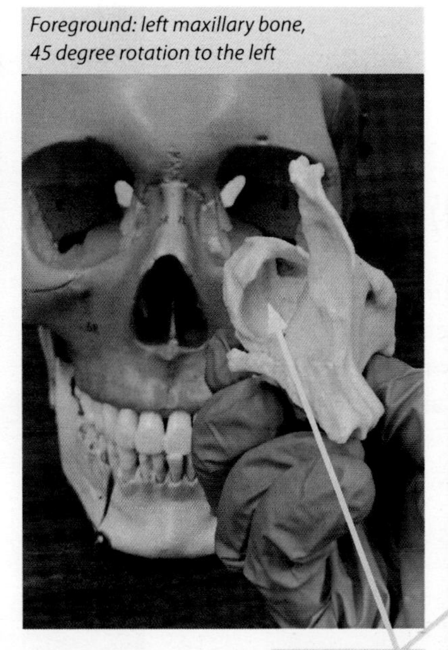

Foreground: left maxillary bone, 90 degree rotation to the left

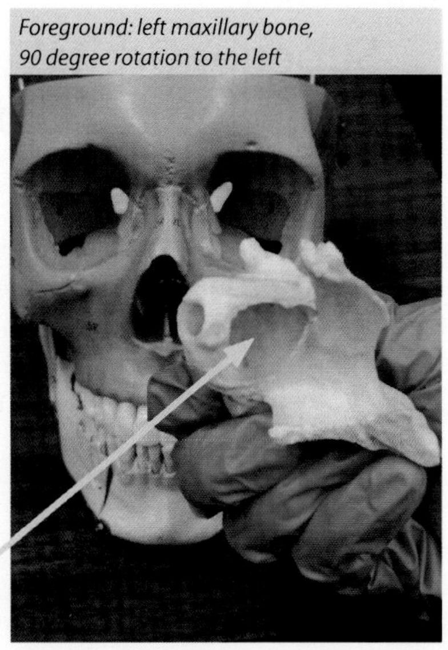

# FORAMINA

## Optic Canal (Sphenoid)

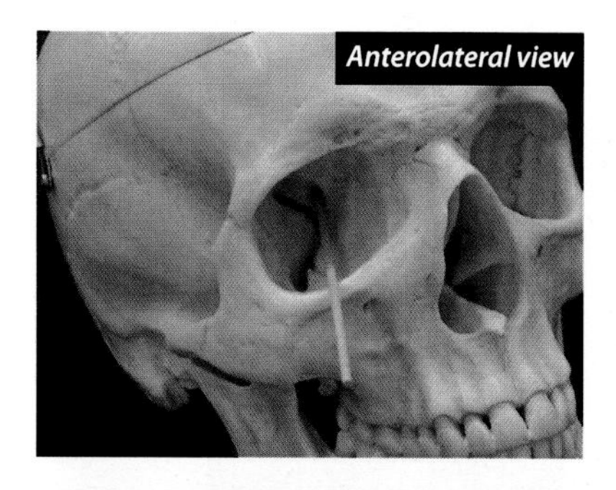
Anterolateral view

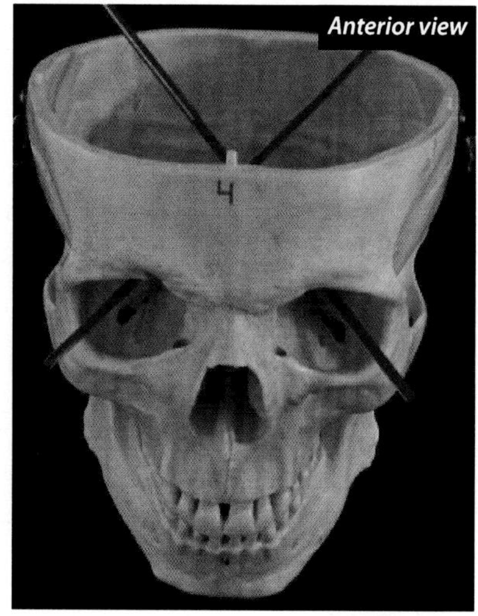
Anterior view

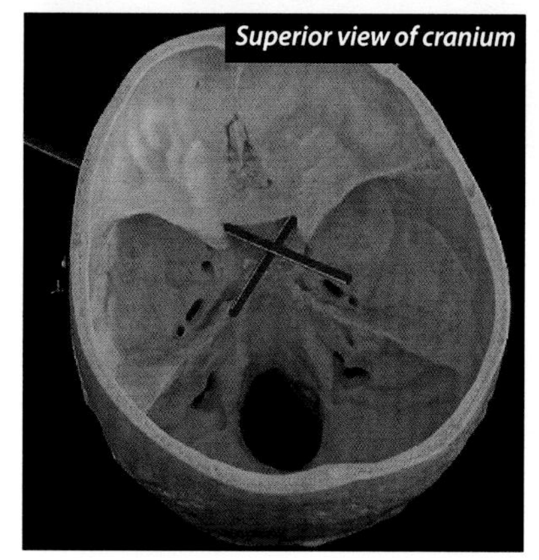
Superior view of cranium

The optic canal is occasionally referred to as the optic foramen. It may seem that there isn't much difference, but remember that the name of the canal suggests it is an opening with appreciable depth. Here we have another Grant thing—the optic canal houses the **optic nerve (II)**. The optic canal is in the lesser wing of the sphenoid bone. Note that if you pass a wooden applicator stick through this canal on each side, the sticks crisscross inside the cranium.

## Superior Orbital Fissure (Sphenoid)

The superior orbital fissure of the sphenoid bone provides passage for the following nerves when moving from the cranium to the orbit: **oculomotor (III)**, **trochlear (IV)**, **ophthalmic ($V_1$)**, and **abducens (VI)**. This fissure separates the lesser wing of the sphenoid from the greater wing of the sphenoid.

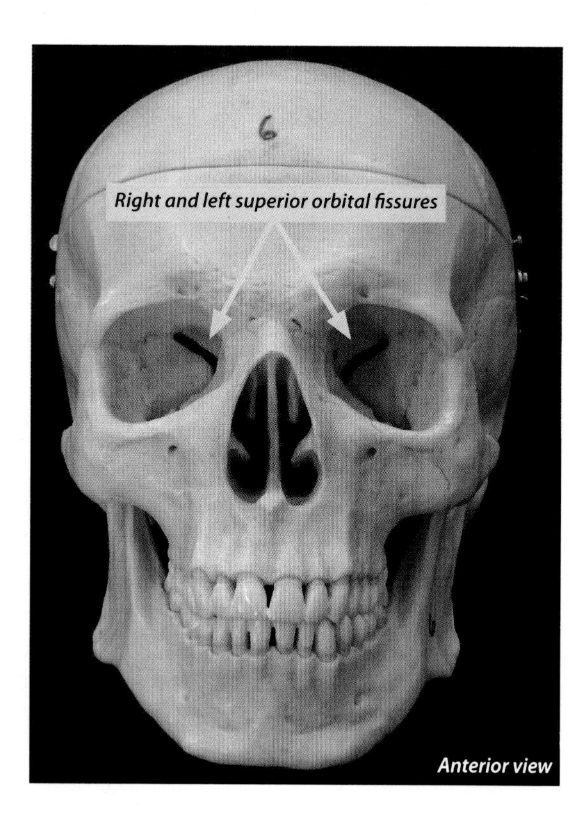

Anterior view

## Inferior Orbital Fissure (Maxilla, Sphenoid, Palatine, Zygomatic)

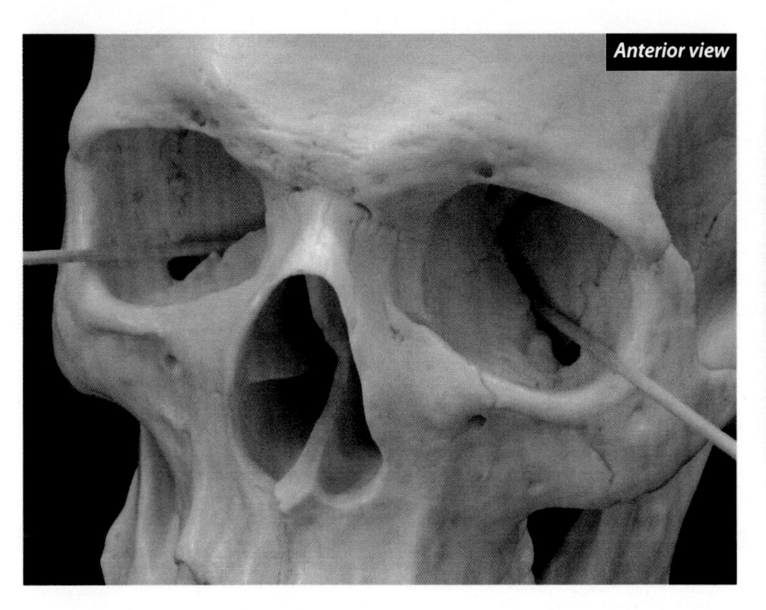


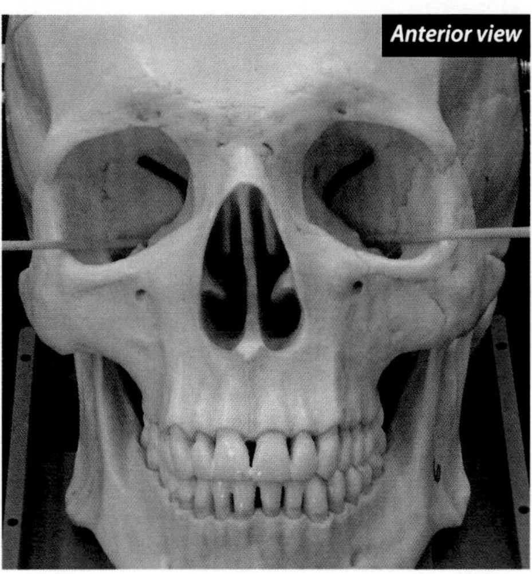


The inferior orbital fissure is formed by the maxilla, the greater wing of the sphenoid, the palatine, and the zygomatic bones. The **maxillary nerve ($V_2$)** passes through this fissure after leaving the foramen rotundum on its way to the infraorbital foramen.

## Lacrimal Canal (Lacrimal)

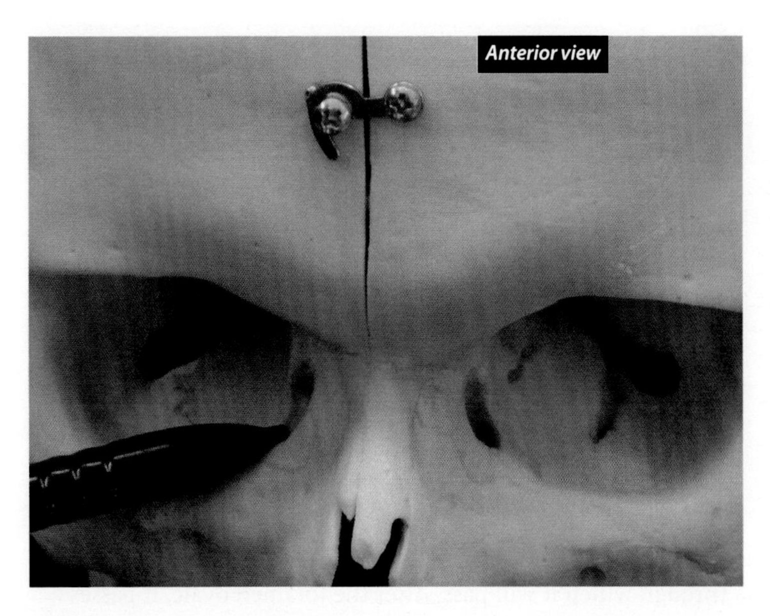


The lacrimal canal holds the lacrimal duct. The duct serves to conduct the fluid from the eye into the nasal cavity. The duct is sometimes called the nasolacrimal duct.

## Foramen Rotundum (Sphenoid)

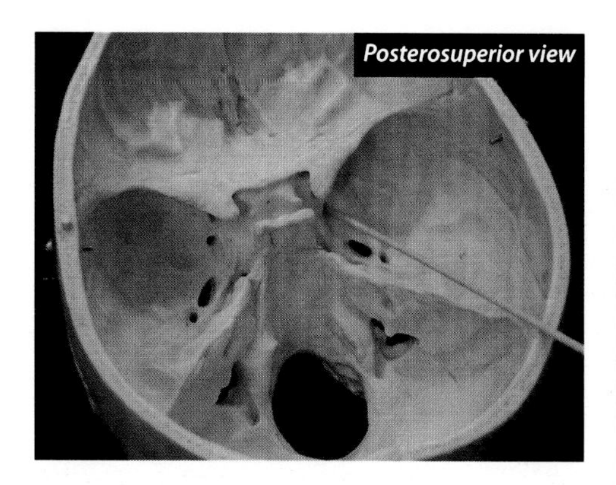


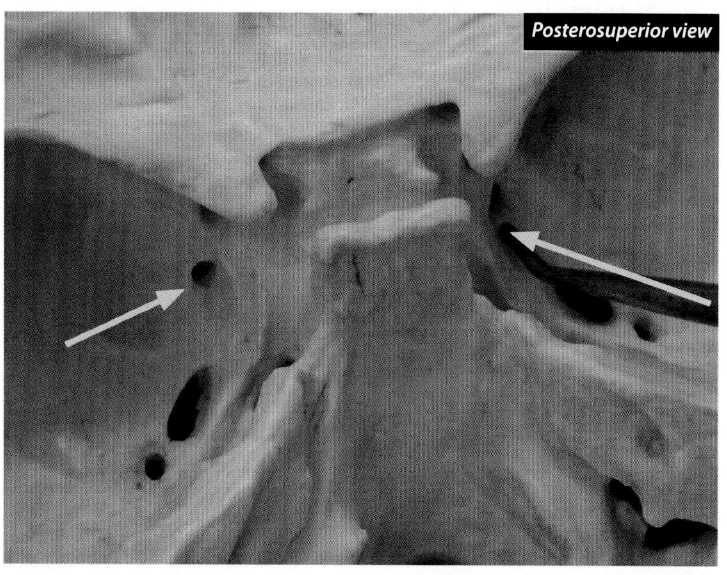


The foramen rotundum of the sphenoid bone houses the **maxillary nerve** ($V_2$) as it passes toward the infraorbital foramen of the maxillary bone. This is the first of the three foramina through which it will pass. After the foramen rotundum, the **maxillary nerve** ($V_2$) passes though the inferior orbital fissure and the infraorbital foramen, respectively.

## Foramen Ovale (Sphenoid)

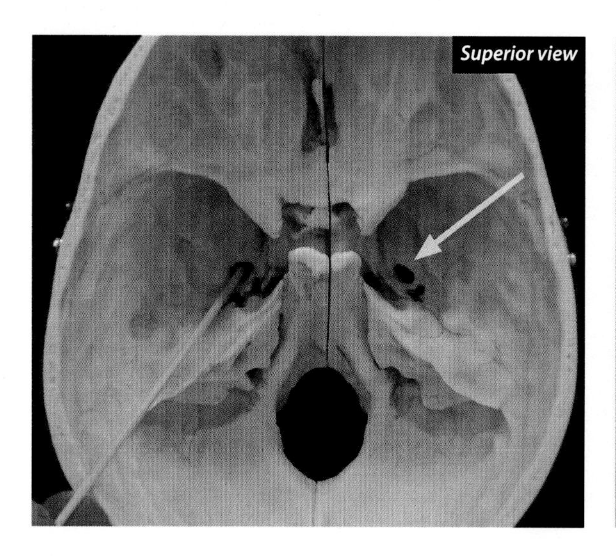


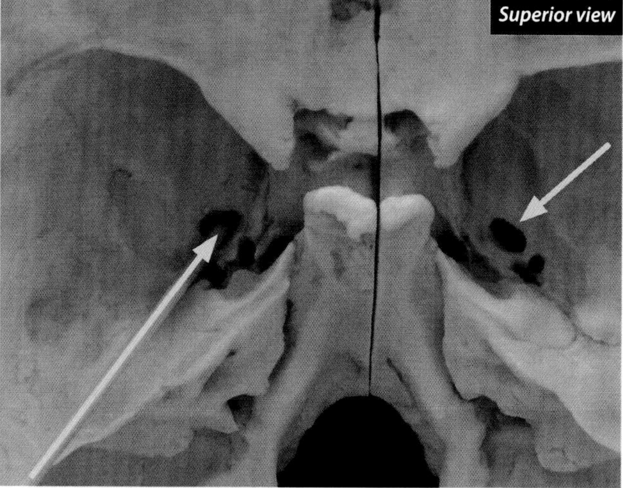


The foramen ovale of the sphenoid bone houses the **mandibular nerve** ($V_3$) as it passes toward the mental foramen. This is the first of the three foramina through which it will pass. After the foramen ovale, it passes through the mandibular foramen and mental foramen, respectively.

## Internal Auditory Meatus (Temporal)

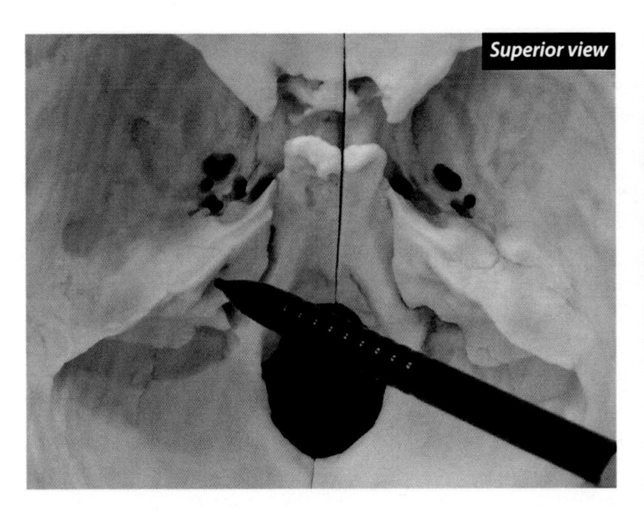


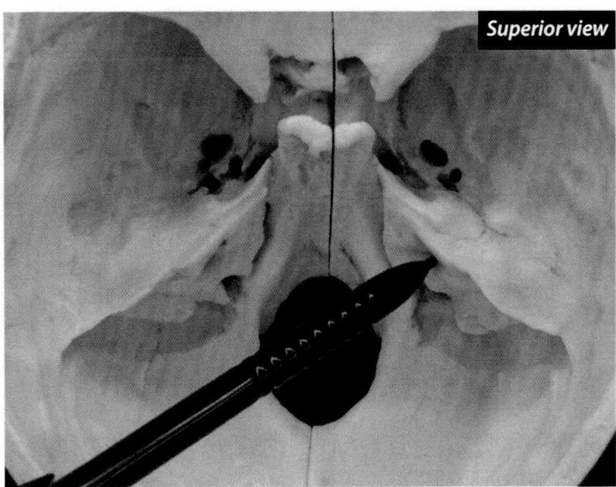


The internal auditory meatus is found on the vertical portion of the petrous ridge of the temporal bone. The **facial nerve (VII)** and the **vestibulocochlear nerve (VIII)** pass into this canal. They separate once inside the temporal bone.

## Jugular Foramen (Temporal and Occipital)

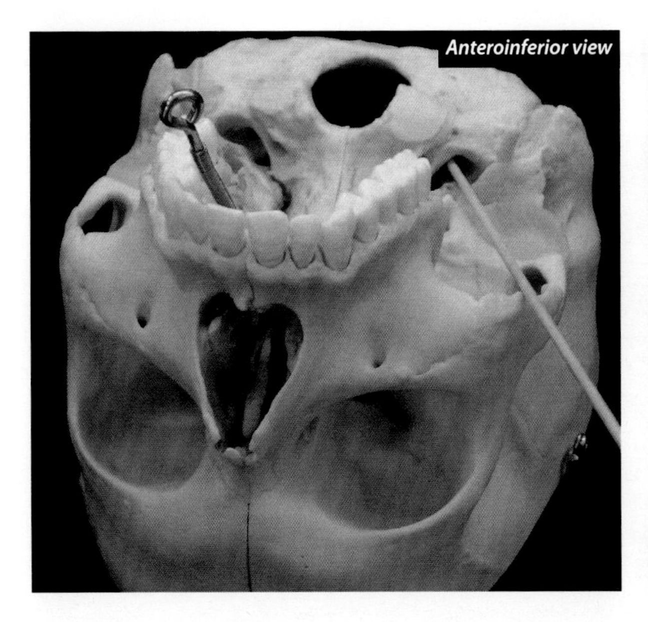


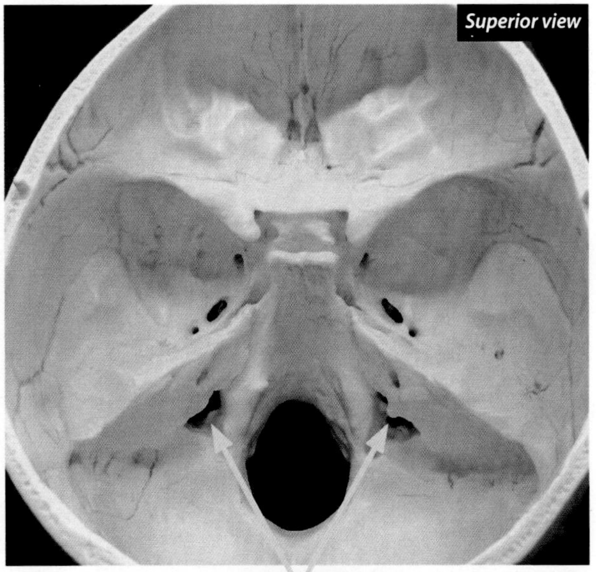


The jugular foramen is between the temporal and occipital bones. The **glossopharyngeal (IX)**, **vagus (X)**, and **spinal accessory (XI) nerves**, as well as the internal **jugular vein**, all pass through this foramen.

## Foramen Magnum (Occipital)

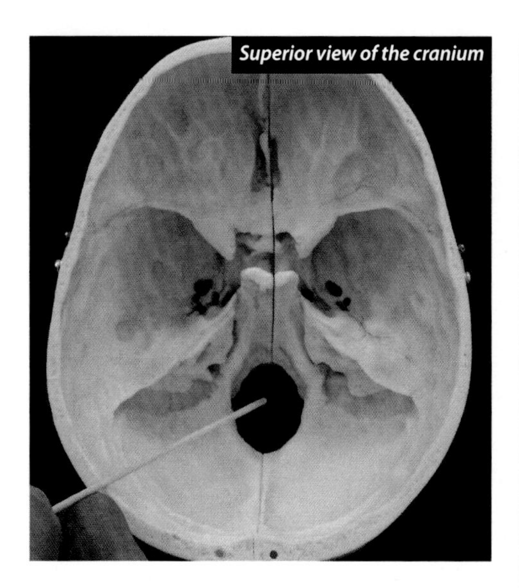


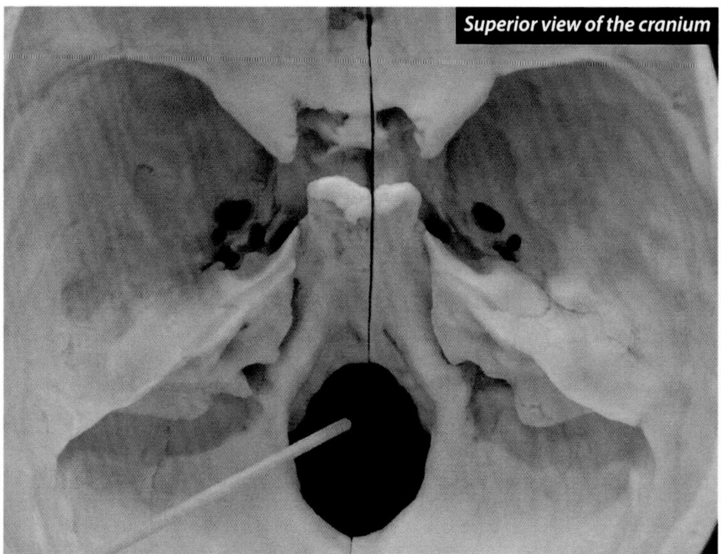


Several structures pass through the foramen magnum. They include the **medulla oblongata** and the meninges that surround it, the ascending portions of the **spinal accessory nerves (XI)**, and the two **vertebral arteries**. The **medulla oblongata** connects to the spinal cord.

## Hypoglossal Canal (Occipital)

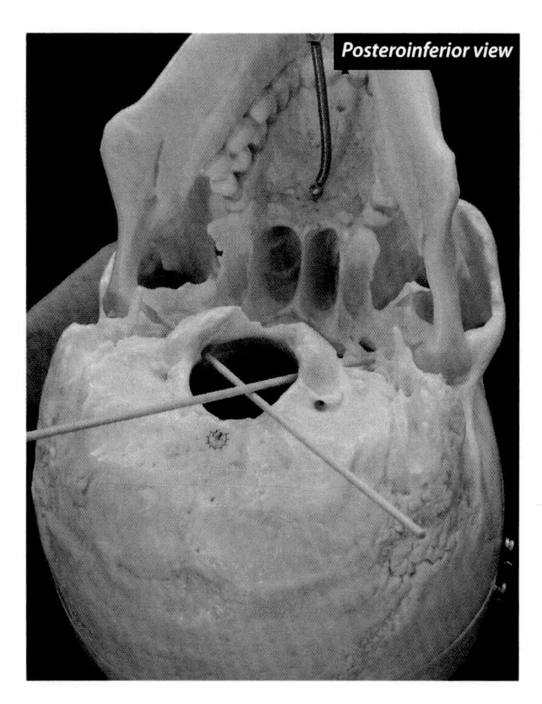


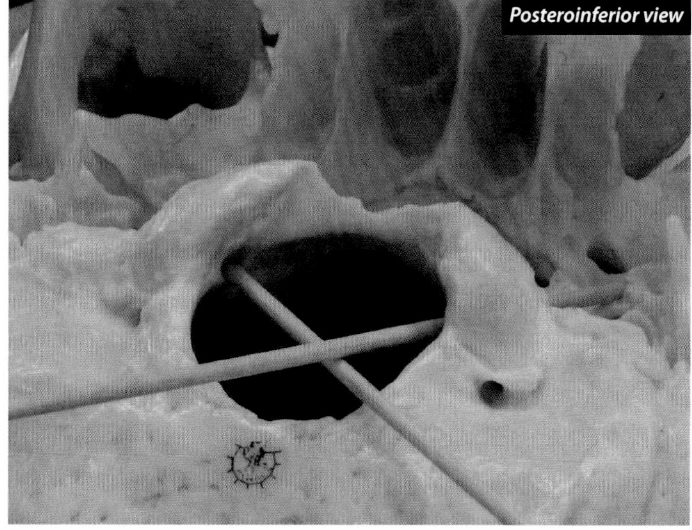


The hypoglossal canal is found in the occipital bone on the lateral surface of the foramen magnum. It provides a passage for the **hypoglossal nerve (XII)**.

## Stylomastoid Foramen (Temporal)

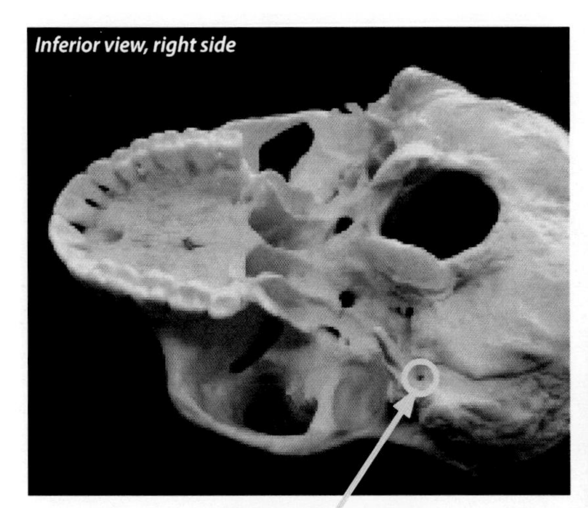


The stylomastoid foramen of the temporal bone surrounds the **facial nerve (VII)** as it passes away from the cranium toward the face. This is the second of two foramina through which it passes. The first is the internal auditory meatus.

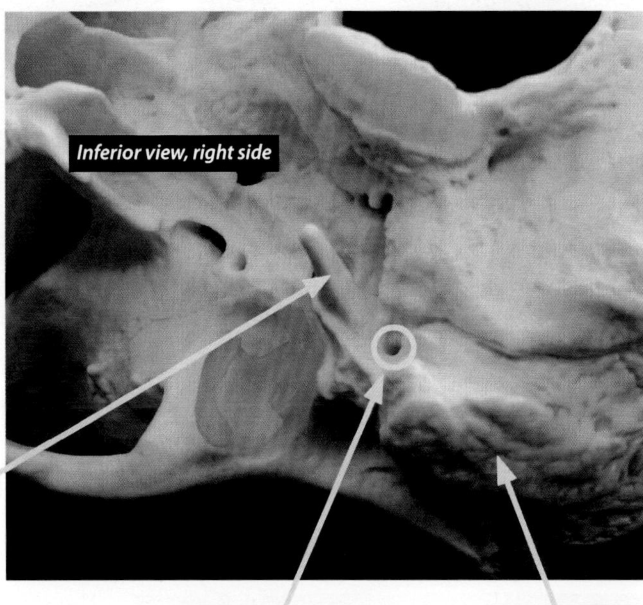


## Mandibular Foramen (Mandible)

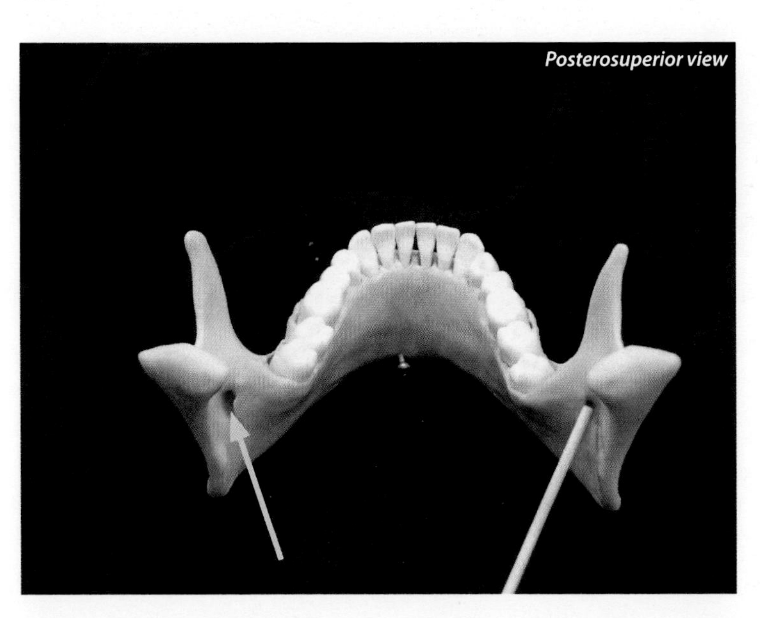


This is the second foramen that the **mandibular nerve ($V_3$)** passes through as it moves toward the mental foramen. Dentists usually try to anesthetize this nerve near this foramen when working on teeth in the mandible.

## External Auditory Meatus (Temporal)

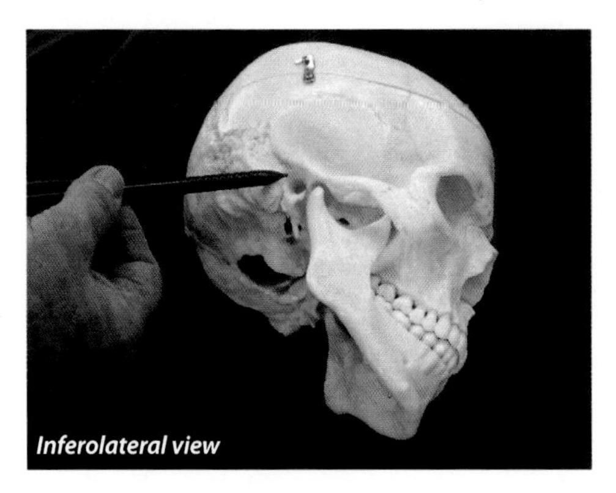


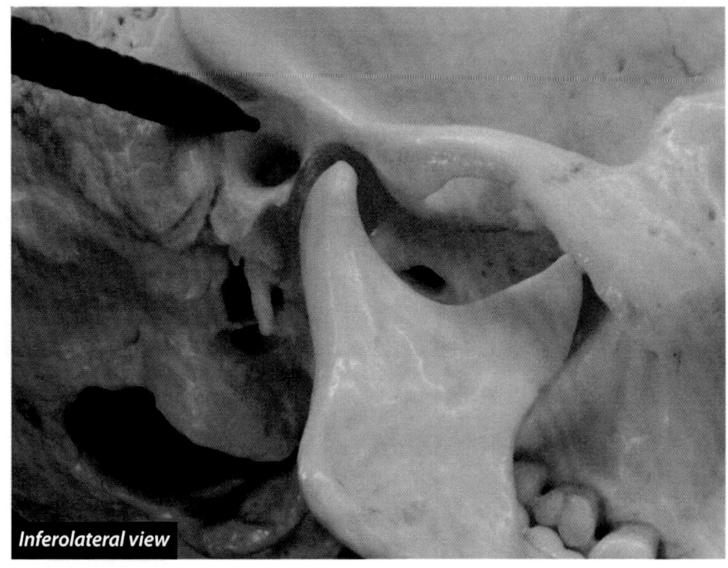


The external auditory meatus is found in the temporal bone. Functionally, it is important because sound enters the ear through this canal and comes in contact with the tympanic membrane (eardrum) at its deep end.

## Supraorbital Foramen (Notch) (Frontal)

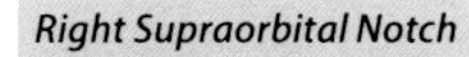


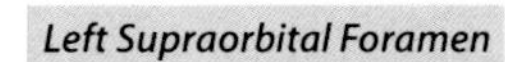


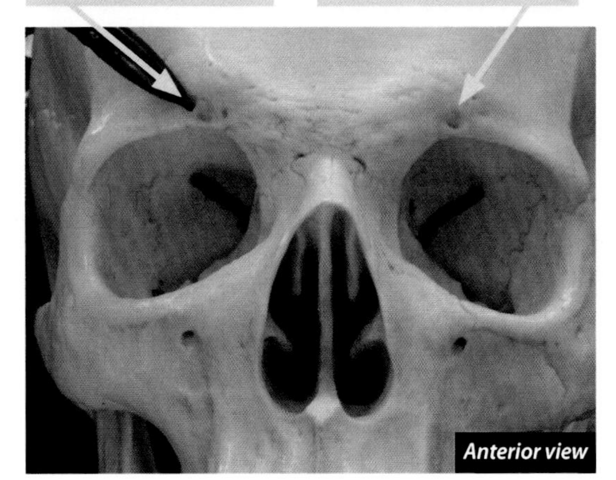


The supraorbital foramen (notch) of the frontal bone is a passageway for the **ophthalmic nerve** ($V_1$) as it moves onto the face. This is the second of two foramina through which it passes. The first was the superior orbital fissure.

## Infraorbital Foramen (Maxilla)

The infraorbital foramen is found in the maxillary bone, just inferior to the orbit. It is the third (and last) foramen through which the **maxillary nerve** ($V_2$) passes to the face.

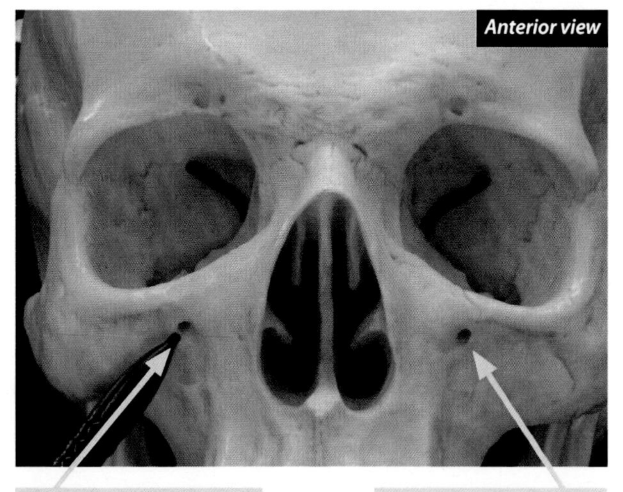


Right Infraorbital Foramen

Left Infraorbital Foramen

## Mental Foramen (Mandible)

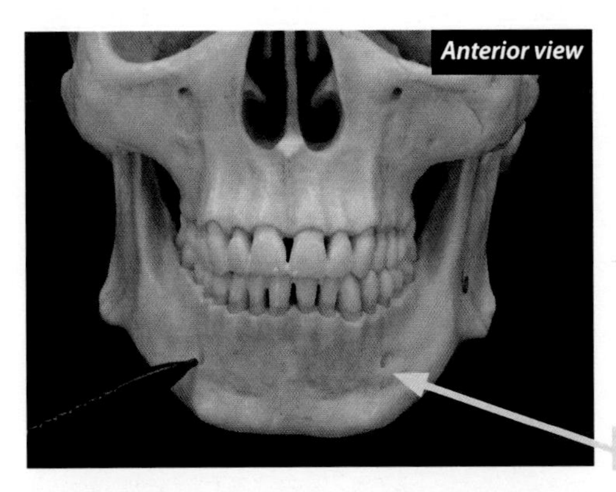


Left (probe) and right (arrow) mental foramina

This is the last foramen that the **mandibular nerve ($V_3$)** passes through as it courses away from the brain. In order, it passes through the foramen ovale, the mandibular foramen, and, finally, the mental foramen. Repetition is your friend.

# OVERVIEW—Foramina of Anterior Surface of Skull

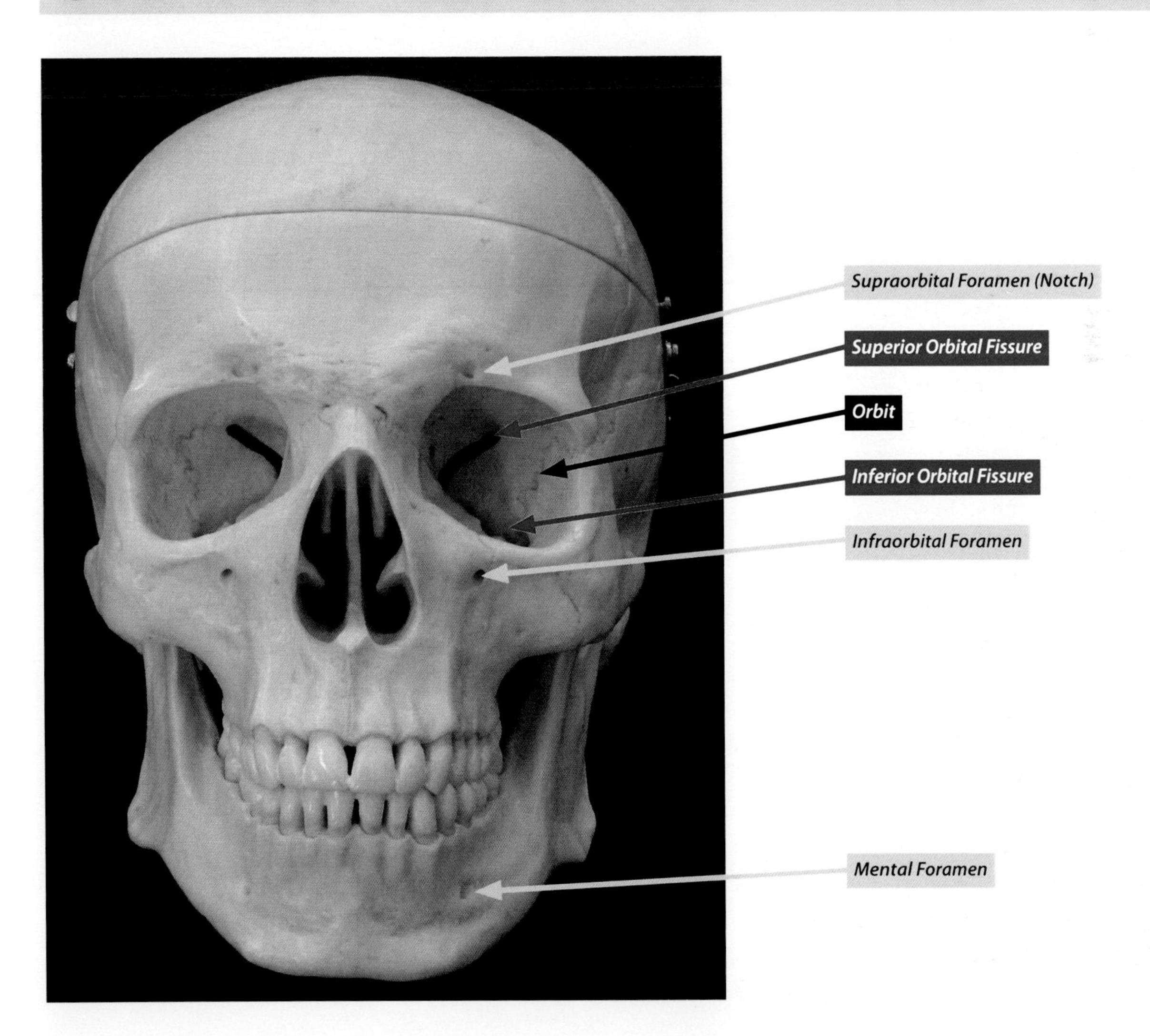

# CRANIAL NERVE OVERVIEW

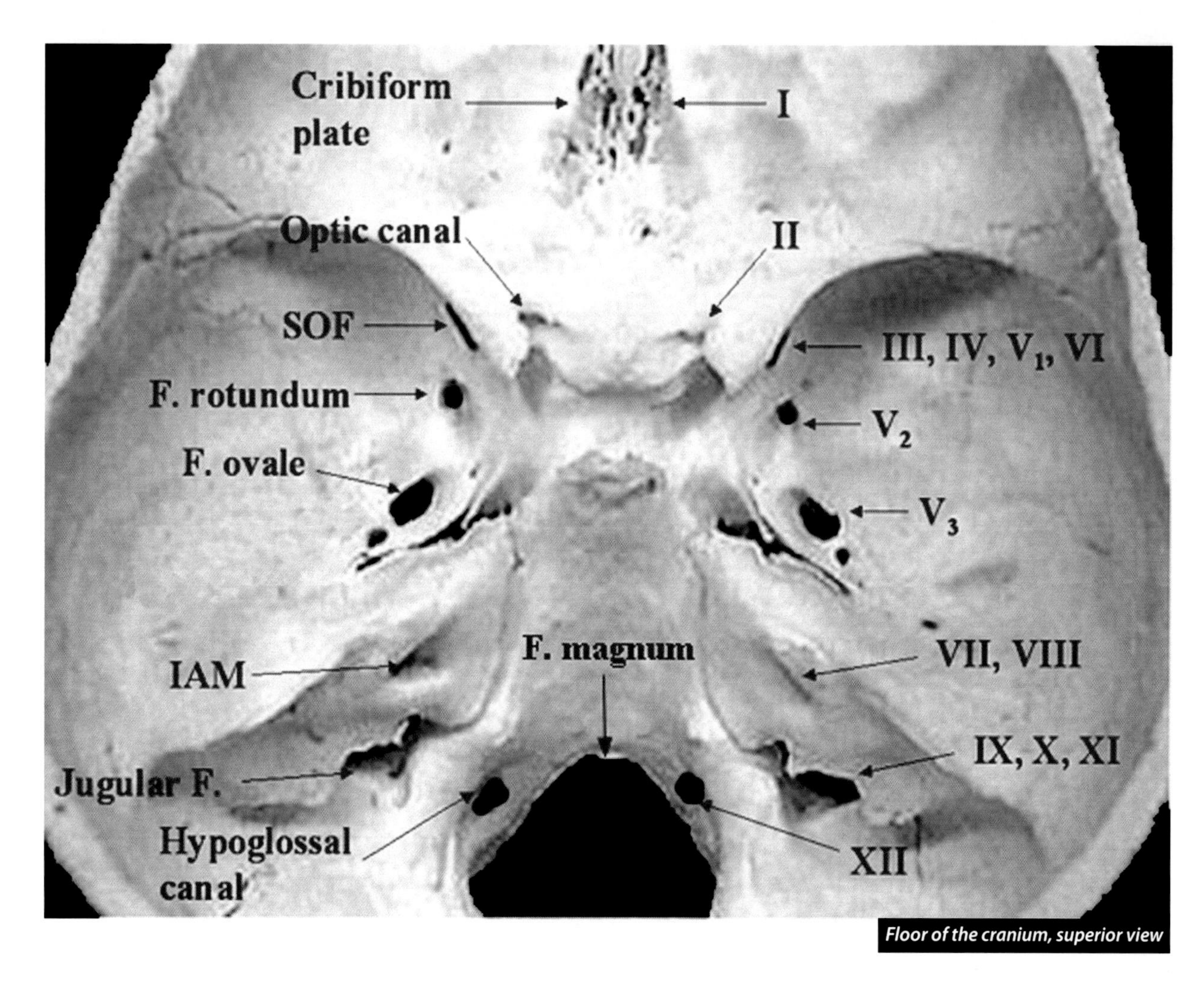


*Floor of the cranium, superior view*

**F. = Foramen**
**SOF = Superior Orbital Fissure**
**IAM = Internal Auditory Meatus**

# CRANIAL NERVES

S – Sensory M—Motor B—Mixed (motor and sensory) P—Parasympathetic E—Eyes/Vision

| # | NAME | FORAMINA / BONE(S) | FUNCTION |
|---|---|---|---|
| I | **Olfactory** | Cribriform plate of **ethmoid** | S—smell |
| II | **Optic** | Optic canal of **sphenoid** | E—S—vision |
| III | **Oculomotor** | Superior orbital fissure of **sphenoid** | E—P—M—muscles of the eyes except lateral rectus and superior oblique; parasympathetic- sphincter of pupil and ciliary muscle of the lens |
| IV | **Trochlear** | Superior orbital fissure of **sphenoid** | E—M—Superior oblique muscle of eye |
| V | **Trigeminal** | | B—see below for details |
| | **$V_1$—Ophthalmic** | Superior orbital fissure of **sphenoid,** then supraorbital foramen of **frontal** | E—S—cornea, nasal mucous membrane, skin of face |
| | **$V_2$—Maxillary** | Foramen rotundum of **sphenoid,** inferior orbital fissure (**sphenoid, maxilla, palatine,** and **zygomatic)** then infraorbital foramen of **maxilla** | S—skin of face , oral cavity, teeth |
| | **$V_3$—Mandibular** | Foramen ovale of **sphenoid,** mandibular foramen of **mandible,** then mental foramen of **mandible** | B—**MOTOR** to muscle of mastication; **SENSORY** to skin of face, teeth, anterior 2/3 of tongue ( general senses, NOT taste) |
| VI | **Abduscens** | Superior orbital fissure of **sphenoid** | E—M—lateral rectus muscle of eye |
| VII | **Facial** | Internal auditory meatus of **temporal,** then stylomastoid foramen of **temporal** | P—B—**MOTOR** to muscle of facial expression; **SENSORY** to anterior 2/3 of tongue (taste); parasympathetic to lacrimal, mandibular and sublingual glands |
| VIII | **Vestibulocochlear** (Statoacoustic, auditory, or acoustic) | | S—see below for details |
| | **Vestibular** | Internal auditory meatus of **temporal** | S—equilibrium |
| | **Cochlear** | Internal auditory meatus of **temporal** | S—hearing |
| IX | **Glossopharyngeal** | Jugular foramen between **temporal and occipital** | P—B—**MOTOR** to stylopharyngeus muscle; **SENSORY** to posterior 1/3 of tongue (general senses and taste), pharynx, branch to carotid sinus; parasympathetic to parotid gland |
| X | **Vagus** | Jugular foramen between **temporal and occipital** | P—B—**MOTOR** to pharynx and larynx; **SENSORY** to pharynx (including taste to epiglottis) and larynx, thoracic and abdominal organs; parasympathetic to thoracic and abdominal viscera |
| XI | **Spinal Accessory** | Exits via Jugular foramen (suture btw. **temporal & occipital),** enters cranium via foramen magnum | M—trapezius and sternocleidomastoid muscles |
| XII | **Hypoglossal** | Hypoglossal canal of **occipital** | M—muscles of the tongue |

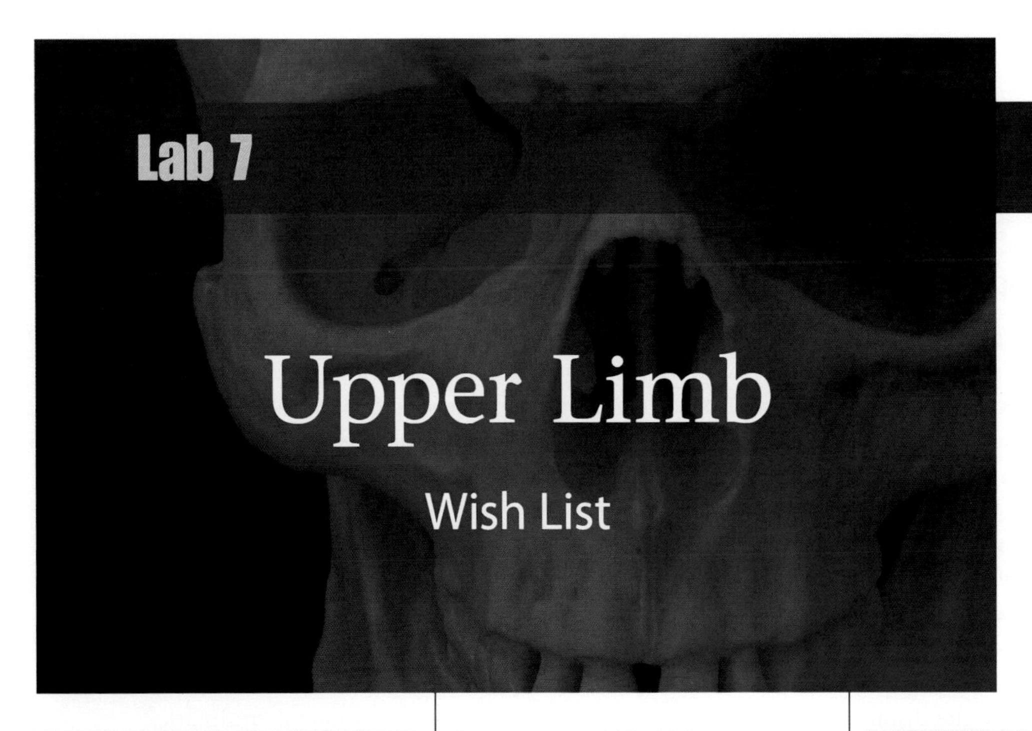

# Lab 7

# Upper Limb

## Wish List


**LAB 7 OVERVIEW, pp. 172–173**

**BONES, pp. 174–189**

- **Scapula, pp. 174–177**
  - Acromion Process, pp. 174–176
  - Supraspinous Fossa, pp. 174, 176
  - Suprascapular Notch, pp. 174–176
  - Spine, pp. 174, 176
  - Infraspinous Fossa, pp. 174, 176
  - Subscapular Fossa, pp. 175–176
  - Coracoid Process, pp. 174–176
  - Metacromion Process, p. 177
  - Glenoid Cavity, pp. 175, 177
  - Infraglenoid Tubercle, p. 177

**Arm, pp. 178–181**

- **Humerus, pp. 178–181**
  - Head, p. 178
  - Bicipital Groove, p. 178
  - Lesser Tubercle, p. 178
  - Greater Tubercle, p. 178
  - Deltoid Tuberosity, p. 179
  - Supracondyloid Foramen (cat), p. 179
  - Lateral Epicondyle, p. 180
  - Coronoid Fossa, p. 180
  - Capitulum, p. 180
  - Trochlea, p. 181
  - Medial Epicondyle, p. 181
  - Olecranon Fossa, p. 181

**Forearm, pp. 182–185**

- **Radius, pp. 182–183**
  - Head, p. 182
  - Neck, p. 182
  - Radial (Bicipital) Tuberosity, p. 182
  - Styloid Process, p. 183
  - Ulnar Notch, p. 183
- **Ulna, pp. 184–185**
  - Olecranon Process, p. 184
  - Trochlear (Semilunar) Notch, p. 184
  - Coronoid Process, p. 184
  - Radial Notch, p. 185
  - Head, p. 185
  - Styloid Process, p. 185
- **Hand, pp. 186–189**
  - Carpals, pp. 186–188
    - Scaphoid, p. 187
    - Lunate, p. 187
    - Triquetral (triangular), p. 187
    - Pisiform, p. 187
    - Trapezium, p. 187
    - Trapezoid, p. 187
    - Capitate, p. 187
    - Hamate, p. 187
  - Metacarpals, pp. 188–189
  - Phalanges, pp. 188–189

**MUSCLES, pp. 190–197**

- **Supraspinatus, pp. 190–191**
- **Infraspinatus, pp. 190–191**
- **Teres Minor, pp. 192–193**
- **Subscapularis, pp. 192–193**
- **Teres Major, pp. 194–195**
- **Brachialis, pp. 194–195**
- **Triceps Brachii (lateral, long and medial heads), pp. 196–197**
- **Anconeus, pp. 196–197**

**NERVES, pp. 198–199**

- **Suprascapular Nerve, p. 198**
- **Radial Nerve, p. 199**

**VESSELS, pp. 199–201**

**ARTERIES, pp. 199–201**

- **Subscapular Artery, p. 199**
- **Suprascapular Artery, p. 200**
- **Transverse Scapular Artery, p. 200**
- **Caudal Humeral Circumflex Artery, p. 201**

**VEINS, p. 201**

- **Cephalic Vein, p. 201**

# LAB 7 OVERVIEW

## 1. Bones of the Upper Limb (Pectoral Appendage)

Lab 7 introduces us to the pectoral appendage (upper limb). We will begin with the bones of the pectoral region and then study the muscles, nerves, and vessels. We will move from proximal to distal.

We have studied the **scapula** before, so most of this is review. You still need to distinguish one side from the other. Observe the **acromion process**, the **supraspinous fossa**, the **suprascapular notch**, the **spine**, the **metacromion process** (quadruped only), the **infraspinous fossa**, the **coracoid process**, the **supraglenoid tubercle**, the **glenoid cavity**, the **infraglenoid tubercle**, and the **subscapular fossa**.

The **head** of the **humerus** forms a ball and socket joint with the **glenoid cavity** of the **scapula**. Lateral to the head is the **greater tubercle**. Anterior to the greater tubercle is the **bicipital (intertubercular) groove**, and anterior to that is the **lesser tubercle**. Along the lateral side, proximal to the middle of the humerus, you will find the **deltoid tuberosity**. Cats have a **supracondyloid foramen** proximal to the distal end of the humerus. Some humans (less than 1 percent) have this landmark, or vestiges of this landmark. At the distal end of the humerus you will observe the most lateral landmark, the **lateral epicondyle**, which is functionally important as it serves as the origin of most of the extensor muscles in the forearm. Medial to the lateral epicondyle you will find the **capitulum (lateral condyle)**, which articulates with the **head** of the **radius** and allows for rotation, as well as for flexion and extension. On the posterior side of the humerus, you will find the **olecranon fossa**. On the anterior, you will find the **coronoid fossa**. Medial to the **captiulum** is the **trochlea (medial condyle)**, which articulates with the trochlear notch of the ulna. This is a hinge joint, allowing for flexion and extension of the elbow. Medial to the trochlear notch you will observe the **medial epicondyle**, which is functionally important as the origin for most of the flexor muscles in the forearm.

The **radius** is the lateral of the two bones in the forearm. The **head** of the radius is at the proximal end where it articulates with the **capitulum** of the **humerus**. Distal to the head is the **neck** of the **radius**. Immediately distal to the **neck** on the anterior side you will observe the **radial (bicipital) tuberosity**, one of the insertions for the **biceps brachii**. The most lateral landmark of the **radius** is the **styloid process**, the insertion for the **brachioradialis**. On the medial surface of the radius, you will find the **ulnar notch**, which forms a pivot joint with the **head** of the **ulna**, allowing for pronation and supination of the forearm (as the radius pivots around the stationary ulna)

The **ulna** is the medial of the two bones in the forearm. The **olecranon process** is the most proximal landmark on the **ulna**. It is on the posterior side and it is the insertion of the **triceps brachii**. On the anterior side of the **olecranon process** is the **trochlear (semilunar) notch**, which forms a hinge joint with the **trochlea** of the **humerus**. Anteriorly, you will observe the **coronoid process** of the **ulna**. Lateral to the trochlear notch is the **radial notch** of the **ulna**. Use this information to distinguish left from right. Place the **trochlear notch** on the anterior side and notice that the radial notch is lateral. The distal end of the ulna is the **head** and it forms a pivot joint with the ulnar notch of the radius. On the posterolateral side of the **ulna** at the distal end is the **styloid process**.

The **carpals** form the wrist. The mnemonic "**S**ome **L**overs **T**ry **P**ositions **T**hat **T**hey **C**an't **H**andle" gives us the first letters for the carpals starting at the lateral carpal of the proximal row and moving medially; then proceeding to the lateral most carpal of the distal row and moving medially. They are: **scaphoid**, **lunate**, **triquetrum (triangular)**, **pisiform**, **trapezium**, **trapezoid**, **capitate**, **hamate**. Note, the proximal carpals form a condyloid joint (at the wrist) with the distal end of the **radius**. The intercarpal joints are considered plane joints, which allow for gliding (sliding) movements.

Distal to the carpals are the **metacarpals**. You will not be responsible for their numbers, but the **first metacarpal** is on the lateral side. The joint between the **trapezium** and the **first metacarpal** is a saddle joint. Distal to the metacarpals are the **phalanges** (thumb and fingers). Again, you will not be held responsible for the specific numbers of the phalanges. You should know both of these sets of bones generically.

## 2. Muscles, Nerves, and Vessels of the Arm and Shoulder

We begin our dissection on the lateral side of the arm. We separate the **lateral head** of the triceps brachii, and once that is complete we transect it. Running along its cranial margin you should observe the **cephalic vein**. As you transect the **medial head** of the **triceps brachii**, you should be especially careful to avoid cutting the **radial nerve**, which is on the deep side of the **lateral head** of the **triceps brachii**. Reflect the **lateral head** of the **triceps** and observe the **brachialis muscle** with the **radial nerve** running over the surface of that muscle. The **radial nerve** bifurcates to a superficial branch that runs onto the surface of the **brachioradialis muscle** as it courses toward the wrist, and to a deep branch that serves the extensor muscles in the forearm.

Notice the **medial head** of the **triceps brachii** that is medial to the **lateral head** of the **triceps brachii**. Also, notice the **long head** of the **triceps brachii**, which is running along the caudal margin of the **lateral head** of the **triceps brachii**. Observe the shiny material that covers the distal portion of the lateral side of the **long head** of the **triceps brachii**. That is the **perimysium**. All skeletal muscles have this feature, which is functionally important as it helps prevent the risk of tearing the muscle cells when they are under a physical load. Distally, you will see a relatively small triangular muscle, the **anconeus**, which extends to the elbow. Proximally you will see an **artery**, the **caudal humeral circumflex artery**, as it emerges from the heads of the **triceps brachii**.

Transect the **spinodeltoid muscle** to expose the small, triangular **teres minor** and part of the **infraspinatus muscle**. Observe the **subscapular artery** as it emerges from between the **teres major** and the **infraspinatus muscle**. Move dorsally and reflect the **acromiotrapezius muscle** to expose the **supraspinatus muscle**. Cut this muscle so that there is a cranial half and a caudal half. Then cut the muscle away from the **supraspinous fossa** of the **scapula** all the way to the **suprascapular notch** to expose the **suprascapular artery** and the **suprascapular nerve** as they pass through that notch. Now pull the **scapula** away from the thoracic wall. Observe the **suprascapular nerve** and the **transverse scapular artery**. Also observe the **subscapularis muscle**, which is on the deep side of the scapula.

# BONES

## Scapula

The scapula is of functional significance as it forms part of the girdle for the pectoral appendage. The large flat surfaces of the scapula serve as attachment areas for many of the muscles that move the arm or stabilize the girdle.

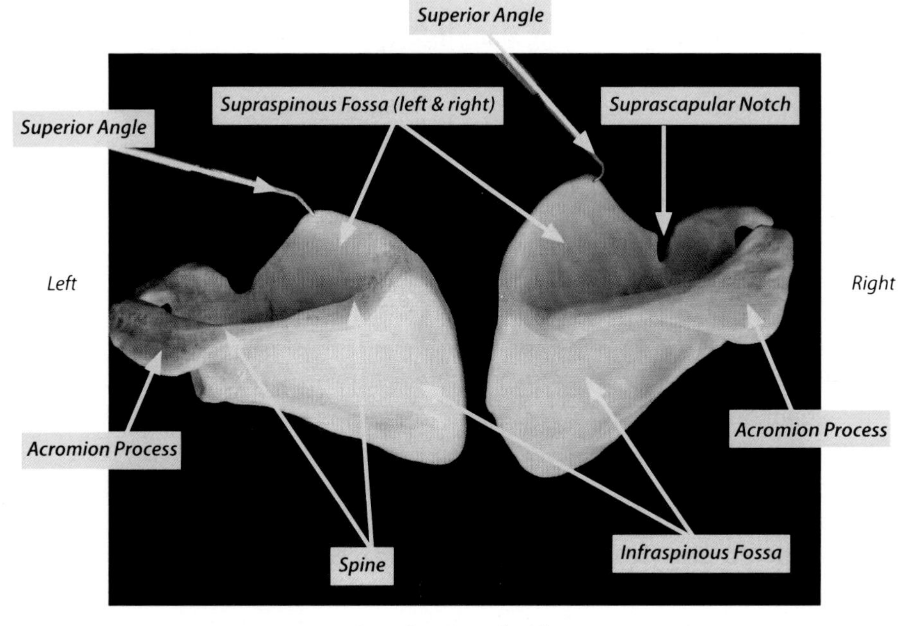


***Scapula—Posterior View***

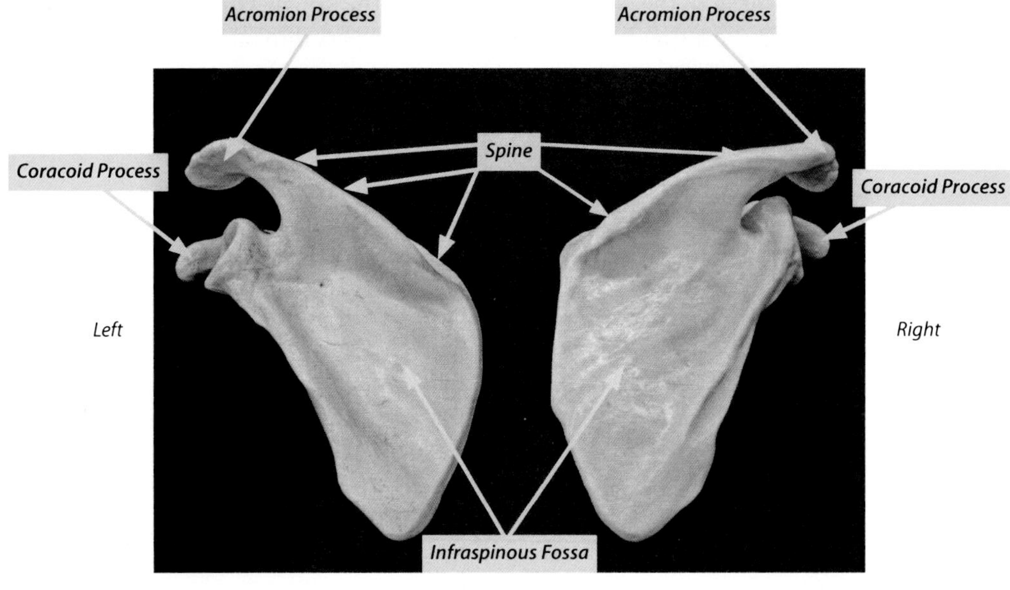


***Scapula—Posterior View***

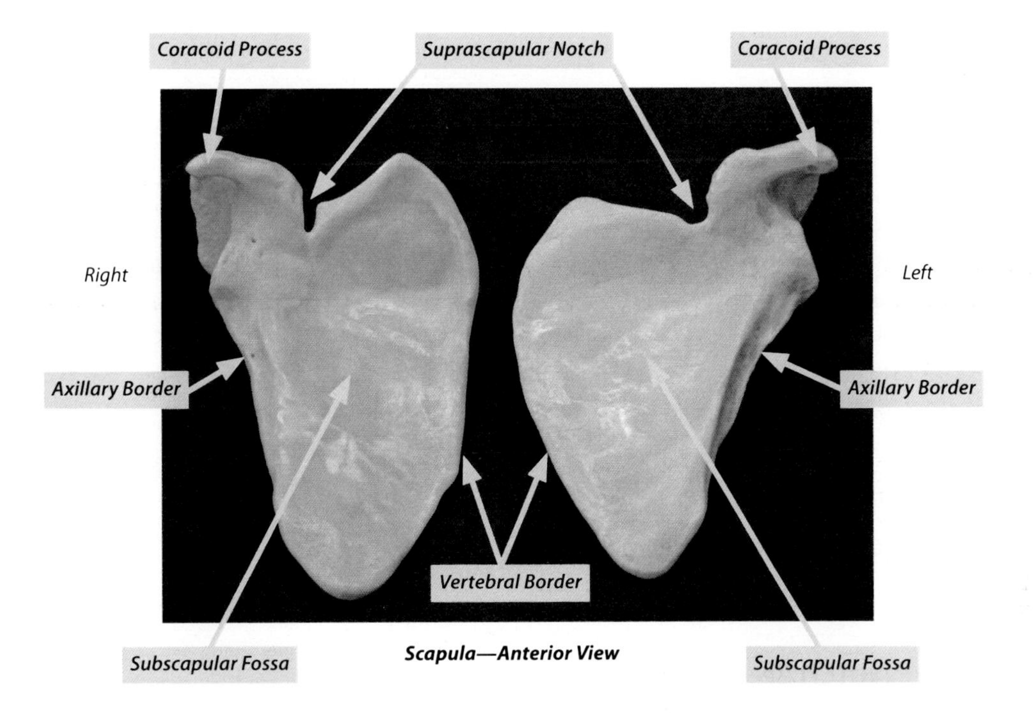


**Scapula—Anterior View**

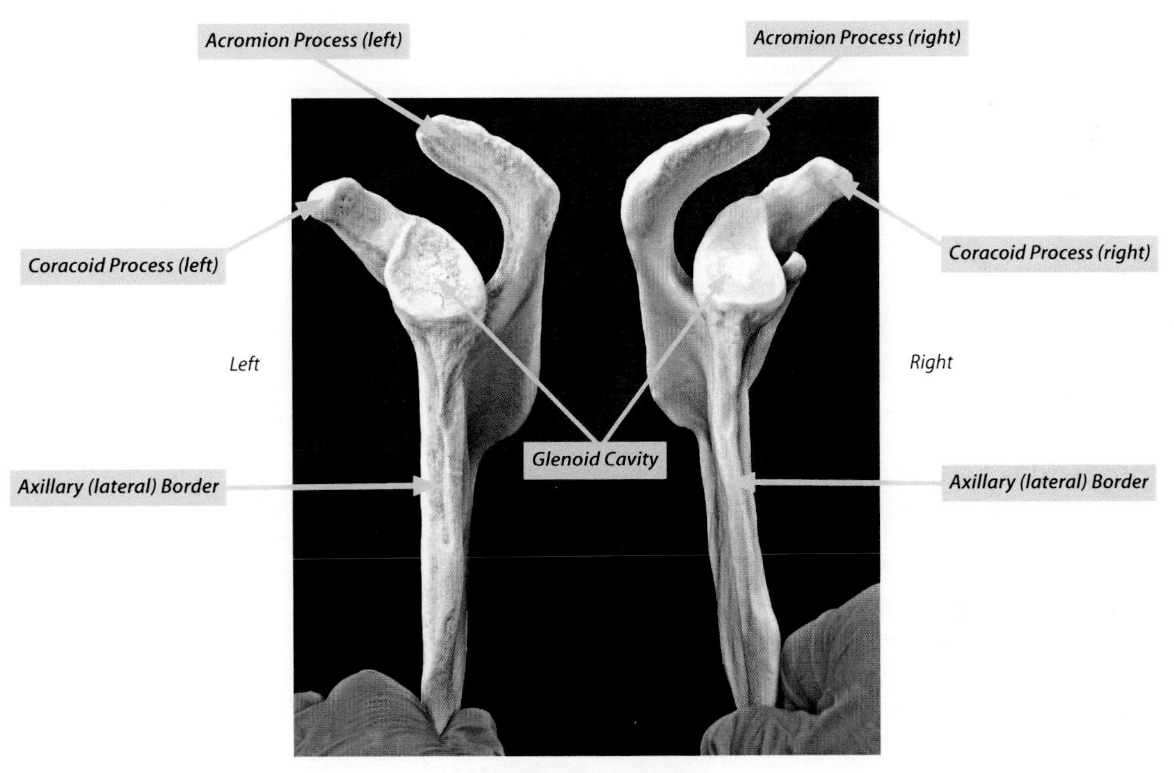


**Scapula—Glenoid View**

# Scapula

## Acromion Process *(see pages 174–175)*

Physically, the acromion process (or just acromion) is a projection of the spine of the scapula. The acromion is an important landmark of the scapula because it articulates with the acromial end (distal end) of the clavicle to form the acromioclavicular joint. This is an unusual plane joint in that it allows for gliding (anteroposterior and vertical) and some axial rotation of the scapula. The joint capsule of this joint is made stronger by two ligaments, the small acromioclavicular ligament on the vertical surface of the bones and the larger coracoclavicular ligament (two parts) between the coracoid process and the inferior surface of the clavicle. The acromion process of the scapula is the insertion for:

1. the superior portion of the trapezius,

and the origin for:

1. the middle portion of the deltoid.

## Supraspinous Fossa *(see page 174)*

The supraspinous fossa of the scapula is the origin for:

1. supraspinatus.

## Suprascapular Notch *(see pages 174–175)*

The suprascapular notch is of functional importance because the **suprascapular nerve** and the **suprascapular artery** pass through this notch. That would make it a Grant, Grant, Grant thing. It doesn't get better than that!

## Spine *(see page 174)*

The spine of the scapula is the insertion for:

1. the middle portion of the trapezius (the superior border of the spine of the scapula) and
2. the inferior portion of the trapezius (the medial third of the spine of the scapula),

and is the origin for:

1. the posterior portion of the deltoid (inferior margin of the spine of the scapula).

## Infraspinous Fossa *(see page 174)*

The infraspinous fossa is the origin for:

1. infraspinatus,
2. teres major, and
3. teres minor.

## Subscapular Fossa *(see page 175)*

The subscapular fossa of the scapula is the origin for:

1. subscapularis.

## Coracoid Process *(see pages 174–175)*

The coracoid process can be palpated in the infraclavicular fossa. The name implies that it looks like a crow's beak, but Dr. J thinks it looks more like Woodstock, Snoopy's friend. The coracoid process of the scapula is the origin for:

1. biceps brachii and
2. coracobrachialis,

and is the insertion for:

1. pectoralis minor.

## Metacromion Process (Cat)

The metacromion process of the cat is the insertion point for the acromiotrapezius muscle and the levator scapulae ventralis (omotransversarius) muscle. Humans do not have a metacromion process.

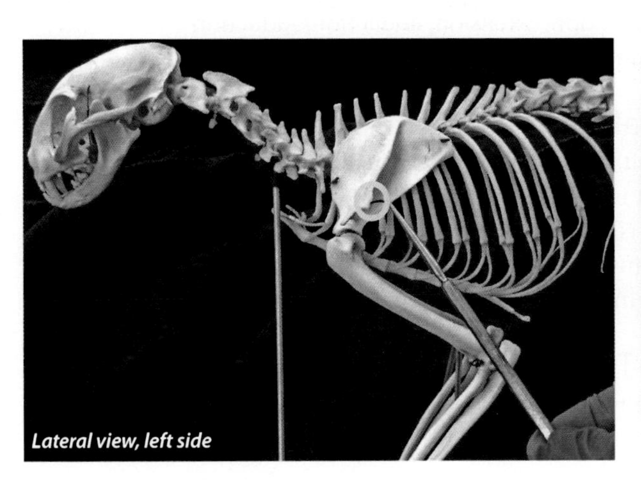

*Lateral view, left side*

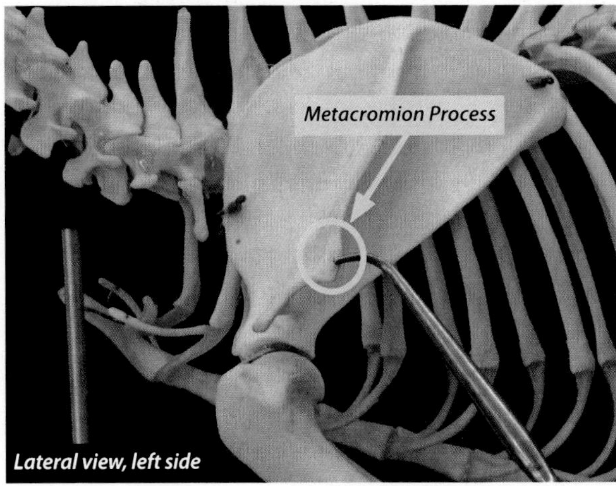


*Lateral view, left side*

## Glenoid Cavity

The glenoid cavity of the scapula articulates with the head of the humerus to form the glenohumeral joint. This is a ball and socket joint (triaxial) allowing for flexion/extension, abduction/adduction (and therefore circumduction), and rotation of the arm.

## Infraglenoid Tubercle

The infraglenoid tubercle of the scapula is the origin for:

1. the long head of the triceps brachii.

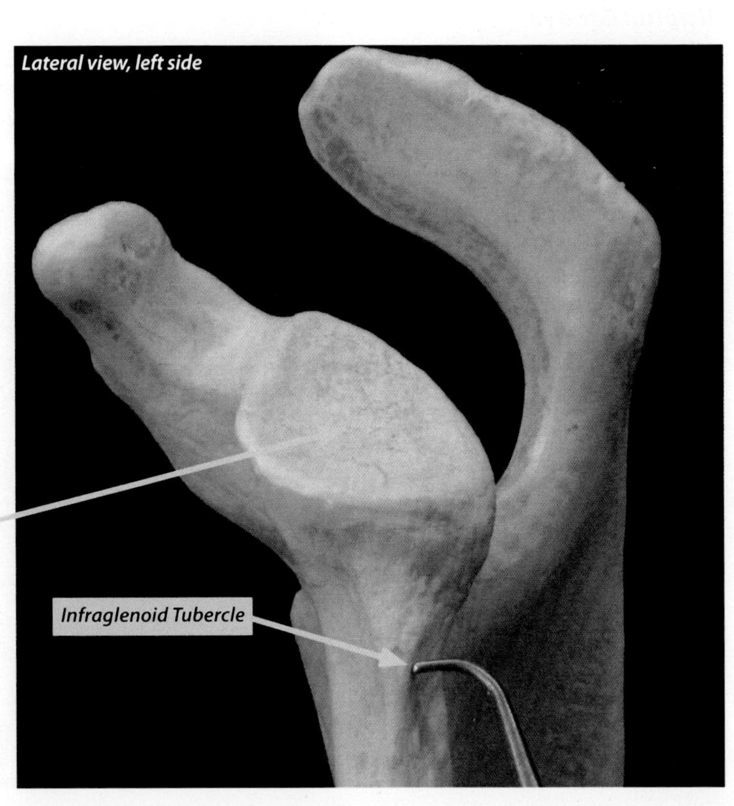


*Lateral view, left side*

## Humerus

### *Head*

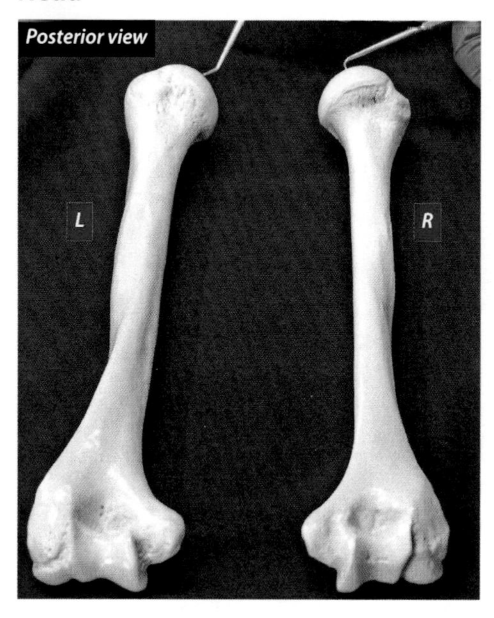


### *Bicipital Groove*

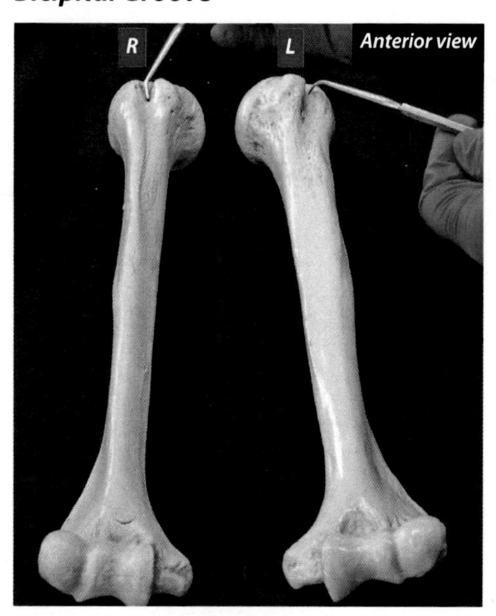


### Head

The head of the humerus articulates with the glenoid cavity of the scapula to form the glenohumeral joint. This is a ball and socket joint (triaxial) in that it allows for flexion/extension, abduction/adduction (and circumduction), and rotation of the arm.

### Bicipital Groove

The bicipital groove is also known as the intertubercular groove. It is located at the proximal end of the intertubercular sulcus. It is of functional importance because the tendon from the long head of the biceps brachii is guided by this groove to its origin on the supraglenoid tubercle of the scapula.

### Lesser Tubercle

The lesser tubercle of the humerus is located anterior to the bicipital groove. It is the insertion for:

1. subscapularis and
2. teres major.

### Greater Tubercle

Three of the four rotator cuff muscles insert on the greater tubercle of the humerus. They are:

1. supraspinatus (superior facet),
2. infraspinatus (posterior aspect), and
3. teres minor (inferior facet on posterior surface).

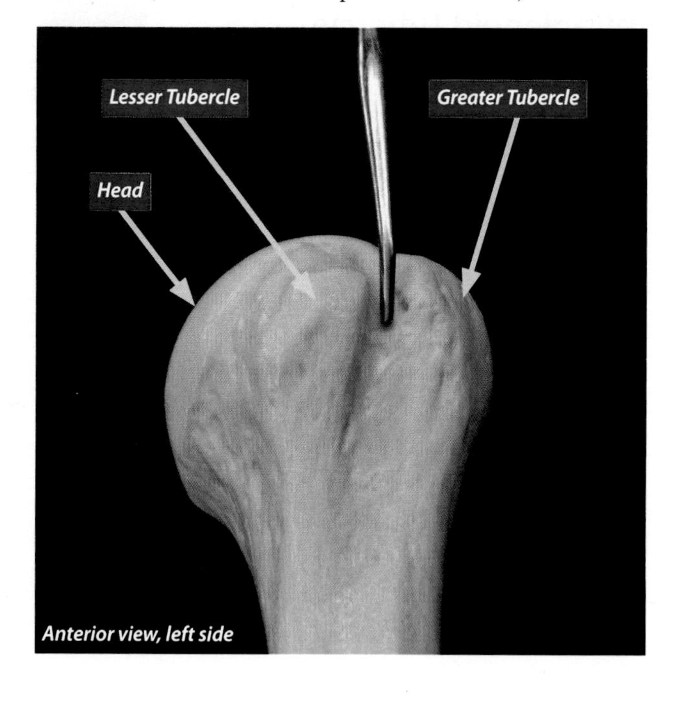

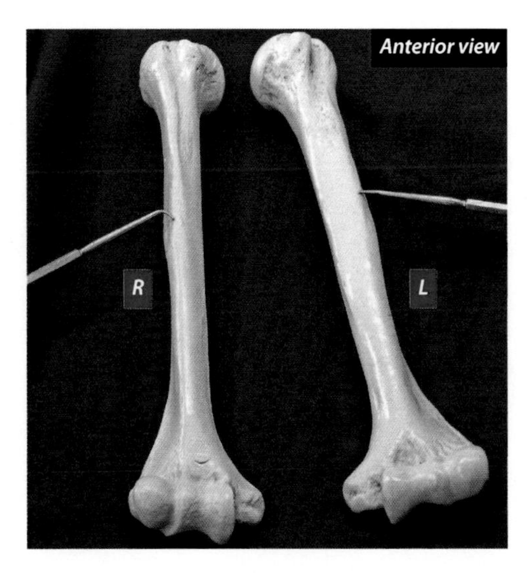


## Deltoid Tuberosity

The deltoid tuberosity of the humerus is of functional importance because it is the insertion for:

1. deltoids (all three heads).

## Supracondyloid Foramen (Humerus—Cat)

The supracondyloid foramen is found on the humerus of the cat, but NOT on the humerus of most humans. Approximately 1 percent of humans have a vestige of this feature that can cause various **median nerve** pathologies. The **brachial artery** and the **median nerve** of the cat pass through this structure and are, therefore, somewhat protected by this arch of bone.

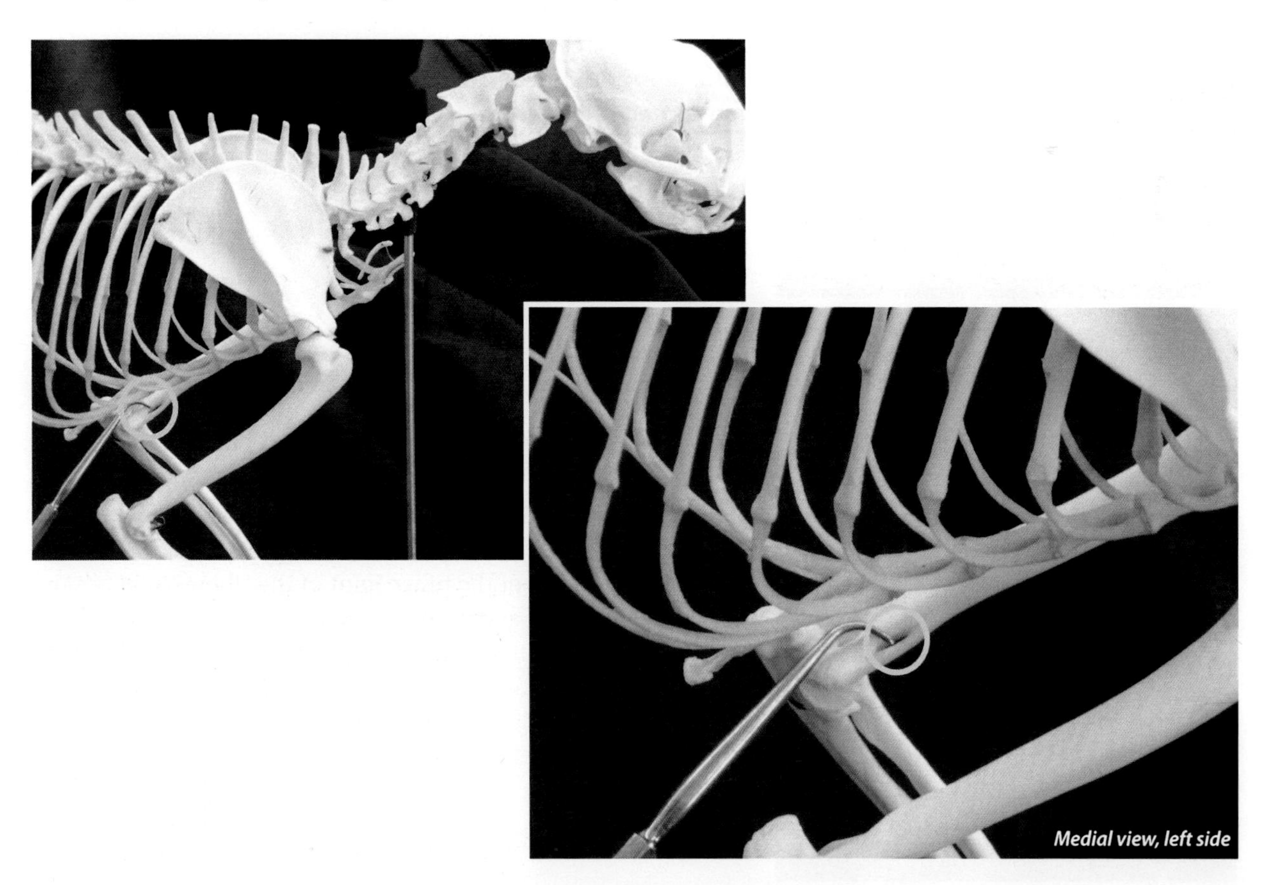

# Humerus

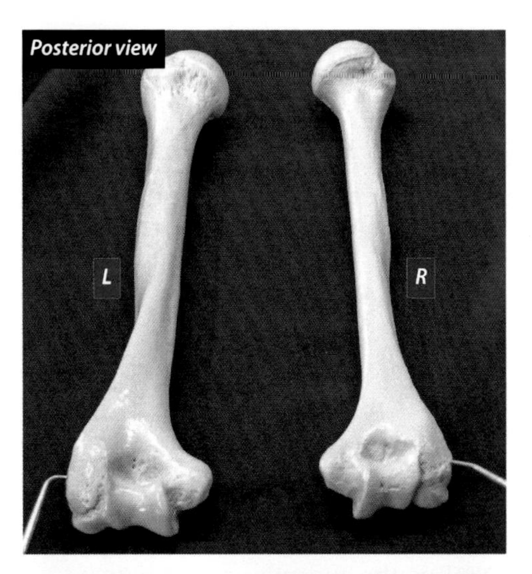


## Lateral Epicondyle

The lateral epicondyle of the humerus is the origin for:

1. extensor carpi radialis brevis,
2. extensor carpi radialis longus,
3. extensor carpi ulnaris (one head),
4. extensor digitorum communis, and
5. anconeus (posterior surface).

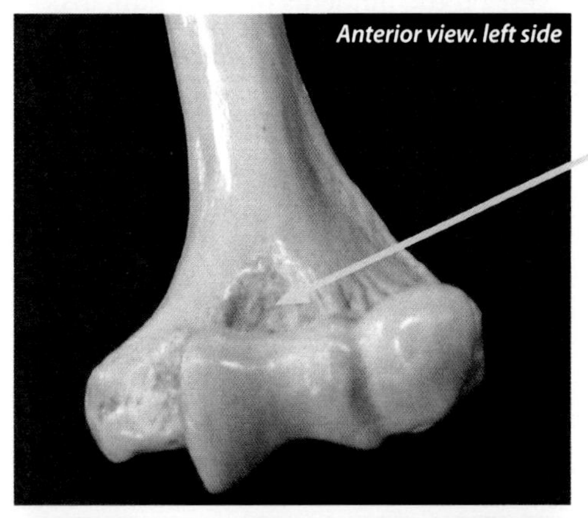


## Coronoid Fossa

The coronoid fossa is a depression on the anterior distal humerus. It receives the coronoid process of the ulna when the forearm is flexed.

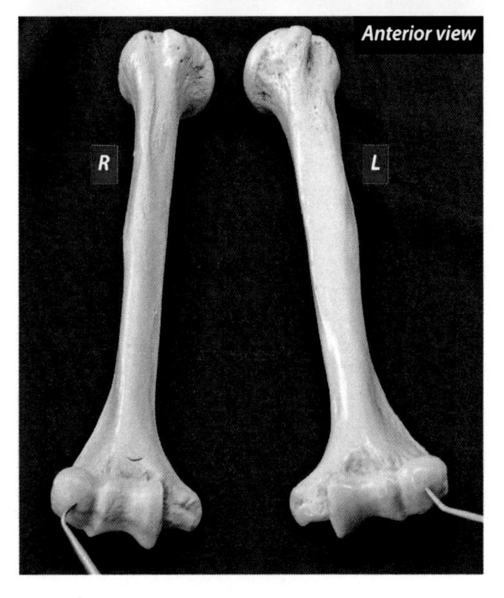


## Capitulum

The capitulum (translated means little head) is also known as the lateral condyle of the humerus. It is an important landmark because it articulates with the head of the radius to form the humeroradial joint. This is an unusual joint in that it forms part of the hinge joint of the elbow while allowing supination and pronation of the forearm (hand) at the proximal radioullnar joint. Dr. J recommends that you call it the capitulum because then the little head of the humerus articulates with the head of the radius, making it a Grant thing.

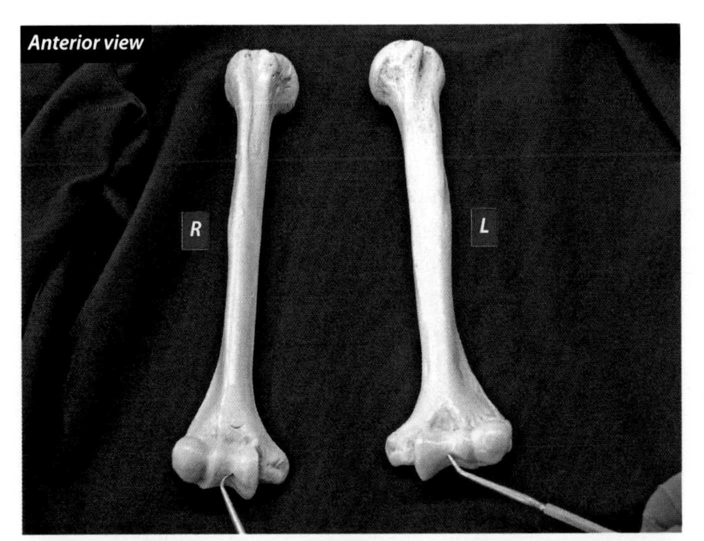


## Trochlea

The trochlea is an important landmark because this is where the humerus articulates with the trochlear notch of the ulna to form the humeroulnar joint. This is a hinge joint that allows for flexion/extension of the forearm. Some refer to the trochlea as the medial condyle. Dr. J recommends that you call it the trochlear, as it articulates with the trochlear notch, which makes it a Grant thing.

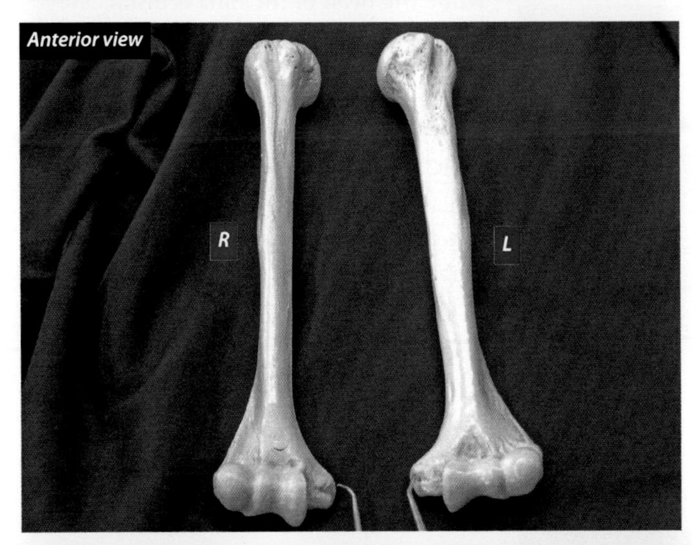


## Medial Epicondyle

The medial epicondyle of the humerus is the origin for:

1. one head of pronator teres,
2. flexor carpi radialis,
3. one head of flexor digitorum superficialis,
4. one head of flexor carpi ulnaris, and
5. palmaris longus.

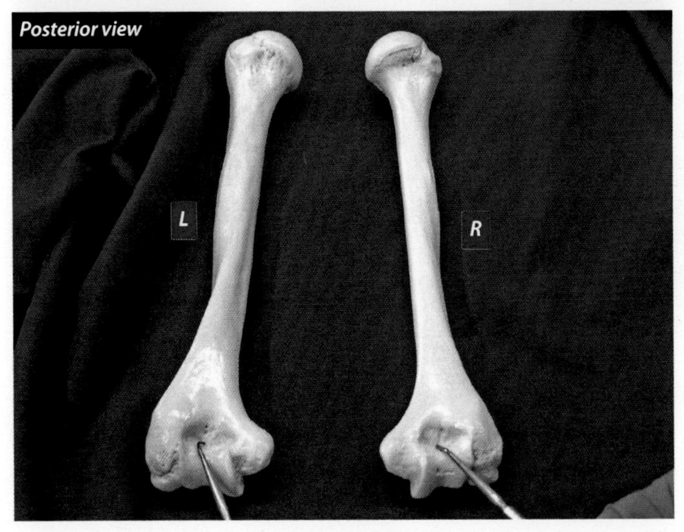


## Olecranon Fossa

The olecranon fossa is a depression on the posterior distal humerus. It receives the olecranon process of the ulna when the forearm is extended.

# Radius

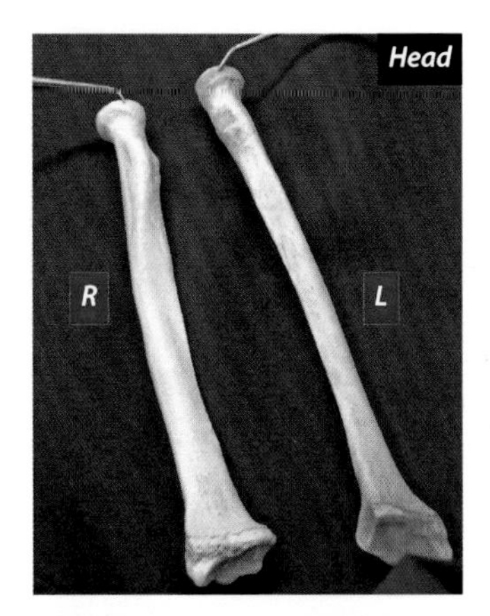


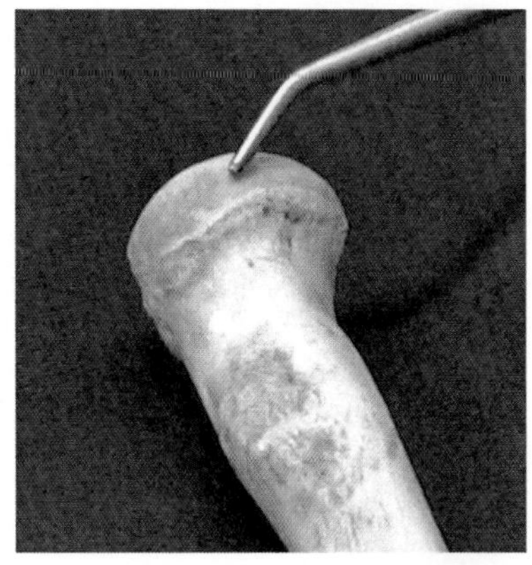

## Head

The head is an important landmark of the radius because it articulates with the capitulum of the humerus to form the humeroradial joint. This is an unusual joint for flexion/extension at the elbow with the humeroulnar joint and supination and pronation of the forearm (hand) at the proximal radioulnar joint. Remember, the ulna does **NOT** rotate. Only the radius rotates. Note that the head of the radius is proximal while the head of the ulna is distal.

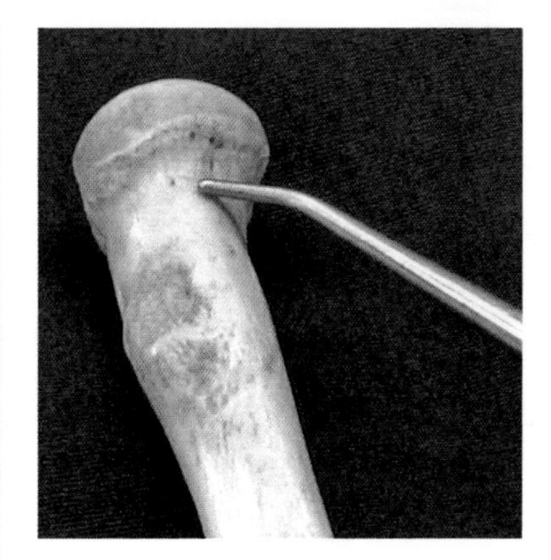

## Neck

The neck of the radius is not very noteworthy in terms of function. It is a narrow, short, cylindrically shaped piece of bone between the radial (bicipital) tuberosity and the head of the radius.

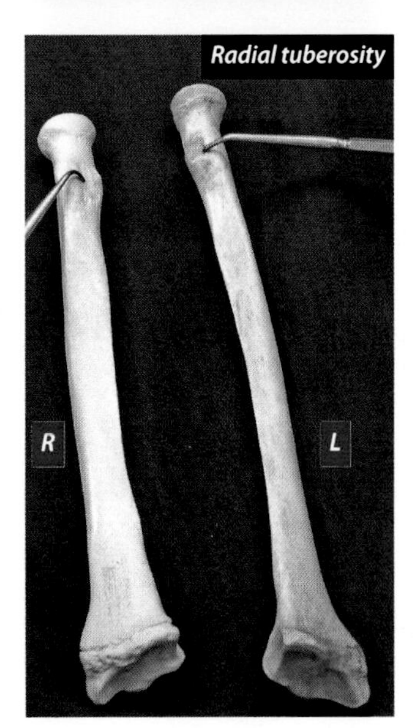


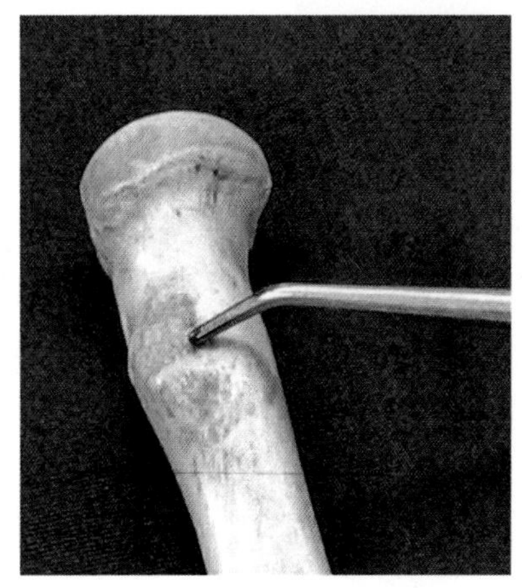

## Radial (Bicipital) Tuberosity

The radial tuberosity of the radius is the insertion for:

1. biceps brachii.

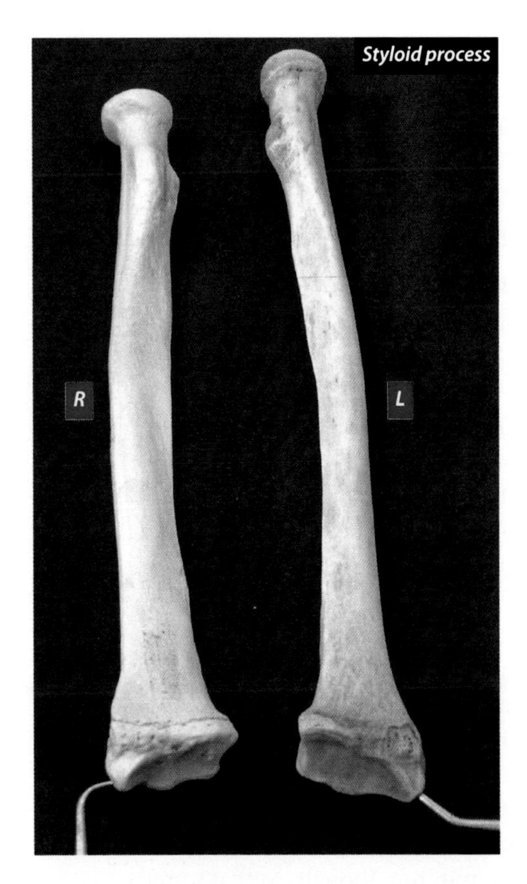


## Styloid Process

The styloid process of the radius is a distal, pointed structure located on the lateral aspect of the wrist (in the anatomical position). It is functionally important because it provides attachment for the radial collateral ligament of the wrist, which supports the wrist laterally. It is also the insertion for:

1. brachioradialis (the lateral side of the styloid process).

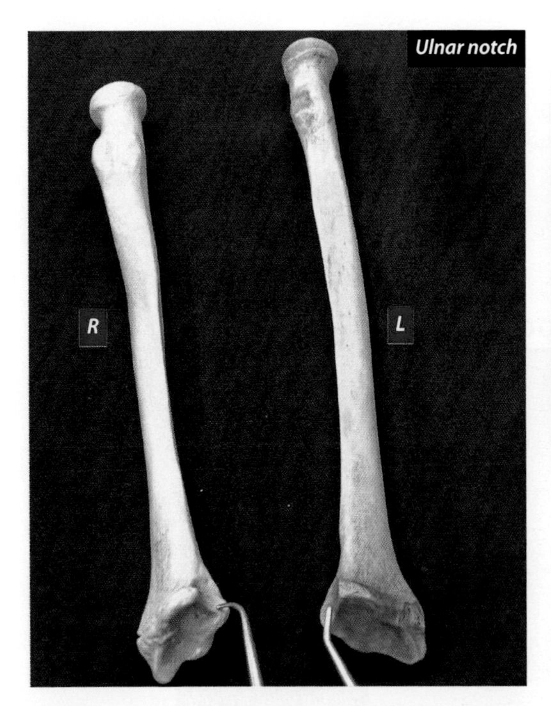


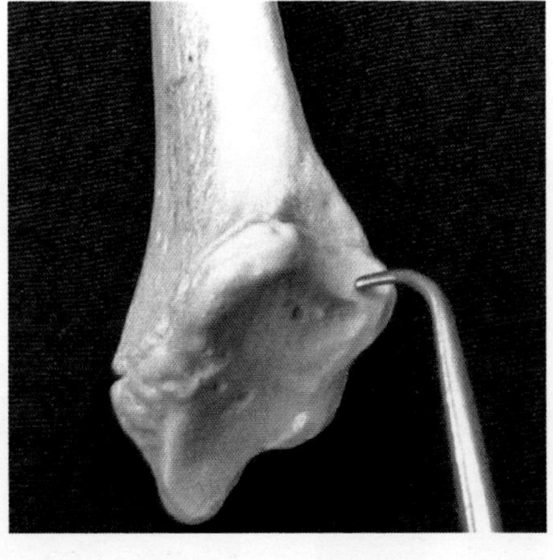

## Ulnar Notch

The ulnar notch is an important landmark of the radius because it is where the head of the ulna articulates with the radius to form the distal radioulnar joint. Both the proximal and distal radioulnar joints are uniaxial pivot joints, allowing for rotation of the radius over the ulna and, therefore, supination and pronation of the forearm (hand). Remember, the ulna does **NOT** rotate. Only the radius rotates.

# Ulna

## Olecranon Process

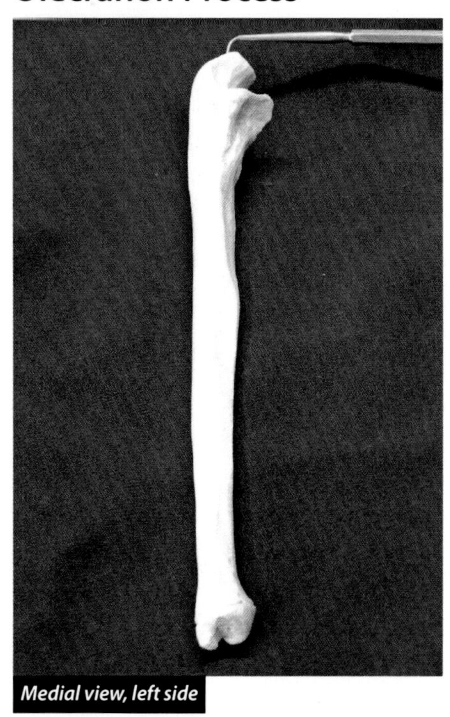
*Medial view, left side*

The olecranon process is a proximal posterior projection of bone that makes up the pointy part of the elbow. It is the insertion for:

1. anconeus (lateral aspect) and
2. triceps brachii (all three heads).

The olecranon process is the origin for:

1. one head of flexor carpi ulnaris.

## Trochlear (Semilunar) Notch

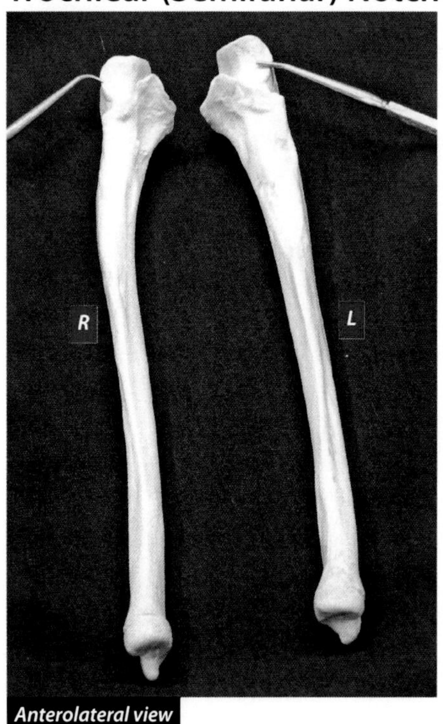

*Anterolateral view*

The trochlear (semilunar) notch is an important landmark of the ulna because it articulates with the trochlea of the humerus to form the humeroulnar joint. This is a (uniaxial) hinge joint allowing for flexion/extension of the forearm. Dr. J recommends that you remember that the trochlea notch articulates with the trochlea, which makes it a Grant thing.

The coro**N**oid process of the ulna is a projection on the anterior proximal ulna that makes up the bottom portion of the trochlear notch. Do not confuse this with the cora**C**oid process of the scapula. One thing that might help is to notice that s**C**apula and cora**C**oid both have the letter "c" in them, while ul**N**a and coro**N**oid both have the letter "n" in them. There is also a coro**N**oid process in the ma**N**dible (jaw), but we'll get to that later (mandible has an "n" in it, too, though, so that's pretty cool). The coronoid process of the ulna is the insertion for:

1. brachialis

and the origin for:

1. one head of pronator teres and
2. one head of flexor digitorum superficialis.

## Coronoid Process

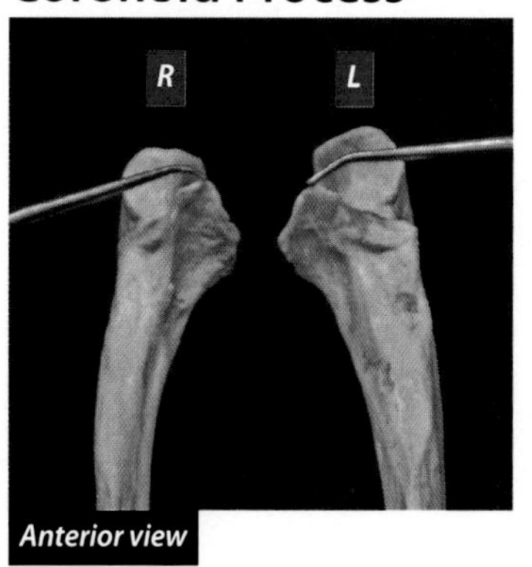

*Anterior view*

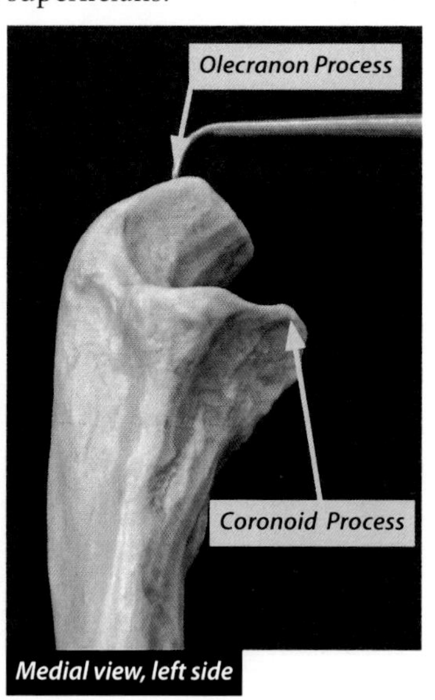

*Medial view, left side*

## Radial Notch

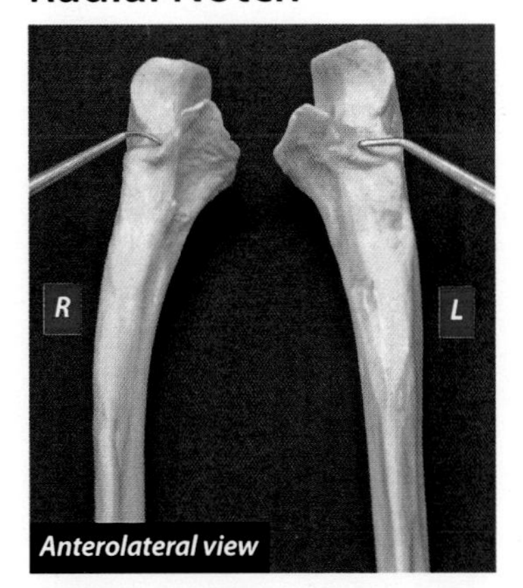


The radial notch is an important landmark of the ulna because it articulates with the head of the radius to form the proximal radioulnar joint. Like the distal radioulnar joint, this is a pivot joint (uniaxial) allowing rotation of the radius over the ulna to provide for supination and pronation of the forearm (hand). Remember, the ulna does **NOT** rotate. Only the radius rotates.

## Head

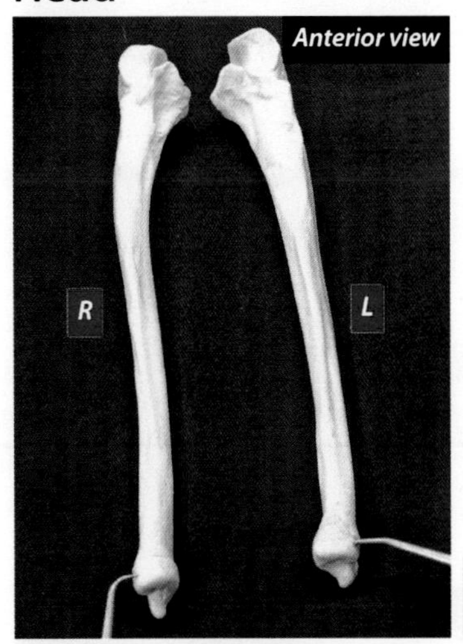


The head is an important landmark of the ulna because it is where the distal ulna articulates with the ulnar notch of the radius to form the distal radioulnar joint. Just like the proximal radioulnar joint, this is a pivot joint (uniaxial) in that it allows for rotation to provide for supination and pronation of the forearm (hand). Only the radius rotates.

*Note that the head of the ulna is distal, while the head of the radius is proximal.*

## Styloid Process

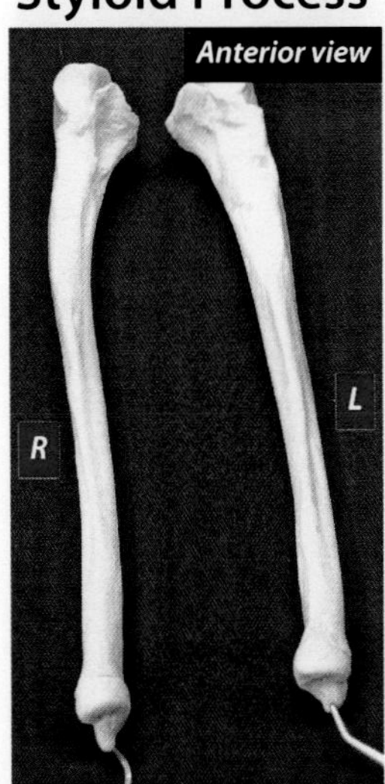


The styloid process of the ulna is a pointed structure found on the distal posteromedial aspect of the ulna. It is functionally important because it provides attachment for the ulnar collateral ligament of the wrist, which supports the wrist medially.

# Hand

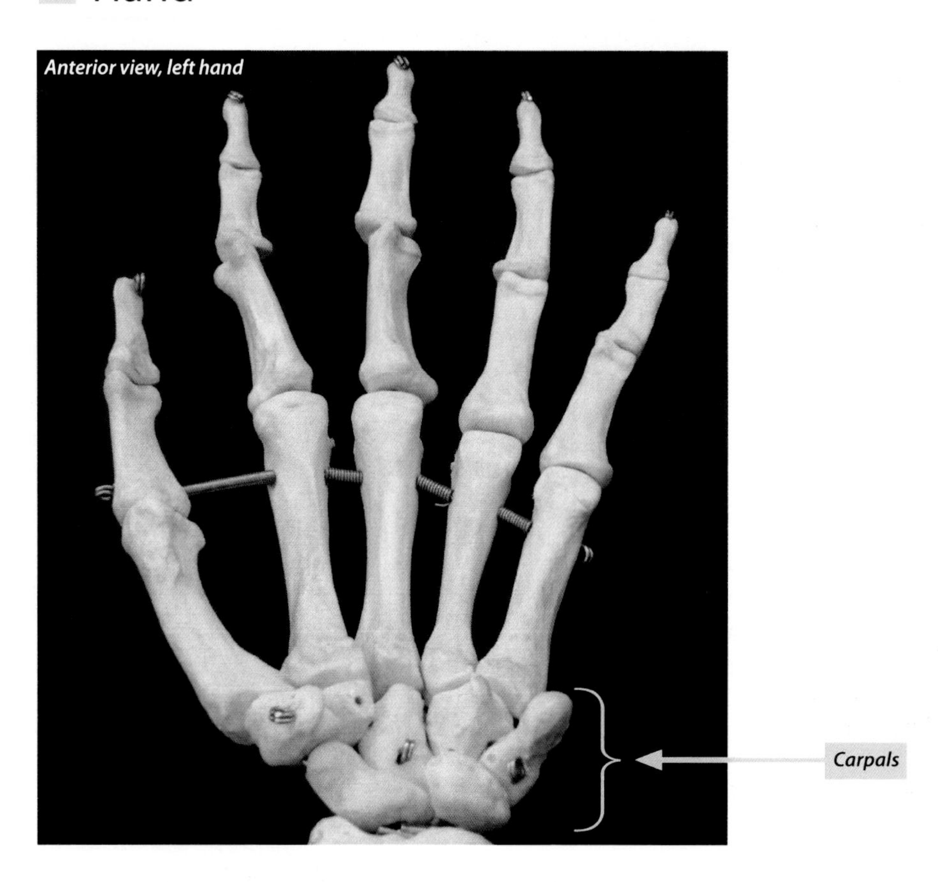


## Carpals—close up

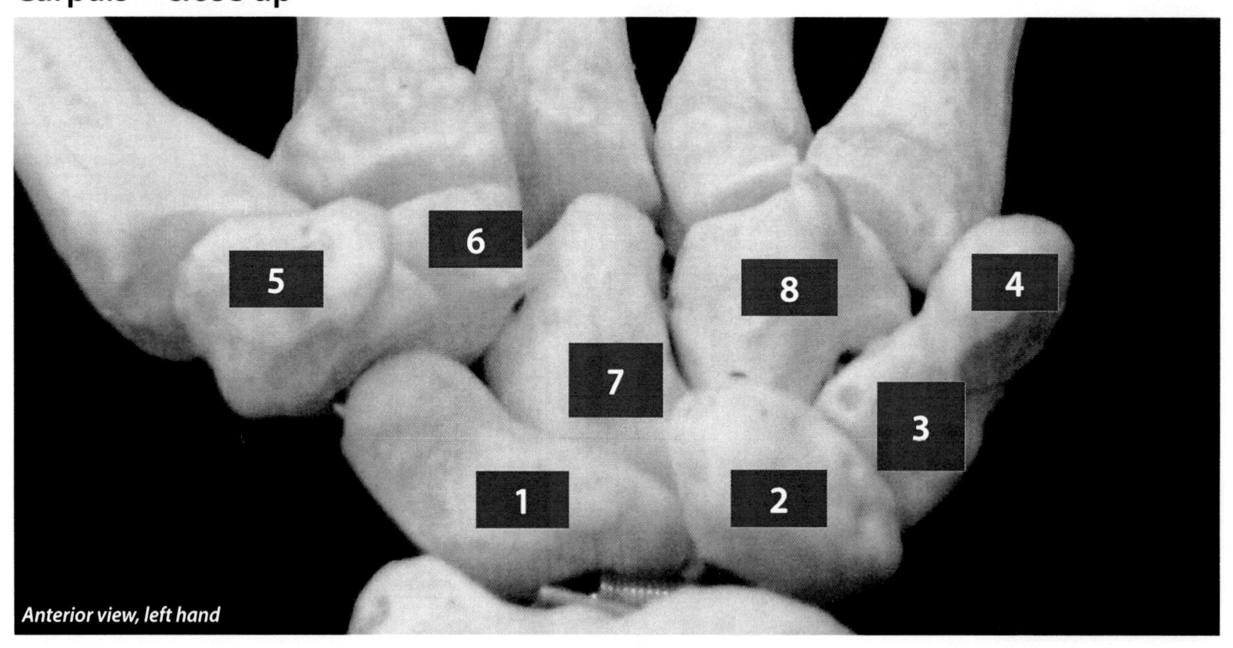

## Carpals

The carpal bones are the small bones of the wrist consisting of two rows of four bones each.

### *1. Scaphoid*

A mnemonic for the carpals is "Some Lovers Try Positions That They Can't Handle!" I am not sure what that means. With the hand in the anatomical position, the mnemonic starts with the most lateral bone of the proximal row and works medially; then it moves to the distal row, again starting with the most lateral bone and working medially. The "S" from "Some" is for the scaphoid. Therefore, it is in the proximal row and is the most lateral carpal of that row. I hope that helps.

### *2. Lunate*

The "L" from "Lovers" is for the lunate. The lunate is in the proximal row and is second from the lateral side.

### *3. Triquetral (Triangular)*

The "T" from "Try" from the above mnemonic is for the triquetral. It is in the proximal row and is the third from the lateral side. Using Dr. J's mnemonic, we see that the first word in the mnemonic is "Try," and the only carpal with the name beginning with "tri" is the triquetral (yup, another Grant thing!). Also, it is the "third" bone in the proximal row and "tri" means three. Yowee—it doesn't get any better than that!

### *4. Pisiform*

The "P" stands for "Positions" from the above mnemonics. It stands for the pisiform (meaning "pea-shaped" in Latin). So it is in the proximal row, and it is fourth from the lateral side (it is the most medial bone of the proximal row). It articulates with one carpal only, the triquetral. The pisiform is actually a sesamoid bone (like the patella) that is found within the flexor carpi ulnaris tendon. Sesamoid bones sometimes develop within tendons at sites that experience friction. In this way, they protect the tendon from excessive wear. The sesamoid bones are named for the street they were first discovered on by Bert and Ernie!

### *5. Trapezium*

The "T" from "That" from the above mnemonic is for the trapezium. It is in the distal row and is the most lateral carpal of that row. So, if you were a Talking Head, you might ask yourself, "There are three bones starting with 't', so how do I know which one is the trapezium?" Well, Dr. J came up with another mnemonic to help here—what a guy! In the distal row, there are two bones with nearly the same name, the trapezium and the trapezoid. They are arranged alphabetically when one moves from lateral to medial. So the trapezium is the more lateral of the two and the trapezoid is the more medial of the two. The trapezium is also the carpal that articulates with the first metacarpal and with the first proximal and distal phalanges (that form the thumb). This is a saddle joint, a unique joint that gives the thumb a great range of motion compared to the other digits.

### *6. Trapezoid*

The "T" from "They" is for the trapezoid. It is in the distal row and is second from the lateral side. As mentioned above, there are two bones with nearly the same name in the distal row, the trapezium and the trapezoid. They are arranged alphabetically when one moves from lateral to medial. Therefore, the trapezium is the more lateral of the two and the trapezoid is the more medial of the two.

### *7. Capitate*

The "C" from "Can't" is for the capitate. Therefore, it is in the distal row and is third from the lateral side.

### *8. Hamate*

The "H" from "Handle" from the above mnemonic is for the hamate. Therefore, the hamate is in the distal row and fourth from the lateral side. It is the most medial bone of the distal row and is the only carpal bone that articulates with two metacarpals (the fourth and the fifth). I hope that helps.

## Hand

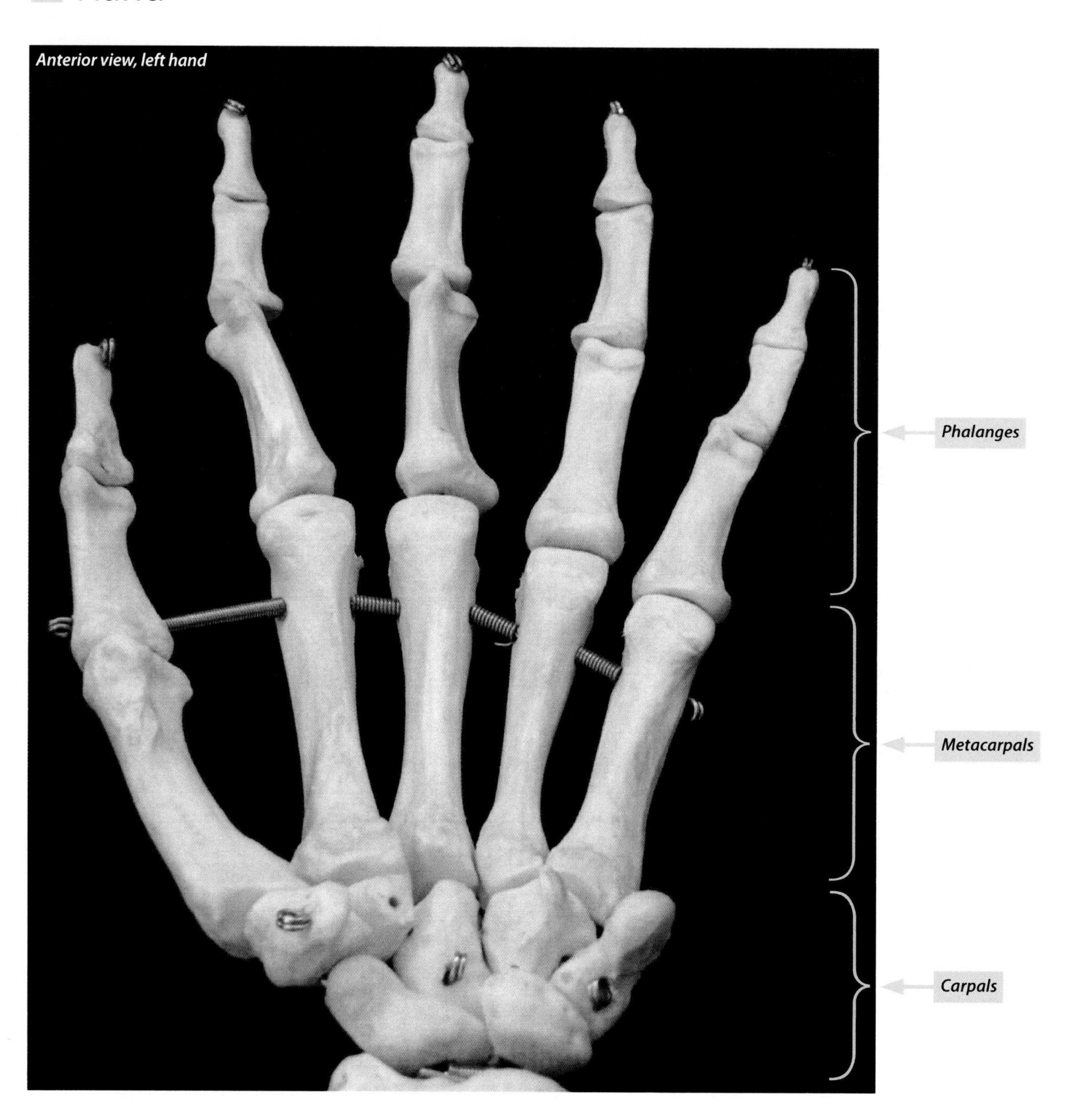

# Hand

## Metacarpals

These are the bones of the hand that extend from the carpals to the phalanges. They are within the hand itself. The second metacarpal is the insertion for:

1. extensor carpi radialis longus (the posterior surface of base of the second metacarpal) and
2. flexor carpi radialis (the base of the second metacarpal).

The third metacarpal is the insertion for:

1. extensor carpi radialis brevis (the base of the third metacarpal) and
2. flexor carpi radialis (the base of the third metacarpal).

The fifth metacarpal is the insertion for:

1. extensor carpi ulnaris (the posterior surface of base of the fifth metacarpal).

## Phalanges (phalanx = singular)

The phalanges are what we commonly refer to as our fingers.
They are the insertion for:

1. extensor digitorum communis (by four tendons, one to each digit) and
2. flexor digitorum superficialis (by four tendons to the middle phalanges of digits 2 to 5).
3. flexor digitorum profundus (by four tendons to the distal phalanges of digits 2 to 5).

# MUSCLES

## Supraspinatus (Rotator Cuff)

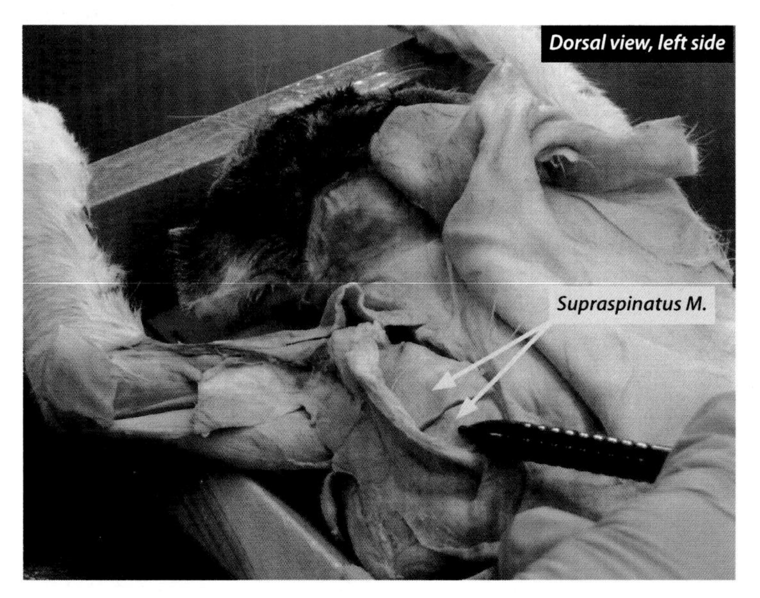


***Human Information:***

**origin:** medial 2/3 of supraspinous fossa of scapula
**insertion:** greater tubercle of humerus and capsule of shoulder joint
**nerve:** suprascapular (C5 and C6)
**action:** ABducts arm

This is one of the four rotator cuff muscles. It is the only rotator cuff muscle that is an abductor of the arm.

## Infraspinatus (Rotator Cuff)

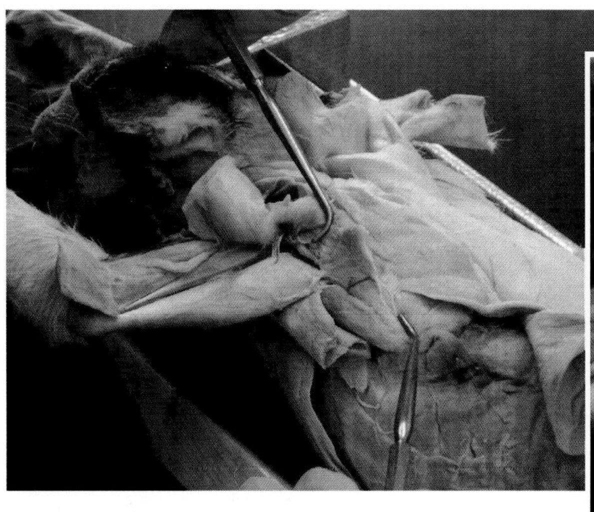

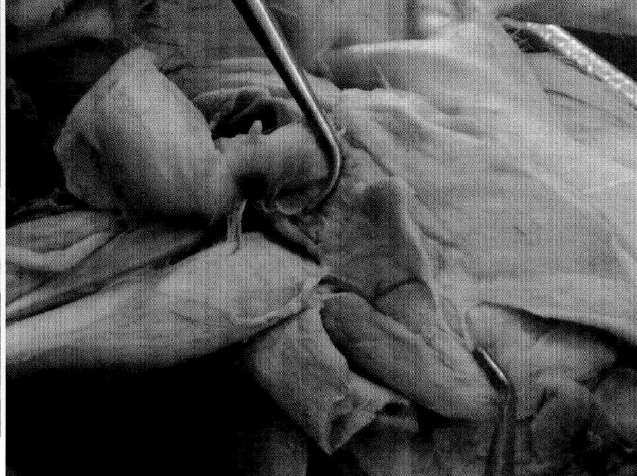

***Human Information:***

**origin:** infraspinous fossa of scapula
**insertion:** posterior aspect of greater tubercle of humerus, and capsule of shoulder joint
**nerve:** suprascapular (C5, C6)
**action:** laterally (externally) rotates arm.

This is one of the four rotator cuff muscles. It has the same action as the teres minor muscle, so Dr. J suggests that you learn these two muscles as a group.

## Supraspinatus (Rotator Cuff)

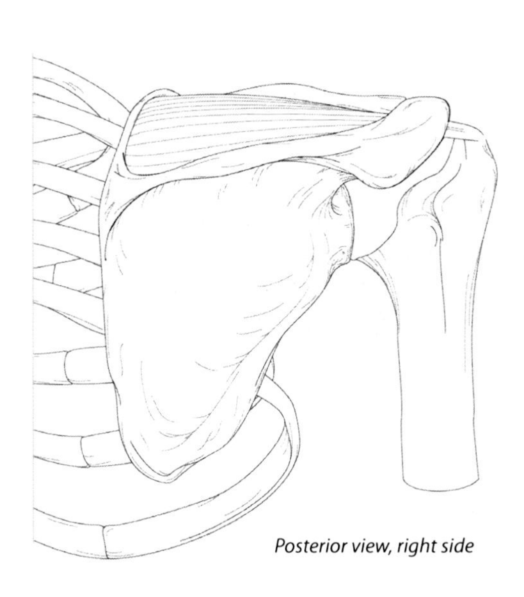

Posterior view, right side

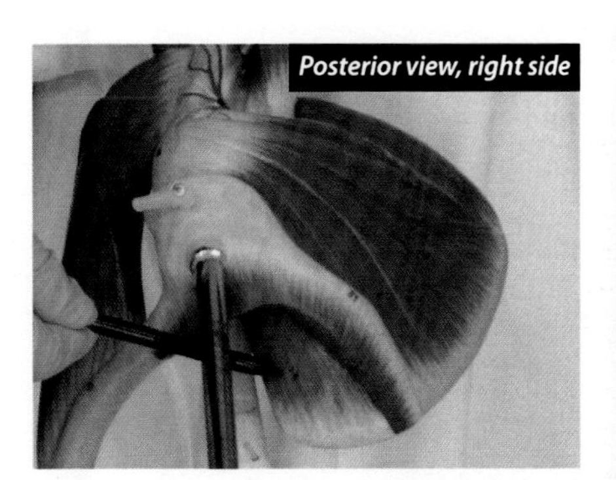


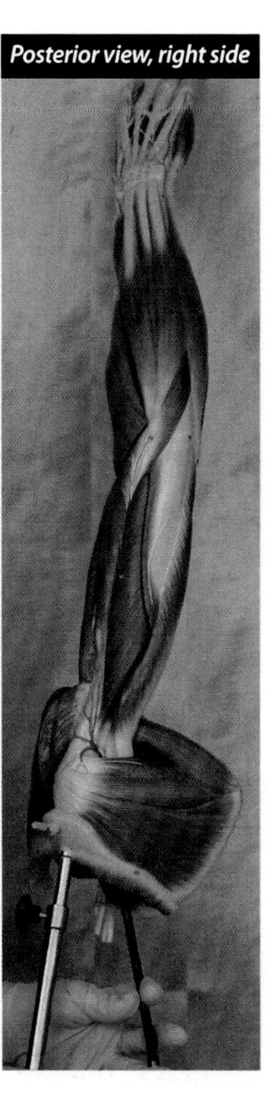


These drawings of the origin and insertion might help you visualize this information (red is the origin, blue the insertion).

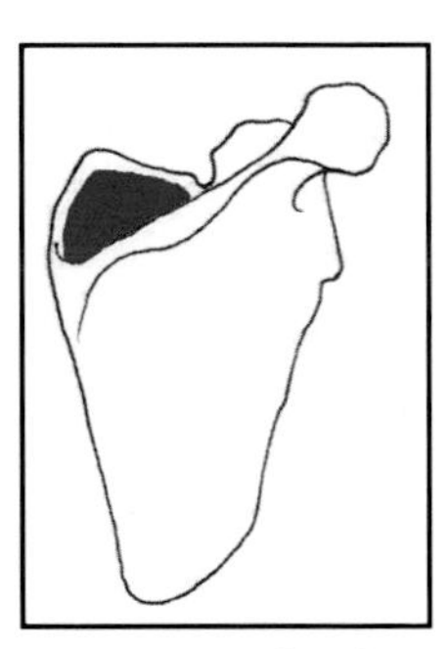

Posterior view, right side

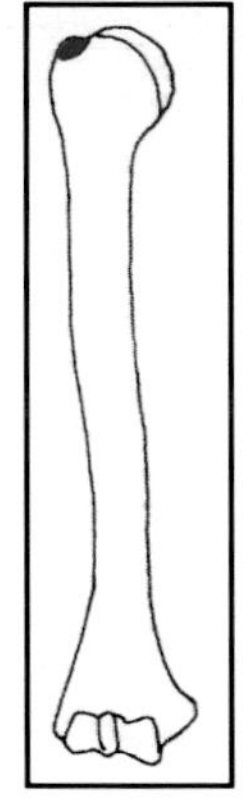

Anterior view, right side

## Infraspinatus (Rotator Cuff)

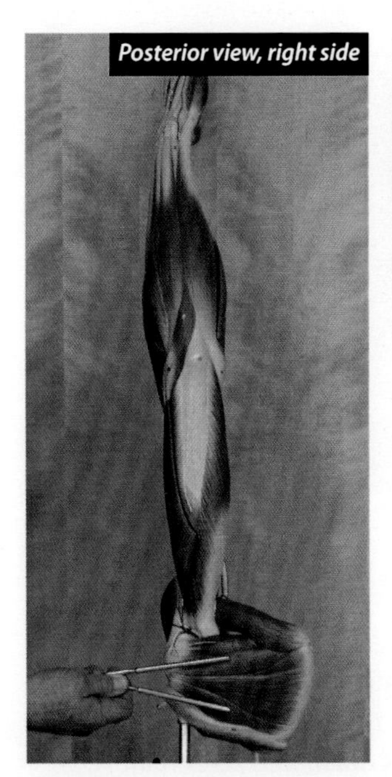


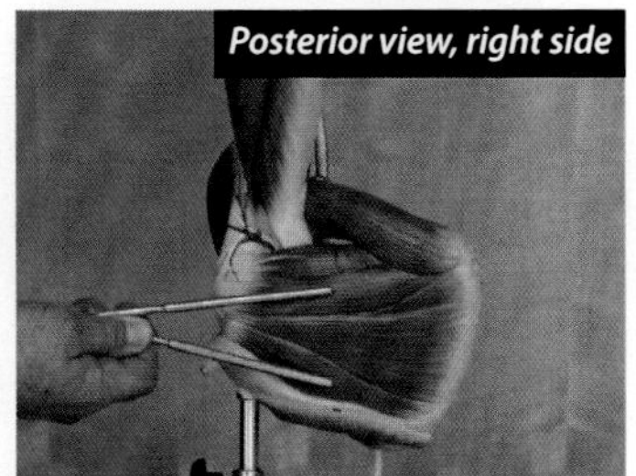


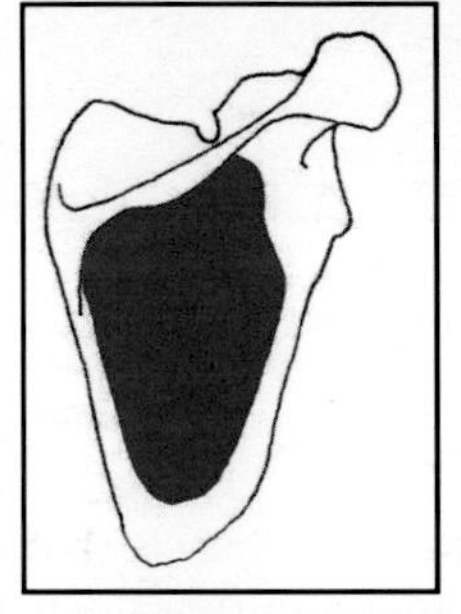

Posterior view, right side

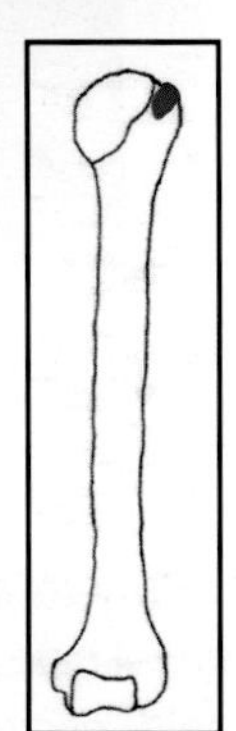

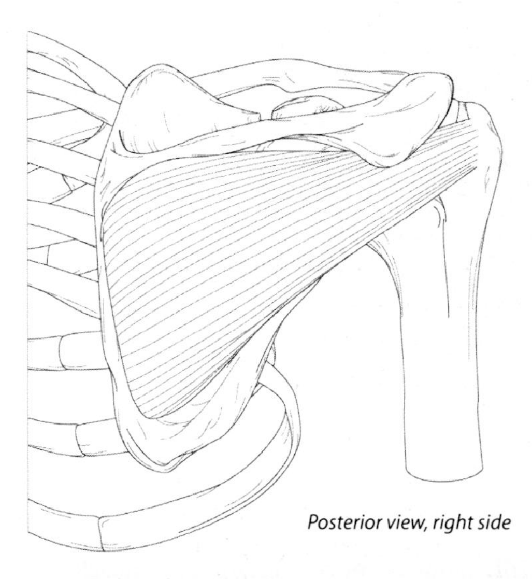

Posterior view, right side

These drawings of the origin and insertion might help you visualize this information (red is the origin, blue the insertion).

## Teres Minor (Rotator Cuff)

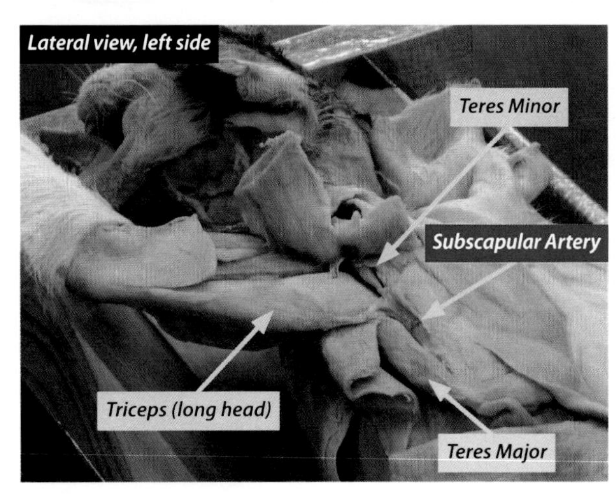


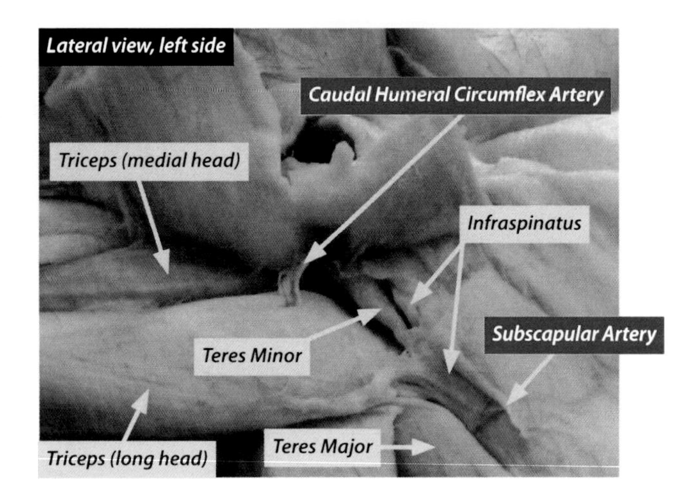


### *Human Information:*

**origin:** lateral (axillary) border of scapula (superior 2/3)
**insertion:** inferior facet on posterior surface of greater tubercle of humerus (inferior to insertion of infraspinatus)
**nerve:** axillary (C5 and C6)
**action:** laterally (externally) rotates arm

This is one of the four rotator cuff muscles. It is sometimes referred to as the little helper of infraspinatus because it has the same action as the infraspinatus muscle and inserts just inferior to it. Therefore, Dr. J suggests that you learn these two muscles as a group. If the **axillary nerve** is damaged and this muscle is no longer active, the individual will exhibit the condition known as head waiter's syndrome, resulting in a medial rotation of the arm.

## Subscapularis (Rotator Cuff)

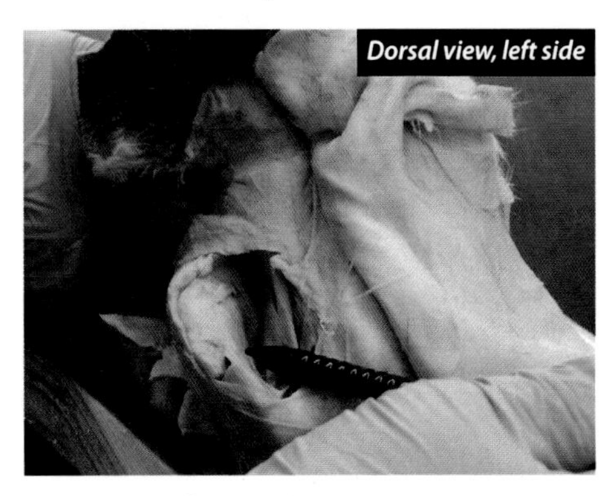


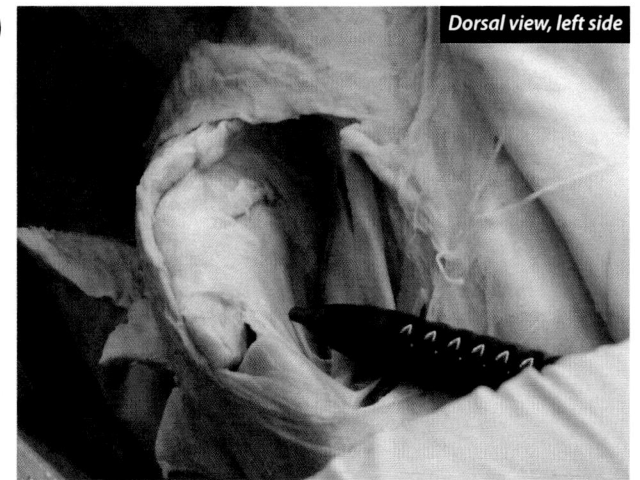


### *Human Information:*

**origin:** subscapular fossa of scapula (entire surface)
**insertion:** lesser tubercle of humerus, and capsule of shoulder joint
**nerve:** upper and lower subscapular nerves (C5, C6)
**action:** medially (internally) rotates arm

This is one of the four rotator cuff muscles. It is the only rotator cuff muscle that medially rotates the arm and that inserts on the lesser tubercle of the humerus. There are four other medial rotators of the arm, but those are not rotator cuff muscles (the anterior deltoid, pectoralis major, latissimus dorsi, and teres major).

## Teres Minor (Rotator Cuff)

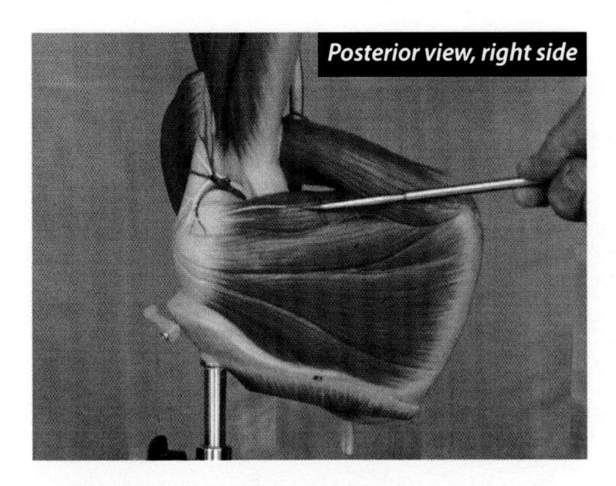


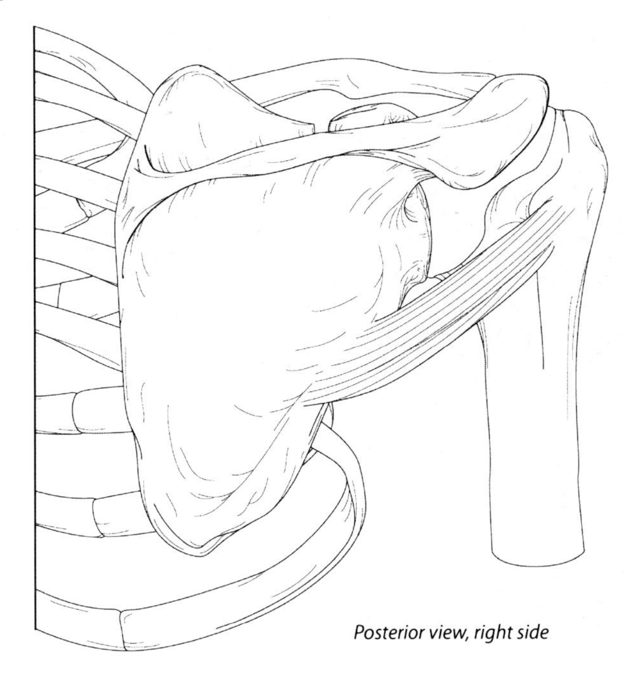
Posterior view, right side

Posterior view, right side

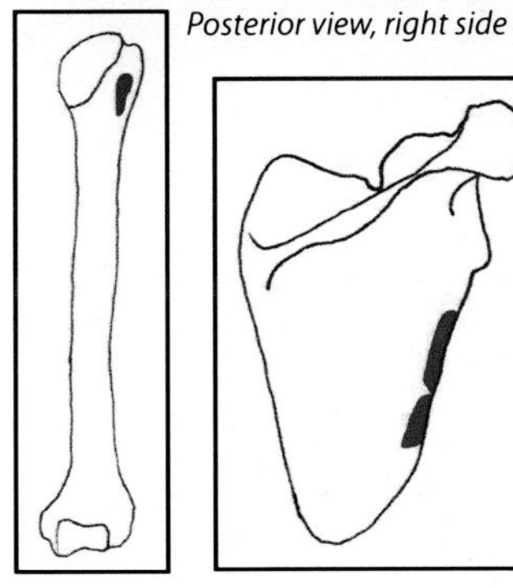

These drawings of the origin and insertion might help you visualize this information (red is the origin, blue the insertion).

## Subscapularis (Rotator Cuff)

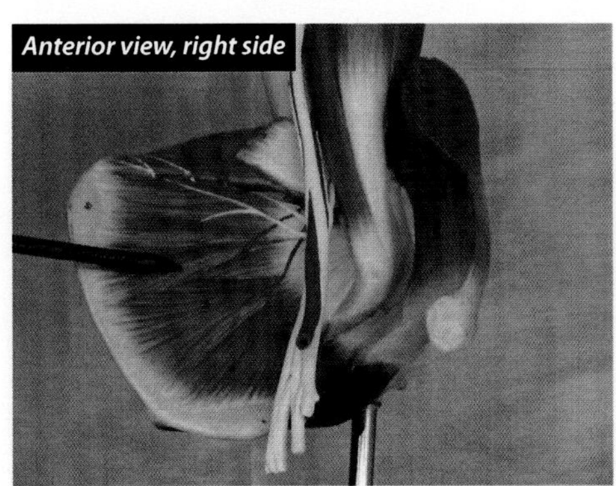


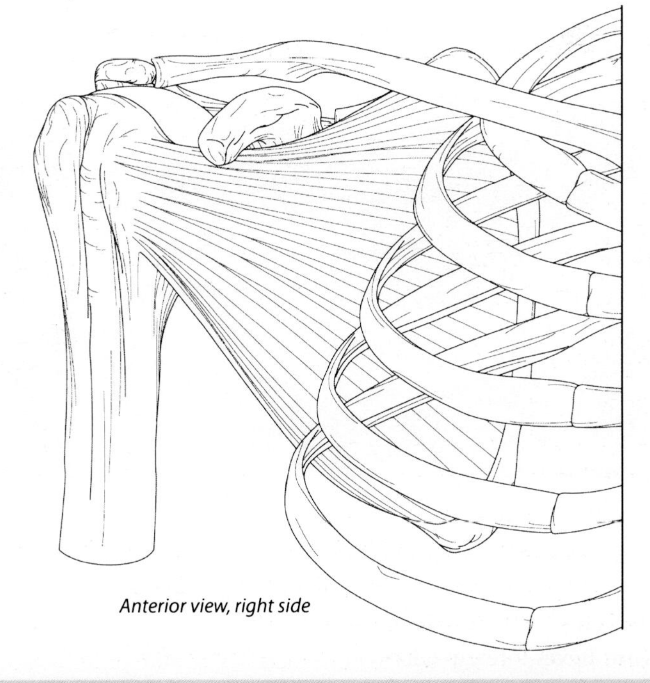
Anterior view, right side

Anterior view, right side

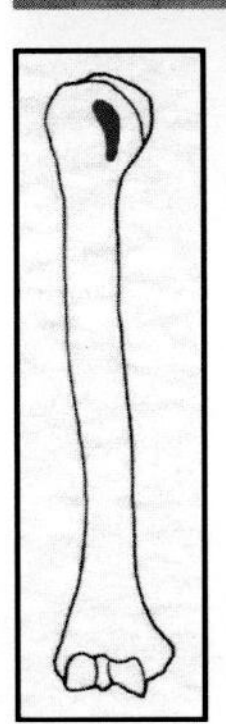

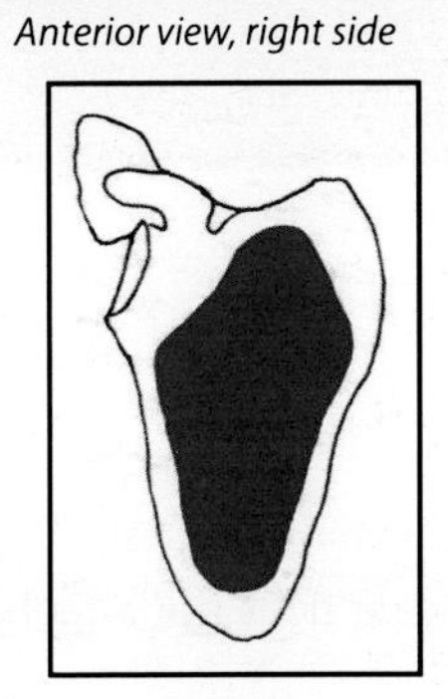

These drawings of the origin and insertion might help you visualize this information (red is the origin, blue the insertion).

## Teres Major

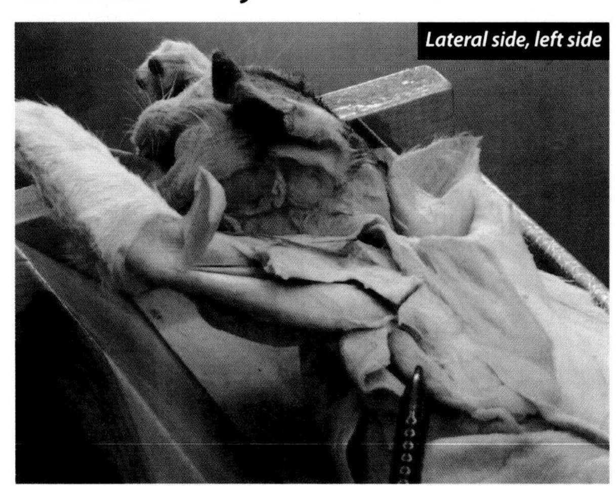


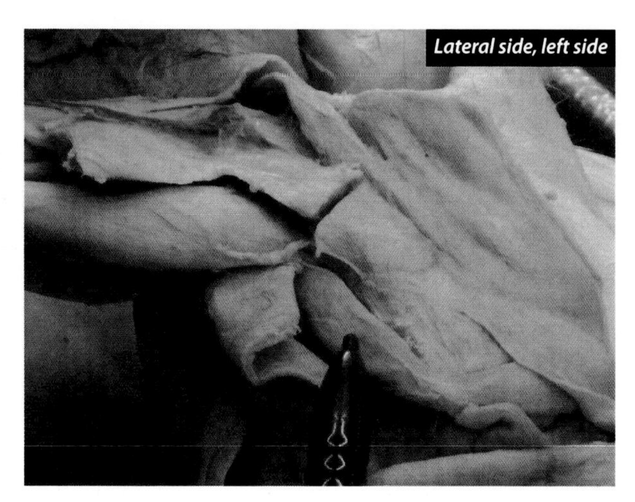


***Human Information:***

**origin:** lateral (axillary) border of scapula (inferior 1/3)
**insertion:** crest of lesser tubercle of humerus, medial to latissimus dorsi tendon and fused with that tendon
**nerve:** lower subscapular (C5 and C6)
**action:** extends, ADducts and medially (internally) rotates the arm

Teres major is one of the five muscles that move the arm. It has the same action as latissimus dorsi. For this reason, it's referred to as the little helper of latissimus dorsi. Therefore, Dr. J suggests that you learn these two as a group.

## Brachialis

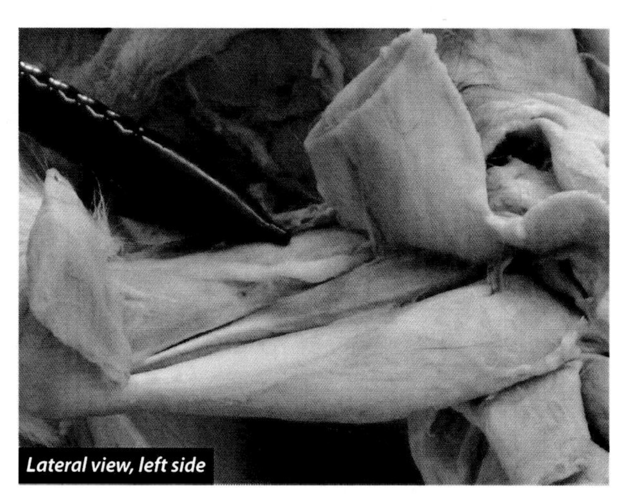


***Human Information:***

**origin:** distal half of the anterior surface of humerus
**insertion:** ulnar tuberosity and coronoid process of ulna
**nerve:** musculocutaneous (C5–C6)
**action:** flexes forearm (elbow)

This is one of the five muscles we will study that move the forearm. Note that since it does not attach to the radius, it does not pronate or supinate the wrist.

## Teres Major

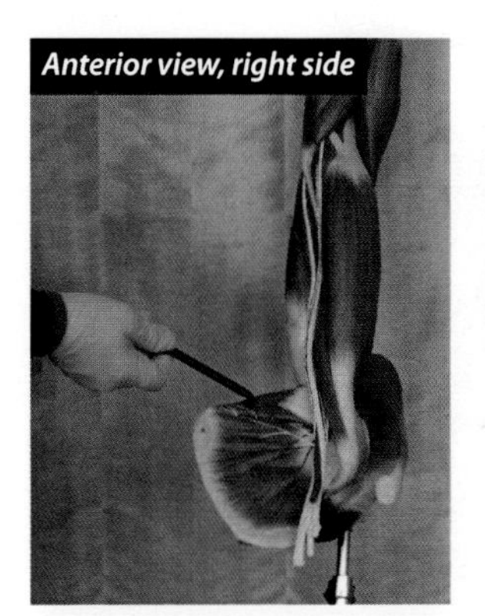


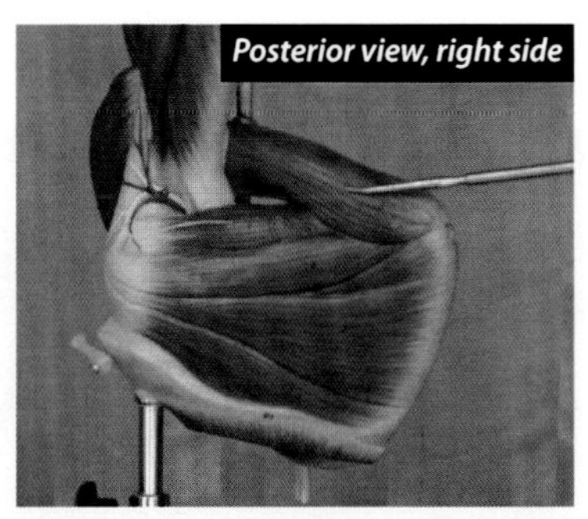


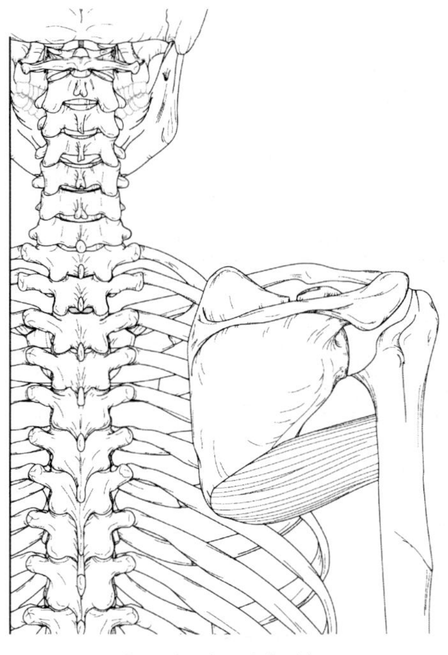
Posterior view, right side

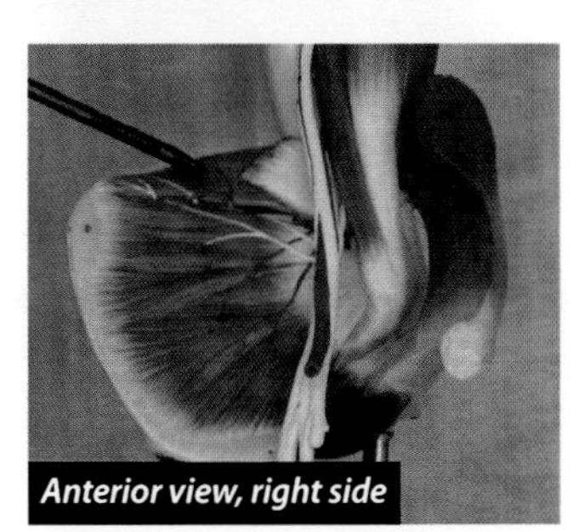


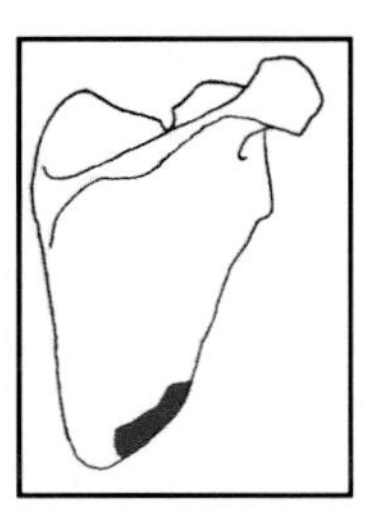
Posterior view, right side

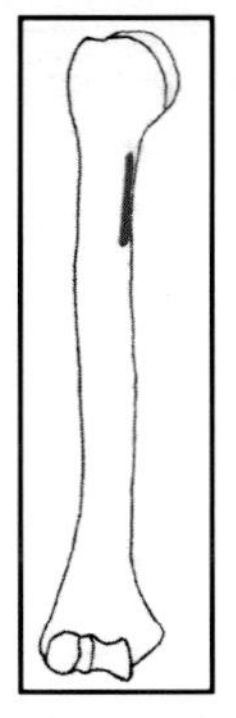
Anterior view, right side

These drawings of the origin and insertion might help you visualize this information (red is the origin, blue the insertion).

## Brachialis

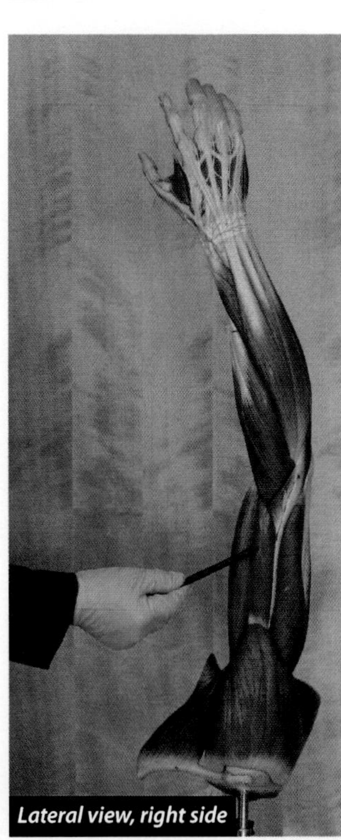


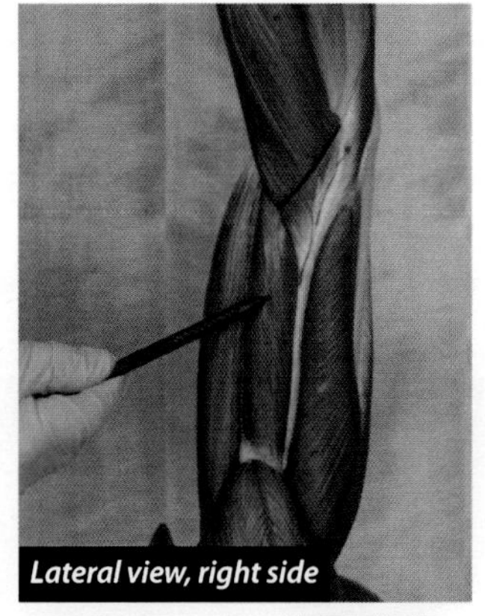


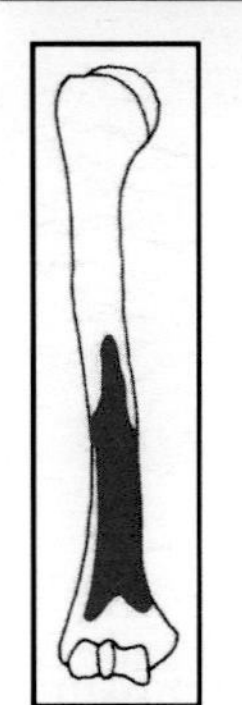
Anterior view, right side

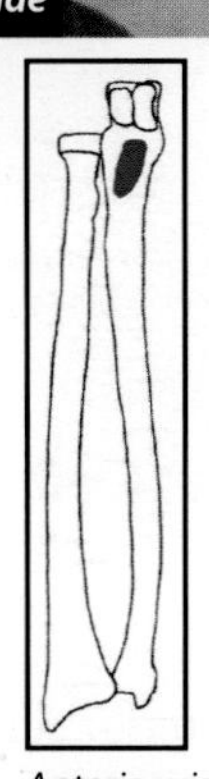
Anterior view, right side

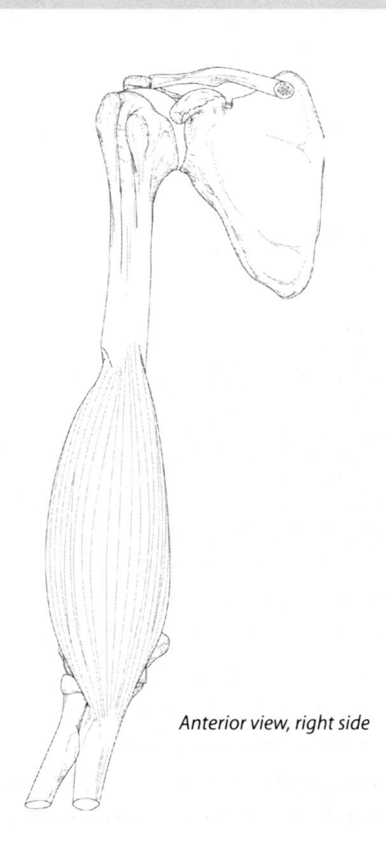
Anterior view, right side

## Triceps Brachii (Lateral, Long, and Medial Heads)

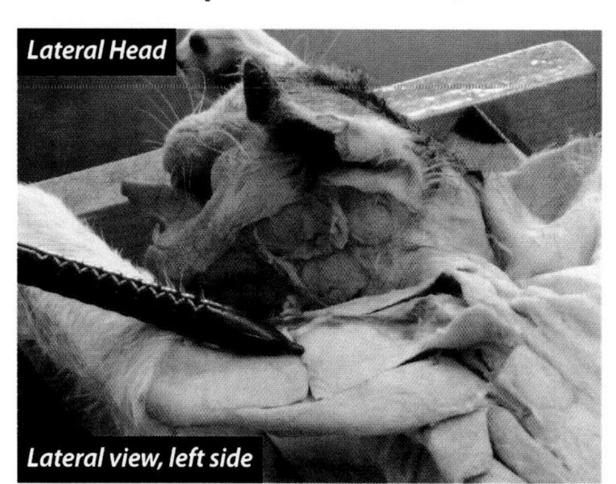
Lateral Head
Lateral view, left side

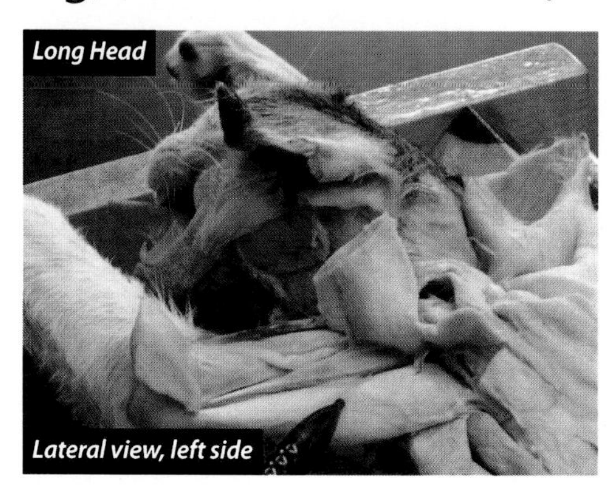
Long Head
Lateral view, left side

Medial Head
Lateral view, left side

*Note: Recent research suggests that the long head is actually innervated by the axillary nerve or posterior cord itself. But you don't need to node this for the exams.*

### *Human Information:*

**Lateral Head**

**origin:** posterior surface of humerus proximal to the spiral groove
**insertion:** olecranon process of ulna
**nerve:** radial (C7–C8)
**action:** extends forearm (elbow)

**Medial Head**

**origin:** posterior surface of humerus
**insertion:** olecranon process of ulna
**nerve:** radial (C7–C8)
**action:** extends forearm (elbow)

**Long Head**

**origin:** infraglenoid tubercle of scapula
**insertion:** olecranon process of ulna
**nerve:** radial (C7–C8)
**action:** extends forearm, extends and adducts arm

## Anconeus

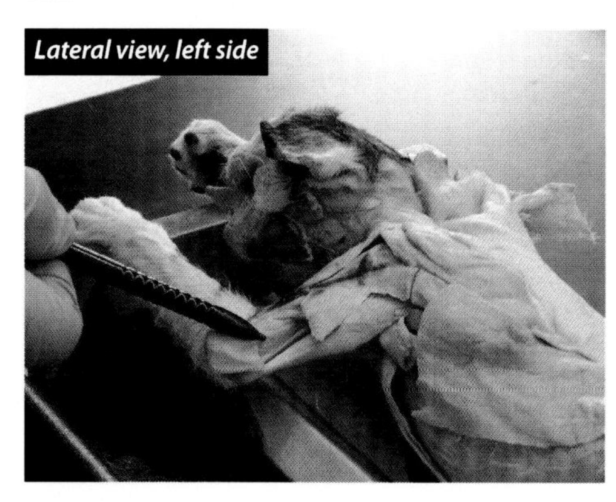
Lateral view, left side

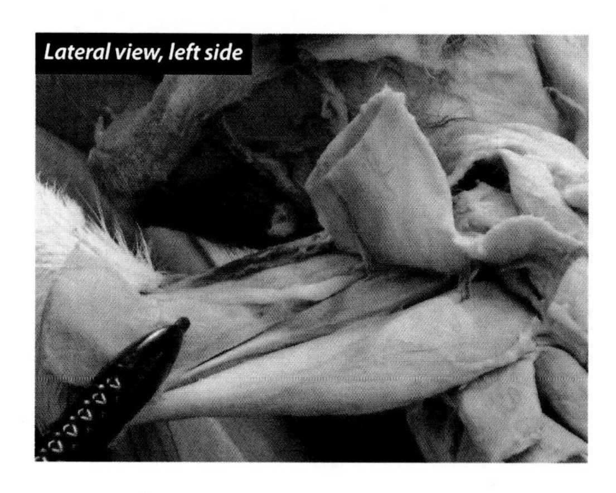
Lateral view, left side

### *Human Information:*

**origin:** posterior surface of lateral epicondyle of humerus
**insertion:** lateral aspect of olecranon process and posterior surface of proximal portion of ulna
**nerve:** radial (C7–C8)
**action:** extends forearm, pronates hand, prevents annular ligament from getting pinched in olecranon fossa

# Triceps Brachii (Lateral, Long, and Medial Heads)

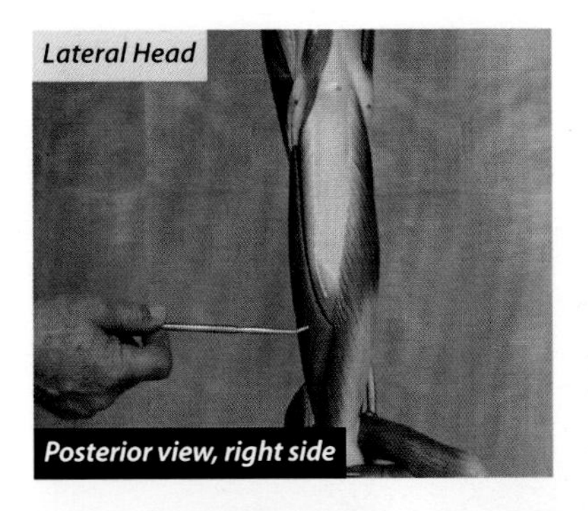


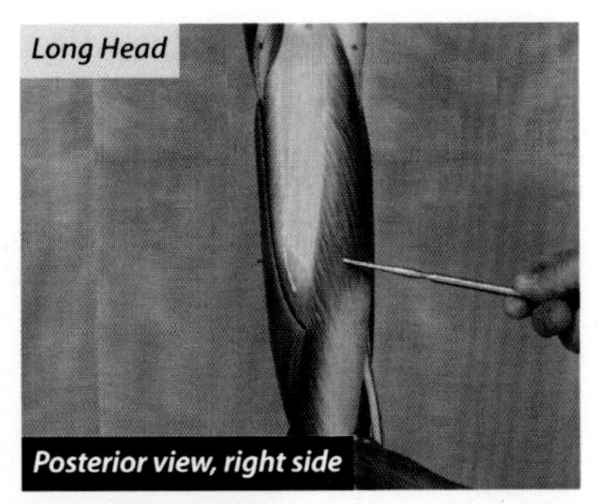


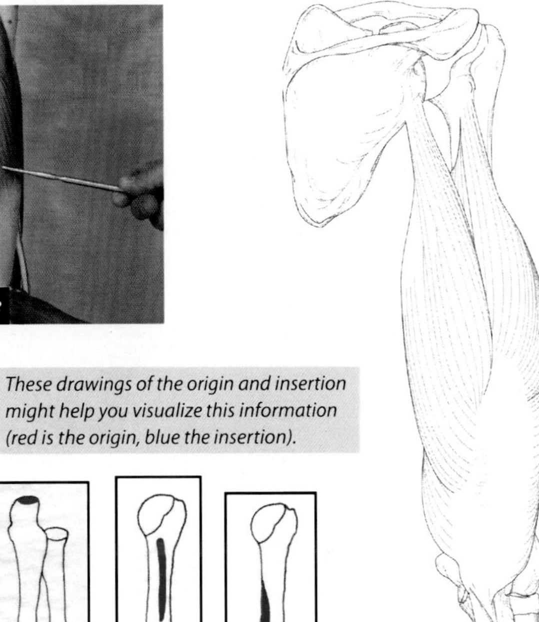
Posterior view, right side

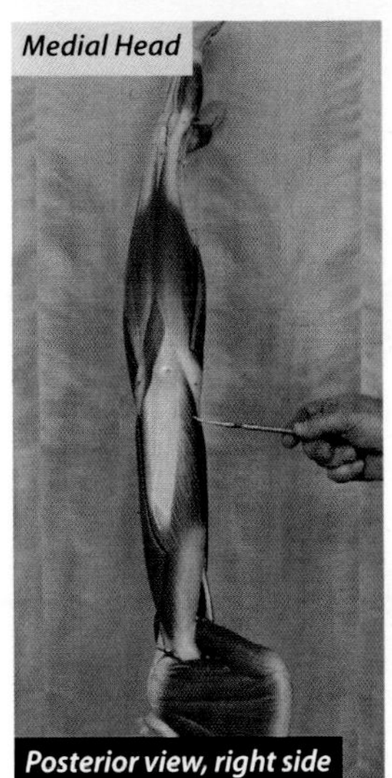


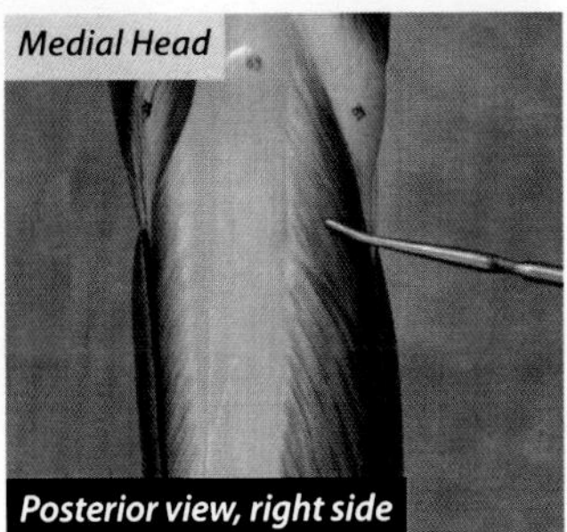


*These drawings of the origin and insertion might help you visualize this information (red is the origin, blue the insertion).*

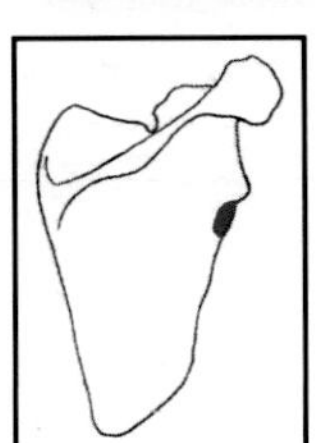
*Long Head*

*Posterior view, right side*

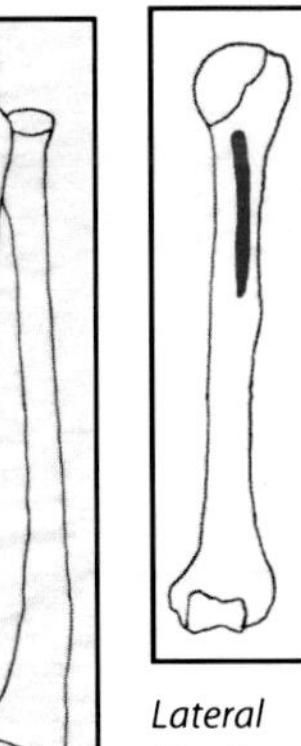
*Lateral Head*

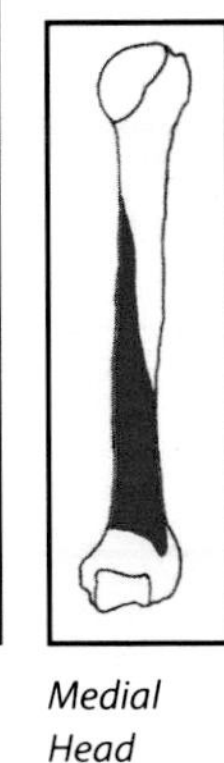
*Medial Head*

The triceps brachii is one of the five muscles of the forearm that we will study. The anconeus and the triceps brachii have the same action, so Dr. J suggests that you learn these as a group. You should note the perimysium (pearlescent pantyhose) on the long head of the triceps brachii. Functionally, this is important as it strengthens the muscle cells and keeps them from tearing. It is found on all skeletal muscles, but it is particularly noticeable in a few muscles, such as this one.

# Anconeus

Anconeus is a synergist of triceps brachii. They both extend the forearm and anconeus provides stability during this motion.

**I hope that before the test you know your gluteals from your anconeus!**

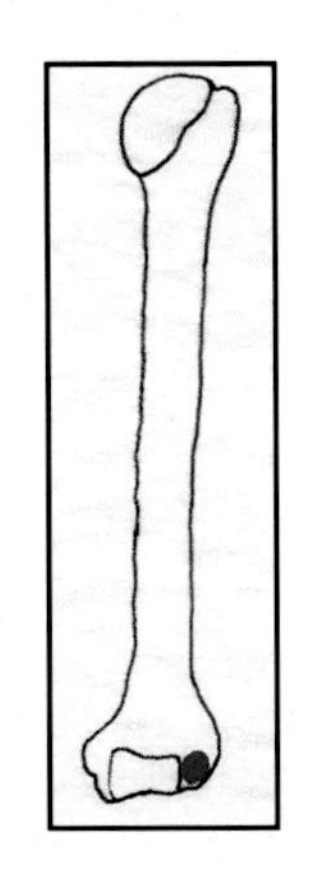

*Posterior view, right side*

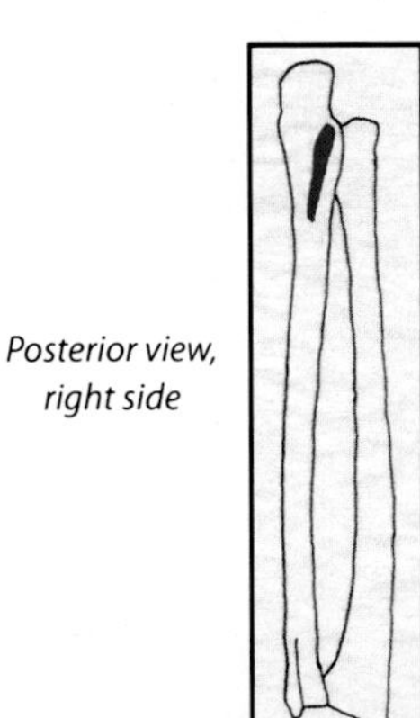

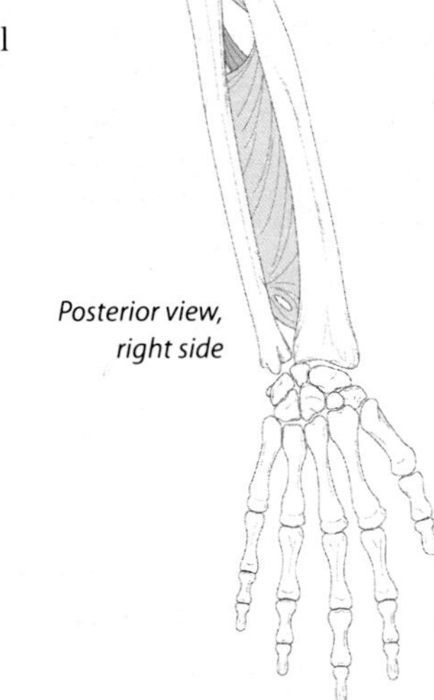
*Posterior view, right side*

# NERVES

## Suprascapular Nerve

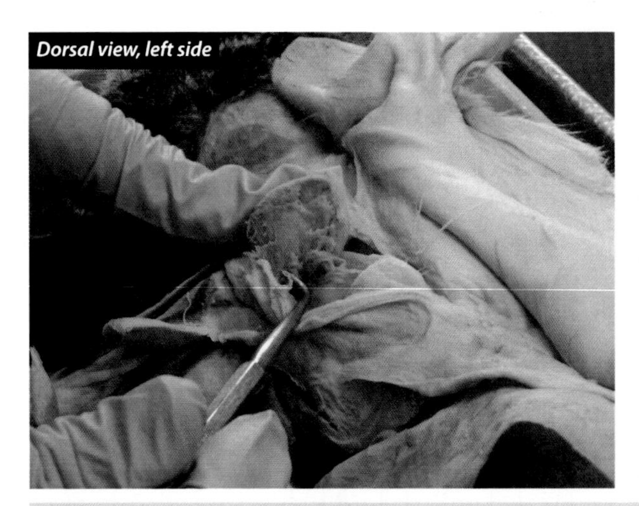


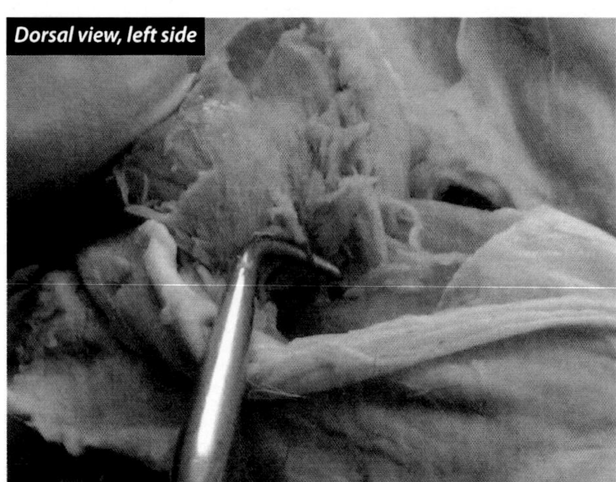


*This is the suprascapular nerve pictured deep to the **supra**spinatus muscle as it runs through the **supra**scapular notch with the **suprascapular artery**.*

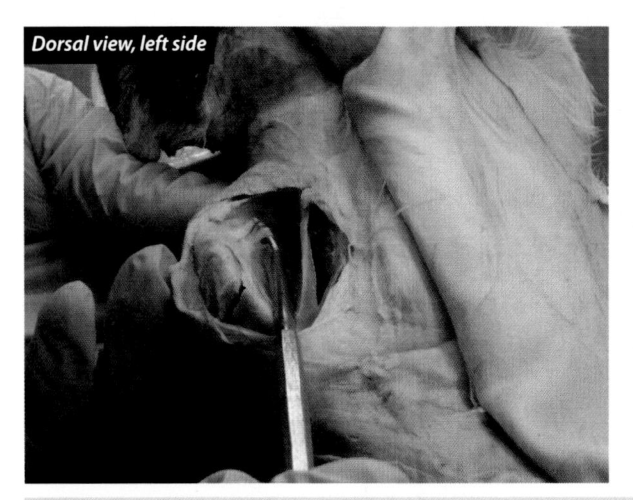


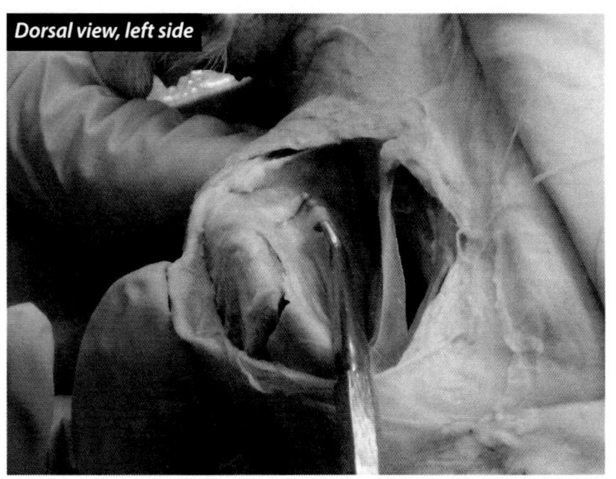


*This is the suprascapular nerve pictured after it has passed through the suprascapular notch. Here it is found running across the **sub**scapularis muscle with the **transverse scapular artery**.*

The **suprascapular nerve** serves the supraspinatus and infraspinatus muscles, as well as the shoulder joint. These are both rotator cuff muscles. The **suprascapular nerve** originates from the **upper trunk** of the brachial plexus (including the **anterior rami** from **spinal nerves C5** and **C6**). The **suprascapular nerve** runs through the **suprascapular notch** with the **suprascapular artery**. That would make it a Grant, Grant, Grant thing.

## Radial Nerve

The **radial nerve** is fondly referred to as the "biggest hugest nervus in the arm" by Dr. J's students. It is a branch of the **posterior cord** containing nerves from the **anterior rami** of **C5**, **C6**, **C7**, **C8**, and **T1**. It serves the posterior compartment of the arm and forearm, as well as the brachioradialis muscle. It controls all the extensor muscles found in the arm and forearm and the supinator muscle (supinator). It also receives sensory input from the posterolateral cutaneous area of the hand. When damaged, it results in the clinical condition called wrist drop.

# VESSELS

## Subscapular Artery

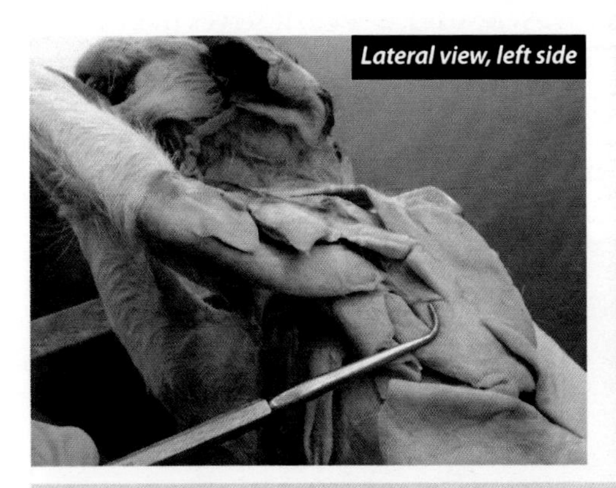


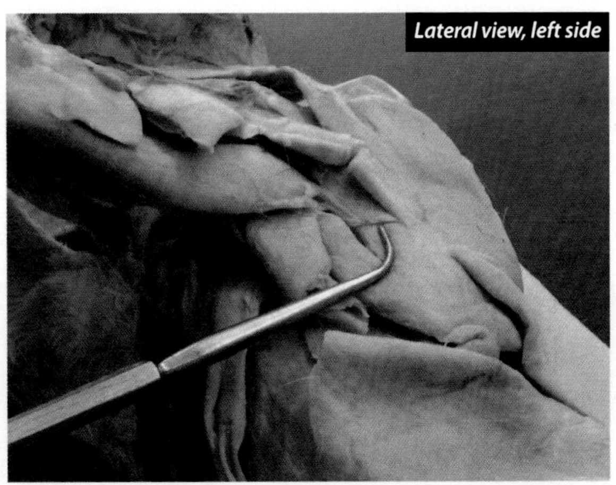


*The **subscapular artery** is pictured here running on the superficial surface of the **infraspinatus muscle**.*

The **subscapular artery** is the largest lateral branch of the **axillary artery**. It marks the end of the **axillary artery** and the beginning of the **brachial artery**. It arises from the distal third of the **axillary artery** on the posterior surface. In the cat, it has two branches, the **caudal humeral circumflex** and the **thoracodorsal**. It continues between the subscapularis and the teres major, and it sends branches to the subscapularis, the long head of the triceps brachii, and the latissimus dorsi. It then continues over the surface of the infraspinatus muscle (where we observe it, in this lab, on the dorsolateral surface of the cat) and serves the infraspinatus, the supraspinatus, the acromiotrapezius, and the spinotrapezius. Occasionally it gives rise to the **cranial humeral circumflex artery**, although that vessel is more often a branch of the **brachial artery**. In humans, the **subscapular artery** serves the posterior wall of the axilla and the posterior scapular region and has just the two terminal branches, the **thoracodorsal artery** and the **circumflex scapular artery**.

# VESSELS

## Suprascapular Artery

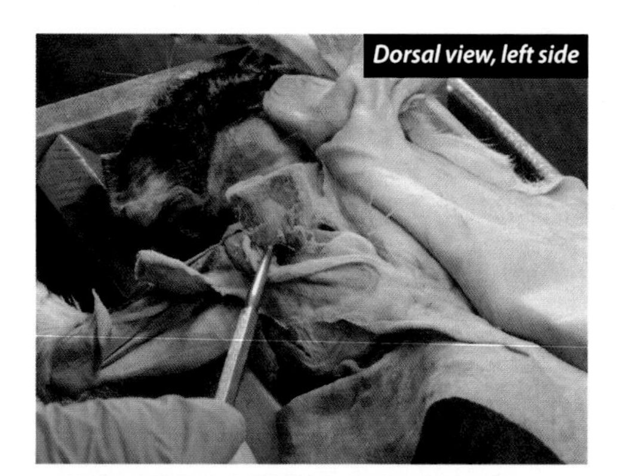


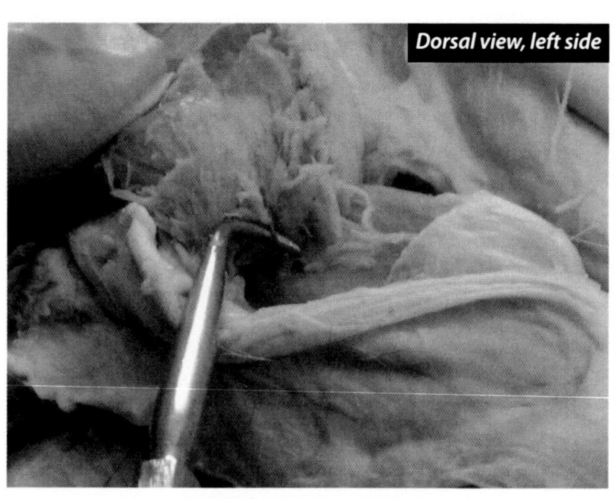


*The **suprascapular artery** is pictured running with the suprascapular nerve through the **suprascapular notch** found deep to the **supraspinatus muscle**.*

The **suprascapular artery** is a branch of the **transverse scapular artery**. It passes through the suprascapular notch with the **suprascapular nerve**, and it sends branches to the supraspinatus and infraspinatus muscles. This makes it a Grant, Grant, Grant thing. In humans, the **suprascapular artery** is a branch of the **thyrocervical trunk**, or it may be a branch of the **subclavian artery**. It passes superior to the suprascapular notch but then runs with the **suprascapular nerve**. It serves the infraspinatus and the supraspinatus, as well as many other structures as it moves toward those muscles.

## Transverse Scapular Artery

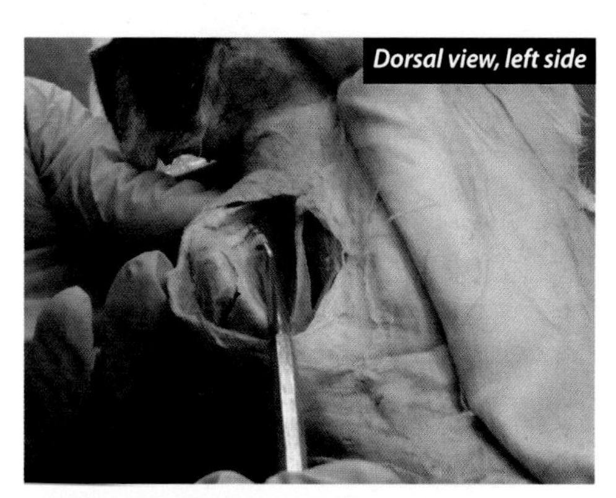


*The **transverse scapular artery** is found running across the **subscapularis muscle** with the suprascapular nerve. This is NOT a Grant thing.*

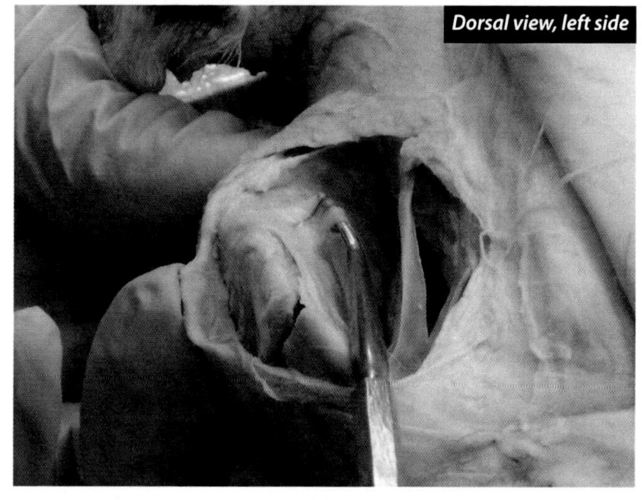


The **transverse scapular artery** is a branch of the **thyrocervical artery**. On the deep side of the scapula, it runs with the **suprascapular nerve**, and it then gives rise to the **suprascapular artery**. There is no comparable artery in humans.

## Caudal Humeral Circumflex Artery

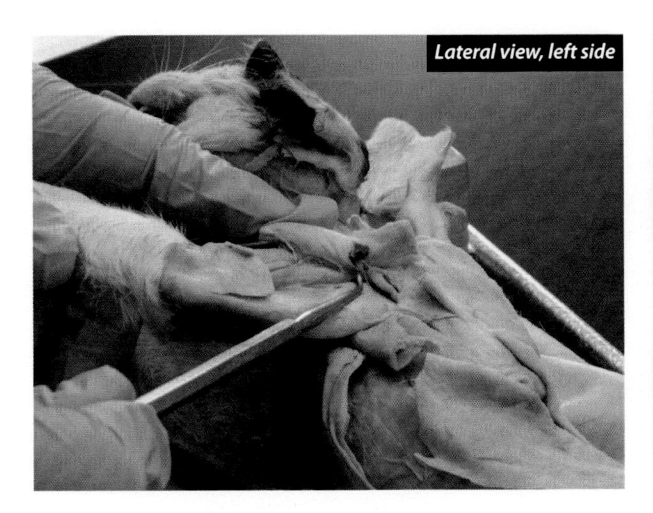


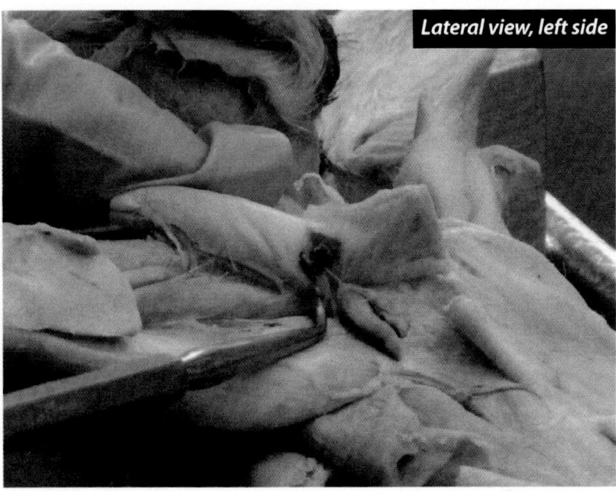


The **caudal humeral circumflex artery** is a branch of the **subscapular artery**. It gets its name from the fact that it passes around the caudal side of the humerus. It forms collateral circulation with the **cranial humeral circumflex artery** in the lateral region of the arm. In the cat, it serves the lateral and long heads of the triceps, the acromiodeltoid, and the spinodeltoid. In humans, it serves the deltoid muscle and the shoulder joint.

## Cephalic Vein

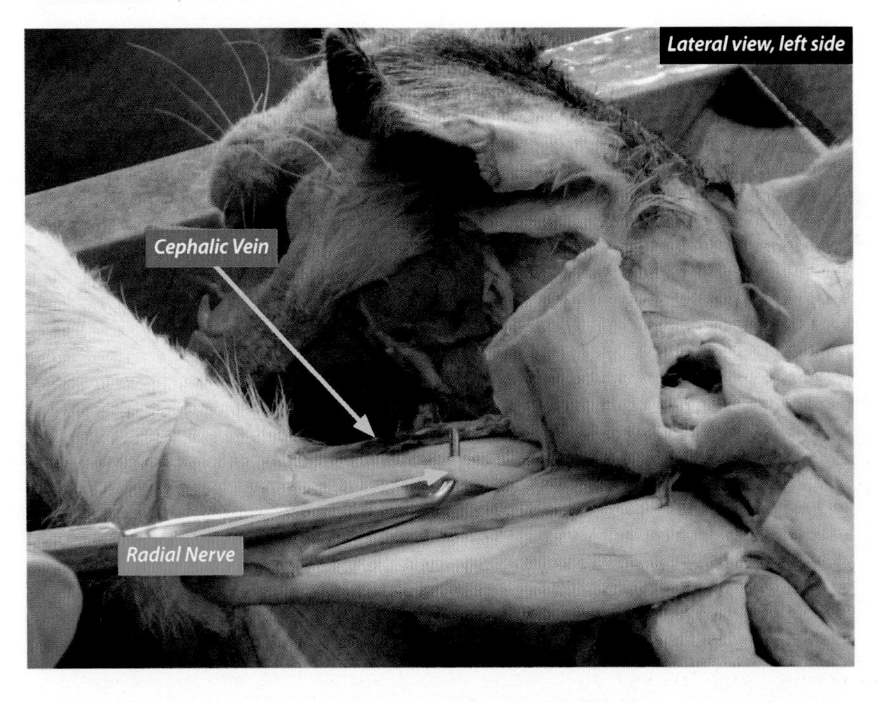


The **cephalic vein** is observed on the superficial, lateral side of the forearm and arm. Inferior to the clavicle, it passes into the axilla where it joins the **axillary vein**. The **axillary vein** passes through the thoracic wall to become the **subclavian vein**. In the forearm, the **cephalic vein** runs with the superficial branch of the **radial nerve** on the surface of the brachioradialis muscle.

Lab 8

# Upper Limbs: Arm and Forearm

## Wish List


**LAB 8 OVERVIEW, pp. 204–206**

**AREA DIAGRAM: NERVES & VESSELS (CAT), p. 207**

**MUSCLES, pp. 208–233**

- Biceps Brachii, pp. 208–209
- Coracobrachialis, pp. 210–211
- Epitrochlearis, p. 211
- Pronator Teres, pp. 212–213
- Flexor Carpi Radialis, pp. 214–215
- Flexor Digitorum Superficialis, pp. 216–217
- Flexor Carpi Ulnaris, pp. 218–219
- Palmaris Longus (human only), p. 220
- Anterior ("Flexor") Compartment Overview, p. 221
- Brachioradialis, pp. 222–223
- Extensor Carpi Radialis Longus, pp. 224–225
- Extensor Carpi Radialis Brevis, pp. 226–227
- Extensor Digitorum Communis, pp. 228–229
- Extensor Carpi Ulnaris, pp. 230–231
- Extensor Digitorum Lateralis (cat), p. 232
- Posterior ("Extensor") Compartment Overview, p. 233

**VESSELS, pp. 234–241**

**ARTERIES, pp. 234–240**

Axilla Overiew: Arteries, p. 234

- Axillary Artery, p. 235
- Ventral Thoracic Artery, p. 235
- Lateral Thoracic Artery, p. 235

Axilla Overiew: Arteries (continued), p. 236

- Subscapular Artery, p. 237
- Caudal Humeral Circumflex Artery, p. 237
- Thoracodorsal Artery, p. 237

Arm Overiew: Arteries, p. 238

- Brachial Artery, p. 239
- Cranial Humeral Circumflex Artery, p. 239
- Deep Brachial Artery, p. 239

Forearm: Arteries, p. 240

- Radial Artery, p. 240
- Ulnar Artery, p. 240

**VEIN, p. 241**

- Cephalic Vein, p. 241

**NERVES, pp. 242–249**

- Cranial Ventral Thoracic Nerve, p. 242
- Lateral Thoracic Nerve, p. 242
- Medial Cord, p. 243
- Lateral Cord, p. 243
- Ventral Nerves, p. 244
- Ulnar Nerve, pp. 244–245
- Median Nerve, pp. 244–245
- Musculocutaneous Nerve, p. 246
- Dorsal Nerves, p. 247
- Posterior Cord, p. 247
- Axillary Nerve, p. 247
- Radial Nerve, p. 248
- Thoracodorsal Nerve, p. 249

**AREA DIAGRAMS, pp. 250–252**

- Brachial Plexus (Anterior), p. 250
- Brachial Plexus (Posterior), p. 251
- Brachial Plexus—Label this diagram, p. 252

**TISSUE, p. 253**

- Carpal Ligament (Flexor Retinaculum, p. 253

**MUSCLE CHARTS, pp. 254–256**

- Muscles of the Scapula, p. 254
- Muscles of the Arm, p. 254–255
- Muscles of the Forearm and Hand, pp. 255–256

# LAB 8 OVERVIEW

This lab involves investigation of the axilla and the paths that the nerves and vessels take down to the forearm. Remember, in nursery school we called the axilla the armpit. A description of the axilla can be found in the PowerPoint and Camtasia for Lecture Quiz 11. Together, these resources will help you navigate and learn the road maps of the upper limb. **Note:** There are three **nerves** and three **arteries** that you are responsible for on the human arm model. They are documented with their feline counterparts in the Lab Dissector Companion.

Before we begin our study of the axilla, we must transect the **epitrochlearis muscle**. It is a cat only muscle. Once that is transected, we will transect the **pectoral muscles** to expose the axillary contents.

## 1. Axilla

The study starts with the axillary sheath, a neurovascular bundle in the axilla. This is similar to the carotid sheath of the neck (**common carotid artery**, the **vagosympathetic trunk**, and the **internal jugular vein**), but it is more complex. The nervous system in this area involves three **cords** (**posterior**, **medial**, and **lateral cords**) and five terminal branches (**axillary**, **radial**, **ulnar**, **median**, and **musculocutaneous nerves**), as well as a few lateral branches. The vascular portion includes the **axillary artery** and several of its branches.

The **axillary artery** and its branches are Grant things; so if you can describe where they go, you can come up with their names. The most proximal branch we study is the **ventral thoracic artery**, which serves the ventral thoracic wall. It will run with the **cranial ventral thoracic nerve**. The intermediate branch is the **lateral thoracic artery**, which serves the lateral thoracic wall with the **lateral thoracic nerve**. The distal branch is the **subscapular artery**. It marks the distal end of the axillary artery and the beginning of the **brachial artery**.

The **subscapular artery** will look like a pitchfork with three tines. The handle of the pitchfork and the central tine are the **subscapular artery**, while the other tines are the lateral branches (the **thoracodorsal artery** and the **caudal humeral circumflex artery**).

The **thoracodorsal artery** is joined by the **thoracodorsal nerve** (**third subscapular nerve**), and they both move caudally to where they serve the **latissimus dorsi muscle**. The **caudal humeral circumflex artery** moves cranially, but it receives its name from the fact that it passes around the caudal side of the **humerus**. It will provide collateral circulation to the lateral portions of the arm with the **cranial humeral circumflex artery**. We will discuss this vessel soon. You will also observe the **axillary nerve** on its way to the teres minor and the deltoids, as it passes between the **teres major** and the **subscapularis muscles** with the **caudal humeral circumflex artery**. The **axillary nerve** can be recognized because it goes deep into the muscles, unlike most of the other nerves we study that travel toward the elbow.

## 2. Arm

Before we continue reviewing the **arteries** in the arm, we will examine the **biceps brachii muscle**, because it is an important landmark in this area. In the cat, this muscle looks like the stem of a chubby lower case "i" (we owe thanks to Jan Stone for this visual). We won't be able to see it on the lateral side of the arm, but we can see it on the medial side. If you run your finger along the length of the **biceps brachii** toward the axilla, you will eventually palpate a bump—that will be the head of the **humerus**, and the **coracobrachialis**

**muscle** covers it. The **coracobrachialis muscle** is small in the cat, but relatively larger in humans. It is like the dot of the lower case "i" (again, thanks to Jan Stone for this visual—I think it is a good one).

The **brachial artery** extends distally from the end of the **axillary artery** (remember, the **axillary artery** terminates at the **subscapular artery**). After this, it becomes the **brachial artery** in the arm, making it a Grant thing.) It will run to the forearm with the **median nerve**. These two structures pass through the **supracondylar foramen** of the **humerus** before reaching the elbow. This structure is found in cats, and in 1 to 2.7 percent of humans it manifests as a vestigial hooked bony process.

There are two branches of the **brachial artery** that we study. The more proximal branch is the **cranial humeral circumflex artery**. It runs to the belly of the **biceps brachii muscle** (as does the **musculocutaneous nerve**, which we will discuss soon). In fact, this is how one can verify which vessel it is. It runs around the cranial side of the **humerus** and forms collateral circulation in the lateral side of the arm with the **caudal humeral circumflex artery**. The **cranial humeral circumflex** sometimes branches from the **subscapular artery**, and may also, in rare cases, branch from the **thoracodorsal artery**. The more distal branch of the **brachial artery** that we will study is the **deep brachial artery**. It is small in cats and runs deep into the arm with the **radial nerve**. The **deep brachial artery** is relatively much larger in humans than it is in cats.

In this area of the arm, the **radial nerve** runs in deep, about midway to the elbow, and then reemerges on the lateral side of the arm, where we saw it in the last lab running deep to the lateral head of the **triceps brachii muscle**, across the **brachialis muscle**. Now we will observe that the **radial nerve** branches superficially and runs on the surface of the **brachioradialis muscle** with the **cephalic vein**, while its deep branch serves the muscles of the posterior compartment of the forearm. You won't be responsible for the deep branch on the lab practical, but you will need to know about it for the written exam.

Now we will focus more on the nerves of the brachial plexus in the arm. There are three **cords**: **posterior**, **medial**, and **lateral**. The **posterior cord** gives rise to two branches, the **radial nerve** and the **axillary nerve**. We have discussed these already. The **medial cord** gives rise to the **ulnar nerve** and half of the **median nerve**. The **ulnar nerve** runs through the arm alone, and then passes on the posterior side of the **medial epicondyle** of the **humerus** before entering the forearm. As we have already noted, the **median nerve** runs with the **brachial artery** in the arm. The **lateral cord** gives rise to the **musculocutaneous nerve** and half of the **median nerve**. The **musculocutaneous nerve** runs along the belly of the **biceps brachii**. This is because it serves the **biceps brachii**, (and also the **coracobrachialis** and the **brachialis**). It does not serve any muscles in the forearm or hand, although it does serve the lateral cutaneous area of the forearm.

## 3. Forearm

In the forearm, we will observe several vessels: the terminal branches of the **brachial plexus** and the muscles they serve. The **ulnar nerve** runs with the **ulnar artery** along the **ulna bone**, then passes on the superficial side of the wrist into the hand. Once in the forearm, the **median nerve** runs with the **radial artery** along the **radius bone**. The **median nerve** then passes deep to the **carpal ligament** (**flexor retinaculum**) into the hand. This is the only terminal branch of the **brachial plexus** that passes through the carpal tunnel. As was mentioned above, we also observe the superficial branch of the **radial nerve** running with the **cephalic vein** in the forearm.

To study the **muscles** of the forearm, I suggest using mnemonics and learning them in order. I will present mnemonics for both the anterior and posterior compartments. You should feel free to make up your own mnemonics if you feel that will work better for you. All the mnemonics I have given you start at the lateral side and work medially. We will hold the cat in the human anatomical position so that the **radius** and the **ulna** are parallel.

For the anterior compartment of the forearm, the cat mnemonic will be "**P**ete **R**ose **S**wings **U**p." The first name is completely different than the others with both the anterior mnemonic and the posterior mnemonic. Pete is for **pronator teres**. The other three start with flexor, and the second words are carpi, digitorum, and carpi. The names with carpi in them include the name of the bone to which the muscle is adjacent. So the next muscle is the **flexor carpi radialis**, then the **flexor digitorum superficialis**, and most medial is the **flexor carpi ulnaris**. In humans, the **palmaris longus**, which is not found in the cat, covers most (but not all) of the **flexor digitorum superficialis**, so the mnemonic becomes "**P**ete **R**ose **S**ometimes **P**umps **U**p."

The mnemonic for the posterior compartment of the cat is "**B**ig **L**arry **B**ird **C**oach **L**oves **U**nicorns." As with the anterior compartment, the first muscle has a name entirely unlike the others. That is **brachioradialis** and it is not actually in the posterior compartment, nor is it an extensor, but it is closer to the posterior compartment than to the anterior compartment. All the remaining muscles start with the word extensor and then carpi, carpi, digitorum, digitorum, and carpi. As with the anterior compartment, the names with carpi in them include the name of the bone to which the muscle is adjacent. So the next muscle is the **extensor carpi radialis longus**, then the **extensor carpi radialis brevis**, then the **extensor digitorum communis** (sometimes called simply **extensor digitorum**), then the **extensor digitorum lateralis** (cat only), and then the most medial is the **extensor carpi ulnaris**. In humans, there is no **extensor digitorum lateralis**, but the other muscles are the same and in the same order, so the mnemonic is "**B**ig **L**arry **B**ird **C**oach **U**nderstands."

# NERVES & VESSELS (CAT)

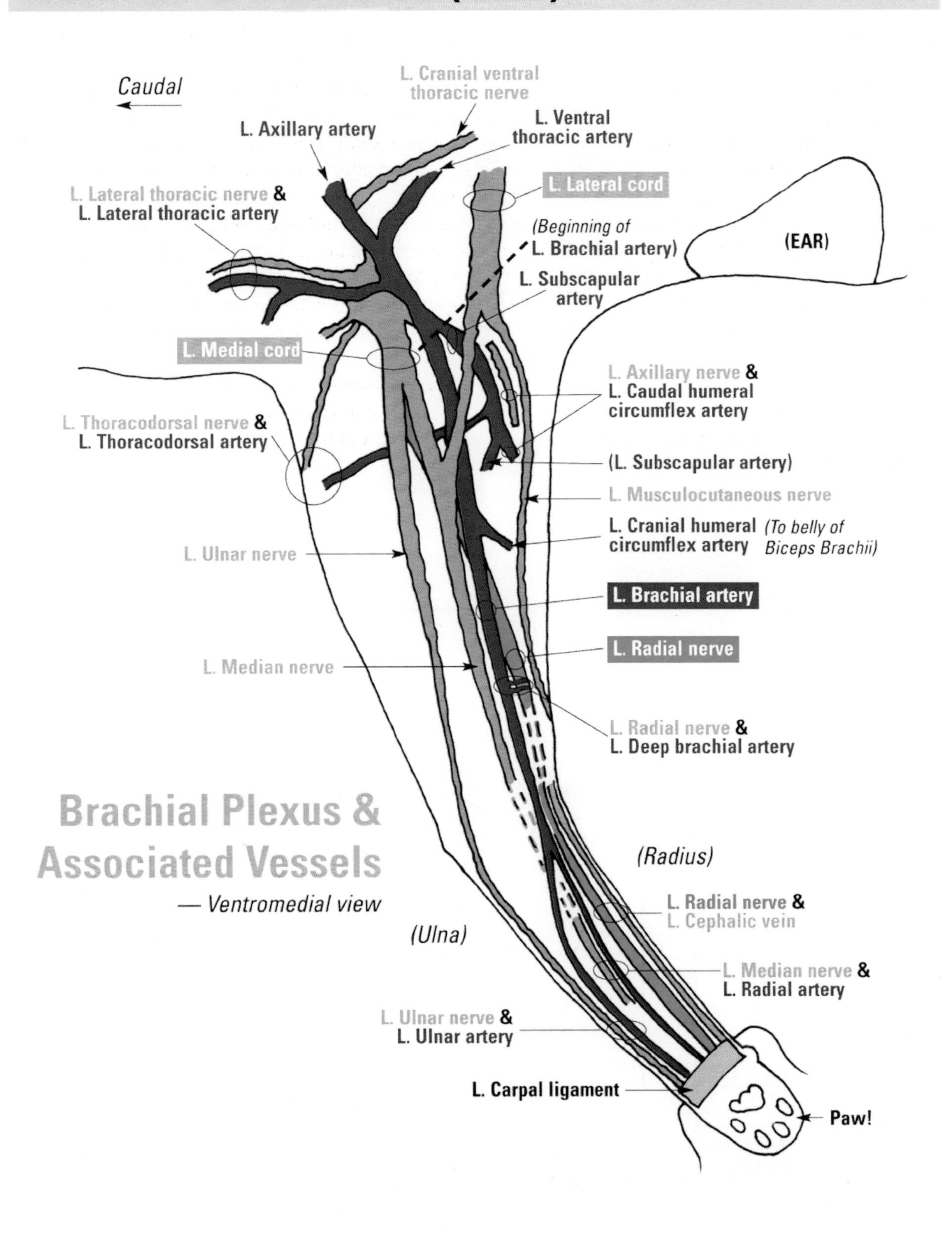

# MUSCLES

## Biceps Brachii

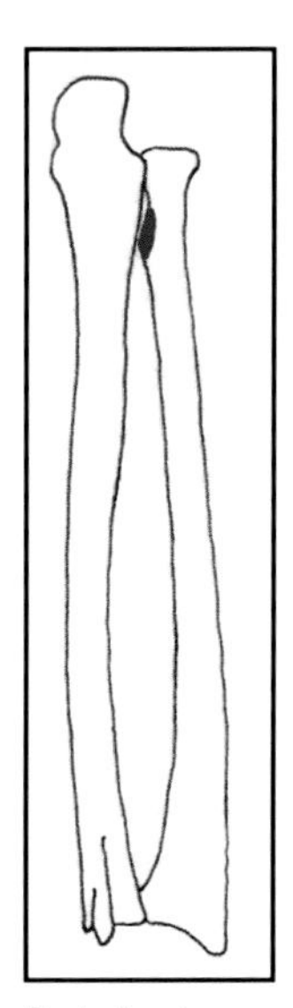

*Posterior view, right side*

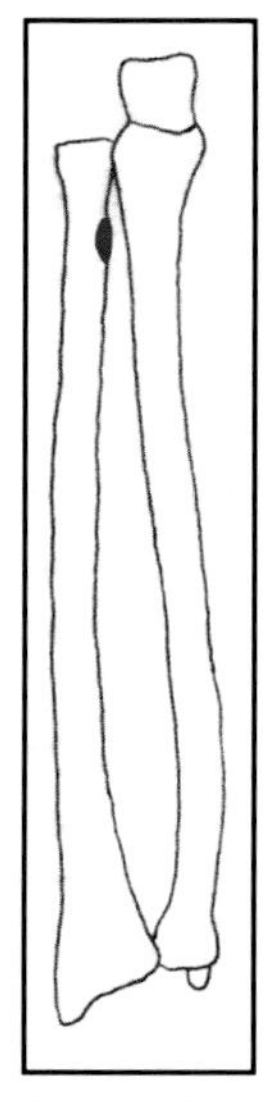

*Anterior view, right side*

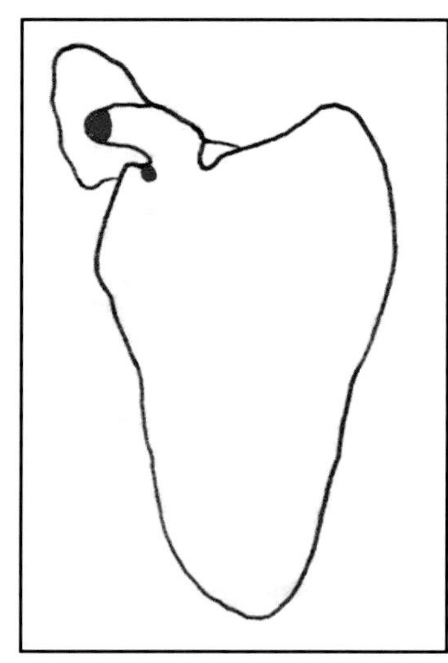

*Anterior view, right side*

*These drawings of the origin and insertion might help you visualize this information (red is the origin, blue the insertion).*

### Human Information:

**origin:** supraglenoid tubercle of scapula (long head) coracoid process of scapula (short head)

**insertion:** radial (bicipital) tuberosity of radius and bicipital aponeurosis

**nerve:** musculocutaneous

**action:** flexes forearm (elbow), supinates forearm and wrist (hand)—both heads; flexes arm (shoulder)—short head only

## Biceps Brachii

Lateral view, right side

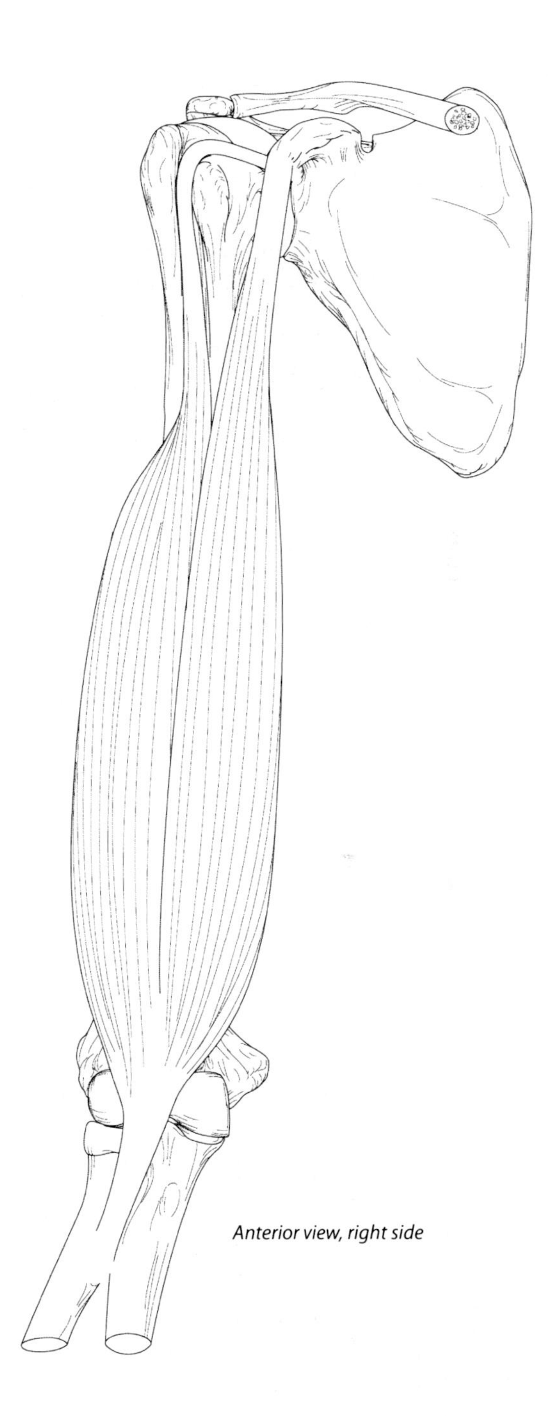
Anterior view, right side

The biceps brachii is one of five muscles that move the forearm for which you are responsible. The others are brachialis, brachioradialis, triceps brachii, and anconeus. Note that supinators insert on the radius. This makes sense because only the radius rotates in the forearm, the ulna does not rotate.

# Coracobrachialis

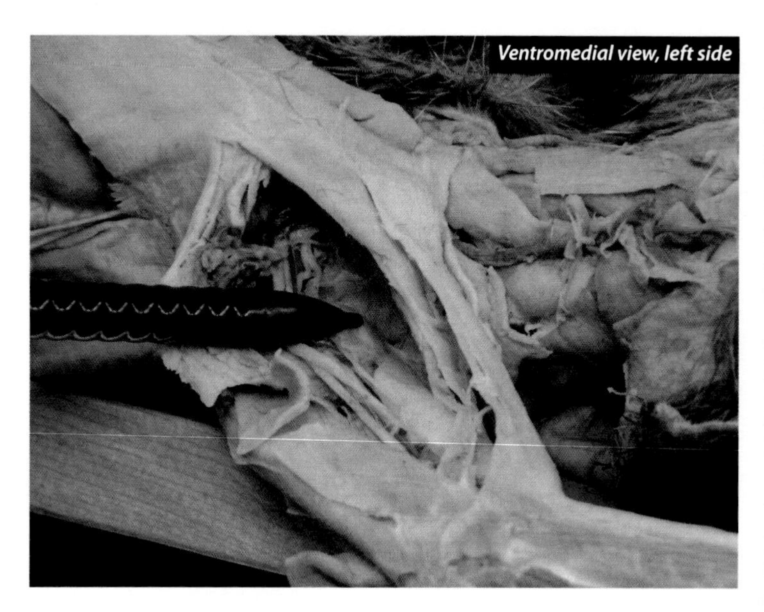
*Ventromedial view, left side*

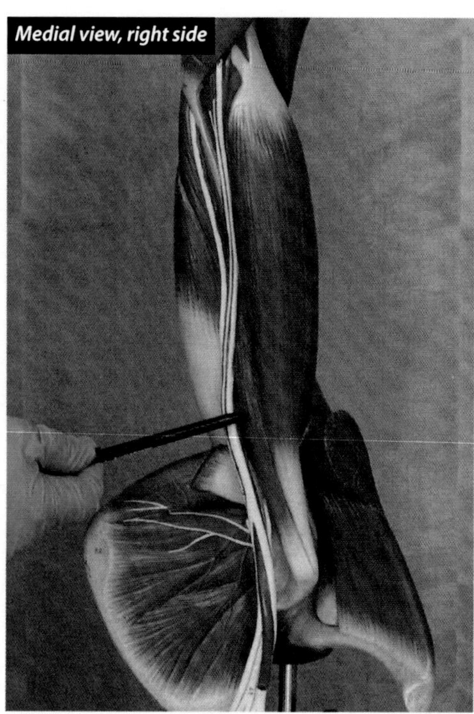
*Medial view, right side*

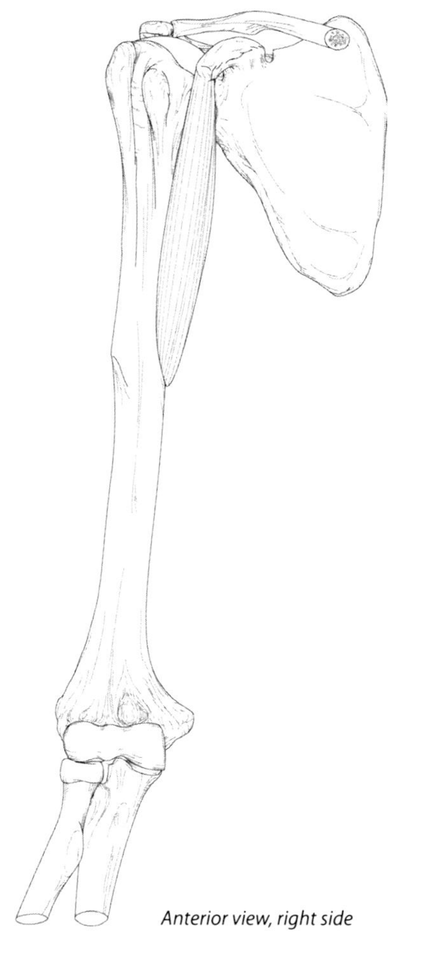
*Anterior view, right side*

The coracobrachialis is one of five muscles that move the arm, for which you are responsible. The others are deltoid, pectoralis major, latissimis dorsi, and teres major. Jan Stone used to tell her students that if one were to view the biceps brachii as the lower portion of the small letter "i," then the dot would be the coracobrachialis. Dr. J likes that, so he tells his students that now. He hopes that helps.

## Coracobrachialis

***Human Information:***
**origin:** coracoid process of scapula
**insertion:** medial humerus opposite deltoid tuberosity
**nerve:** musculocutaneous
**action:** flexes and adducts arm; stabilizes shoulder joint

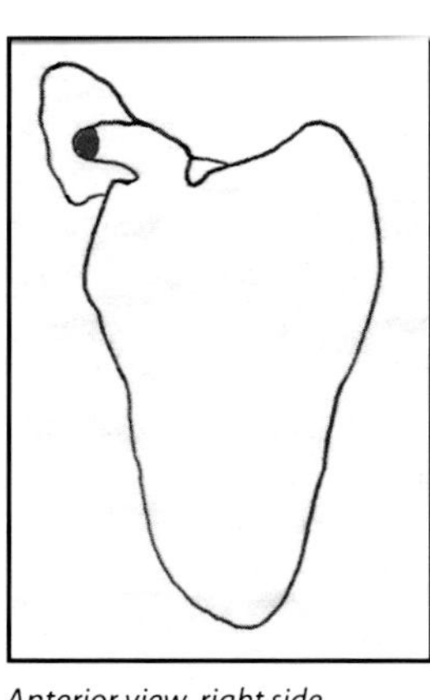

*Anterior view, right side*

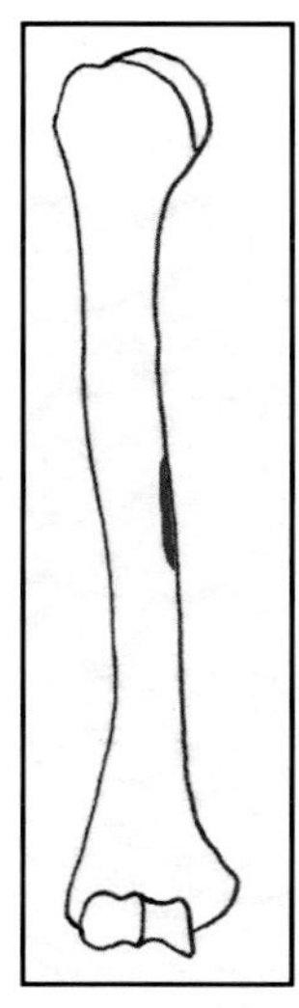
*Anterior view, right side*

## Epitrochlearis (Cat)

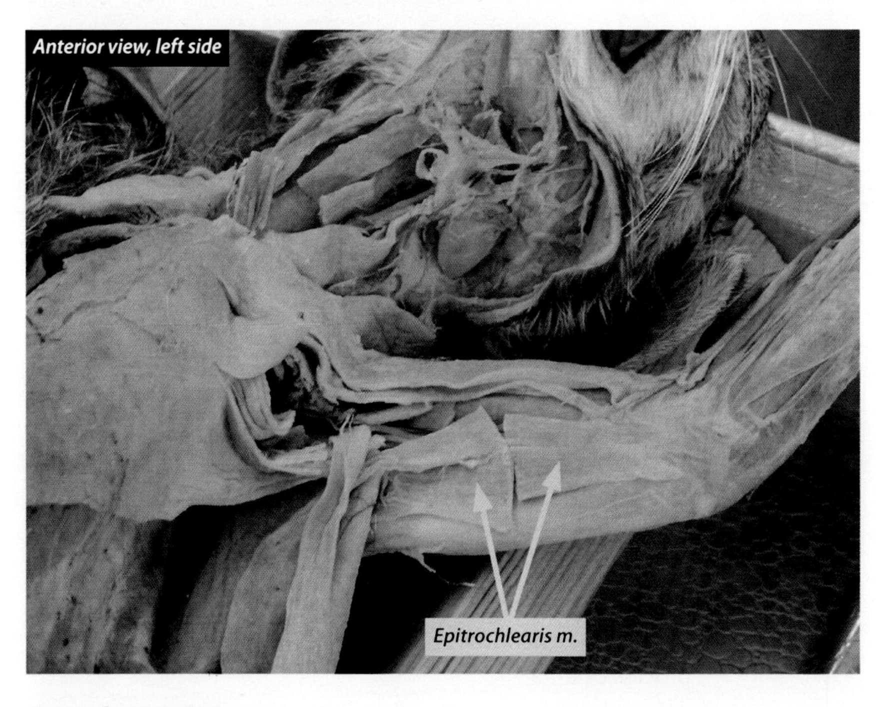


***Cat Information:***
**origin:** lateral or outer surface of the ventral border of latissimus dorsi; often fibers are attached to teres major and pectoralis minor
**insertion:** olecranon process of ulna
**nerve:** radial
**action:** extends of forearm

**Note that this muscle is found only in cats.**

# Pronator Teres

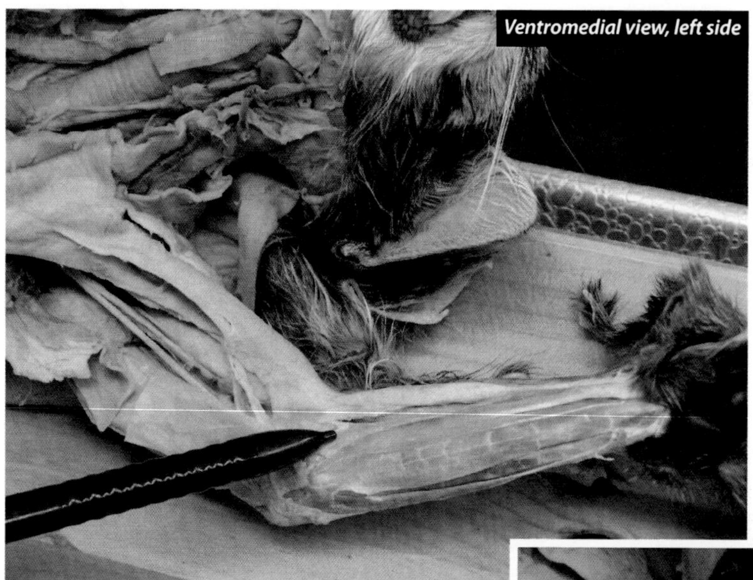


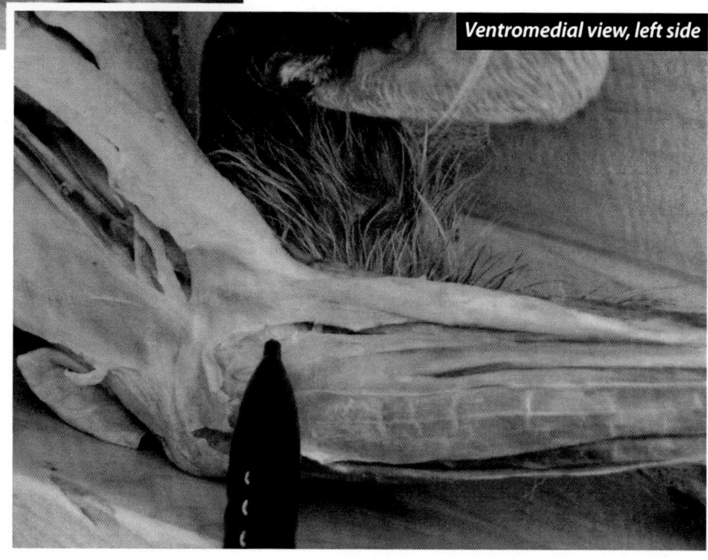


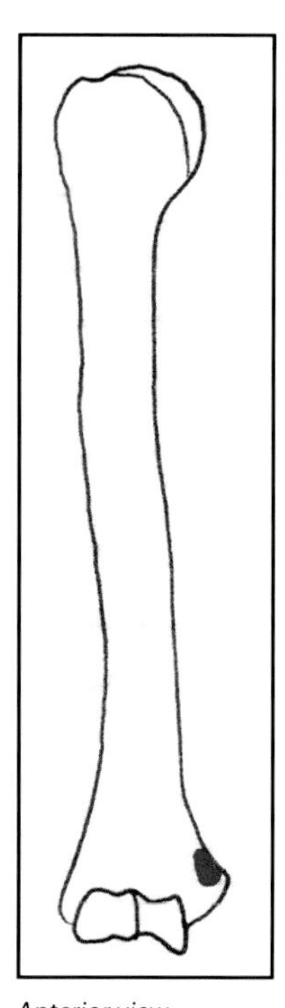

*Anterior view, right side*

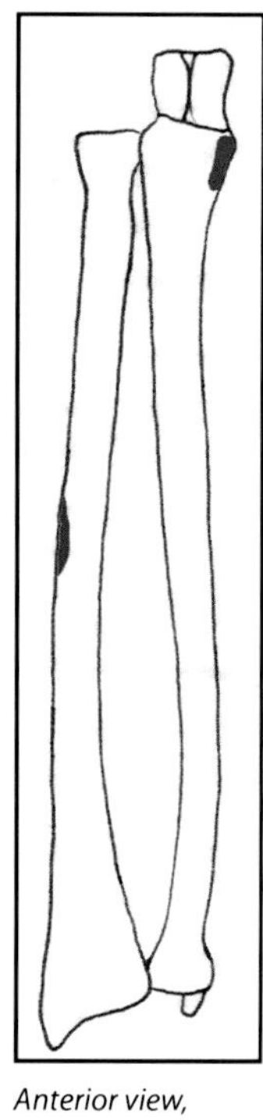

*Anterior view, right side*

*These drawings of the origin and insertion might help you visualize this information (red is the origin, blue the insertion).*

### *Human Information:*

**origin:** medial epicondyle of humerus, medial side of coronoid process of ulna

**insertion:** lateral aspect of radius at its midpoint

**nerve:** median

**action:** pronates forearm and wrist (hand), assists elbow flexion

# Pronator Teres

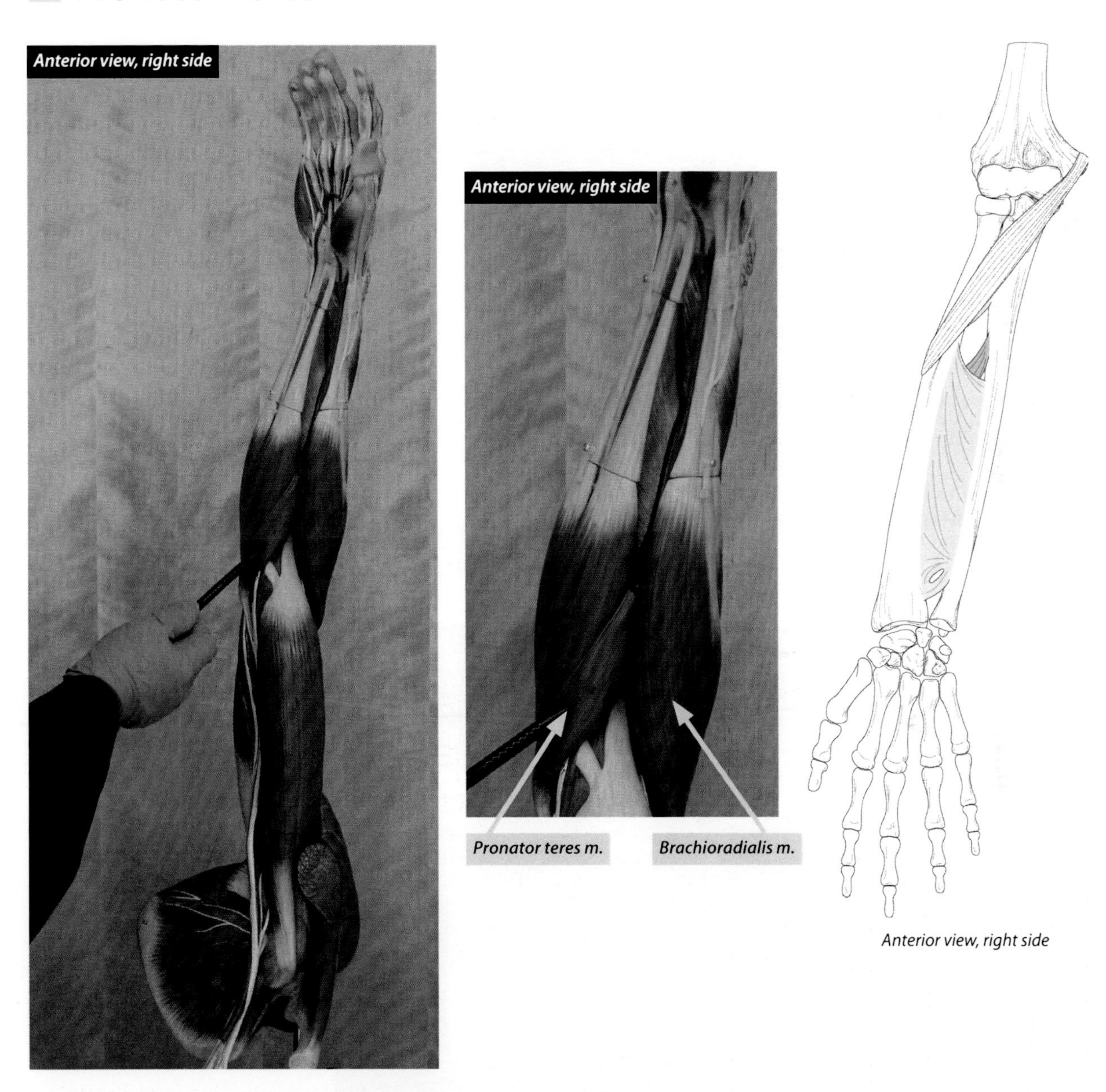


Anterior view, right side

The pronator teres is part of the anterior compartment of the forearm. The origin of this muscle is by way of the common flexor tendon. The common flexor tendon attaches to the medial epicondyle of the humerus, and it also serves as the origin for flexor carpi radialis, palmaris longus, flexor digitorum superficialis (humeroulnar head), and flexor carpi ulnaris. This is one of two pronator muscles. Note that pronators insert on the radius. This makes sense because only the radius rotates in the forearm; the ulna does not rotate. It has the same action as pronator quadratus, so Dr. J suggests you learn these two muscles as a group. Dr. J often tells his students that this muscle looks like a teardrop. It is the "Pete" of the cat mnemonic "Pete Rose Swings Up" and of the human mnemonic "Pete Rose Sometimes Pumps Up." Note that these mnemonics are slightly different due to the addition of the palmaris longus, which the cat does not have.

# Flexor Carpi Radialis

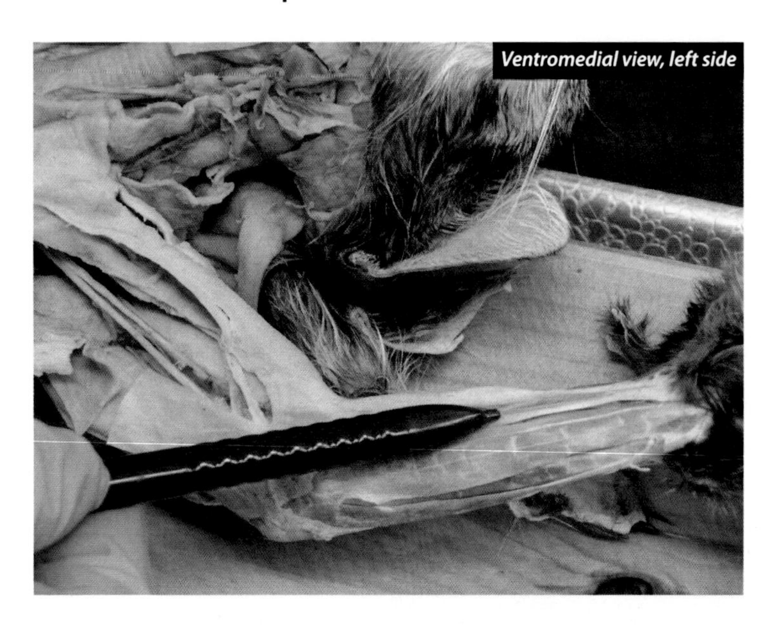


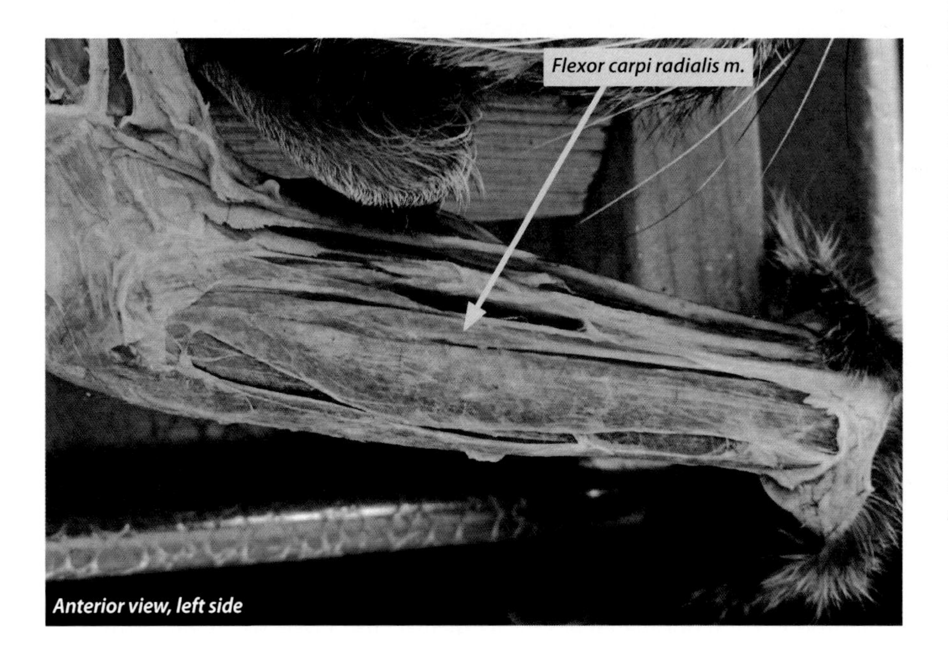


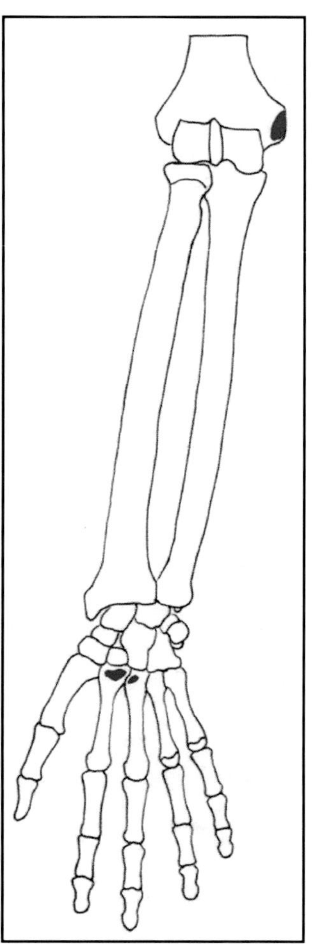

*Anterior view, right side*

### *Human Information:*

**origin:** medial epicondyle of the humerus at common flexor tendon
**insertion:** bases of the second and third metacarpals
**nerve:** median
**action:** flexes and abducts of wrist (hand)

# Flexor Carpi Radialis

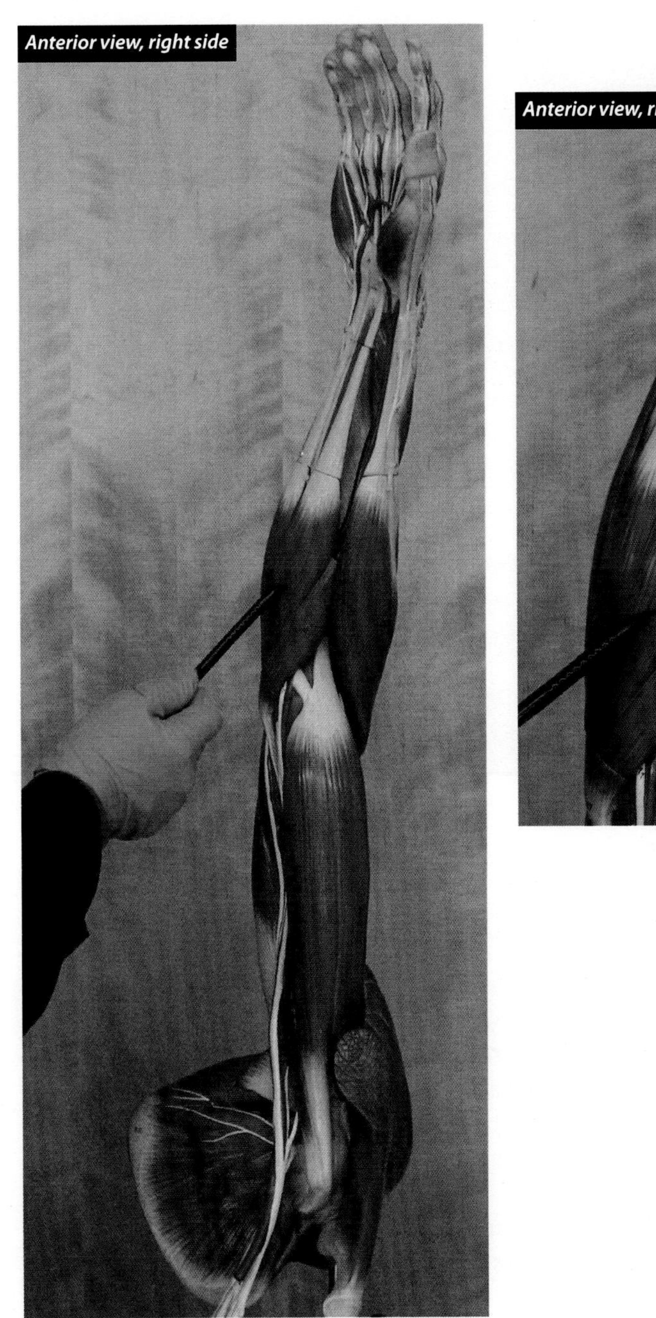

*Anterior view, right side*

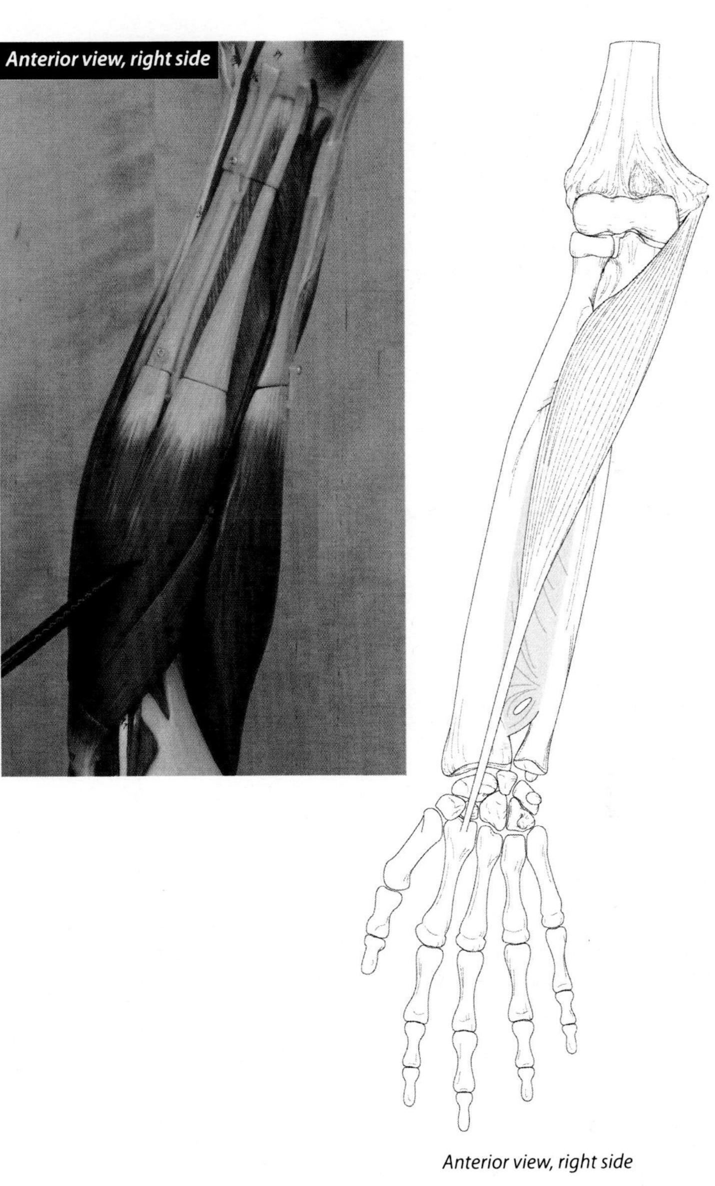

*Anterior view, right side*

*Anterior view, right side*

The origin of flexor carpi radialis is by way of the common flexor tendon. The common flexor tendon attaches to the medial epicondyle of the humerus, and it also serves as the origin for pronator teres, palmaris longus, flexor digitorum superficialis (humeroulnar head), and flexor carpi ulnaris muscles. It is the "Rose" of the cat mnemonic "Pete Rose Swings Up" and of the human mnemonic "Pete Rose Sometimes Pumps Up." Note that these mnemonics are slightly different due to the addition of the palmaris longus, which the cat does not have.

## Flexor Digitorum Superficialis

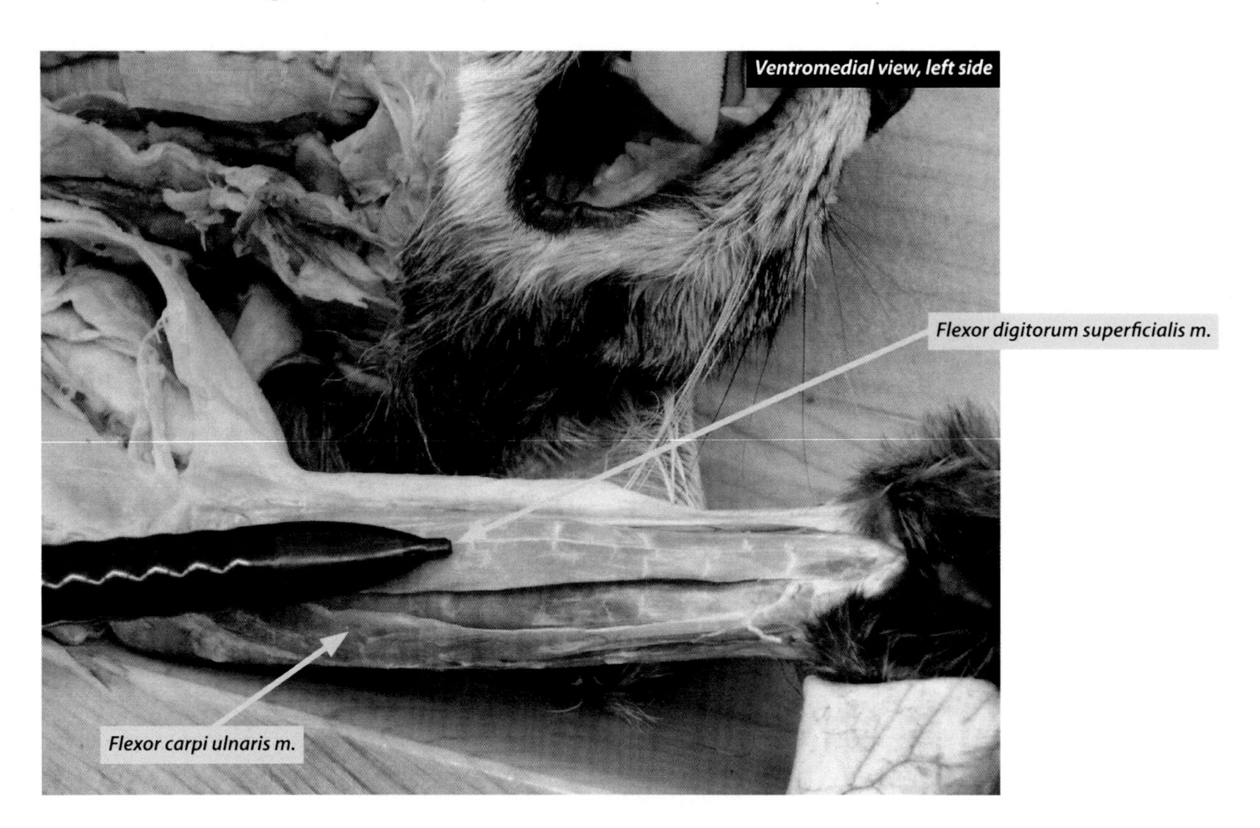


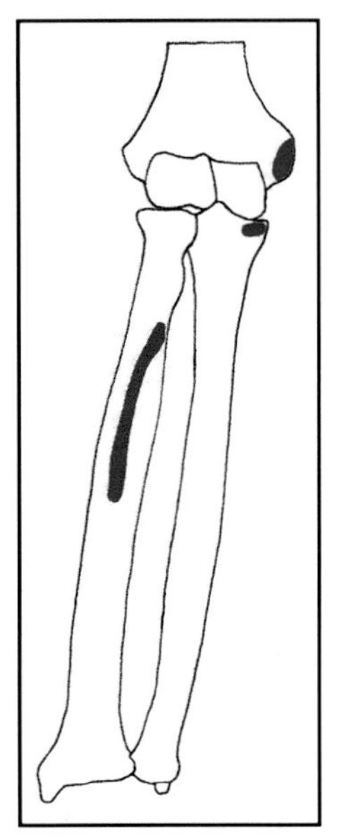

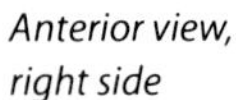
*Anterior view, right side*

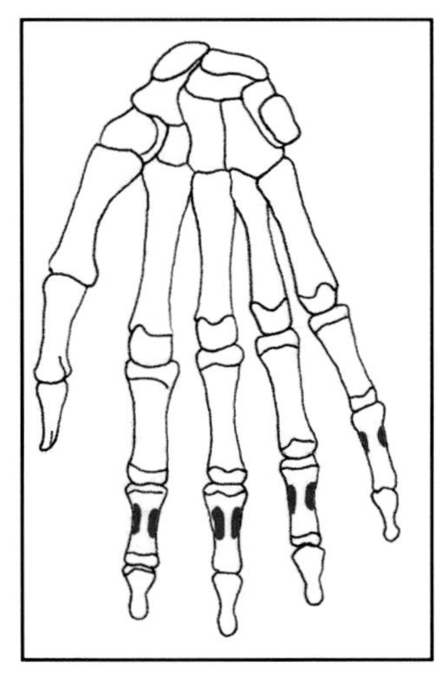

***Human Information:***

**origin:** medial epicondyle of humerus, coronoid process of ulna, shaft of radius

**insertion:** by four tendons to middle phalanges of digits 2 to 5

**nerve:** median

**action:** flexes of wrist (hand) and digits 2 to 5

*These drawings of the origin and insertion might help you visualize this information (red is the origin, blue the insertion).*

# Flexor Digitorum Superficialis

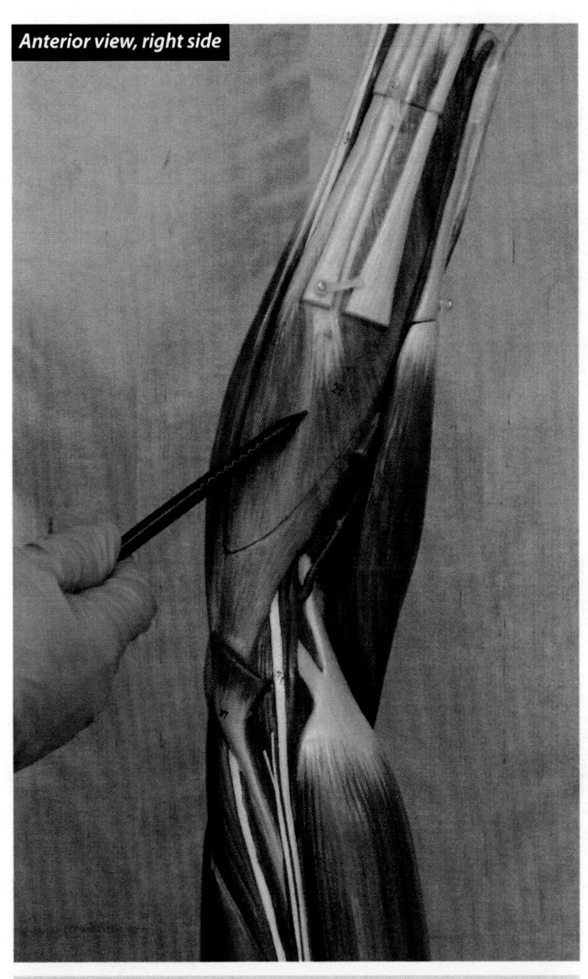


*Note: in this picture palmaris longus was removed to reveal flexor digitorum superficialis.*

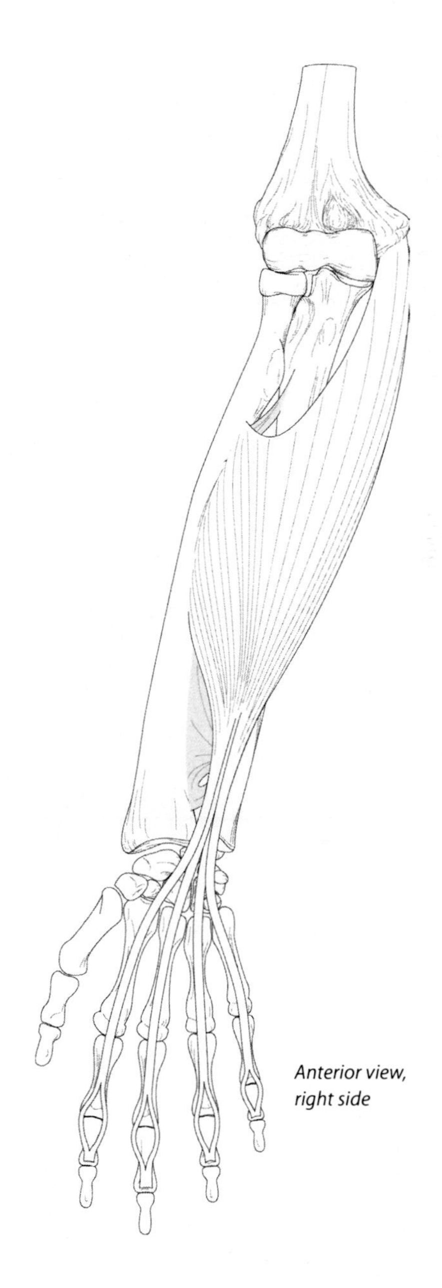

*Anterior view, right side*

In humans, this muscle is deep to palmaris longus. The origin of the humeroulnar head of flexor digitorum superficialis is by way of the common flexor tendon. The common flexor tendon attaches to the medial epicondyle of the humerus, and it also serves as the origin for pronator teres, flexor carpi radialis, palmaris longus, and flexor carpi ulnaris. It has the same action as flexor digitorum profundus, so Dr. J suggests you learn these two muscles as a group for the lecture exam and quiz. It is the "Sometimes" of the human mnemonic "Pete Rose Sometimes Pumps Up." Note that this mnemonic is slightly different from the cat mnemonic due to the addition of the palmaris longus, which the cat does not have. In the cat mnemonic it is the "Swings" of "Pete Rose Swings Up."

# Flexor Carpi Ulnaris

*Ventromedial view, left side*

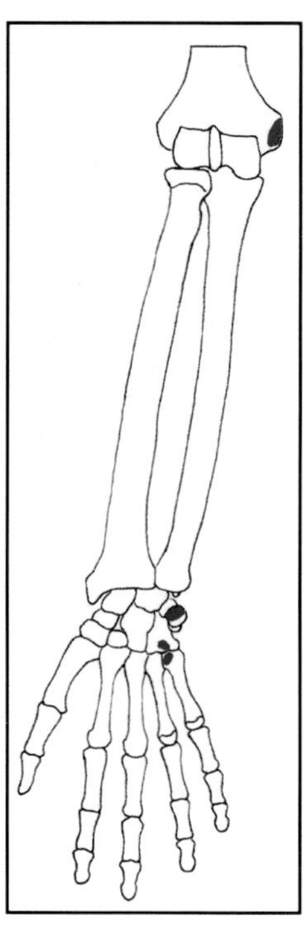
*Anterior view, right side*

*These drawings of the origin and insertion might help you visualize this information (red is the origin, blue the insertion).*

### *Human Information:*

**origin:** medial epicondyle of the humerus at common flexor tendon (two heads)
**insertion:** palmar surface of pisiform, hook of the hamate, and base of fifth metacarpal
**nerve:** ulnar
**action:** flexes and adducts wrist (hand)

# Flexor Carpi Ulnaris

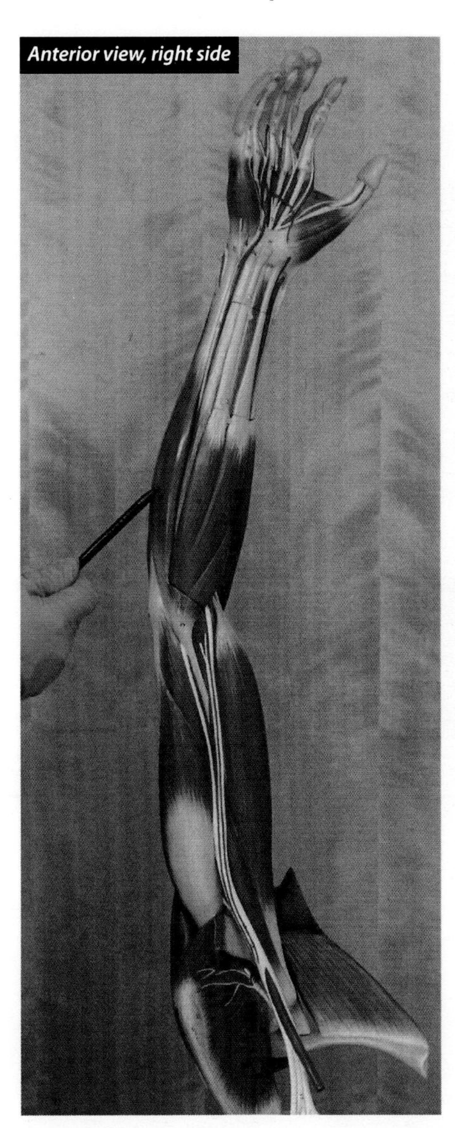
Anterior view, right side

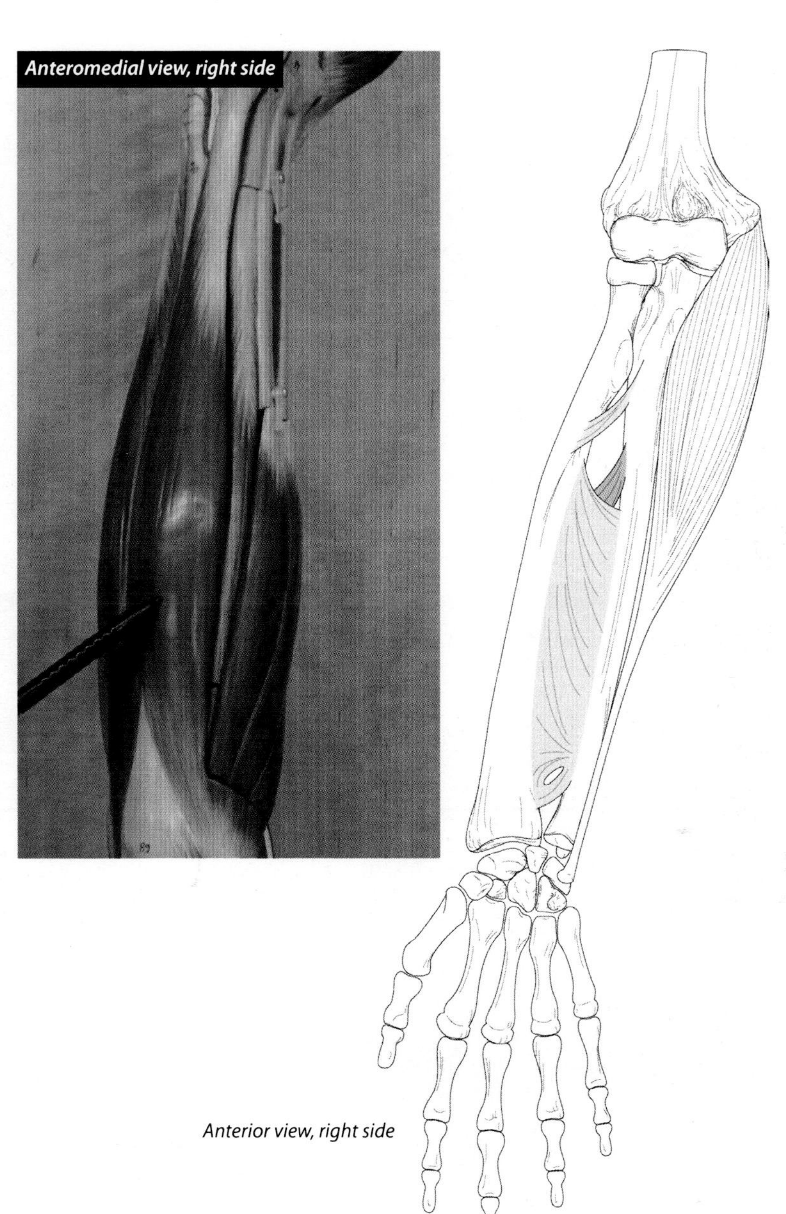
Anteromedial view, right side

Anterior view, right side

The flexor carpi ulnaris is unusual as a flexor muscle because it is served by the **ulnar nerve**, while most muscles in the anterior compartment of the forearm are served by the **median nerve**. This makes it a Grant, Grant, Grant, Grant thing: ulna bone, **ulnar nerve**, **ulnar artery**, and flexor carpi ulnaris. LOMG! The origin of this muscle is by way of the common flexor tendon. The common flexor tendon attaches to the medial epicondyle of the humerus, and it also serves as the origin for pronator teres, flexor carpi radialis, palmaris longus, and flexor digitorum superficialis (humeroulnar head). It is the "Up" of the human mnemonic "Pete Rose Sometimes Pumps Up" and the cat mnemonic "Pete Rose Swings Up."

## Palmaris Longus

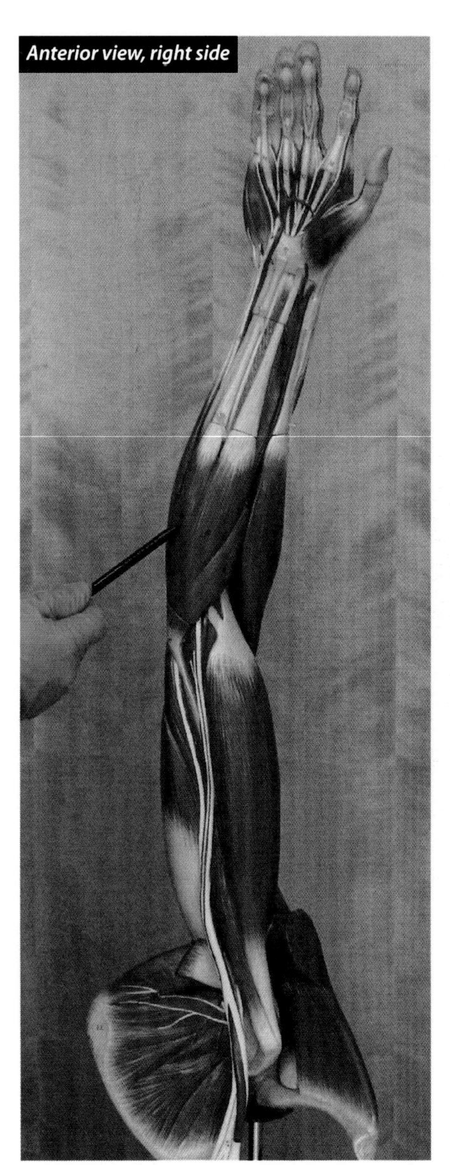


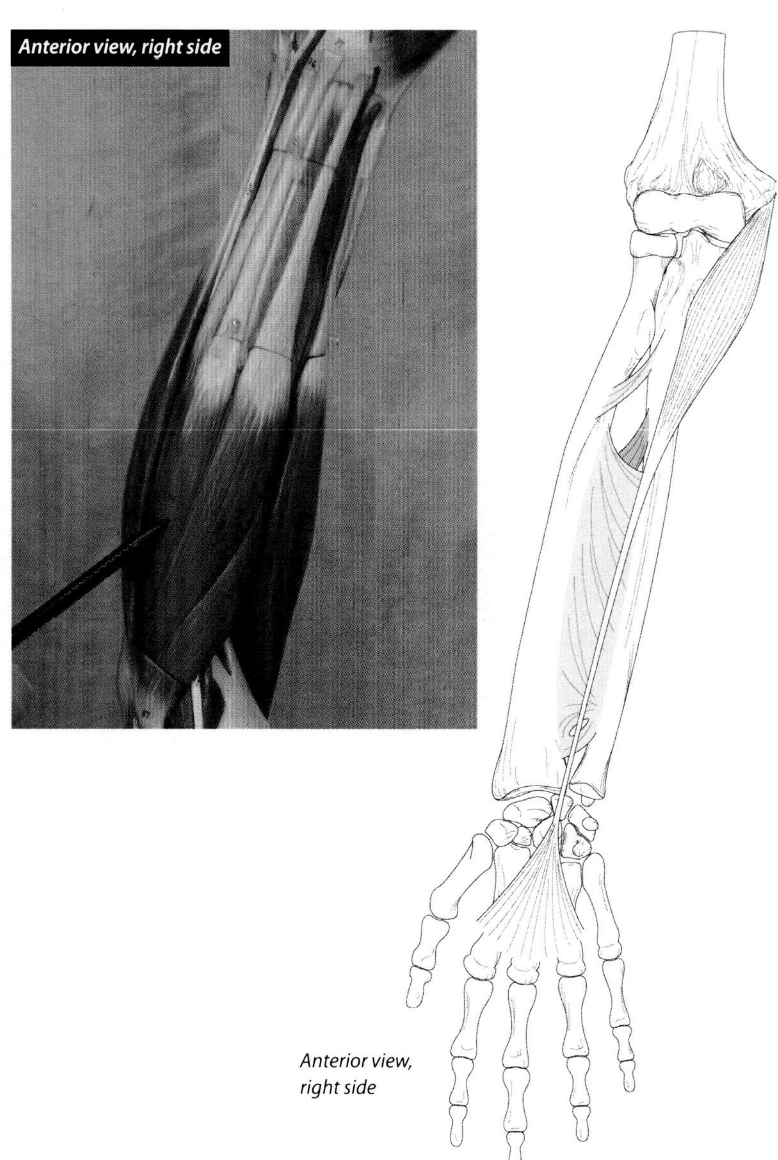


Anterior view, right side

### Human Information:

**origin:** medial epicondyle of humerus
**insertion:** palmar aponeurosis of hand
**nerve:** median
**action:** flexes wrist (hand)

Palmaris longus is part of the anterior compartment of the forearm, and it is superficial to flexor digitorum superficialis. It is the "Pumps" of the human mnemonic "Pete Rose Sometimes Pumps Up." Note that this mnemonic is slightly different from the cat mnemonic due to the addition of this muscle, which the cat does not have. Approximately 10 percent of the human population does not have this muscle either; it may be absent bilaterally or unilaterally. If an individual requires hand surgery and does have a palmaris longus, its tendon is often harvested because its contribution to hand flexion is weak.

# ANTERIOR ("FLEXOR") COMPARTMENT OVERVIEW

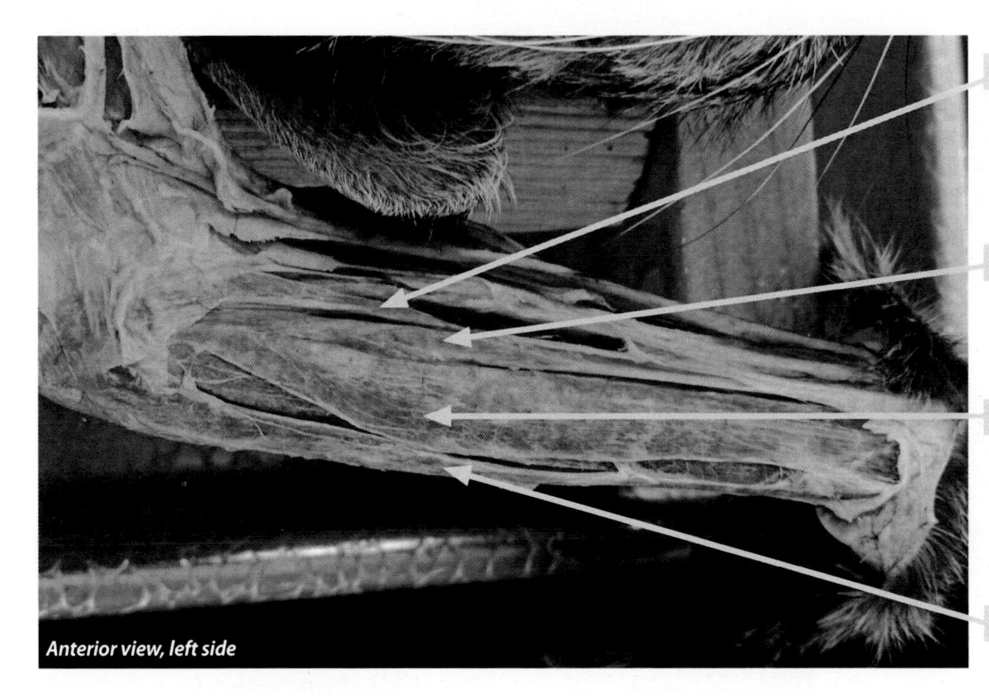


Anterior view, left side

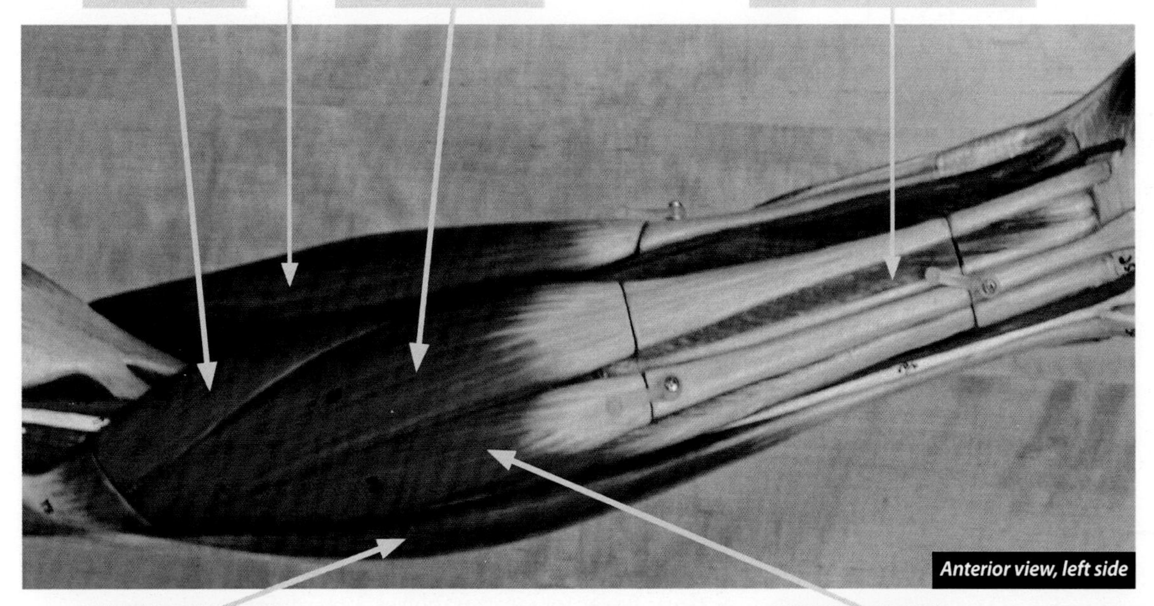


Anterior view, left side

# Brachioradialis

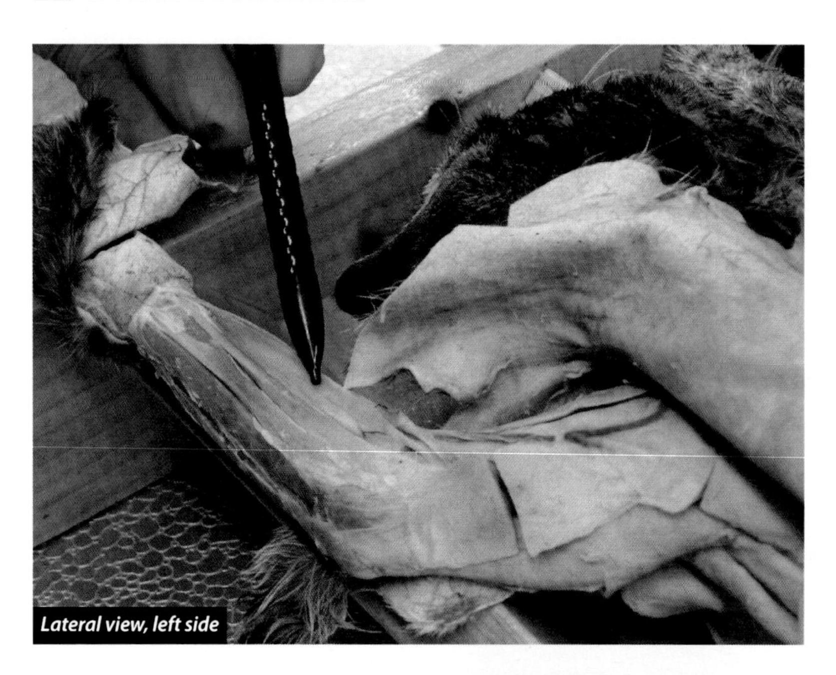
Lateral view, left side

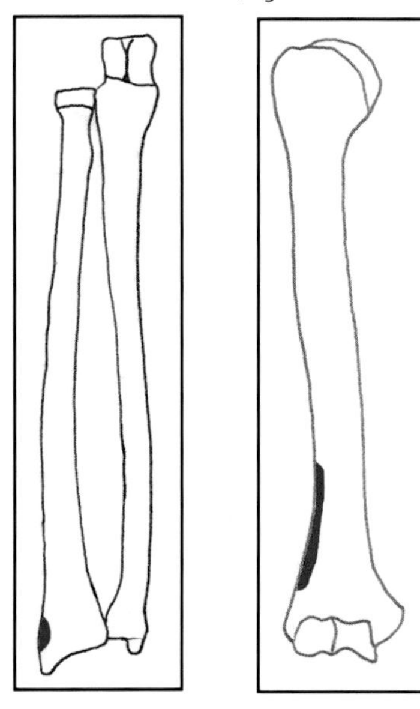
Anterior view, right side

These drawings of the origin and insertion might help you visualize this information (red is the origin, blue the insertion).

Lateral view, left side

### *Human Information:*

**origin:** supracondylar ridge of humerus
**insertion:** lateral styloid process of radius
**nerve:** radial
**action:** flexes forearm (stabilizes forearm for small, quick, movements—"hammering muscle")

## Brachioradialis

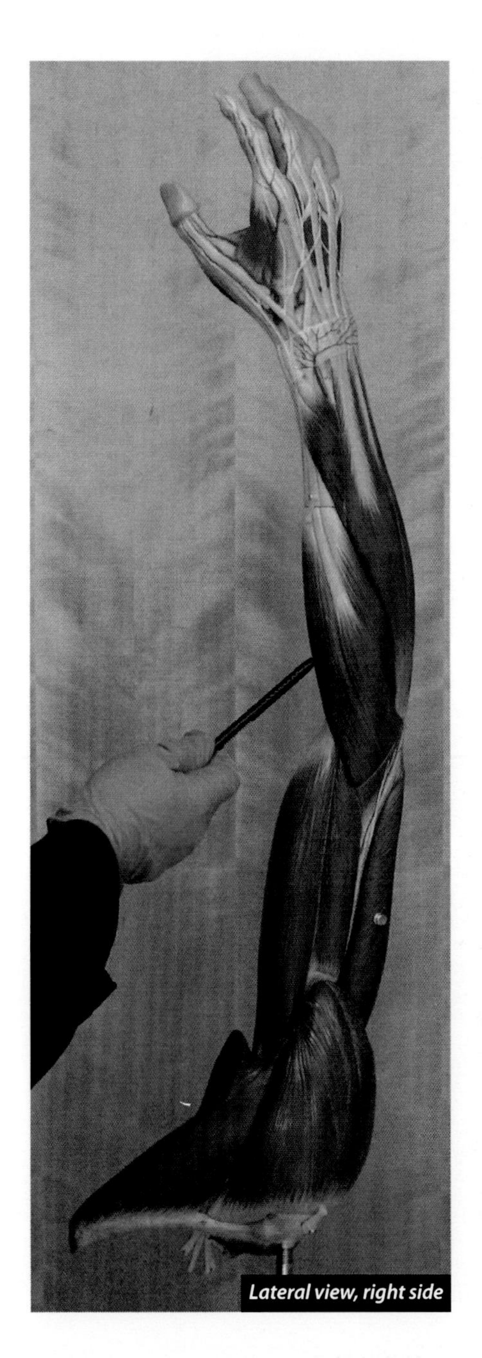
Lateral view, right side

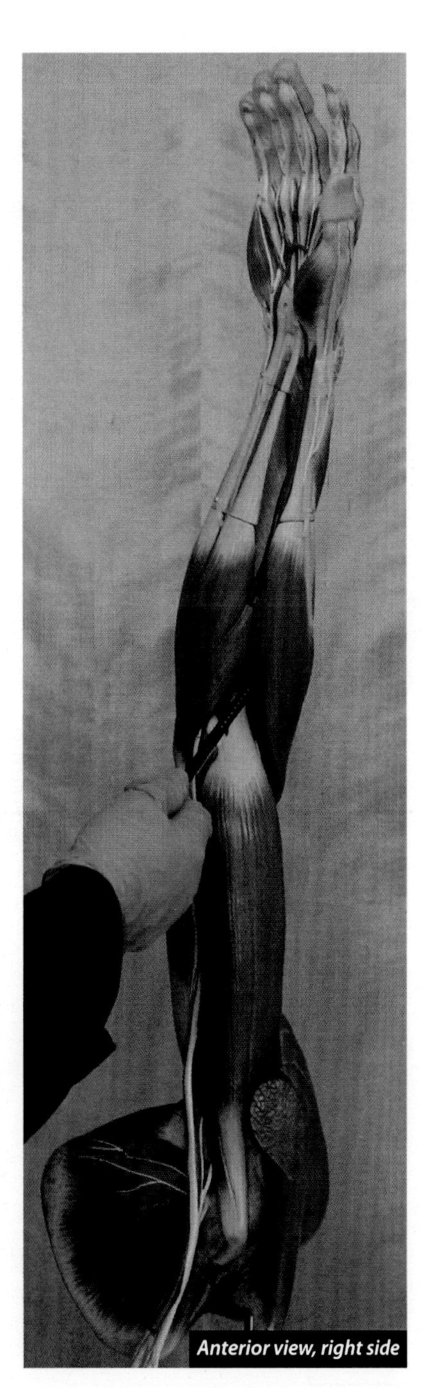
Anterior view, right side

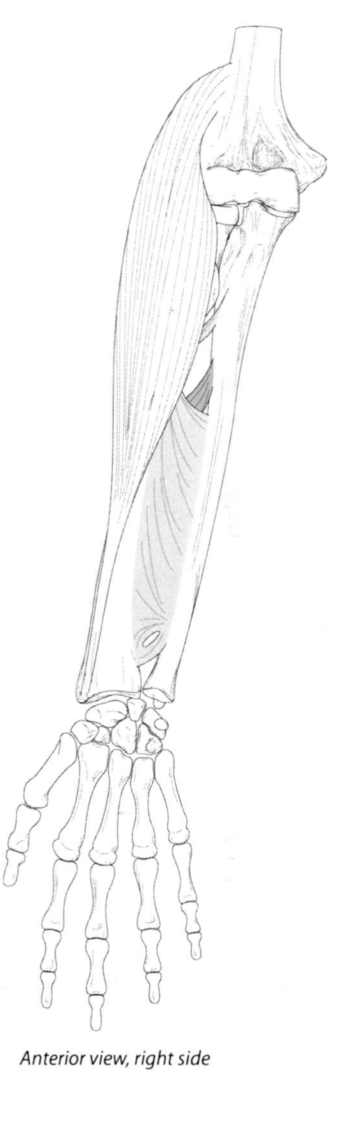
Anterior view, right side

The brachioradialis is the only flexor we study that is served by the **radial nerve**. It is one of five muscles that move the forearm, for which you are responsible. The others are biceps brachii, brachialis, triceps brachii, and anconeus. It is the "Big" of the human mnemonic "Big Larry Bird Coach Understands" and the cat mnemonic "Big Larry Bird Coach Loves Umbrellas."

# Extensor Carpi Radialis Longus

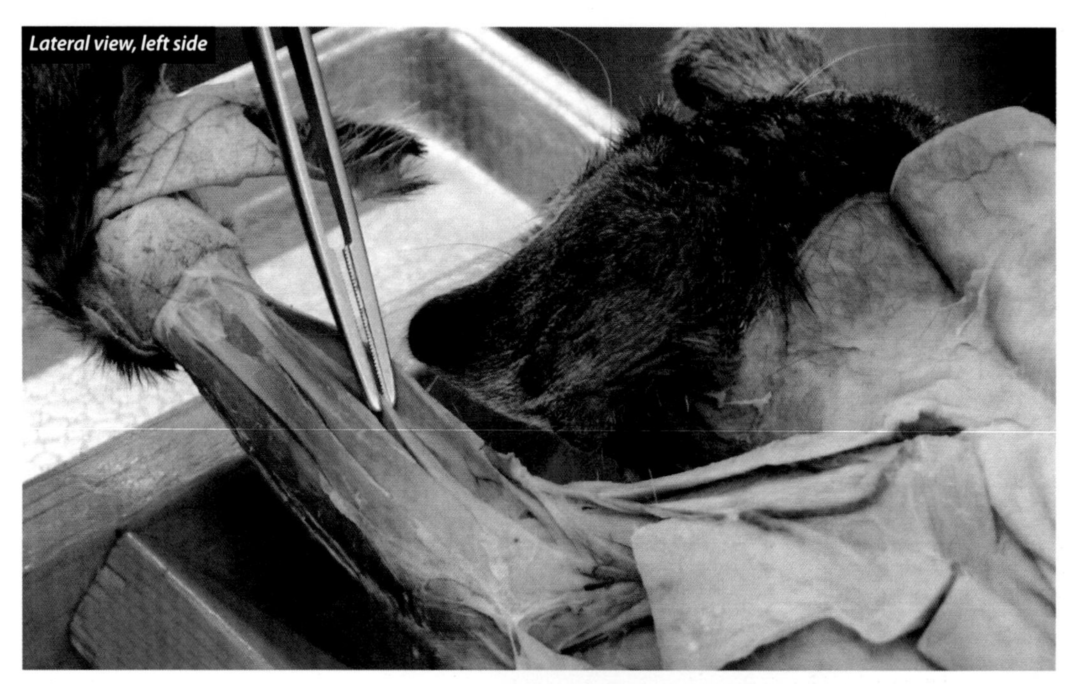


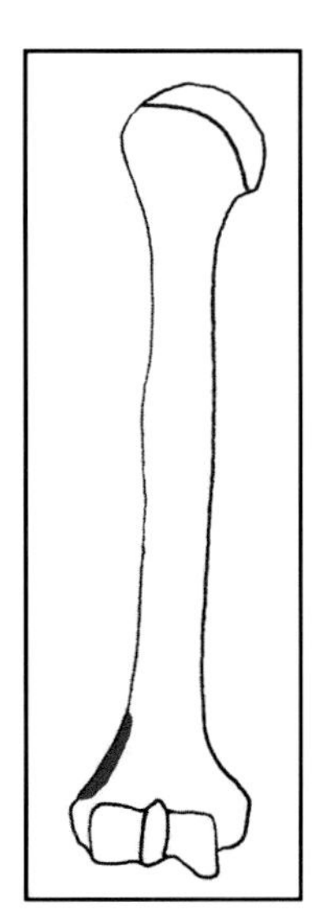
Anterior view, right side

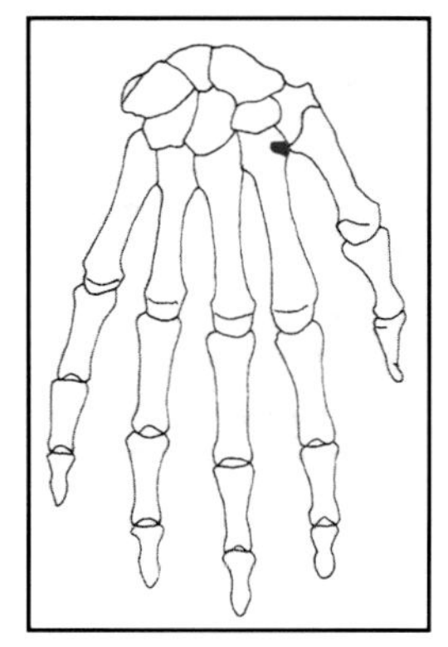
Posterior view, right side

These drawings of the origin and insertion might help you visualize this information (red is the origin, blue the insertion).

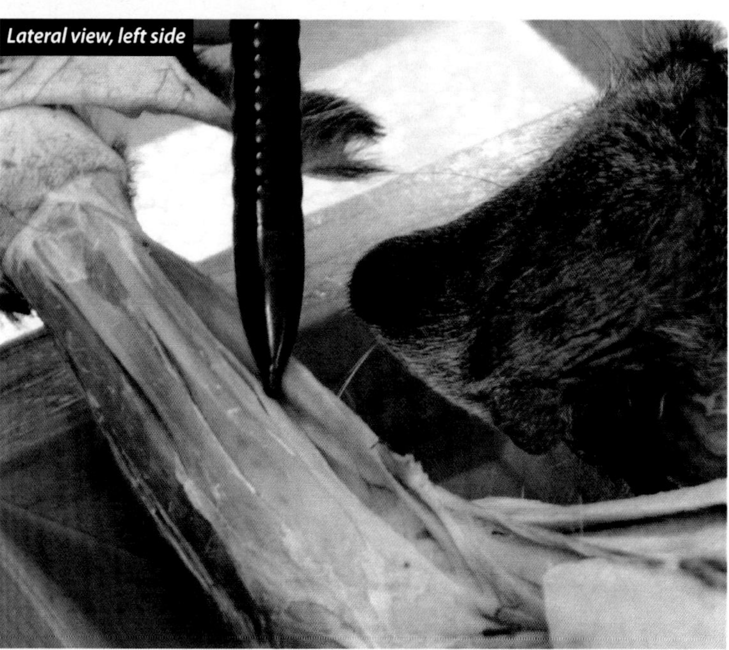


***Human Information:***

**origin:** lateral supracondylar ridge of humerus
**insertion:** posterior surface of base of second metacarpal
**nerve:** radial
**action:** extends and abducts wrist (hand)

## Extensor Carpi Radialis Longus

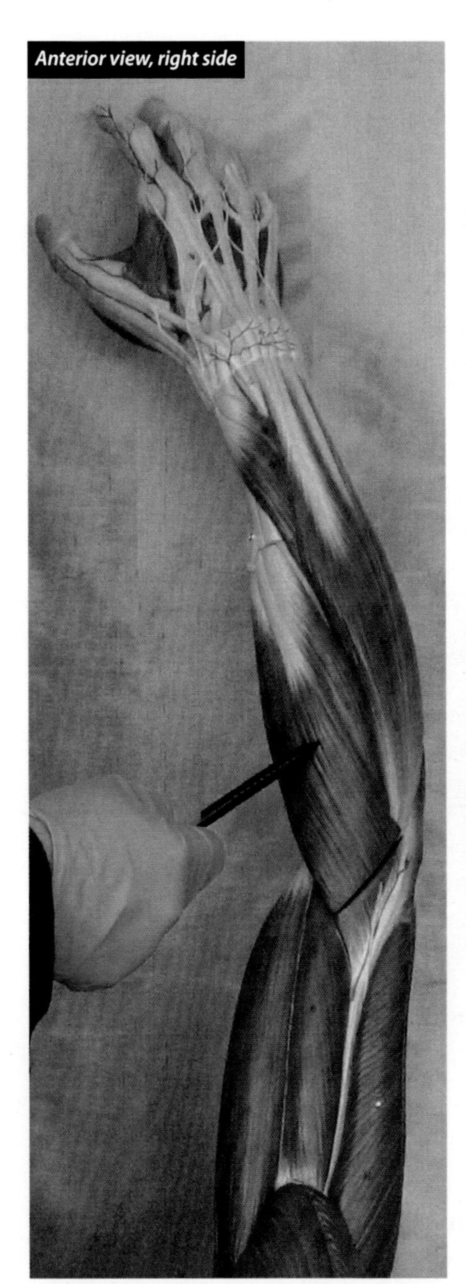

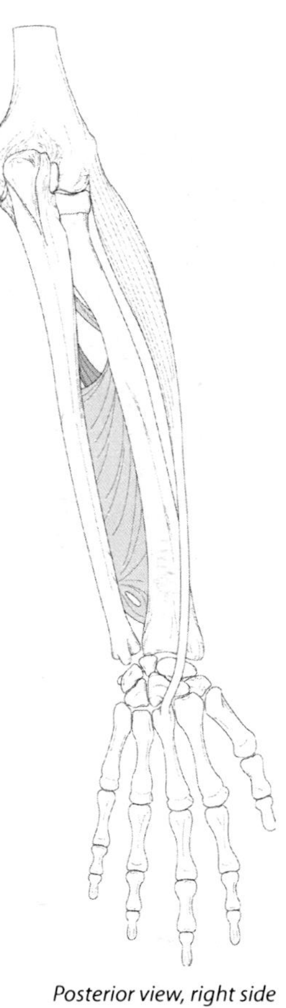

Posterior view, right side

Extensor carpi radialis longus has the same action as the extensor carpi radialis brevis. Dr. J suggests you learn those two muscles as a group. This muscle is the "Larry" of the mnemonic of the human mnemonic "Big Larry Bird Coach Understands" and the cat mnemonic "Big Larry Bird Coach Loves Umbrellas."

# Extensor Carpi Radialis Brevis

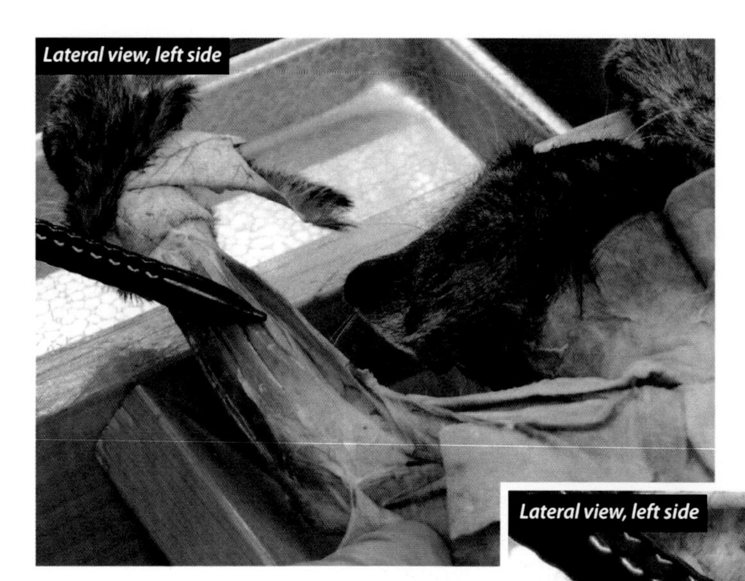
*Lateral view, left side*

*Lateral view, left side*

*These drawings of the origin and insertion might help you visualize this information (red is the origin, blue the insertion).*

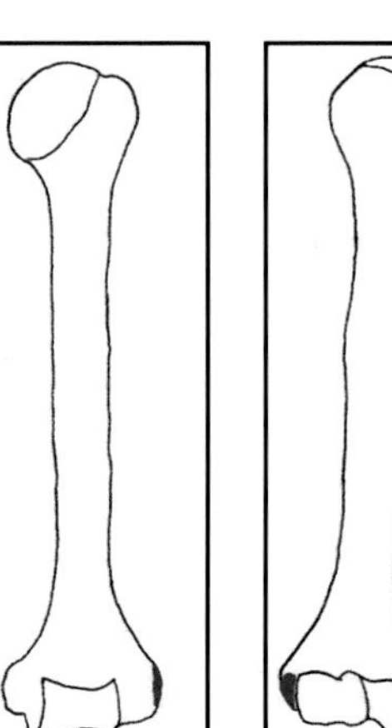
*Posterior view, right side*

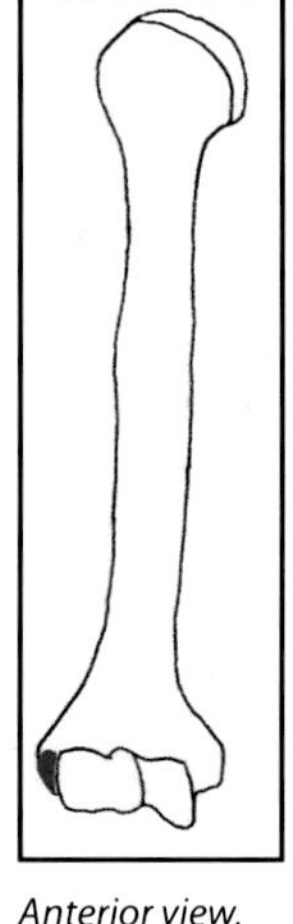
*Anterior view, right side*

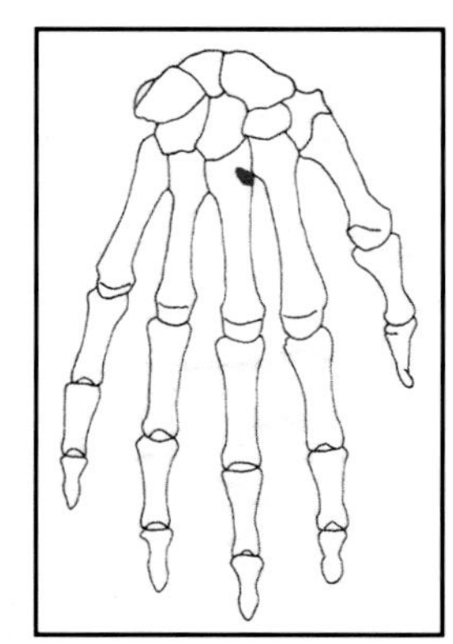
*Posterior view, right side*

### *Human Information:*

**origin:** lateral epicondyle of humerus

**insertion:** posterior surface of base of third metacarpal

**nerve:** radial

**action:** extends and abducts wrist (hand)

## Extensor Carpi Radialis Brevis

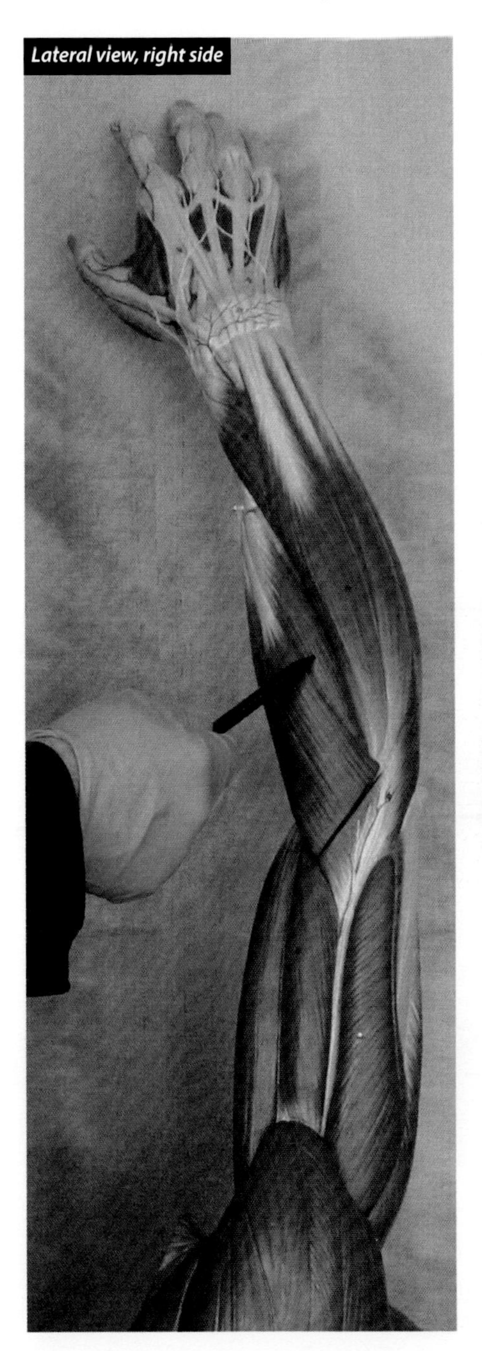
Lateral view, right side

Lateral view, right side

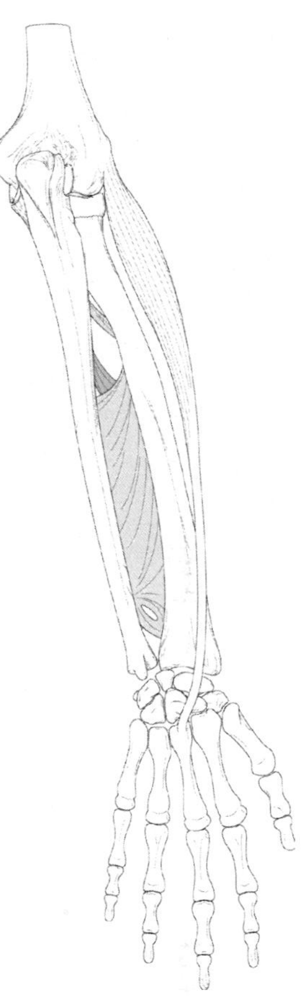
Posterior view, right side

Extensor carpi radialis brevis has the same action as extensor carpi radialis longus. Dr. J suggests you learn those two muscles as a group. The origin of extensor carpi radialis brevis is by the common extensor tendon that is shared with extensor digitorum communis, extensor digiti minimi (which we will not be studying), and extensor carpi ulnaris muscles. This muscle is the "Bird" of the mnemonic of the human mnemonic "Big Larry Bird Coach Understands" and the cat mnemonic "Big Larry Bird Coach Loves Umbrellas."

# Extensor Digitorum Communis

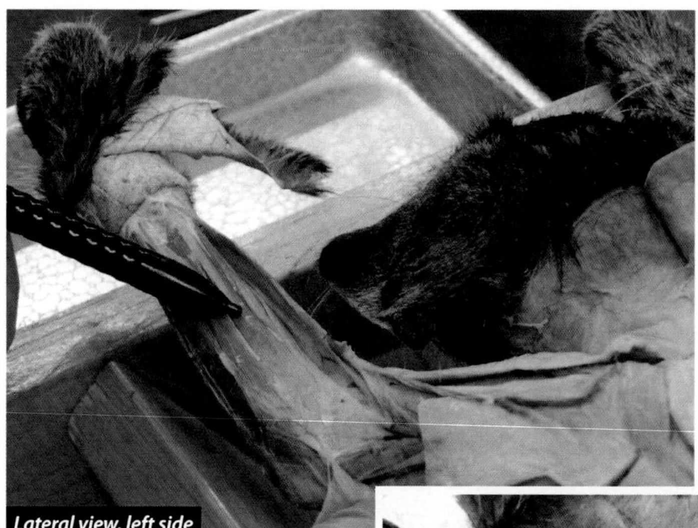
Lateral view, left side

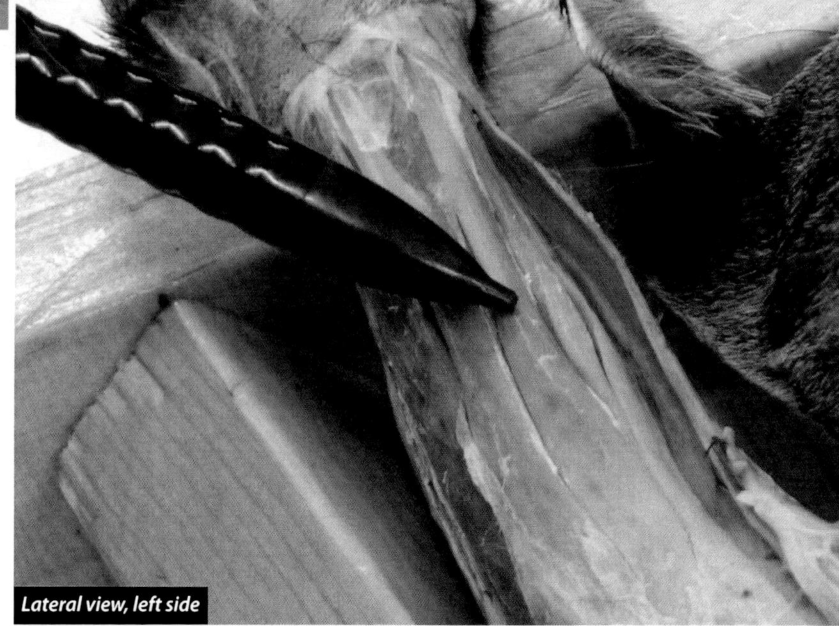
Lateral view, left side

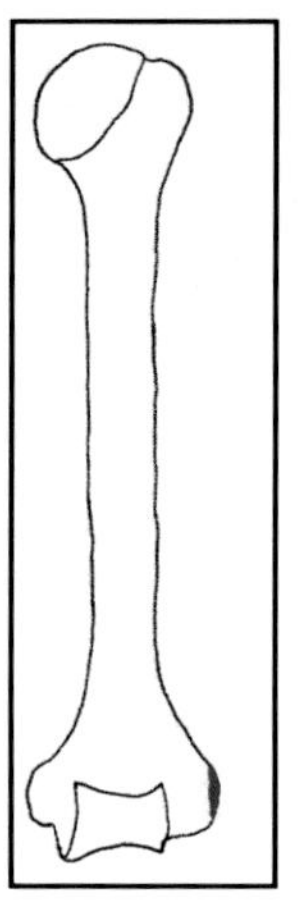
*Posterior view, right side*

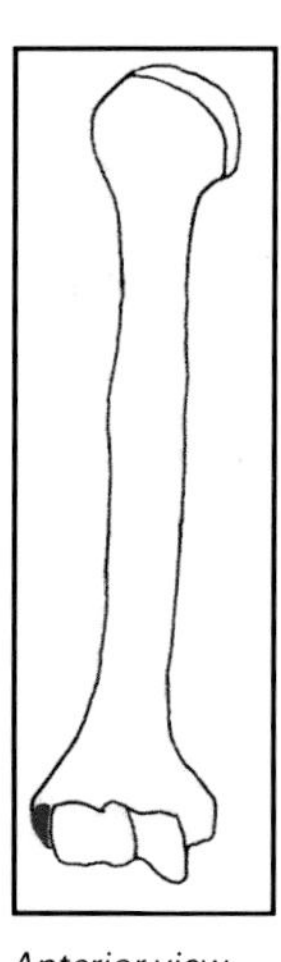
*Anterior view, right side*

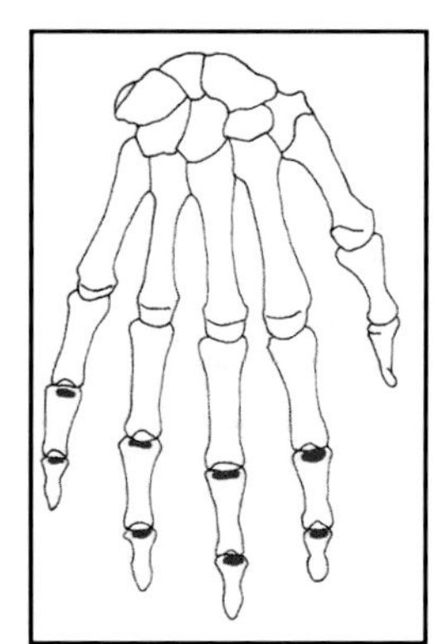
*Posterior view, right side*

***Human Information:***

**origin:** lateral epicondyle of humerus at the common extensor tendon
**insertion:** by four tendons, one to each digit
**nerve:** radial
**action:** extends wrist (hand) and digits 2 to 5

*These drawings of the origin and insertion might help you visualize this information (red is the origin, blue the insertion).*

# Extensor Digitorum Communis

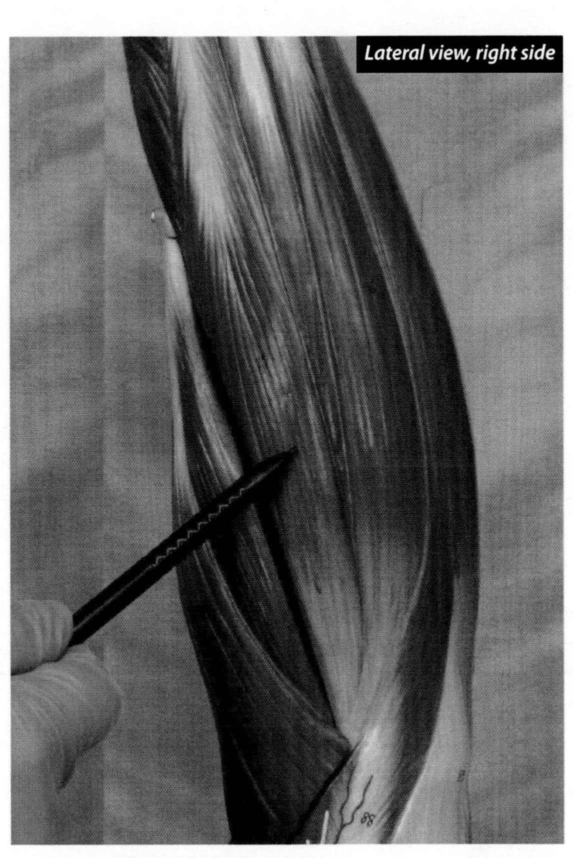


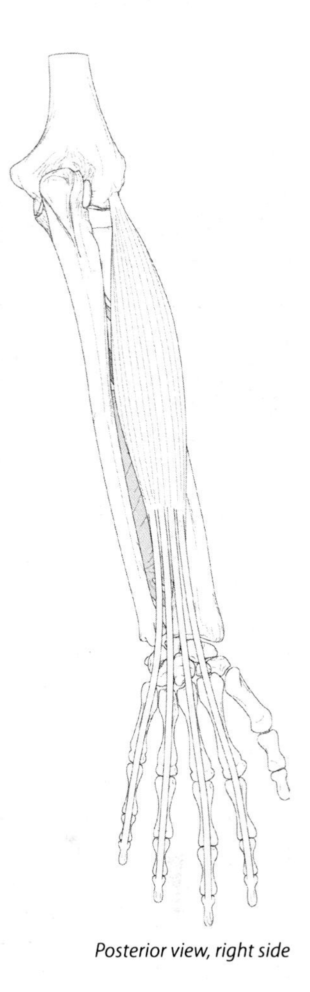

Posterior view, right side

Extensor digitorum communis is sometimes simply called extensor digitorum. The origin of extensor digitorum communis is by the common extensor tendon that is shared with extensor carpi radialis brevis, extensor digiti minimi (that we will not be studying), and extensor carpi ulnaris muscles. This muscle is the "Coach" of the mnemonic of the human mnemonic "Big Larry Bird Coach Understands" and the cat mnemonic "Big Larry Bird Coach Loves Umbrellas."

# Extensor Carpi Ulnaris

*Lateral view, left side*

*Lateral view, left side*

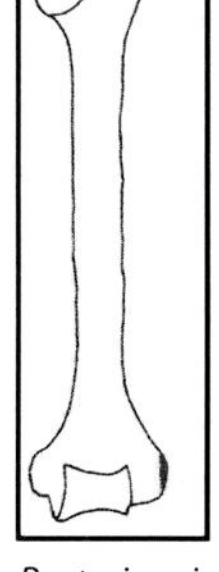

*Posterior view, right side*

*Anterior view, right side*

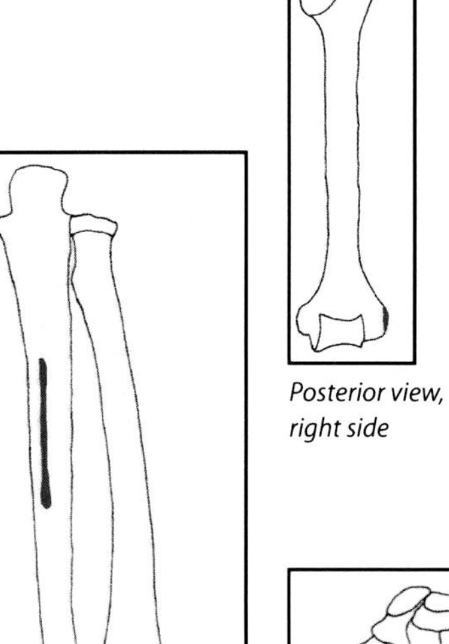

*Posterior view, right side*

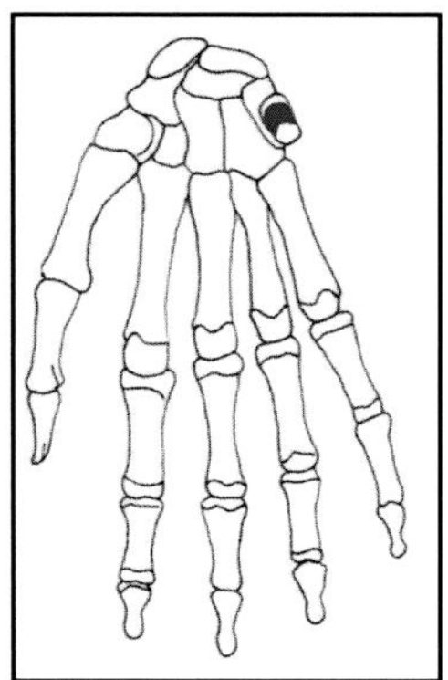

*Anterior view, right side*

### *Human Information:*

**origin:** lateral epicondyle of humerus at common extensor tendon; posterior border of ulna

**insertion:** posterior surface of base of fifth metacarpal

**nerve:** radial

**action:** extends and adducts of wrist (hand)

*These drawings of the origin and insertion might help you visualize this information (red is the origin, blue the insertion).*

## Extensor Carpi Ulnaris

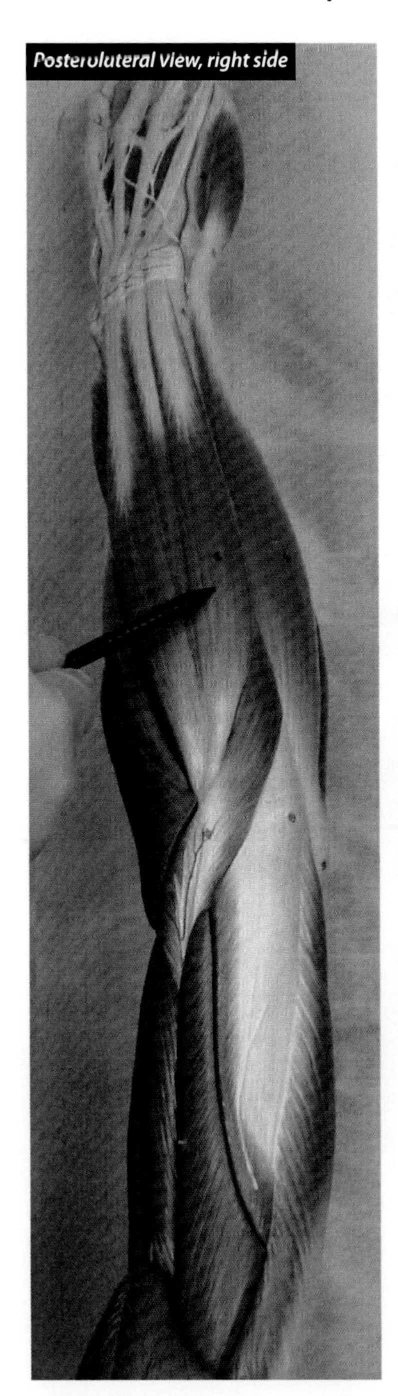


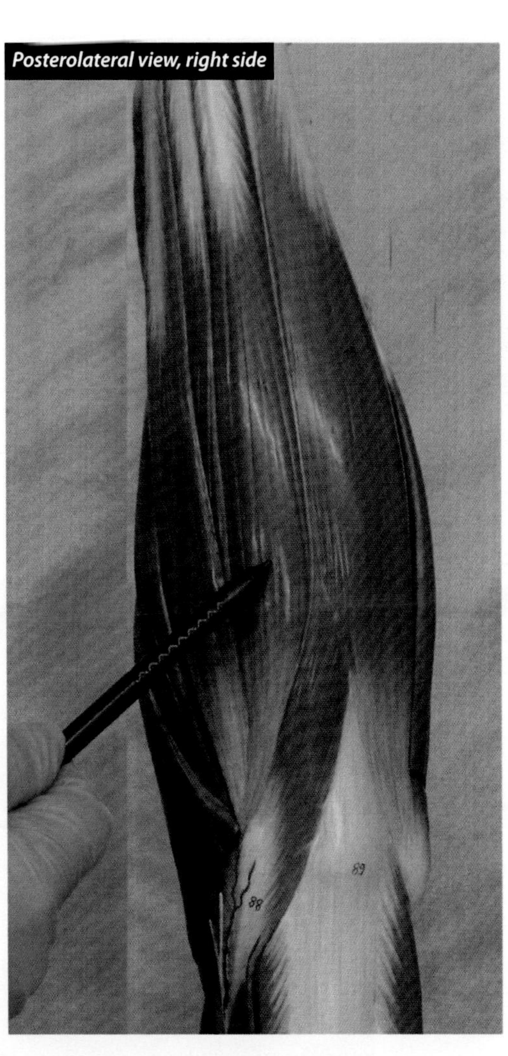


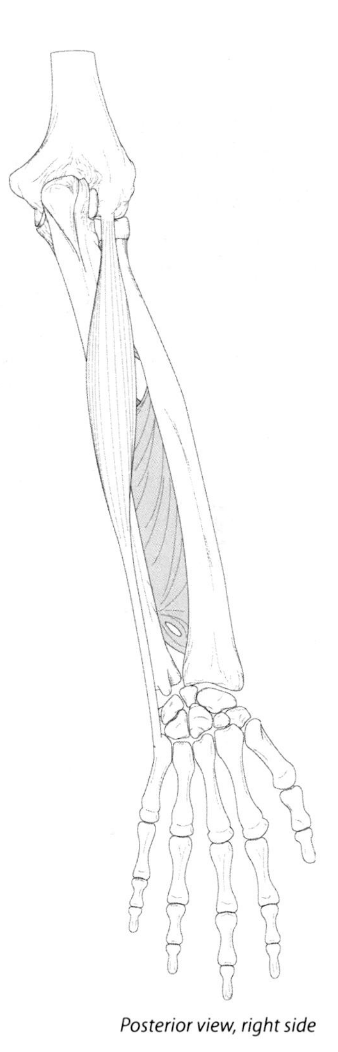

*Posterior view, right side*

The origin of extensor carpi ulnaris is by the common extensor tendon that is shared with extensor carpi radialis brevis, extensor digitorum communis, and extensor digiti minimi (which we will not be studying). This muscle is the "Understands" of the human mnemonic "Big Larry Bird Coach Understands" and the "Umbrellas" of the cat mnemonic "Big Larry Bird Coach Loves Umbrellas."

# Extensor Digitorum Lateralis (Cat)

*Lateral view, left side*

*Lateral view, left side*

### *Cat Information:*

**origin:** lateral supracondyloid ridge of humerus distal to the origin of the extensor digitorum communis

**insertion:** the three tendons on the ulnar side join the tendons of the extensor digitorum communis at their insertions

**nerve:** radial

**action:** extends four principal digits

Extensor digitorum lateralis is only found in the cat. This muscle is the "Loves" of the mnemonic "Big Larry Bird Coach Loves Umbrellas."

# POSTERIOR ("EXTENSOR") COMPARTMENT OVERVIEW

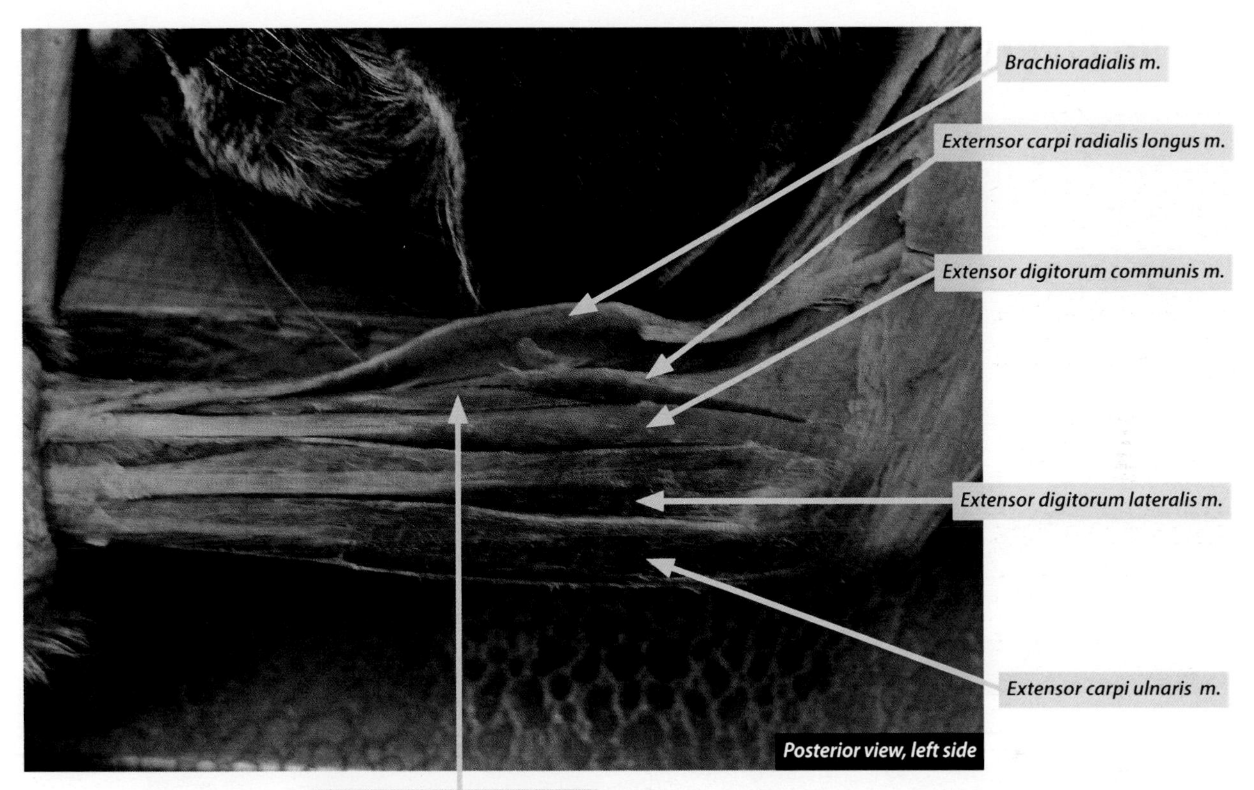

# VESSELS

## Axilla Overview: Arteries

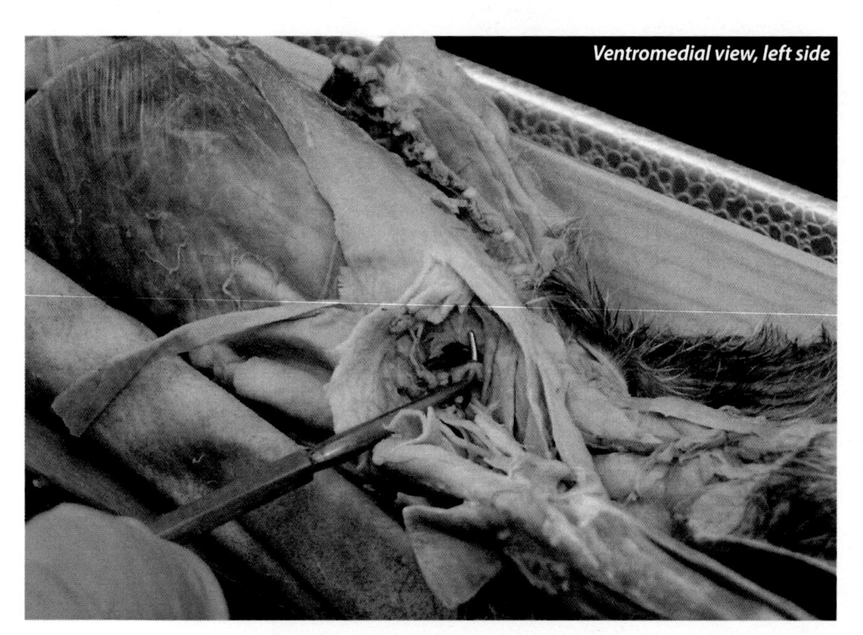


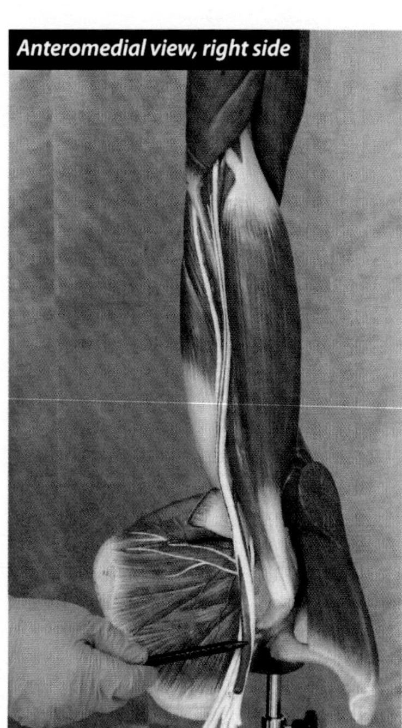


*Axillary Artery*

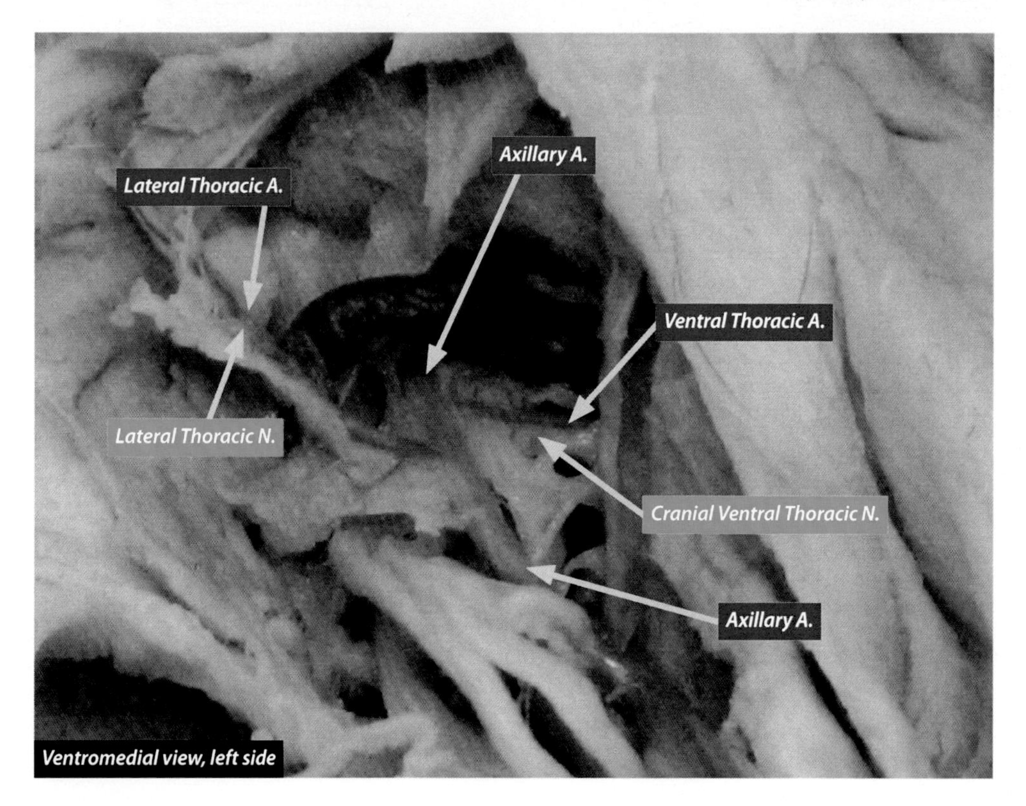

## Axillary Artery

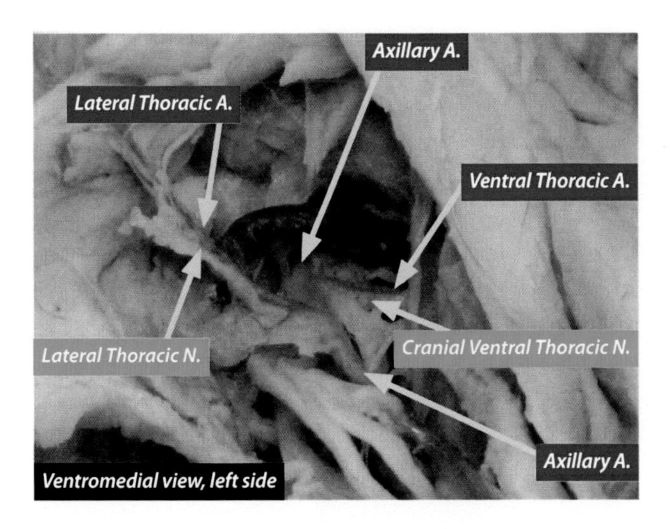


The **axillary artery** begins where the **subclavian artery** terminates and ends where the **brachial artery** begins. It has three branches that we will study in the cat: the **ventral thoracic artery**, the **lateral** (**long**) **thoracic artery**, and the **subscapular artery**. The beginning of the **subscapular artery** is often used to delineate the distal end of the **axillary artery**.

## Ventral Thoracic Artery

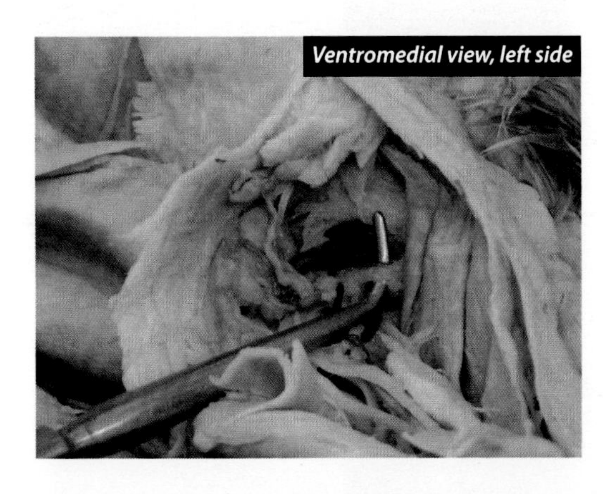


The **ventral thoracic artery** is the first lateral branch of the **axillary artery**. It serves the media ends of the latissimus dorsi and the pectoralis muscles. It runs with the **cranial ventral thoracic nerve**. In humans, this vessel is called the **superior thoracic artery**, and it originates from the anterior surface of the proximal third of the **axillary artery**. It serves the anterior and medial walls of the axilla.

## Lateral Thoracic Artery

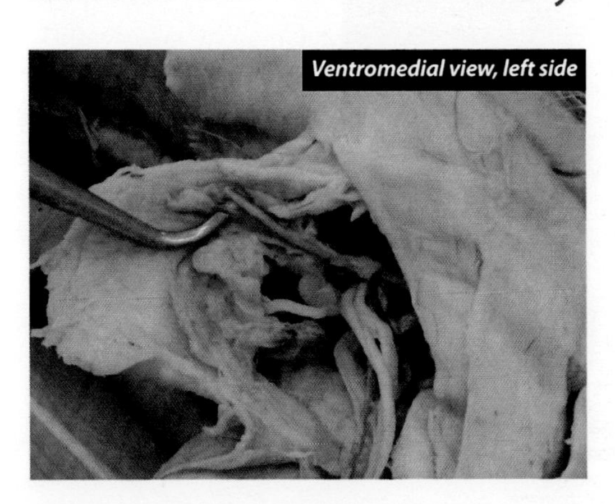


The **lateral thoracic** (or **long thoracic**) **artery** is a branch of the middle third of the **axillary artery**. It gets its name from the fact that it passes to the lateral thoracic wall, yet another Grant thing. In humans, it serves the anterior and medial walls of the axilla. In women, it also has branches that serve the mammary gland.

# Axilla Overview: Arteries (continued)

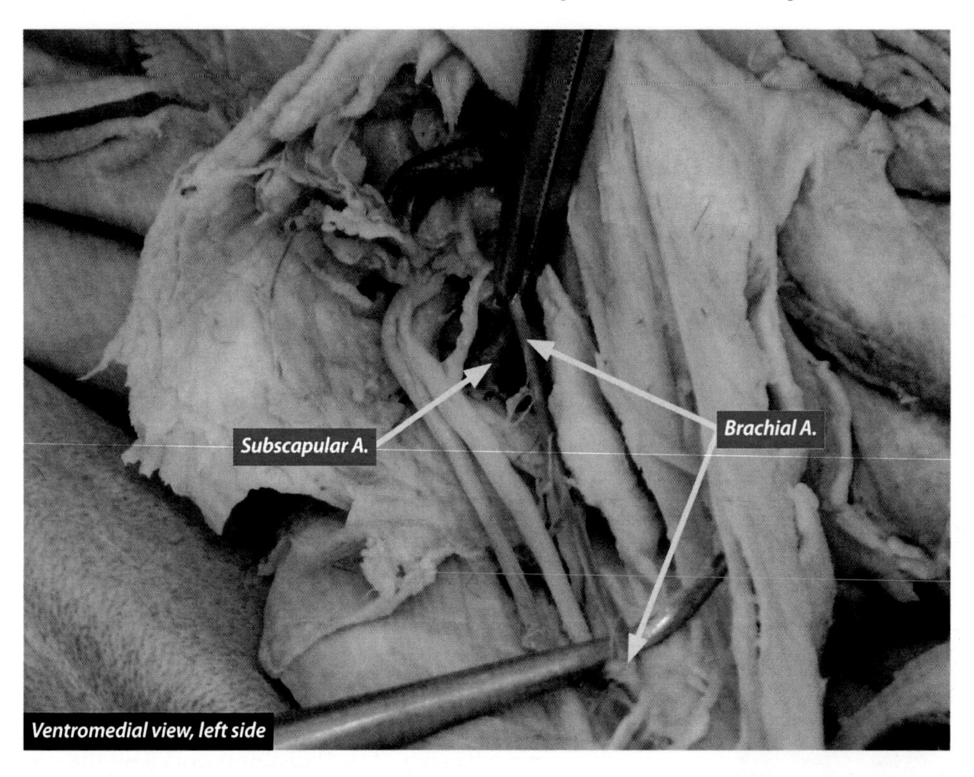


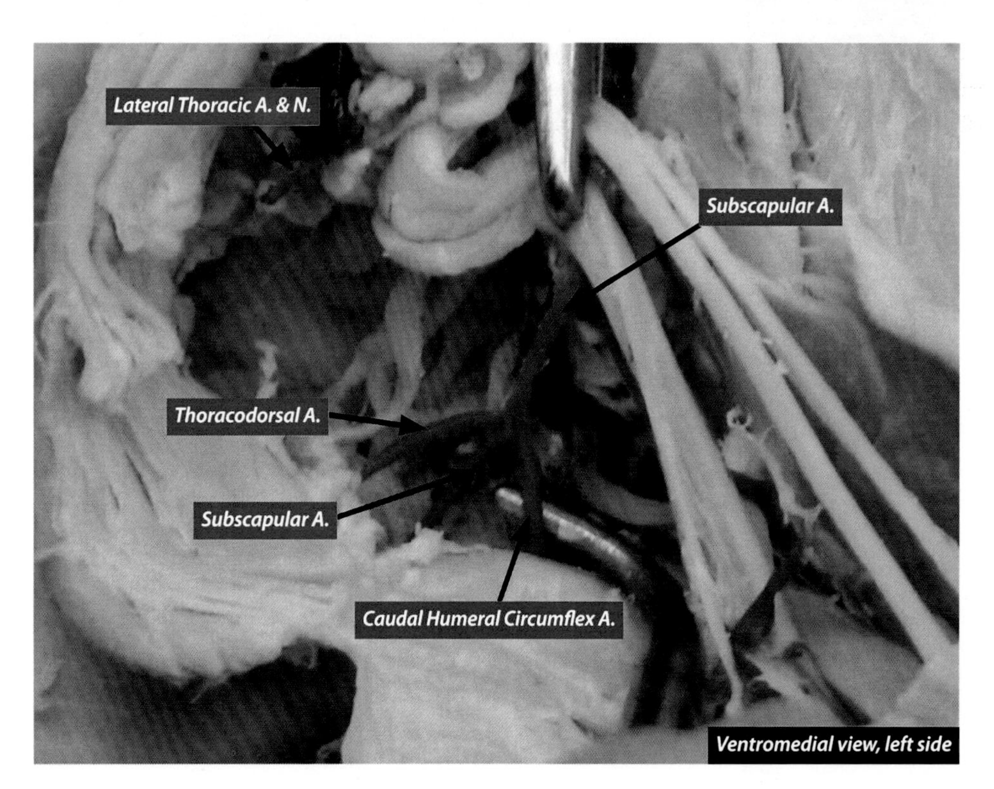

## Subscapular Artery

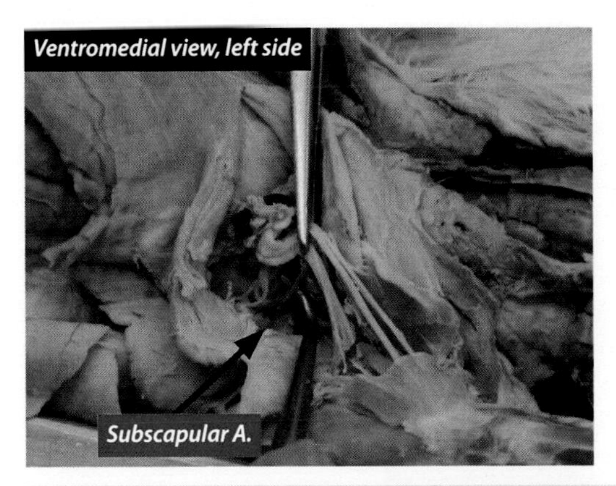


The subscapular a. is the handle and middle tine of the pitchfork.

The **axillary artery** begins at the end of the **subclavian artery** and ends with the beginning of the **brachial artery**. It has three branches that we will study in the cat: the **ventral thoracic artery**, the **lateral (long) thoracic artery**, and the **subscapular artery**. The beginning of the **subscapular artery** is often used to delineate the end of the **axillary artery**.

## Caudal Humeral Circumflex Artery

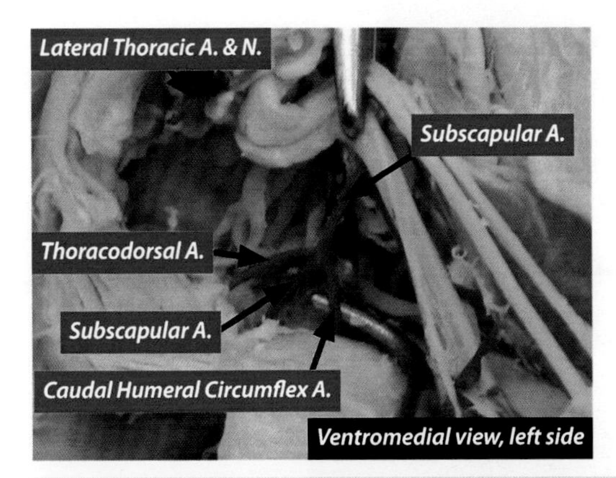


The caudal humeral circumflex a. is the cranial tine of the pitchfork.

The **caudal humeral circumflex artery** is a branch of the **subscapular artery**. It gets its name from the fact that it passes around the caudal side of the humerus. It forms collateral circulation with the **cranial humeral circumflex artery** in the lateral region of the arm. In the cat, it serves the lateral and long heads of the triceps, acromiodeltoid, and spinodeltoid. In humans, it serves the deltoid muscle and the shoulder joint.

## Thoracodorsal Artery

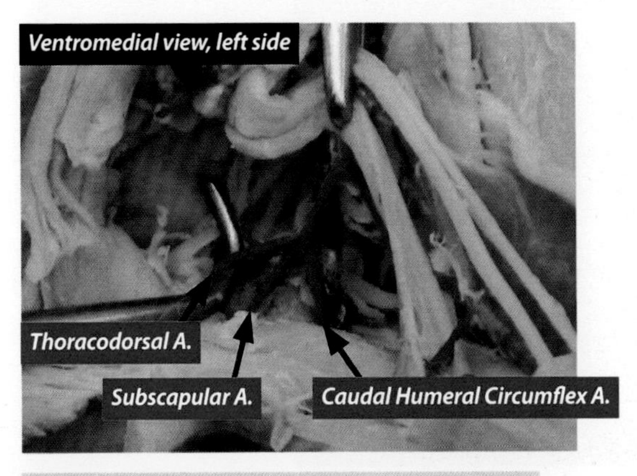


The thoracodorsal a. is the caudal tine of the pitchfork.

The **thoracodorsal (dorsal thoracic) artery** is a branch of the **subscapular artery**. We saw this artery back in laboratory exercise 2, where we found it on the deep side of the humeral end of the latissimus dorsi with the nerve of the same name. In the cat, it serves the teres major, the epitrochlearis, and the latissimus dorsi. In humans, it serves the medial and posterior walls of the axilla.

# *Arm Overview: Arteries*

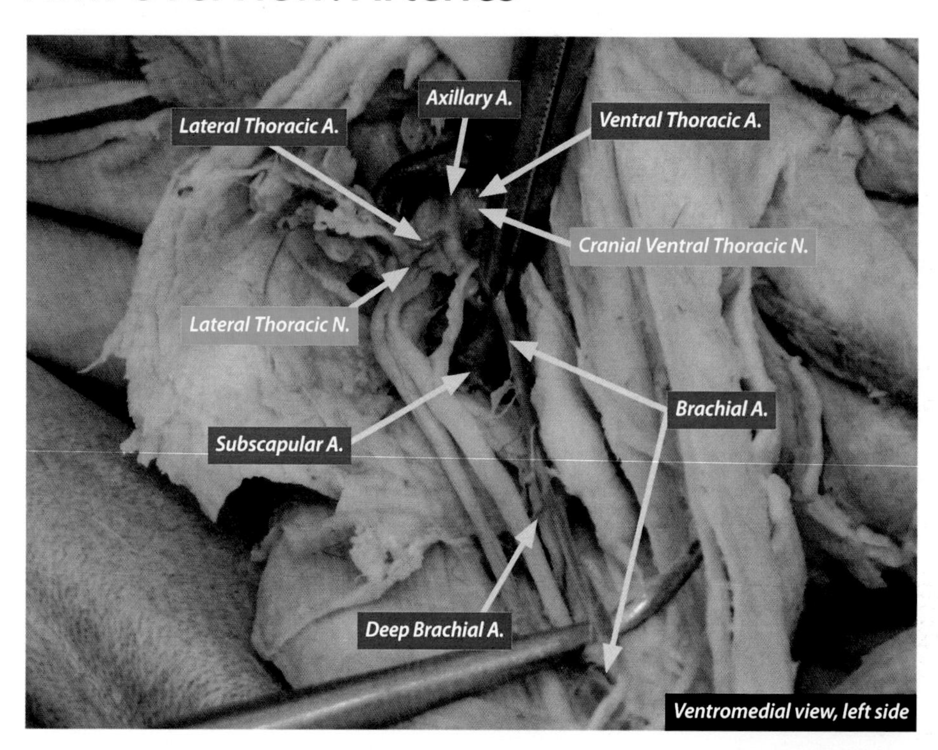


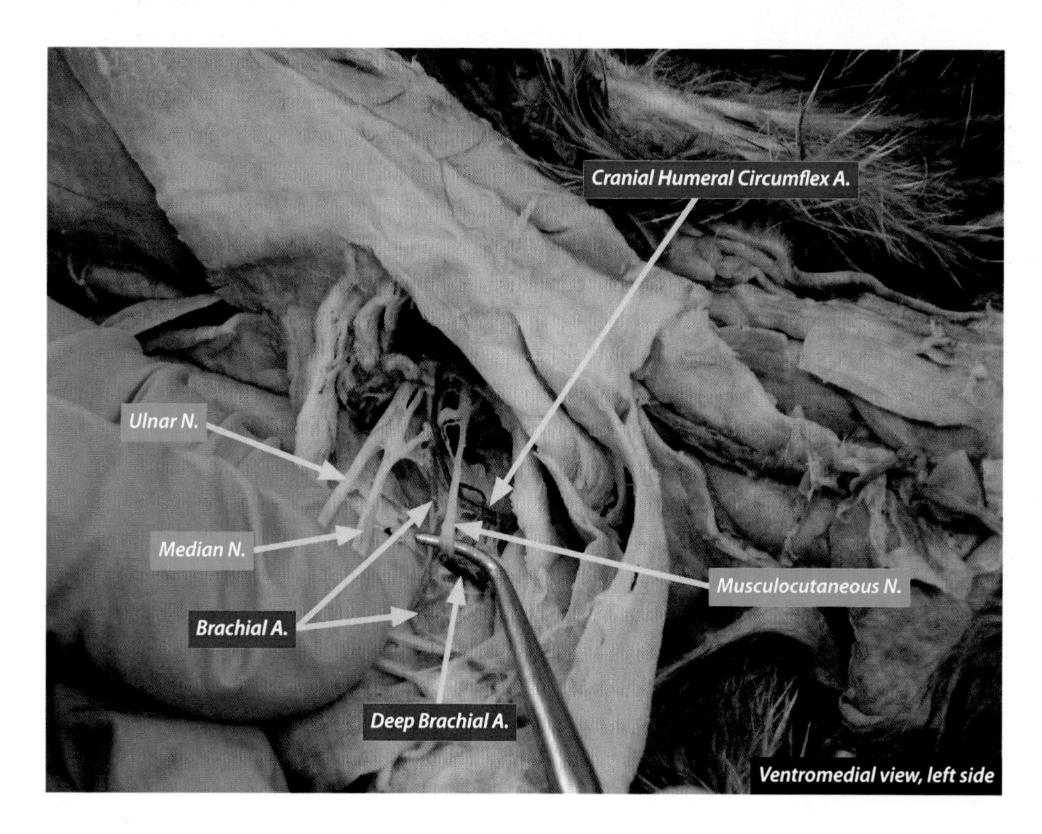

## Brachial Artery

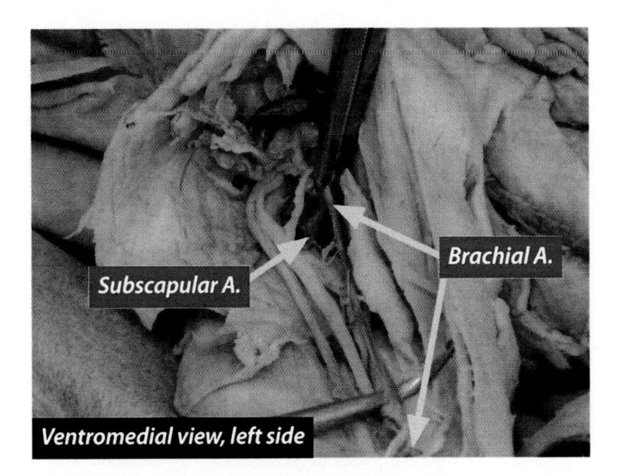


The **brachial artery** begins at the distal end of the **axillary artery** where the **subscapular artery** branches from the **axillary artery**. The **brachial artery** ends near the neck of the radius where it branches into the **ulnar** and **radial arteries**. It provides the main arterial blood supply to the arm, and it has two branches that we will study in the cat: the **cranial humeral circumflex artery** and the **deep brachial artery**. It runs with the **median nerve** in the arm, and in the cat both of these structures run through the supracondyloid foramen of the humerus. Approximately 1 percent of humans have a vestige of the supracondyloid foramen, which can cause various median nerve pathologies. It normally passes deep to the bicipital aponeurosis in the cubital fossa (so it is not at risk when blood is drawn there). A brachial pulse can be taken where the **brachial artery** passes through the medial bicipital groove.

## Cranial Humeral Circumflex Artery

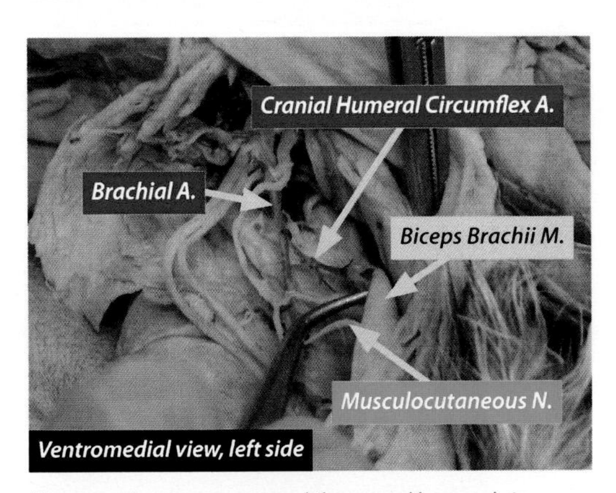


*The musculocutaneous n. and the cranial humeral circumflex a. will pass into the belly of the biceps and will move with it if you lift this muscle gently. The brachial and deep brachial arteries wilL not. The deep brachial a. is obscured by the probe in this picture.*

The **cranial humeral circumflex artery** is usually a branch of the **brachial artery**, although I have seen it branch from the **subscapular artery**, the **caudal humeral circumflex artery**, and the **thoracodorsal artery**. One can tell that it is the **cranial humeral circumflex** by where it goes—the belly of the biceps brachii muscle—rather than where it comes from. It gets its name from the fact that it passes around the cranial side of the humerus. It forms collateral circulation with the **caudal humeral circumflex artery** in the lateral region of the arm. In the cat, it serves the biceps brachii, as well as the head of the humerus. In the human, it branches from the lateral side of the distal third of the **axillary artery** and serves the head of the humerus and the glenohumeral (shoulder) joint.

## Deep Brachial Artery

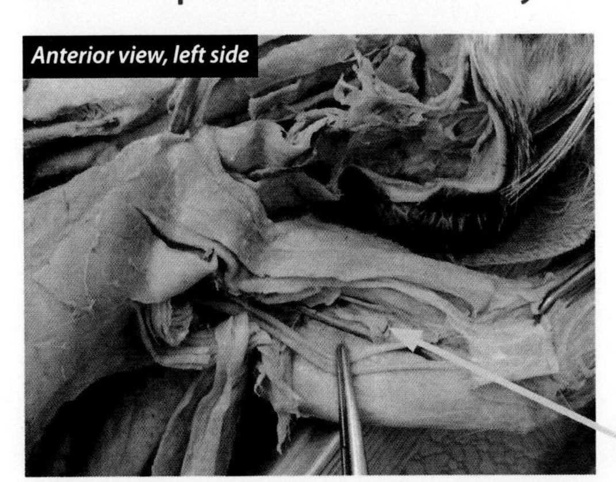


The **deep brachial artery** (the **deep artery of the arm** or the **profunda brachial artery**) is a small branch, or, more often, two or three small branches of the **brachial artery**. It runs deep into the arm with the **radial nerve**. In humans, it serves the posterior compartment of the arm and forms anastomoses with the **posterior humeral circumflex artery**. In the cat, it serves the triceps brachii, the epitrochlearis, and the latissimus dorsi.

# *Forearm: Arteries*

## Radial Artery

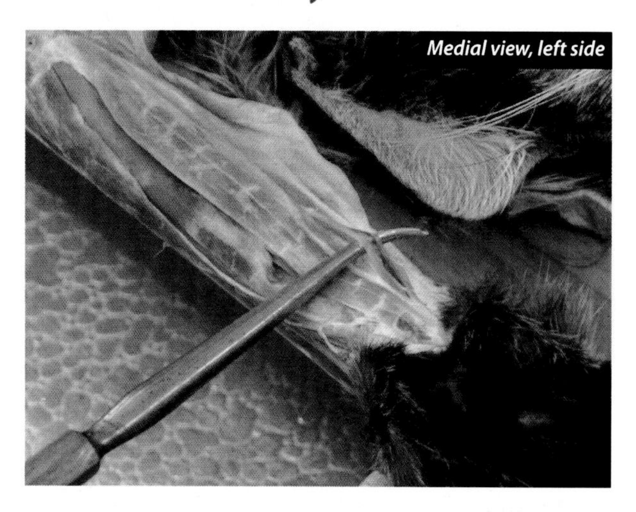

The **radial artery** is a branch of the **brachial artery**. Its origin is in the forearm, and it runs with the **median nerve** on the lateral side of the forearm (which is the side closest to the cat's ear in the left-hand picture). It is usually found deep and medial to flexor carpi radialis. Dr. J often refers to it as a "Grant, Grant, Lincoln thing" because we find the radius bone, the **radial artery**, and the **median nerve** together. This vessel is of particular importance because it is where we detect one's radial pulse.

## Ulnar Artery

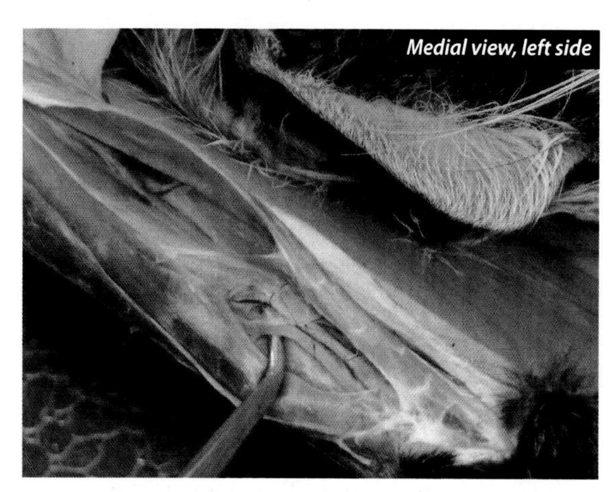


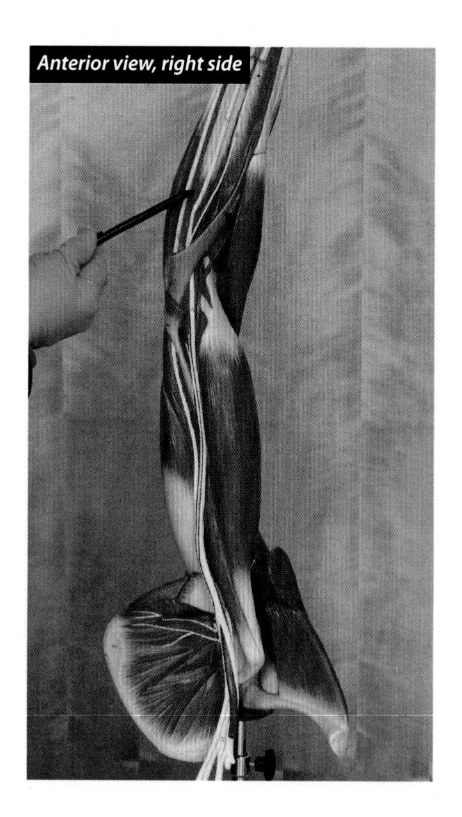


The **ulnar artery** is a branch of the **brachial artery**. Its origin is in the forearm, and it runs with the **ulnar nerve** on the medial side of the forearm. It is usually found deep and lateral to the flexor carpi ulnaris muscle. Dr. J often refers to it as a "Grant, Grant, Grant thing" because we find the ulna bone, the **ulnar artery**, and the **ulnar nerve** together.

# VESSELS

## Cephalic Vein

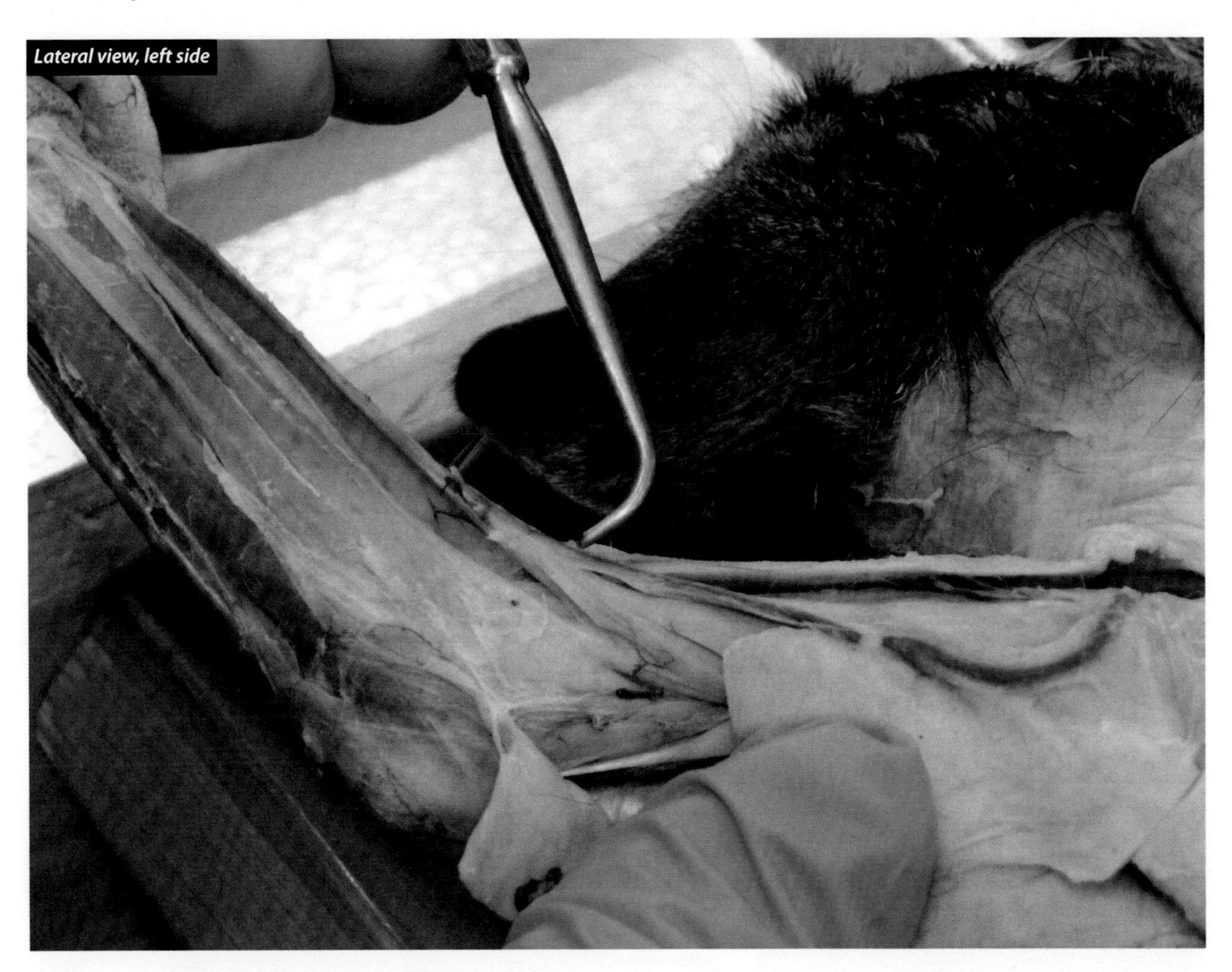


The **cephalic vein** is observed on the superficial, lateral side of the forearm and arm. We saw it for the first time way back in the second laboratory exercise. Inferior to the clavicle, it passes into the axilla where it joins the **axillary vein**. The **axillary vein** passes through the thoracic wall to become the **subclavian vein**. In the forearm, it runs with the **radial nerve** on the surface of the brachioradialis muscle.

# NERVES

## Cranial Ventral Thoracic Nerve

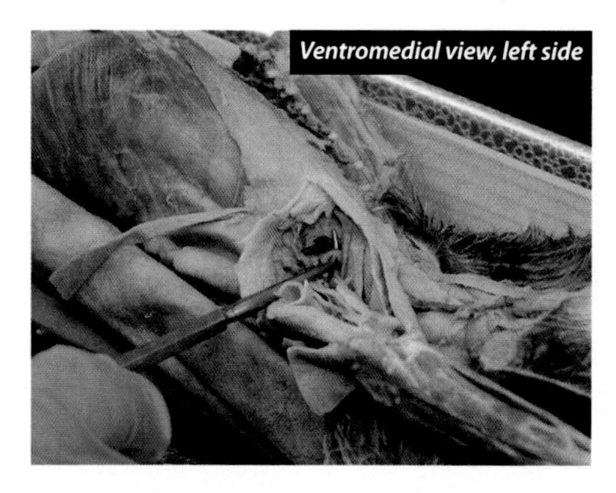


The **cranial ventral thoracic nerve** runs with the **ventral thoracic artery** to the ventral thoracic wall. Here is another example of a Grant thing.

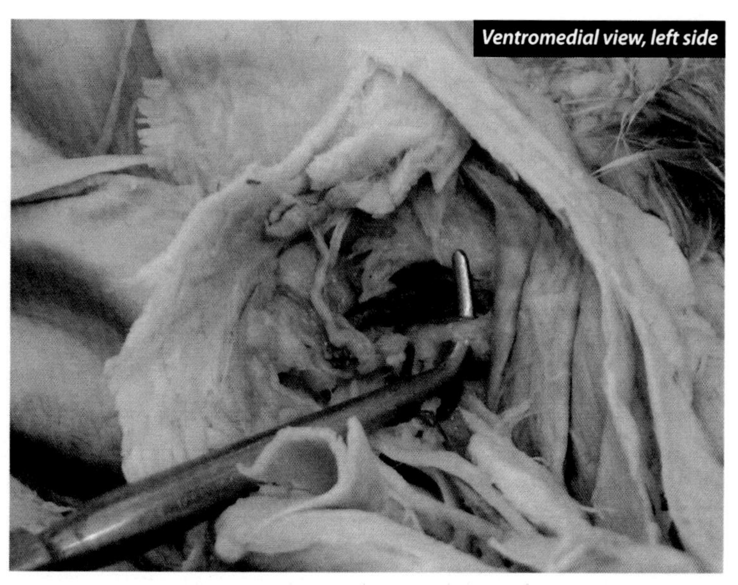


## Lateral Thoracic Nerve

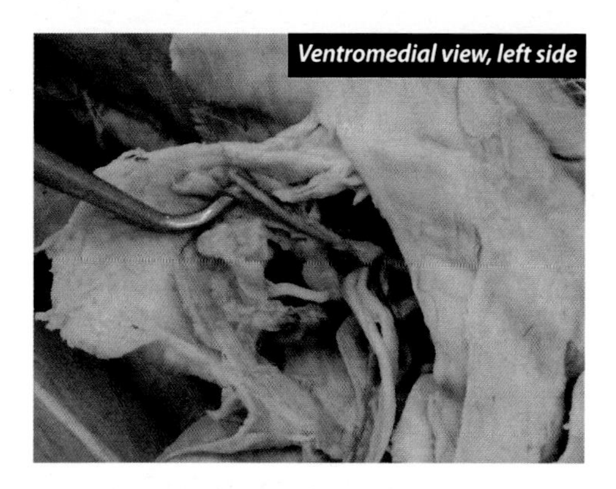


The **lateral thoracic nerve** runs with the **lateral thoracic artery** to the lateral thoracic wall. That fact makes it a Grant, Grant, Grant thing.

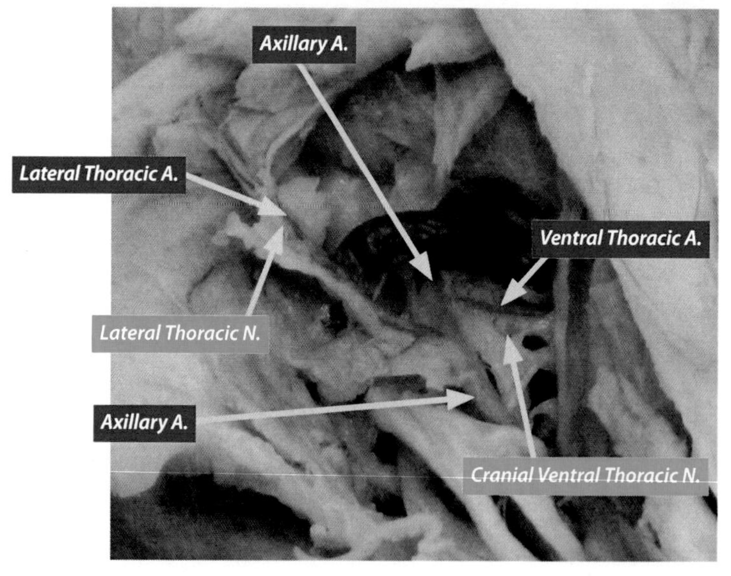

## Medial Cord

The **medial cord** is formed from the **anterior division** of the **lower trunk** of the **brachial plexus**. It contains neurons from the **anterior rami** of **C8** and **T1**, and it gives rise to the **ulnar nerve**, as well as to part of the **median nerve**. It gets its name from its position relative to the middle third of the **axillary artery**. Notice that it forms the top, left side (your left) of the "M" as you look at it (the cranial end from the cat's anatomical position).

## Lateral Cord

The **lateral cord** is formed from the **anterior divisions** of the **upper** and **middle trunks** of the **brachial plexus**. It contains neurons from the **anterior rami** of **C5**, **C6**, and **C7**. It gives rise to the **musculocutaneous nerve**, as well as to part of the **median nerve**. It gets its name from its position relative to the middle third of the **axillary artery**. Notice that it forms the top right side of the "M" as you look at it (the cranial end from the cat's anatomical position).

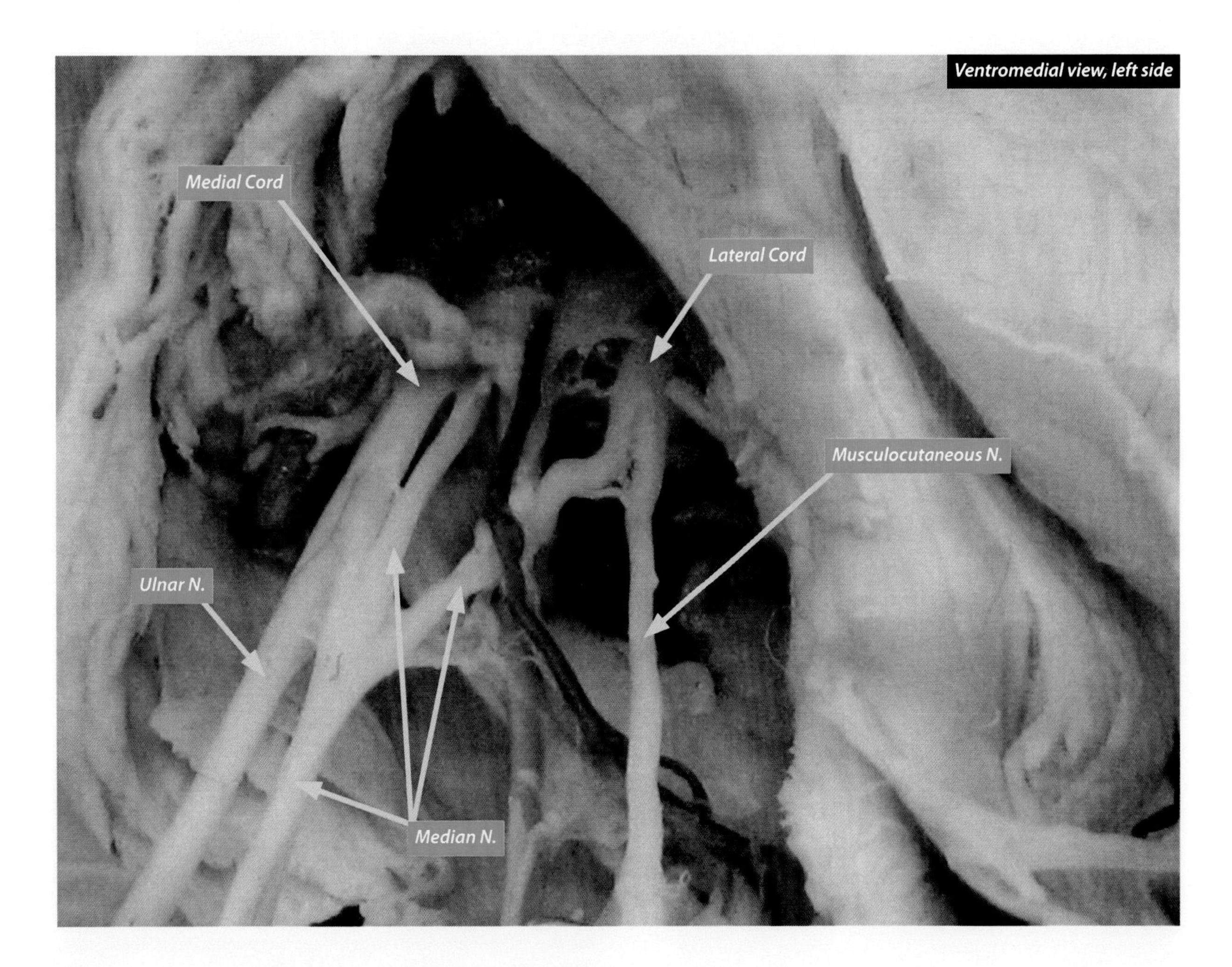

# VENTRAL NERVES

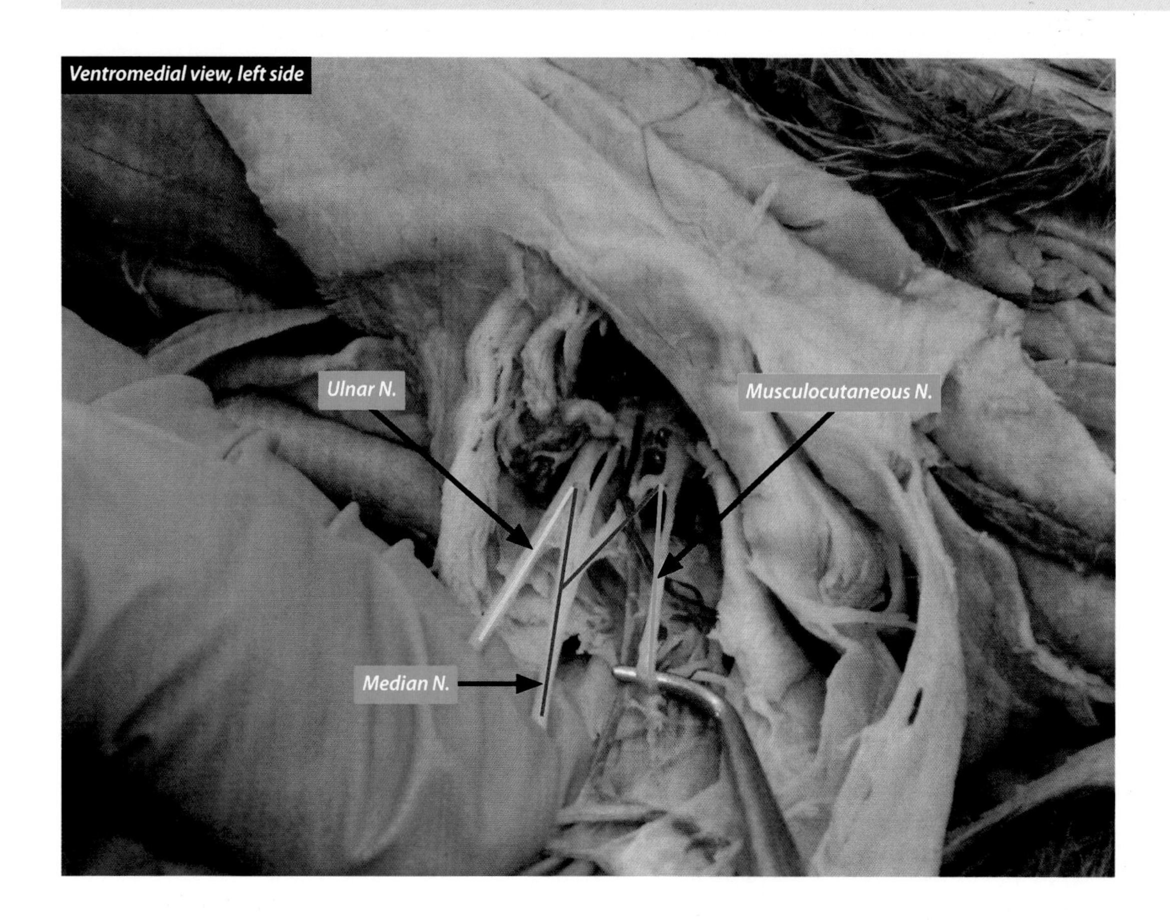


## Ulnar Nerve

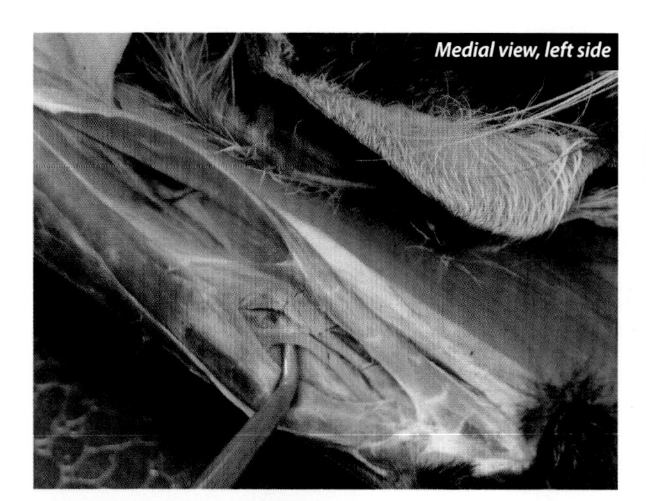


*Ulnar nerve and **artery** in the forearm.*

## Median Nerve

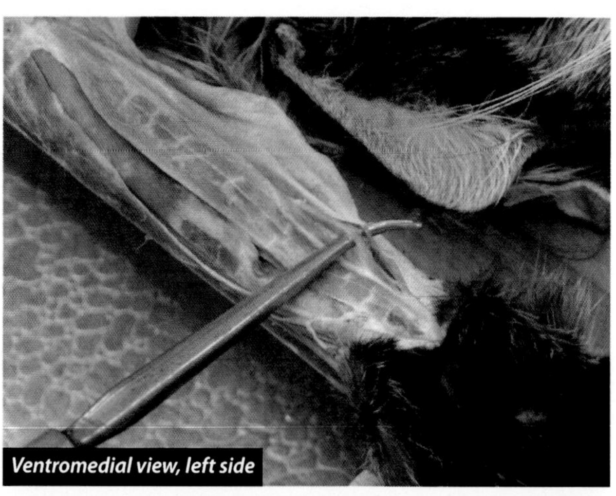


*The median nerve runs with the **radial artery** in the forearm of the cat; the radial nerve runs with the **cephalic vein**. This is not a Grant thing.*

## Ulnar Nerve

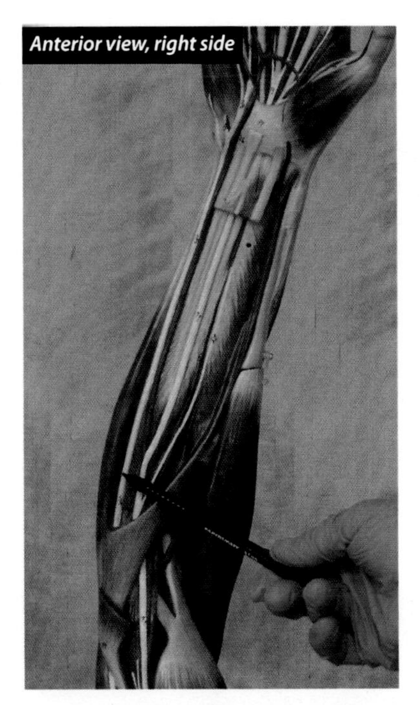


As with the **radial nerve**, in Lab 8 we will observe the **ulnar nerve** in two areas of the pectoral appendage. The **medial cord** gives rise to the **ulnar nerve** in the axilla. It contains nerves from the **anterior rami** of **C8** and **T1** that come directly from the **anterior divisions** of the **brachial plexus**, as well as the **anterior ramus** of **C7** from communication with the **median nerve**. We find it first running by itself on the medial surface of the arm. In the forearm, it joins the **ulnar artery** and runs along the ulna bone. That would make it a "Grant, Grant, Grant thing!" LOMG! It passes into the hand on the anterior medial portion of the wrist just deep to the skin and superficial to the flexor retinaculum. Because of that, it is often damaged when the wrist is cut. Although the **ulnar nerve** does not innervate any muscles in the arm, it innervates 1.5 muscles in the forearm (flexor carpi ulnaris and half of flexor digitorum profundus) and fifteen out of twenty muscles of the hand (except most of those controlling the thumb). It also receives sensory input from the posterior and anterior medial cutaneous area of the hand, the little finger, and the medial half of the ring finger. Damage to the **ulnar nerve** results in the clinical condition called "claw hand."

## Median Nerve

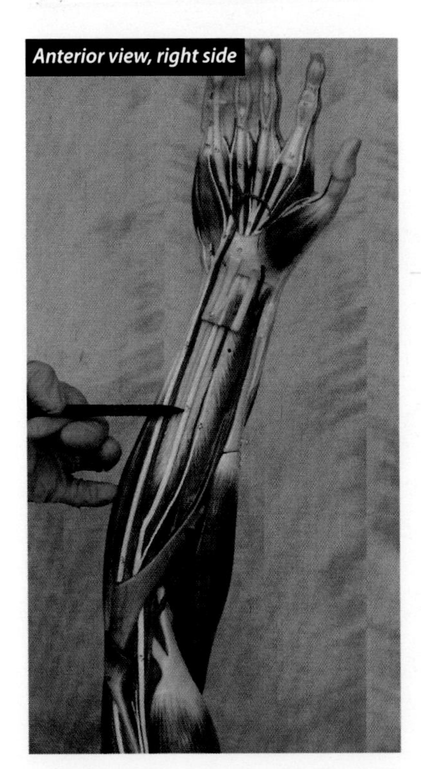


The **median nerve** runs with the **brachial artery** in the arm. It passes through the supracondyloid foramen of the cat with the **brachial artery**, and then in the forearm it runs with the **radial artery**. It also passes through the carpal tunnel into the hand. The **median nerve** is formed from branches of the **medial** and **lateral cords**. It contains nerves from the **anterior rami** of **C6**, **C7**, **C8**, and **T1** that come directly from the **anterior divisions** of the **brachial plexus**, as well as the **anterior ramus** of **C5** from communication with the **musculocutaneous nerve**. It serves the anterior compartment of the forearm by controlling the two pronator muscles and all the flexor muscles except 1.5. It also controls most muscles of the thumb. The **median nerve** also receives sensory input from the anterolateral cutaneous area of the hand, the lateral half of the ring finger, and the middle and index fingers. Because it runs deep to the flexor retinaculum, inflammation deep to the flexor retinaculum results in carpal tunnel syndrome. When damaged, it results in the clinical condition called "ape hand."

## Musculocutaneous Nerve

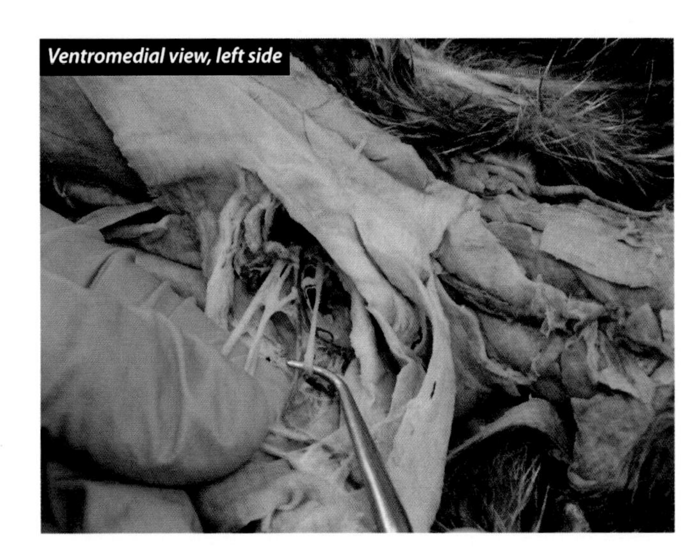


The **musculocutaneous nerve** is formed from the **lateral cord**. It contains nerves from the **anterior rami** of **C5**, **C6**, and **C7**. It serves the anterior compartment of the arm, and it also receives sensory input from the lateral cutaneous area of the forearm.

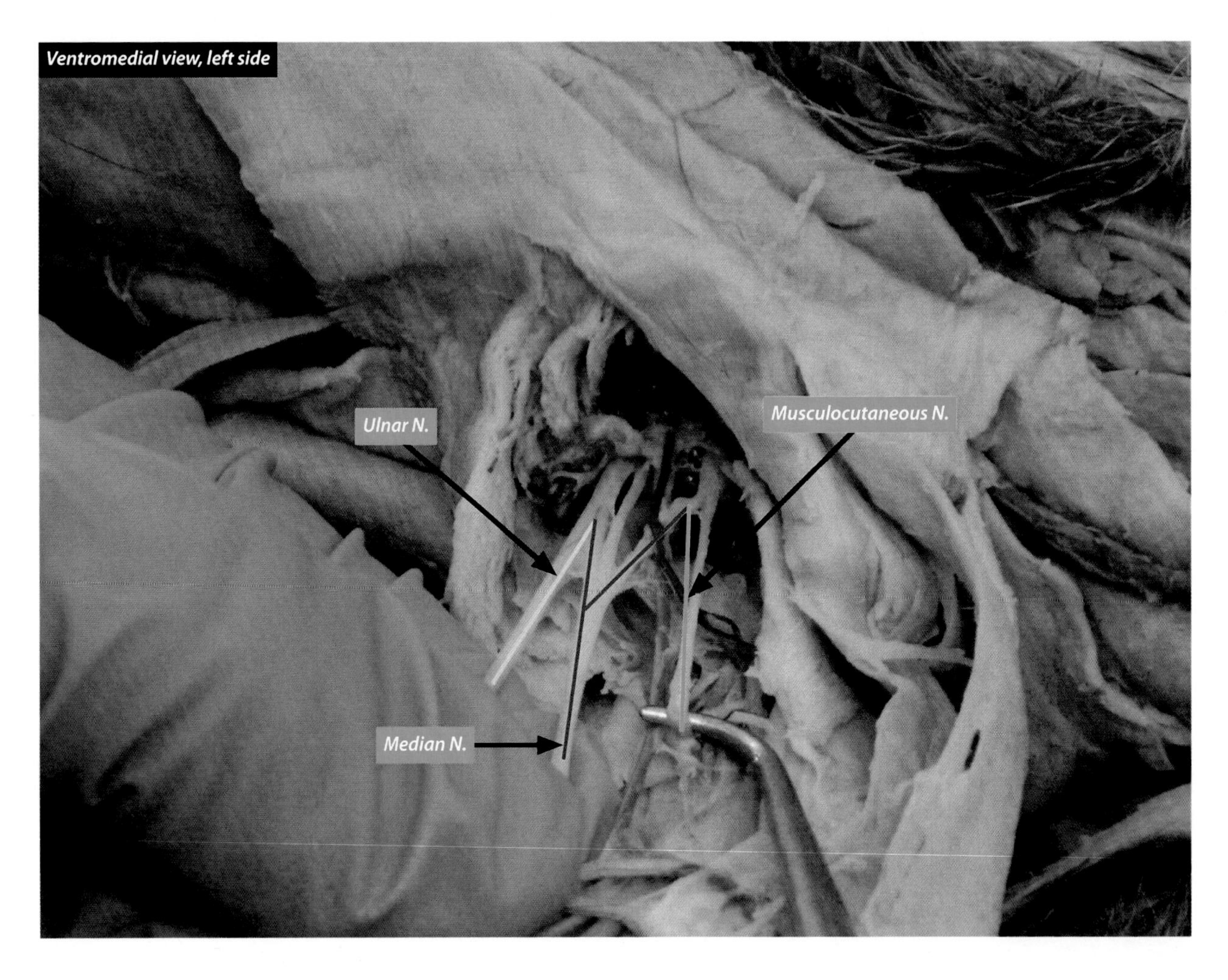

# DORSAL NERVES

## Posterior Cord

The **posterior cord** is formed from the **posterior divisions** of the **upper**, **middle**, and **lower trunks** of the **brachial plexus**. It contains neurons from **anterior rami** of **C5**, **C6**, **C7**, **C8**, and **T1**. It gives rise to the **radial** and **axillary nerves** and gets its name from its position relative to the middle third of the **axillary artery**. This nerve is deep in the axillary sheath and is difficult to expose without doing massive amounts of cat terrorism.

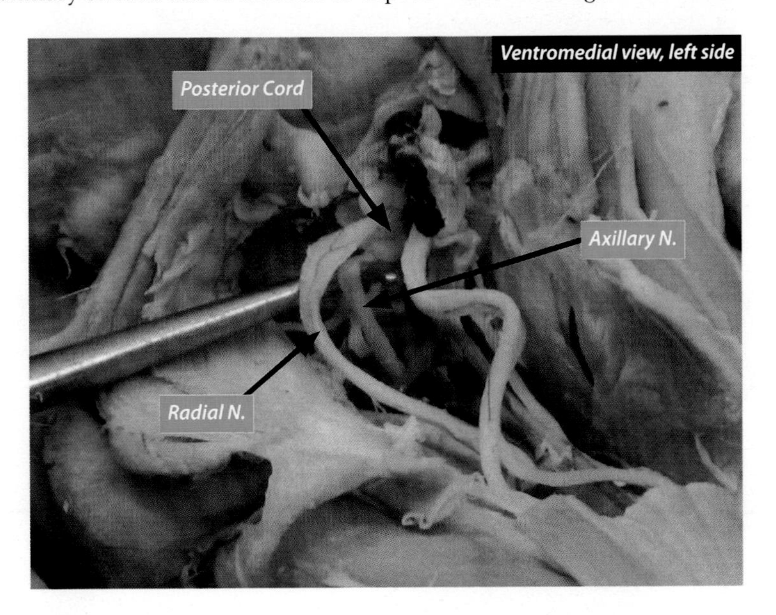


## Axillary Nerve

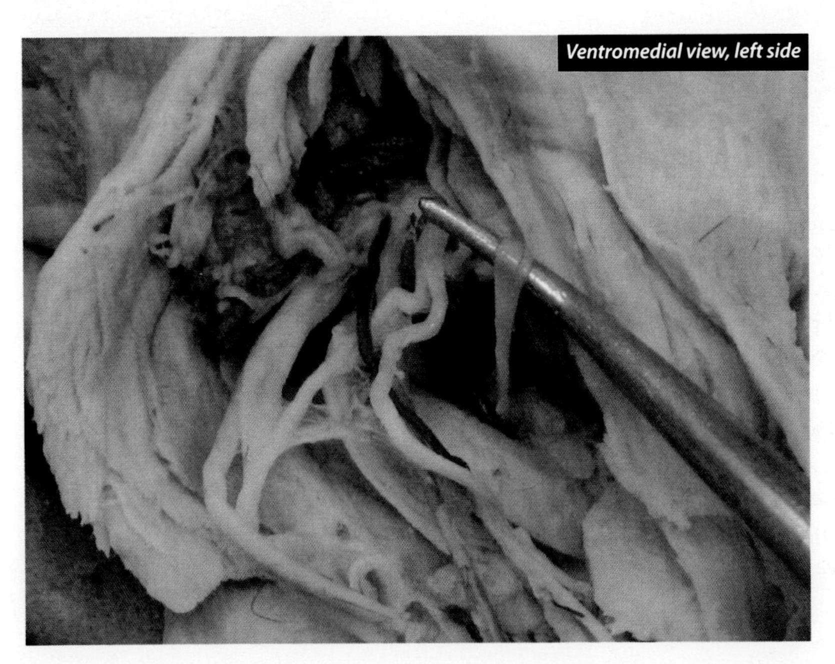


The **axillary nerve** is a branch of the **posterior cord** of the **brachial plexus** and it contains neurons from the **anterior rami** of **C5** and **C6**. It can be seen in the axilla adjacent to the **caudal humeral circumflex artery**. It runs between the subscapularis and teres major muscles and then reappears on the lateral surface of the arm, where it serves the deltoids (deltoid in humans) and teres minor. It also has cutaneous branches to the shoulder joint and to the skin that covers the distal half of the deltoid muscle in humans.

## Radial Nerve

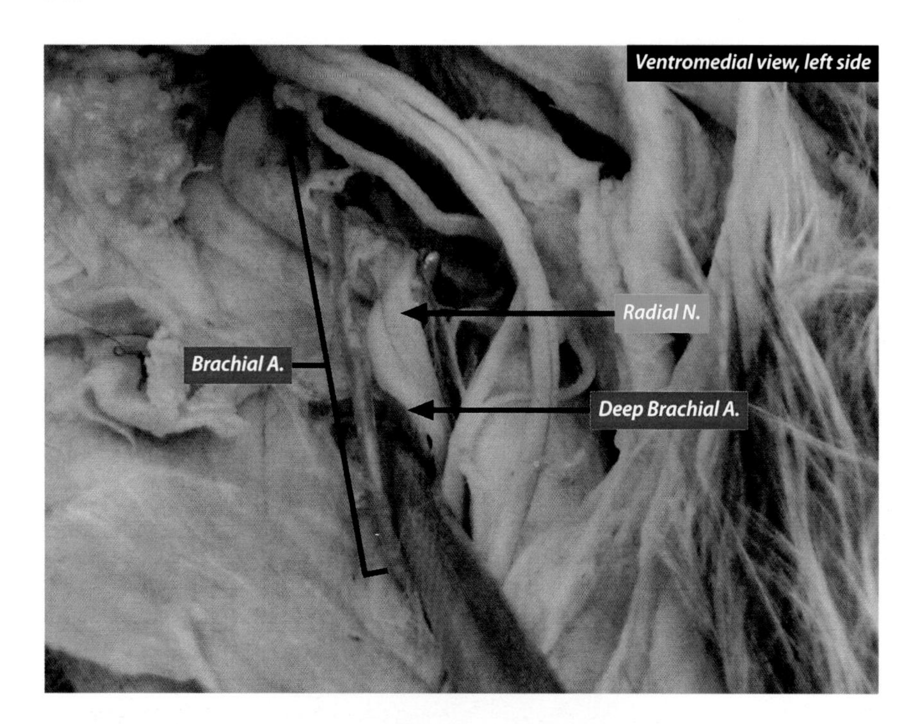


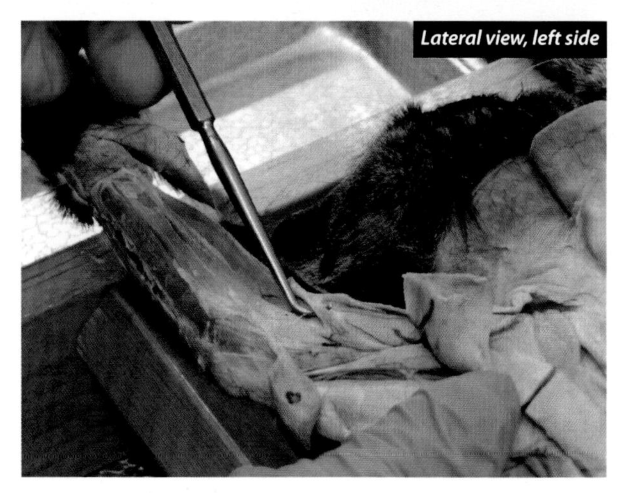


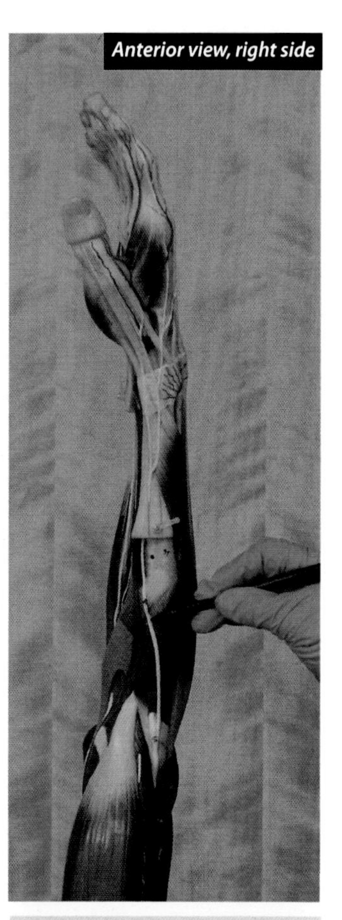


*In this picture of the human arm model, we find the* ***radial nerve*** *deep to brachioradialis in the posterior compartment of the forearm.*

With our work in Lab 8, we will observe the **radial nerve** in two areas of the pectoral appendage. First, we will find it on the medial side of the arm. Here, Dr. J's students fondly refer to it as the "**biggest hugest nervus in the armus**." It is a branch of the **posterior cord** containing nerves from the **anterior rami** of **C5**, **C6**, **C7**, **C8**, and **T1**. It serves the posterior compartment of the arm. The second place we will find it is where it has become the **superficial branch** of the **radial nerve** as it passes subcutaneously over the surface of the brachioradialis with the **cephalic vein** toward the wrist. The **radial nerve** serves the posterior compartment of the forearm, as well as the brachioradialis muscle. Therefore, it controls all the extensor muscles found in the arm and forearm, the supinator muscle, and brachioradialis. It also receives sensory input from the posterolateral cutaneous area of the hand. When damaged, it results in the clinical condition called "wrist drop."

## Thoracodorsal Nerve

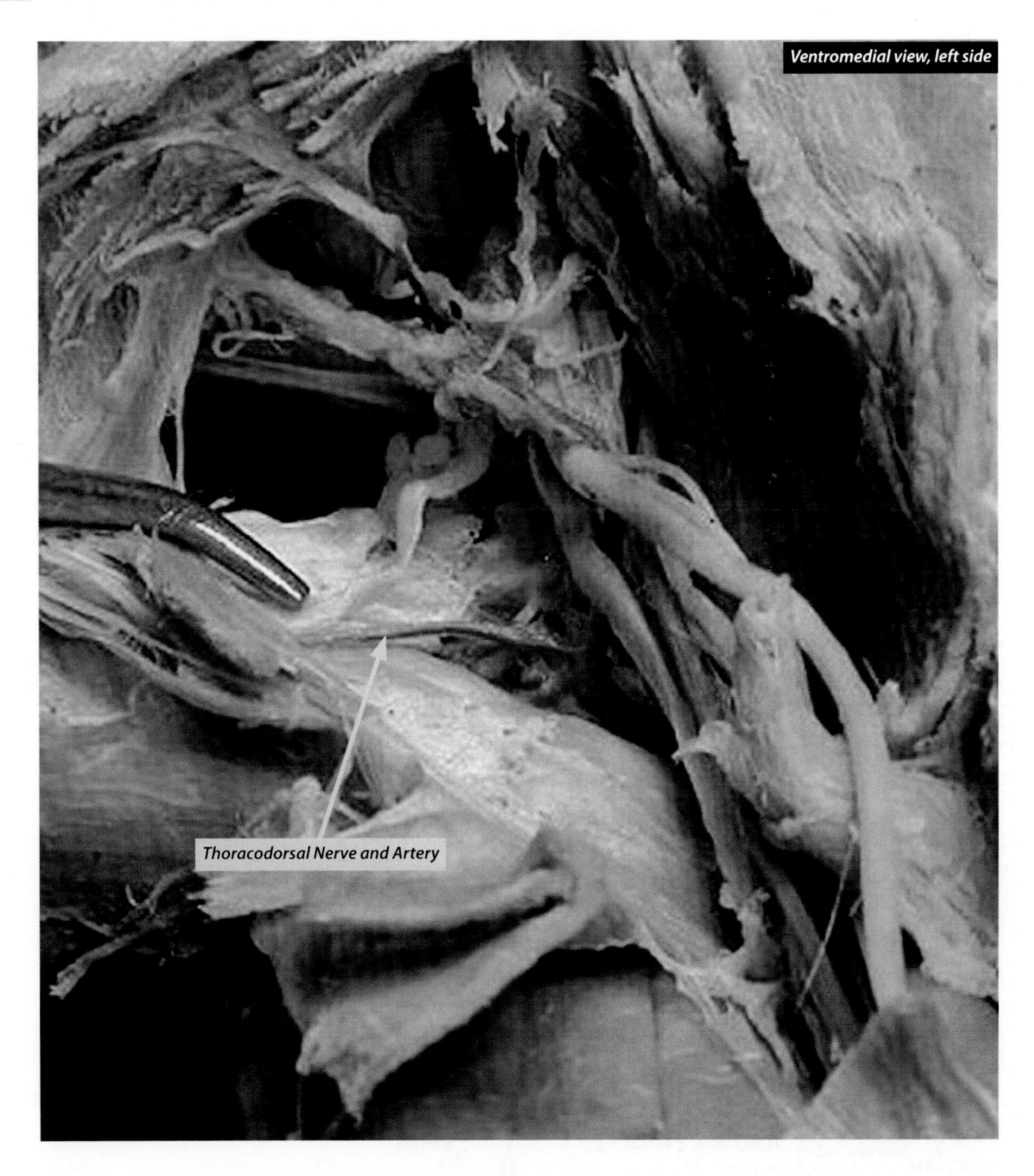


The **thoracodorsal nerve** is also called the **third subscapular nerve**. It serves the latissimus dorsi muscle. We will find it on the deep side of latissimus dorsi at its humeral end, running with the **thoracodorsal artery**. There are actually three **subscapular nerves**. The **upper** and **lower subscapular nerves** both innervate the subscapularis muscle. The **lower subscapular nerve** also innervates the teres major muscle. All of these **subscapular nerves** are lateral branches of the **posterior cord** of the **brachial plexus**.

# AREA DIAGRAMS

## Brachial Plexus, Anterior

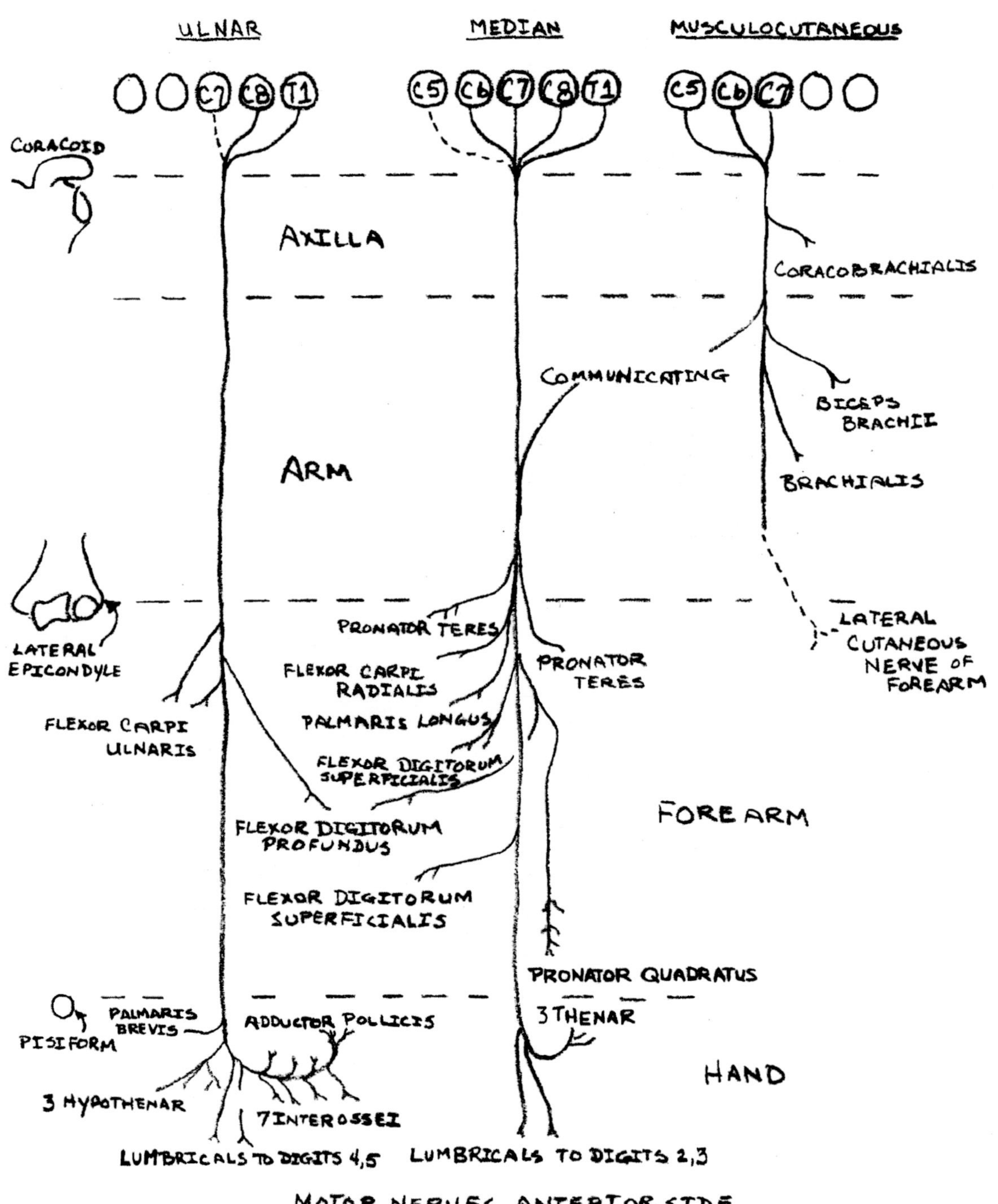

# AREA DIAGRAMS

## Brachial Plexus, Posterior

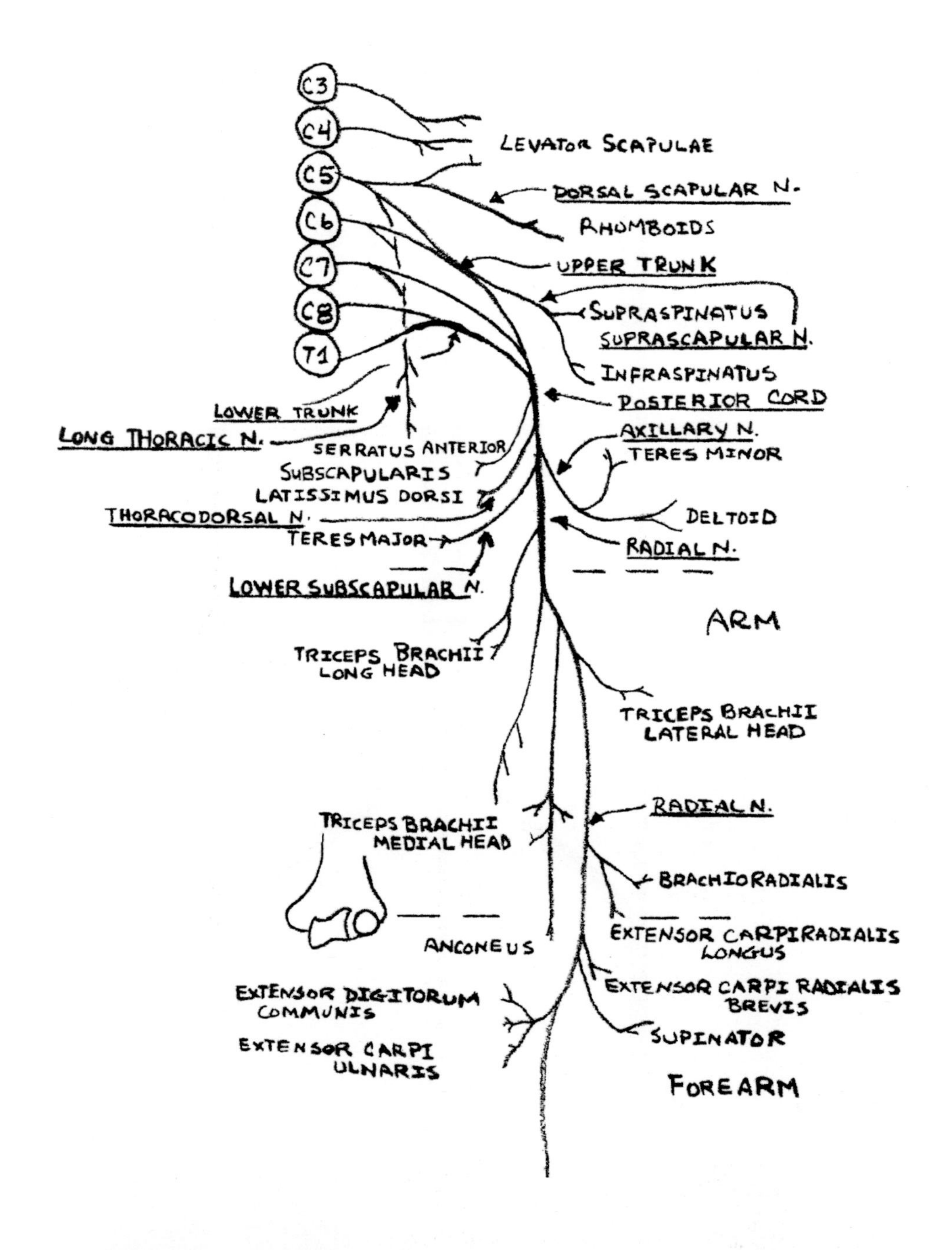

## Brachial Plexus—Label this diagram

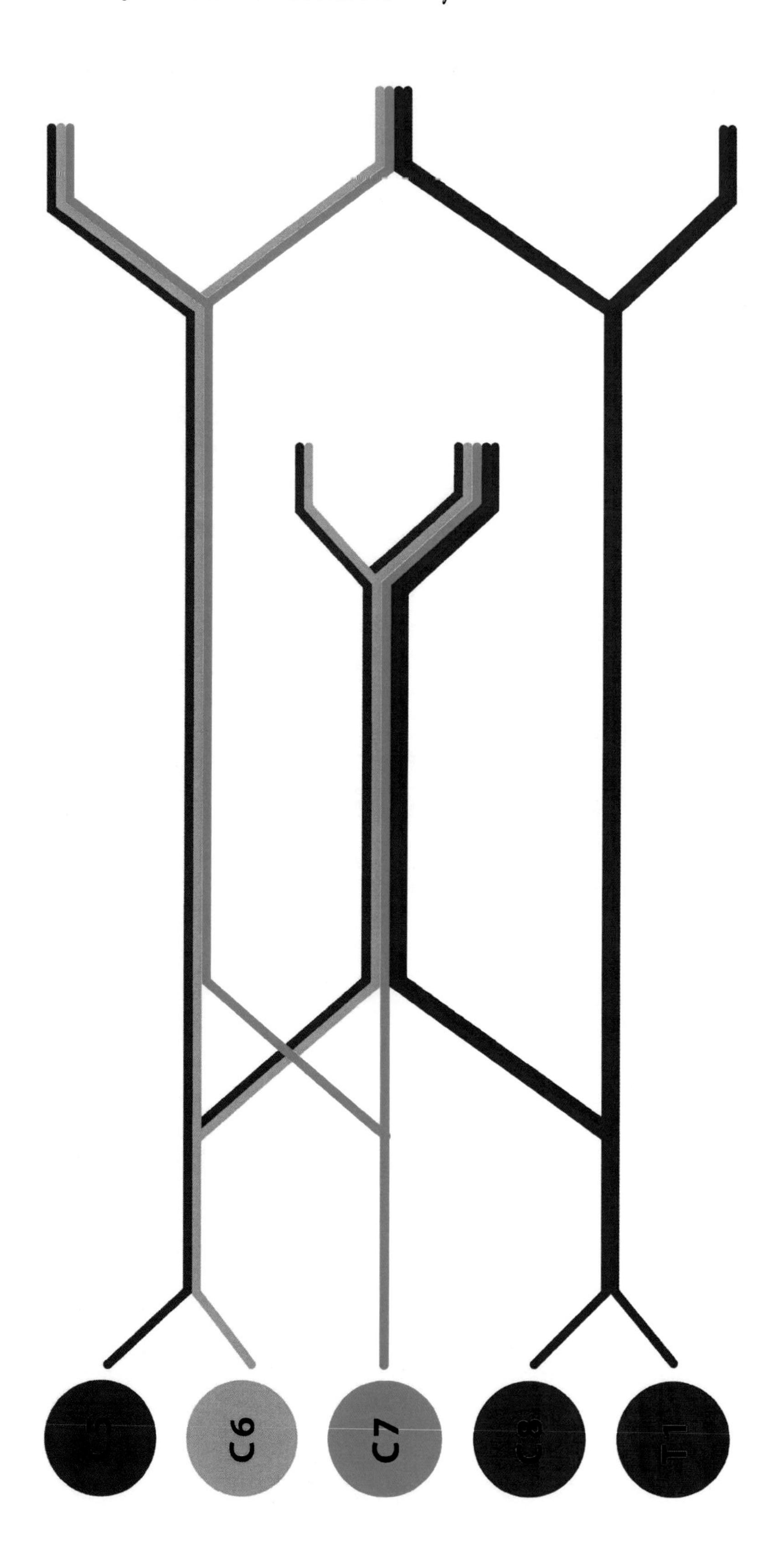

# TISSUE

## Carpal Ligament (Flexor Retinaculum)

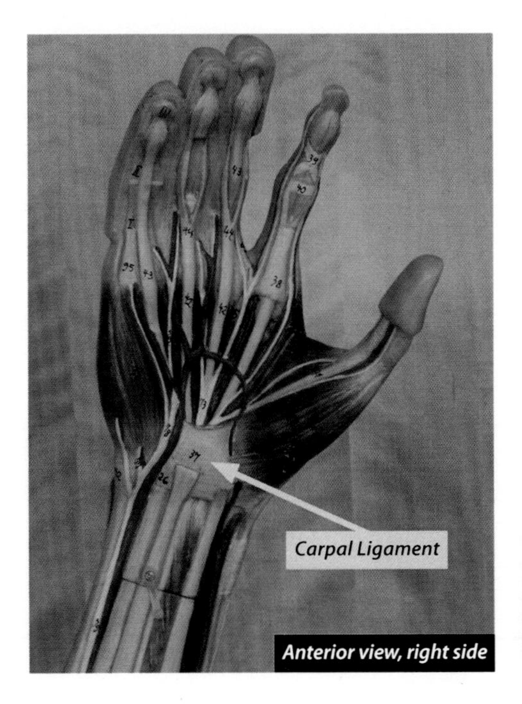


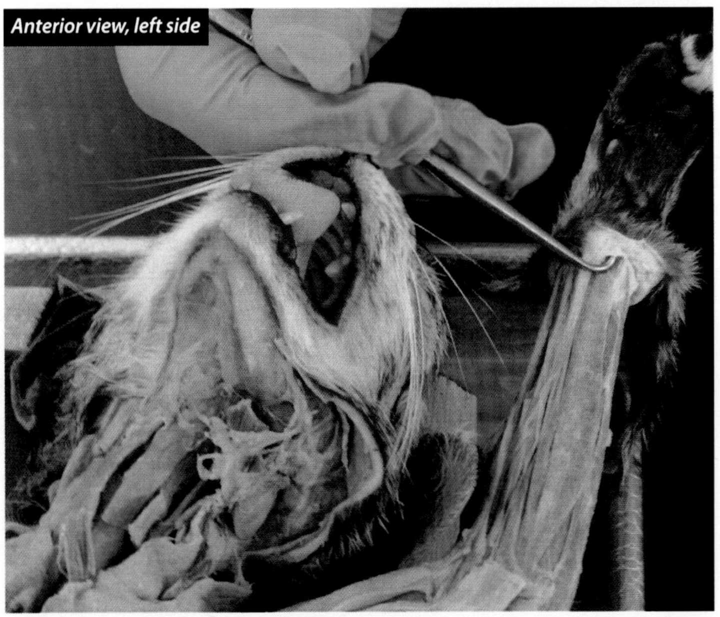


The flexor retinaculum of the wrist is a thickening of the deep fascia (ligament) that wraps across the anterior side of the wrist. It forms the roof of a tunnel anterior to the carpal bones called the carpal tunnel, which passes superficially to all of the flexor tendons except that of palmaris longus. Medially, it attaches to the hook of the hamate and the pisiform bone, and laterally it attaches to the trapezium and the tubercle of the scaphoid. Functionally, it is important because it prevents the tendons from "bowing." The **median nerve** is the only nerve that passes through the carpal tunnel. The impingement of this nerve may lead to a condition called "carpal tunnel syndrome."

# MUSCLE CHARTS

## Muscles of the Scapula

**Stabilize the Scapula—All insert on the scapula**

| Muscle/Nerve | Origin | Action |
|---|---|---|
| **Trapezius**—<br>Spinal accessory nerve (XI) | Ligamentum nuchae and C7–T12 | Elevates, retracts (ADducts), rotates, and depresses the scapula. |
| **Levator Scapulae (Ventralis)**—<br>Dorsal scapular nerve and the anterior rami of C3 and C4 (lateral branches of cervical spinal nerves 3–5) | C1–C4 | Elevates the scapula, flexes neck laterally (lateral side bending), and rotates the scapula. |
| **Rhomboids**—<br>Dorsal scapular nerve | C7–T5 | Retracts (ADducts), elevates, and rotates the scapula |
| **Pectoralis Minor**—<br>Medial pectoral nerve (lateral branches of cervical spinal nerves C7, C8, and T1) | Ribs 3–5 | Depresses, protracts (ABducts), and rotates the scapula. |
| **Serratus Anterior (Ventralis)**—<br>Long thoracic nerve | Ribs 1–9 | Protracts (ABducts) and rotates the scapula. |

## Muscles of the Arm

**"Rotator Cuff" (SITS) – All insert on the humerus and originate on the scapula**

| Muscle/Nerve | Origin | Action |
|---|---|---|
| **Supraspinatus**—<br>Suprascapular nerve | Supraspinous fossa of scapula | ABducts the arm (humerus) |
| **Infraspinatus**—<br>Suprascapular nerve | Infraspinous fossa of scapula | Laterally rotates the arm (humerus) |
| **Teres Minor**—<br>Axillary nerve | Lateral (axillary) border of scapula (superior 2/3 ) | |
| **Subscapularis**—<br>Upper and Lower subscapular nerves (C5 and C6) | Subscapular fossa of scapula | Medially rotates the arm (humerus) |

# MUSCLE CHARTS

**Move the arm—All insert on the humerus**

| Muscle/Nerve | Origin | Action |
|---|---|---|
| **Deltoid**—<br>Axillary nerve | Lateral 1/3 of the clavicle, acromion, and spine of the scapula | Prime flexor and ABdctor of the arm (humerus). Extends and hyperextends the arm (humerus); laterally and medially rotates the arm (humerus). |
| **Pectoralis Major**—<br>Medial and Lateral Pectoral nerves (C5–T1) | Sternum, costal cartilage of ribs 1–6 (or 7), clavicle, and aponeurosis of the external oblique muscle | ADducts, flexes, and medially rotates the arm (humerus), sternal portion extends the arm (humerus). |
| **Latissimus Dorsi** —<br>Thoracodorsal nerve | Iliac crest, T7–T12 , ribs 10–12, and the lumbar fascia | Extends, hyperextends, ADducts, and medially rotates the arm (humerus). |
| **Teres Major**—<br>Lower Subscapular nerve | Lateral (axillary) border of scapula (inferior 1/3) | Extends, ADducts, and medially rotates the arm (humerus). |
| **Coracobrachialis**—<br>Musculocutaneous nerve | Coracoid process of the scapula | Flexes and ADducts the arm (humerus); stabilizes the shoulder joint. |

## Muscles of the Forearm and Hand

| Muscle/Nerve | Origin and Insertion | Action |
|---|---|---|
| **PRONATORS** | | |
| **Pronator Teres**—<br>Median nerve | **Origin**: medial epicondyle of the humerus and coronoid process of the ulna.<br>**Insertion**: lateral aspect of the radius at its midpoint | Pronates (medially rotates) the wrist (hand); assists elbow flexion. |
| **Pronator Quadratus**—<br>Median nerve | **Origin**: distal ¼ of the ulna.<br>**Insertion**: distal ¼ of the radius. | Pronates (medially rotates) the wrist (hand). |
| **SUPINATORS** | | |
| **Supinator**—<br>Radial nerve | **Origin**: lateral epicondyle of the humerus and posterio-lateral proximal ulna.<br>**Insertion**: anterior proximal radius. | Supinates (laterally rotates) the wrist (hand). |
| **Biceps Brachii**—<br>Musculocutaneous nerve | **Origin**: supraglenoid tubercle and coracoid process of the scapula.<br>**Insertion**: radial tuberosity of the radius and bicipital aponeurosis. | Flexes the forearm and supinates (laterally rotates) forearm and wrist (hand); short head flexes the arm (humerus). |

# MUSCLE CHARTS

## Muscles of the Forearm and Hand (continued)

| Muscle/Nerve | Origin and Insertion | Action |
|---|---|---|
| **FLEXORS** | | |
| **Biceps Brachii—** Musculocutaneous nerve | **Origin**: Supraglenoid tubercle and coracoid process of the scapula **Insertion**: radial tuberosity of radius and bicipital aponeurosis | Flexes the forearm and supinates (laterally rotates) forearm and wrist (hand); short head flexes the arm (humerus). |
| **Brachialis—** Musculocutaneous nerve | **Origin**: Anterior surface of the humerus **Insertion**: ulna | Flexes the forearm. |
| **Brachioradialis—** Radial nerve | **Origin**: Supracondylar ridge of the humerus **Insertion**: radius | Flexes the forearm (stabilizes forearm for small, quick, movements—"hammering muscle"). |
| **EXTENSORS** | | |
| **Triceps Brachii—** Radial nerve | **Origin**: Infraglenoid tubercle and shaft of the humerus **Insertion**: ulna | Extends the forearm; long head extends and adducts arm (humerus). |
| **Anconeus—** Radial nerve | **Origin**: Lateral epicondyle of the humerus **Insertion**: ulna | Extends the forearm. |

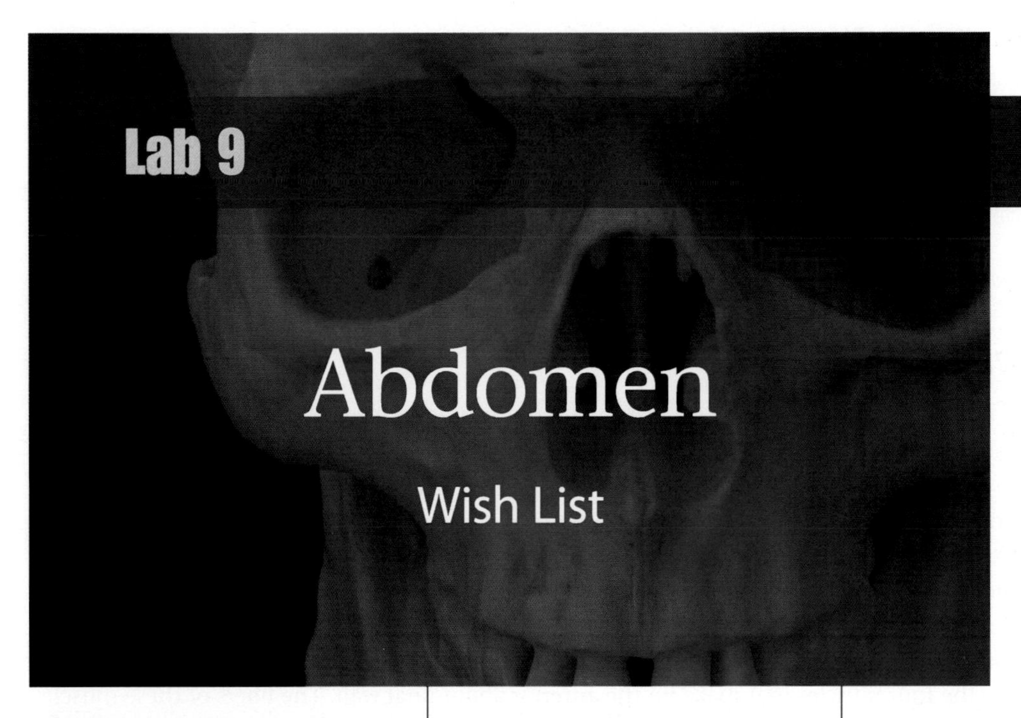

Lab 9

# Abdomen

## Wish List


**LAB 9 OVERVIEW, pp. 258–260**

**MUSCLES, pp. 261–270**

- **Linea Alba, p. 261**
- **Rectus Abdominis, pp. 262, 263**
- **External Abdominal Oblique, pp. 262, 263**
  - Aponeurosis of External Abdominal Oblique, p. 263
- **Internal Abdominal Oblique, pp. 264, 265**
  - Aponeurosis of Internal Abdominal Oblique, p. 265
- **Transverse Abdominis, pp. 264, 265**
  - Aponeurosis of Transverse Abdominis, p. 265
- **Quadratus Lumborum, pp. 266, 267**
- **Psoas Major, pp. 266, 267**
- **Psoas Minor, pp. 268, 269**

**ORGANS, pp. 268–286**

- **Diaphragm, pp. 268–270**
  - Central Tendon, p. 269
  - Crura (singular = Crus), p. 270
- **Billary Ducts (Quack), pp. 270–271**
  - Common Bile Duct, p. 270
  - Cystic Duct, p. 271
  - Hepatic Duct, p. 271
- **Liver, pp. 272–275**
  - Left Lateral Lobe, p. 273
  - Left Medial Lobe, p. 273
  - Quadrate Lobe, p. 273
  - Right Medial Lobe, p. 273
  - Right Lateral Lobe, p. 275
  - Caudate Lobe, p. 275
  - Papillary Process of Caudate Lobe, 275
- **Lesser Omentum, p. 276**
- **Esophagus, p. 276**
- **Stomach, p. 277**
- **Greater Omentum, p. 277**
- **Spleen, p. 278**
- **Pancreas, pp. 278–279**
- **Small Intestine, pp. 279–280**
  - Duodenum, p. 279
  - Jejunum, p. 280
  - Ileum, p. 280
- **Mesentery** (Mesenteric Ligament), p. 281
- **Large Intestine, pp. 281–282**
  - Cecum, p. 281
  - Colon, p. 282
  - Rectum, p. 282
- **Urinary Bladder, p. 283**
- **Ureter, p. 283**
- **Urethra, p. 283**
- **Peritoneum, p. 284**
- **Kidney, pp. 284–285**
  - Calyx, p. 285
  - Cortex, p. 285
  - Medulla, p. 285
  - Pyramid, p. 285
- **Suprarenal (Adrenal) Gland, p. 286**

**NERVES, pp. 287–288**

- Lumbar Nerves, p. 287
- Lateral Cutaneous Nerve, p. 287
- Femoral Nerve, p. 288
- Genitofemoral Nerve, p. 288

**HUMAN ABDOMEN—OVERVIEW, pp. 289–291**

# LAB 9 OVERVIEW

Lab 9 is a study of the abdomen. We will examine the abdominal wall and viscera (internal organs).

## 1. Abdominal Wall

We will start our work by removing the skin on the right side of the abdomen. Then we will make a cut through the abdominal wall approximately ¼ inch (close to the width of a pencil) to the right of the midline. We do this to avoid the **lina alba** so that we may observe that structure later. Start the incision at the diaphragm and proceed caudally to the pelvis, taking care not to dig deeply into the abdomen with the scissors. You may begin this portion of the lab yourself or wait for Dr. J to make the cut for you. Once that cut is complete, you should make a second and third cut along the margin of the diaphragm to the right and left sides. Stop cutting when you are close to the vertebral border on each side.

The flap of abdominal wall that is formed on the right side by these incisions can now be delaminated (peeled apart) to expose the four muscles that make up the anterior abdominal wall. The fibers of these muscles crisscross, adding strength to the abdominal wall. Three of the four muscles also have their own distinctly-shaped aponeuroses, which are collectively referred to as the **rectus sheath**. They will be discussed in turn.

The most superficial of the abdominal wall muscles is the **external abdominal oblique**. Note that its muscle fibers move in a caudal direction as they approach the **linea alba**. We have studied the **external intercostal muscle** in lecture. The **external abdominal oblique** is an extension of the same muscle sheet. The **aponeuroses** of the **external abdominal oblique** are shaped like a bell.

Deep to the **external abdominal oblique** we find the **internal abdominal oblique** muscle. It makes up the intermediate layer of oblique muscles, and its fibers move in a cranial direction as they approach the **linea alba**. The **internal abdominal oblique** is an extension of the same muscle sheet that forms the **internal intercostal muscle**. Its **aponeuroses** are shaped like a "V" that widens as it approaches the costal margin.

The deepest of the three oblique muscles is the **transverse abdominis**. Its fibers move caudally in cats, but they are nearly transverse in humans. The **transverse abdominis** is part of the same muscle sheet that forms the **innermost intercostal muscles**. Its **aponeuroses** are shaped like a narrow oval.

The fourth and final anterior abdominal muscle is the **rectus abdominis**. It runs from the **pubic crest** and **pubic symphysis** to the costal cartilages of ribs five through seven. It is enclosed within the **aponeuroses** of the **internal abdominal oblique** like meat in a pita pocket. Remember, the three abdominal aponeuroses are called the **rectus sheath**.

Before we continue on to the viscera, observe the shiny material on the deep side of the abdominal wall. This is the **parietal peritoneum**.

## 2. Abdominal Viscera

Deep to the ventral abdominal wall you will once again observe the **diaphragm**, but this time from the caudal side. This is an opportunity to observe the **central tendon**, which is the **diaphragm's** central insertion point. Because this tendon is relatively transparent, like a domed, circular window, you should be able to place a probe on the cranial side of it and then see that probe when you look toward the thoracic cavity from the caudal side of it. We will also observe the **left crus** of the **diaphragm** a little later in this dissection.

Moving caudally, we observe the **liver**. The cat **liver** has six lobes, while the human liver has four. If you move counterclockwise from left to right, these lobes are the: **left lateral lobe**, **left medial lobe**, **quadrate lobe**, **right medial lobe**, **right lateral lobe**, and **caudate lobe**. We will observe the **papillary process** of the **caudate lobe** between the **stomach** and the **left medial lobe** of the **liver**. We will also see the **gall bladder** between the **quadrate lobe** and the **right medial lobe** of the **liver**. Later, in the next lab, we will observe the **biliary ducts** when we clean up some of the abdominal vessels that are adjacent to them.

Caudal to the liver, we observe the **lesser omentum**, **stomach**, and **greater omentum** (which covers the **small intestine** like a blanket and looks like a mozzarella cheese-impregnated hairnet). The **omenta** (plural) are primarily composed of epithelial tissue and are part of the visceral peritoneum. Note the **gastrohepatic ligament**, which is part of the **lesser omentum** and runs between the lesser curvature of the **stomach** and the **liver**. It is a typical ligament, as it is two layers thick. There will be communication between the **greater omentum** and the **lesser omentum** by way of the **epiploic foramen**, which we will observe later between the dorsal side of the **pancreas** and **duodenum** and the ventral surface of the **right lateral lobe** of the **liver**.

You may tease the **greater omentum** away from the **small intestines** but do not remove it. Note that resting on the left side of the **greater omentum** is the **spleen**. The shape of a cat's **spleen** is relatively flat, often like a small, narrow pancake. The human **spleen** is oval-shaped and lies superior to the **transpyloric line** (**TPL**—this is the subject of lecture material).

As we move deep to the **greater omentum**, we will observe the **intestines. Chyme** passes from the stomach to the **duodenum**. It is the proximal portion of the **small intestine** that is attached to the **pylorus** of the **stomach**. The orientation of the **duodenum** is different in cats than in humans. In the cat, the **duodenum** is oriented in a cranial/caudal direction with the **right lobe** of the **pancreas** adjacent to it. The human **duodenum** reflects and moves transversely to the left, taking the right lobe of the **pancreas** with it.

Attached to the caudal end of the **duodenum** is the **jejunum**, which is highly convoluted (coiled). Here we will make a point to observe the **mesentery** (**mesenteric ligament**). This two-layered sheet of peritoneum is functionally important because it prevents the intestine from tying itself in knots. You can see the **intestinal arteries**, the **intestinal veins**, and the **lymphatic vessels** pass between the two layers of the **mesentery**. The **jejunum** attaches to the **ileum**, which is relatively short and attaches to the **cecum** (**caecum**) at the **ileocecal junction**.

The **cecum** marks the end of the small intestine and is considered the beginning of the **large intestine**, which includes the **cecum**, the **colon**, the **rectum**, and the **anal canal**. The **cecum** is found in the right iliac region of the abdomen. Its name comes from the Latin word meaning "blind," because it is a blind pouch. It also curls like the toes of the Keebler Elves' slippers, which is a good way to remember it. The cat lacks an **appendix**, but humans have an **appendix** that is attached to the **cecum**.

The **large intestine** presents us with the anatomical conundrum of the cat: If it is shaped like a question mark, why do we call it a **colon**? If you find an answer, please send it to Dr. J. The **colon** courses to the left side and then moves caudally to the **rectum**. Ventral to the **rectum** you will find the **urinary bladder**. We will wait until we work on Lab 11 (the reproductive organs) to see the **urethra**.

We now will move to the dorsal abdominal wall, removing adipose tissue from the *left side only*. (It is very important that we not disturb the right side, because we will dissect that side during Lab 11 to expose the more delicate structures of the reproductive system.) We will start by observing the left **kidney**. You will notice that on its ventral side there is **peritoneum**, which extends out onto the dorsal abdominal wall, but does not surround the entire **kidney**. This physical arrangement means that the **kidney** is **retroperitoneal**. Any time an organ has **peritoneum** on the ventral side only, it is recognized as being **retroperitoneal**. This is a useful description because it tells us that the organ is located in the dorsal portion of the abdomen.

I will dissect the **kidney** with a partial coronal cut, so that you will be able to open it to observe its features. The outer layer is the **cortex**, which is usually lighter in color. The **cortex** surrounds the **pyramid of the medulla**, which is usually darker than the **cortex**. The **pyramid** releases urine into the **calyx**, a central, funnel-shaped structure. In turn, the **calyx** will drain into the **pelvis** of the **kidney** and that will drain into the **ureter**. You can observe the **ureter** as it leaves the **kidney** as it courses to the posterior side of the **urinary bladder**. The **ureter** is usually very white in color.

Medial and cranial to the cat **kidney** you will observe the **suprarenal gland**, which may also be called the **adrenal gland**. The human **adrenal glands** sit directly on the superior portion of each kidney. Slightly cranial and lateral to the **suprarenal gland** on the dorsal wall of the abdomen, we will uncover the **left crus** of the **diaphragm**.

Next we will observe structures along the dorsal wall of the abdomen. Actually, we will begin just cranial to the diaphragm, where we will observe the **quadratus lumborum muscle**. This muscle is also in the abdomen, dorsal to the two psoas muscles, but it would be better not to remove them to see it there. Please observe the **psoas minor** and **psoas major muscles** caudal and medial to the **kidney**. **Psoas minor** will be medial and ventral to **psoas major**.

Adjacent to the **adrenolumbar artery** and its branches, we will observe the **medial** and **lateral lumbar nerves**. They can be very challenging to find, as they blend in with the muscle fibers of the dorsal wall. Adjacent to the **deep iliac circumflex artery** we will find the **lateral cutaneous nerve** which runs obliquely (caudal and lateral). Moving further in a caudal direction we should observe the **external iliac artery** and be able to isolate the **genitofemoral nerve** running adjacent to that. This nerve is very small in diameter. Lastly, we will find the **femoral nerve** within the **psoas major muscle**, proximal to where it emerges in the thigh.

****NOTE:** During this lab you should observe a dissected pig **kidney** (also prepared with a coronal cut). You can see the **cortex**, but rather than having a single **pyramid**, the **medulla** is composed of ten or more **pyramids**. Instead of having a single **calyx**, there is a **minor calyx** associated with each **pyramid**. The **minor calyces** drain into **major calyces**. These in turn drain into the **pelvis** of the **kidney**. The human **kidney** is more similar to the pig **kidney** because it has between eight and ten **pyramids**.

# MUSCLES

## Linea Alba

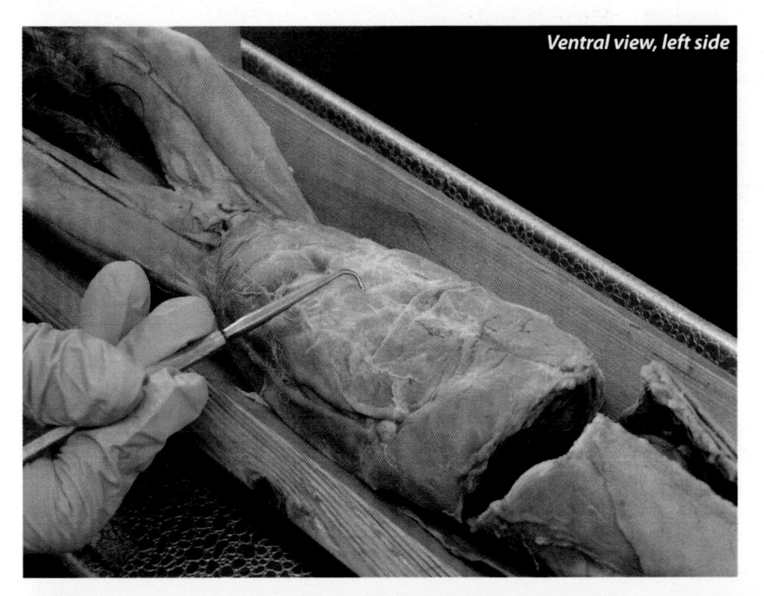


The linea alba (white line) is where the aponeuroses of the anterior abdominal muscles fuse with each other to form a single tendinous band. It runs along the anterior midline from the xiphoid process to the pubic symphysis, and it contains the umbilicus. The linea alba has surgical significance in that it does not contain muscle or much vascular tissue; therefore, it is often cut in surgery to avoid muscle damage. It is also the insertion of the:

1. external abdominal oblique,
2. internal abdominal oblique (by aponeurosis), and
3. transverse abdominis.

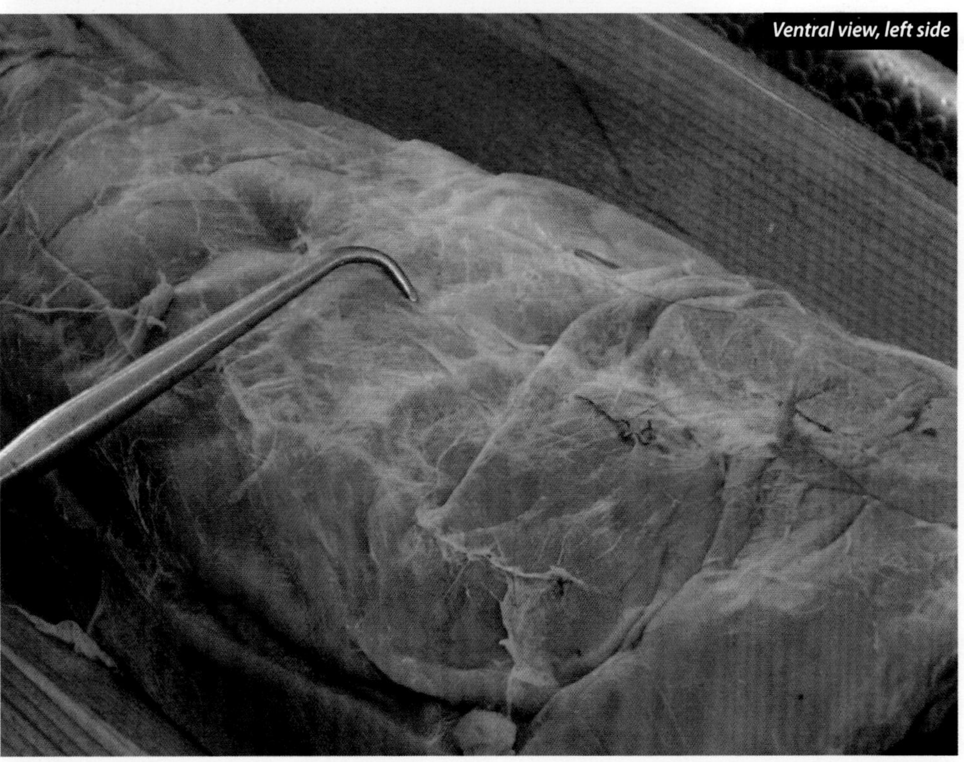

## Rectus Abdominis

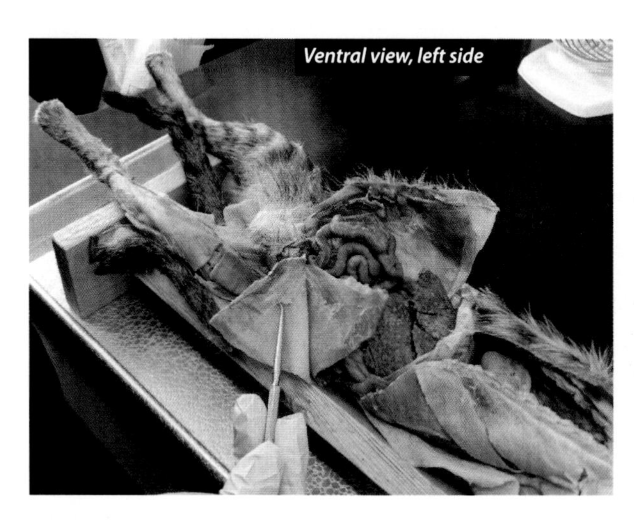


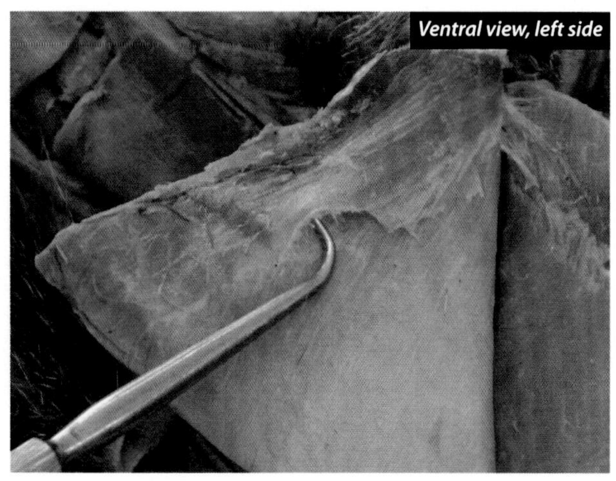


***Human Information:***
**origin:** pubic tubercle, pubic symphysis, and pubic crest
**insertion:** costal cartilages of ribs 5–7; xiphoid process
**nerve:** anterior rami of inferior thoracic spinal nerves (T7–T12)
**action:** abdominal press, flexion of vertebral column (trunk flexion)

## External Abdominal Oblique

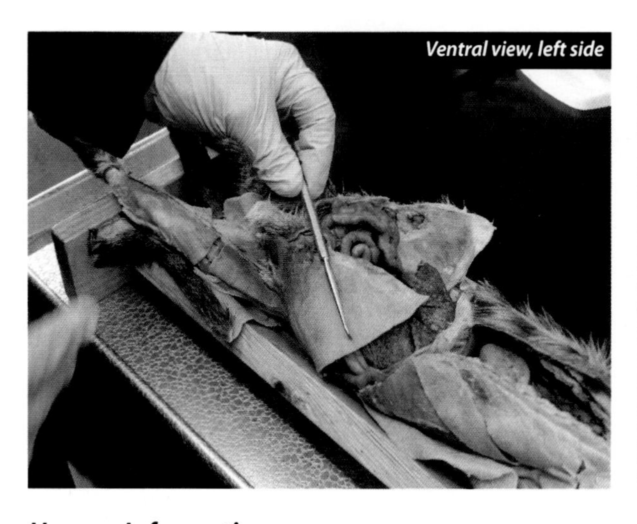


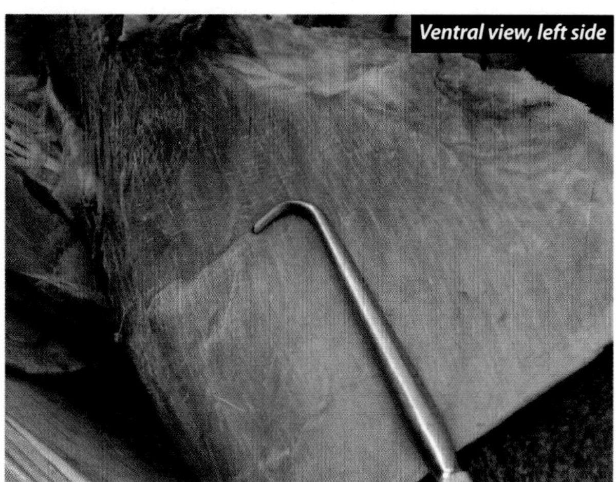


***Human Information:***
**origin:** external lateral surface of inferior eight ribs
**insertion:** linea alba and anterior half of iliac crest
**nerve:** intercostal (8–12), iliohypogastric, and ilioinguinal
**action:** bilaterally: abdominal press, assists vertebral column (trunk) flexionunilaterally: contralateral vertebral column (trunk) rotation and sidebending

## Rectus Abdominis

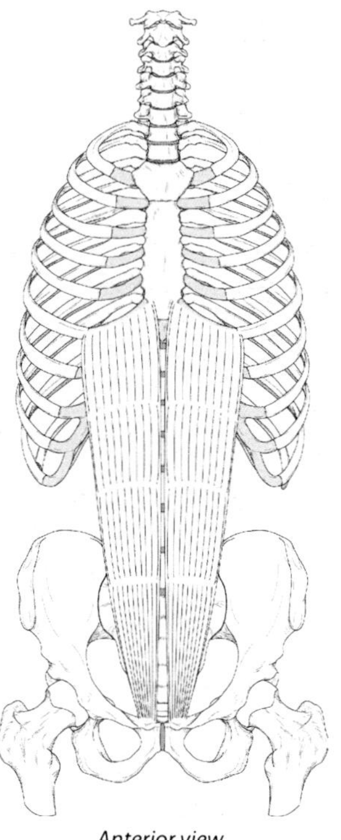
*Anterior view*

The rectus abdominis is often incorrectly known as the six pack muscle. Actually, it is an eight pack muscle. The four divisions on each side result from three tendinous intersections of this muscle. The muscle is enclosed within the aponeurosis of the internal abdominal oblique, much like sandwich meat in a pita pocket. Functionally, it is important because it participates in the abdominal press. It is sometimes used in breast reconstruction surgery. The rectus abdominis in the cat is sometimes referred to as the bacon muscle by Dr. J's students. This is because it looks suspiciously like bacon. In fact, Dr. J suspects that this may be where bacon comes from. A few students refer to it as one of the Oscar Mayer muscles, the other being the rectus femoris, which looks a lot like a cocktail wiener.

## External Abdominal Oblique

The external abdominal oblique muscle is continuous with the external intercostal muscle. Its fibers point in an inferior direction as it moves anteriorly until it inserts on the linea alba via its aponeurosis. The external abdominal oblique is the most superficial of the muscles of the abdominal wall. It is superficial to the internal abdominal oblique. The abdominal muscles have a number of actions, but collectively they aid in the abdominal press, which is very important for many abdominal functions.

### Aponeurosis of the External Abdominal Oblique

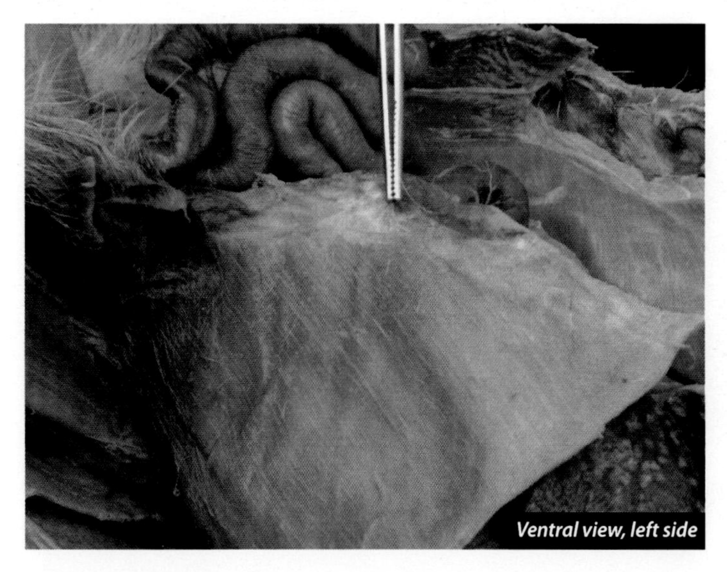
*Ventral view, left side*

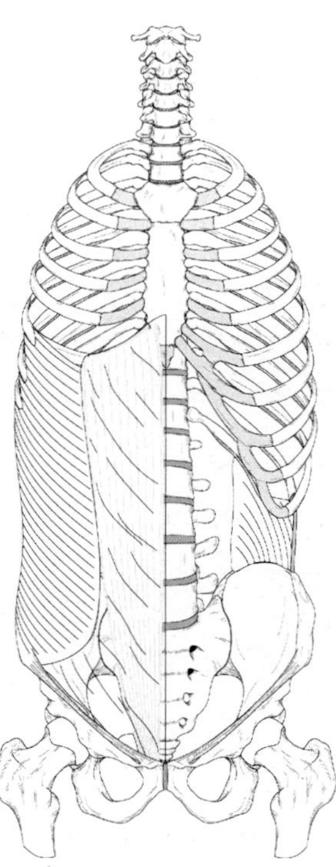
*Anterior view, right side*

The aponeurosis for the external abdominal oblique muscle is shaped a little like a bell. It is milky in appearance and flares laterally as it approaches the inferior portion of the abdomen. Remember that an aponeurosis is a broad, flat tendon.

## Internal Abdominal Oblique

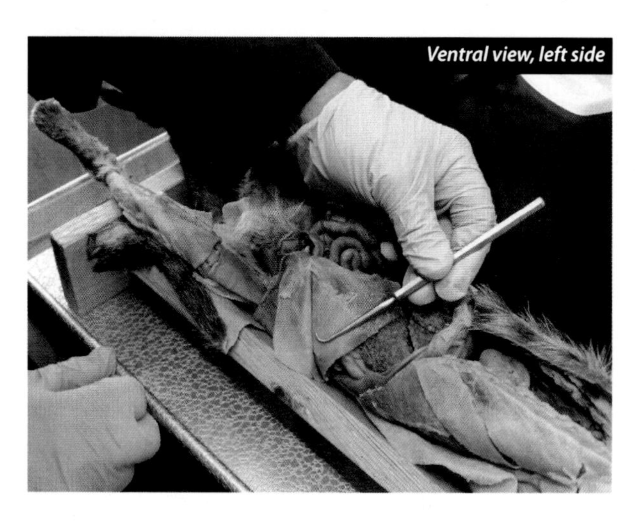


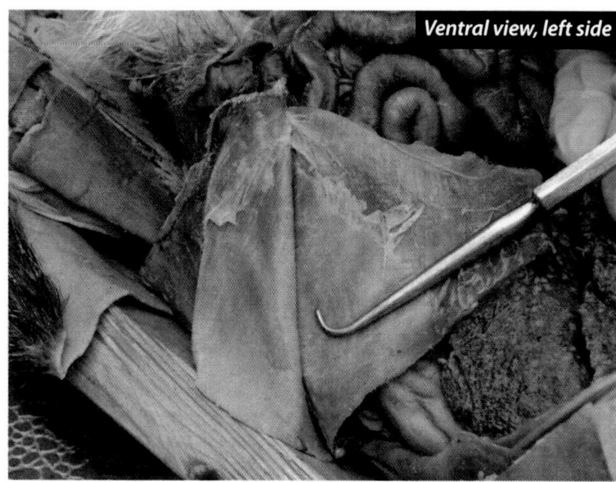


***Human Information:***

**origin:** thoracolumbar fascia; iliac crest between origins of external and transverse abdominis; lateral 2/3 of the inguinal ligament
**insertion:** inferior border of the inferior three or four ribs; aponeurosis on linea alba; pubic crest and pectineal line
**nerve:** anterior rami of inferior six thoracic spinal nerves (T7–T12); iliohypogastric, and ilioinguinal nerves
**action:** bilaterally: abdominal press, assists vertebral column (trunk) flexion; unilaterally: ipsilateral vertebral column (trunk) flexion (sidebending) and rotation

## Transverse Abdominis

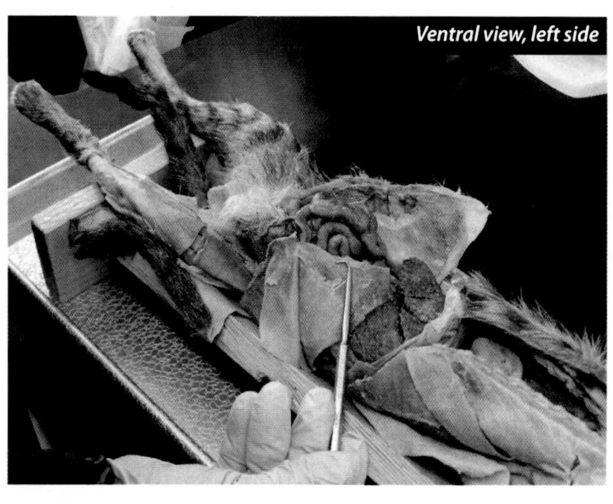


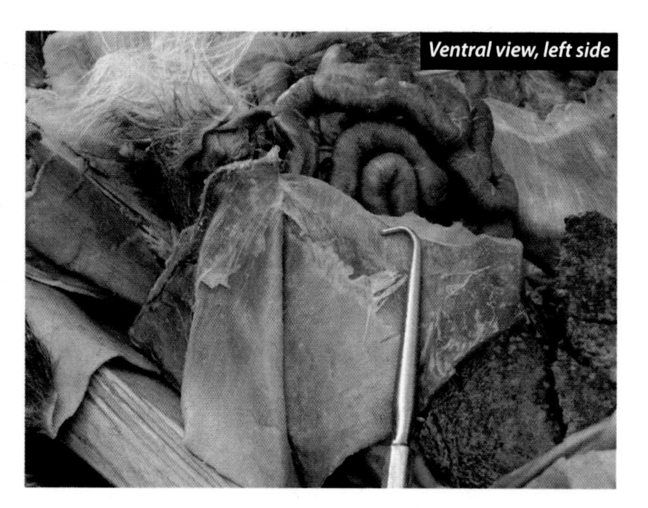


***Human Information:***

**origin:** deep surface of costal cartilages of ribs 7–12; lateral third of inguinal ligament; thoracolumbar fascia; medial lip of iliac crest
**insertion:** linea alba; pubic crest and pectineal line
**nerve:** anterior rami of T7–T12; iliohypogastric and ilioinguinal nerves
**action:** abdominal press

# Internal Abdominal Oblique

The internal abdominal oblique muscle is continuous with the internal intercostal muscle. Its fibers point in a superior direction as it moves anteriorly until it inserts on the linea alba via its aponeurosis. It is deep to the external abdominal oblique and superficial to the transverse abdominis muscle. Its aponeurosis encapsulates the rectus abdominis. Part of this muscle is pulled into the spermatic fascia in the male, and that portion is known as the cremaster muscle. The abdominal muscles have a number of actions, but collectively they aid in the abdominal press, which is very important for many abdominal functions.

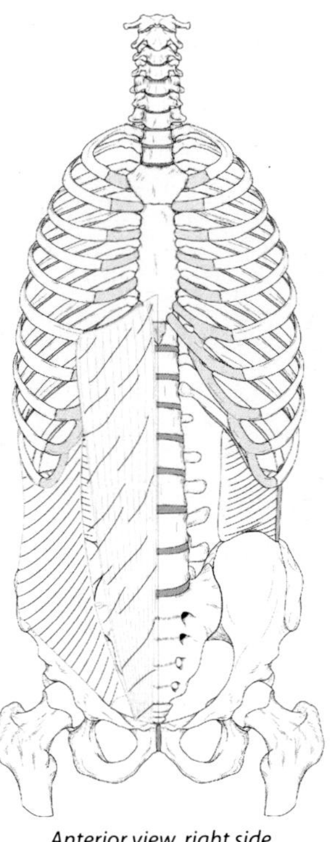
*Anterior view, right side*

## Aponeurosis of the Internal Abdominal Oblique

The aponeurosis for the internal abdominal oblique muscle is shaped a little like a "V," getting wider at the cranial end and narrower at the caudal end. It is milky in appearance. Remember that an aponeurosis is a broad, flat tendon.

# Transverse Abdominis

The transverse abdominis is the deepest of the three abdominal muscles, making up the majority of the abdominal wall and residing deep to the internal abdominal oblique. The transverse abdominis muscle is continuous with the innermost intercostal muscle. In the cat, its fibers move caudally as they approach the midline, while in the human they are nearly transverse.

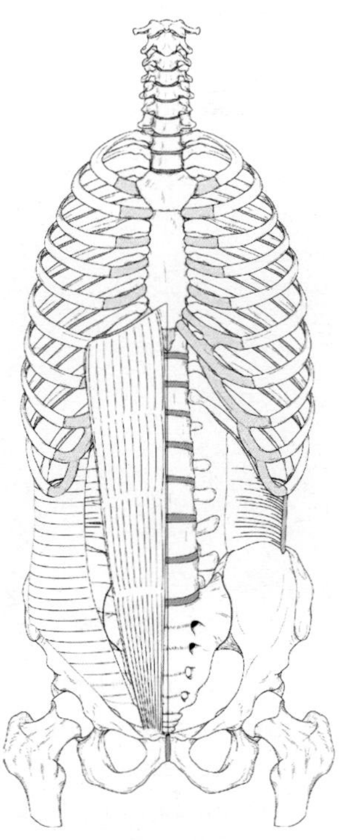
*Anterior view, right side*

## Aponeurosis of the Transverse Abdominis

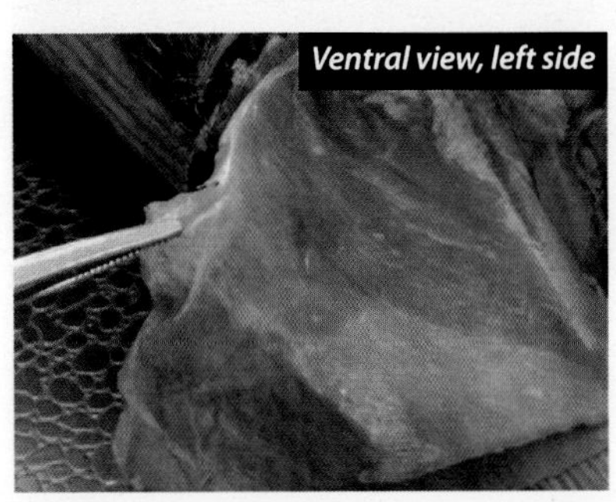

*Close-up of aponeurosis of transverse abdominis*

The aponeurosis for the transverse abdominis muscle is shaped a little like a narrow oval. It is usually relatively clear. Remember that an aponeurosis is a broad, flat tendon.

## Quadratus Lumborum

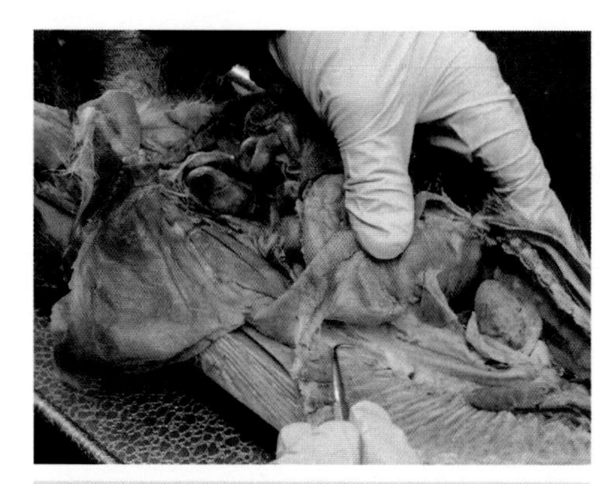

*Ventral view, left side, cranial to diaphragm, and lateral to the vertebral column.*

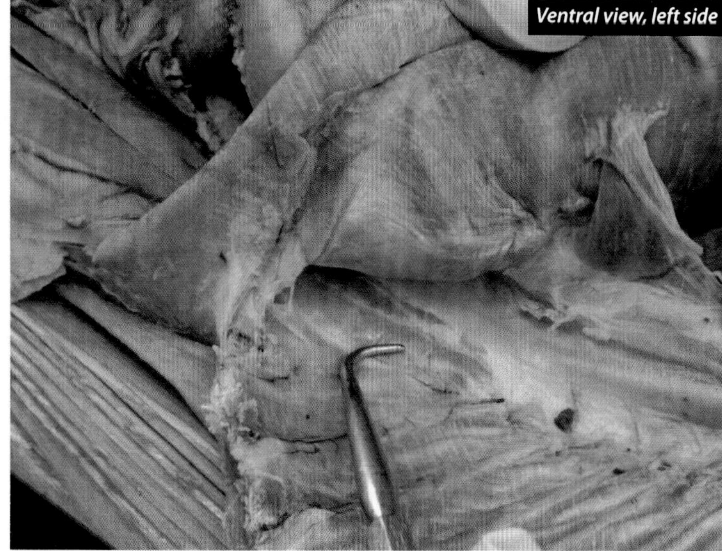

Ventral view, left side

***Human Information:***

**origin:** transverse process of L5, iliolumbar ligament, and iliac crest

**insertion:** transverse processes of L1 to L5 and inferior border of rib 12

**nerve:** anterior rami of T12, and L1–L4

**action:** depresses and stabilizes rib 12, some lateral flexion of vertebral column (trunk sidebending), elevates hip

## Psoas Major

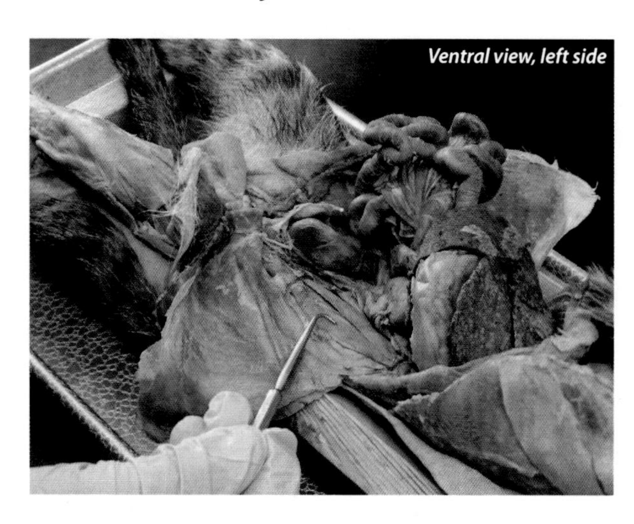

Ventral view, left side

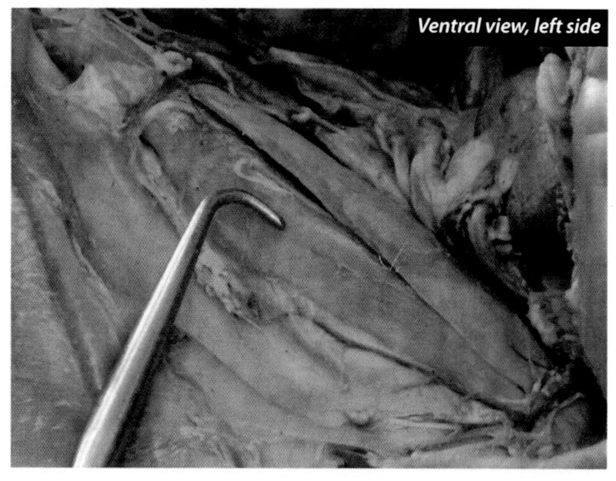

Ventral view, left side

***Human Information:***

**origin:** transverse processes, bodies, and intervertebral discs of L1–L5 and T12 vertebrae

**insertion:** lesser trochanter of femur

**nerve:** anterior rami of L1–L3 vertebrae

**action:** prime flexor of thigh, flexor of trunk, lateral flexor of vertebral column

## Quadratus Lumborum

The quadratus lumborum can be seen in two places in the cat. It can be seen in the thoracic cavity running along the vertebral border, and it can be found dorsal to the psoas major in the abdomen. In humans, it would not be seen in the thoracic cavity.

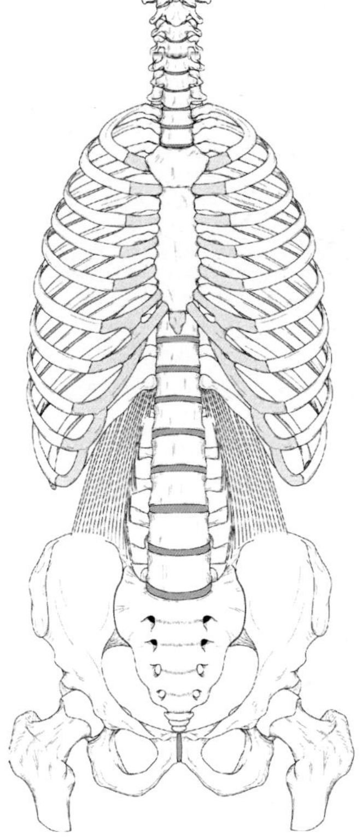

*Anterior view*

## Psoas Major

The psoas major is of particular importance because it merges with the iliacus to form the iliopsoas, which is the prime flexor of the thigh. It is known as the tenderloin behind the meat counter of markets. Both of the psoas muscles were named for a famous anatomist that later became Superman's girlfriend, Psoas Lane!

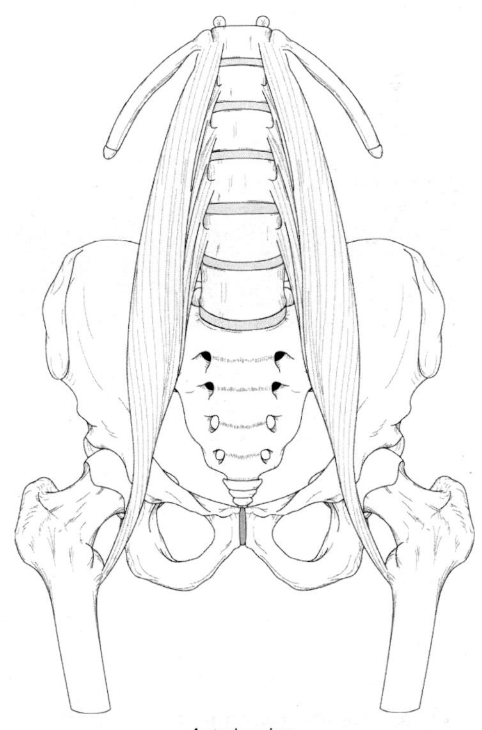

*Anterior view*

## Psoas Minor

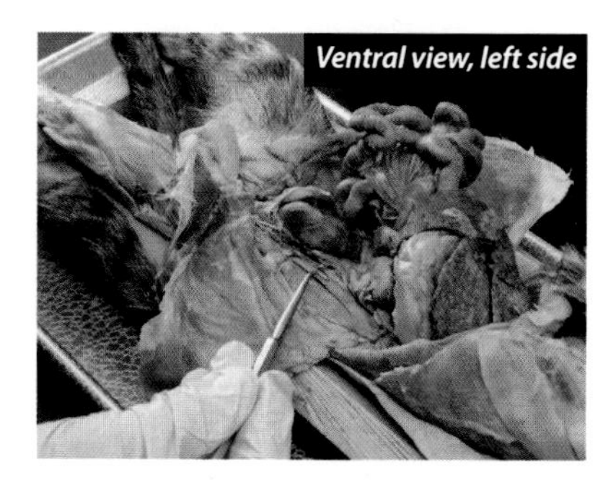


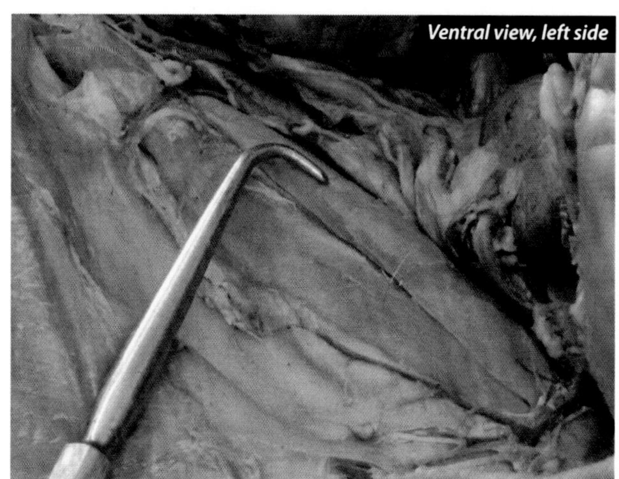


***Human Information:***
**origin:** lateral surface of T12 and L1 vertebral bodies and intervertebral discs between them
**insertion:** pectineal line of pelvic brim and iliopubic eminence
**nerve:** anterior rami of L1
**action:** weak flexion of lumbar region of vertebral column

## Diaphragm

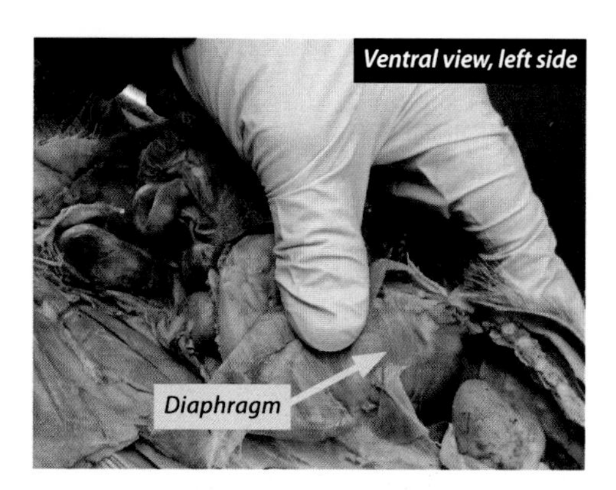


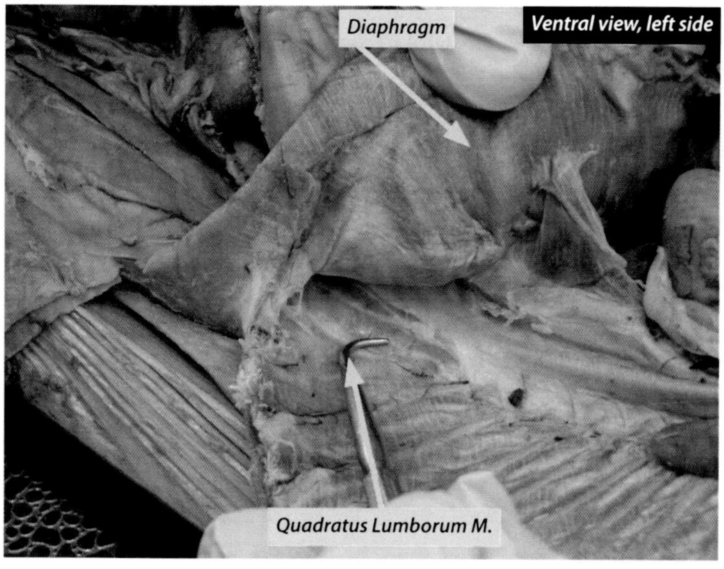


The diaphragm is a dome-shaped organ made primarily of skeletal muscle that forms an anatomical barrier between the thoracic cavity and the abdominal cavity. Functionally, it is very important because it is responsible for most breathing at rest. In humans, the origin of the diaphragm is along its outer margin—the deep surface of the bony thorax, the costal cartilage of ribs 7 to 12, the sternum, and the lumbar vertebrae 1 to 3. Its insertion is medial, at the central tendon. When one inspires (inhales), the central insertion point moves inferiorly toward the origin at the outer margin, thereby flattening the diaphragm and compressing the abdominal viscera. When one expires (exhales), the diaphragm relaxes and the compressed organs expand, which assists in pushing the diaphragm back up to its original dome shape.

## Psoas Minor

The psoas minor is missing in about 40 percent of the population. Both of the psoas muscles were named for a famous anatomist that later became Superman's girlfriend, Psoas Lane!

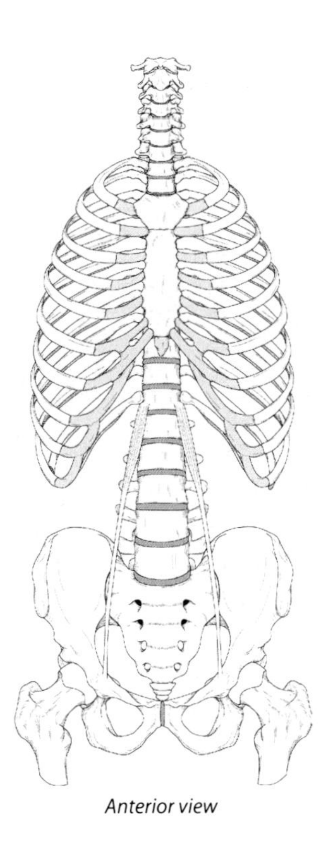

Anterior view

## Diaphragm

### Central Tendon

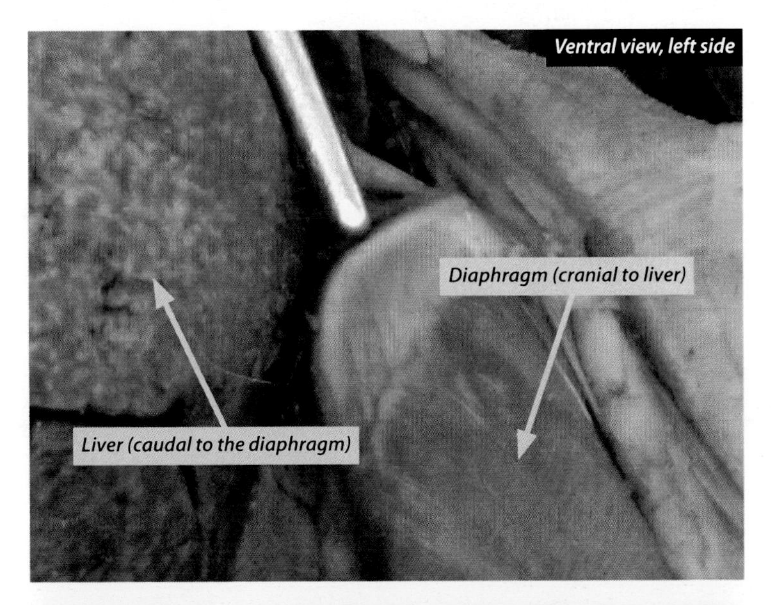


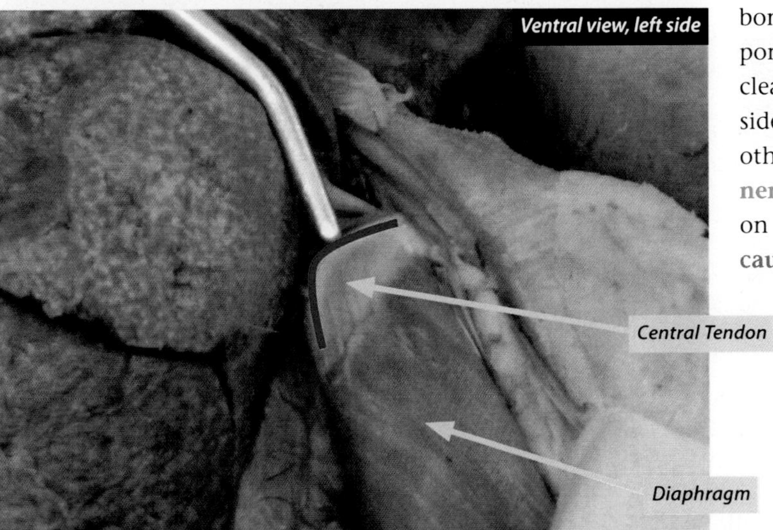


The central tendon is the insertion of the diaphragm. It is an aponeurosis and unusual in that it does not attach to a bone, but rather connects the central portions of the diaphragm. It is relatively clear, and if a probe is placed on one side, the probe can be seen from the other side. Remember that the **phrenic nerve** passes through the central tendon on both sides, as well as through the **caudal vena cava** on the right.

# Diaphragm

## Crura (singular = Crus)

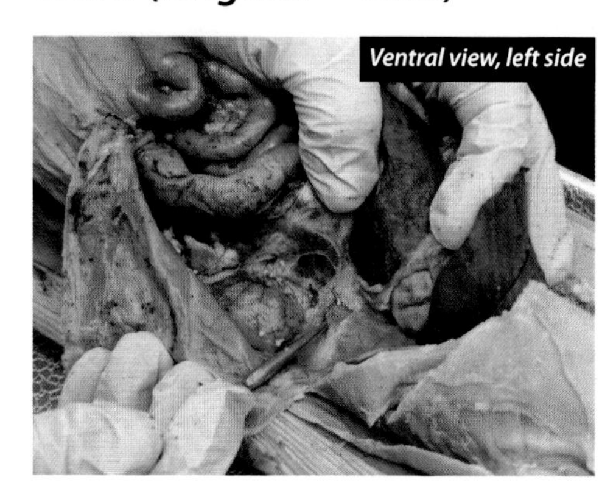


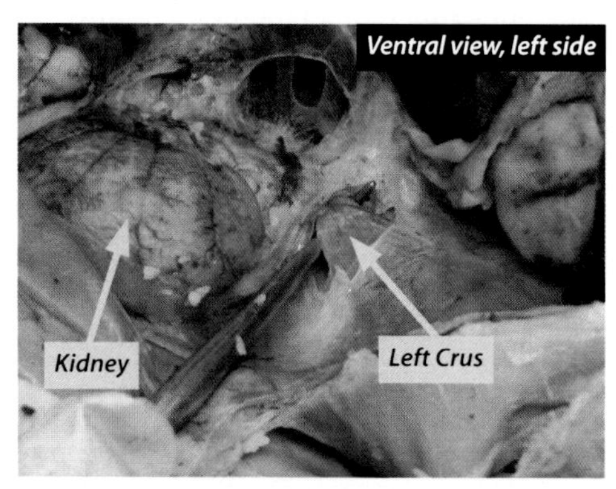


We talked about the crura back in the material covered on the first written exam. They are musculotendinous extensions of the diaphragm. Now we finally will see one. Dr. J wants you to clean up on the cat's left side. Then we will be able to observe the left crus. In the human, the right crus secures the diaphragm to the bodies of lumbar vertebrae 1, 2, and 3. The left crus secures the diaphragm to the bodies of lumbar vertebrae 1 and 2. Later in the course we will study the crura of the penis. Each crus was named for a famous movie star—the left crus was named for Tom (it is shorter), and the right crus was named for Penelope.

# ORGANS

# Biliary Ducts (Quack)

## Common Bile Duct

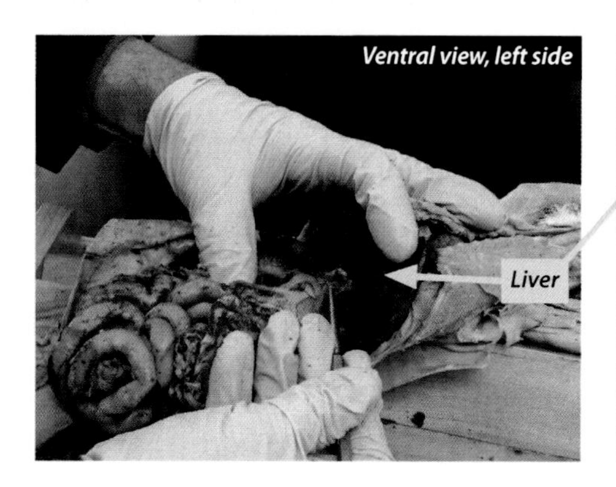


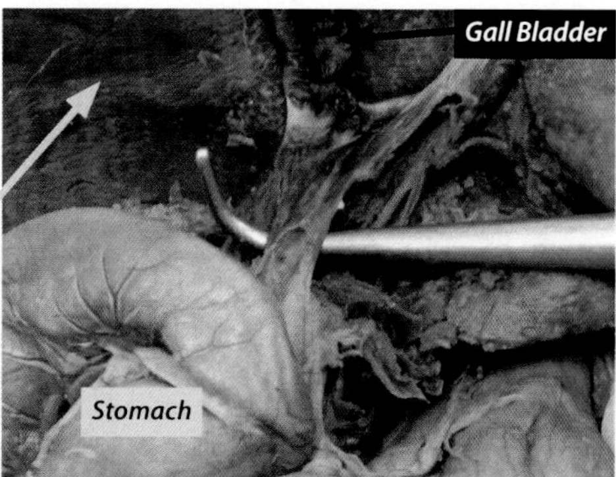


The common bile duct can also be called the hepatopancreatic duct. It carries bile from the liver and pancreatic secretions from the pancreas to the duodenum. You will not see it in the cat, but there is an ampulla that projects into the duodenum and has a sphincter at the end of it. The ampulla is called the ampulla of Vater or the hepatopancreatic ampulla. The sphincter at the end of the ampulla of Vater can be called the sphincter of Oddi or the hepatopancreatic sphincter. You might have guessed that the ampulla of Vater is named for Darth Vater, an anatomist that later joined the forces of the Empire!

## Cystic Duct

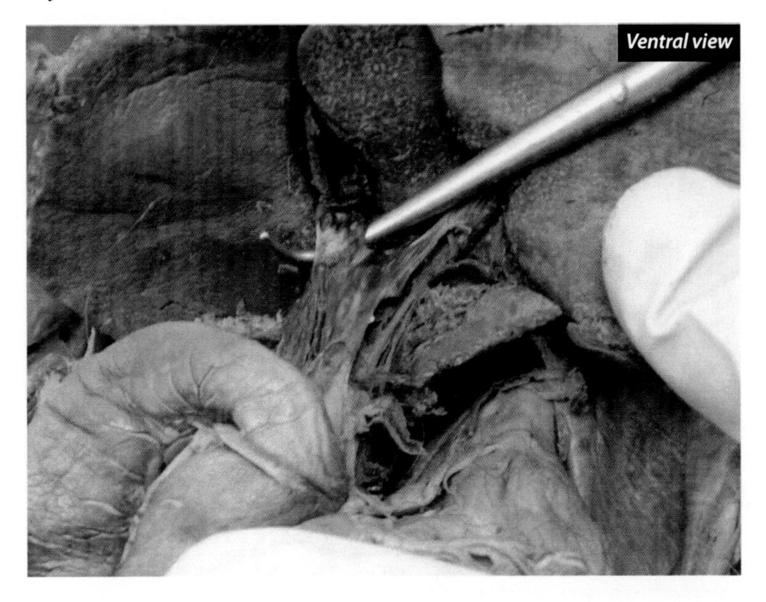


In the picture, the cystic duct (quack) is on the cat's right side and the hepatic duct is on the cat's left side. This is intended to help you orient one relative to the other.

## Hepatic Duct

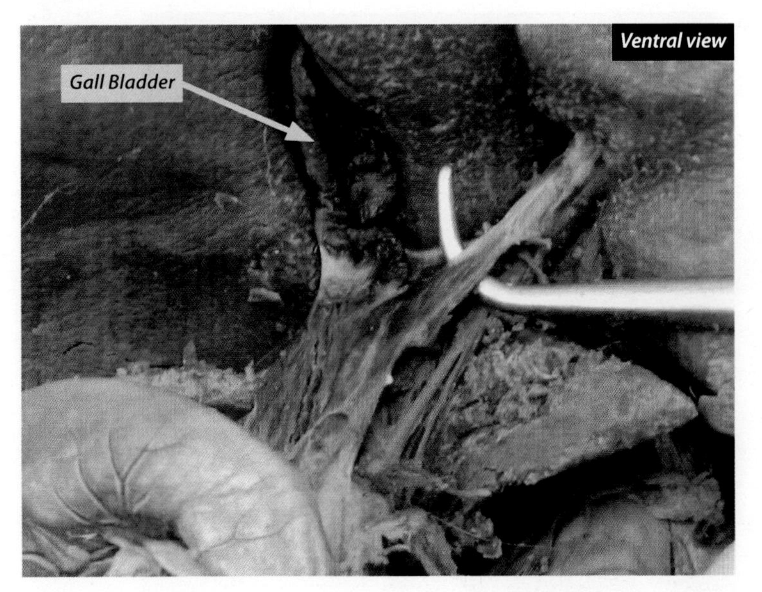


In the picture, the cystic duct (quack) is on the cat's right and the hepatic duct (quack) is on the cat's left side. This is intended to help you orient one relative to the other. The hepatic duct conducts bile from the liver to the common bile duct. The cystic duct conducts bile to and from the gall bladder.

# Liver

*Probe is pointing to the left lateral lobe of the liver.*

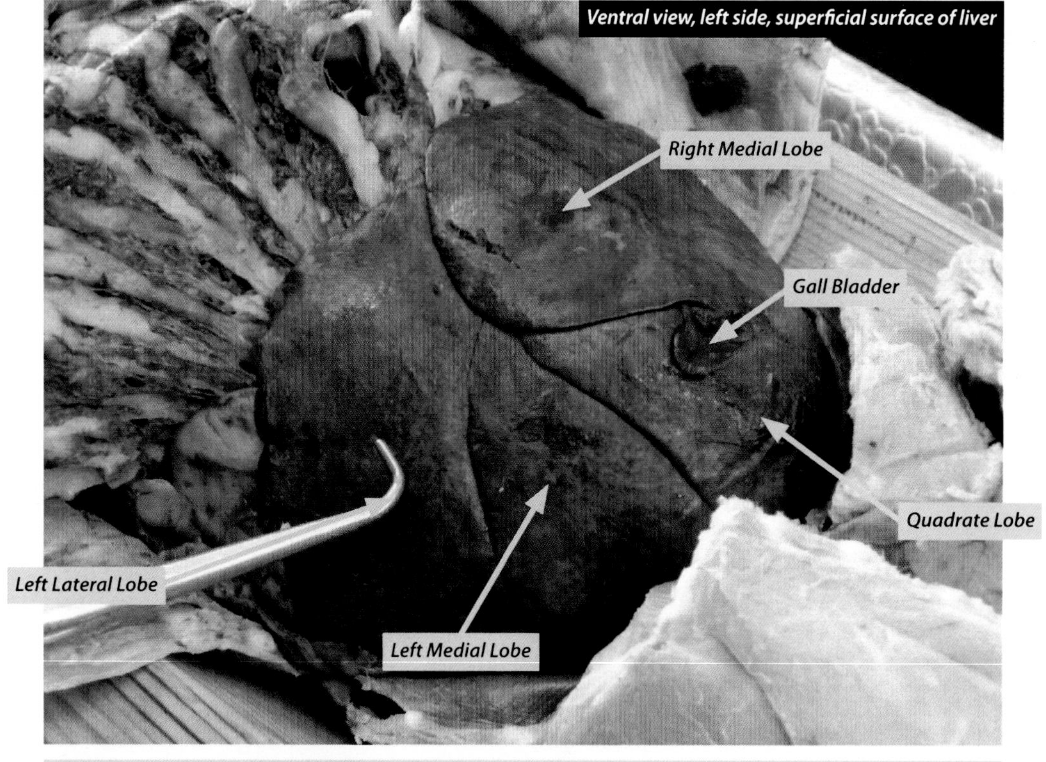


*NOTE: the right lateral lobe, caudate lobe, and papillary process of caudate lobe cannot be seen in these pictures.*

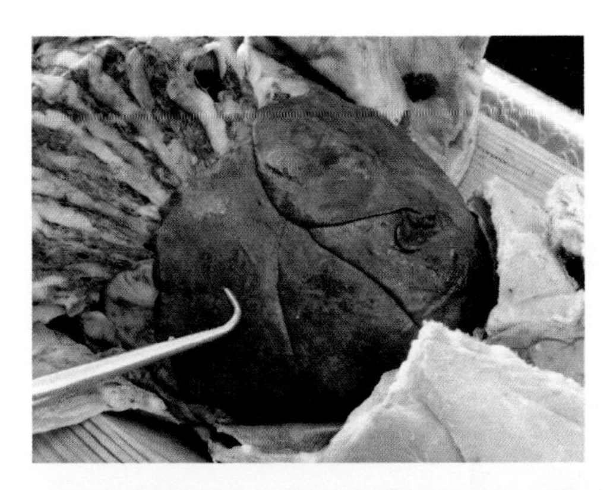

## Left Lateral Lobe of the Liver

The left lateral lobe of the liver is caudal to the left medial lobe. It is the larger of the two lobes on the left side.

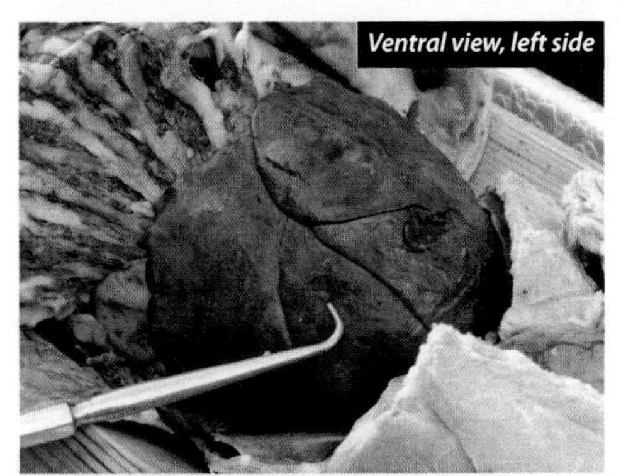
Ventral view, left side

## Left Medial Lobe of the Liver

The left medial lobe of the liver is cranial to the left lateral lobe. It is the smaller of the two lobes on the left side.

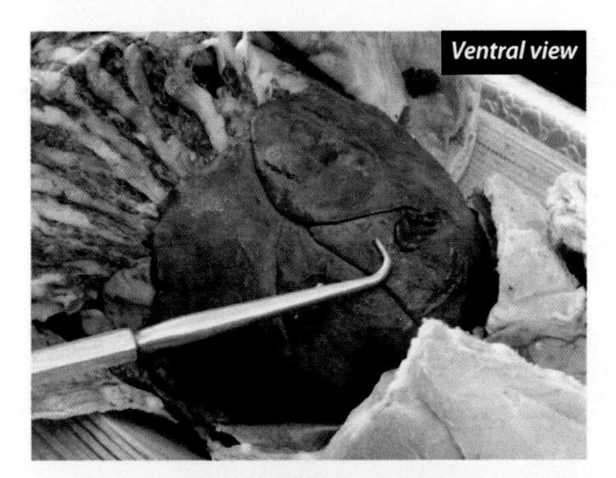
Ventral view

## Quadrate Lobe of the Liver

The quadrate lobe of the liver is only found on the right side. It is between the gall bladder and the left medial lobe. It gets its name because it is sometimes four-sided in appearance.

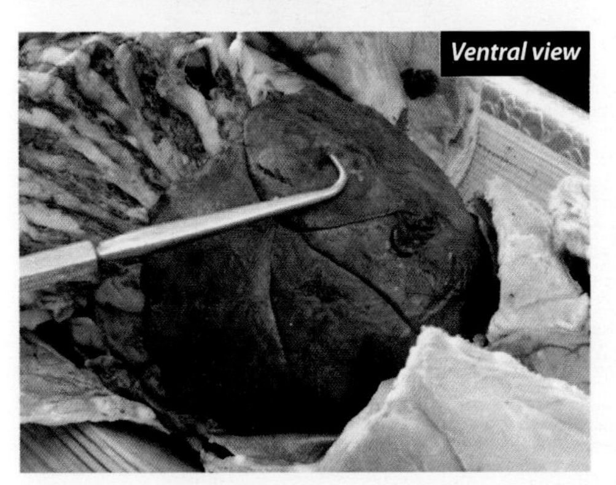
Ventral view

## Right Medial Lobe of the Liver

The right medial lobe of the liver is cranial to the right lateral lobe and lateral to the quadrate lobe. It is larger than the right lateral lobe.

# Liver

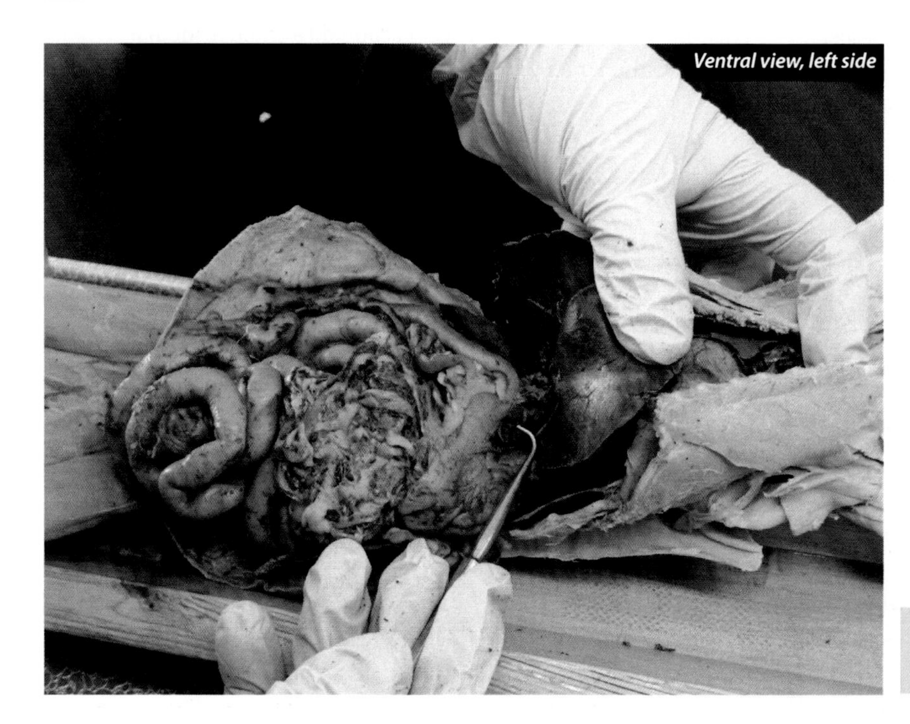


*In this picture the probe is pointing to the papillary process of the caudate lobe.*

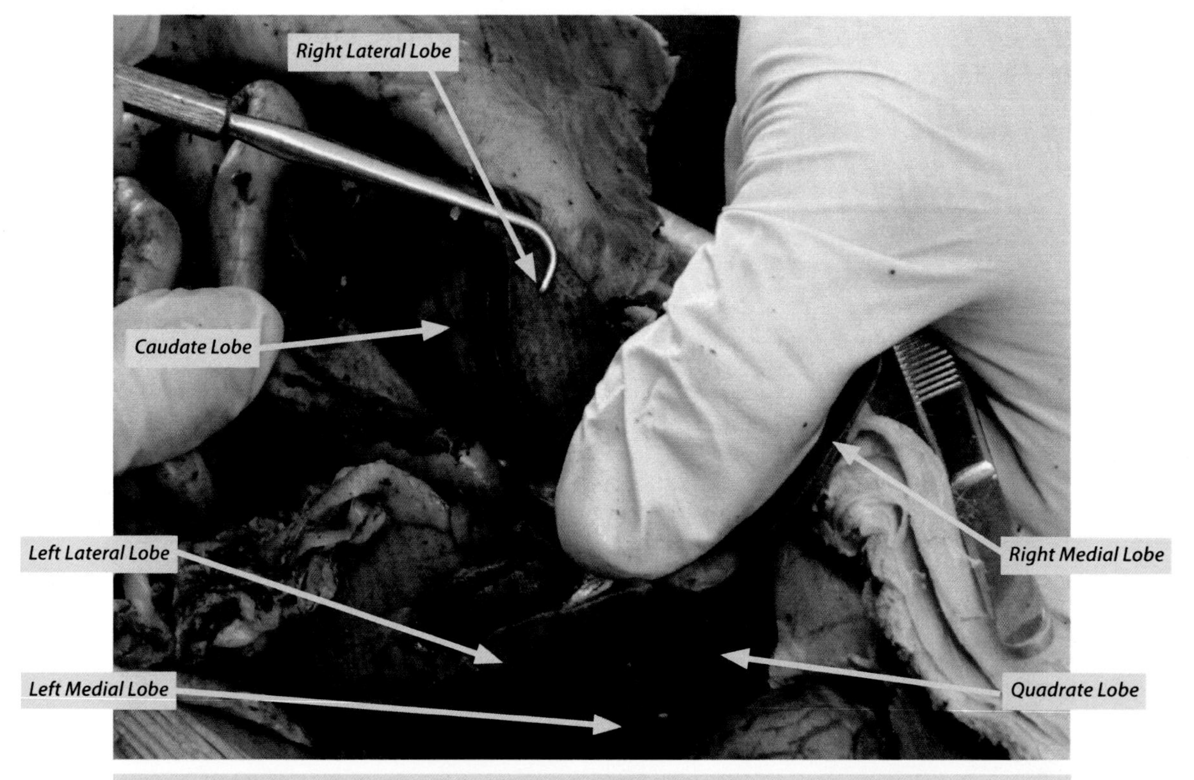


*In this deep view, Dr. J is reflecting the right medial lobe of the liver and his thumb is on the papillary process of the caudate lobe of the liver.*

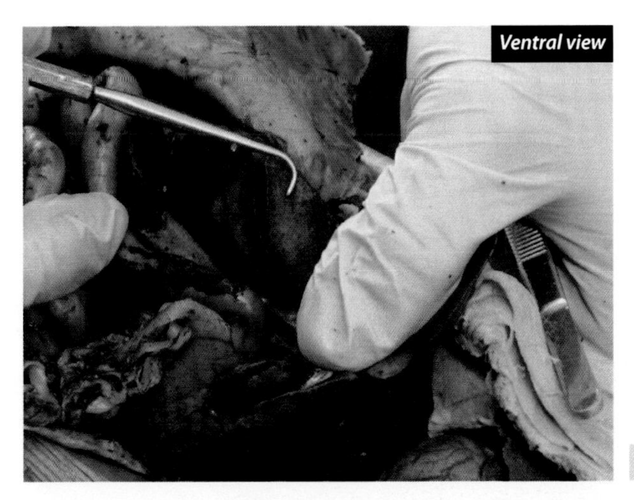


## Right Lateral Lobe of the Liver

The right lateral lobe of the liver is caudal to the right medial lobe and cranial to the caudate lobe. It is smaller than the right medial lobe.

*Probe is on the right side.*

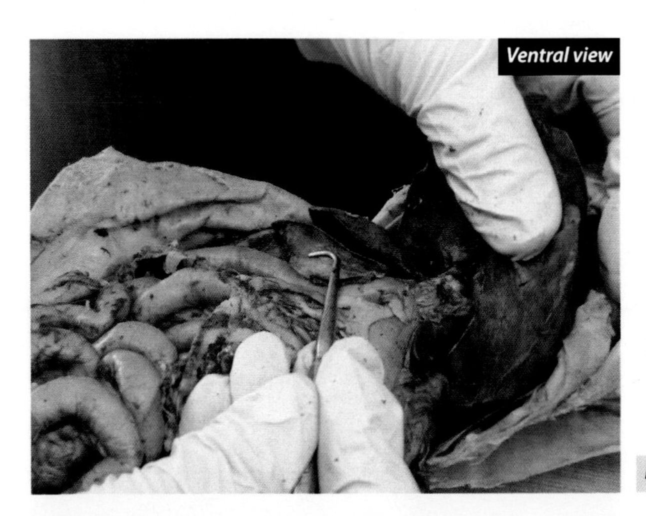


## Caudate Lobe of the Liver

The caudate lobe of the liver is only found on the right side. The largest portion is caudal to the right lateral lobe of the liver. Note that the caudate lobe can be seen in two places, one being the papillary process of the caudate lobe.

*Probe is on the right side.*

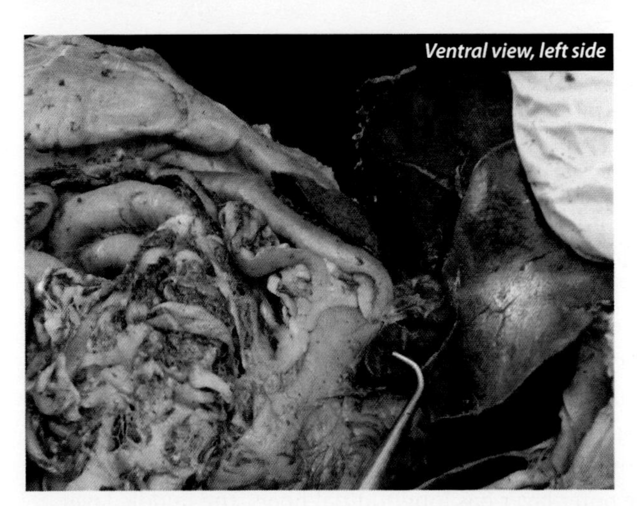


## Papillary Process of Caudate Lobe

The papillary process of the caudate lobe of the liver is only found on the right side. It is a relatively small portion of the caudate lobe, and it is found dorsal to the lesser omentum (gastrohepatic ligament) and to the left of the common bile duct and the **hepatic artery**. It is one of the two portions of the caudate lobe of the liver that can be tagged.

## Lesser Omentum

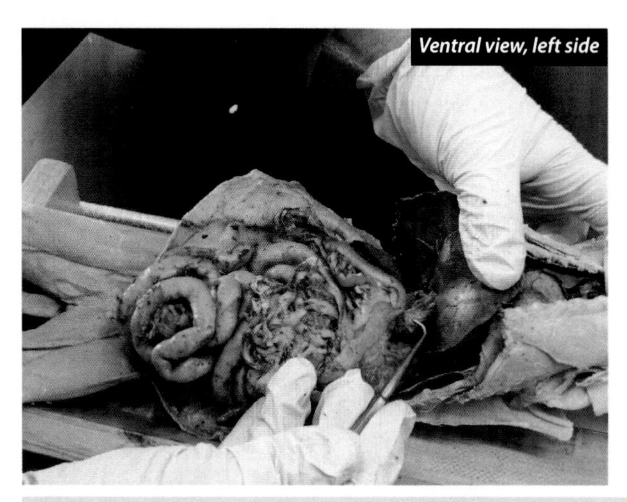


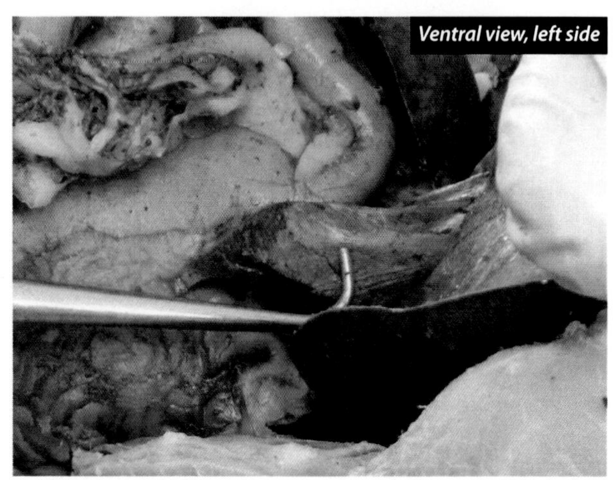


*Note: the papillary process of the caudate lobe of the liver is deep to the lesser omentum in these pictures.*

The lesser omentum is part of the peritoneum and is located primarily between the stomach, duodenum, liver, and diaphragm. It forms the gastrohepatic and hepatoduodenal ligaments. The lesser omentum is significantly smaller than the greater omentum. Most of it will be destroyed when you clean this area to study the abdominal vessels of Lab 10.

## Esophagus

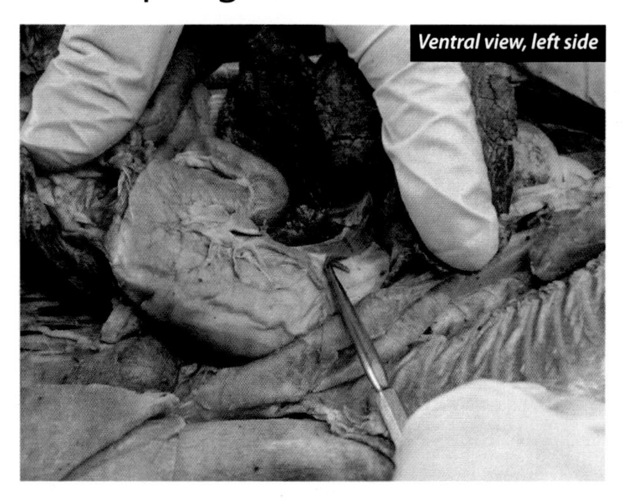


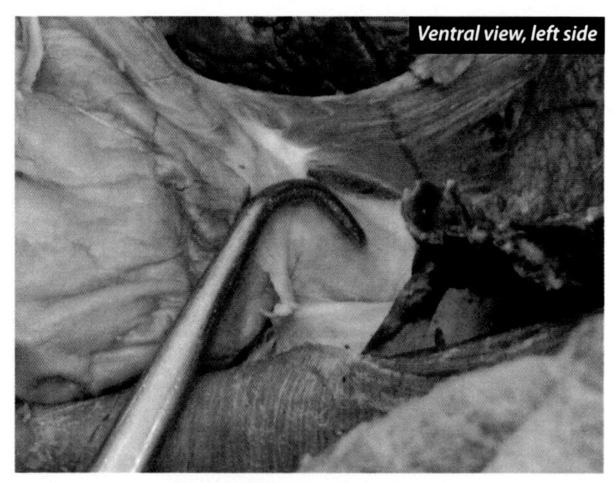


The esophagus begins at the inferior end of the pharynx. It passes through the cervical region, then through the thoracic cavity, then through the diaphragm at the level of thoracic vertebra T-10 body at mid-inhalation, and on to the stomach. Remember that it is surrounded by the crura at the diaphragm. At the junction of the stomach, we find the gastroesophageal sphincter (the cardiac sphincter). The esophagus has rugae in its walls that allow for expansion without substantial increases in pressure. Food is pushed through the esophagus by peristaltic contractions, and passage is facilitated by mucal secretions. Although the walls are made of muscle very much like skeletal muscle at the superior end, by the time one reaches the middle third of the esophagus, he or she would find three layers of smooth muscle. These three layers are consistent for much of the remainder of the gastrointestinal tract. The outer layer has longitudinal fibers, the middle layer oblique fibers, and the inner layer transverse fibers. The esophagus is superior to the transpyloric line (TPL).

## Stomach

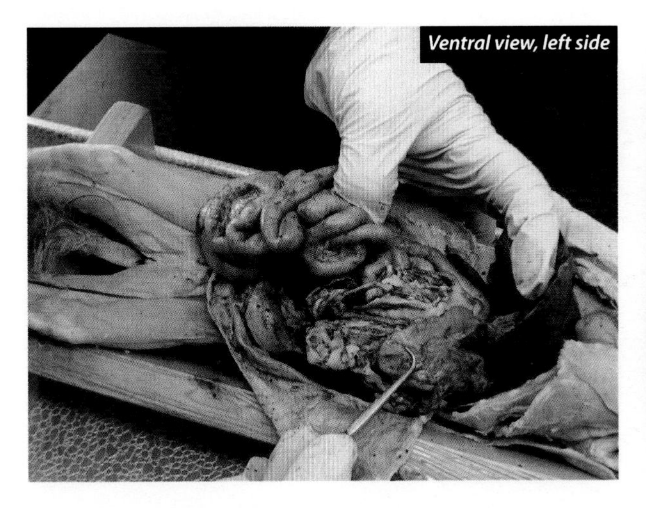


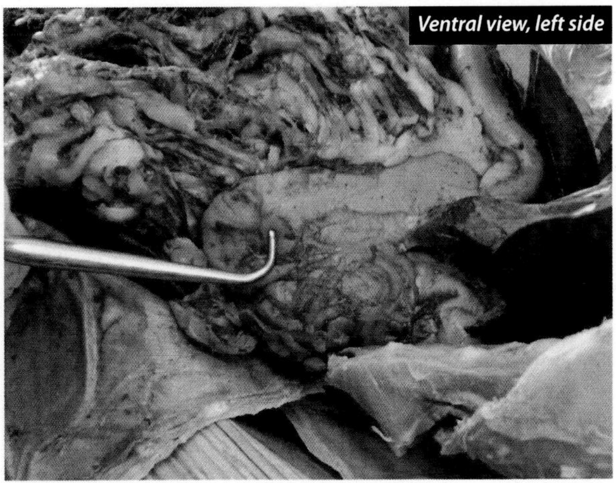


The stomach is a stomach-shaped organ that exists between the esophagus and the duodenum of the small intestine. The walls of the stomach are heavily muscled with three layers of smooth muscle. It has rugae in its walls, making it possible for them to stretch to hold food without increasing pressure. The stomach serves in several capacities. One function is to store food until the duodenum is ready to receive it. A second function is to mix the food with acid and enzymes to begin the digestive process. There is a pacemaker in the fundus of the stomach so that contractions occur automatically, usually beginning about half an hour before the normal meal time.

## Greater Omentum

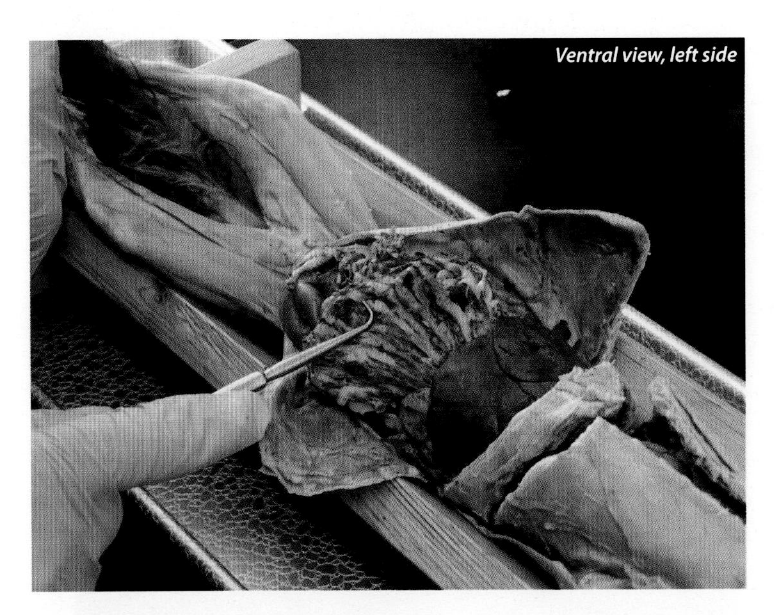


The greater omentum is part of the peritoneum. It is attached to the greater curvature of the stomach and to part of the duodenum. It is two layers thick and therefore it forms a bursa. It contains adipose tissue and that, coupled with the serous fluids it produces, acts as lubricants for the abdominal organs. It also can seal off areas that become infected, thereby reducing the spread of the infection. It produces macrophage cells that help remove debris from the omental cavity. Many of Dr. J's students think it looks suspiciously like a mozzarella cheese-impregnated hairnet!

## Spleen

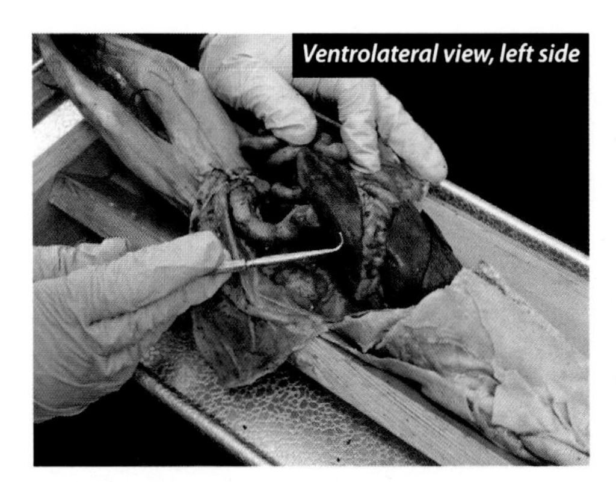


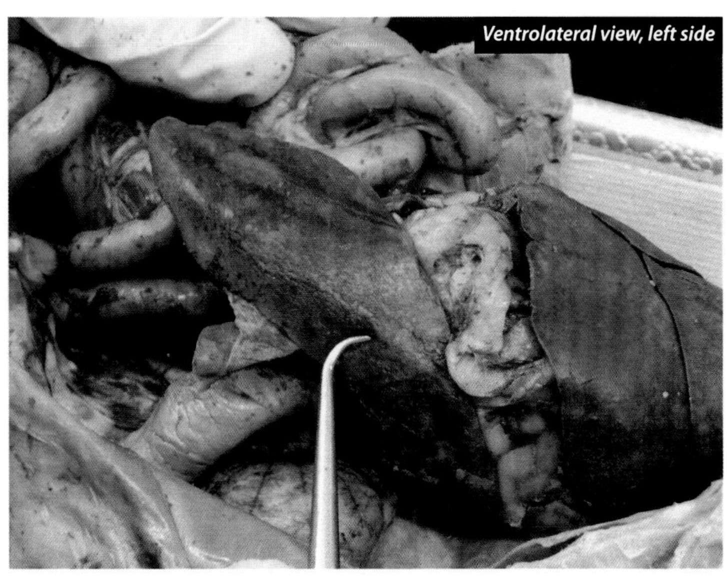


The spleen produces red blood cells before birth. It is an important part of the lymphatic system. The spleen can produce some white blood cells (lymphocytes and monocytes). It is the primary organ that removes old red blood cells, old platelets, and debris from the blood. If the spleen is removed, the liver takes over as the site of red blood cell removal. The spleen is also an area where blood accumulates. This is important in the case of blood loss, as this extra blood volume (350 to 550 ml) can be reflexively driven into general circulation to help restore the volume of actively circulating blood, and thereby help maintain adequate blood pressure to ensure normal functioning of the circulatory system. The spleen in humans is located along the midaxillary line, deep to ribs 9, 10 and 11. It is superior to the transpyloric line; thus it is in the left hypochondriac region. It is ovoid or spherical in shape. In the cat, we see that it strongly resembles Johnnycake, and in fact Dr. J suspects this may be where Johnnycakes come from!

## Pancreas

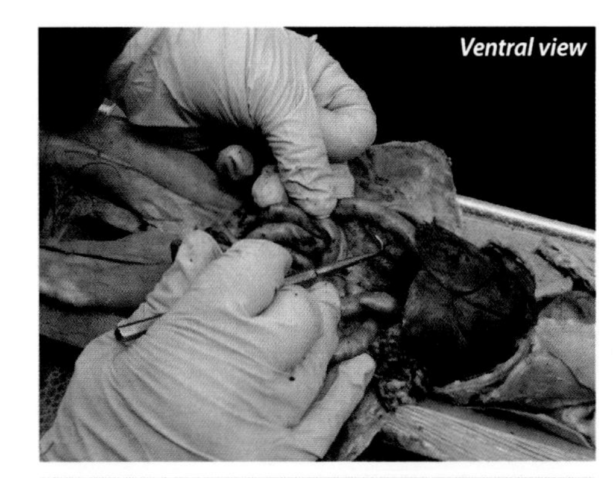


*Right lobe runs cranial to caudal and is found adjacent to the duodenum. Probes are on the right side.*

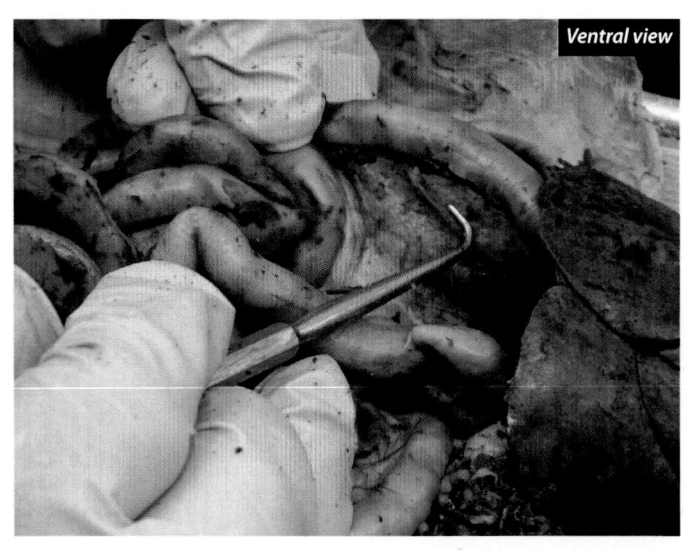

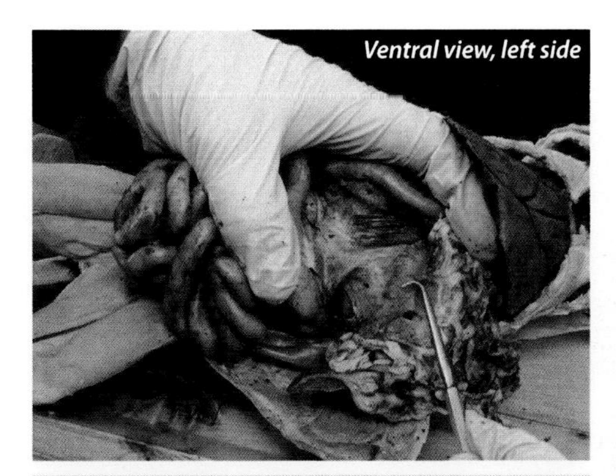


*Left lobe is transverse and is found caudal to the greater omentum.*

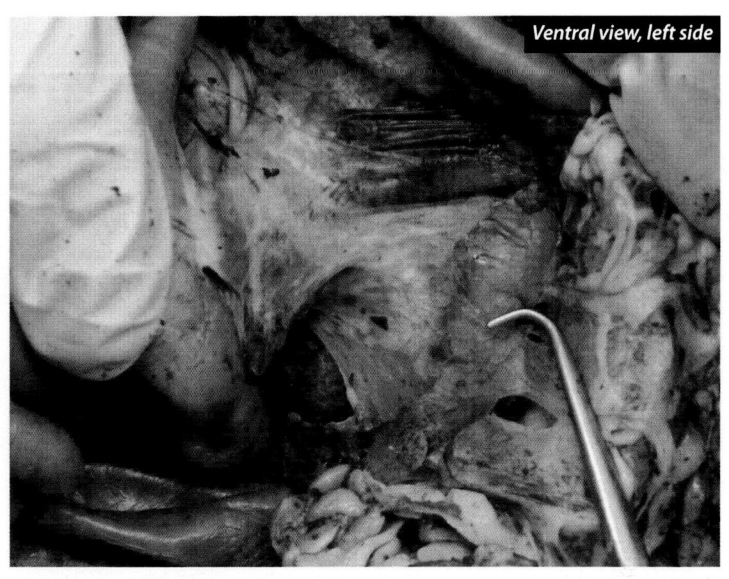


The pancreas is a bilobate mixed gland, both endocrine and exocrine. Its exocrine secretions include bicarbonates and ten digestive enzymes. These pass into the common bile duct, which in turn empties into the duodenum. The endocrine secretions include insulin and glucagon. These hormones are of particular importance in carbohydrate metabolism. The right lobe runs in a cranial/caudal orientation adjacent to the duodenum. The left lobe is transverse and extends from the duodenum on the right side to the spleen on the left side. The pancreas is retroperitoneal. It looks remarkably like Hamburger Helper®; in fact, this may be where Hamburger Helper® comes from!

# Small Intestine

## Duodenum

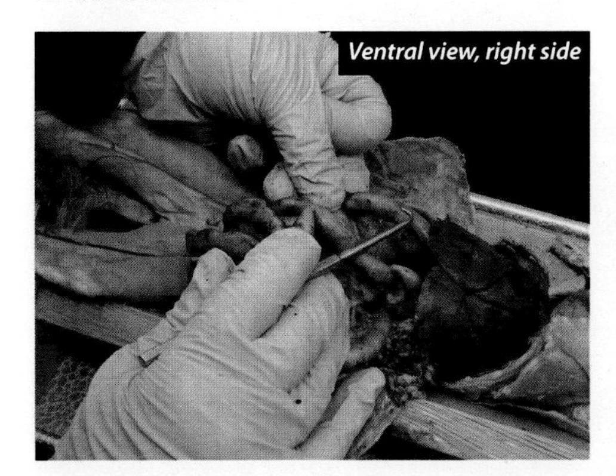


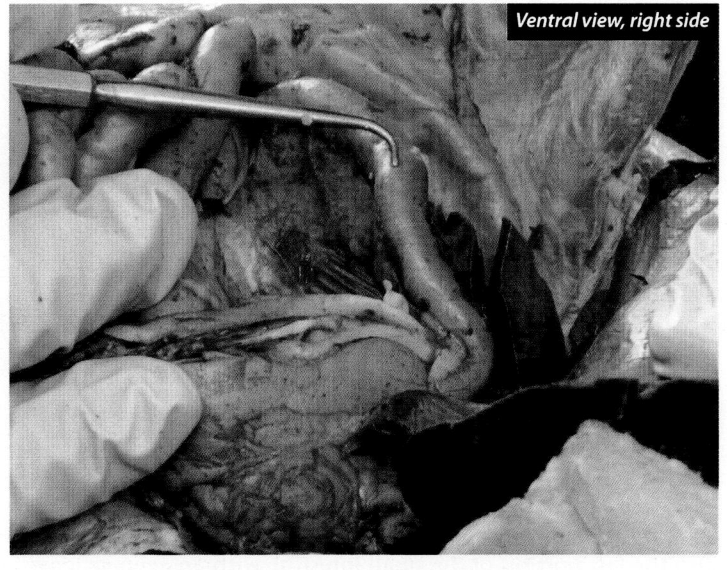


The duodenum is the first portion of the small intestine. The chyme enters the duodenum from the stomach and material leaves the duodenum to enter the jejunum. In a human, the duodenum is about ten inches in length. In the cat, the duodenum travels in a caudal direction from the pylorus of the stomach, while in the human it reflects and moves transversely toward the left side. In both, it travels adjacent to the pancreas. It receives the common bile duct (the hepatopancreatic duct) and the accessory pancreatic duct. The duodenum is retroperitoneal and is at the level of the transpyloric line (TPL).

## Jejunum

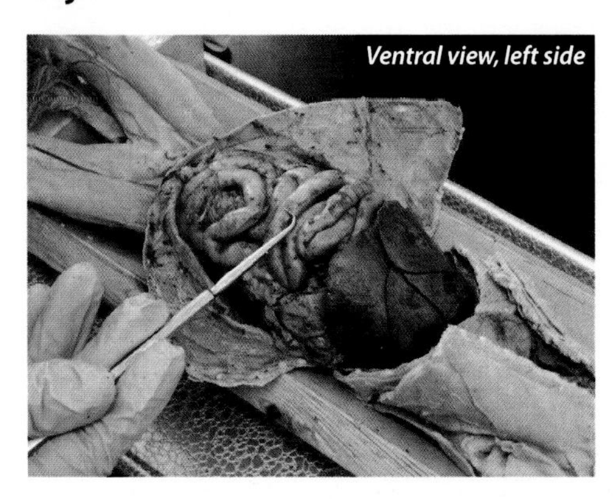


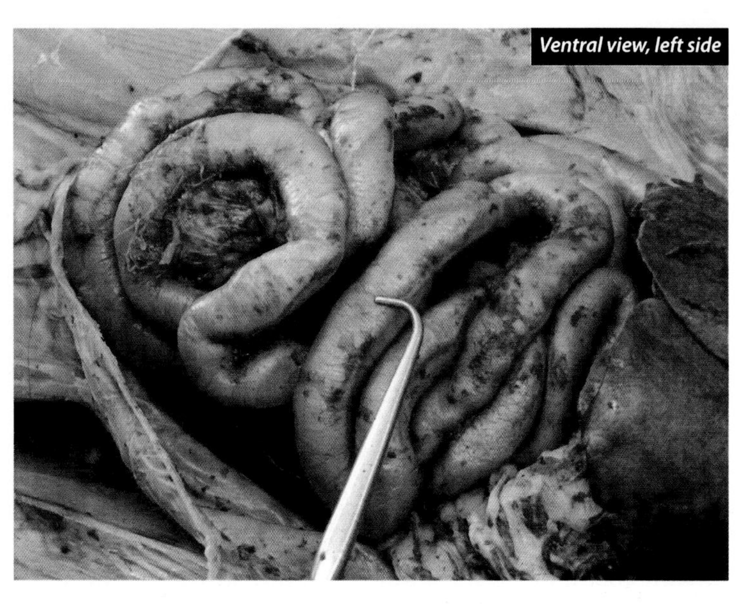


The jejunum of the cat makes up most of the small intestine and is highly coiled. The jejunum is the middle portion of the small intestine and has an average length of eight feet in humans. It receives material from the duodenum, and the material that leaves enters the ileum. The plicae circulares are more pronounced, and the villi are larger in the jejunum than in the ileum. It is largely in the left hypochondriac and left lumbar regions of the abdomen. Many students come to anatomy believing that Jejunum comes after MeMayum. They are wrong, it comes after the duodenum!

## Ileum

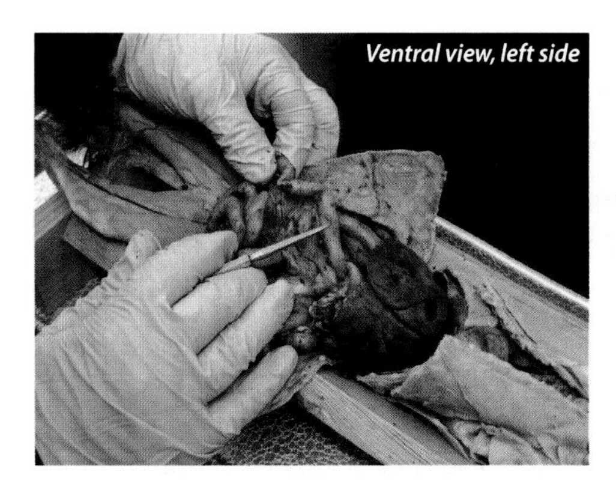


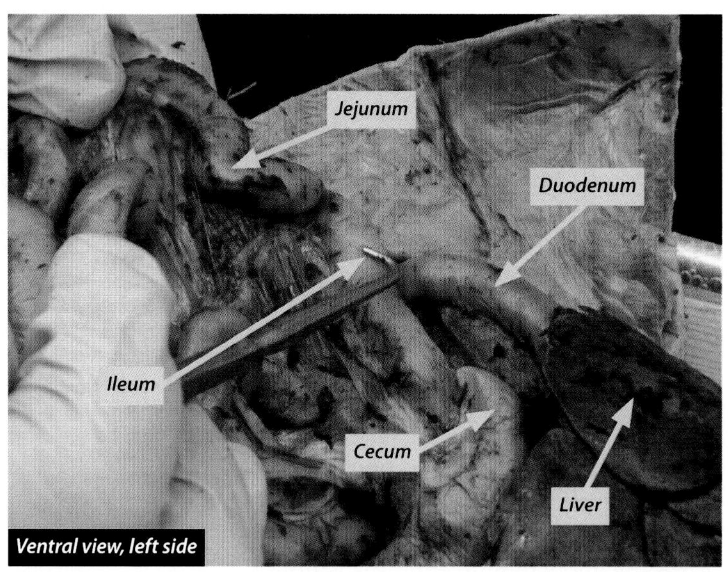


Spelling counts with the ileum! There is also an ilium that is a bone and part of the os coxa. The ileum of the cat is less coiled than the jejunum. It is approximately five inches long and joins the colon at the ileocecal (ileocolic) junction. The human ileum is also the last portion of the small intestine with an average length of eleven feet. It receives material from the jejunum and releases material into the colon by way of the ileocecal sphincter. The ileum has thinner walls than the jejunum. It is mostly in the right lumbar and right iliac regions of the human abdomen. It was named for the famous Cuban refugee boy, Ileum Gonzales, who was an anatomist in Cuba before coming to the United States.

## Mesentery

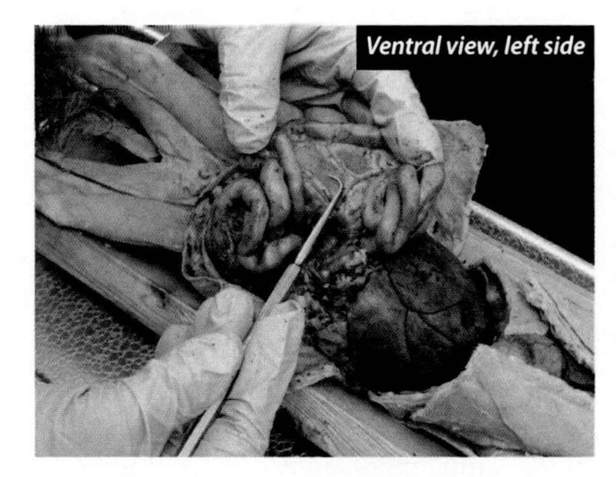


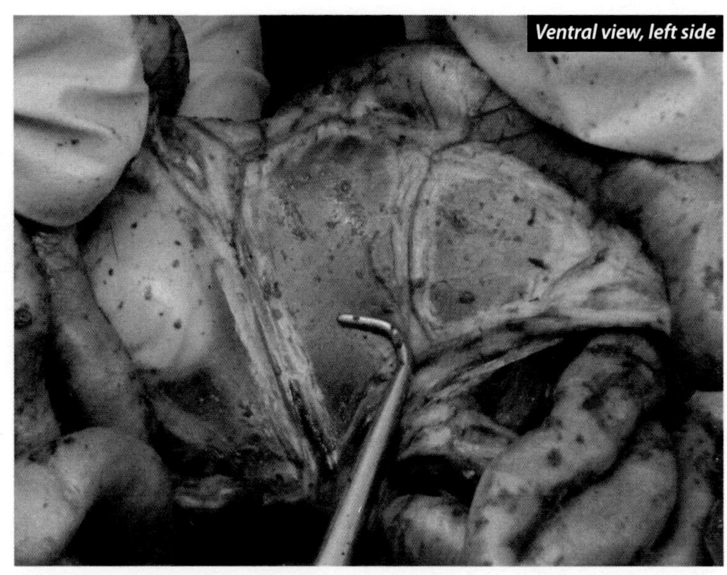


The mesentery is also known as the mesenteric ligament. It is a double layer of peritoneum that comes out around the small intestine. The arteries, veins, nerves, and lymphatics that come to and from the small intestine pass between these two layers. The mesentery is also of great importance because it prevents the small intestine from tying itself in knots. If the intestine does this, the person normally dies. This is what contributed to the death of Maurice Gibb, one of the *Bee Gees*. He toured as a lead singer for *Puff Diddy* too.

## Large Intestine

### Cecum

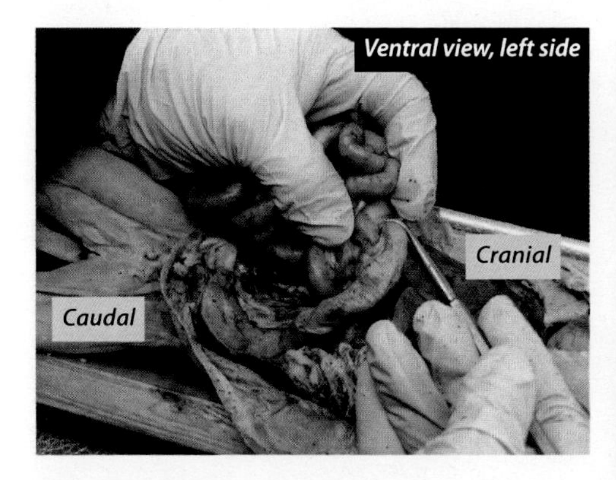


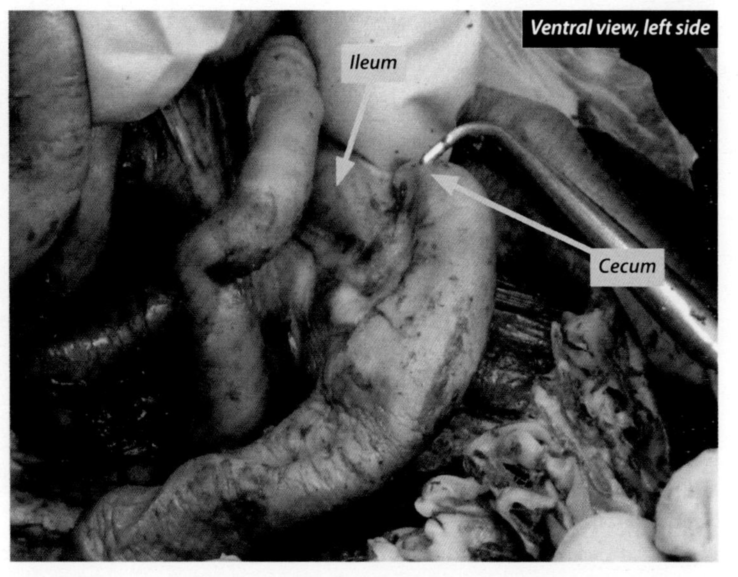


The cecum (caecum) is a blind sac, about 2.5 inches long, and located at the inferior end of the ascending colon. In the human, the appendix is attached to it. In the cat, it looks like the pope's slippers. Remember—cecum and you shall findem.

# Large Intestine

## Colon

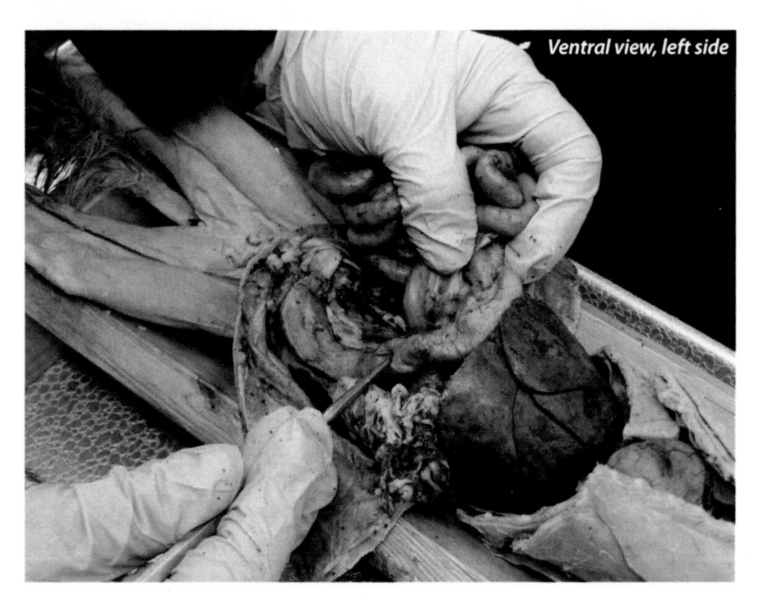


The colon is actually the anatomical conundrum of the cat. Why do we call it the colon when it looks like a question mark? Let Dr. J know if you come up with an answer. The colon is a major portion of the large intestine, functioning as a storage place for fecal material until that material is discharged by a process called elimination. The large intestine absorbs vitamins K and B12 produced by bacteria that inhabit it, and it is also important in the supplemental absorption of water and electrolytes. Be clear, though, that MOST absorption of water and electrolytes occurs in the small intestine. The shape of a human large intestine is quite different from that of the cat. In the human, there is a cecum, an ascending colon, a transverse colon, a descending colon, a sigmoid colon, a rectum, and an anal canal. In total, the human large intestine is about five feet long.

## Rectum

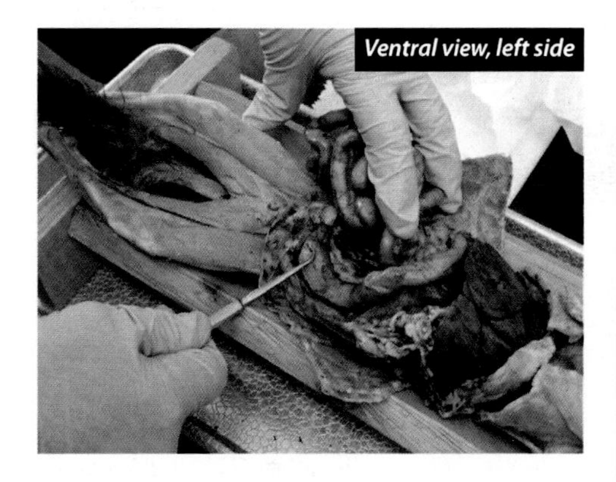


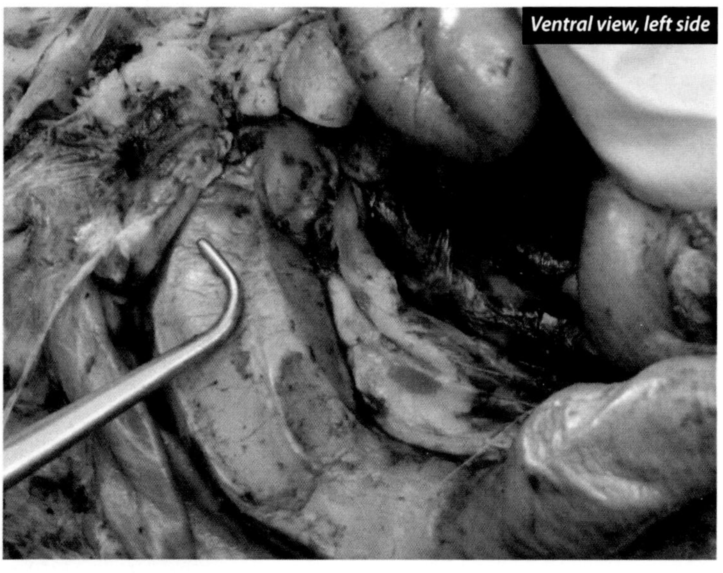


The rectum is a portion of the large intestine that is between the sigmoid colon and the anal canal. It is about five inches in length. The junction between the sigmoid colon and the rectum of a human is at the level of sacral vertebra 3. The rectum is of medical significance because a number of structures that are anterior to it can be palpated from within it. These structures include the prostate of the male and the cervix of the female.

## Urinary Bladder

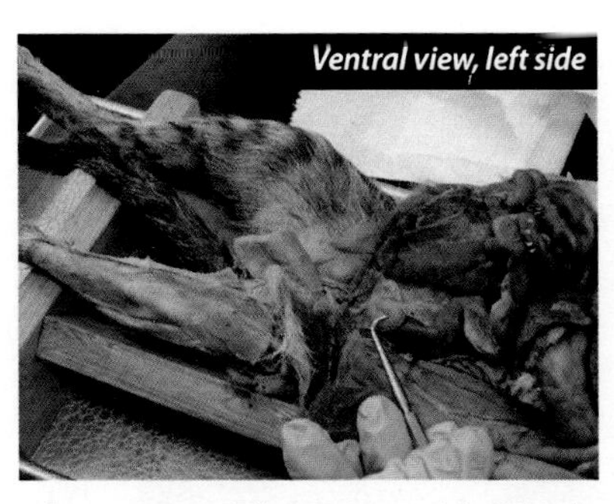
Ventral view, left side

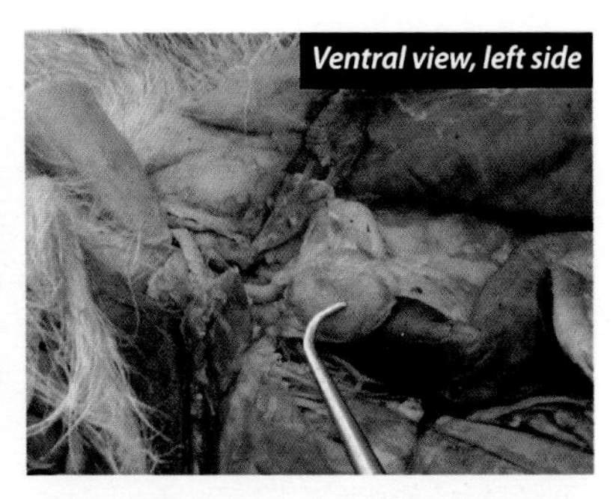
Ventral view, left side

The urinary bladder receives urine from the two ureters and stores it until it is discharged to the outside by means of the urethra. NOTE that it does **NOT** form the urine; it stores the urine. Urine is formed by the kidneys. The urinary bladder is retroperitoneal. Because the urinary bladder holds fluids, it must be able to expand without building up too much pressure. The urinary bladder has rugae that help it do this. The rugae are longitudinal folds similar to pleats in a skirt. They were named for Mr. October, Rugae Jackson, who was an anatomist before taking up baseball.

## Ureter

Ventral view, dorsal portion of abdomen, left side

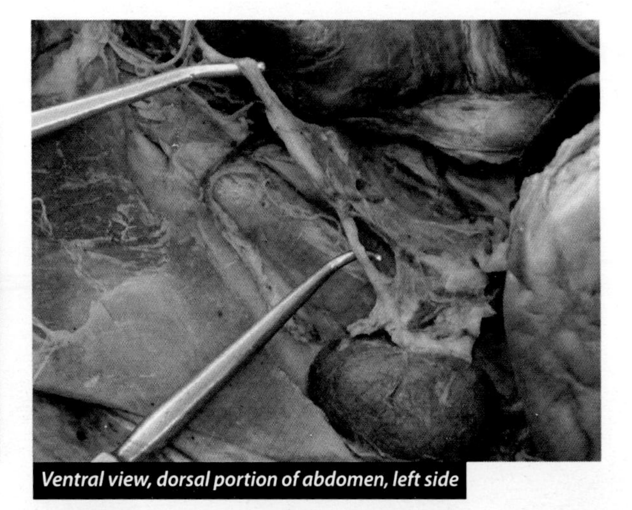
Ventral view, dorsal portion of abdomen, left side

This is the duct that transports urine from the kidney to the urinary bladder. Spelling is important for this term as it is close to the spelling of urethra, which is another structure. There is only one urethra, whereas there are two ureters, one on each side.

## Urethra

This is the duct (quack) that transports urine from the urinary bladder to the outside. It was named for Urethra Franklin, otherwise known as Lady Soul. Remember there is only one Urethra Franklin and there is only one urethra. Spelling is important for this term as it is close to the spelling of ureter. That is another structure, and there are two of them.

## Peritoneum

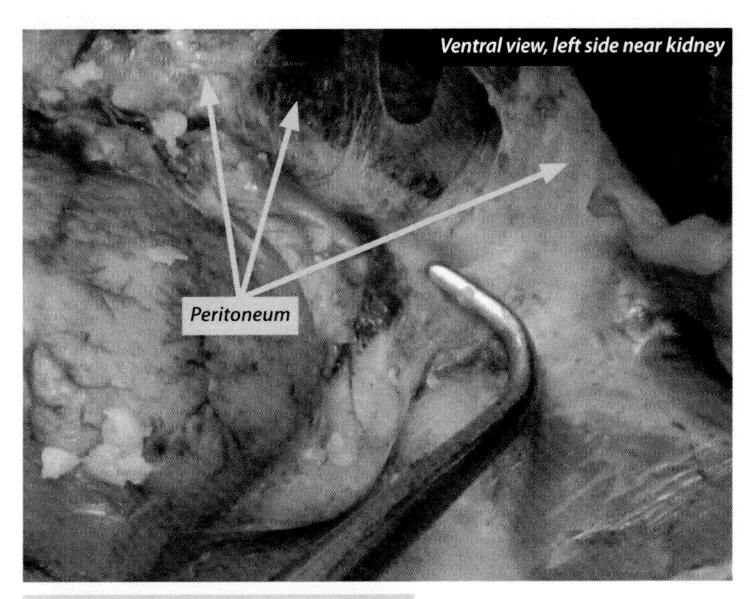


The probe is pointing to the suprarenal gland.

The peritoneum forms two abdominal organs, the greater omentum and the lesser omentum, which consist primarily of epithelial cells. They secrete a serous fluid that lubricates and reduces heat buildup. It covers the abdominal wall (the parietal peritoneum), as well as the organs (the visceral peritoneum). The mesentery is peritoneum, as is the broad ligament of the uterus. The pleura of the thoracic cavity is much like the peritoneum.

## Kidney

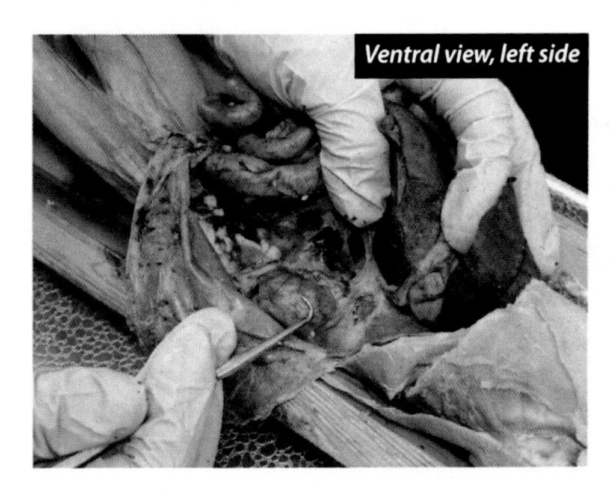


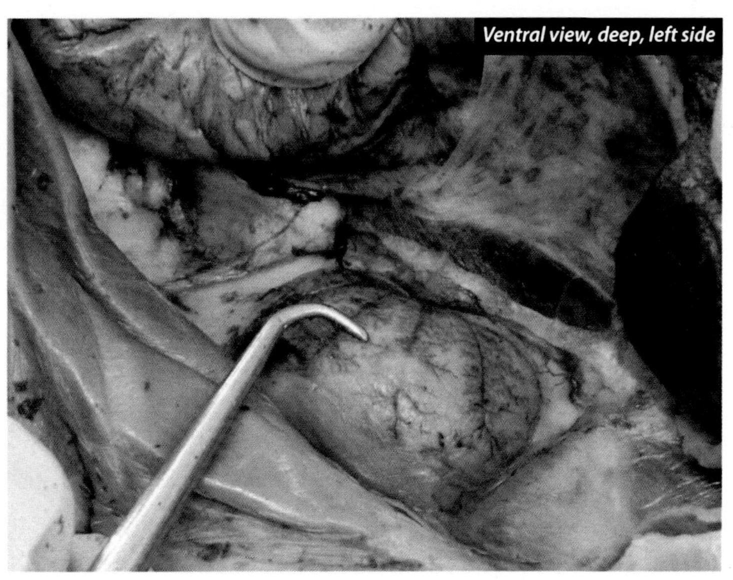


The kidney is a kidney shaped organ (LOMG) and is the major excretory organ of the abdomen. It has portal circulation. It is also retroperitoneal. In the cat, the right kidney is cranial to the left kidney, while in humans the right kidney is inferior to the left kidney.

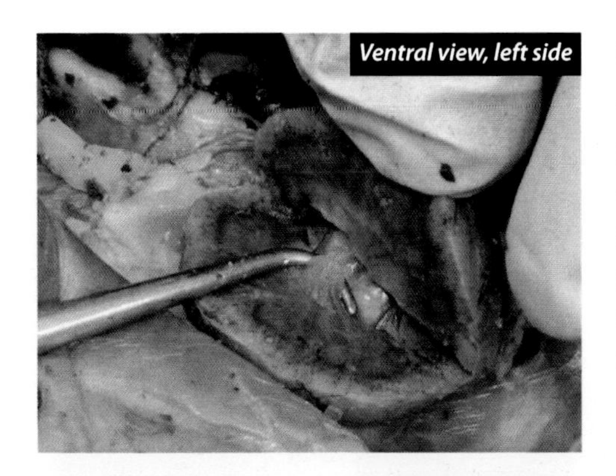


## Calyx

The calyx of the kidney is actually a funnel or cup-shaped space. In the lab specimens of the pig kidney, this space is often filled with yellow latex. There is a calyx for each pyramid in humans and in pigs, while there is only one calyx in the cat since the medulla does not have subdivisions. In humans and pigs, the calices drain into the renal pelvis. In cats, the calyx leads to the renal pelvis.

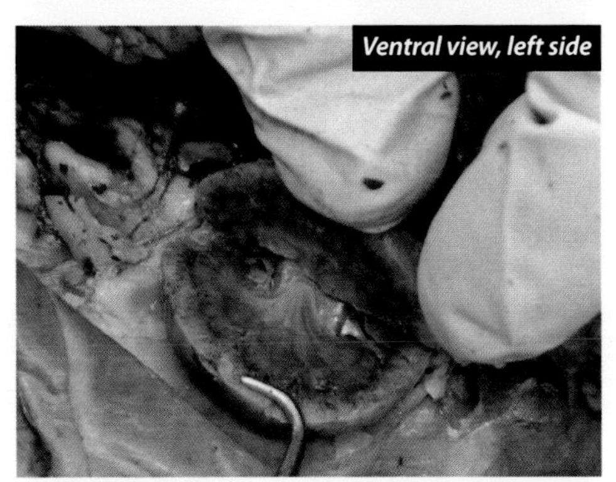


## Cortex

Cortex is a term that refers to the outer layer of a structure. The cortex of the kidney is where the glomerulus, Bowman's capsule, and part of the tubular system of the nephron are found. It is usually a lighter color than the medulla.

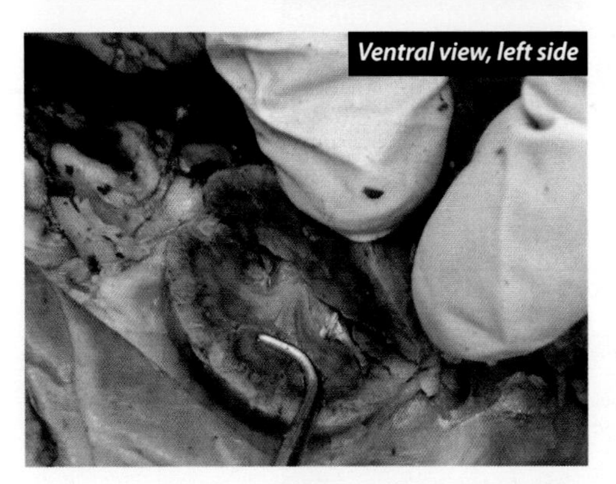


## Medulla

Medulla is a term that refers to the inner portion of a structure. The medulla of the kidney is where the Loop of Henle is found and where the tubule system empties its contents into the calyx. At this point, the content of the tubules is urine. In humans and in pigs, the medulla is broken down into triangular divisions called pyramids. The apex of the pyramid points toward the calyx it empties into. Normally a human will have eight to ten pyramids while pigs usually have ten or more pyramids in each kidney. The cat has a single piece medulla that is sometimes called a pyramid. The medulla is usually darker than the cortex.

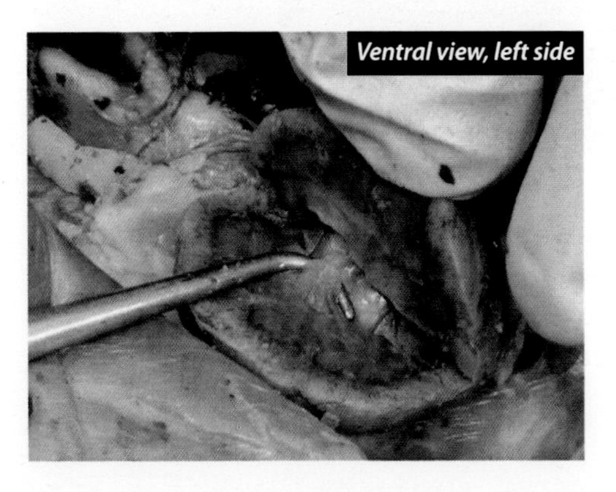


## Pyramid

In humans and in pigs, the medulla is broken down into triangular divisions called pyramids. The apex of the pyramid points toward the calyx into which it empties. Normally a human will have eight to ten pyramids while pigs usually have ten or more pyramids in each kidney. The cat has a single piece medulla that is sometimes called a pyramid. The pyramids of the medulla are usually darker than the cortex. Dr. J thinks they look more like chocolate kisses since they are circular in cross section.

## Suprarenal (Adrenal) Gland

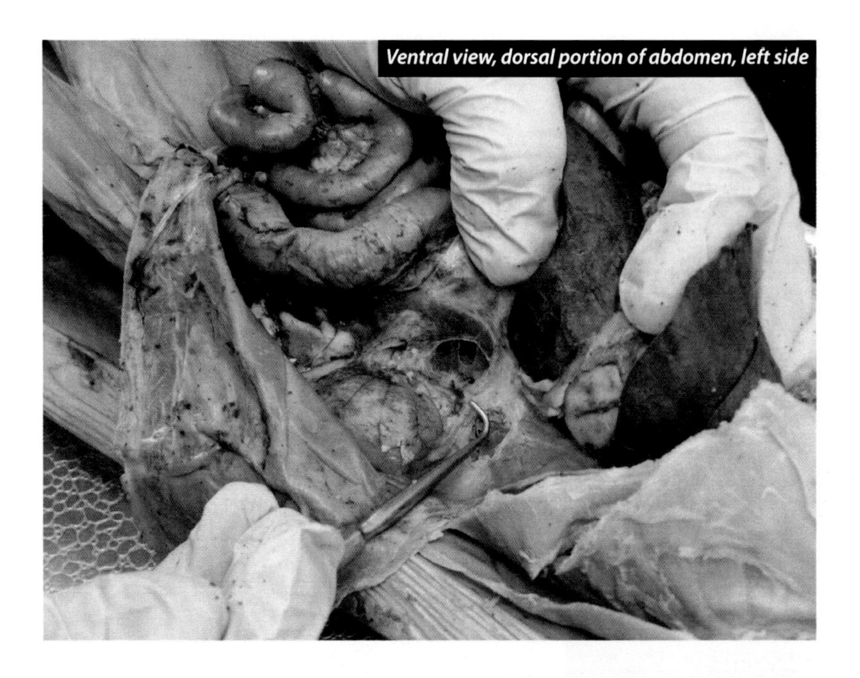

*Ventral view, dorsal portion of abdomen, left side*

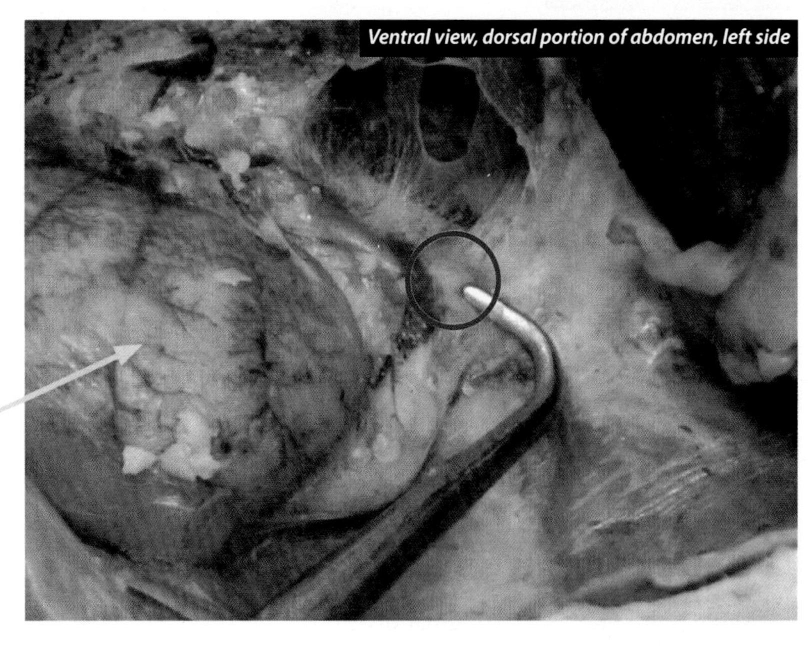


*Ventral view, dorsal portion of abdomen, left side*

The adrenal glands are endocrine glands. There is an outer layer, the cortex, and an inner layer, the medulla. The cortex secretes adrenal steroids (including sex hormone precursors—female and male in both sexes), aldosterone, and cortisol. The medulla produces epinephrine and norepinephrine. The adrenal glands of humans are on the superior surface of the kidneys, whereas in the cat they are about 1 cm cranial and medial to the kidneys. They are retroperitoneal.

# NERVES

## Lumbar Nerves

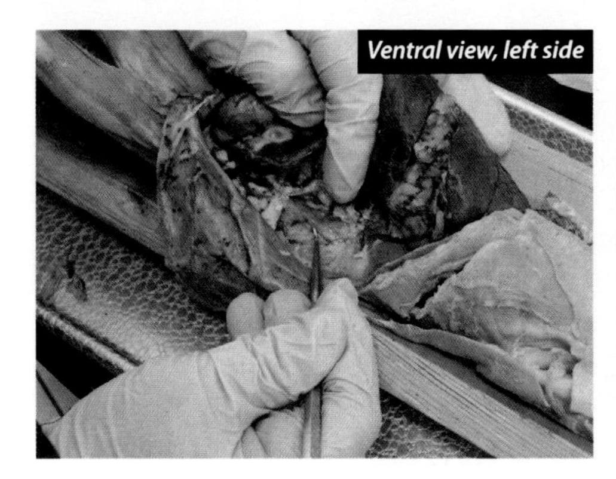


The medial and lateral lumbar nerves are branches of the ventral rami of the third lumbar nerve. They can be found dorsal to the kidney and lateral to the psoas major muscle.

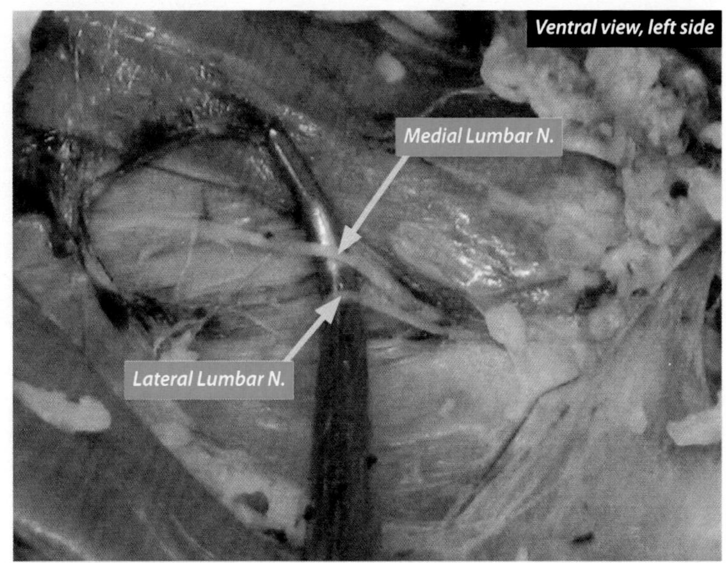


Ventral view, deep to kidney, left side. Dr. J has pulled left kidney aside with his left hand.

## Lateral Cutaneous Nerve

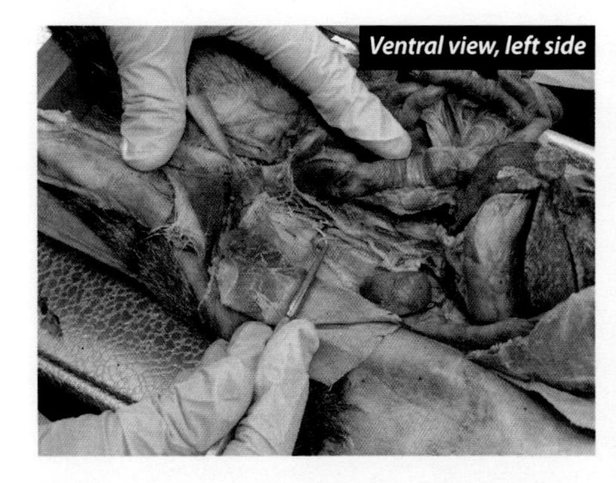


The lateral cutaneous nerve is usually found close to the deep **iliac circumflex artery** on the surface of the psoas major. Usually it is the diameter of dental floss in the cat. It passes through the internal and external abdominal oblique muscles and innervates the posterolateral gluteal skin.

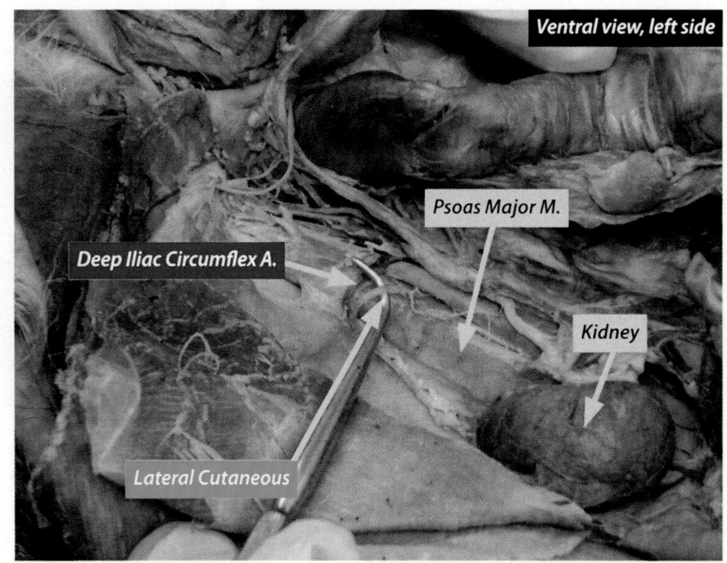

## Femoral Nerve

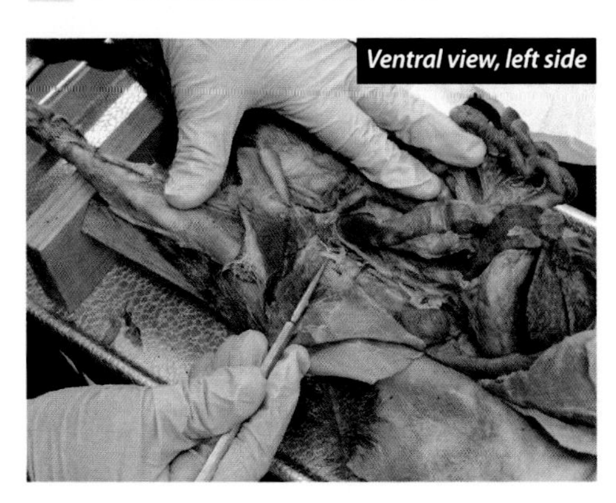


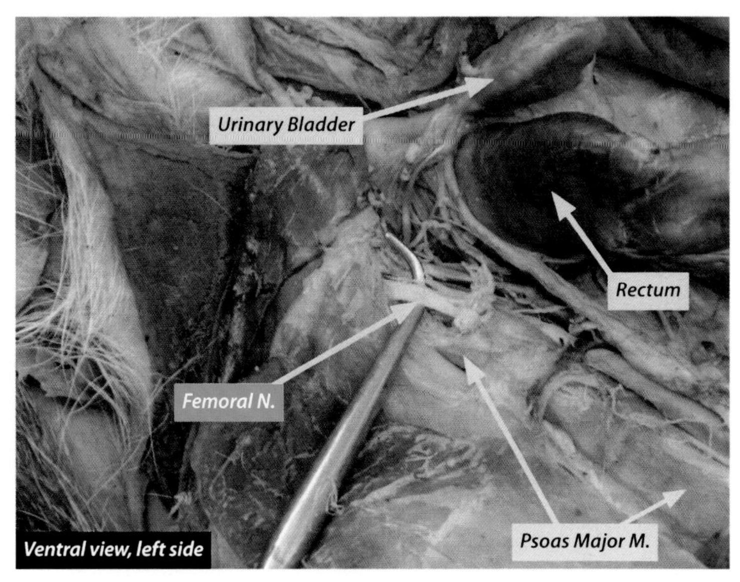


The origin of the femoral nerve is the anterior rami of L2–L4. The femoral nerve emerges from the psoas major muscle just before it passes deep to the inguinal ligament. It enters the thigh lateral to the **femoral artery**. It serves the iliacus, and once in the thigh it divides and serves the pectineus muscle as well as the muscles of the anterior compartment. It gives rise to the saphenous nerve. It also has a sensory function to the skin of the anterior surface of the thigh, and via the saphenous nerve, to the medial surface of the leg. In the cat, the femoral nerve is about the size of number 9 vermicelli. The genitofemoral nerve is smaller and found medial to the femoral nerve, as you will see in the pictures below. This information is included to help you orient one relative to the other. Note: We have labeled the picture below to reflect this.

## Genitofemoral Nerve

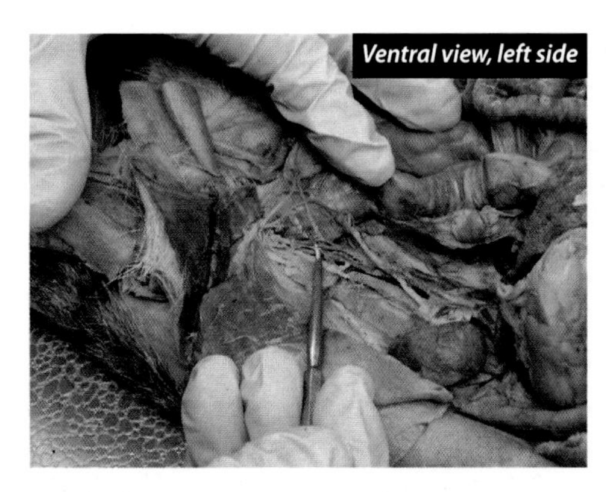


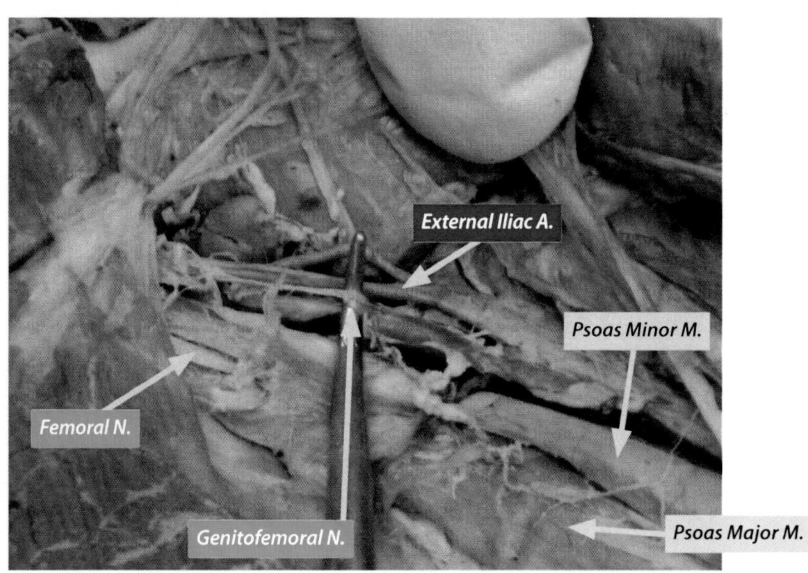


The genitofemoral nerve is usually found adjacent to the **external iliac artery** in the cat. It bifurcates into a genital branch and a femoral branch. The genital branch in men innervates the cremaster muscle, and some fibers continue on the superior skin of the scrotum. In females, the fibers travel with the round ligament of the uterus to the inguinal canal and then on to the labium majus and the skin of the mons pubis. The femoral branch passes into the femoral sheath and then innervates the skin on the anterior proximal portion of the thigh. Because of this, it is sometimes affectionately known to Dr. J's anatomy students as the "heebeegeebee nerve." That's because when someone rubs the proximal portion of the thigh, the person experiences the heebeegeebees. It is recommended that you inform a person that you are going to do this before performing this experiment; otherwise you may end up on the late news! LOMG!

# HUMAN ABDOMEN—OVERVIEW

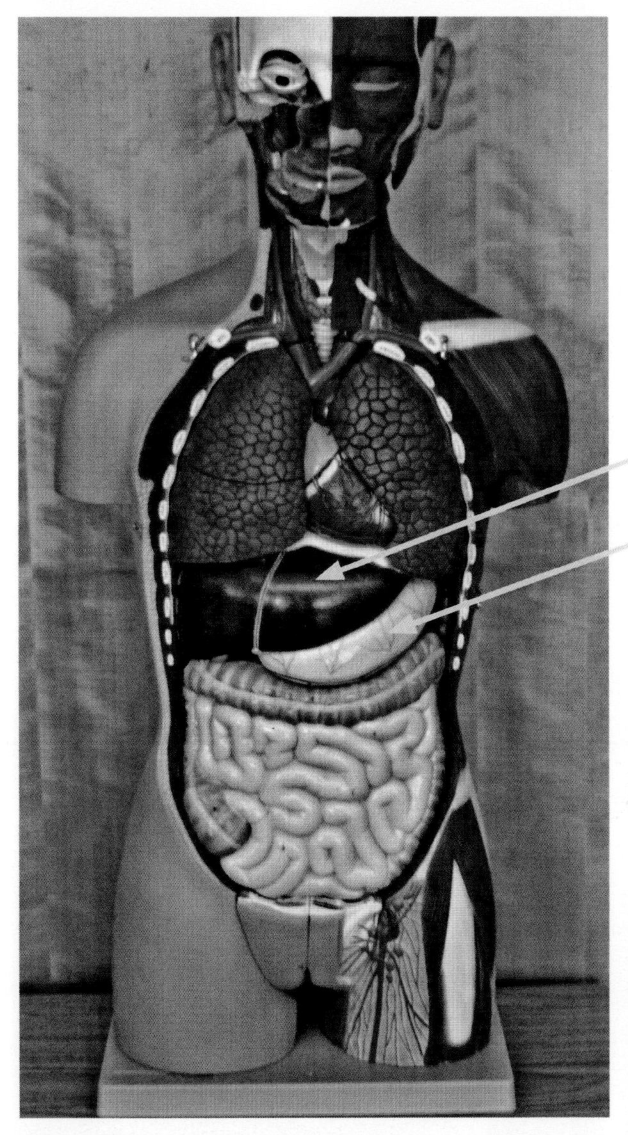

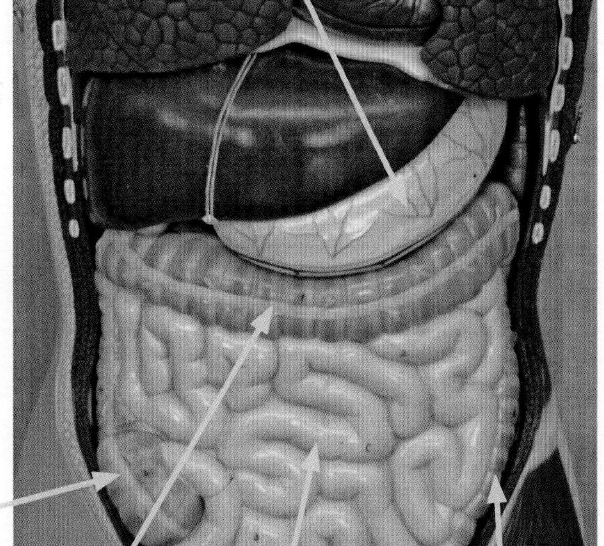

# HUMAN ABDOMEN—OVERVIEW

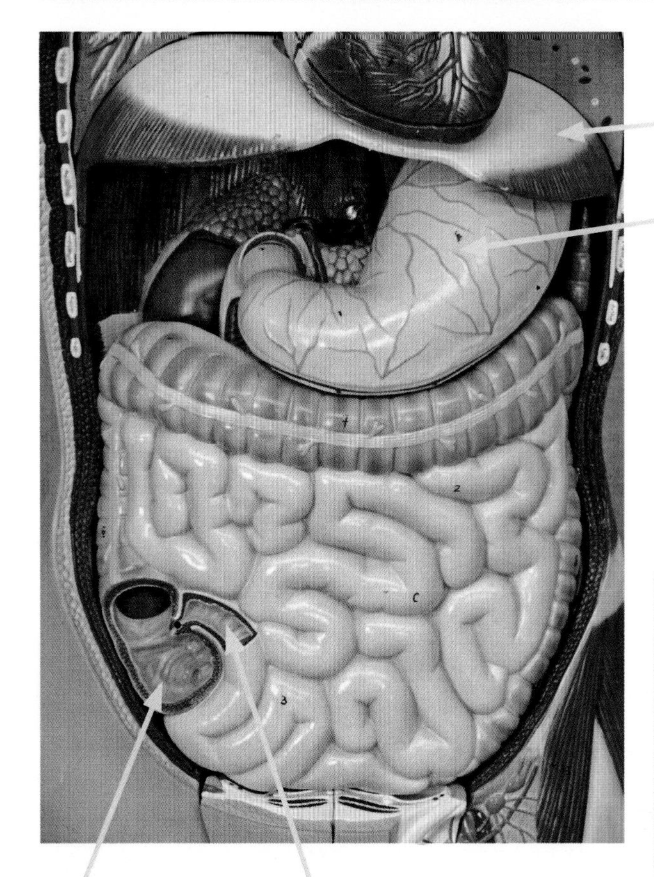


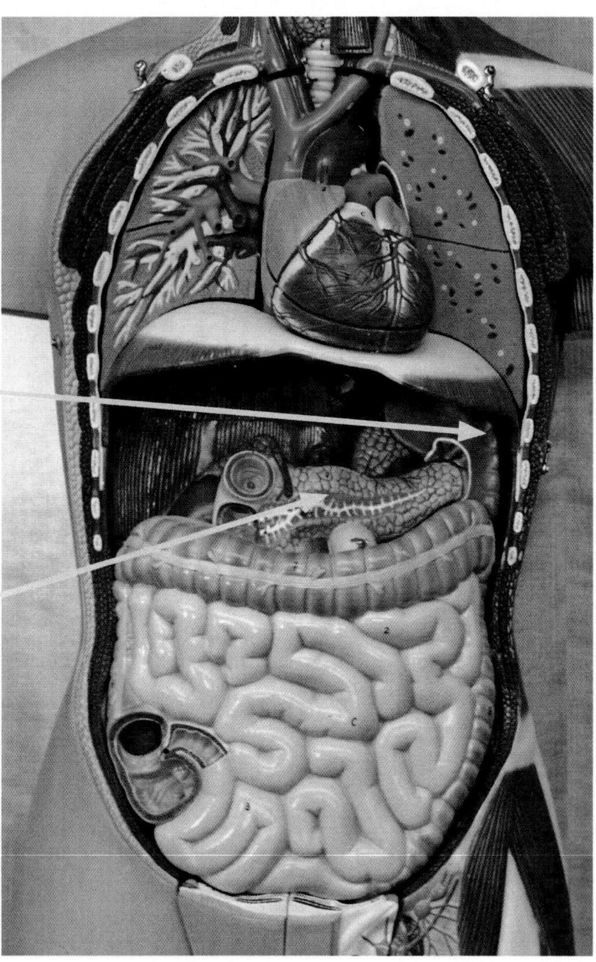

# HUMAN ABDOMEN—OVERVIEW

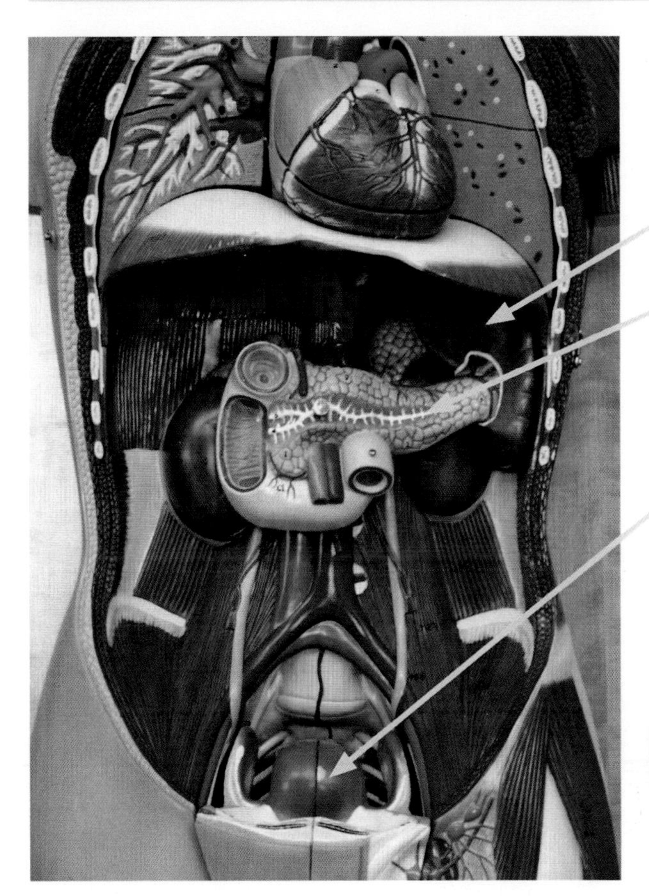


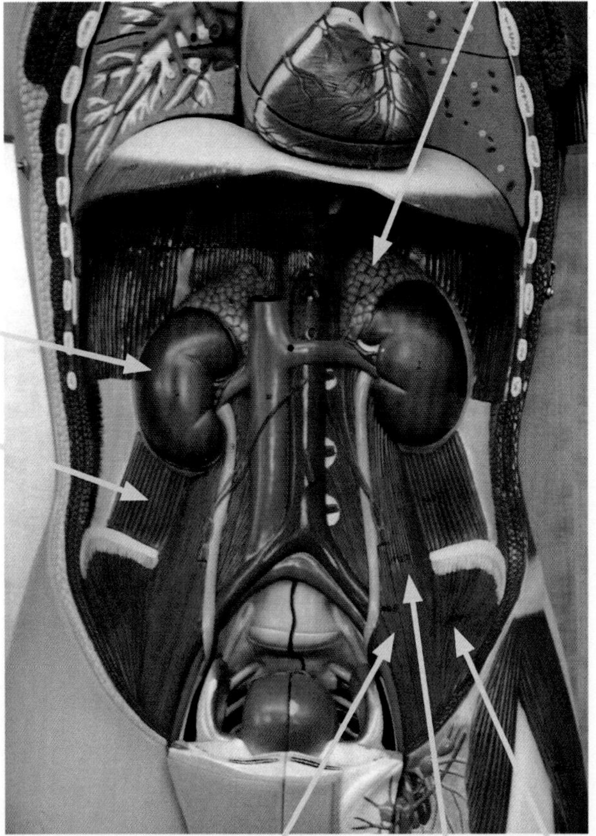


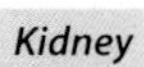

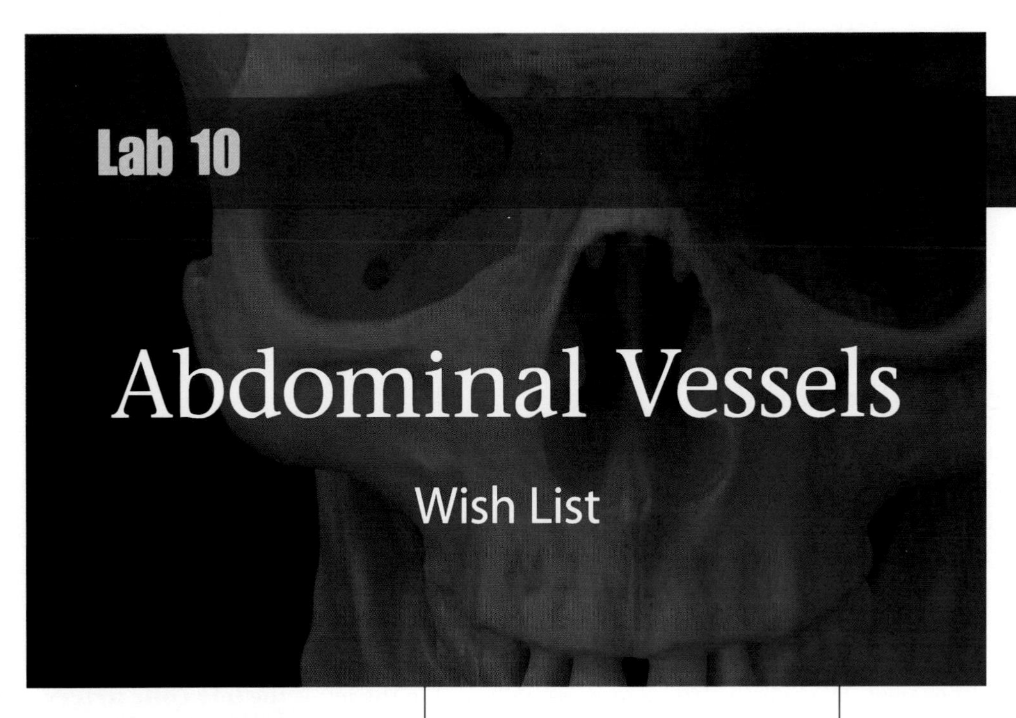


# Abdominal Vessels

## Wish List


**LAB 10 OVERVIEW, pp. 294–296**
- **Abdominal Vessel Diagram, p. 297**

**NERVES, p. 298**
- Obturator Nerve, p. 298
- Lumbosacral Cord, p. 298

**VESSELS, pp. 297, 299–329**

**ARTERIES** and **VEINS**
- Caudal Vena Cava, p. 299

**HEPATIC PORTAL VEIN, p. 299**
- **Vessel Plate—Ventral Branches of Abdominal Aorta: View 1, p. 300**
- **Vessel Plate—Ventral Branches of Abdominal Aorta: View 2, p. 301**
- **Abdominal (Descending) Aorta, p. 302**
- **Ventral Branches of Abdominal Aorta, pp. 302–312**
  - Celiac Artery Trunk, p. 302
  - • Hepatic Artery, pp. 302, 305
  - • Left Gastric Artery, p. 303
  - • Splenic Artery, p. 304
    - –Cranial Splenic Artery, p. 304
      - • Left Gastroepiploic Artery, p. 304
    - –Caudal Splenic Artery, p. 304
  - Hepatic Artery (continued), p. 305
  - • Cystic Artery, p. 305
  - • Gastroduodenal Artery, p. 305
    - –Right Gastric Artery, p. 306
    - –Right Gastroepiploic Artery, pp. 306–307
    - –Cranial Pancreaticoduodenal Artery, p. 308
  - Cranial Mesenteric Artery, p. 308
- **Vessel Plate—Ventral Branches of Abdominal Aorta: View 3, p. 309**
  - • Middle Colic Artery, p. 310
  - • Caudal Pancreaticoduodenal Artery, p. 310
  - • Ileocecal (Ileocolic) Artery, p. 311
  - • Intestinal Arteries, p. 311
  - Caudal Mesenteric Artery, p. 312
  - • Left Colic Artery, p. 312
  - • Cranial Rectal (Hemorrhoidal) Artery, p. 312
- **Vessel Plate—Lateral Branches of Abdominal Aorta: S/He Cat, p. 313**
- **Lateral Branches of the Abdominal Aorta, pp. 314–316**
  - Adrenolumbar Artery, p. 314
  - Renal Artery and Vein, p. 314
  - Gonadal Arteries and Veins, p. 315
  - • Internal Spermatic Artery and Vein, p. 315
  - • Ovarian Artery and Vein, p. 315
  - Deep Iliac Circumflex (Deep Circumflex Iliac) Artery, p. 316
  - External Iliac Artery, p. 316
- **Dorsal Branches of the Abdominal Aorta, p. 317**
  - Lumbar Arteries p. 317
- **Vessel Plate—Pelvic Vessels, p. 318**
- **Pelvic Vessels Overview, p. 319**
- **Pelvic Vessels, pp. 320–323**
  - Common Iliac Artery, p. 320
  - Median Sacral Artery, p. 320
  - Internal Iliac (Hypogastric) Artery, p. 321
  - • Umbilical Artery, p. 321
  - • Cranial Gluteal Artery, p. 322
  - • Caudal Gluteal Artery, p. 322
  - • Internal Pudendal Artery, p. 323

**ADDITIONAL VESSEL PLATES, pp. 324–326**

**PRACTICE QUIZ AND ANSWERS, pp. 327–329**

# LAB 10 OVERVIEW

Lab 10 focuses on the study of the **abdominal aorta** and its branches. Particular attention will be given to the organs these vessels serve and to five abdominal **anastomoses**.

## 1. Ventral Branches of the Abdominal Aorta

We begin our observations by placing the cat on its right side. We will reflect the organs to expose the three major ventral branches of the **abdominal aorta**. The most cranial is the **celiac trunk**, and it courses ventrally. Caudal, and in very close proximity to the **celiac trunk**, is the **cranial mesenteric artery**. It is directed at about a 45° angle toward the ventral side and has a substantial diameter. The most caudal ventral branch of the abdominal aorta, the **caudal mesenteric artery**, will be found just cranial to the two **external iliac arteries**. Often the ventral end of the **caudal mesenteric artery** will lean toward the caudal end of the cat as though the artery were under acceleration. It usually looks like a capital Y with two terminal branches of its own. The branch that courses cranially and to the left is the **left colic artery**, serving the left side of the **colon**. The caudally directed branch of the **caudal mesenteric artery** is the **cranial rectal artery** and it serves the cranial end of the **rectum**. As you can see, there are many Grant things in the abdomen!

Next we will move cranial to observe the **celiac trunk** and its branches. The most dorsal branch of the celiac trunk is the **hepatic artery**. This thick vessel usually is directed obliquely to the right side and it is sometimes difficult to see because the **left gastric artery** covers it. We will see the **hepatic artery** again as the left hand side of the "X." As its name implies, it serves the **liver**. Just ventral to the **hepatic artery** will be the **left gastric artery**. The **left gastric artery** will be seen again on the left side of the lesser curvature of the **stomach**; and, as its name implies, it serves the **stomach**. It is also important because it forms functional **anastomoses** with the **right gastric artery**. What remains of the **celiac trunk** is the **splenic artery**. The **splenic artery** moves ventral and bifurcates into a **cranial splenic artery** and a **caudal splenic artery**. These two branches serve the **spleen**. The **left gastroepiploic artery** is usually a branch of the **cranial splenic artery**. Although we will not observe this vessel, it has functional importance because it forms **anastomoses** with the **right gastroepiploic artery**, thereby providing collateral circulation to the greater curvature of the **stomach** and the **greater omentum**.

Place the cat on its back and move it so that the edge of the cradle is at the level of the cat's **diaphragm**. Then bend the cranial end of the cat dorsally and this will open up the space between the **liver** and the **stomach**. Reflect the **papillary process of the caudate lobe of the liver** to the left to expose the hepatic artery. You should observe that it courses obliquely to the right as it approaches the **liver**. Then it bends back to the left side and in doing so it forms the left side of the "X." The **hepatic artery** has two branches at the "waist" of the "X." The cranial branch is the **cystic artery**, which serves the **gall bladder**. The caudal branch is the **gastrohepatic artery**. As already noted, we can also observe the **left gastric artery** from this view as it arrives at the left side of the lesser curvature of the **stomach**.

The **gastrohepatic artery** will have three branches that we study. The first branch is the **right gastric artery** that serves the right side of the **stomach** along the lesser curvature. You are not required to find this small artery, but you will need to know that it forms an **anastomosis** with the **left gastric artery** on the lesser curvature of the **stomach**. The **gastrohepatic artery** then bifurcates into a **right gastroepiploic artery** and a **cranial pancreaticoduodenal artery**. The **right gastroepiploic artery** will be found in the **greater omentum** as it follows the greater curvature of the **stomach** toward the left side. It will form an **anastomosis** with the **left gastroepiploic artery**, as mentioned above. It is usually about one quarter of an inch from the **stomach** and it serves both the **stomach** and the **greater omentum**. The **cranial pancreaticoduodenal artery** will be found at the cranial end of the **pancreas** and the **duodenum**. It usually runs alongside of the **cranial pancreaticoduodenal vein**. It serves the **pancreas** and the **duodenum** and forms an **anastomosis** with the **caudal pancreaticoduodenal artery**. This is a good opportunity to see the **caudal pancreaticoduodenal artery** as it enters the caudal end of the **pancreas** and **duodenum**. This vessel is a branch of the **cranial mesenteric artery**, which is the next vessel we will investigate. Note that I have the vein in yellow. This vessel is part of the **hepatic portal system** and, if it has dye in it, the dye will be yellow.

We will study the branches of the **cranial mesenteric artery**. First we will observe the **middle colic artery**. I suggest that you find the middle of the **colon** and reflect it to the right side. Be careful **NOT** to pull the **colon** caudally! If you do that, you will most likely rip off all of the vessels going to or coming from the **colon**. If you are holding the middle of the **colon**, you should see the **middle colic artery** running toward your fingers. This vessel serves the **colon** and forms **anastomoses** with the **left colic artery**, which is a branch of the **caudal mesenteric artery**. You already observed the **caudal pancreaticoduodenal artery**. It is usually the second branch off the **cranial mesenteric artery**. Next we will find the **ileocecal** (**ileocolic**) **artery** as it runs over the **ileocecal (ileocolic) junction**. We then can observe the terminal branches of the **cranial mesenteric artery**, the **intestinal arteries**, by holding the **jejunum** away from the cat and inspecting the **mesentery (mesenteric ligament)**. The **intestinal arteries**, **intestinal veins**, and **lymphatic vessels** all pass through the double layered **mesentery**.

The branches of the **caudal mesenteric artery** are not very large. The **left colic artery**, as already noted, serves the left side of the **colon** and is functionally important because it forms an **anastomosis** with the **middle colic artery**. The **cranial rectal artery** serves the cranial end of the **rectum**.

## 2. Lateral Branches of the Abdominal Aorta

Reflect the **gastrointestinal organs** toward the right side to expose the dorsal abdominal wall on the left side. We will be able to study the lateral branches of the **abdominal aorta** on the cat's left side from this vantage point. Caudal to the **diaphragm** and cranial to the **kidney**, you will find the **left adrenolumbar artery**. It runs laterally and then caudally, to where it forms **anastomoses** with the **left deep iliac circumflex artery**, which is also a lateral branch of the **abdominal aorta**. Caudal to the **left adrenolumbar artery** is the **left renal artery**, which serves the **kidney**. Moving caudally, we will find the **left gonadal artery**. Every cat you see will either be a male or a female. Therefore, **gonadal artery** will not be a correct answer on the third practical. If the cat is a male, the vessel will be the **left internal spermatic** (**testicular**) **artery**, and if it is a female, it will be the **left ovarian artery**. These vessels leave the aorta at the same location in males and females because of the common area of origin for the **gonads**. The fourth lateral branch as we move caudally is the **left deep iliac circumflex** (**deep circumflex iliac**) **artery**. This vessel moves laterally and then cranially to where it forms **anastomoses** with the **left adrenolumbar artery**. The most caudal lateral branch of the **abdominal aorta** in the cat is the **left external iliac artery**. It leaves the abdomen and courses into the thigh. It ends with the **left deep femoral artery**, which will be studied in Lab 12. We will also see that the arrangement of the **iliac arteries** is slightly different in humans.

Before moving into the study of the dorsal branches and the pelvic vessels, we should note an anomaly to the usual bilateral symmetry of the lateral branches. We have described the five major lateral branches of the **abdominal aorta**. The same branches exist in nearly the same positions on the right side. However, we find that although there are **veins** that correspond to each **artery** on the right side, and that each of these

**veins** enters the **caudal vena cava** separately, on the left there is a mistake. On the left, the **gonadal vein** continues cranially and joins the **renal vein** to enter the **caudal vena cava** as a trunk. The same mistake is found in humans. This will be discussed in lab.

## 3. Dorsal Branches of the Abdominal Aorta

If you gently pull the **aorta** away from the **vertebral column**, you will see a number of branches running in a dorsal direction. Those are the **lumbar arteries** and they serve the **spinal cord** and the muscles of the dorsal abdominal wall.

## 4. Pelvic Vessels

You will find a single vessel continuing from the point where the aorta gives rise to the two **external iliac arteries**. This single vessel is the **common iliac artery**. It will give rise to a **left** and **right internal iliac artery**.

As the **left internal iliac artery** courses away from the **common iliac artery**, the first branch moves medially to the **urinary bladder**. This is what remains of the **left umbilical artery**. You should remember that the **umbilical artery** originally carried **mixed fetal blood** to the **placenta**. Next is a lateral branch, the **left cranial gluteal artery**. It effectively pulls the **left internal iliac artery** laterally, and it is situated very close to the **left obturator nerve**. This nerve passes through the obturator foramen and then serves the medial compartment of the thigh. The **left internal iliac artery** then courses back medially to where it ends with a medial branch, the **left internal pudendal** (**middle hemorroidal**) **artery**, which serves the rectum, and a lateral branch, the **left caudal gluteal artery**. Dorsal to the **left obturator nerve**, within the pelvis, you should observe the **left lumbosacral cord**. This forms part of the **sciatic nerve** that serves the posterior compartment of the thigh.

After the two **internal iliac arteries** branch, the **common iliac artery** becomes the **medial sacral artery**.

## Abdominal Vessels

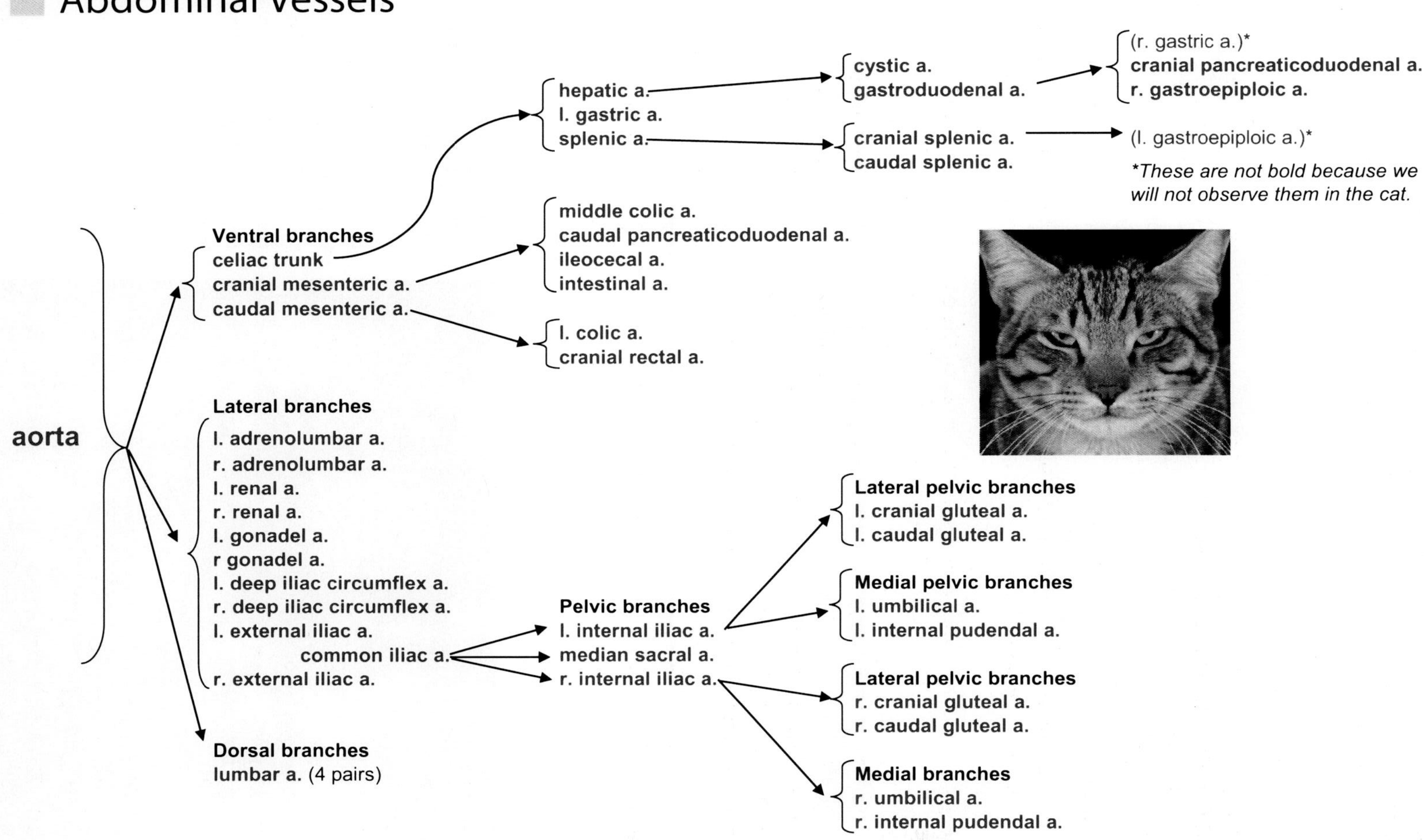

# NERVES

## Obturator Nerve

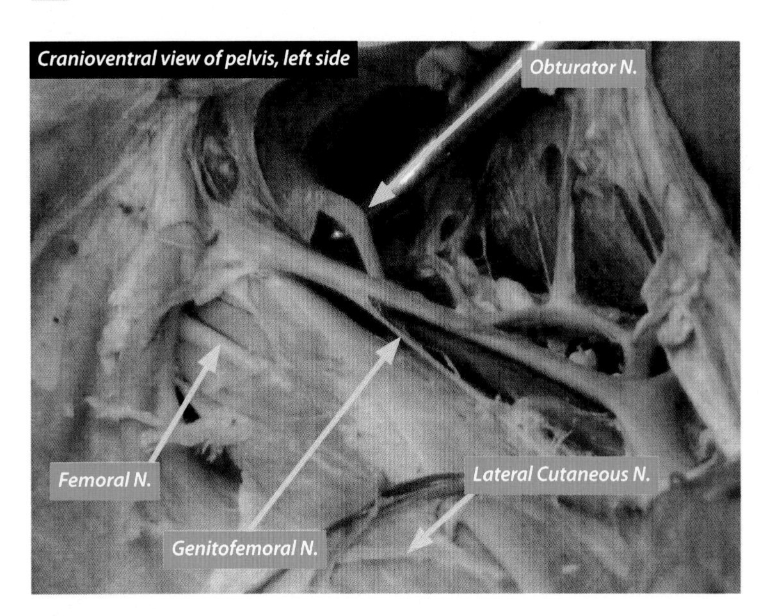


The **obturator nerve** runs close to the **cranial gluteal artery** where it branches off the **internal iliac artery**. It innervates most of the adductor muscles and the skin on the medial side of the thigh. The **obturator nerve** serves:

1. Adductor Femoris (magnus et brevis),
2. Adductor Longus and
3. Gracilis

This **nerve** gets its name from the Jim Croce song "**Obturator**—Will you help me place this call?" and the Sade song "**Smooth Obturator**." Wow.

## Lumbosacral Cord

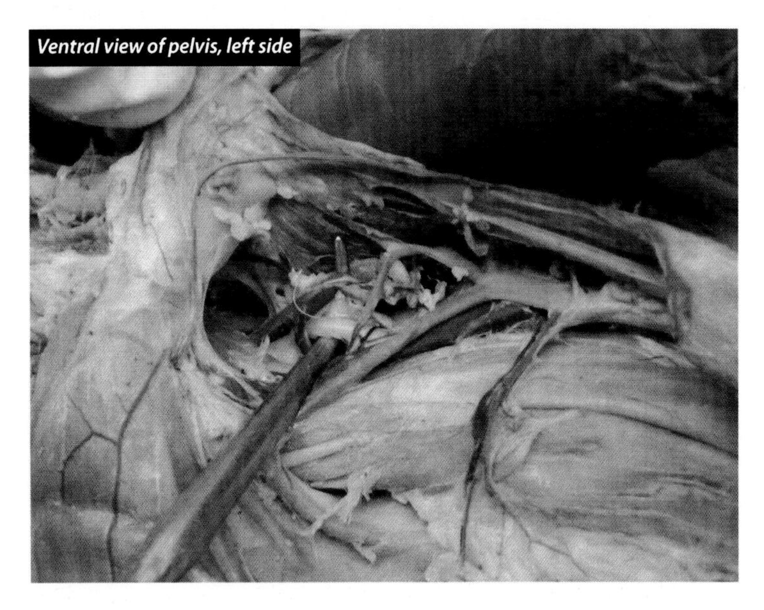


The **lumbosacral cord** (**trunk**) is formed from part of the **anterior ramus** of **L4** and the entire **L5 anterior ramus**. It becomes part of the **sacral plexus**. Branches of the **sacral plexus** include the **pudendal nerve**, the **gluteal nerves**, and the **sciatic** (**ischiatic**) **nerve**. The **lumbosacral cord** is known to many of Dr. J's students as the "**biggest hugest nervus in the pelvis**!"

# VESSELS

## Caudal Vena Cava

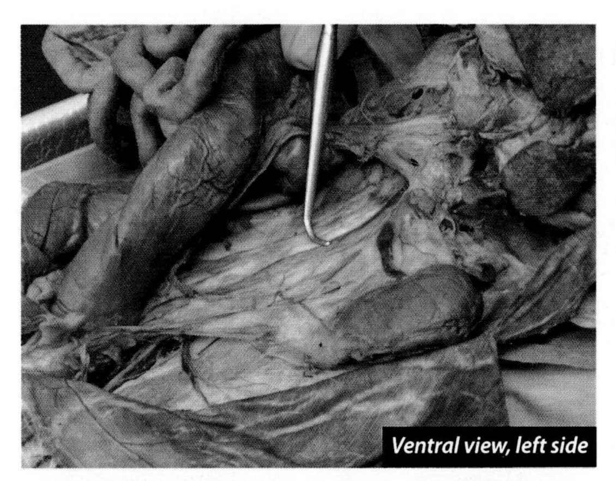


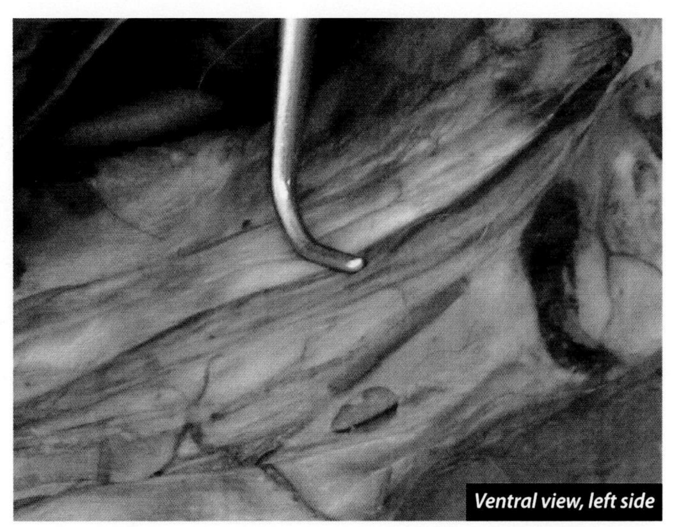


The **caudal vena cava** is the vessel that brings all the blood caudal to the diaphragm back to the right atrium of the heart. In the cat and the human, it forms at the union of the two **common iliac veins**, which is at the level of the L5 vertebra in humans. It then proceeds toward the heart, passing through the central tendon of the diaphragm in the process. You will remember that it is slightly to the right side of the midline and passes through the diaphragm at the level of the T8 body at mid-inhalation. The **caudal vena cava** is of particular importance in the adult, as some of this blood has been to the liver, where nutrient levels have been adjusted. Some of the blood has also been to the kidneys, where it was modified to reduce many wastes and where the pH, electrolyte, and water content were adjusted.

# HEPATIC PORTAL VEIN

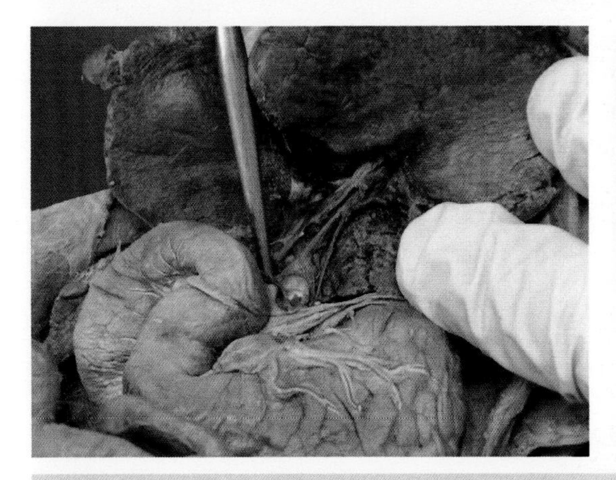

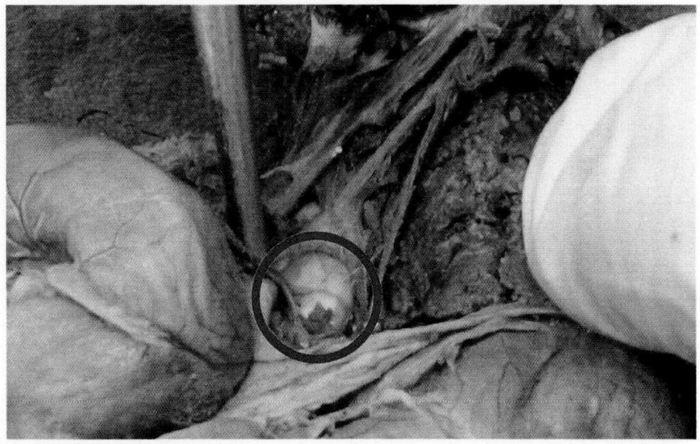

*These photos show the ventral view with cranial end of the cat at the top of the picture.*

The **hepatic portal vein** (**portal vein**) will be **yellow** in the cats where it has dye in it. It is of functional importance because it is the one vessel that receives blood from the ENTIRE gastrointestinal system. It brings blood from those organs to the liver ("save the liver") where it can be processed and altered for distribution to the body.

# VESSELS

## Ventral Branches of Abdominal Aorta

### View #1—Includes branches of the celiac trunk and splenic artery

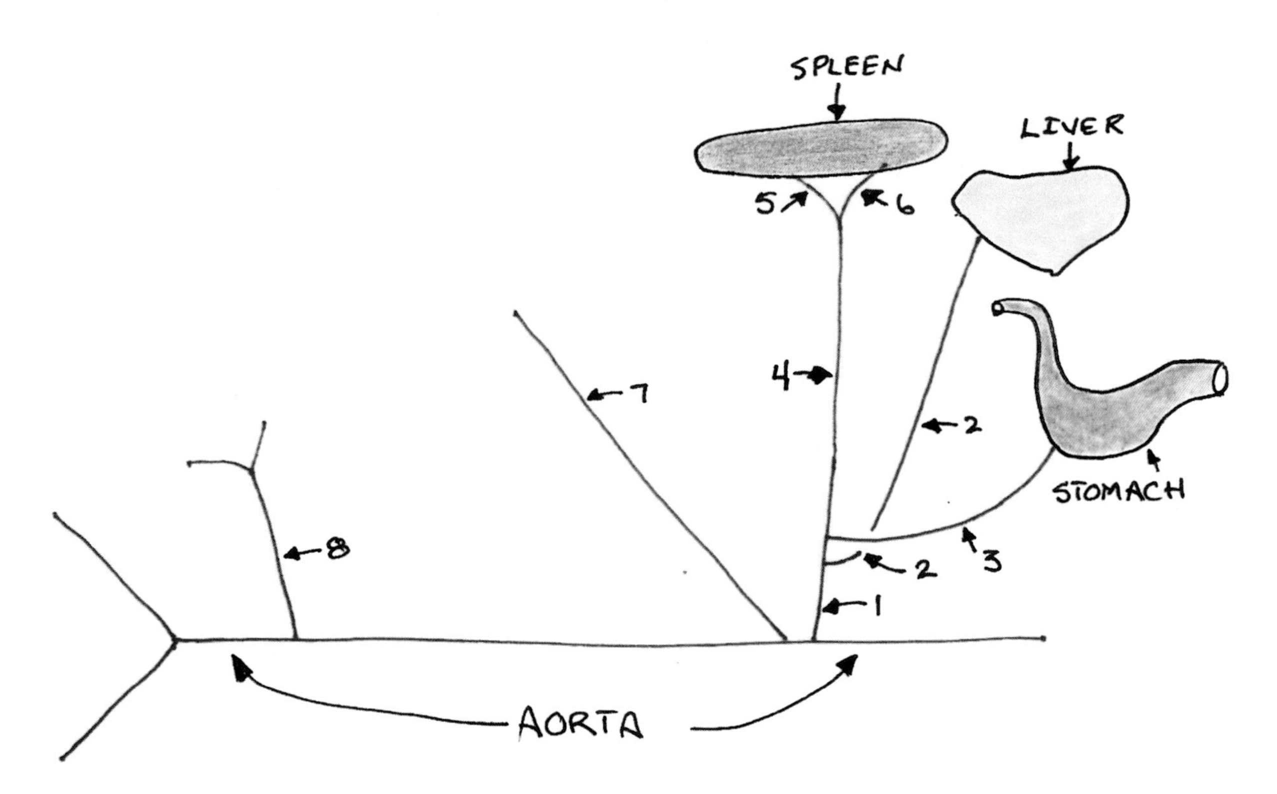


1. **Celiac Trunk**
2. **Hepatic Artery**
3. **Left Gastric Artery**
4. **Splenic Artery**
5. **Caudal Splenic Artery**
6. **Cranial Splenic Artery**
7. **Cranial Mesenteric Artery**
8. **Caudal Mesenteric Artery**

# VESSELS

## Ventral Branches of Abdominal Aorta

### View #2—Includes branches of the celiac trunk, splenic, hepatic, and gastroduodenal arteries

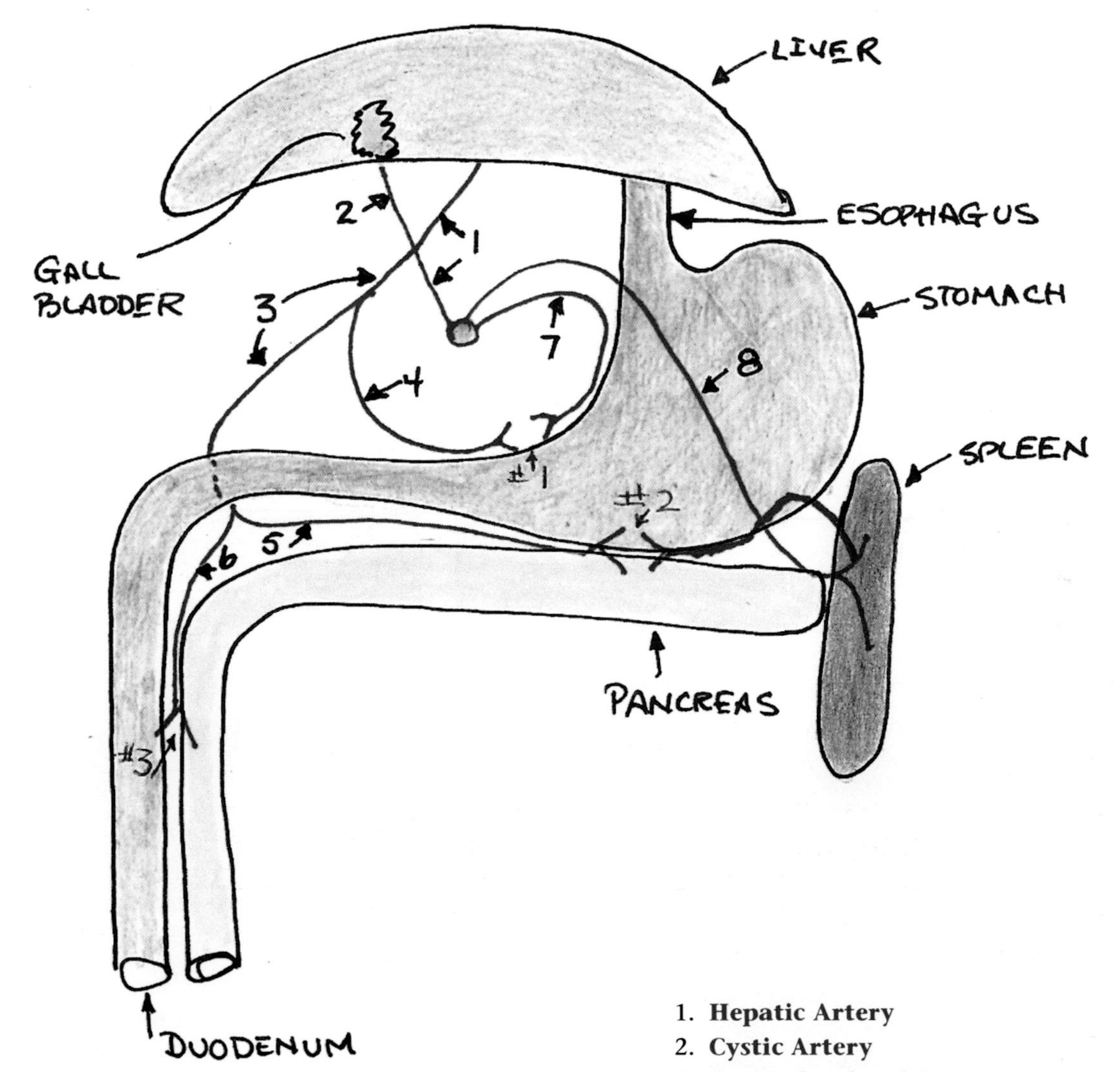


1. **Hepatic Artery**
2. **Cystic Artery**
3. **Gastroduodenal Artery**
4. **Right Gastric Artery**
5. **Right Gastroepiploic Artery**
6. **Cranial Pancreaticoduodenal Artery**
7. **Left Gastric Artery**
8. **Splenic Artery**

### Anastomoses

**#1. Right** and **Left Gastric Arteries**
**#2. Right** and **Left Gastroepiploic Arteries**
**#3. Cranial** and **Caudal Pancreaticoduodenal Arteries**

# Abdominal (Descending) Aorta

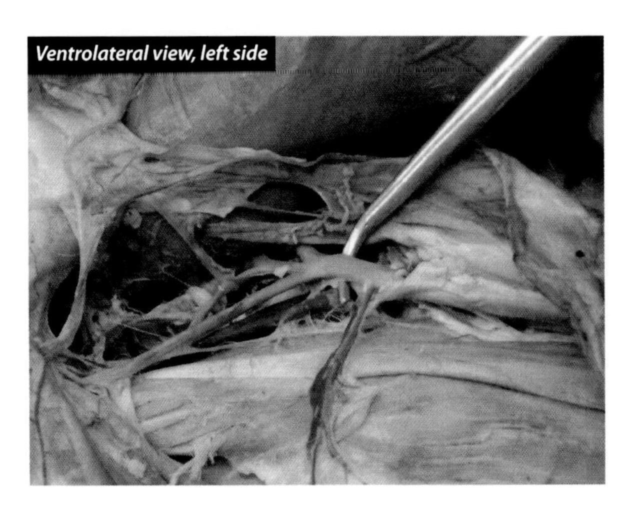


The **abdominal (descending) aorta** has three major ventral branches and five major lateral branches on each side that we will study. The ventral branches starting at the cranial end in the cat are the **celiac trunk**, the **cranial mesenteric artery**, and the **caudal mesenteric artery**. Again, beginning at the cranial end, the lateral branches found on both sides are the **left** and **right adrenolumbar arteries**, the **left** and **right renal arteries**, the **left** and **right gonadal** (either **internal spermatic** or **ovarian**) **arteries**, the **left** and **right deep iliac circumflex arteries**, and the **left** and **right external iliac arteries**. In a human, the **abdominal aorta** passes along the vertebral border, through the diaphragm at the level of the T12 body, and terminates at the level of the L4 body, where it gives rise to the **left** and **right common iliac arteries**. This picture has been taken at the caudal end of the **aorta**.

# Ventral Branches of the Abdominal Aorta

## Celiac Artery (Trunk)

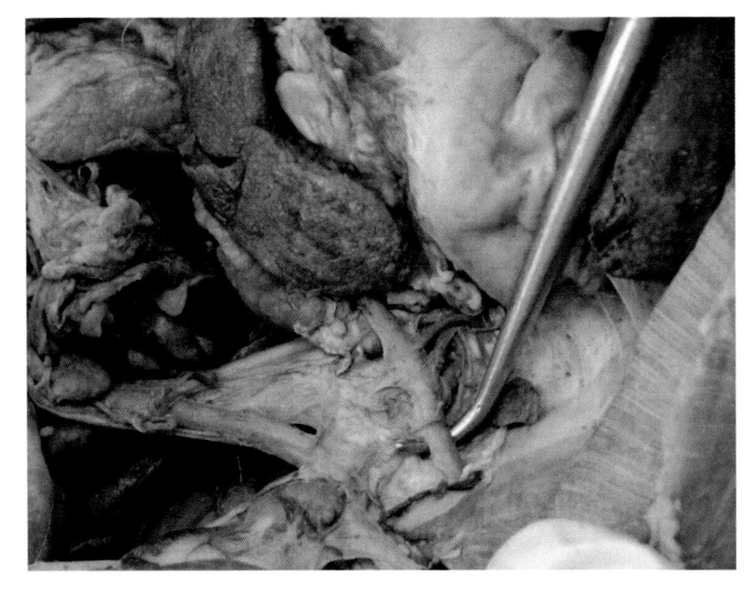

The **celiac artery (trunk)** is the first anterior (ventral) branch of the **abdominal aorta**. It gives rise to the **hepatic artery (common hepatic artery)**, the **left gastric artery**, and the **splenic artery**. It has a substantial diameter in most cats.

## Hepatic Artery

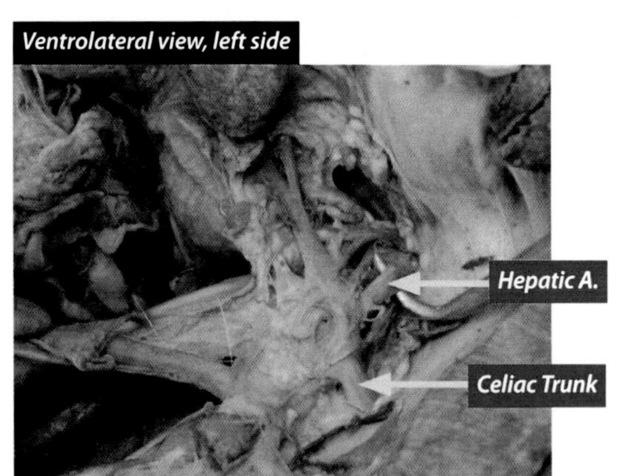


*In these pictures the hepatic artery can be seen branching off the celiac trunk.*

In this view we see the **hepatic artery** for the first time where it branches from the celiac trunk. On page 305 we will see it again where it forms an "X" with its branches, the **cystic artery** and the **gastroduodenal artery**.

# Ventral Branches of the Abdominal Aorta

## Left Gastric Artery

Ventrolateral view, left side

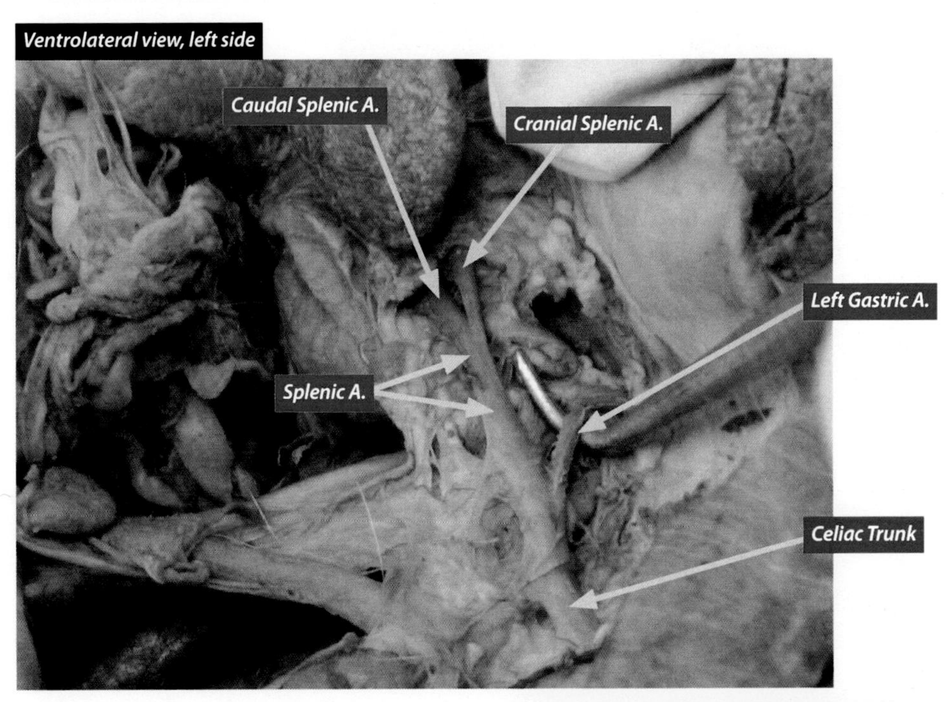


Superficial view, both pictures

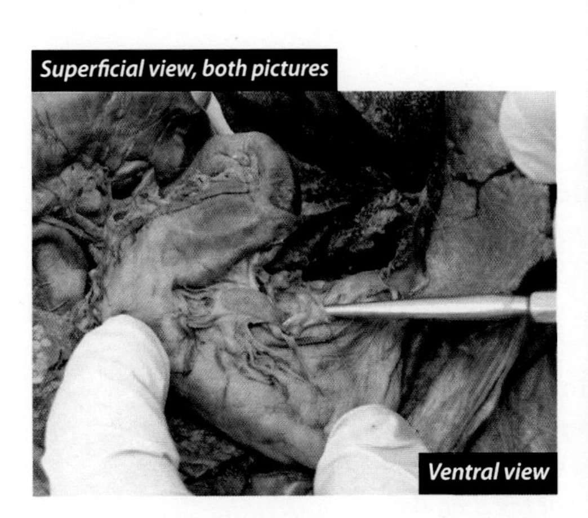


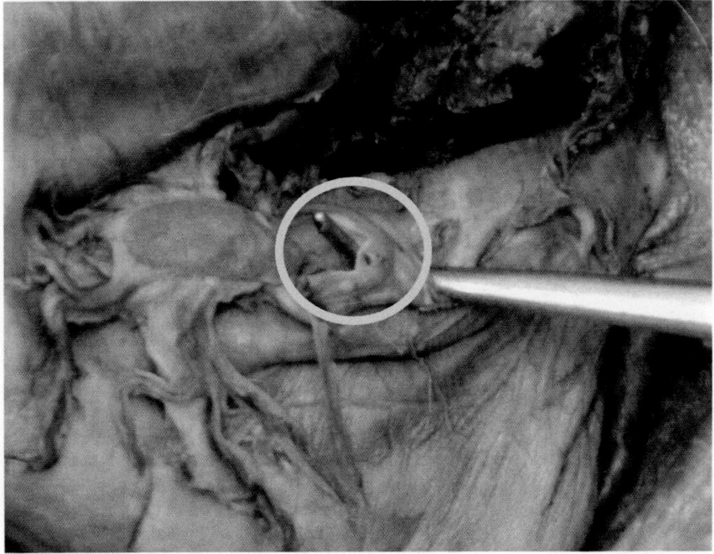

The **left gastric artery** is usually the second branch of the **celiac trunk** in the cat and is the smallest branch of the **celiac trunk** in humans. It joins the lesser curvature of the stomach from the left side and forms an anastomosis with the **right gastric artery**. In humans, it also gives rise to esophageal branches that serve the abdominal portion of the esophagus. It can be tagged in two places in the cat: where it branches from the **celiac trunk** and where it joins the lesser curvature of the stomach. These places are shown in the above pictures.

# Ventral Branches of the Abdominal Aorta

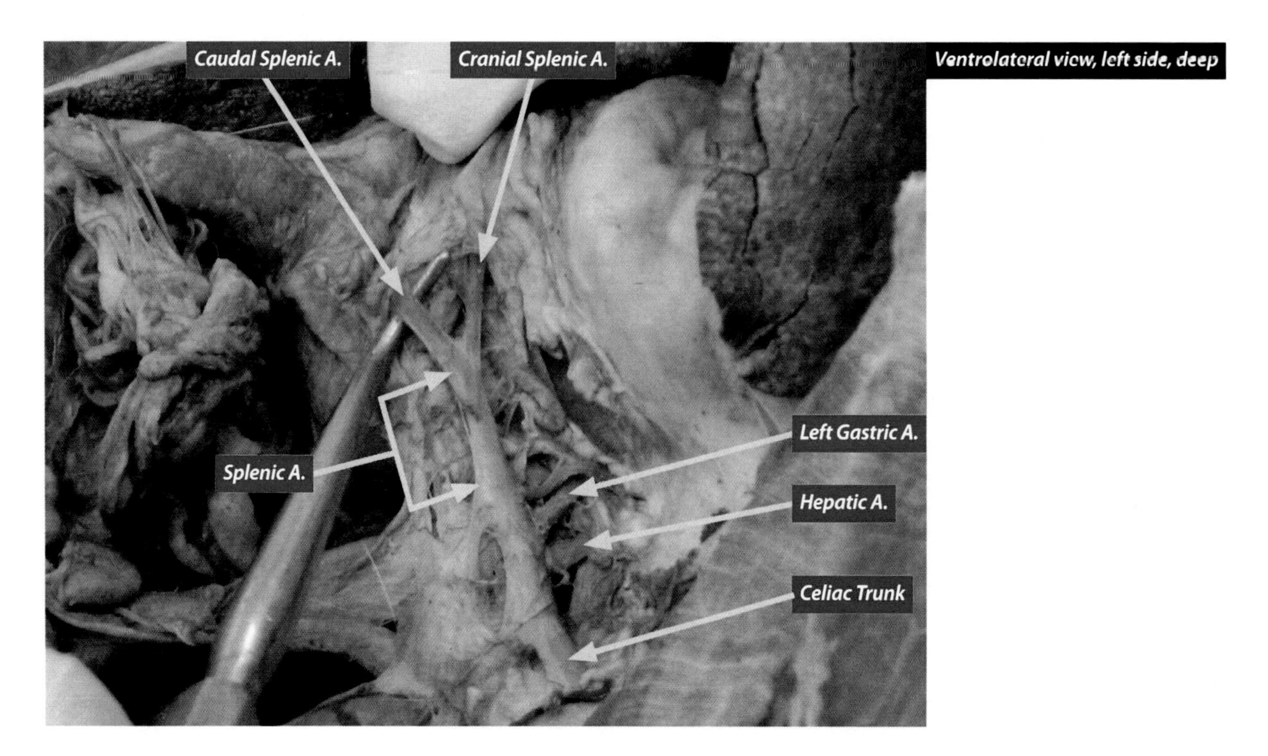


## Splenic Artery

The **splenic artery** is the terminal branch of the **celiac trunk**. It gives rise to the **cranial** and **caudal splenic arteries** that serve the spleen. In humans, it is the largest branch of the **celiac trunk**; and, in addition to serving the spleen, some of its branches also serve the pancreas, the greater omentum, and the stomach.

## Cranial Splenic Artery

The **cranial splenic artery** is one of two branches of the **splenic artery** that we will study. Simply, as with the **caudal splenic artery**, it is a Grant thing. It supplies blood to the cranial end of the spleen.

## Left Gastroepopploic Artery

The **left gastroepiploic artery** is usually a branch of the **cranial splenic artery**, while the **right gastroepiploic artery** is a branch of the **gastroduodenal artery**. As the name implies, the **left** and **right gastroepiploic arteries** serve the greater curvature of the stomach and the greater omentum.
*NOTE: You will not see the left gastroepiploic artery in the cat, so I will not tag it on the practical.*

## Caudal Splenic Artery

The **caudal splenic artery** is one of two branches of the **splenic artery** that we will study. Simply, it is a Grant thing. It supplies blood to the caudal end of the spleen.

# Ventral Branches of the Abdominal Aorta

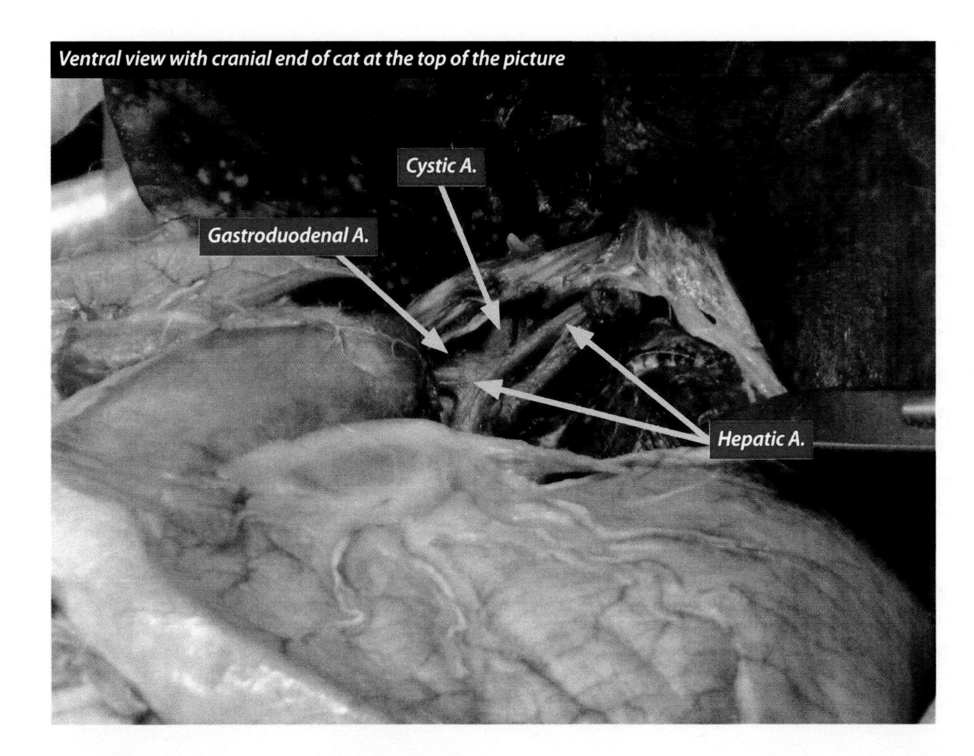


*In this picture the **hepatic artery** can be seen making up the entire left side of the "X."*

## Hepatic Artery (continued)

Notice the "X" in this view. The **hepatic artery** makes up the entire left side of the "X." It is a branch of the **celiac trunk (artery)**. It has two major branches in the cat that we will study: the **cystic artery** and the **gastroduodenal artery**. In humans, this vessel is called the **common hepatic artery**, and it courses to the right side, where it gives rise to the **hepatic artery proper** and to the **gastroduodenal artery**. The **hepatic artery** proper then continues to the right where it gives rise to the **right** and **left hepatic arteries**.

## Cystic Artery

The **cystic artery** is on the cranial right side of the "X." It serves the gall bladder and is a branch of the **hepatic artery**. In humans, the **cystic artery** is a branch of the **right hepatic artery**. When you see the word "cystic," you should remember that it refers to the gall bladder.

## Gastroduodenal Artery

The **gastroduodenal artery** is on the caudal right side of the "X." It is a branch of the **hepatic artery**. In the cat, the three main branches of the **gastroduodenal artery** are (1) the **right gastric artery**, (2) the **cranial pancreaticoduodenal artery**, and (3) the **right gastroepiploic artery**. Because of this, it serves four organs: the stomach, the greater omentum, the pancreas, and the duodenum. In humans, the **gastroduodenal artery** gives rise to the **supraduodenal artery**, the **right gastroomental artery**, and the **superior pancreaticoduodenal artery**.

# Ventral Branches of the Abdominal Aorta

## Right Gastric Artery

The **right gastric artery** is the first branch of the **gastroduodenal artery** in the cat. In humans, it is a branch of the **common hepatic artery**. It joins the lesser curvature of the stomach on the right side and forms an anastomosis with the **left gastric artery** in both cats and humans. The name is another Grant thing, telling us that it serves the stomach. *It is so small in the cat that I will not tag it on a practical because it is likely to be destroyed by overzealous students almost immediately upon inspection.*

## Right Gastroepiploic Artery

The **right gastroepiploic artery** forms functional **anastomoses** with the **left gastroepiploic artery**. The **right gastroepiploic artery** is a branch of the **gastroduodenal artery** while the **left gastroepiploic artery** is usually a branch of the **cranial splenic artery**. As the name implies, the **right gastroepiploic artery** serves the greater curvature of the stomach and the greater omentum. Some authors call it the **right gastro-omental artery**. *NOTE: You will not see the **left gastroepiploic artery** in the cat, so I will not tag it on the practical.*

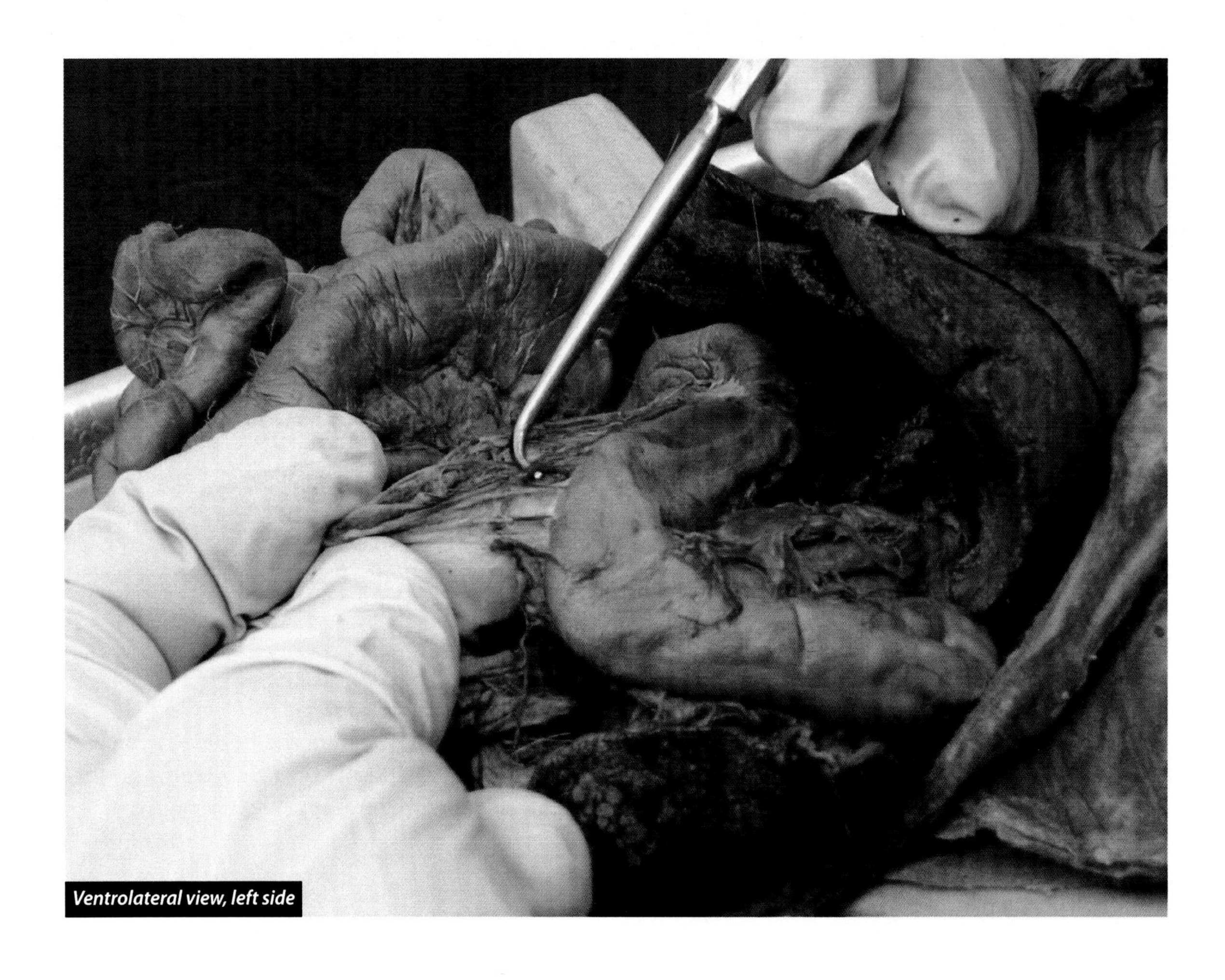

Ventrolateral view, left side

# Ventral Branches of the Abdominal Aorta

## Right Gastroepiploic Artery

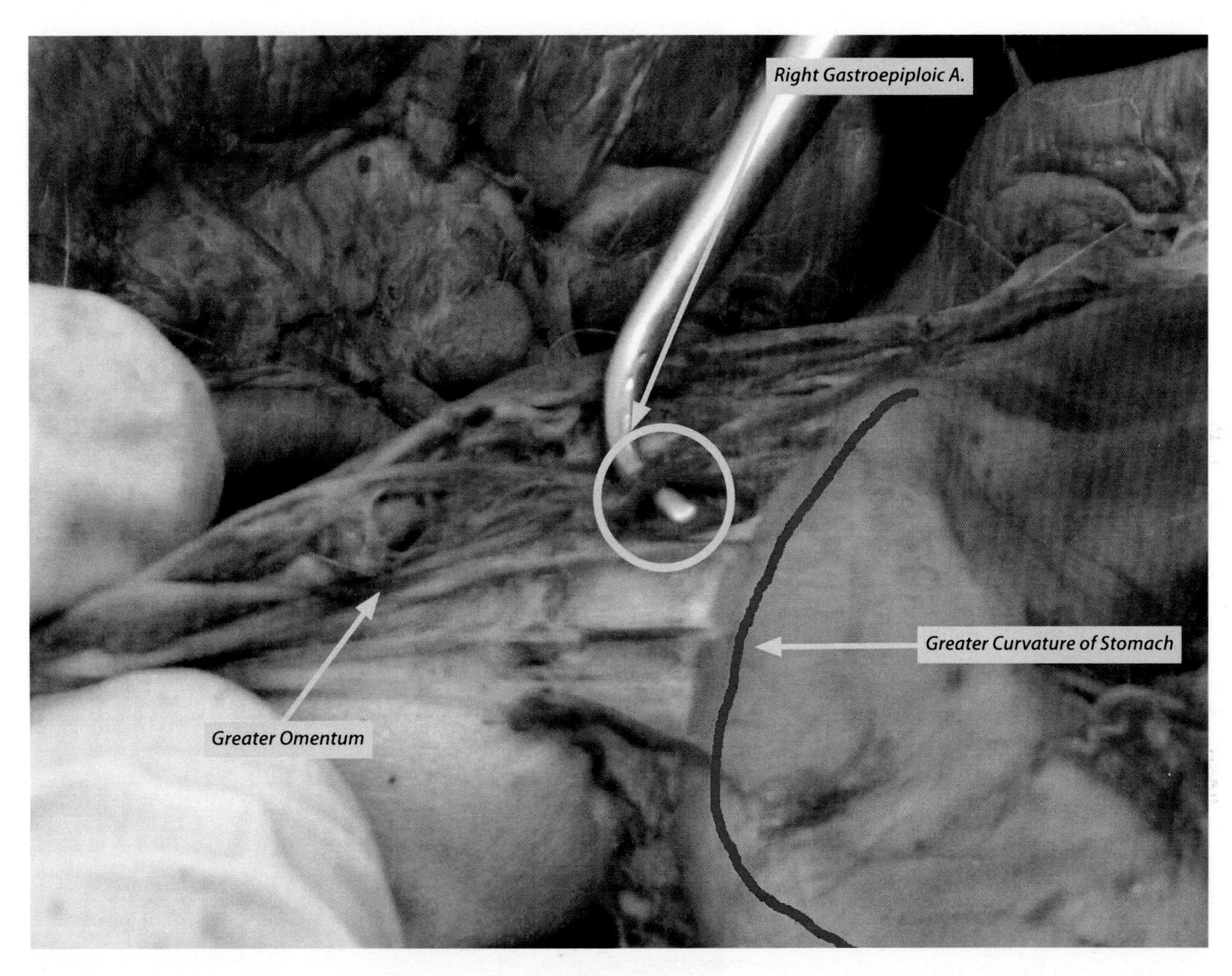

# Ventral Branches of Abdominal Aorta

## Cranial Pancreaticoduodenal Artery

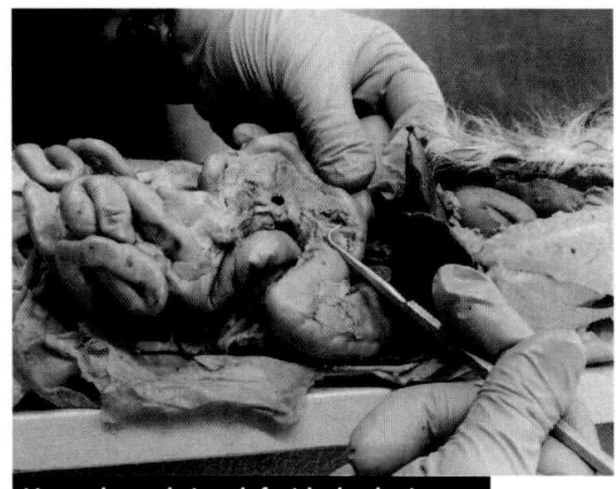

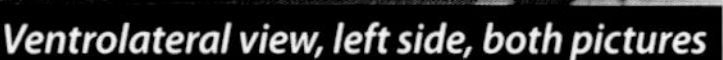

*Ventrolateral view, left side, both pictures*

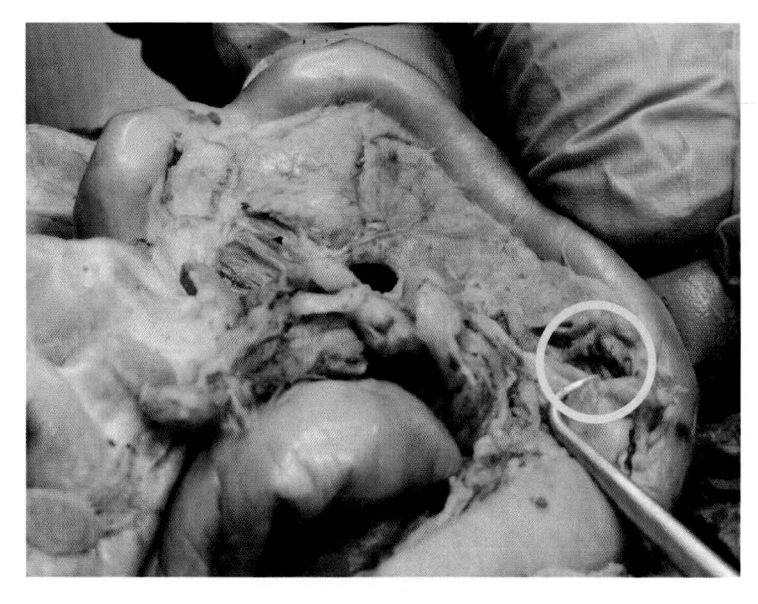

The **cranial pancreaticoduodenal artery** is a branch of the **gastroduodenal artery**. It forms one of the five major abdominal **anastomoses** with the **caudal pancreaticoduodenal artery**. As always, the best place to look for it is where it goes (and its name tells you that). Explore the cranial end of where the pancreas and duodenum run together. In humans, it would be called the **superior pancreaticoduodenal artery**. It divides into anterior and posterior branches, which would supply blood to the head of the pancreas and duodenum and form **anastomoses** with comparable branches of the **inferior pancreaticoduodenal artery**.

## Cranial Mesenteric Artery

The **cranial mesenteric artery** is the second ventral branch of the aorta. It moves obliquely in a caudal direction toward the ventral midline and usually has a large diameter. It gives rise to the **middle colic artery**, the **caudal pancreaticoduodenal artery**, and the **ileocolic** (**ileocecal**) **artery**, and then it terminates in the **intestinal arteries**. In humans, the **superior mesenteric artery** is credited with supplying blood to the abdominal mid-gut with the branches we find in the cat, as well as to the **right colic artery**.

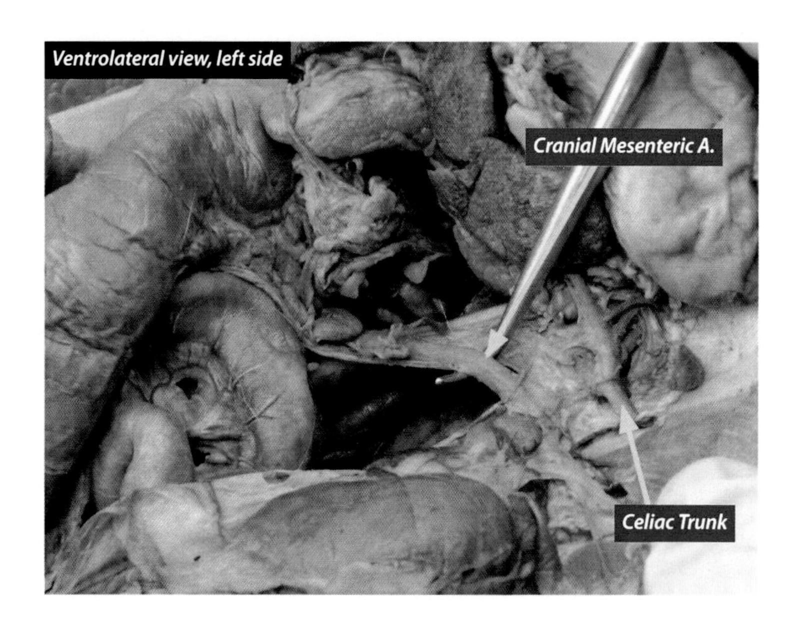


*Ventrolateral view, left side*

# Ventral Branches of Abdominal Aorta

## View #3—Includes branches of the cranial mesenteric and caudal mesenteric arteries

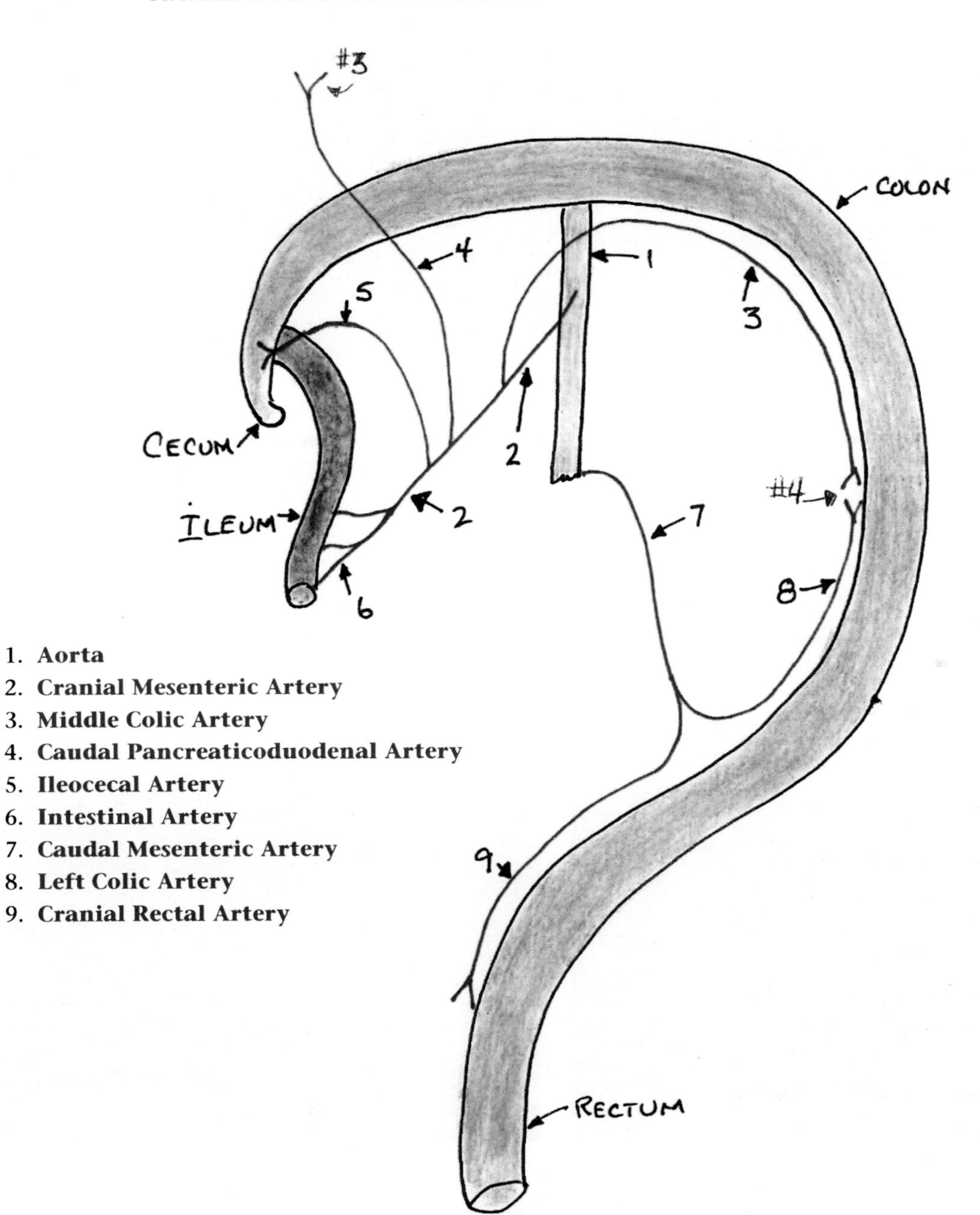


1. **Aorta**
2. **Cranial Mesenteric Artery**
3. **Middle Colic Artery**
4. **Caudal Pancreaticoduodenal Artery**
5. **Ileocecal Artery**
6. **Intestinal Artery**
7. **Caudal Mesenteric Artery**
8. **Left Colic Artery**
9. **Cranial Rectal Artery**

### Anastomoses

**#3. Cranial** and **Caudal Pancreaticoduodenal Arteries**

**#4. Left** and **Middle Colic Arteries**

## Middle Colic Artery

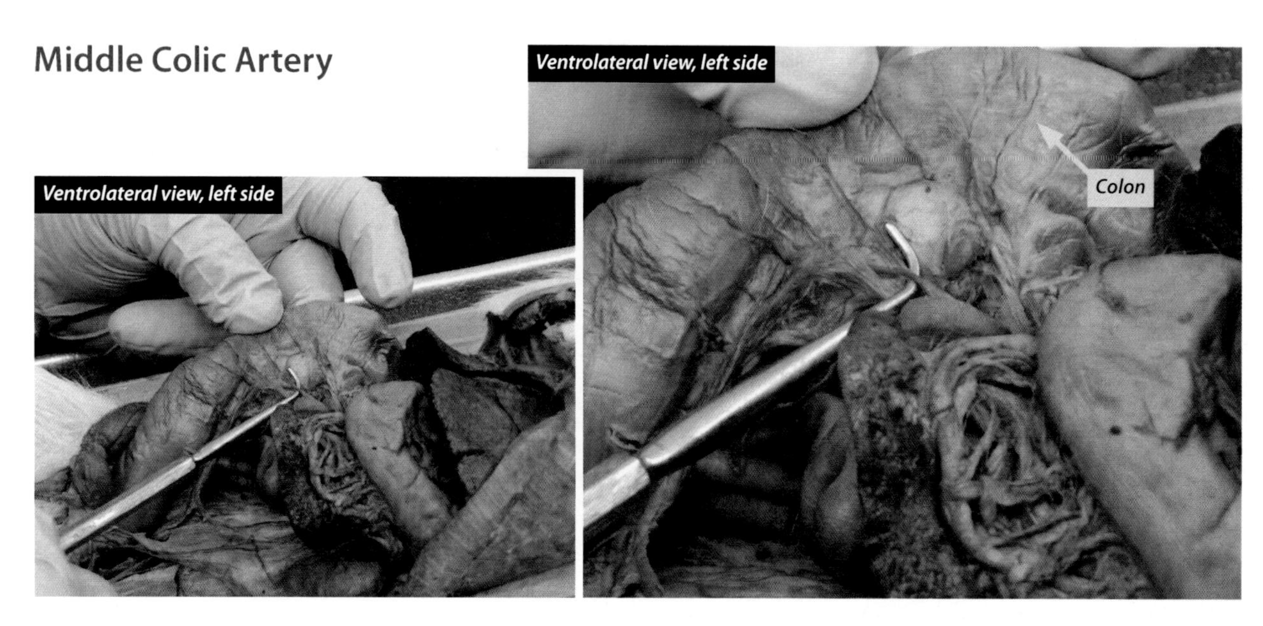


The **middle colic artery** is usually the first branch of the **cranial mesenteric artery**. It forms functional **anastomoses** with the **left colic artery**, which is a branch of the **caudal mesenteric artery**. In humans, this vessel bifurcates. Its left branch **anastomoses** with the **left colic artery** (as it does in the cat) while its right branch **anastomoses** with the **right colic artery**. Dr. J believes this is one of the easiest arteries to find in the cat, as it is a Grant thing. Go to the middle of the colon and CAREFULLY reflect it to the cat's right side. The **middle colic artery** should be right there going to the middle of the colon. A word of caution—if you pull the colon toward the caudal end of the cat, this vessel, among others, will most likely be broken. Please don't do that.

## Caudal Pancreaticoduodenal Artery

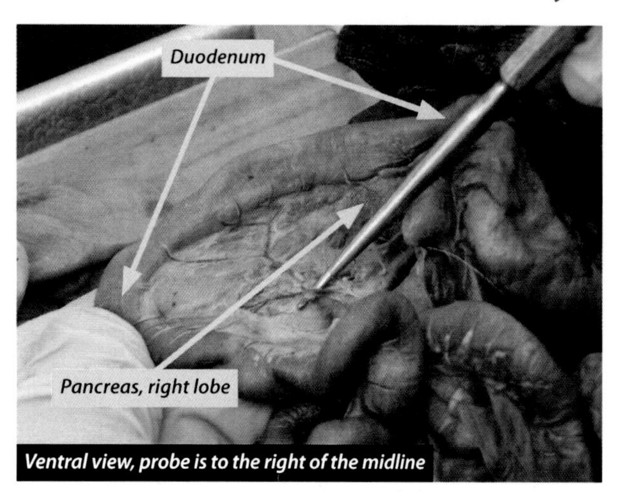


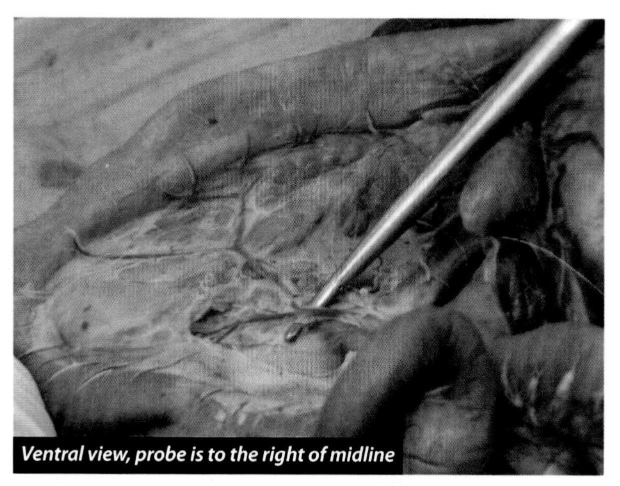


The **caudal pancreaticoduodenal artery** is usually the second branch of the **cranial mesenteric artery** of cats, while it is the first branch of the **superior mesenteric artery** of humans. It forms one of the five major abdominal **anastomoses** with the **cranial pancreaticoduodenal artery** in the cat. In humans, it divides into two branches that form **anastomoses** with the branches of the **superior pancreaticoduodenal artery**, and this network serves the pancreas and the duodenum. This is on both the left and right sides. It is found, as the name suggests, at the caudal end of the duodenum and at the right lobe of the pancreas (yikes—another Grant thing!).

## Ileocecal (Ileocolic) Artery

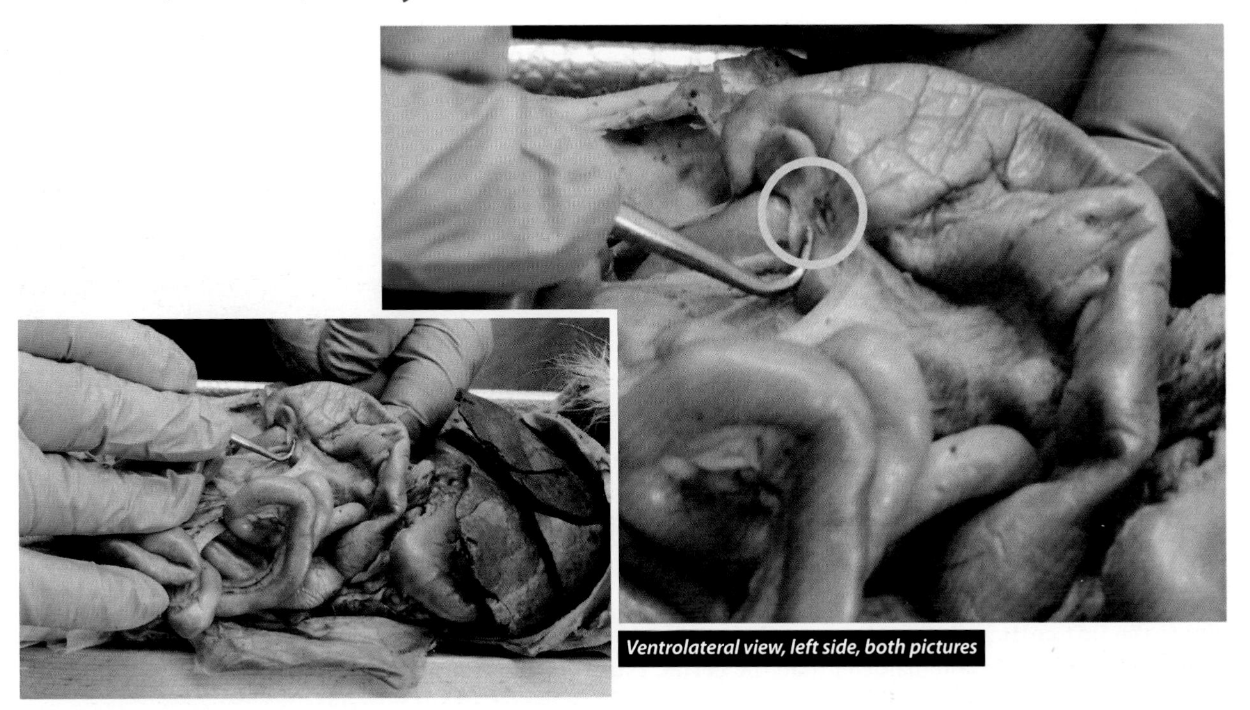

Ventrolateral view, left side, both pictures

The **ileocecal (ileocolic) artery** is another Grant thing! It serves the area around the ileocolic (ileocecal) junction. It is usually the third branch of the **cranial mesenteric artery**. Branches of this artery supply blood to the cecum, the appendix, and the ascending colon in humans.

## Intestinal Arteries

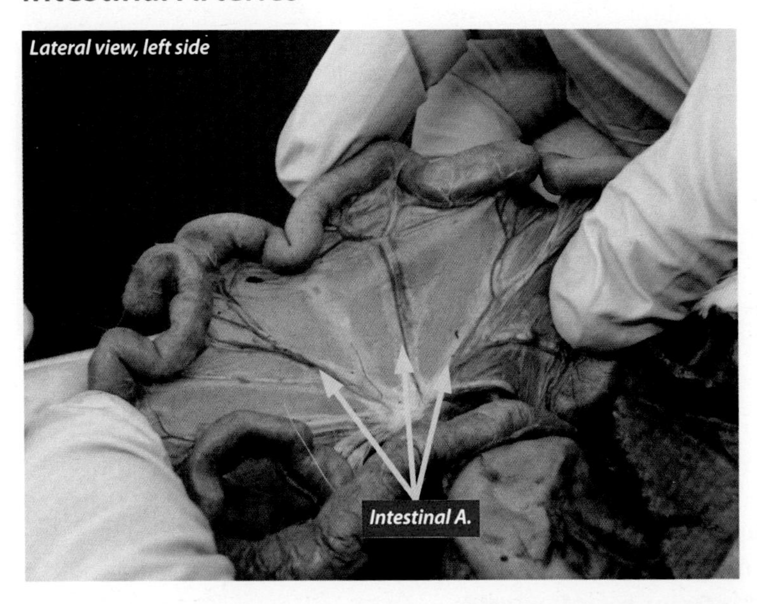


Lateral view, left side

The **intestinal arteries** pass through the mesenteric ligament to the intestines. Functionally, this is very important as it keeps the intestines and the structures serving them from becoming tied in knots. The intestinal arteries are the terminal branches of the **cranial mesenteric artery**.

## Caudal Mesenteric Artery

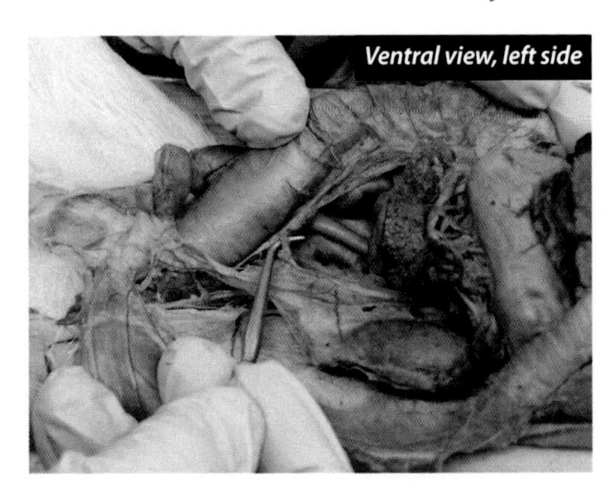


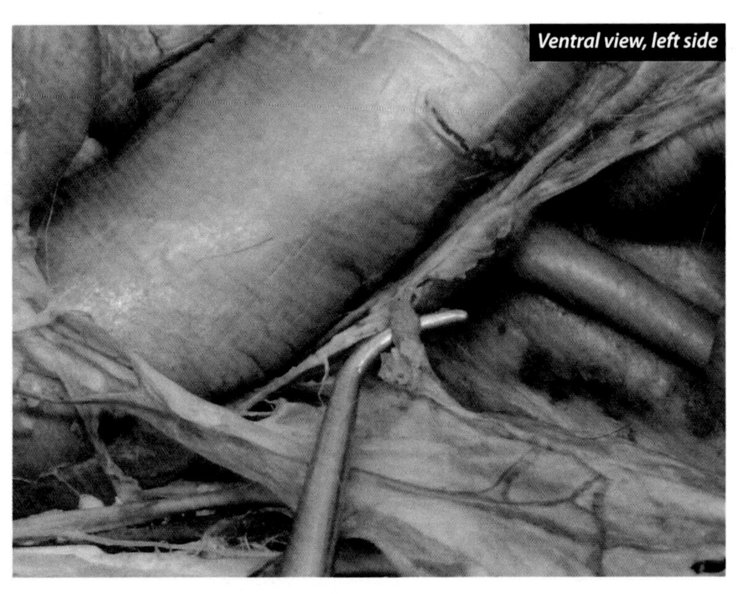


The **caudal mesenteric artery** is the third ventral branch of the **abdominal aorta**. In humans, it branches from the **aorta** at the level of the L3 body. It gives rise to the **left colic** and **cranial rectal arteries** in the cat, and in humans it gives rise to the **left colic** and **superior rectal**, as well as to several **sigmoid arteries**. It is usually significantly smaller both in length and diameter than the **celiac trunk** or the **cranial mesenteric arteries**.

### *Left Colic Artery*

The **left colic artery** is one of two branches of the **caudal mesenteric artery** in the cat. It passes to the left side of the colon and forms collateral circulation with the **middle colic artery**. In humans, it bifurcates into ascending and descending branches. The descending branch supplies blood to the descending colon and forms collateral circulation with the **first sigmoid artery**. The ascending branch passes toward the left kidney and then supplies blood to the superior portion of the descending colon, as well as to the transverse colon. As in the cat, this vessel forms an **anastomosis** with the **middle colic artery**.

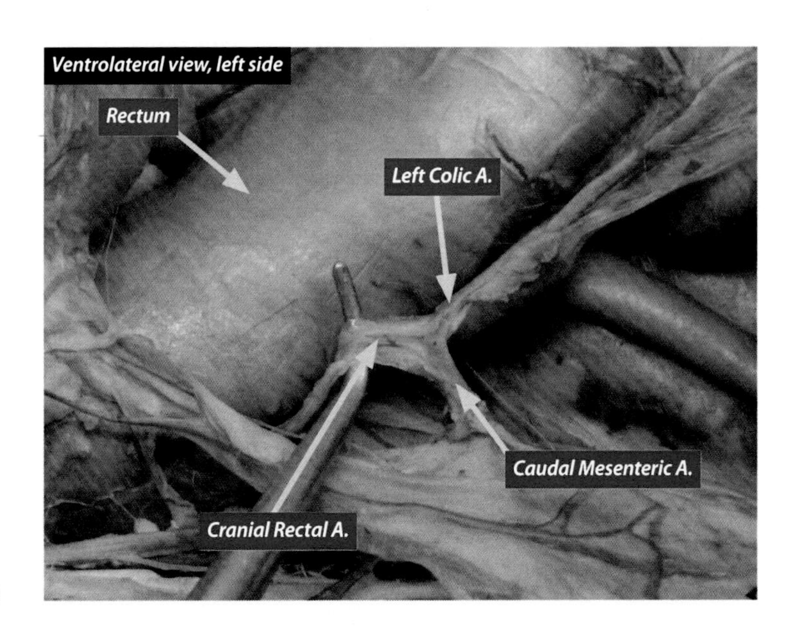


### *Cranial Rectal (Hemorrhoidal) Artery*

The **cranial rectal (hemorrhoidal) artery** is a branch off of the **caudal mesenteric artery** of the cat. It passes caudally along the rectum of the cat and forms collateral circulation with other vessels in the pelvic region. In humans, it would be called the **superior rectal (hemorrhoidal) artery**. It runs along the rectum, divides into two branches that run along the lateral sides of the rectum, and eventually forms pelvic **anastomoses** with the **middle rectal** and **inferior rectal arteries**.

## Lateral Branches of the Abdominal Aorta

### S/He Cat

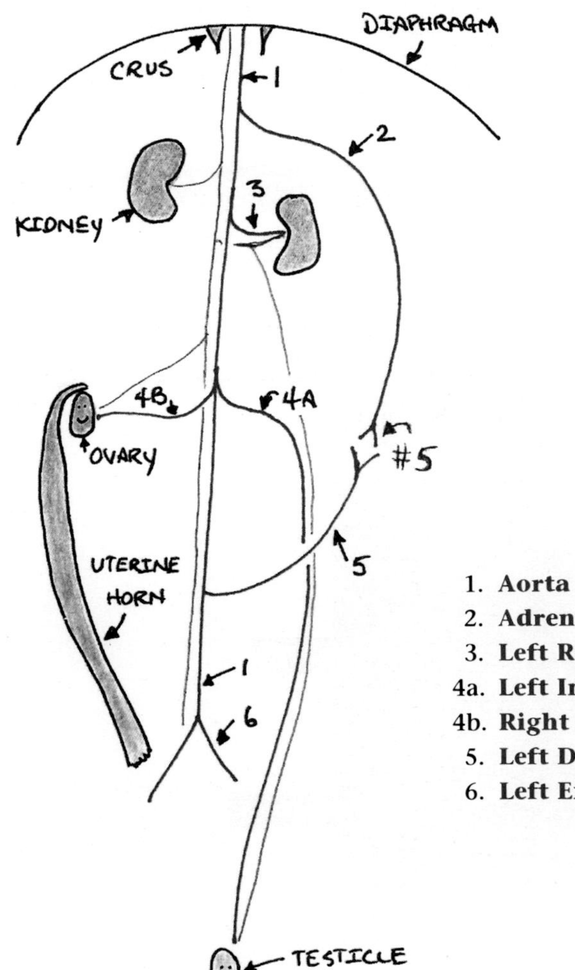


1. **Aorta**
2. **Adrenolumbar Artery**
3. **Left Renal Artery**

4a. **Left Internal Spermatic Artery**

4b. **Right Ovarian Artery**

5. **Left Deep Iliac Circumflex Artery**
6. **Left External Iliac Artery**

### Anastomoses

**#5. Adrenolumbar** and **Deep Iliac Circumflex Arteries**

# Lateral Branches of Abdominal Aorta

## Adrenolumbar Artery

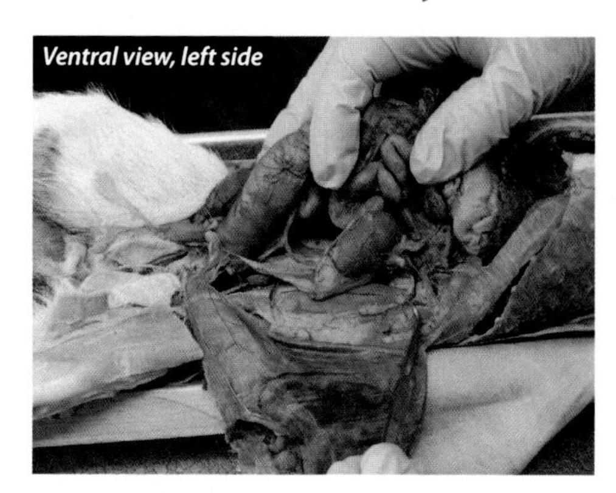


The **adrenolumbar artery** (**left** and **right**) are the first lateral branches of the **abdominal aorta**. They form one of the five major abdominal **anastomoses** with the **deep iliac circumflex artery** (**left** and **right**). The **anastomoses** are on the posterior (dorsal) wall of the abdomen.

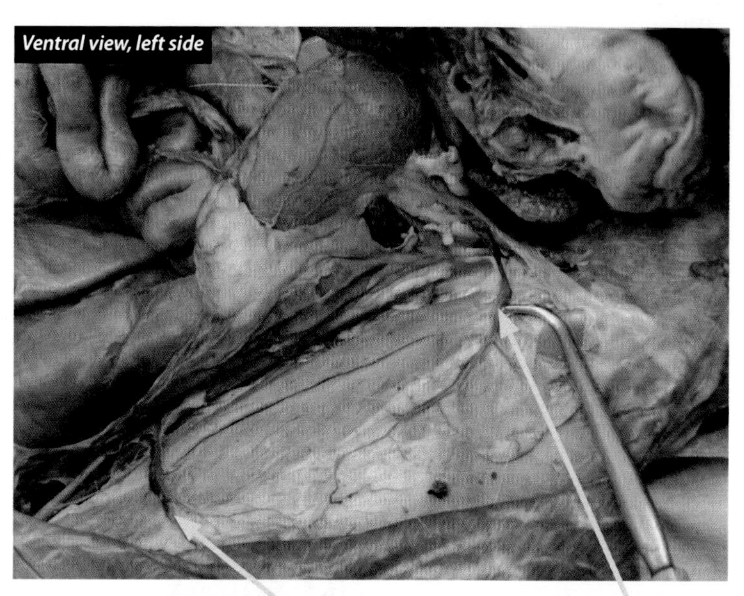


## Renal Artery and Vein

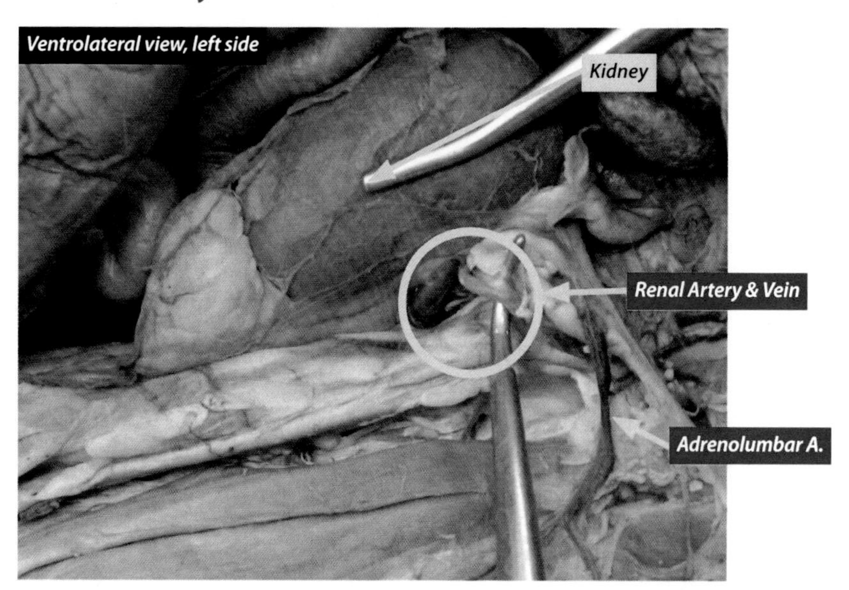


The **renal artery** and **vein** are Grant things—they serve the kidney. The **renal artery** is the second lateral branch of the **abdominal aorta** that we will study. Note that on the right side the **gonadal vein** joins the **caudal vena cava** separately from the **renal vein**. On the left side, the **gonadal vein** joins the **renal vein**, and this trunk then enters the **caudal vena cava**. This is the case for humans and cats of both sexes.

## Gonadal Arteries and Veins

### *Internal Spermatic Artery and Vein*

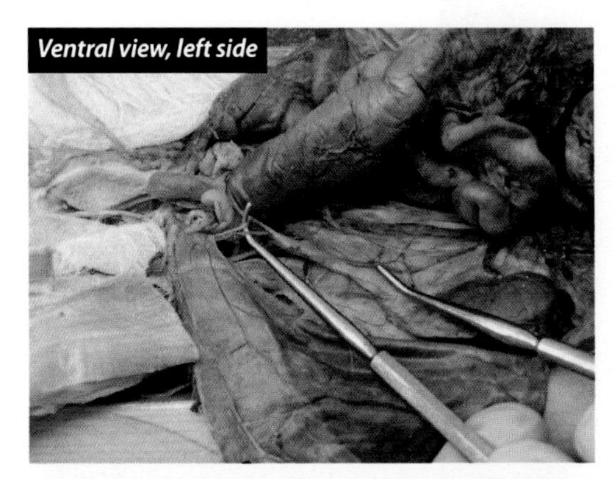


*Observe in the pictures that the two probes indicate the length of these vessels. At the caudal end, the probe is also lifting the vas deferens as it emerges from the spermatic cord and travels dorsal to the bladder.*

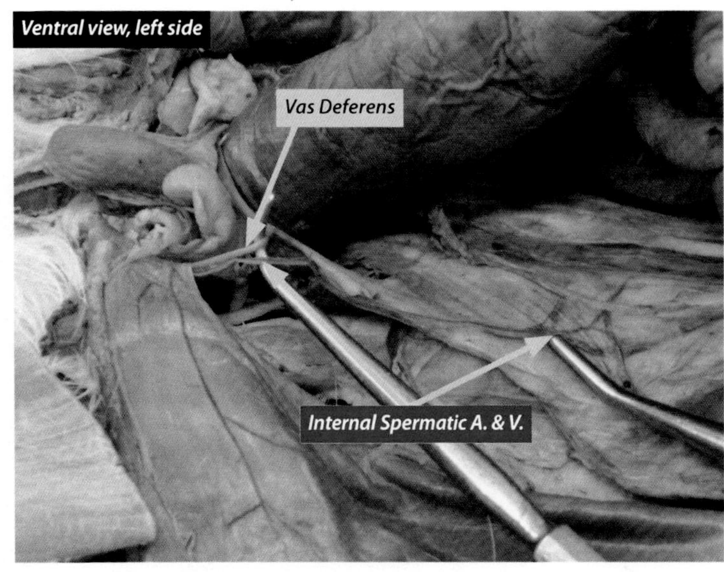


This **artery** is very small and very long. Please be careful not to break it. The **internal spermatic (testicular) artery** is the third lateral branch of the **abdominal aorta** that we will study. You will note that it branches from the **aorta** in about the same area that the **ovarian artery** branches from the **aorta**, just caudal to the **kidney**. The obvious question is why it would leave the **aorta** there when the **testicle** is significantly caudal to that. The answer is that the **testicle** begins its development in the abdomen and descends to the scrotum, taking its nerve and vascular supply with it. In humans, the **internal spermatic (testicular) artery** branches from the aorta at the level of the L2 vertebral body.

### *Ovarian Artery and Vein*

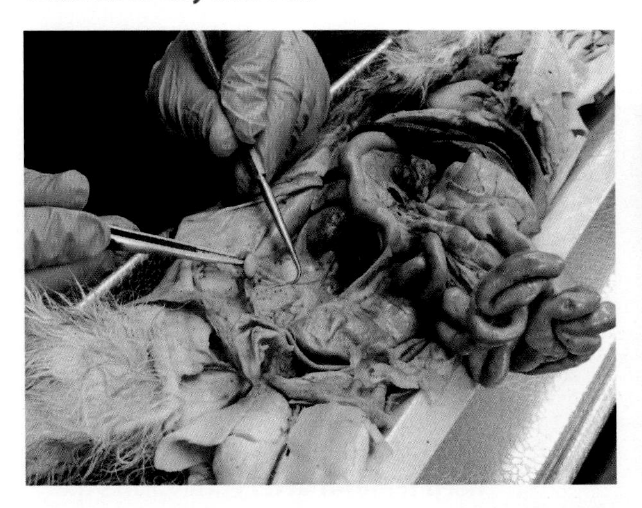

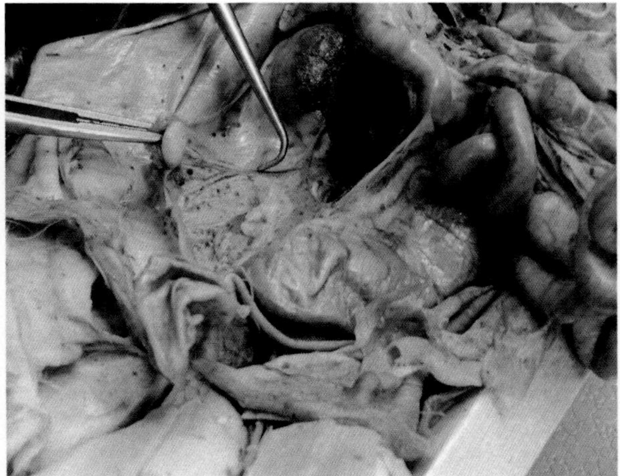

The **ovarian artery** and **vein** serve the **ovary**, as their names imply. The **ovarian artery** is the third lateral branch of the abdominal aorta that we will study. In humans, it passes inferiorly and laterally on the anterior surface of the psoas major muscle on its way to the **ovary**. Note that on the right side the **gonadal vein** joins the **caudal vena cava** separately from the **renal vein**. On the left side, the **gonadal vein** joins the **renal vein**, and this trunk then enters the **caudal vena cava**. This is the case for humans and cats of both sexes.

## Deep Iliac Circumflex (Deep Circumflex Iliac) Artery

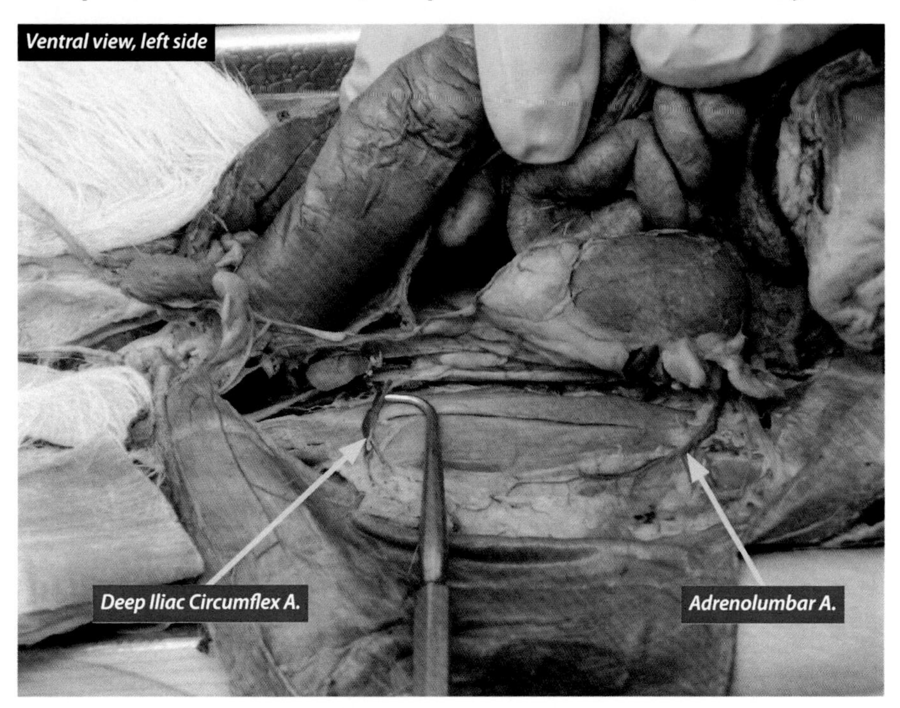


The **deep iliac circumflex artery** (or **deep circumflex iliac artery**) is the fourth lateral branch of the **abdominal aorta** that we study. It forms one of the five major abdominal **anastomoses** with the **adrenolumbar artery**. This vessel is located on the left and right sides, on the posterior (dorsal) wall of the abdomen between the **gonadal artery** and the **external iliac artery**.

## External Iliac Artery

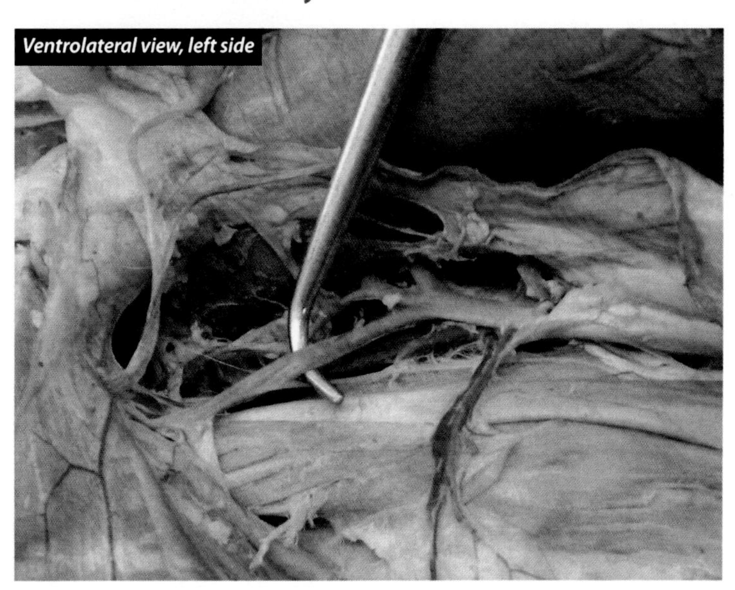


The **external iliac artery** of the cat is the last lateral branch of the aorta, while in a human it is one of the two main branches of the **left** and **right common iliac arteries**, which are pelvic vessels. The **external iliac artery** passes deep to the inguinal ligament into the thigh. The first and only branch of each **external iliac artery** that we will study is the **deep femoral artery**. This marks the end of the **external iliac artery** and the beginning of the **femoral artery** (which is essentially a continuation of the **external iliac artery**).

# Dorsal Branches of the Abdominal Aorta

## Lumbar Arteries

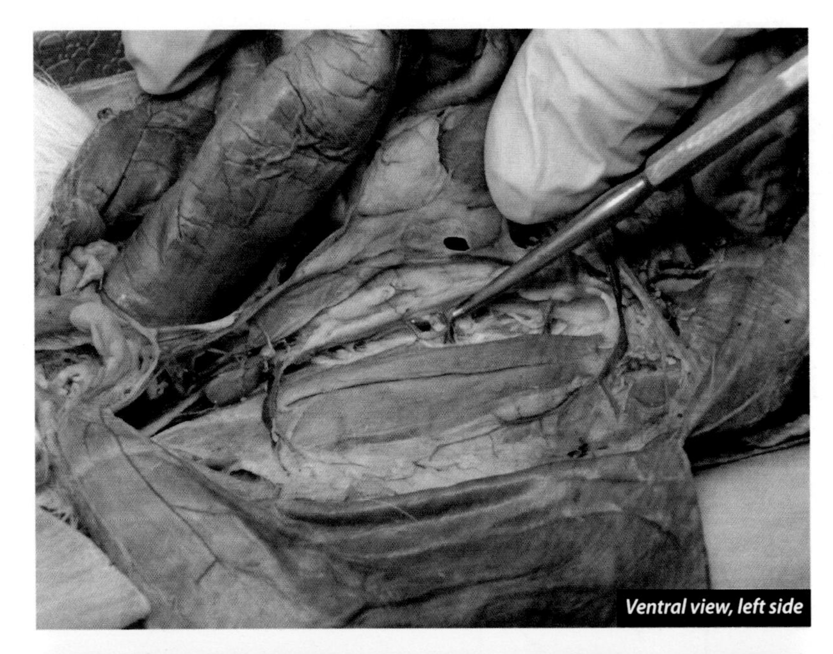
Ventral view, left side

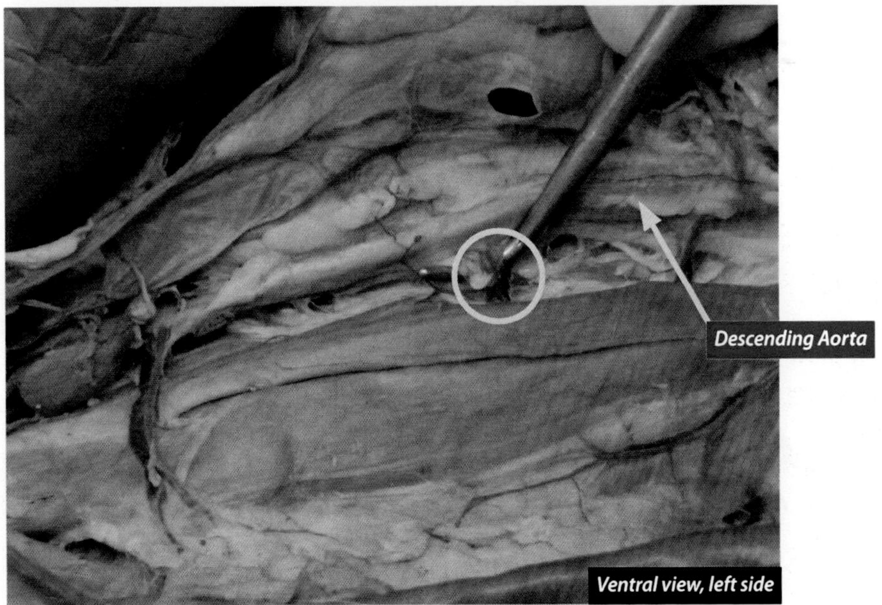

Ventral view, left side

Normally there are four pairs of **lumbar arteries** that branch from the posterior surface of the **abdominal aorta**. They pass posteriorly and laterally to the bodies of the lumbar vertebrae, serving the abdominal wall and branches to the spinal cord. They are most easily seen in the cat by lifting the **aorta** ventrally so that they are exposed where they exit the **aorta's** dorsal surface.

# Pelvic Vessels

## Includes branches of the common iliac and internal iliac arteries

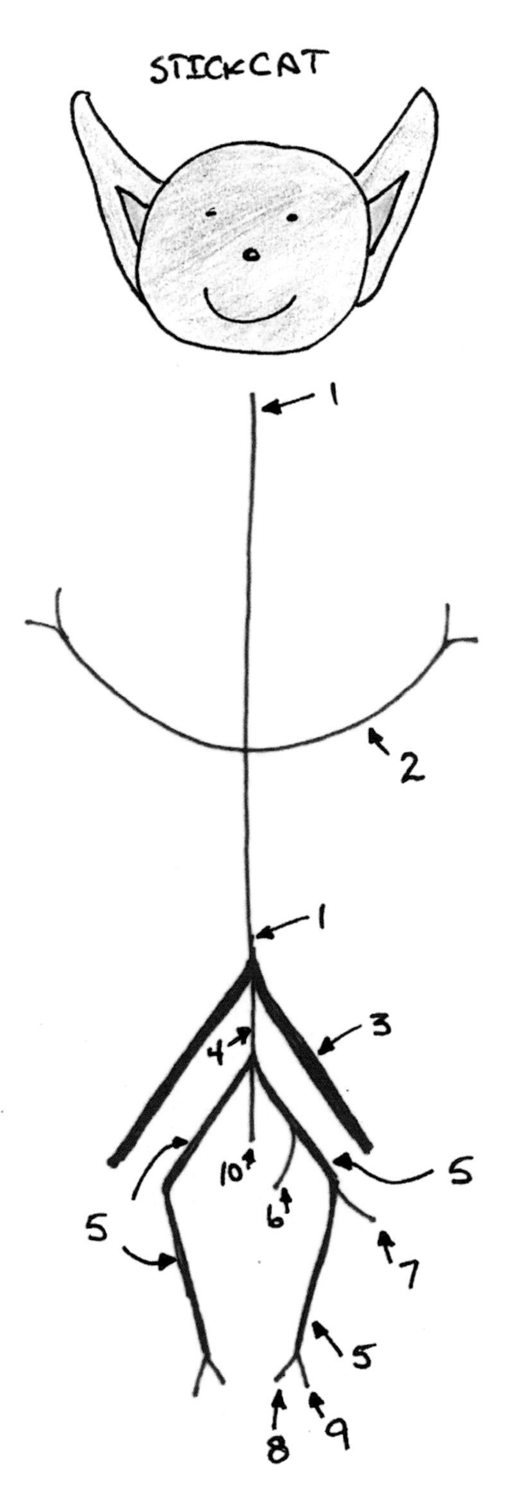


1. **Aorta**
2. **Deep Iliac Circumflex Artery**
3. **Left** and **right Exernal Iliac Artery**
4. **Common Iliac Artery**
5. **Left** and **right Internal Iliac Arteries**
6. **Left Umbilical Artery**
7. **Left Cranial Gluteal Artery**
8. **Left Internal Pudendal Artery**
9. **Left Caudal Guteal Artery**
10. **Median Sacral Artery**

# Pelvic Vessels (Overview)

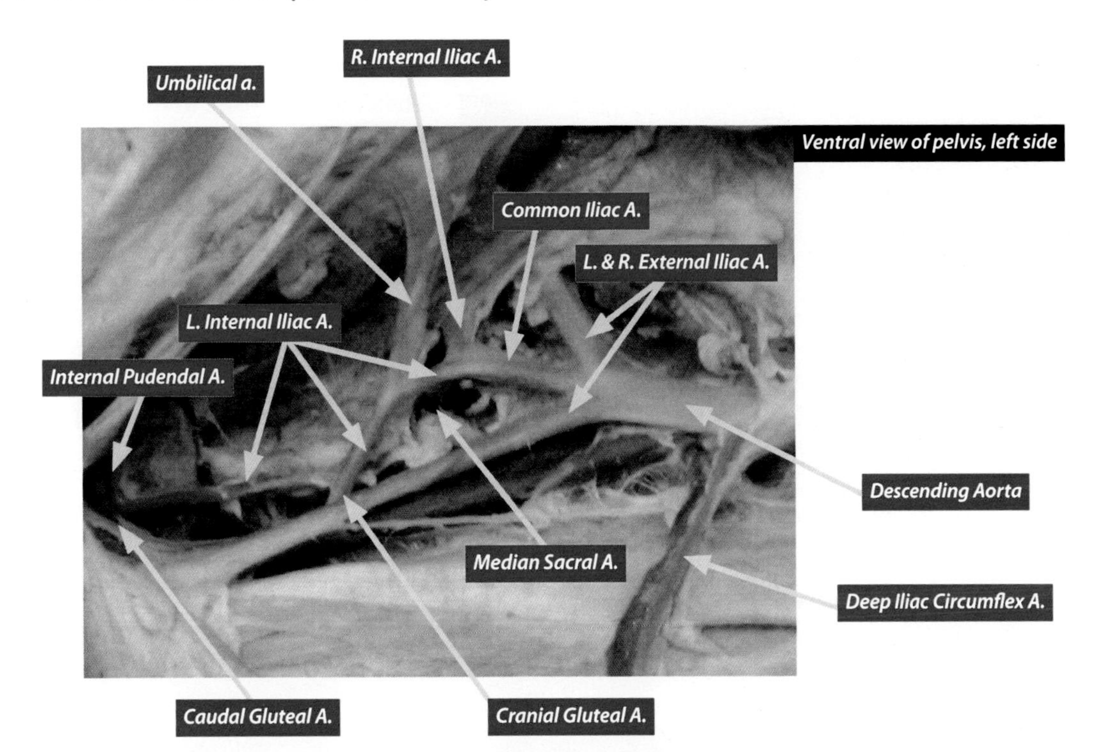


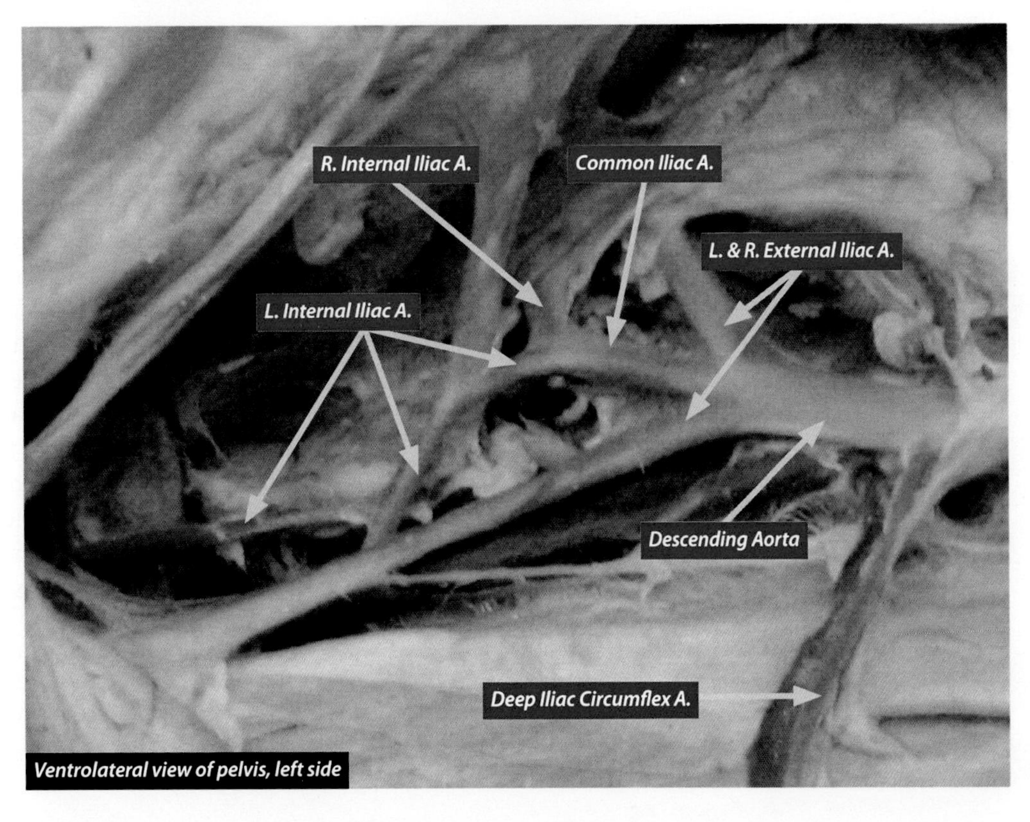

# Pelvic Vessels

## Common Iliac Artery

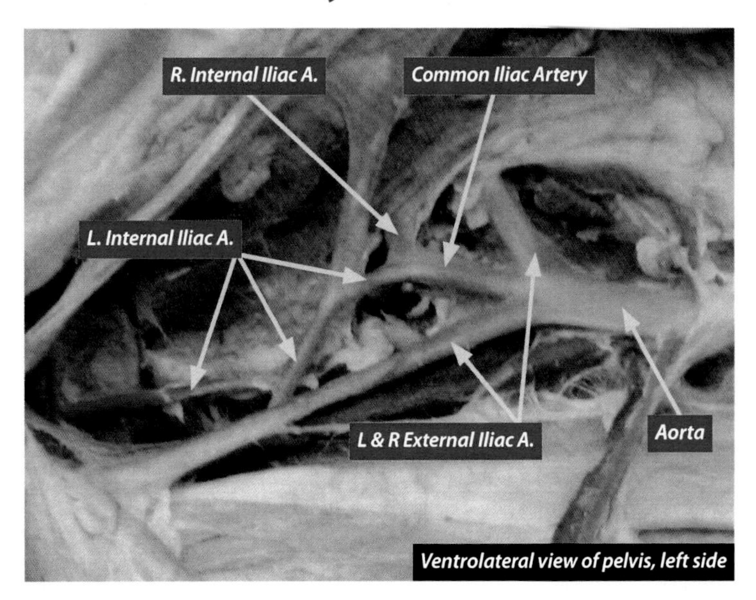


The **common iliac artery** is usually short and located between the **external iliac arteries** and the **median sacral artery** of cats. It gives rise to the **left** and **right internal iliac arteries** of the cat. Occasionally there is no **common iliac artery** in the cats. In humans, there are **left** and **right common iliac arteries** that branch from the aorta at the level of the L4 body; and, in turn, they give rise to the **external** and **internal iliac arterie**s on both sides.

## Median Sacral Artery

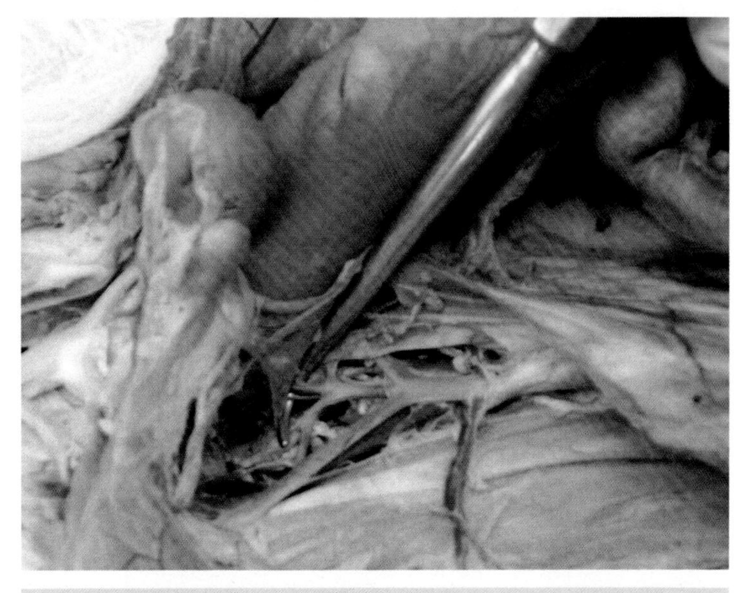

*In this photo, Dr. J is pulling the **median sacral a.** up with the probe to make it more visible. See the photo on the next page to get an idea of its natural position.*

In the cat, the **median sacral artery** begins where the **left** and **right internal iliac arteries** branch from the **common iliac artery**. It serves the sacral region and tail and is sometimes called the **caudal artery**. In humans, the **median sacral artery** branches from the posterior surface of the **aorta** just superior to L4, which marks the terminus of the **aorta**. It then descends through the pelvic inlet, passing along the anterior surface of the sacrum and coccyx. The **median sacral artery** gives rise to the inferior most **lumbar arteries** and forms two **anastomoses** with pelvic vessels.

## Internal Iliac (Hypogastric) Artery

The **left** and **right internal iliac arteries** are lateral branches of the **common iliac artery** of the cat. We will study four main branches of this artery: (1) the **umbilical artery** (medial branch), (2) the **cranial gluteal artery** (lateral branch), (3) the **internal pudendal** or **middle hemorrhoidal artery** (medial branch), and (4) the **caudal gluteal artery** (lateral branch). In humans, there are **left** and **right common iliac arteries**, and each gives rise to an **internal iliac artery** and an **external iliac artery**. The **internal iliac artery** has an anterior and posterior trunk. These vessels supply blood to the posterior pelvic and abdominal walls, and to the gluteal region, the pelvic viscera, the adductor region of the thigh, the perineum, and, in the fetus, the placenta. *(See pictures on left)*

## Umbilical Artery

As you can see, this is not the biggest vessel in the cat. The **umbilical artery** is the first medial branch of the **internal iliac artery**. Functionally, it is important because one of its branches serves the urinary bladder. In the fetus, it also carried blood to the placenta.

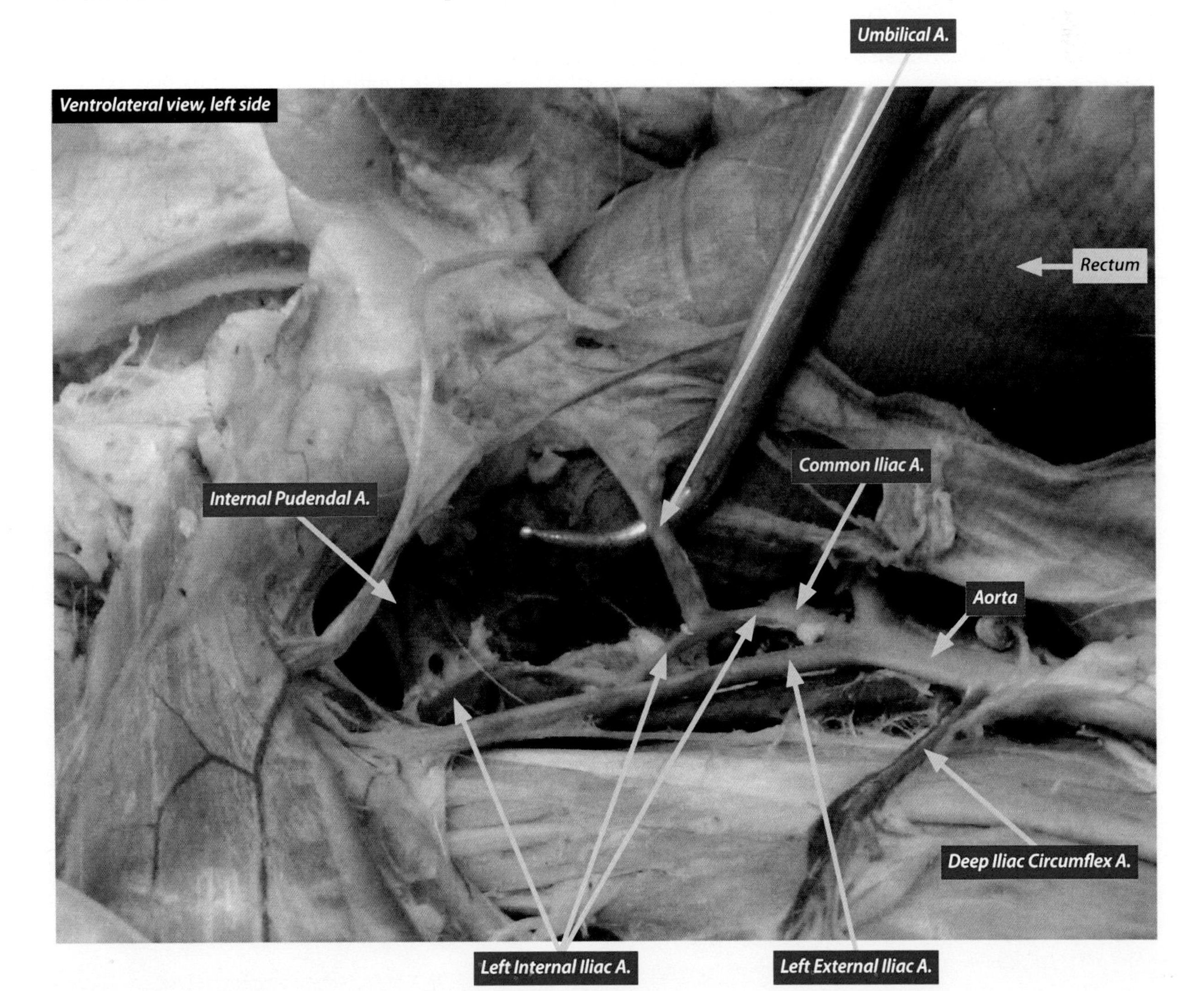

## Cranial Gluteal Artery

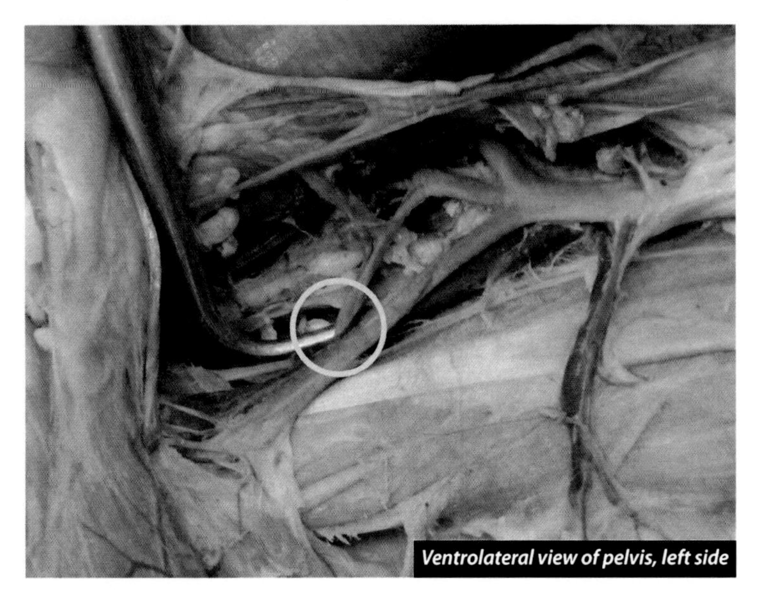

Ventrolateral view of pelvis, left side

The **cranial gluteal artery** is the first lateral branch of the **internal iliac artery**, both on the left and right sides of the cat. This occurs in the pelvis of the cat. This artery will form collateral circulation with the **caudal gluteal artery**. In humans, it would be called the **superior gluteal artery**. It leaves the pelvic cavity by way of the greater sciatic foramen superior to the piriformis muscle. It supplies the gluteus maximus muscle and adjacent muscles. It also supplies blood to the hip joint. The **cranial gluteal artery** also forms collateral circulation with the **medial** and **lateral femoral arteries**. The greater sciatic foramen is an opening between the greater sciatic notch, the sacrotuberous ligament, and the sacrospinous ligament.

## Caudal Gluteal Artery

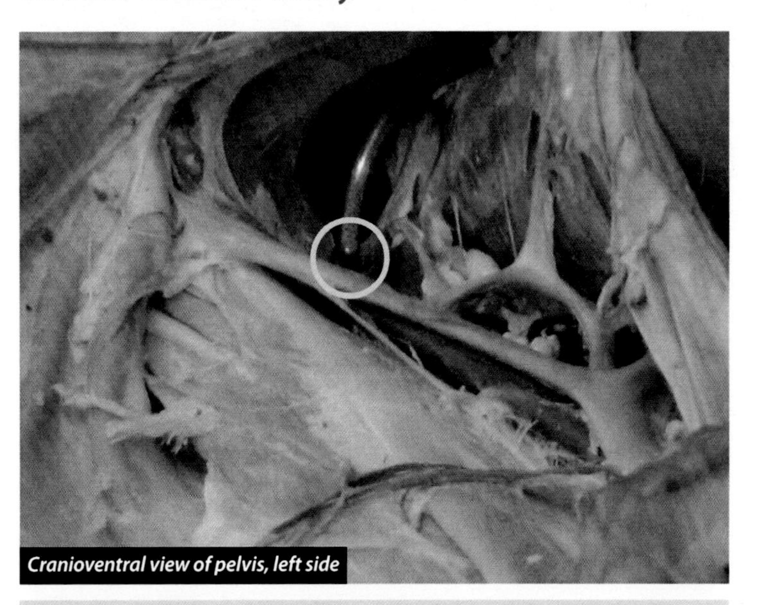

Cranioventral view of pelvis, left side

*The **caudal gluteal artery** courses laterally and the **internal pudendal artery** courses medially—the probe is touching both arteries in these pictures.*

The **caudal gluteal artery** is the second lateral branch of the **internal iliac artery**, both on the left and the right sides of the cat. This occurs in the pelvis of the cat, as did the **cranial gluteal artery**. In humans, it would be called the **inferior gluteal artery**. It leaves the pelvic cavity by way of the greater sciatic foramen inferior to the piriformis muscle. It supplies muscles in the area and passes into the posterior thigh where it forms collateral circulation with branches of the **femoral artery**. It also supplies blood to the **sciatic nerve**.

## Internal Pudendal Artery

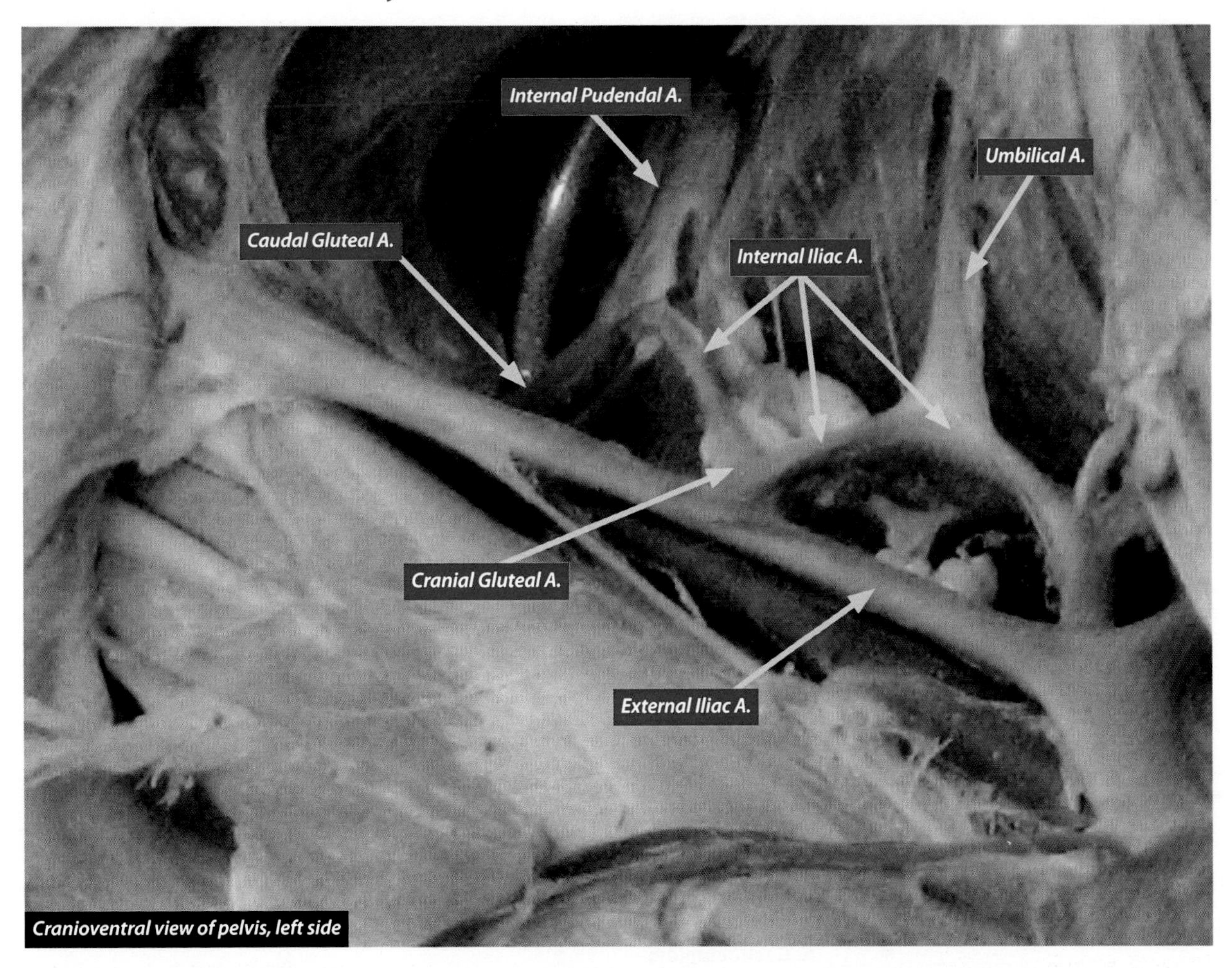


Cranioventral view of pelvis, left side

The second (and terminal) medial branch of the **internal iliac artery** that we will study in the cat is the **internal pudendal artery**. It is sometimes called the **middle hemorrhoidal artery** in the cat. In humans, the **internal pudendal artery** is a branch of the **internal iliac artery**, as it is in the cat. It supplies blood to the inferior half of the anal canal and the perineum, as well as to the penis in the male and the labia and clitoris in the female.

# ADDITIONAL VESSEL PLATES

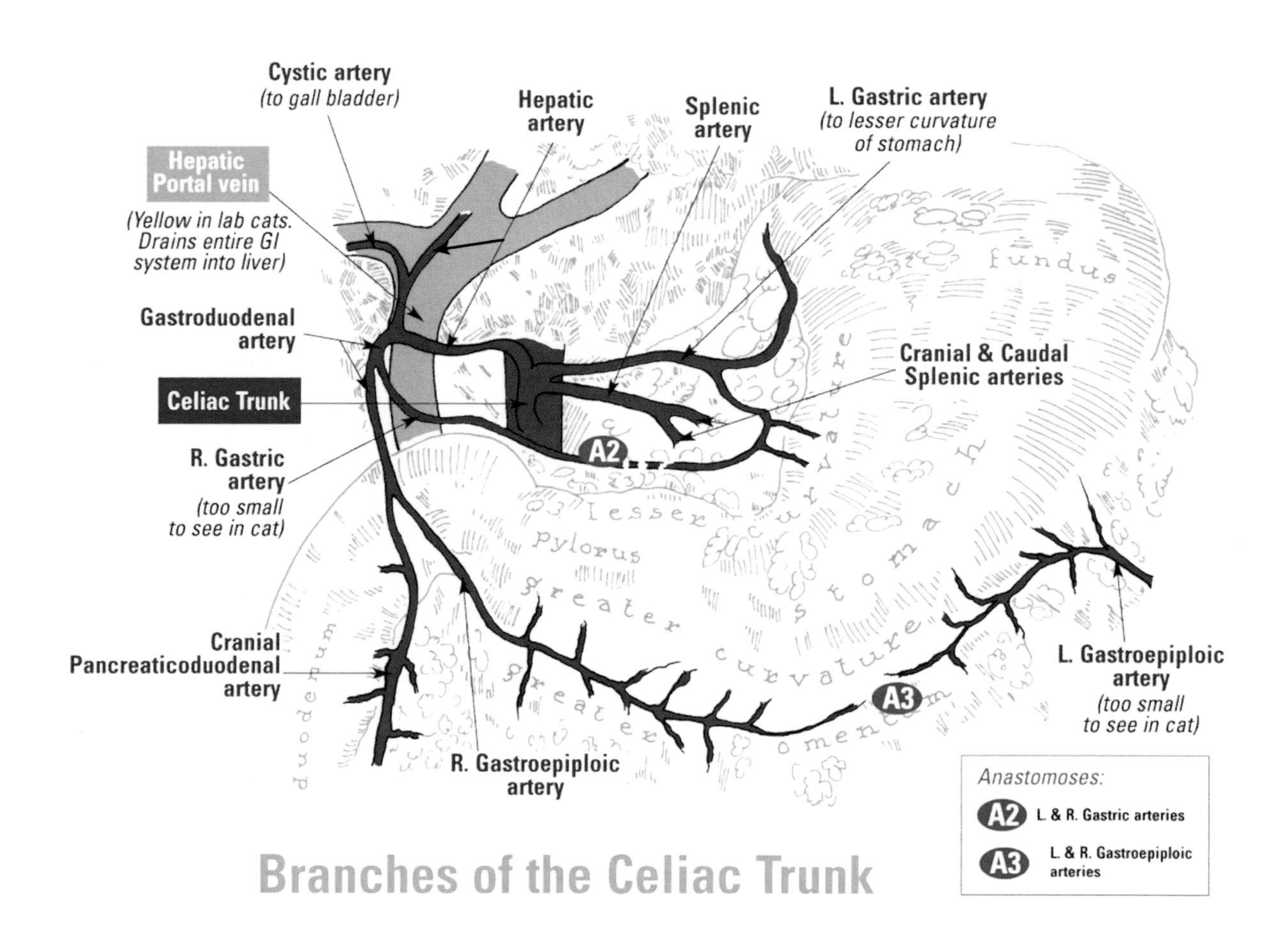


Branches of the Celiac Trunk

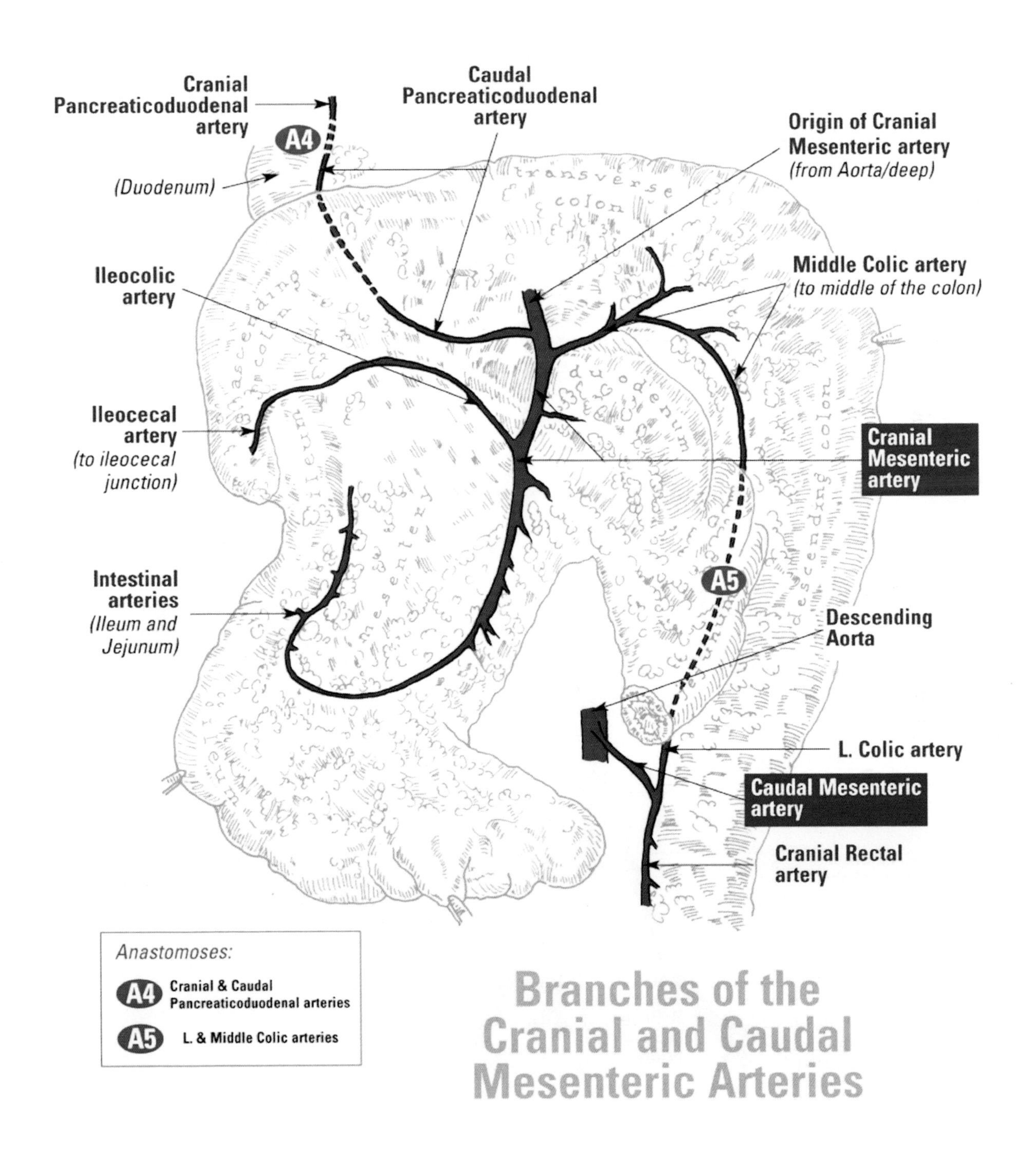


## Branches of the Cranial and Caudal Mesenteric Arteries

# VESSEL PLATES

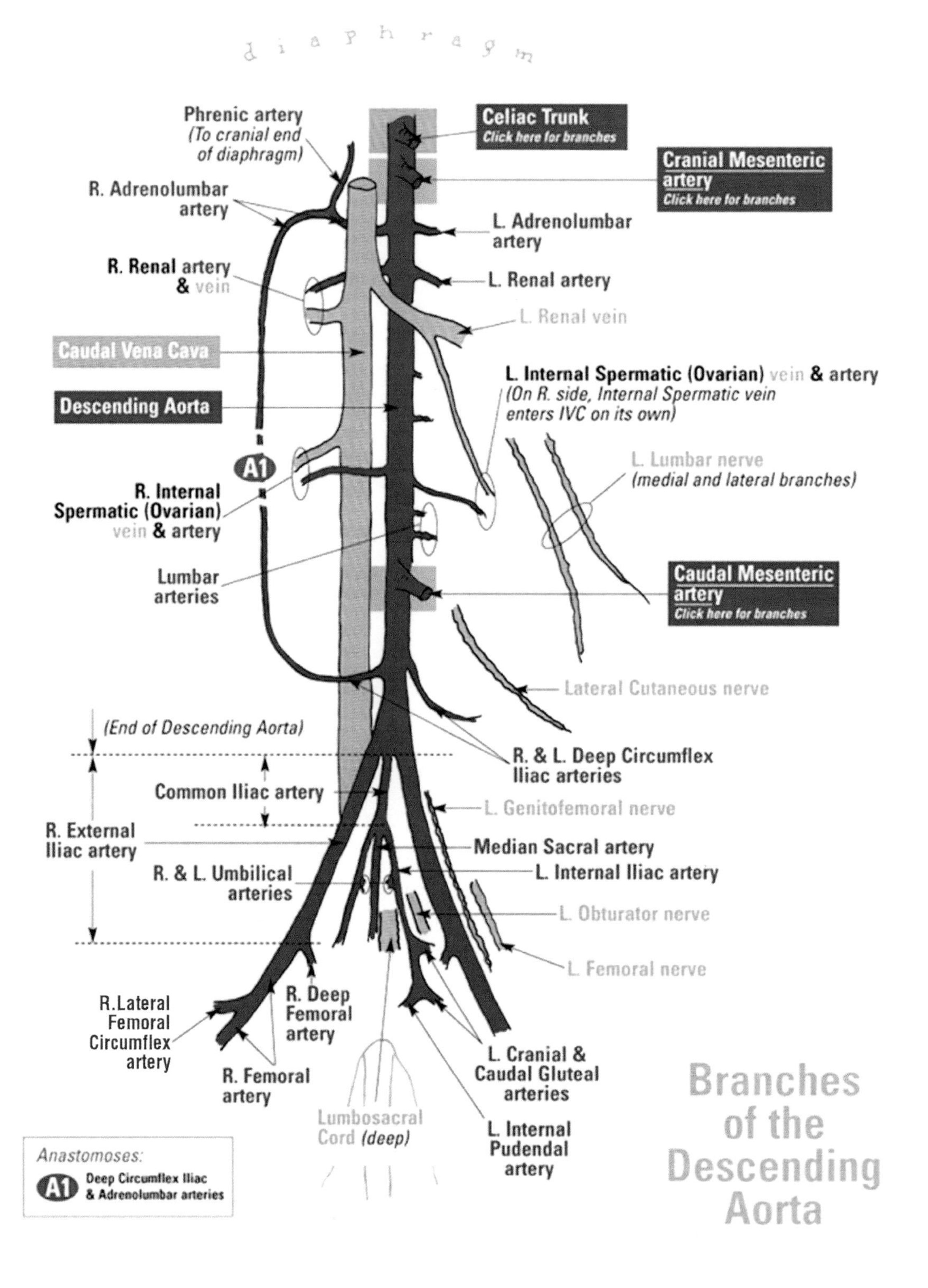

# PRACTICE QUIZ

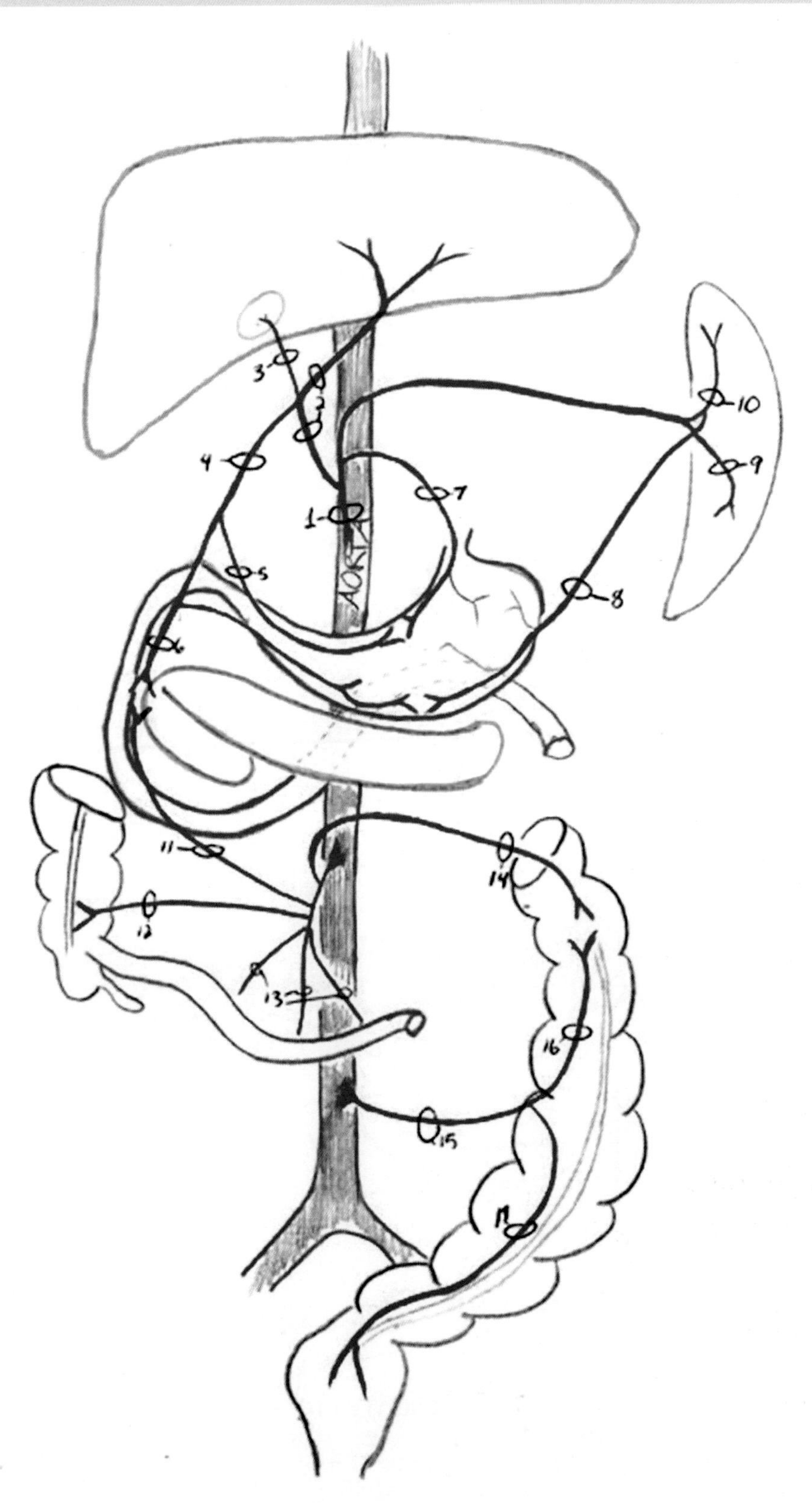

# PRACTICE QUIZ ANSWERS

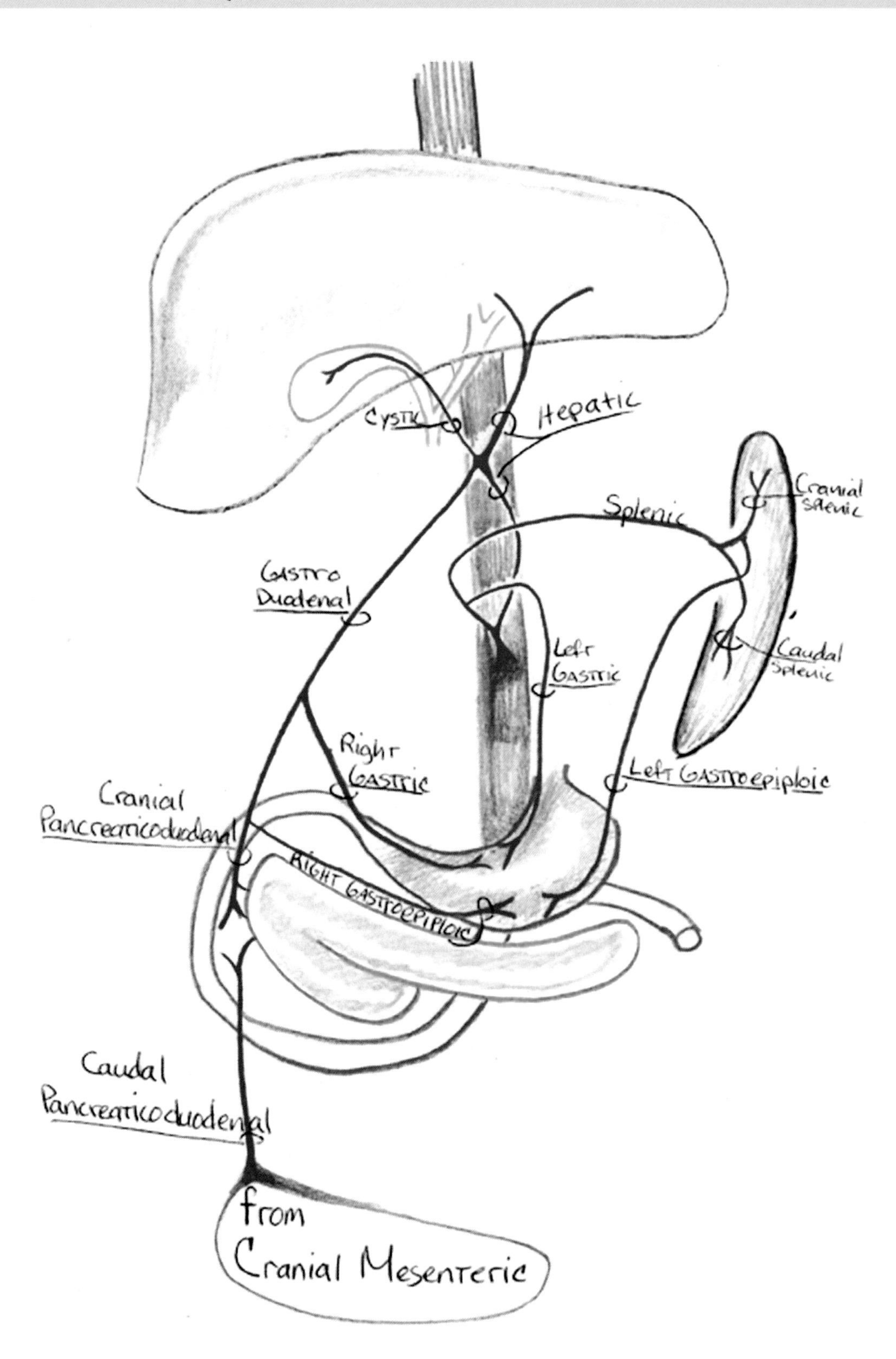

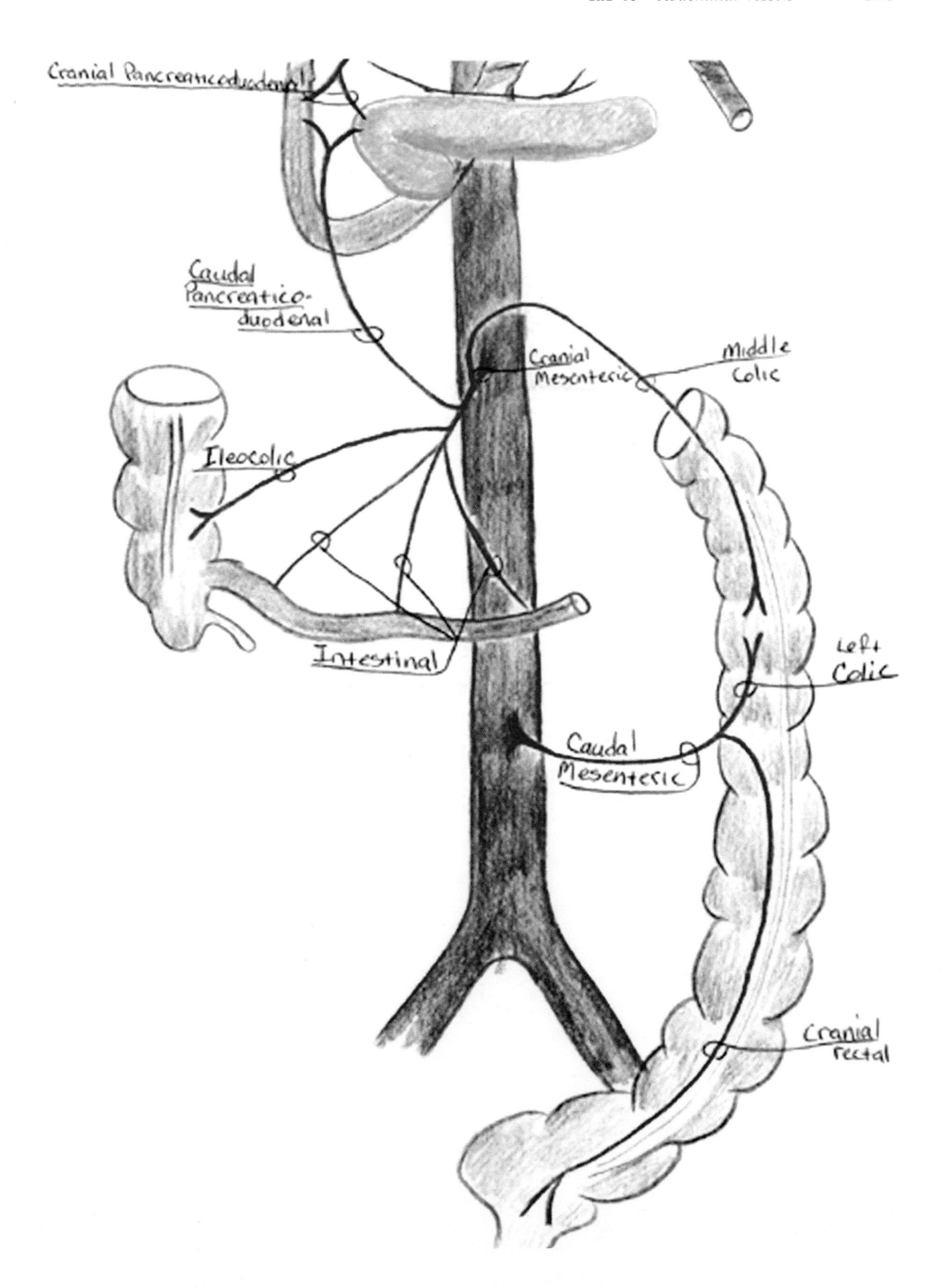
Cranial Pancreaticoduodenal
Caudal Pancreatico-duodenal
Cranial Mesenteric
Middle Colic
Ileocolic
Intestinal
Left Colic
Caudal Mesenteric
Cranial rectal

# Lab 11

# Pelvis and Reproductive Organs

## Wish List


**LAB 11 OVERVIEW, pp. 332–334**

**BONES, pp. 335–348**

- **Os Coxae, pp. 335–338, 347**
  - Iliopectineal Line (Arcuate Line), p. 336
  - Acetabulum, p. 336
  - Obturator Foramen, p. 336
  - Pelvic Inlet (Brim), p. 337
  - Pelvic Outlet, p. 337
- **Ilium, pp. 339–342**
  - Iliac Crest, p. 339
  - Sacroiliac Articulation/Auricular Surface, p. 339
  - Iliac Fossa, p. 339
  - Anterior Superior Iliac Spine (ASIS), p. 340
  - Anterior Inferior Iliac Spine (AIIS), p. 340
  - Posterior Superior Iliac Spine (PSIS), p. 341
  - Posterior Inferior Iliac Spine (PIIS), p. 341
  - Greater Sciatic Notch, p. 342
  - Iliac Tubercle, p. 342
- **Ischium, pp. 343–344**
  - Ischial Spine, p. 343
  - Lesser Sciatic Notch, p. 344
  - Ischial Tuberosity, p. 344
- **Pubis, pp. 345–347**
  - Pubic Crest, p. 345
  - Pubic Tubercle (Pubic Spine), p. 345
  - Pubic Arch (Subpubic Angle), p. 346
  - Symphysis Pubis (Symphysis Pubis), p. 347
- **Sacrum, p. 348**
  - Sacral Promontory, p. 348
  - Sacral Canal, p. 348
    - Anterior Sacral Foramen, p. 348
    - Posterior Sacral Foramen, p. 348

**MUSCLES, p. 349**

- **Levator Ani (Pelvic Diaphragm), p. 349**
- **Cremaster Muscle, p. 349**

**REPRODUCTIVE ORGANS, pp. 350–365**

- **Female, pp. 350–357**
  - Ovaries, p. 350
    - Tunica Albuginea (no female pic), p. 350
  - Infundibulum, p. 350
  - Uterine Tube (Fallopian Tube, Oviduct), p. 351
  - Urethra (Female Cat), p. 352
  - Uterus, p. 353
    - Uterine Horn (Horn of the Uterus), p. 353
    - Body of the Uterus, p. 353
  - Broad Ligament of the Uterus, p. 354
  - Round Ligament of the Uterus, p. 355
  - Vestibule, p. 356
    - Rugae (Vestibule), p. 356
  - Vagina, p. 357
- **Male, pp. 358–365**
  - Urethra (Male Cat), p. 358
  - Prostate, p. 358
  - Vas Deferens, p. 359
  - Bulbourethral Glands, p. 360
  - Penis, p. 360
    - Crus of the Penis, p. 361
    - Glans Penis, p. 361
  - Spermatic Cord, p. 362
  - Testes, p. 362
    - Fascial Sac (Tunica Vaginalis), p. 363
    - Tunica Albuginea, p. 363
    - Seminiferous Tubules, p. 364
    - Vas Efferens, p. 364
  - Epididymis (Head, Body, and Tail), p. 365

# LAB 11 OVERVIEW

Lab 11 provides us with an opportunity to study the reproductive system and the pelvic region. We will begin with the bones of the pelvic region and then work on the primary reproductive organs and the accessory reproductive organs.

## 1. Bones of the Pelvic Region

The walls of the **pelvis** are formed by the two **os coxae** and by the **sacrum** and **coccyx**. We will study each in turn.

The **os coxa** is formed by the fusion of three bones. Superiorly, we find the **ilium** (note the spelling as there is an ileum that is part of the small intestine); inferiorly, there is the **ischium**, and anteriorly the **pubis**. The superior most landmark of the os coxa is the **iliopectineal line** (**arcuate line**), which forms the anterior portion of the **pelvic brim**. The **pelvic brim** forms the edge of the **pelvic inlet**. The **acetabulum** is also known as the hip socket and upon close inspection you should be able to see all three bones in this landmark. Anterior and slightly inferior to the acetabulum is the **obturator foramen**. The most inferior feature of the os coxae is the **pelvic outlet**. The **pelvic outlet** of the male (and prepubescent female) is a narrow oval shape, while the **pelvic outlet** of the post-pubescent female is round.

The **ilium** is the largest of the three bones. Superiorly, there is an **iliac crest**, a very important point of attachment for many muscles. Anteriorly, it ends with the **anterior superior iliac spine** (**ASIS**), and posteriorly it ends with the **posterior superior iliac spine** (**PSIS**). Slightly anterior to the middle of the iliac crest on its lateral side is the **iliac tubercle**, also a point of attachment for muscles. On the deep side of the ilium you find the **iliac fossa**, a large flat surface that resembles the subscapular fossa. It is also functionally similar in that it is the site for a large origin of a muscle. The **auricular surface** is also known as the **sacroiliac articulation**. This is a Grant thing because this is where the ilium and sacrum articulate. The **anterior inferior iliac spine** (**AIIS**) is slightly superior to the acetabulum. It is the origin of the rectus femoris and the iliofemoral ligament. The **posterior inferior iliac spine** (**PIIS**) is slightly anterior to the inferior portion of the sacroiliac articulation. Anterior to the posterior inferior iliac spine, you will find the **greater sciatic notch**. It is on the posterior border of the **os coxa** between the **posterior inferior iliac spine** and the **ischial spine**.

The **ischium** is inferior and posterior to the ilium and posterior to the pubis. The superior most landmark of the ischium is the **ischial spine**. You will find this anterior and inferior to the **greater sciatic notch**. The **lesser sciatic notch** is inferior to the **ischial spine**. The **ischial tuberosity** is the inferior most landmark of the **ischium**, as well as the inferior most landmark of the **os coxa**.

The **pubis** is anterior and inferior to the **ilium** and anterior to the **ischium**. The **pubic crest** is the superior most landmark of the pubis, and it forms the medial to the **pubic tubercle** (**pubic spine**). The **pubic angle** (**subpubic angle**) is formed by the anterior rami of the pubic and ischial bones. The **symphysis pubis** (**pubic symphysis**) is an amphiarthrosis between the **pubic bones**. You will only observe this on the articulated pelvic girdles.

The **sacrum** forms the majority of the posterior wall of the pelvic cavity. At the superior anterior edge of the first sacral vertebra is the **sacral promontory**. The **sacral canal** is a continuation of the vertebral foramen of the lumbar vertebrae. At this point there is no spinal cord, but there are spinal nerves. We will also observe the **anterior** and **posterior sacral foramina**.

The **pelvic outlet** is a diamond shaped area if one were to connect the inferior end of the pubic symphysis to the ischial tuberosities (which are posterior and lateral to the pubic symphysis) and then the **coccyx** (which is posterior and medial to the ischial tuberosities).

## 2. Dissection of the Feline Pelvis

There will be slightly different ways to prepare the cat, depending on the gender. For female cats, we cut the skin along the midline from the abdominal region to the pelvis. Then we palpate the symphysis pubis (pubic symphysis) and carefully cut through the symphyseal cartilage from the cranial to the caudal end. Firmly spread the thighs laterally, being careful not to tear the **levator ani** (part of the **pelvic diaphragm**) away from the os coxa. Cut the skin to the right lateral side of the vestibular opening. Then insert the scissors into the **vestibule** and cut approximately one half to three quarters of an inch toward the **vagina** along the lateral side of the vestibule. You want this to be a lateral cut because the **urethra** intersects with the vagina in a ventral-dorsal orientation, and if you cut along the ventral side, you make it impossible to observe the junction of these two structures. You may be able to observe the **rugae** of the vestibule. The female cat is now ready to study.

Male cats are prepared slightly differently from the females. Begin with a midventral cut through the skin, being especially careful not to cut the **spermatic cord** that will be just deep to the skin and lateral to the midsternal line on both sides. Then move to the caudal end of the scrotum on the left side and make an incision. Carefully cut along the ventral side of the scrotum toward the midsternal cut that you already made. Remove the **testicle** from the scrotum and gently separate it from the other tissues. Repeat this procedure on the right side. Once both testicles have been freed, palpate the pubic symphysis and carefully cut through the symphyseal cartilage from the cranial to the caudal end. Firmly spread the thighs laterally, being careful not to tear the levator ani away from the os coxa.

Remove the prepuce of the **penis** by starting at the urethral opening and cutting toward the **shaft** of the penis. Cut around the prepuce to expose the **glans penis**. Removing the penis without doing cat terrorism is challenging at the very least. Move laterally with a probe. Be cautious if you use scissors; but if you do, use one side of the scissors and move smoothly laterally until the **crus of the penis** is released. Repeat on the other side. Once the penis is free, pull it away from the rectal area. Clean the area around the bulb of the penis to expose the crura and the **bulbourethral glands**. At this point you should be able to see both crura and both bulbourethral glands. Moving cranially, follow the urethra to where it widens. This wide area is the **prostate gland**. Stop removing the penis at this point.

## 3. Structures of the Female Reproductive System

The female cat has an internal vestibule that extends from the junction of the urethra and vagina to the outside. Therefore, it is a common urogenital sinus. Carefully insert the sharp end of the probe into the urethra. You will have difficulty identifying the vagina as it is very narrow. Reflect the urinary bladder to expose the **body of the uterus**. Notice that the body looks like the head of the cow from Walt's Roast Beef. LOMG, who would have thought?

Proceeding cranially you will see where the **uterine horns** (**horns of the uterus**) converge to form the body. Continuing cranially, you will observe the **infundibulum**, which forms a raincoat hood-like structure that surrounds the **ovary**. This is a loose association. The deep side of the infundibulum has ciliated epithelial cells that cause a current that sweeps the ova into the **oviduct** (**fallopian tube** or **uterine tube**). This arrangement has potential to become problematic and that will be discussed in class. You should also observe the **ovarian artery** that is directed toward the ovary.

Also notice a thin structure that runs from the horn of the uterus to the internal inguinal ring. That structure is the **round ligament of the uterus**. There is also a **round ligament of the ovary**, but you will not be responsible for it. The male has a homologous structure called the gubernaculum testis. You will find a sheet of peritoneum, the **broad ligament of the uterus**, extending between the abdominal wall and the horn of the uterus.

## 4. Structures of the Male Reproductive System

Observe the bull testicle. It should be prepared for you. Reflect the **fascial sac** (**tunica vaginalis**) and note the shiny inner surface that is in part what remains of the peritoneum. Technically, the tunica vaginalis is just the peritoneum, but we will use this term interchangeably with fascial sac. Note that if you lift the testicle by using the fascial sac, the testicle remains attached to the fascial sac. That attachment is the **gubernaculum testis** (**scrotal ligament**), which is what remains of the embryonic structure called the gubernaculum.

You can see the **epididymis** on the dorsal surface of the testicle. It has a **head** (cranially), a **body**, and a **tail** (caudally). Inside the epididymis is a highly coiled **duct of the epididymis**. You will not be responsible for seeing this, but it will be a subject in the lecture. In a human, that duct is approximately 20 feet long! Surrounding the testis you will see a leathery layer, the **tunica albuginea**. This name is used for other layers that cover other organs in the body.

The testicle should be cut in a sagittal plane. Reflect the two halves and you will be able to observe the mass that includes the **seminiferous tubules** and the **rete testes**. You will not be able to distinguish specific tubules, as they are microscopic. However, running from the caudal to the cranial end near the center of the testicle you will be able to see the **vas efferens** (**vas efferentia** is the pleural form), which transports the spermatozoa from the rete testes to the head of the epididymis. In the head of the epididymis, they converge to form the duct of the epididymis.

Leaving the tail of the epididymis, you will find the **vas deferens**, which transports the spermatozoa through the spermatic cord into the body cavity, where it intersects with the **seminal duct** to form the **ejaculatory duct**. You are not responsible for the path of the vas deferens beyond where it loops over the ureter, but you should be able to see where it leaves the **internal spermatic artery** close to the **internal inguinal ring**, inside the abdomen. You should observe the **cremaster muscle** in the spermatic cord. It is formed when fibers of the **internal abdominal oblique** are pulled down into the spermatic cord as the testicle descends to the scrotum. The cremaster muscle is functionally important, as it helps with thermoregulation of the testicle by contracting when the testicle is cold. This contraction brings the testicle into a position near the body wall to help warm it up. When it is sufficiently warm, the cremaster muscle relaxes and the testicle descends back to the scrotum. This action is supplemented by the **dartos muscle** (smooth muscle) of the scrotum.

We have already prepared the cat's testicles and penis for review. Follow the spermatic cord from the testicle to the external inguinal ring. Inside the body cavity, you will see the internal spermatic artery separate from the vas deferens. Notice that the vas deferens passes over the ureter and then descends on the dorsal side of the urinary bladder toward the prostate gland. The other male structures were described in the section that instructs you on how to prepare the cat for lab.

# BONES

## Os Coxa

The os coxa is also known as the pelvic bone or the innominate bone. It is actually three fused bones: the pubis, the ilium, and the ischium. At birth, they are still three bones joined by cartilage, and they fuse when the individual is about sixteen to eighteen years old. The plural is os coxae. The ilium is the superior most of the three, the pubis is the most anterior, and the ischium is the most inferior.

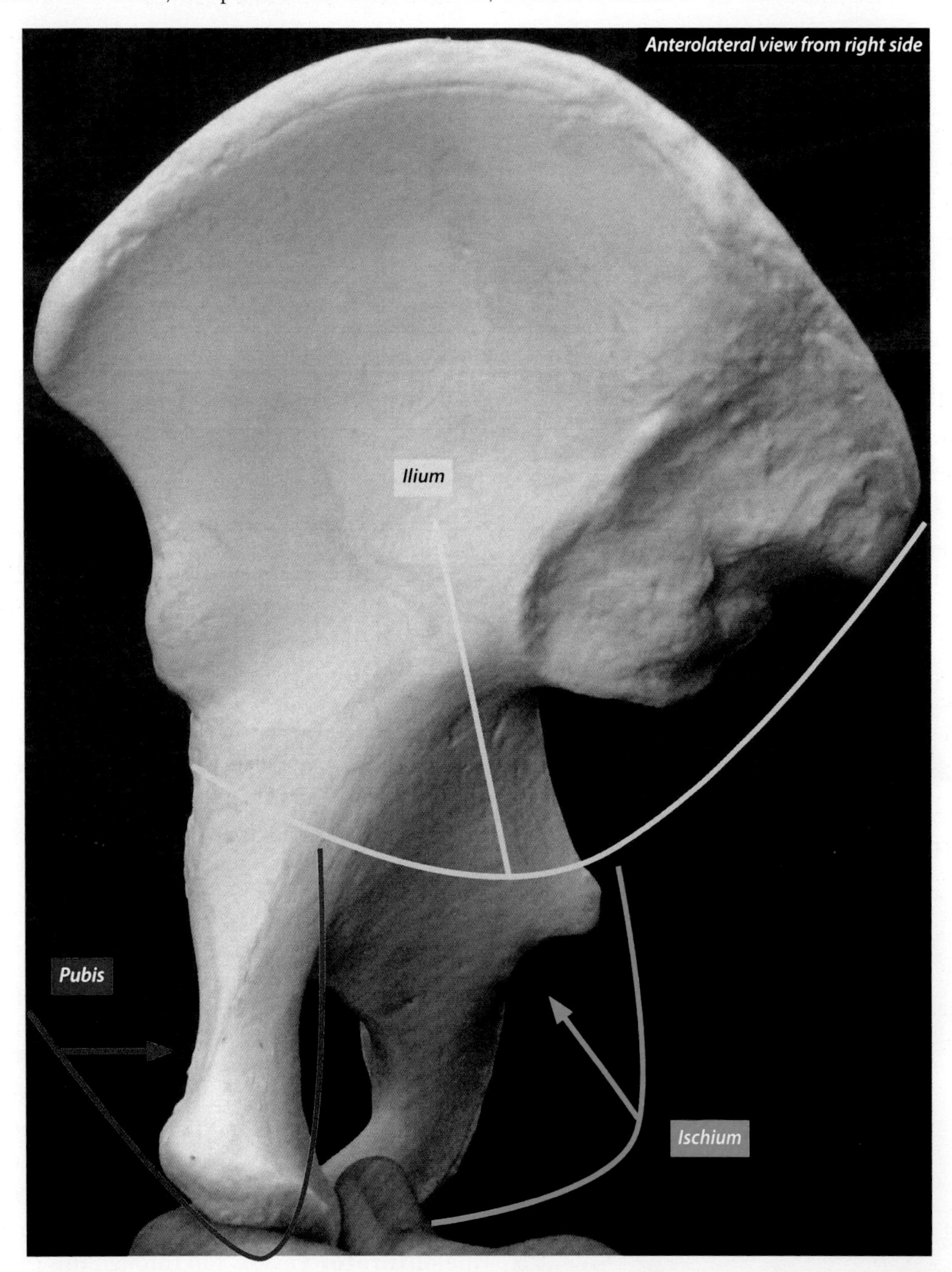

# Os Coxae

## Iliopectineal Line

*Anterolateral view from right side*

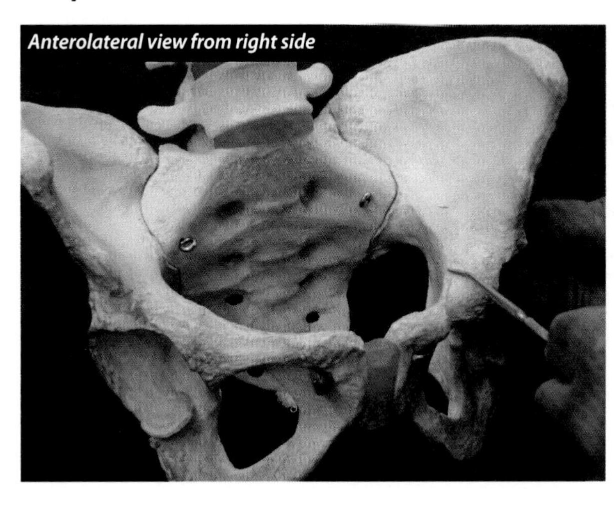

*Anterolateral view from right side*

The iliopectineal line divides the false pelvis from the true pelvis. It forms part of the pelvic inlet or pelvic brim. Part of it is on the pubic bone just posterior to the pubic symphysis, and part is on the ilium extending posteriorly toward the sacral promontory. It is also called the arcuate line.

The iliopectineal line is the insertion for part of the:

1. internal abdominal oblique.

## Acetabulum

*Inferolateral view, rigtht side*

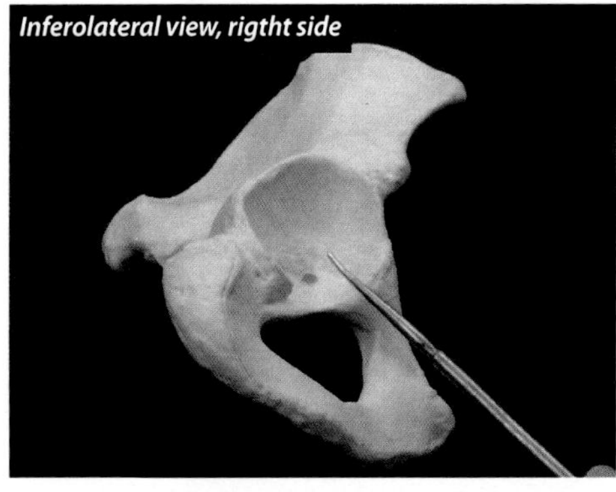

When we were in nursery school, we called the acetabulum the hip joint. It is on the lateral side of the os coxa. The ilium, the ischium, and the pubis all form part of the acetabulum. This is a ball and socket joint where the head of the femur articulates with the acetabulum. It allows for flexion, extension, abduction, adduction, and rotation. The name translates to mean "little vinegar cup." I am not sure who named it that, but we should be looking into what they were smoking at the time!

## Obturator Foramen

*Inferolateral view, right side*

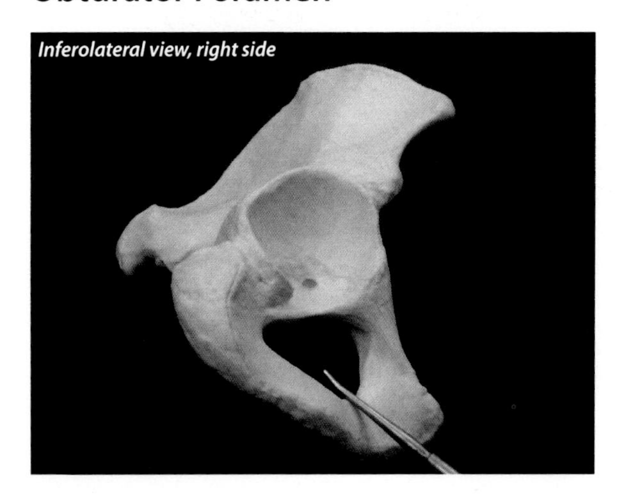

Two of the three bones that make up the os coxa contribute to the obturator foramen. They are the pubic bone and the ischium. Most of the foramen is closed by the obturator membrane, which is a flat membrane made of connective tissue occupying the inferior portion of the foramen. The obturator canal is small, but it remains open, allowing for communication between the lower limb and the pelvis. The **obturator nerve** and the **obturator vessels** run through this canal.

# Os Coxae

## Pelvic Inlet (Brim)

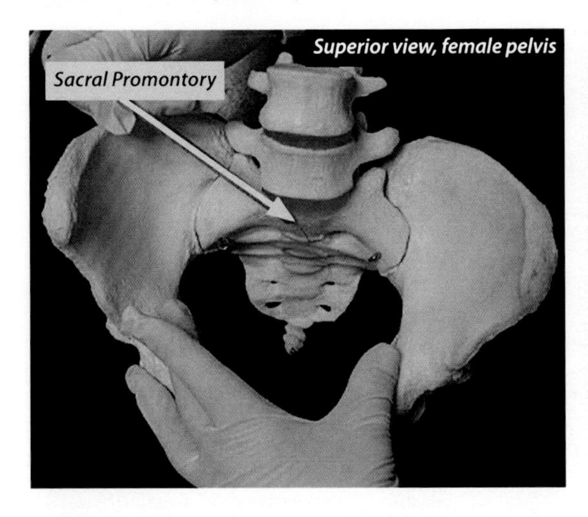


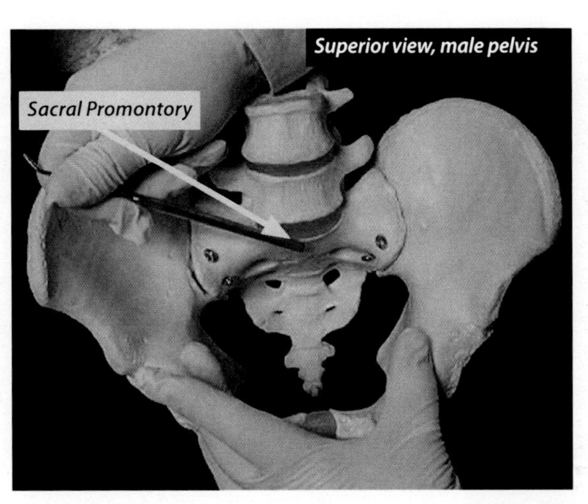


The pelvic brim or inlet of a prepubescent female and a normal male will be heart-shaped. This has led some people to conclude that a male's heart is in his pelvis! The pelvic brim or inlet of a normal post-pubescent female is oval, being wider than it is deep. The change comes about from hormones, and it is functionally important because it facilitates childbirth.

## Pelvic Outlet

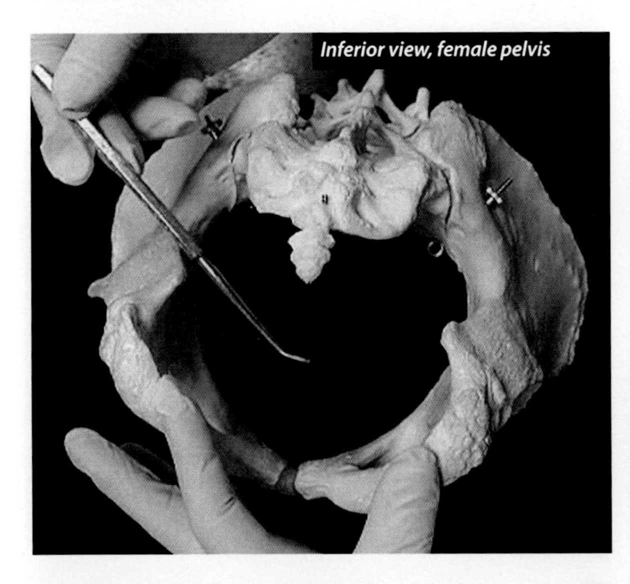


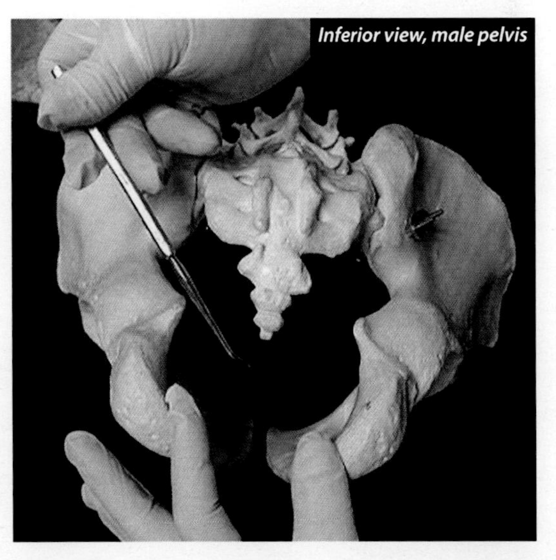


The pelvic outlet of a prepubescent female and a normal male will be a narrow oval shape. The pelvic outlet of a normal post-pubescent female is round. The change comes about from hormones, and it is functionally important because it facilitates childbirth.

# Os Coxa

## Overview

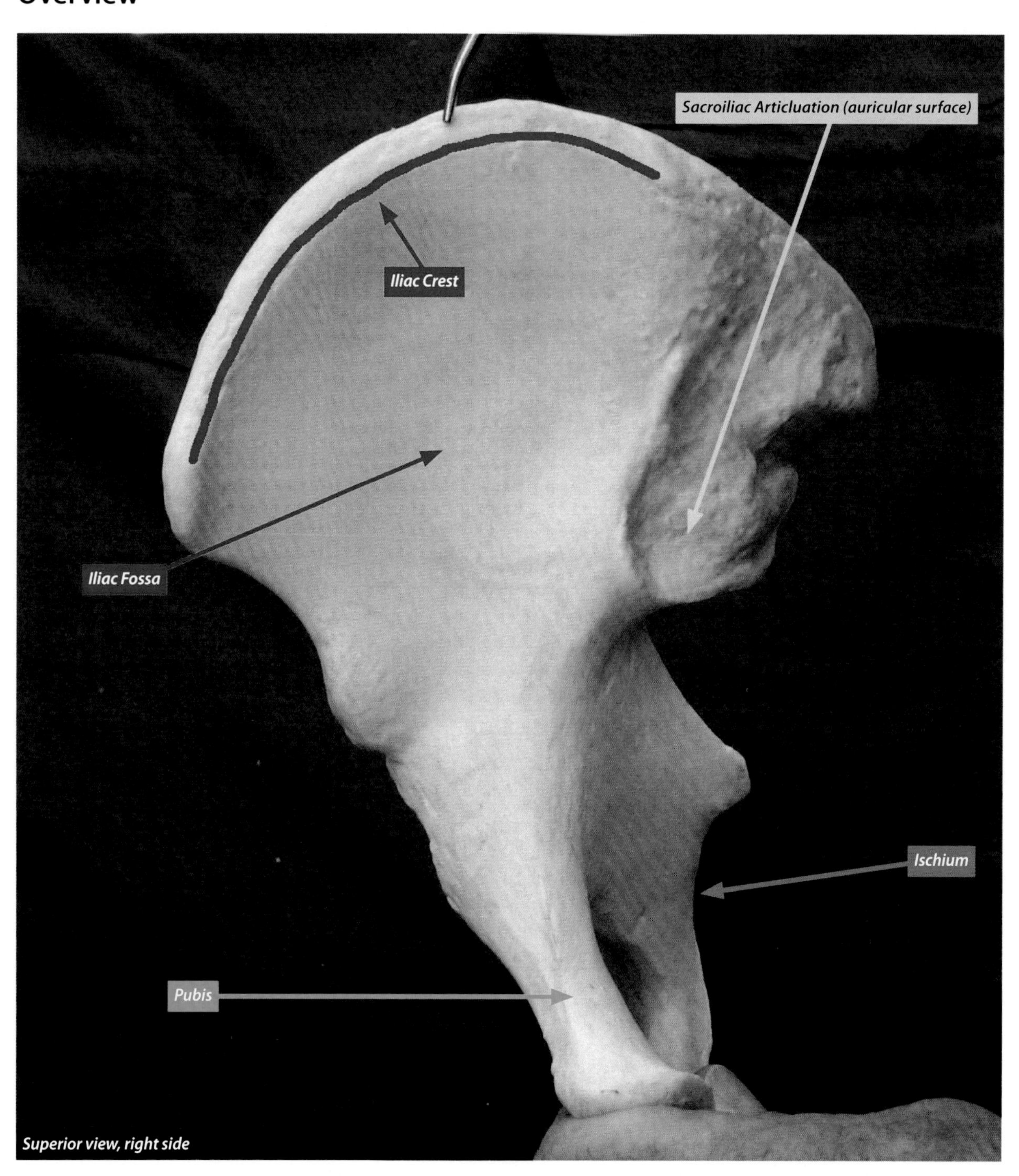

# Ilium

## Iliac Crest (Ilium)

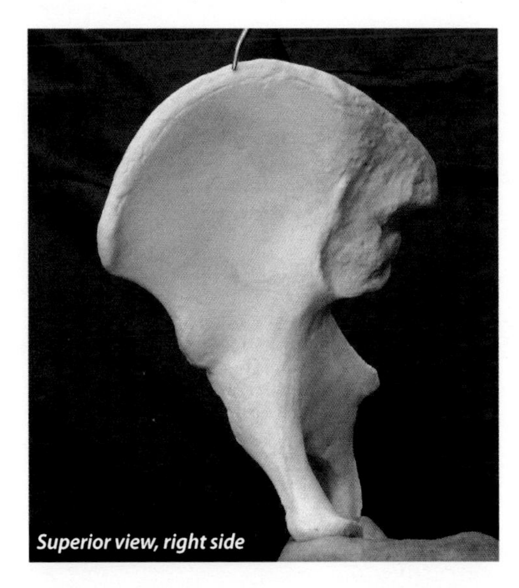
*Superior view, right side*

The iliac crest extends from the anterior superior iliac spine (ASIS) posteriorly to the posterior superior iliac spine (PSIS). The iliac crest is the insertion for the:

1. external abdominal oblique (anterior half of the iliac crest),

and the origin for:

1. gluteus maximus (posterior portion),
2. internal abdominal oblique,
3. latissimus dorsi (posterior portion),
4. quadratus lumborum,
5. tensor fasciae latae (anterior portion of outer lip),
6. transverse abdominis.

## Sacroiliac Articulation/Auricular Surface

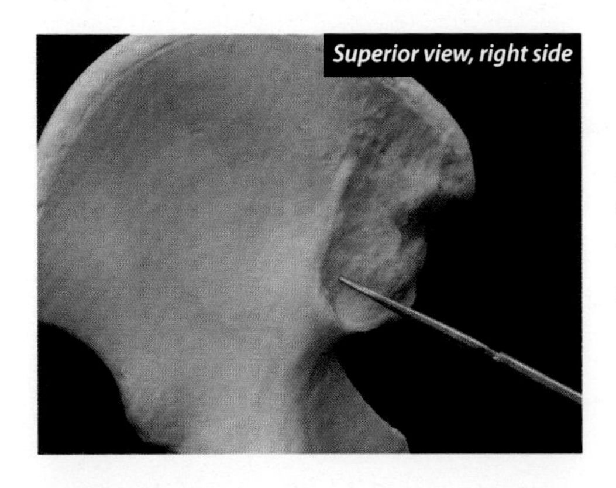
*Superior view, right side*

Auricle translates to mean "little ear." So the auricular surface looks a little like a little ear. This is where the sacrum and the ilium articulate. Therefore, it is also referred to as the sacroiliac articulation.

## Iliac Fossa

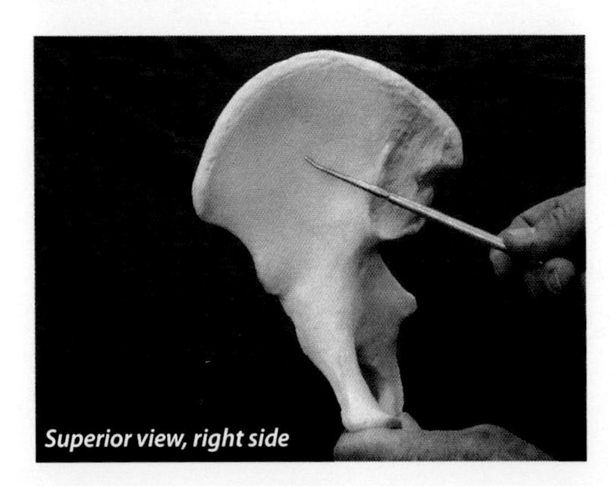
*Superior view, right side*

This landmark is located on the deep side of the ilium and is in some ways similar to the subscapular fossa of the scapula. The iliac fossa is the origin for:

1. iliacus.

# Ilium

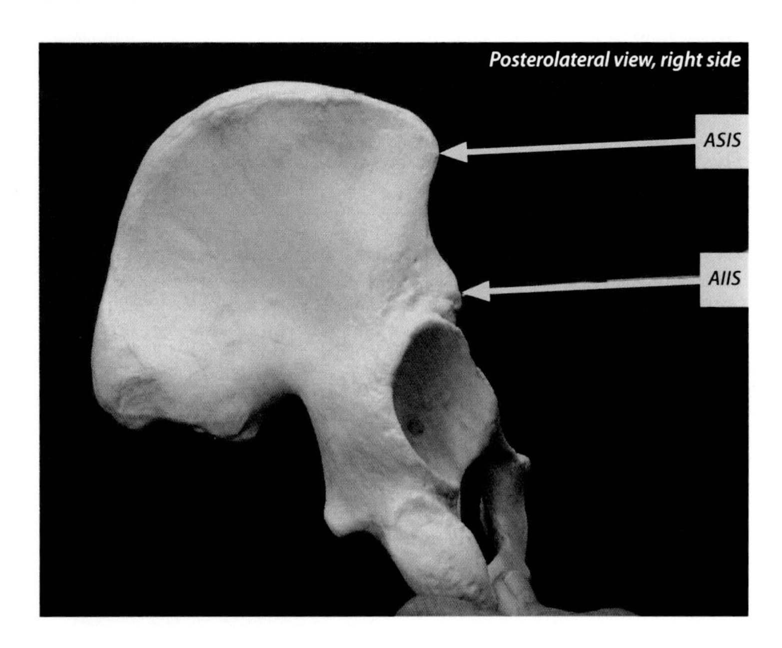


## Anterior Superior Iliac Spine (ASIS)

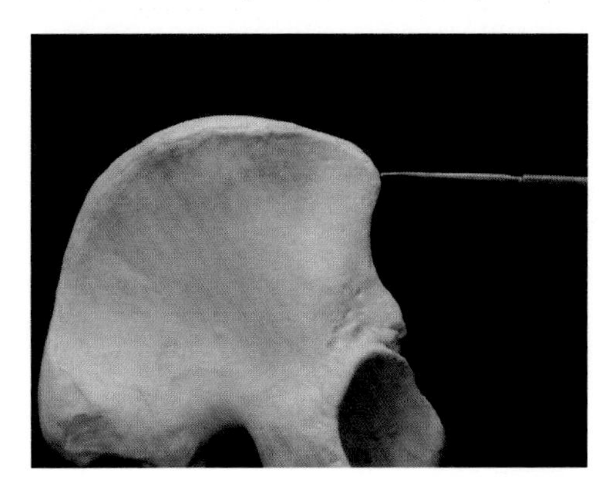

The superior attachment of the inguinal ligament is the anterior superior iliac spine. It is the most anterior point of the iliac crest. The anterior superior iliac spine is the origin for:

1. tensor fasciae latae (slightly superior and posterior on the lateral surface),
2. sartorius.

## Anterior Inferior Iliac Spine (AIIS)

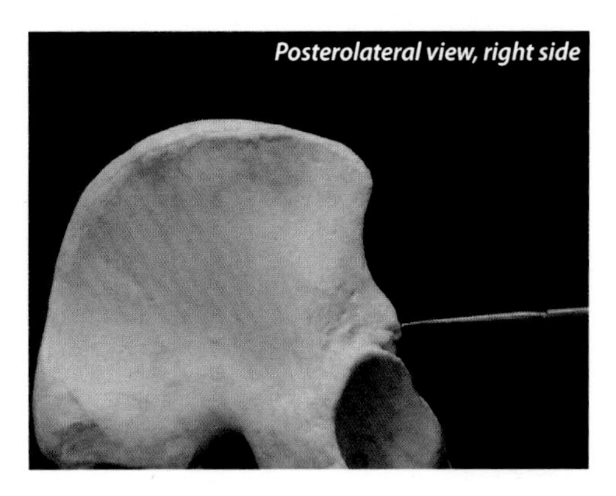


The anterior inferior iliac spine is just inferior to the origin of the inferior gluteal line. That line curves across the ilium and ends near the posterior margin of the acetabulum. The anterior inferior iliac spine is the origin for:

1. rectus femoris.

# Ilium

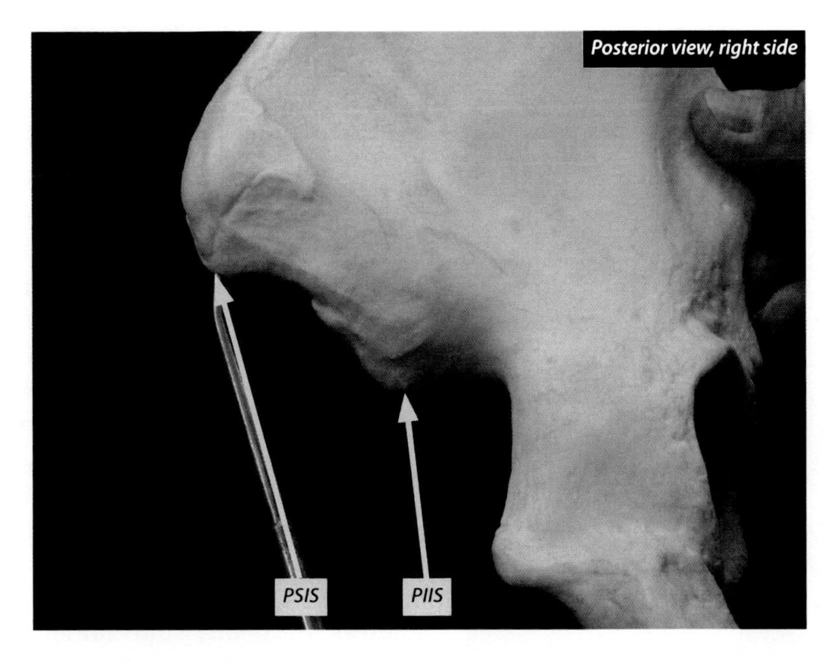


## Posterior Superior Iliac Spine (PSIS)

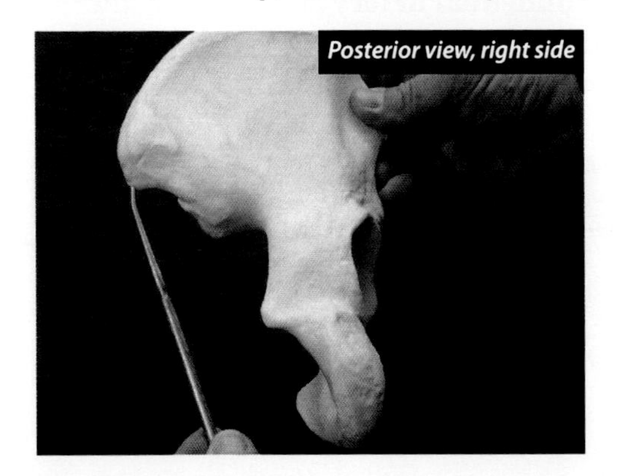


The posterior superior iliac spine (PSIS) is the most posterior point of the iliac crest. This is at the level of the sacral vertebra 2 spinous process, and it marks the inferior extent of the subarachnoid space. This also marks the middle of the sacroiliac joint. It is normally two to three fingers' width lateral to the midline. It is of medical significance in that one centimeter inferolateral to this is where a physician would insert a needle to obtain material for a bone marrow biopsy. Such a sample could be used to determine if the person had leukemia or other blood disorders. This landmark is recognizable as a pair of "dimples" on the posterior pelvis. I have been told by a number of female students that they are particularly easy to see on Mel Gibson in a scene from Lethal Weapon. Wow.

## Posterior Inferior Iliac Spine (PIIS)

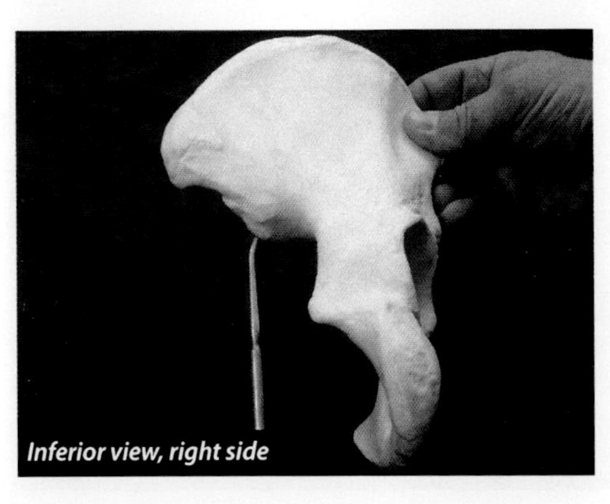


The posterior inferior iliac spine (PIIS) serves as part of the attachment for the sacrotuberous ligament. This strong and relatively inflexible ligament extends from the posterior inferior iliac spine, the lateral part of the sacrum and coccyx, to the ischial tuberosity. Because of the nature of this ligament, it forms part of the perimeter of the pelvic outlet. Functionally, it is of great significance because, with the sacrospinous ligament, it prevents the sacrum from rotating and breaking away from the os coxae. Because of the force from the lumbar vertebrae pushing inferiorly and anteriorly on the sacral promontory, it would rotate forward and inferiorly were it not for these two ligaments.

# Ilium

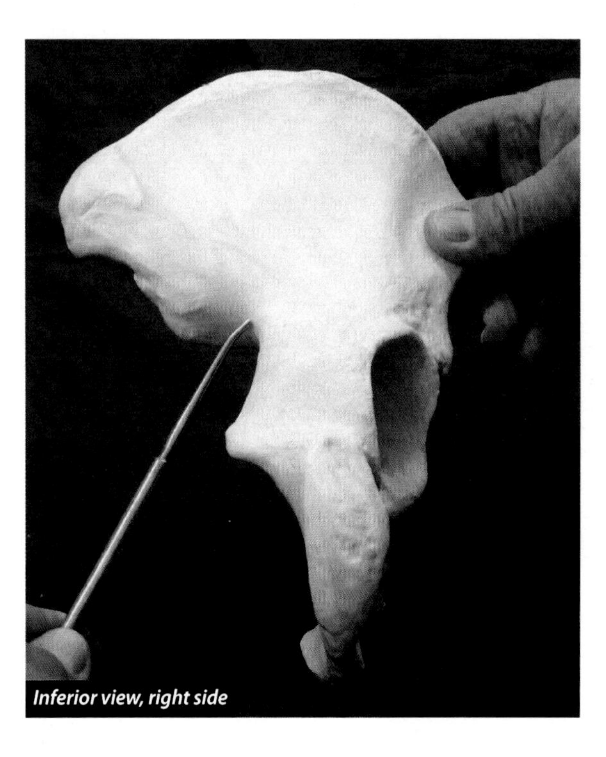
*Inferior view, right side*

## Greater Sciatic Notch

The greater sciatic notch is on the posterior border of the os coxa between the posterior inferior iliac spine and the ischial spine. It forms part of the greater sciatic foramen. The remainder of the foramen is formed by sacrospinous and sacrotuberous ligaments, and the spine of the ischium. The greater sciatic foramen is superior to the pelvic diaphragm. It allows for communication between the lower limb and the pelvic cavity. The piriformis muscle passes through the greater sciatic foramen, but you will not be responsible for this muscle. This landmark is also significant because, in addition to the nerves serving the quadratus femoris and the obturator internus muscles, the **sciatic**, **pudendal**, **posterior femoral cutaneous**, **superior** and **inferior gluteal nerves** pass through it on their course to the thigh. The **internal pudendal artery** and **vein**, and the **superior** and **inferior gluteal arteries** and **veins**, also pass through this foramen.

## Iliac Tubercle

The iliac tubercle is the origin for:

1. tensor fasciae latae (the anterior portion of outer lip).

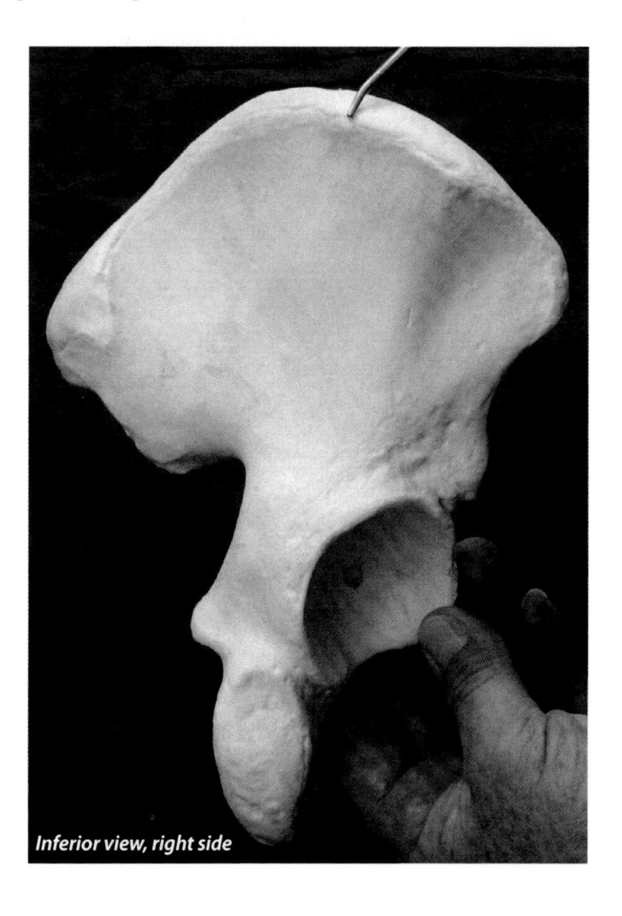
*Inferior view, right side*

***Posterior View of Right Os Coxa (Innominate Bone)***

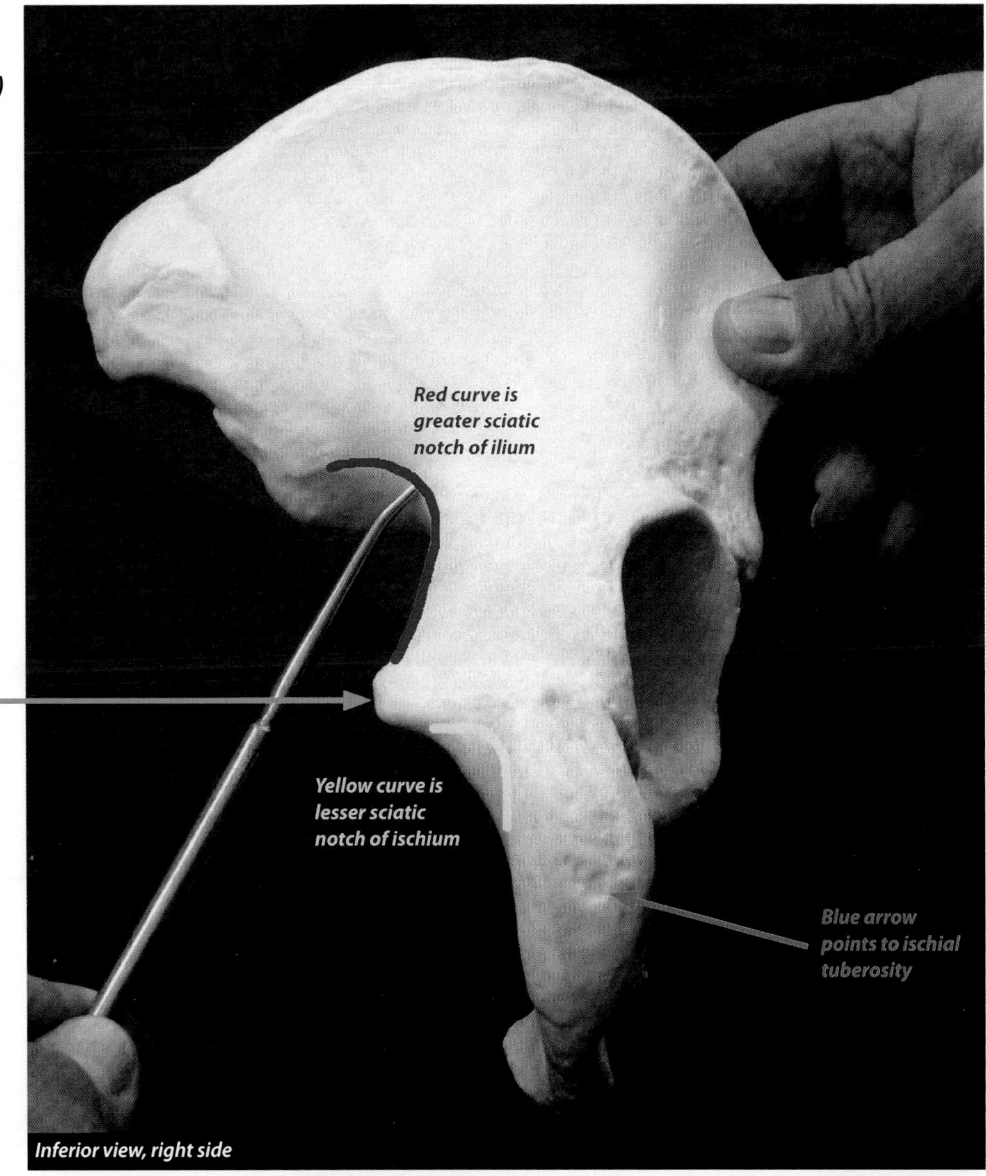


# Ischium

## Ischial Spine

The ischial spine is the attachment for the sacrospinous ligament. The sacrospinous ligament is a triangularly shaped ligament that is strong, attaching the coccyx and sacrum to the ischial spine. With the sacrotuberous ligament, the ischial spine forms the boundaries for the lesser sciatic foramen. Functionally, it is of great significance because, with the sacrotuberous ligament, it prevents the sacrum from rotating and breaking away from the os coxae. Because of the force from the lumbar vertebrae pushing inferiorly and anteriorly on the sacral promontory, it would rotate forward and inferiorly if it were not for these two ligaments.

# Ischium

## Lesser Sciatic Notch

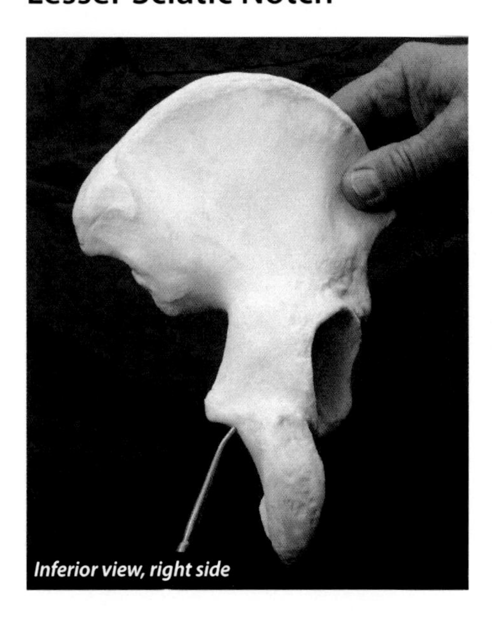


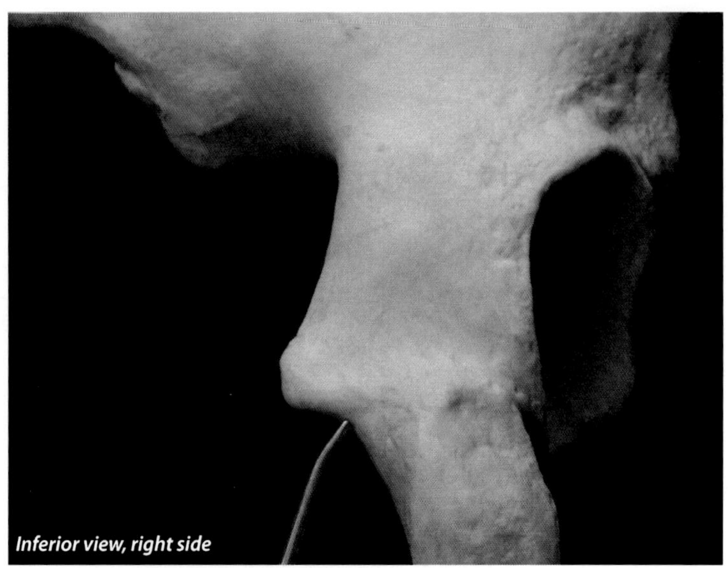


The lesser sciatic notch of the ischium forms part of the wall of the lesser sciatic foramen. The remainder of the wall is formed by the ischial spine and by the sacrotuberous and sacrospinous ligaments. The lesser sciatic foramen is inferior to the pelvic diaphragm and the ischial spine. This foramen provides communication between the perineum and the gluteal region. The **pudendal nerve**, the **internal pudendal vessels**, and the nerve that serves the obturator internus muscle pass through this foramen to reach the perineum. Also, the tendon of the obturator internus muscle passes through this foramen to reach the gluteal region of the lower limb.

## Ischial Tuberosity (Ischium)

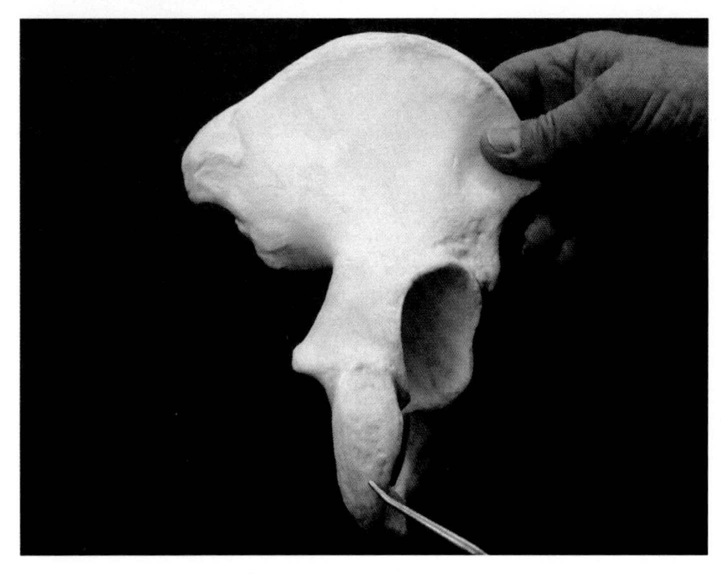

Remember that it is your ischial tuberosity that you sit on! The ischial tuberosity is the origin for the:

1. adductor magnus,
2. biceps fermoris,
3. semimembranosus (superior and lateral aspect of ischial tuberosity), and
4. semitendinosus.

## Pubis

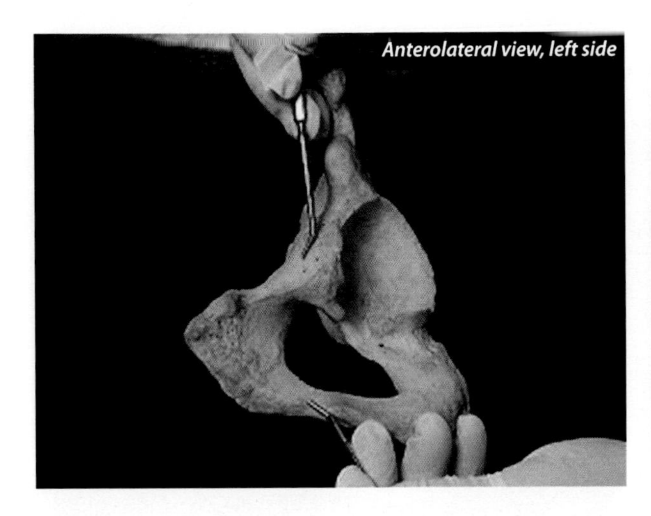


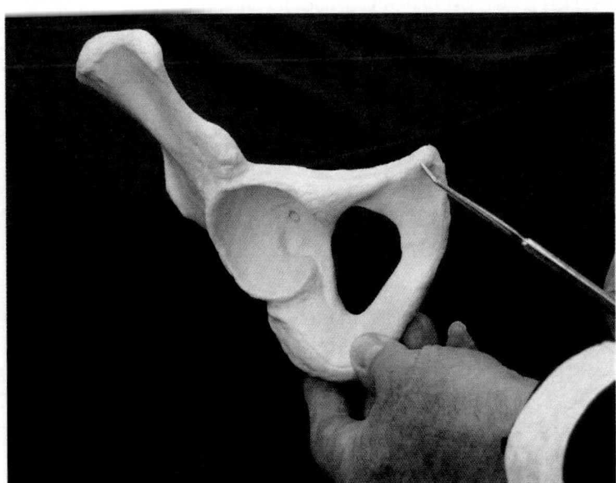

The picture on your left shows two synarthroses. The one toward the top is between the pubis and the ilium. The one toward the bottom is between the pubis and the ischium. In the picture to your right, the end of the probe is just to your left of where the pubic symphysis would be. The pubis is the origin for the:

1. adductor magnus,
2. adductor brevis,
3. adductor longus, and
4. pectineus.

### Pubic Crest

The pubic crest forms the anterior potion of the pelvic inlet. It is immediately superior and slightly medial to the pubic tubercle. The pubic crest is the insertion for part of the:

1. internal abdominal oblique.

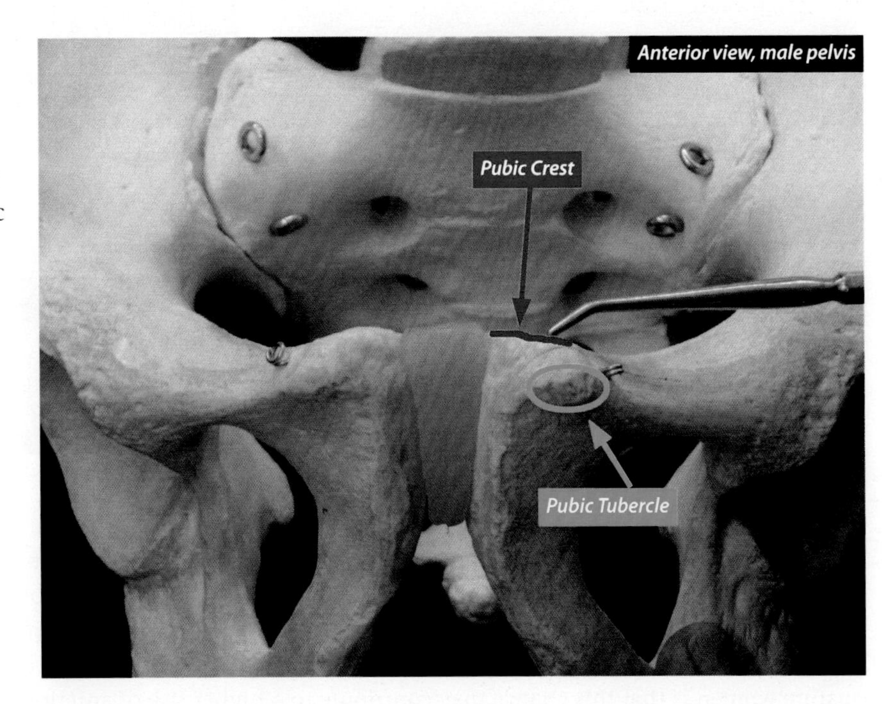


### Pubic Tubercle

The pubic tubercle is of significance because it is the place where the inferior end of the inguinal ligament attaches. The pubic tubercle is the origin for:

1. adductor longus.

# Pubis

## Pubic Arch (Subpubic Angle)

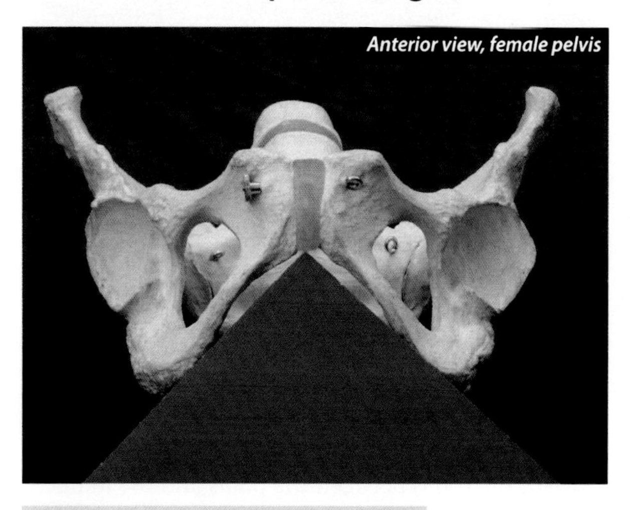


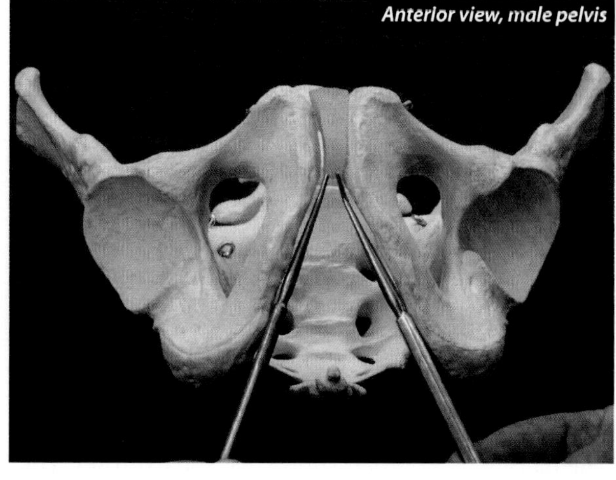


*A red piece of paper is included in this picture to demonstrate the angle, which is close to 90°—usually 80° to 85° in post-pubescent females.*

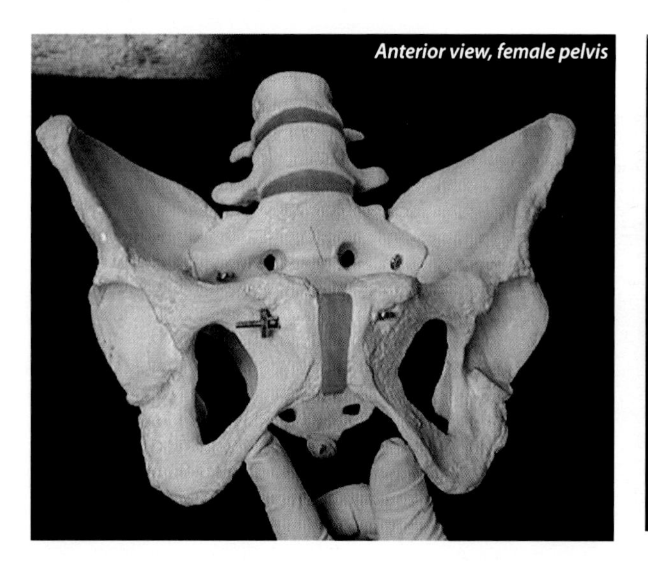


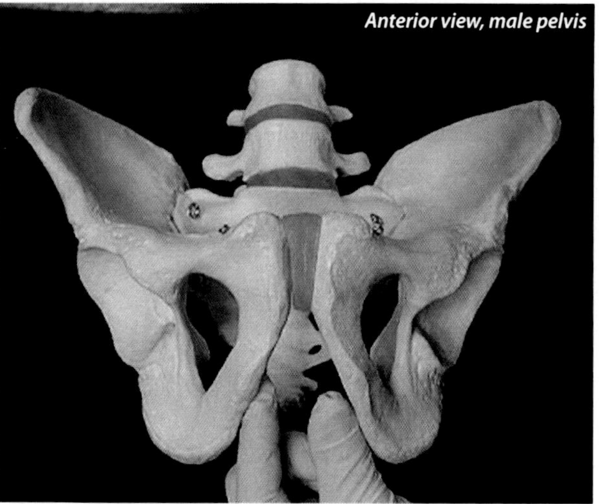


The adult male pelvis is on your right. The adult female pelvis is on your left. The pubic angle of a prepubescent female and a normal male will be a narrow 50° to 60°. The pubic angle of a normal post-pubescent female is relatively wide, 80° to 85° (not all women have this change). The change comes about from hormones, and it is functionally important because it facilitates the birth of children. The downside of this for mature women is that this change may contribute to a higher risk of anterior cruciate ligament (ACL) damage due to the resulting change of geometry of the knee joint. This may also be useful at parties when trying to determine whether Pat is a man or a woman!

# Pubis

## Symphysis Pubic

The symphysis pubis (pubic symphysis) is an important fibrocartilaginous joint, as it allows for slight motion during walking, thereby reducing the risk of damage to the pelvis. In females, it may become even more flexible during the birth of their children, which facilitates the delivery of the baby. This change comes about from hormonal influences on the joints.

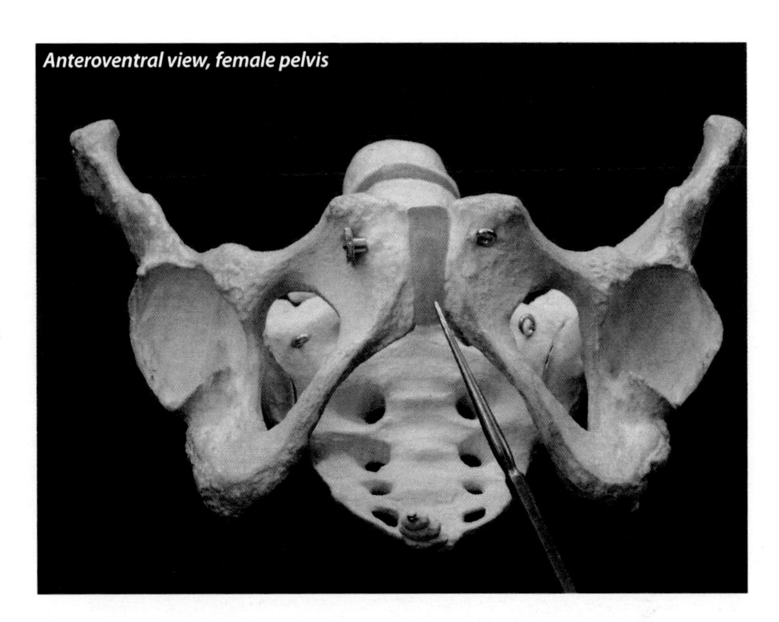


# Os Coxae

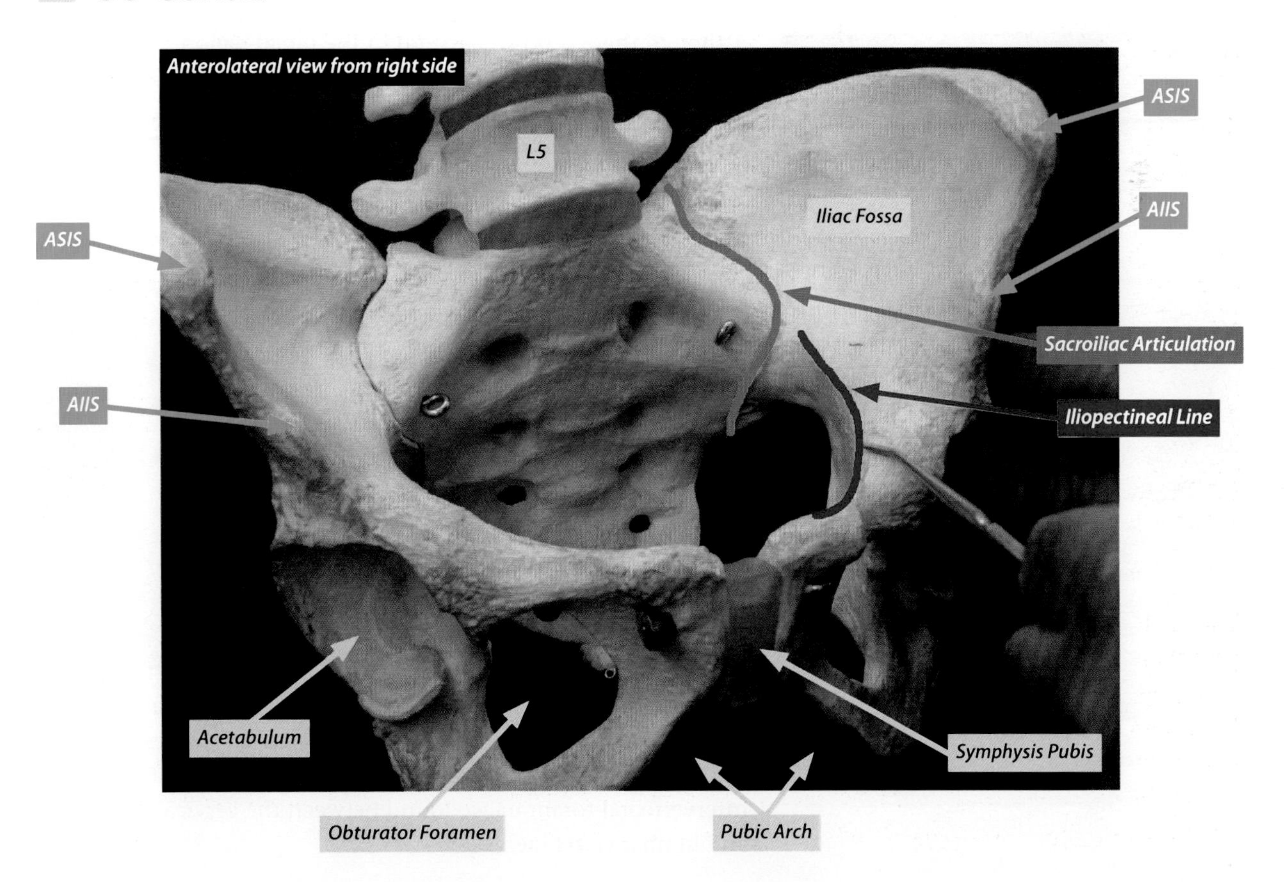

# Sacrum

## Sacral Promontory

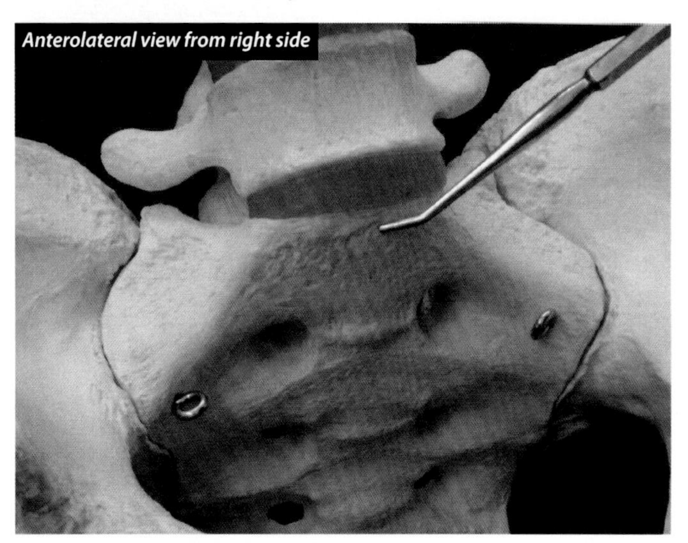
Anterolateral view from right side

Note that the sacral promontory is the anterior, superior edge of the body of the first sacral vertebra. It projects into the pelvic inlet differently in adult males and females. It forms the posterior portion of the pelvic inlet. The center of gravity for a human is approximately one centimeter posterior to this landmark.

## Sacral Canal (Sacrum)

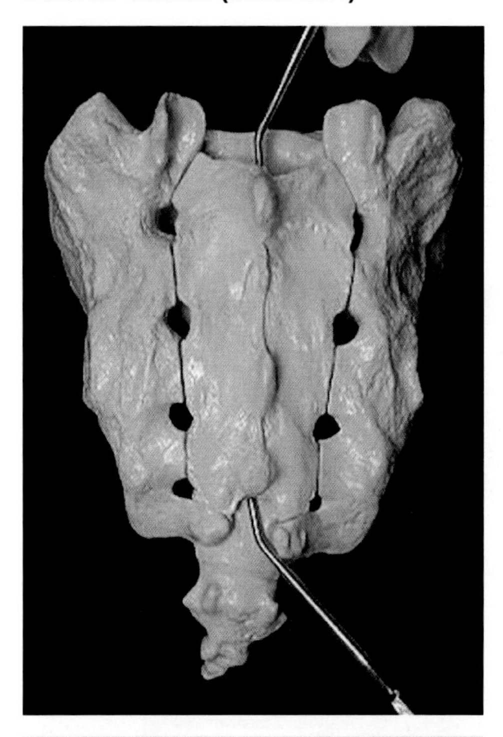
*Posterior view. Note that the probe in this photograph is where the spinal nerves would be found.*

Although there is no spinal cord in the sacral region, there are spinal nerves passing through the sacral canal. Some exit by way of the anterior and posterior sacral foramina, and some continue on to the coccygeal region.

### *Anterior Sacral Foramen*

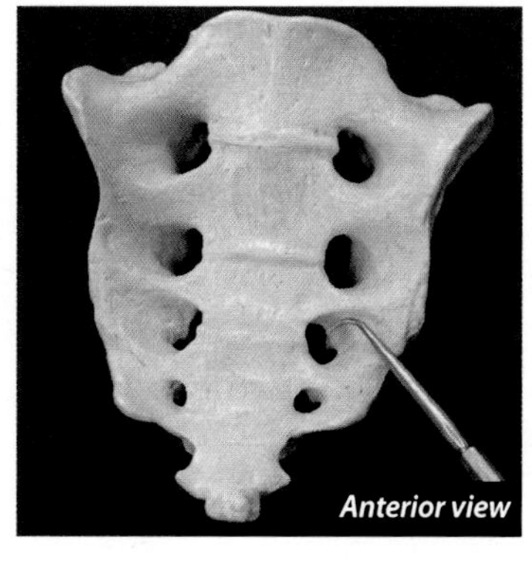
Anterior view

### *Posterior Sacral Foramen*

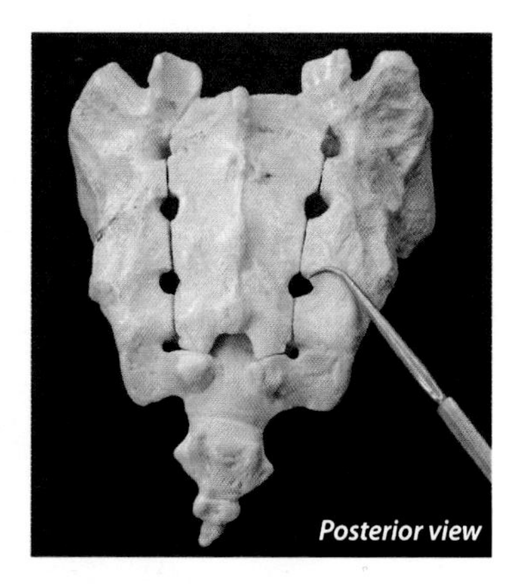
Posterior view

In the sacrum, we find both anterior and posterior foramina where the spinal nerves (anterior and posterior rami respectively) leave the sacral canal. These foramina are comparable to the intervertebral foramina we found between the cervical, thoracic, and lumbar vertebrae.

# MUSCLES

## Levator Ani (Pelvic Diaphragm)

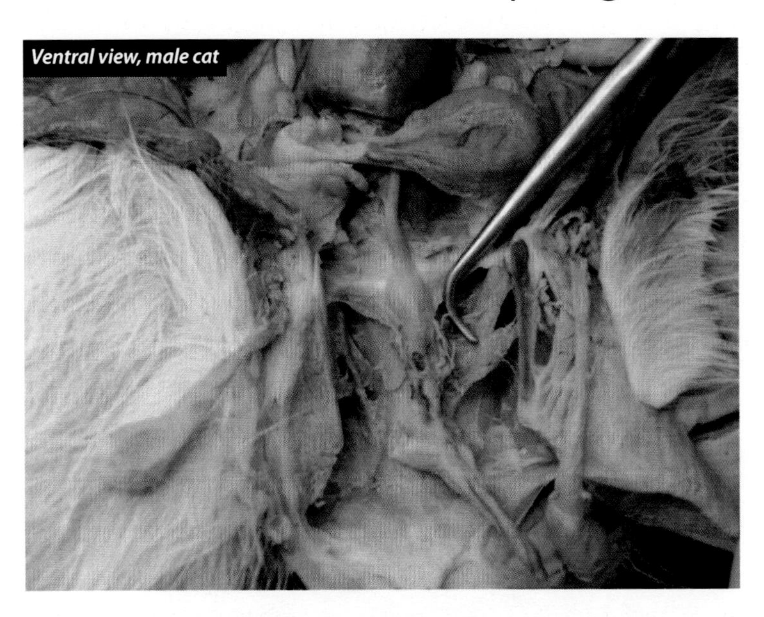
*Ventral view, male cat*

The levator ani is primarily skeletal muscle and is, therefore, under conscious control. It forms the anterior portion of the pelvic diaphragm and makes up the majority of it. You should remember that the coccygeus muscle is also a small part of the pelvic diaphragm and this muscle is posterior to the levator ani. This diaphragm extends across the diamond shape of the pelvic outlet. Some anatomists consider the external sphincter ani to be part of the levator ani, while others consider it a separate structure. The levator ani provides minimal support to the urogenital organs. It also supports pelvic viscera, lifts the anal canal, and forms a sphincter for the vagina. We will consider it one muscle. However, technically it is made up of three muscles, the iliococcygeus, the pubococcygeus, and the puborectalis. You will not be responsible for these names (whew!). This is one of the muscles that is active during Kegel exercises.

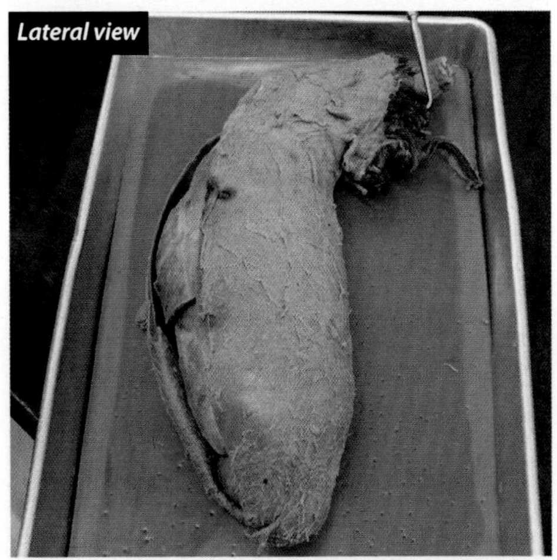
*Lateral view*

## Cremaster Muscle

The cremaster muscle is functionally important because it helps in the thermoregulation of the testicle. Ideally, testicular temperature should be 3° to 4° Celsius cooler than body temperature. If the testicle gets too cold, the cremaster muscle contracts and pulls it closer to the body to help warm it. Conversely, if the testicle gets too warm, the cremaster muscle relaxes, allowing the testicle to move away from the body where it can cool off. Remember, the cremaster muscle was formerly part of the internal abdominal oblique; therefore, it is skeletal muscle. Also, note how suspiciously the cremaster looks like corned beef.

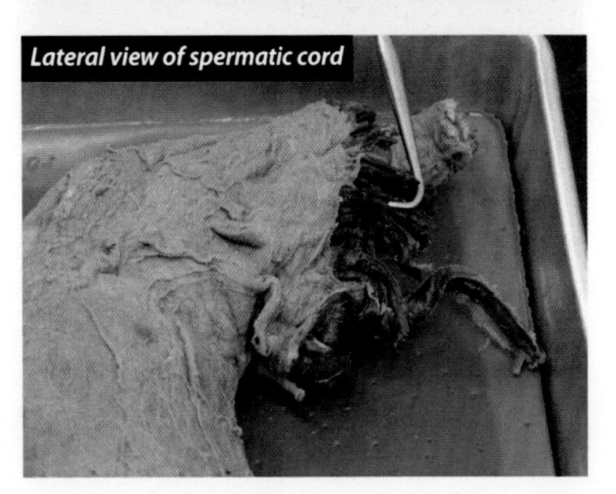
*Lateral view of spermatic cord*

# REPRODUCTIVE ORGANS

## Female

### Ovaries

The ovaries are the primary sex organs of the female. They are homologous to the testicles. In humans, they are found lateral to the uterus. They lie against the lateral wall of the pelvis and are held in place by the mesovarium of the broad ligament (peritoneum). There is a suspensory ligament of the ovary that is part of the broad ligament, as well as by the round ligament of the ovary. The round ligament of the ovary attaches it to the wall of the uterus; this is a remnant of the gubernaculum. The ovary releases ova after the onset of puberty and until menopause. In the cat, the ovary is wrapped within the infundibulum.

#### *Tunica Albuginea* (no female picture—see entry under male)

This is the covering over the testicles and ovaries. It is made of white collagenous fibers. The same name is used elsewhere in the body—for example, for the sclerotic coat of the eyeball and the covering of the spleen.

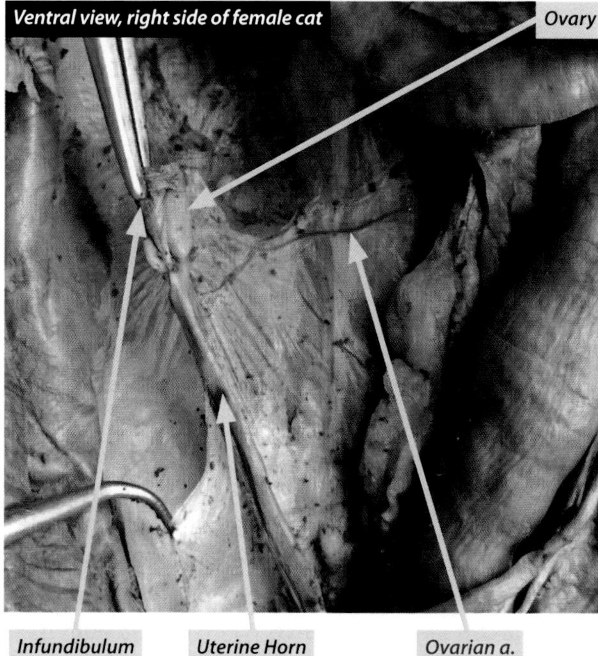


### Infundibulum

The infundibulum is an accessory sex organ. It has embryological origins from the Mullerian duct. It is part of the uterine tube (the fallopian tube or oviduct) and is shaped somewhat like a funnel. It is lined with cilia that produce a current that usually sweeps the ovum into the uterine tube. Dr. J particularly likes the following analogy: The infundibulum is similar to the hood of a raincoat, covering the ovary just as the hood covers one's head. Another analogy he uses is of a catcher's mitt and a baseball. The two objects are not sealed together. We find the uterine tube on the deep side of the infundibulum.

# REPRODUCTIVE ORGANS

## Female

### Uterine Tube (Fallopian Tube, Oviduct)

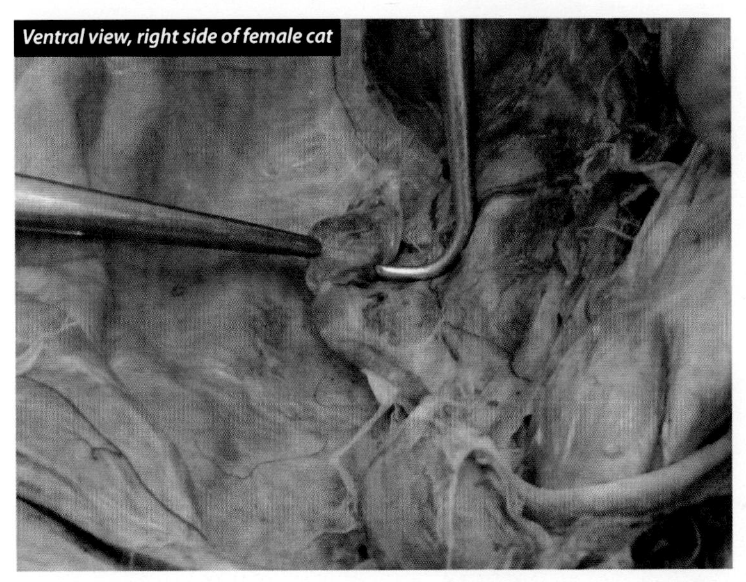


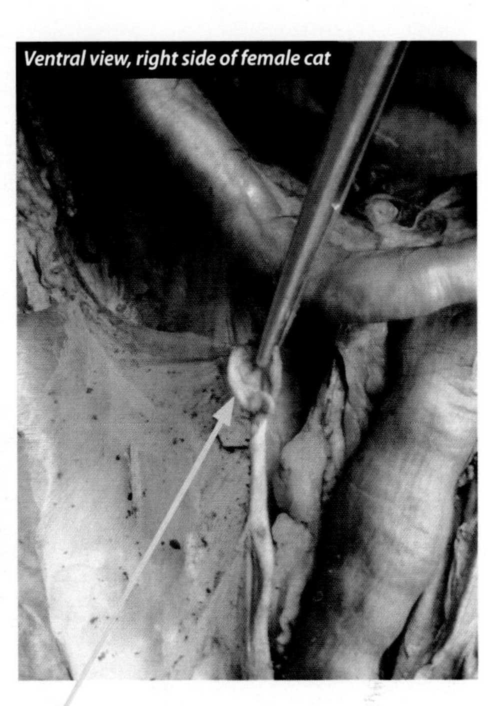


The uterine tube may also be referred to as the fallopian tube or oviduct. The uterine tube is an accessory sex organ. It has embryological origins from the Mullerian duct. In the human, it is about 4 inches long (10 cm) and is found in the superior border of the broad ligament of the uterus. Functionally, it is important because it receives the ovum from the ovary, and fertilization normally occurs in the uterine tube's distal third portion, which is called the ampulla. At its lateral extremity, we find the infundibulum. At its medial border, the uterine tube connects to the uterus in humans, while in cats it connects to the horn. The uterine tube is lined with a ciliated epithelium that propels the ovum or zygote toward the uterus. In addition to the cilia, peristaltic waves of smooth muscle contractions in the wall of the uterine tube assist transport. The uterine tube is supported by the mesosalpinx, which is part of the broad ligament of the uterus (peritoneum).

# REPRODUCTIVE ORGANS

## Female

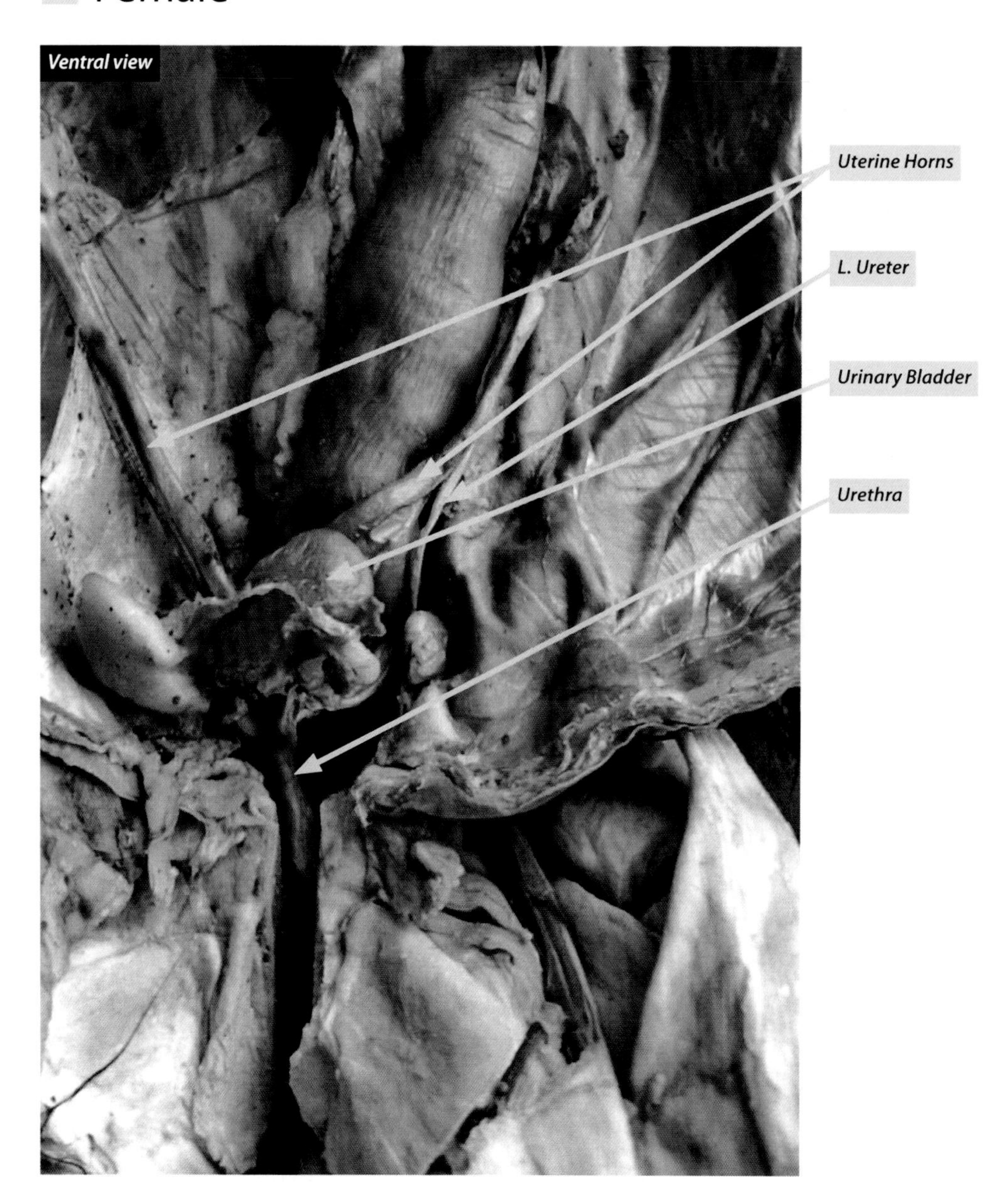


### Urethra (Female Cat)

In humans, the urethra and vagina exit at separate points. In the cat, the vagina and urethra join to form the vestibule, a common urogenital sinus.

# REPRODUCTIVE ORGANS

## Female

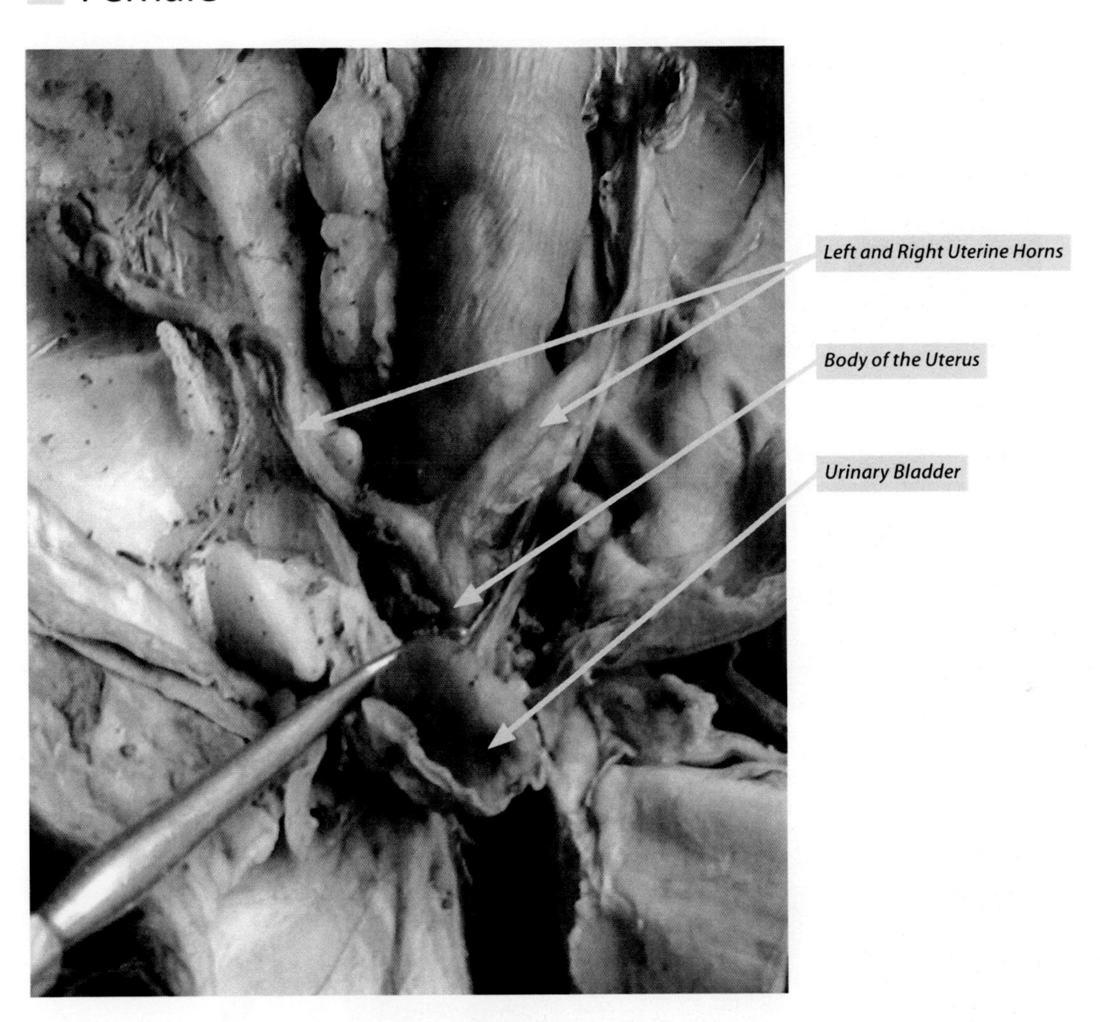


### Uterine Horn (Horn of the Uterus)

The paired horns (cornua) of the uterus are structures found in species that have litters rather than single births. They extend laterally and cranially from the body of the uterus to the infundibulum. The uterine tubes connect to the lumen of the uterine horns on each side. The lumen of the horn provides a space for the developing fetuses, much like peas in a pod.

### Body of the Uterus

The body of the uterus is caudal and medial to the two uterine horns, and cranial to the cervix in the cat. Humans do not have uterine horns. The body makes up the major portion of the uterus. It is found inferior to the entrance of the uterine tubes (fallopian tubes, oviducts). Further inferiorly it narrows to be continuous with the cervix.

# REPRODUCTIVE ORGANS

## Female

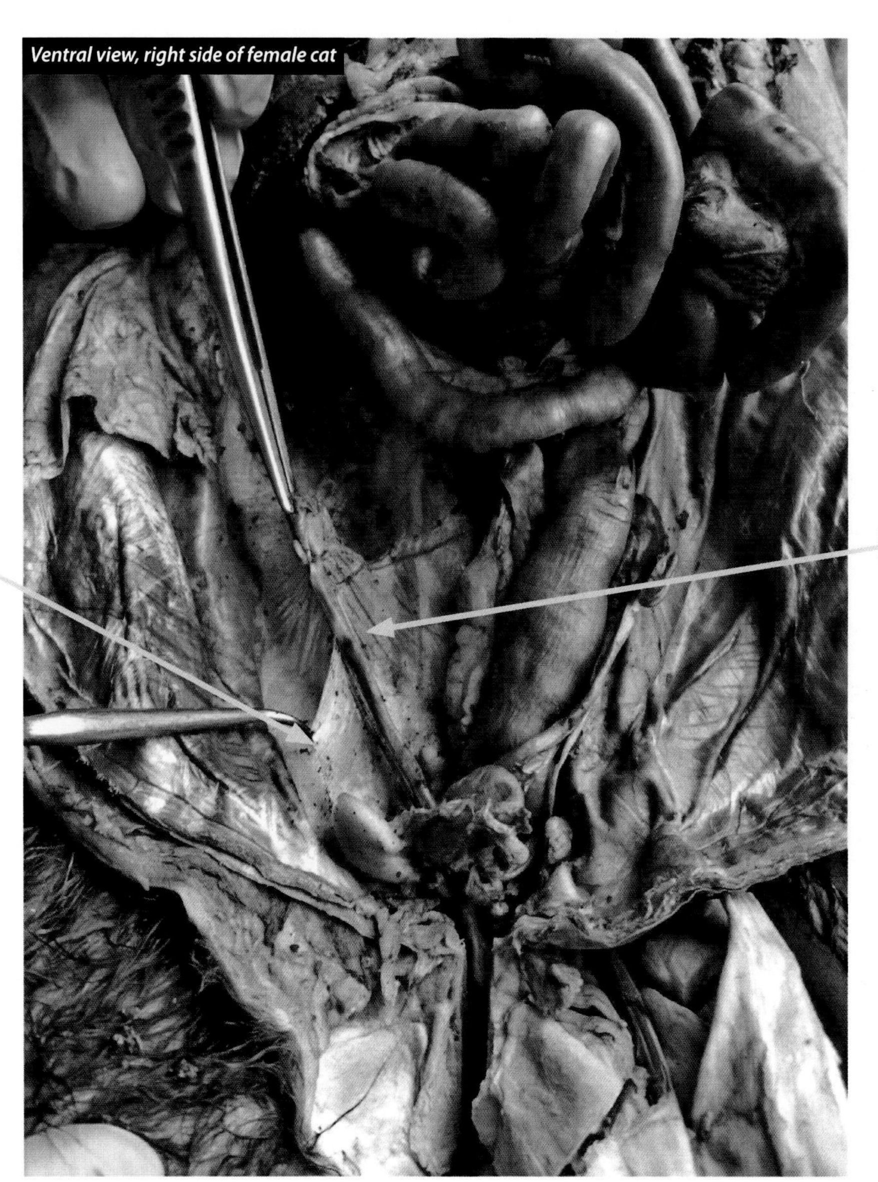


### Broad Ligament of the Uterus

The broad ligament is peritoneum. It provides some support for the uterus, the uterine tubes, and the ovaries. It has three sections: the mesometrium between the uterus and the dorsal wall of the body, the mesoalpinx between the uterine tube and the dorsal body wall, and the mesovarium between the ovary and the dorsal body wall.

# REPRODUCTIVE ORGANS

## Female

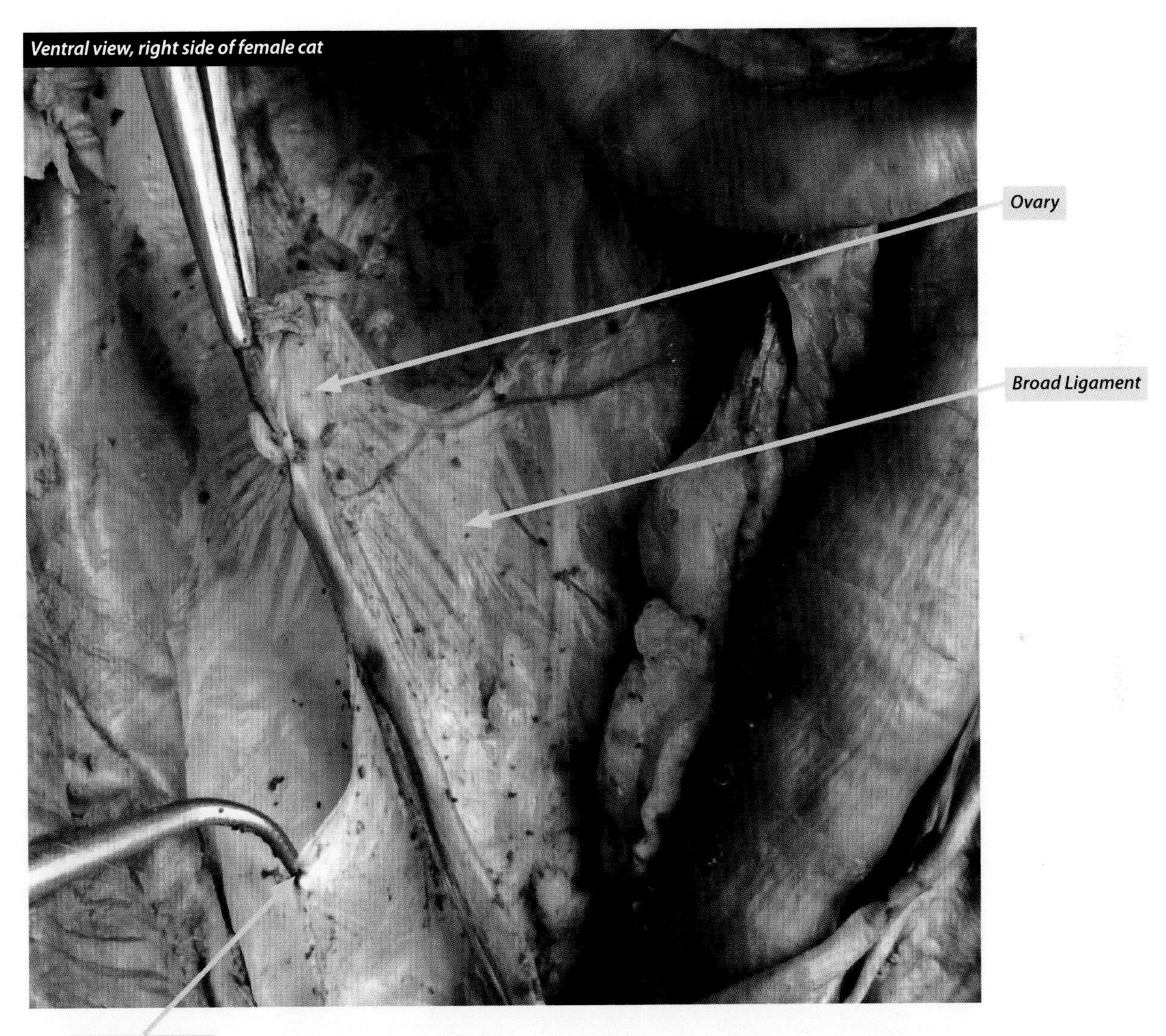


### Round Ligament of the Uterus

In the female, the remnants of the gubernaculum make up the round ligaments of the ovary and the uterus. The round ligament of the uterus passes through the mesometrium and the inguinal canal, and eventually it anchors the uterus to the labium majus. This structure is **NOT** part of the broad ligament of the uterus. The broad ligament is peritoneum, and it is medial to the horn of the uterus. Remember that in the male the remnant of the gubernaculum connects the testis to the fascial sac.

# REPRODUCTIVE ORGANS

## Female

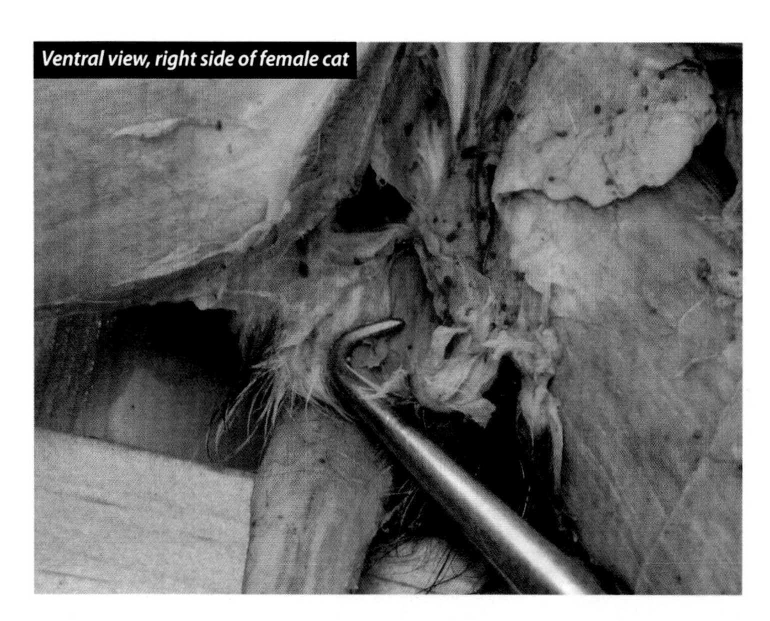
Ventral view, right side of female cat

### Vestibule

The vestibule extends from the union of the urethra and vagina to the outside. It is considered a urogenital sinus. In the human, it is external, surrounding the vagina and the urethral opening.

***Rugae (Vestibule)***

These are longitudinal folds in the walls of a tubular organ. For example, we find rugae in the stomach, the urinary bladder, the uterus, the vagina, and the vestibule, as well as in other places. Functionally, they are important because they allow for stretch without significantly increasing pressure on their contents. The above pictures are of the vestibule. They are important here so that babies aren't born looking like Gumby! They were named for Mr. October, Rugae Jackson, who was an anatomist before he played baseball.

Ventral view, right side of female cat

# REPRODUCTIVE ORGANS

## Female

### Vagina

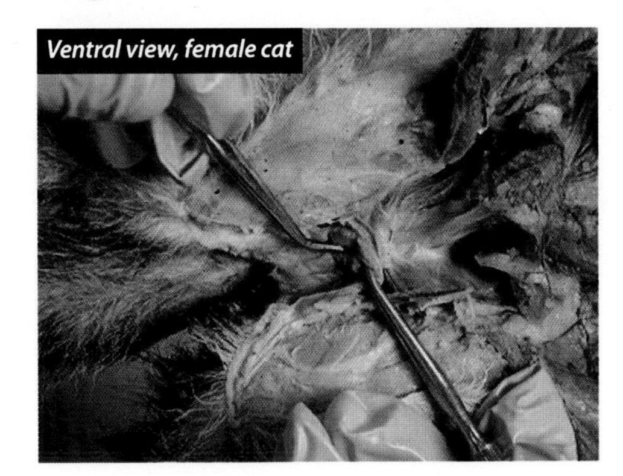
*Ventral view, female cat*

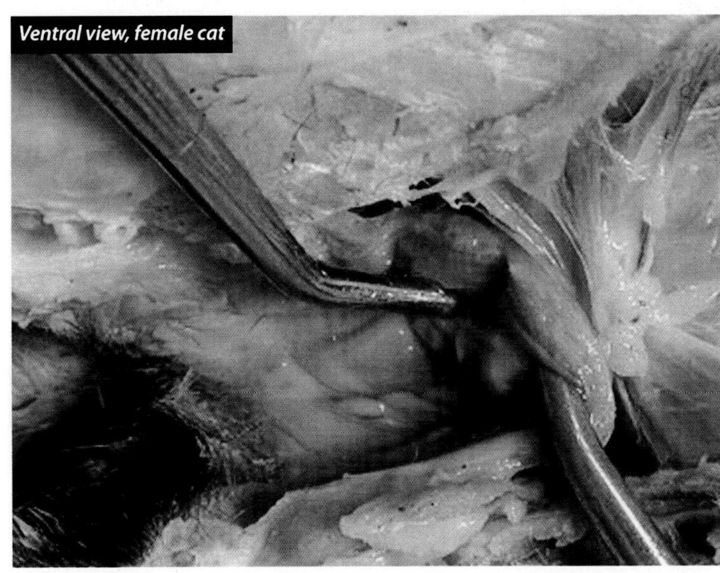
*Ventral view, female cat*

The vagina is a structure unique to females. It is an accessory sex organ in that it receives the penis during intercourse and is where semen is normally deposited. Part of it is derived from the urogenital sinus of the embryo. It is described by many authors as a potential space and is approximately three inches in length. Its walls have rugae that help in expansion without a huge increase in pressure during childbirth. It is an excretory duct for the uterus as well as being part of the birth canal. The inferior half is in the perineum, while the superior half is in the pelvis. The epithelial cells lining the vagina release glycogen into the lumen. This carbohydrate undergoes bacterial fermentation, which produces lactic acid. The lactic acid discourages harmful bacteria, but it also is harmful to spermatozoa. Note that the probe on your right is inserted into the urethra. There is a major difference between the vagina of a human and the vagina of a cat. In the human, the vagina exits separately to the outside from the urethra. In a cat, the vagina and the urethra meet to form an internal vestibule, which is also called a urogenital sinus. The vestibule of an adult human is external. The probe on your left is pointing at the caudal end of the vagina. The vagina extends from that point cranially to the cervix.

# REPRODUCTIVE ORGANS

## Male

### Urethra (Male Cat)

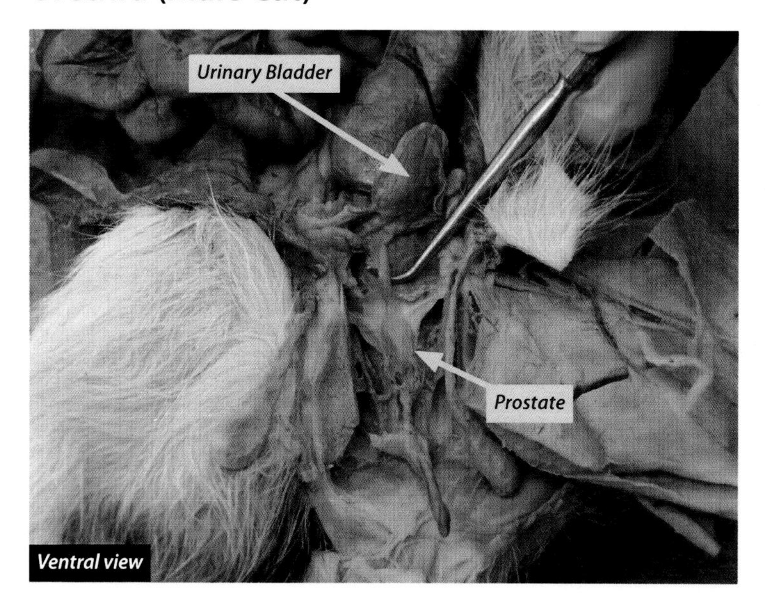


This is the duct (quack!) that transports urine from the urinary bladder to the outside. In males, it is a common urogenital duct, while in females it simply transports urine. It was named for Urethra Franklin, otherwise known as Lady Soul. Remember there is only one Urethra Franklin and there is only one urethra. Spelling is important for this term, as it is close to the spelling of ureter, which is another structure.

### Prostate

The prostate gland is an accessory sex gland, which surrounds the urethra and releases its secretions into the urethra during ejaculation, thereby contributing to the makeup of the semen. The prostate's secretion makes up about one-third of the volume of the semen and includes materials that aid in the motility of the spermatozoa. The prostate forms from the wall of the urethra when hormones act on that wall during development.

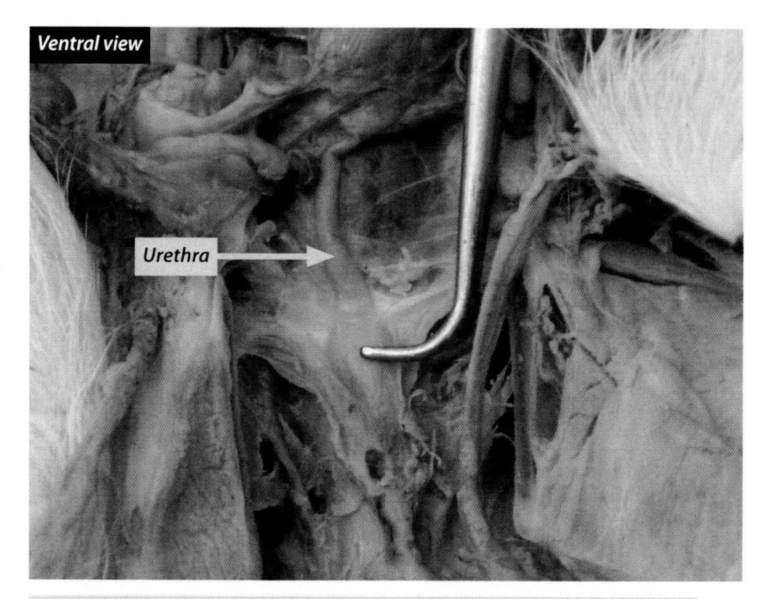


*The prostate is the enlarged area at the tip of the probe. It surrounds the urethra.*

# REPRODUCTIVE ORGANS

## Male

### Vas Deferens

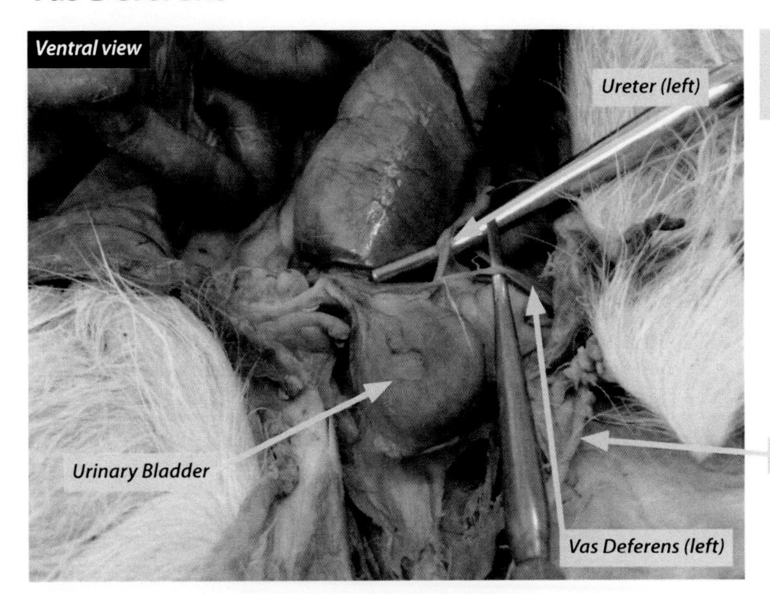


*One probe is lifting a ureter while the other is lifting the vas deferens as it emerges from the spermatic cord and courses behind the bladder.*

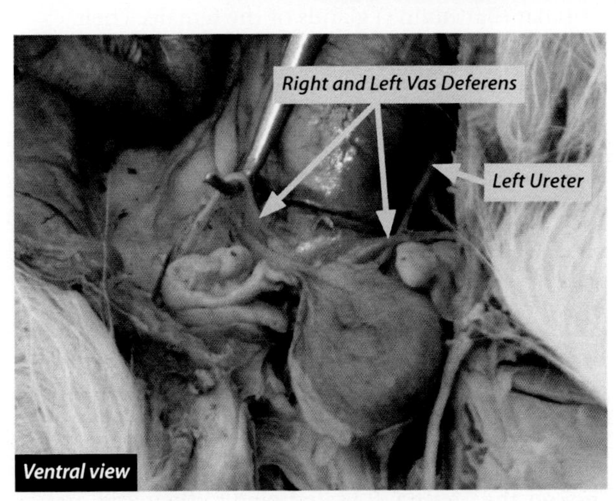


The vas deferens is one of the male accessory sex organs. It conducts spermatozoa from the epididymis to the ejaculatory duct. You can see it as it passes within the cat's spermatic cord—it zigzags back and forth, much like ribbon candy. In fact, this may be where ribbon candy comes from! It is about eighteen inches long in humans and includes smooth muscle in its walls. The vas deferens passes from the testicle through the spermatic cord, through the inguinal canal, over the ureter, and then down the posterior side of the urinary bladder where it meets the seminal duct. It is usually the structure that is severed during male sterilization procedures.

# REPRODUCTIVE ORGANS

## Male

### Bulbourethral Glands

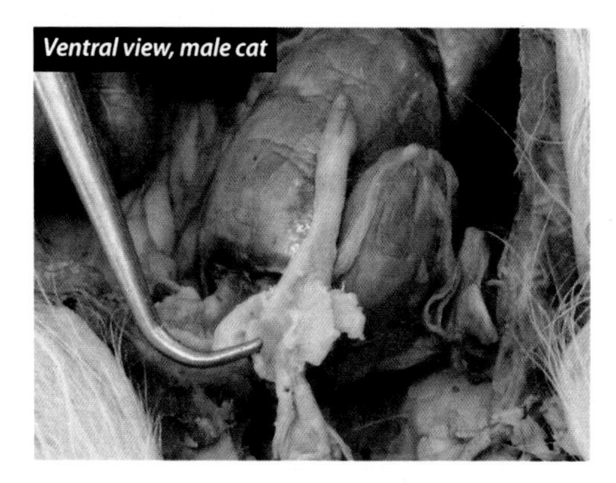


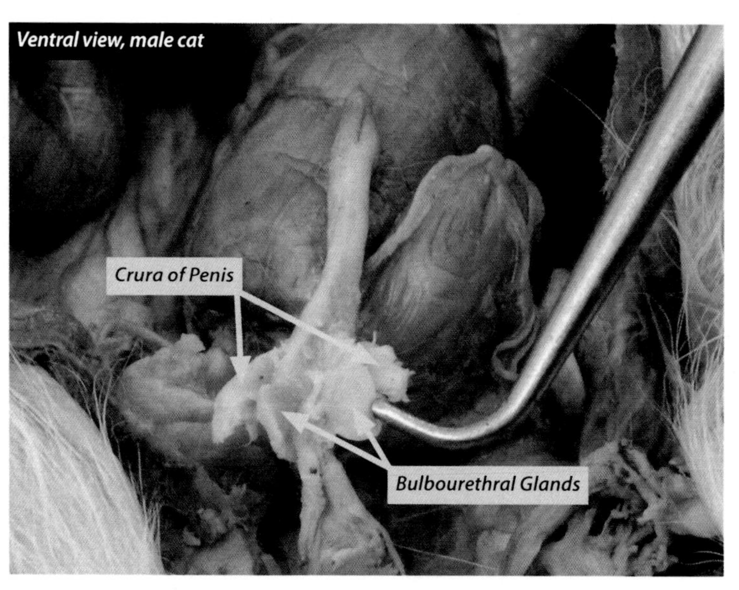


The bulbourethral gland (Cowper's gland) is an accessory sex gland, which releases its secretions into the urethra during ejaculation, thereby contributing to the makeup of the semen. The glands are about the size of a peanut and are homologous to the greater vestibular (Bartholin's) glands of the female. Their secretions are primarily mucus and are released as a result of erotic stimulation. The mucus is important as it lubricates the urethra and thereby enhances ejaculation, and it also neutralizes any traces of urine in the urethra. The urine is usually acidic and could potentially be harmful to the spermatozoa. Note the crura to your left and your right of the bulbourethral glands. As a literary note, the bulbourethral glands were named for the famous Hobbit, Bulbo Baggins.

### Penis

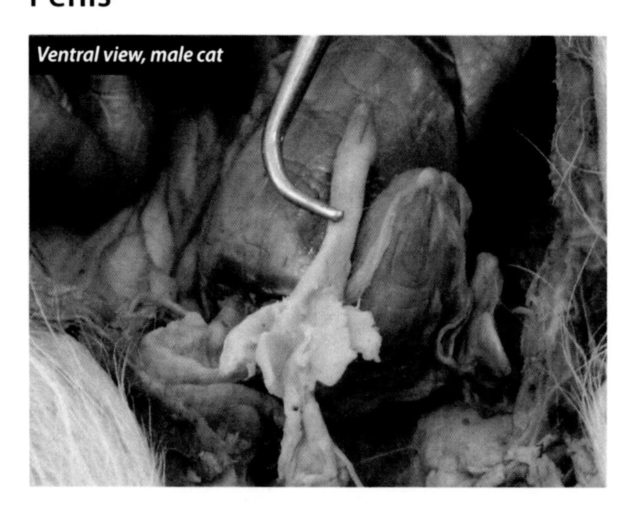


The penis is an accessory sex organ. Proximally, it begins with the root of the penis, then the shaft (or body), and finally the glans penis. It is made up of three erectile cylinders, two corpora cavernosa, and the corpus spongiosum. The corpus spongiosum houses the urethra. It is anchored to the pubic arch by the ischiocavernosus muscles that also attach to the crura. Functionally, it aids in the transfer of semen from the male to the vagina of the female during heterosexual intercourse. The anatomical position of the penis is in the erect state. Dr. J is not sure who decided that, probably Plato.

# REPRODUCTIVE ORGANS

## Male

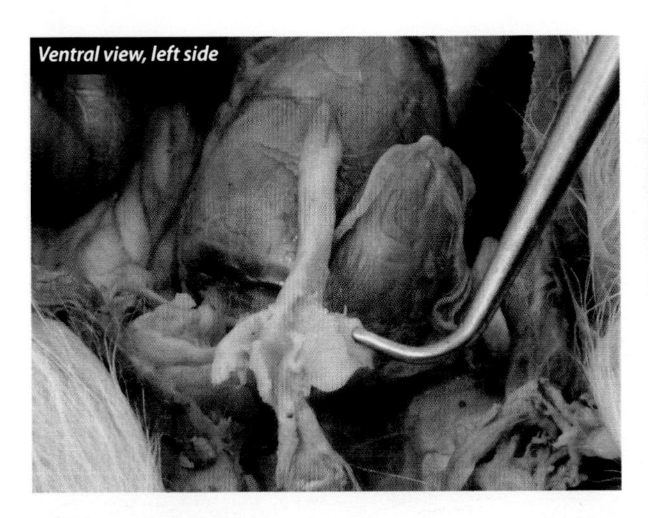


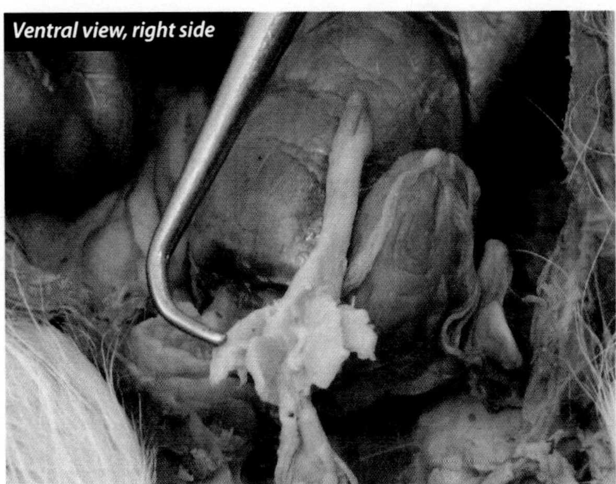


### *Crus of the Penis*

The crura of the penis are part of the root and the proximal ends of the corpus cavernosa. Each crus is covered on its outer surface by the ischiocavernosus muscle, and they serve to anchor the penis to the pubic arch. The right crus was named for Tom and the left was named for Penelope.

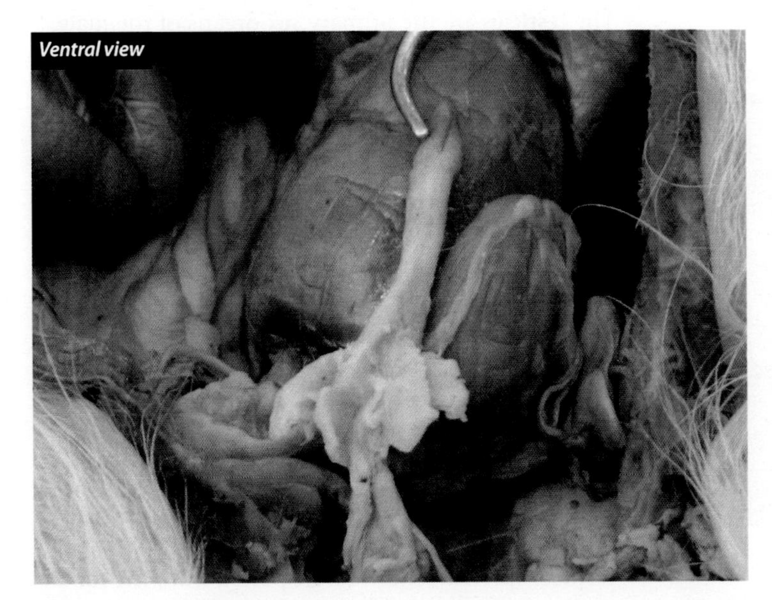


### *Glans Penis*

Check out the description of the penis on page 360. The glans penis is at the distal end of the penis. It is a continuation of the corpus spongiosum and is covered by the prepuce (foreskin) unless the individual has had a circumcision. The opening is the terminal end of the urethra, which serves as a common urogenital opening in the male.

# REPRODUCTIVE ORGANS

## Male

### Spermatic Cord

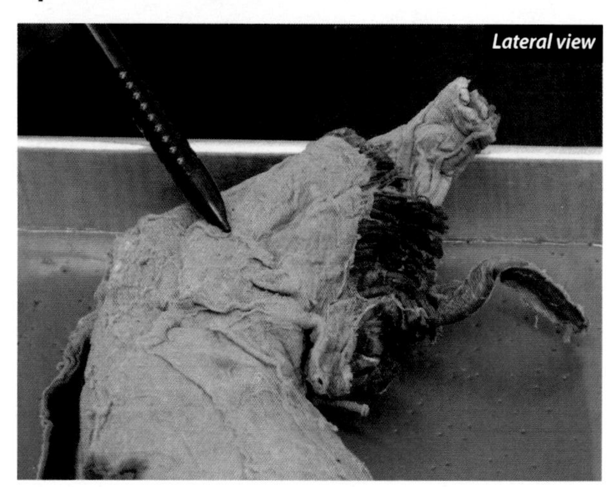


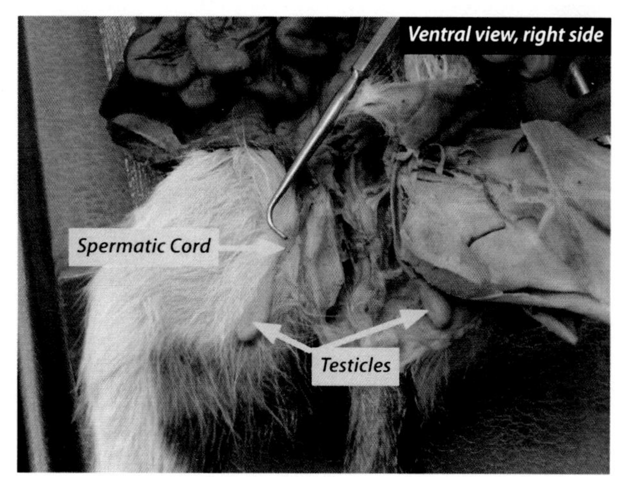


The spermatic cord is the structure extending from the fascial sac surrounding the testicle to the external inguinal ring. It is covered with spermatic fascia and includes the vas deferens, the internal spermatic **artery** and **vein**, the lymphatics, and a branch of the **genitofemoral nerve**. Incorporated in the spermatic fascia is the cremaster muscle, which helps in thermoregulation for the testicle. Originally, the cremaster muscle was part of the internal abdominal oblique muscle.

### Testes

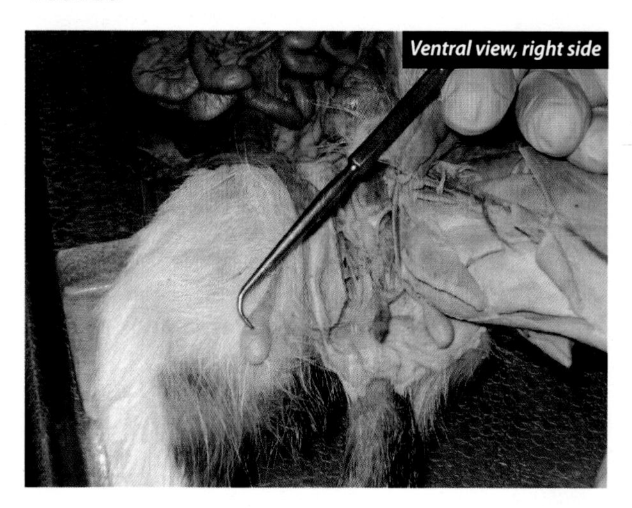


The testicles are the primary sex organs of the male. They are homologous to the ovaries. They produce gametes as well as hormones. Thus, they are mixed glands. The production of the gametes is the result of meiosis. In the male, meiosis begins prior to birth but then stops before it is complete. The completion of the meiosis occurs after the onset of puberty. The testicles normally descend from the body cavity into the scrotum, where they are 3° to 4° Celsius cooler than body temperature. This is essential in order for them to produce viable spermatozoa. The testicle is covered by the tunica albuginea. It is contained within the fascial sac (or tunica vaginalis) and is connected to the sac by what remains of the gubernaculum. The seminiferous tubules, the rete testes, and the vas efferens are contained within the testicle. Spermatozoa are transported from the tubules of the testicle to the epididymis. The epididymis is located on the posterior surface of the testicle.

# REPRODUCTIVE ORGANS

## Male

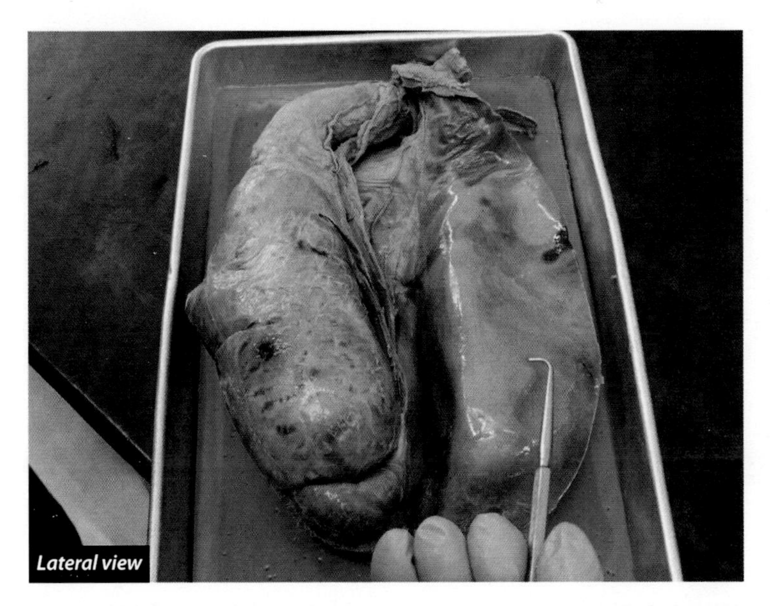
Lateral view

### *Fascial Sac (Tunica Vaginalis)*

The fascial sac or tunica vaginalis (translated this means "ensheathing coat") is formed from the aponeuroses of the abdominal muscles when the testis descends through the inguinal canal on its way to the scrotum. In addition to the aponeuroses, peritoneum is included in the fascial sac. In fact, technically, the tunica vaginalis is composed of peritoneum, even though these terms are used interchangeably. Within the fascial sac is a cavity containing fluid. Occasionally, when the testis becomes inflamed, fluid builds up in this cavity also. A piece of material connects the testis to the fascial sac—this is the remnant of the gubernaculum (scrotal ligament). Remember that the remnants of the gubernaculum in the female make up the round ligaments of the ovary and uterus.

### *Tunica Albuginea*

The tunica albuginea is the covering of the testicle and ovary. It is made of white collagenous fibers. The same name is used elsewhere in the body—for example, the sclerotic coat of the eyeball and the covering of the spleen.

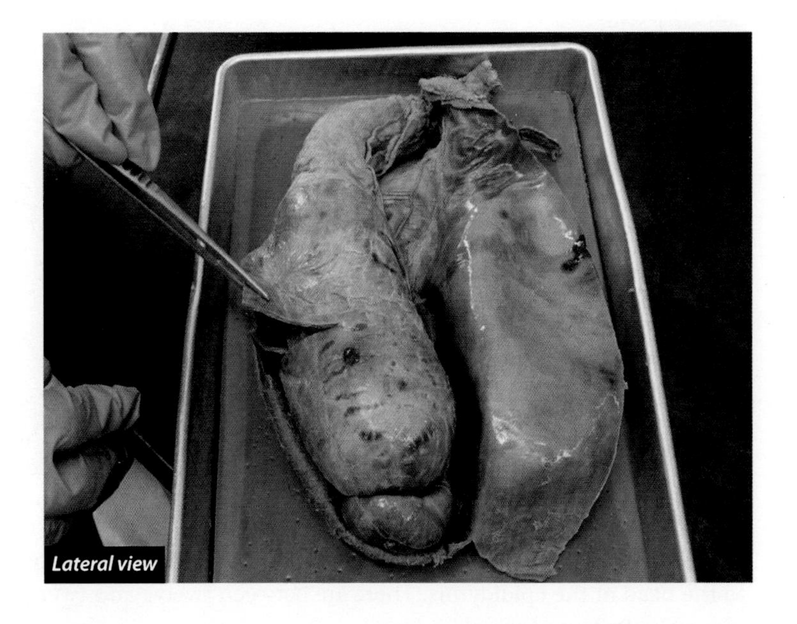
Lateral view

# REPRODUCTIVE ORGANS

## Male

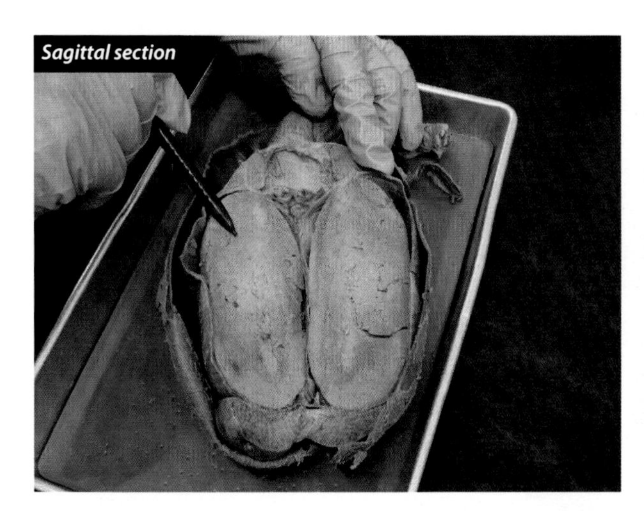


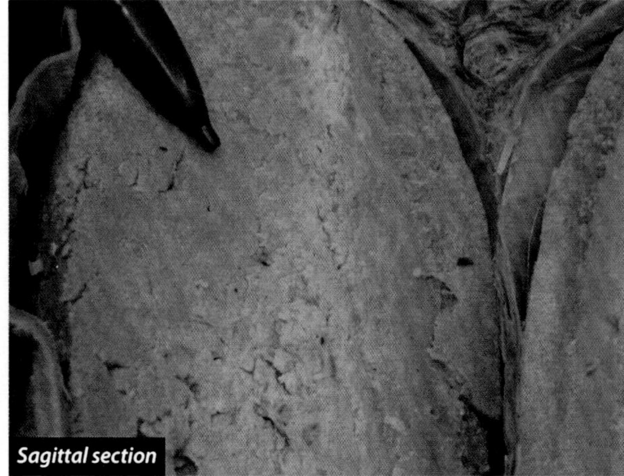


### *Seminiferous Tubules*

The seminiferous tubules are coiled microscopic tubes that produce spermatozoa. Although the initial phases of spermatogenesis begin during embryonic development, the actual production of spermatozoa is delayed until puberty. This production is controlled by two hormones, a follicle-stimulating hormone (FSH) and testosterone. Once formed, the spermatozoa move into the rete testes.

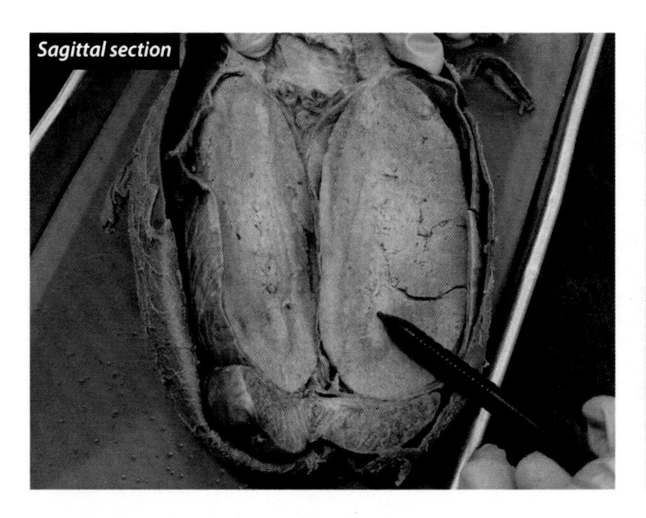


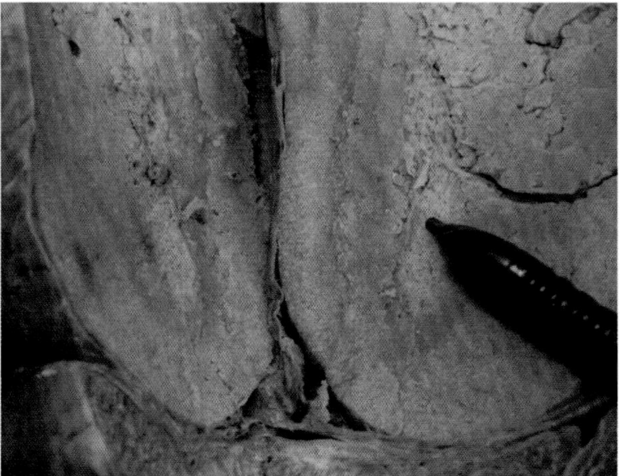

### *Vas Efferens*

The vas efferens (ductus efferens) receives spermatozoa from the rete testes of the testicle and transports them to the head of the epididymis. There are between twelve and twenty in each testicle, and they, in turn, drain into a single duct of the epididymis.

# REPRODUCTIVE ORGANS

## Male

### Epididymis (Head, Body, and Tail)

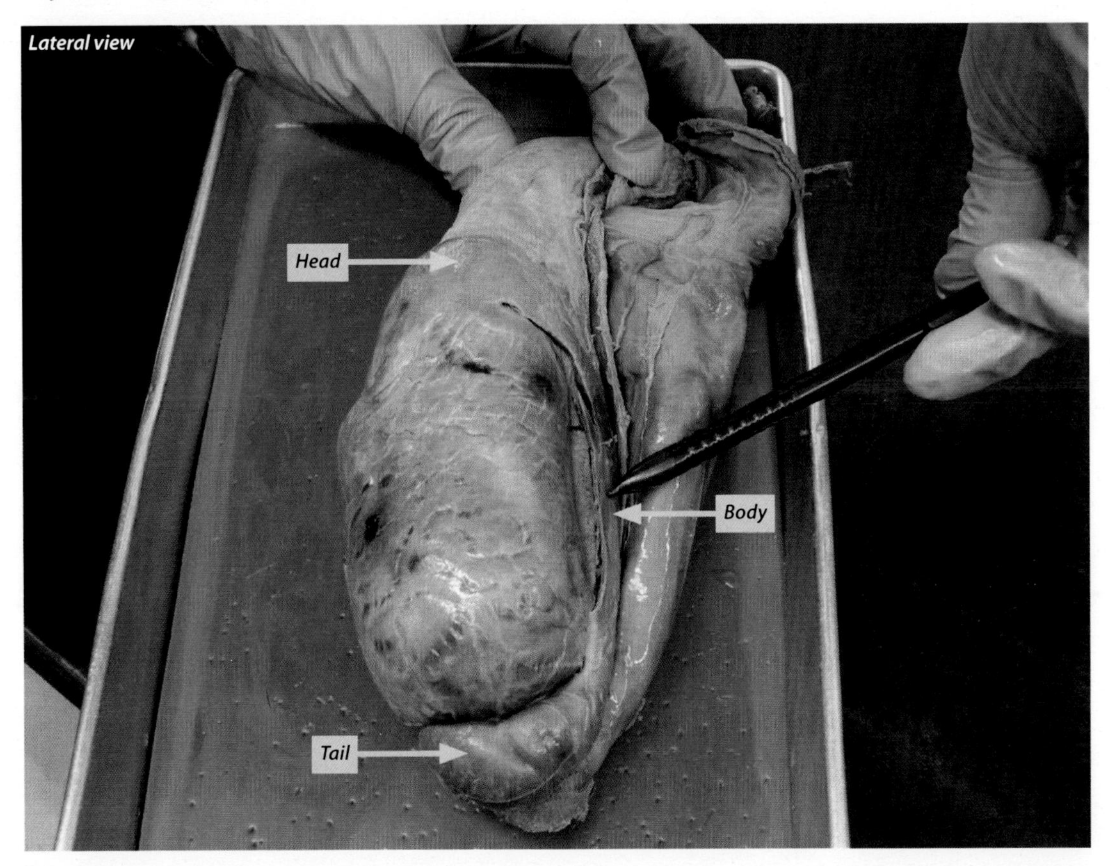


The epididymis is an accessory sex organ. It has embryological origins from the Wolffian duct. It is found posterior to the testis. At the superior end is the broad end called the head of the epididymis, followed inferiorly by the body (the central portion of the epididymis), and then the tail of the epididymis. The vas efferens enters the head of the epididymis and empties into the duct of the epididymis.

The coiled tubes that make up the epididymis are about 20 feet long! Surrounding the tube is connective tissue. The tube emerges as the vas deferens from the tail of the epididymis and begins its course toward the spermatic cord and the external inguinal ring.

The length of the tube is important as a storage place for spermatozoa, and it is in this tube that they mature (which includes developing a flagellum and changing their shape). The spermatozoa spend about 20 days in there. The duct of the epididymis is also important because it absorbs testicular fluids and may add substances to nourish the spermatozoa. Smooth muscle in the walls of the epididymis contracts during ejaculation, causing the spermatozoa to move into the vas deferens. If there is no ejaculation, the spermatozoa stay there for up to several months, after which they are phagocytized by epithelial cells in the duct.

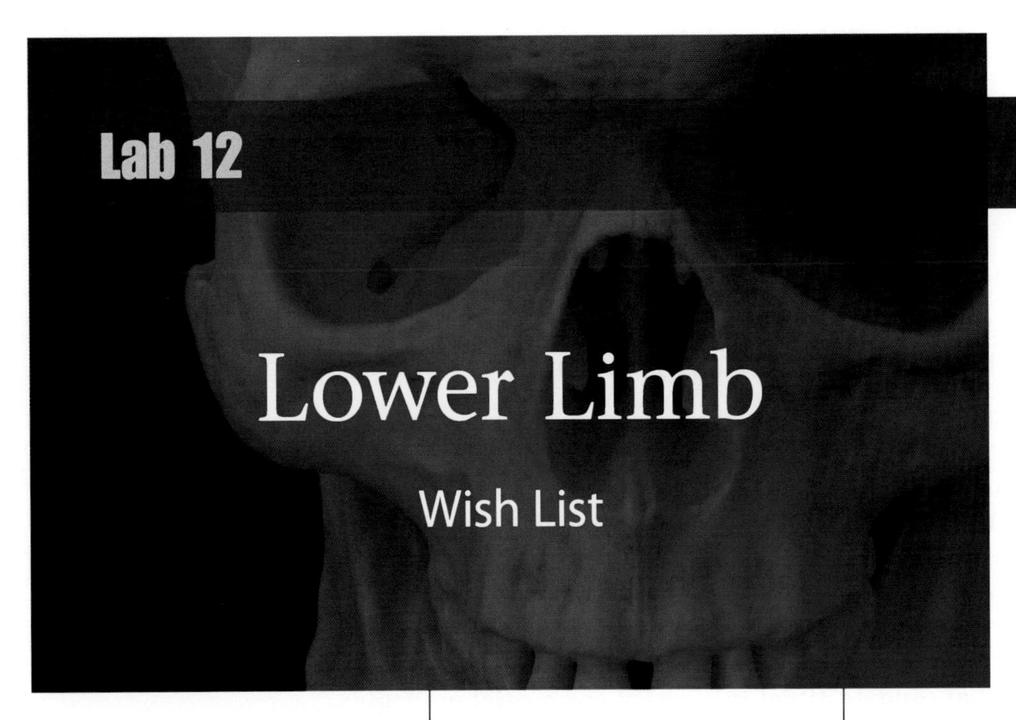


# Lower Limb

## Wish List


**LAB 12 OVERVIEW, pp. 368–369**

**BONES, pp. 370–381**

- **Anterior Femur (Overview), p. 370**
- **Posterior Femur (Overview), p. 371**
- **Femur, pp. 370–374**
  - Head, p. 372
  - Neck, p. 372
  - Greater Trochanter, p. 372
  - Lesser Trochanter, p. 372
  - Trochanteric Fossa, p. 372
  - Linea Aspera, p. 373
  - Lateral Condyle, p. 373
  - Intercondylar Notch, p. 373
  - Medial Condyle, p. 373
  - Lateral Epicondyle, p. 374
  - Patellar Surface, p. 374
  - Medial Epicondyle, p. 374
- **Leg and Ankle (Overview), p. 375**
- **Patella, p. 376**
- **Fibula, pp. 376–377**
  - Head, p. 377
  - Lateral Malleolus, p. 377
- **Tibia, pp. 378–379**
  - Medial Malleolus, p. 378
  - Lateral Condyle, p. 379
  - Medial Condyle, p. 379
  - Tibial Tuberosity, p. 379
- **Tarsals (Specific Bones), p. 380**
  - Calcaneus, p. 380
  - Talus, p. 380
- **Metatarsals, p. 381**
- **Phalanges, p. 381**

**MUSCLES, pp. 382–413**

Muscles of the Hip and Thigh, pp. 382–411

- **Caudofemoralis, p. 382**
- **Tensor Fasciae Latae, pp. 382–383**
  - Fascia Lata, p. 382
- **Gluteals, pp. 384–385**
  - Gluteus Maximus, pp. 384–385
  - Gluteus Medius, pp. 384–385
- **Hamstrings, pp. 386–391**
  - Biceps Femoris, pp. 386–387
  - Semitendinosus, pp. 388–389
  - Semimembranosus, pp. 390–391
- **Adductors, pp. 392–399**
  - Gracilis, pp. 392–393
  - Adductor Femoris (Magnus et Brevis), pp. 394–395
  - Adductor Longus, pp. 396–397
  - Pectineus, pp. 398–399
- **Iliopsoas, pp. 400–401**
- **Sartorius, pp. 402–403**
- **Quadriceps, pp. 404–411**
  - Rectus Femoris, pp. 404–405
  - Vastus Lateralis, pp. 406–407
  - Vastus Medialis, pp. 408–409
  - Vastus Intermedius, pp. 410–411

Muscles of the Leg, pp. 412–413

- **Gastrocnemius, pp. 412–413**
- **Calcaneal (Achilles) Tendon, pp. 412–413**

**NERVES , pp. 414–415**

- Sciatic (Ischiadicus) Nerve, p. 414
- Common Peroneal (Fibular) Nerve, p. 414
- Tibial Nerve, p. 414
- Femoral Nerve, p. 415
- Saphenous Nerve, p. 415

**VESSELS, pp. 416–418**

**ARTERIES, pp. 416–417**

- **External Iliac Artery, p. 416**
- **Deep Femoral Artery, p. 416**
- **Femoral Artery, p. 416**
- **Lateral Femoral Circumflex Artery, p. 416**
- **Muscluar Artery, p. 416**
- **Saphenous Artery, 417**

**VEINS, pp. 416–417**

- Femoral Vein, p. 416
- Saphenous Vein, p. 417

**MUSCLE CHARTS, pp. 419–420**

# LAB 12 OVERVIEW

In Lab 12, we will study the lower limb. This will include information on **bones**, **muscles**, **nerves**, and **vessels**.

## 1. Bones of the Lower Limb (Pelvic Appendage)

We will start proximally and work toward the distal end of the appendage:

The **femur** is the strongest and usually the longest bone in the human body. You will identify the **head**, **neck**, **greater trochanter**, **trochanteric fossa**, **lesser trochanter**, **linea aspera**, **lateral condyle**, **intercondylar notch**, **medial condyle**, **lateral epicondyle**, **patella surface**, and **medial epicondyle**.

The **patella** should be relatively easy for you, as all you need to distinguish is the side the bone is from.

The **fibula** has an "l" in its name and you can remember it because it is the lateral bone and the word lateral begins with an "l." You should be able to identify the **head** and **lateral malleolus** of the fibula.

The **tibia** does **NOT** have an "l" in it. It is medial and medial does **NOT** begin with an "l." Please find the **lateral condyle**, **medial condyle**, **tibial tuberosity**, and **medial malleolus**.

Next you should examine the **tarsal** bones. Pay special attention to the **talus** and the **calcaneus (calcaneum)**, as these are the only tarsals whose names you will need to know. As with the patella, you will only be required to identify the side the bone is from. The other tarsals include the navicular, the medial cuneiform, the intermediate cuneiform the lateral cuneiform, and the cuboid.

The five **metatarsals** are distal (anterior) to the tarsals. You will only need to know these generically.

Note that the numbers and arrangement of the **phalanges** of the foot are similar to those of the phalanges of the hand. There are two phalanges for the big (great) toe and three for each of the other (lesser) toes.

## 2. Lateral Side of the Thigh

Place the cat on its ventral surface and remove the skin from the thigh and leg. Be careful not to cut the vessels that are on the deep side of the skin on the medial portion of the thigh. I suggest that you find the thigh that had dye injected into the vessels. There should be a cut in the skin there. Extend that cut distally to the ankle and then cut all the way around the ankle, again being careful not to cut the vessels. Also extend the cut to the pelvis on the medial side of the appendage. Then reflect the skin toward the pelvis until the entire leg and thigh are exposed. Next separate the **biceps femoris muscle** from the **caudofemoralis muscle**. They are closely associated. Be especially careful of the tendon from the **caudofemoralis**. The **caudofemoralis** is a cat-only muscle. Slip the probe along the deep side of the **biceps femoris** so that you can transect the muscle without transecting the **sciatic nerve**. Once transected, reflect the **biceps femoris** and observe the **sciatic (ischiadicus) nerve**. It bifurcates as it approaches the knee. The medial branch is the **tibial nerve** and it goes into the belly of the **gastrocnemius muscle**. The lateral branch is the **common peroneal nerve** and it runs out to the lateral side of the knee.

Cranial to **caudofemoralis** near its origin, you will find a thin muscle, the **gluteus maximus**. This muscle goes across the lateral side of the hip in the cat, while it goes across the posterior side of the hip in humans. Therefore, it is primarily an abductor of the thigh for the cat and primarily an extensor of the thigh for the human. Carefully cut along the cranial edge of the **gluteus maximus** and remove the fascia that is cranial to that margin. This will expose the **gluteus medius muscle**.

Observe the tensor fascia latae muscle. This relatively thin muscle is superficial to the lateral portion of the quadriceps muscle. It has an aponeuroses called **fascia latae** and it extends toward the knee. Transect this tendon, being careful not to transect the tendon of the **caudofemoralis**. Dr. J likes to draw attention to the fact that this muscle looks like an asymmetric Valentine heart with its apex being proximal and close to the hip joint.

On the lateral side of the thigh, you will find some adipose tissue on the deep side of the **biceps femoris** at the distal end of the muscle. Remove that to help expose the nerves and the muscles. You will see three muscles on the caudolateral side of the thigh. The most caudal is the **semitendinosus muscle**. Cranial to that is the **semimembranosus muscle**, and cranial to that is the **adductor femoris muscle**. You can use "Some Sailors Admire" for a mnemonic for those three muscles. We will observe these same three muscles in the same order on the medial side of the thigh. There are three muscles in the posterior compartment of the thigh, which collectively are called the hamstring muscles. They include the **biceps femoris**, the **semitendinosus**, and the **semimembranosus** (from lateral to medial).

## 3. Medial Side of the Thigh, Posterior Side of the Leg

We will transect the **gracilis**, which is one of the muscles of the medial compartment of the thigh. It is superficial to the other medial compartment muscles, as well as to two of the hamstring muscles. On the medial side, the most caudal muscle is the semitendinosus. Progressing cranially from the **semitendinosus** will be the **semimembranosus**, the **adducutor femoris (magnus et brevis)**, the **adductor longus**, and the **pectineus**. The mnemonic some students use for this group is **S**ome **S**ailors **A**dmire **A**ttractive **P**arrots. A note for caution! You will see that the adductor femoris has two heads (magnus and brevis). **PLEASE DO NOT SEPARATE THESE HEADS!** If you do, it will lead to confusion if your cat is used on the third practical exam.

We will transect the **sartorius** muscle. It is cranial to the gracilis muscle. This is a thin muscle and it covers the medial view of the quadriceps muscles. Reflect this muscle to expose the **iliopsoas** muscle. This muscle is tubular in shape and is nearly the diameter of a wooden pencil. It is formed by the union of the **iliacus muscle** with the **psoas major.** Notice that it is on the lateral side of the neurovascular bundle while the medial compartment muscles are on the medial side of that bundle. You should be able to observe the **femoral nerve** as it emerges from the **iliopsoas**. Close to this, the **saphenous nerve** branches from the **femoral nerve**, and it runs with the **femoral artery** and the **femoral vein** toward the knee. Trace the **femoral artery** proximally to where it begins. This will be where the **deep femoral artery** branches medially from the **external iliac artery** as it passes into the thigh from the abdomen. Trace the **femoral artery** distally from its origin and you will observe the **lateral femoral circumflex artery** as it branches laterally from the **femoral artery**. Continue distally to where the **muscular artery** will branch medially from the **femoral artery**. This will mark the beginning of the **saphenous artery**. The **saphenous vein** will carry blood proximally to the **femoral vein**. At the beginning of the **saphenous artery** we have a neurovascular bundle that is a Grant thing: **saphenous artery**, **saphenous vein**, **saphenous nerve**. There is no comparable artery in humans to the **saphenous artery** of cats.

We will now examine the quadriceps (four heads) muscles. Reflect the **sartorius** and the **tensor fascia latae**. Deep to these muscles you will find the quadriceps. Notice the uncanny resemblance to a miniature hot dog in a bun, even more so once the **rectus femoris muscle** has been transected. The lateral portion of the bun is the **vastus lateralis**. It is the largest of the quadriceps heads. On the medial side of the quadriceps, we find the **vastus medialis muscle**. The hot dog in the middle of the bun is the **rectus femoris muscle**. You will need to transect this muscle to observe the **vastus intermedius muscle**, which is between the **vastus medialis** and the **vastus lateralis**, and deep to the **rectus femoris**.

We now move distally to the leg. There will only be one muscle for you to know in the leg, the **gastrocnemius**. It is on the posterior side of the leg and it has two heads. You should observe its tendon, the **calcaneal tendon (tendo calcaneus)**.

# BONES

## Anterior Femur (Overview)

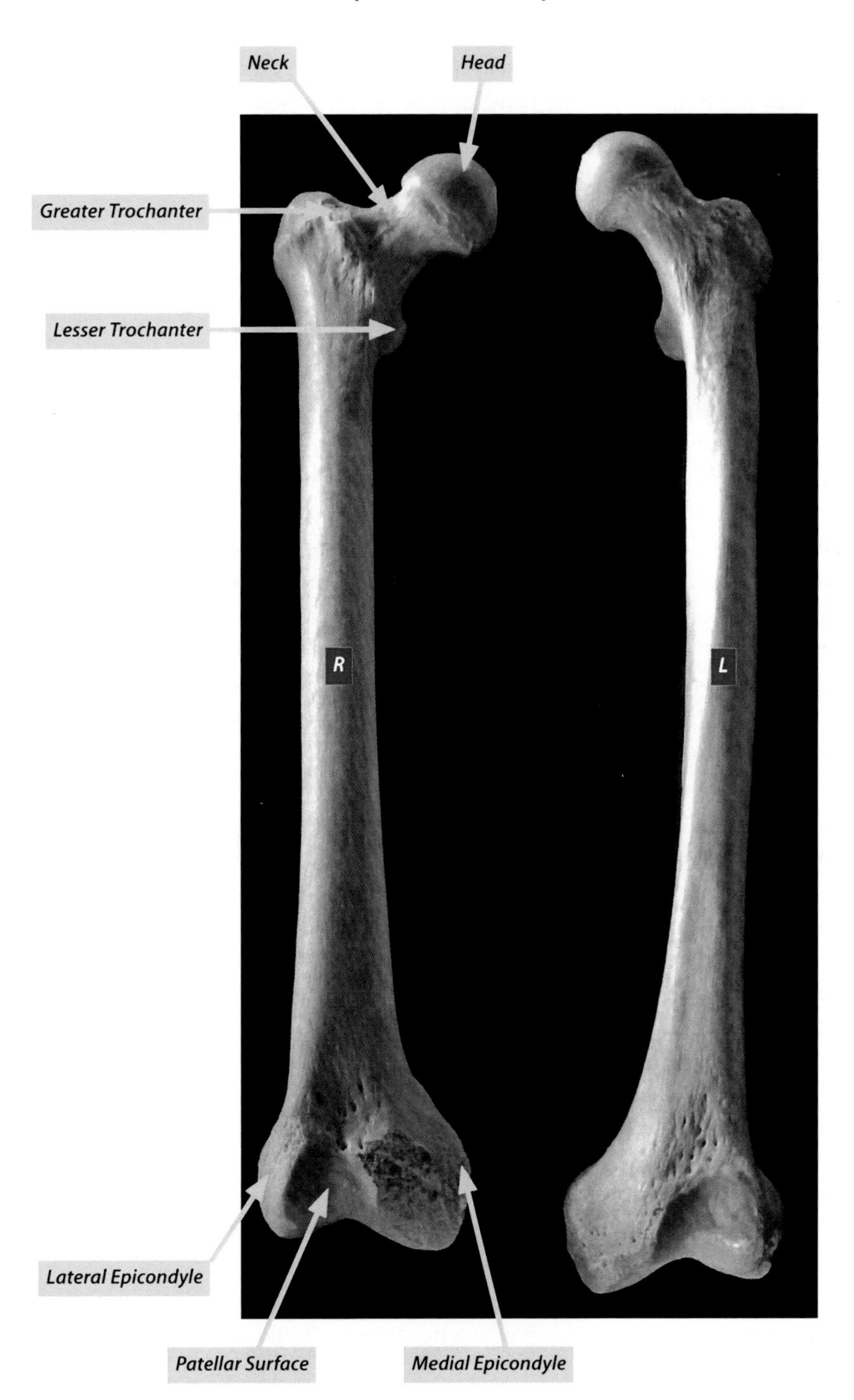


Dr. J has a suggestion to tell the left femur from the right. Place the patellar surface at the distal end (it is on the anterior surface). Then look at the head—it is pointing medially. He hopes that helps.

# Posterior Femur (Overview)

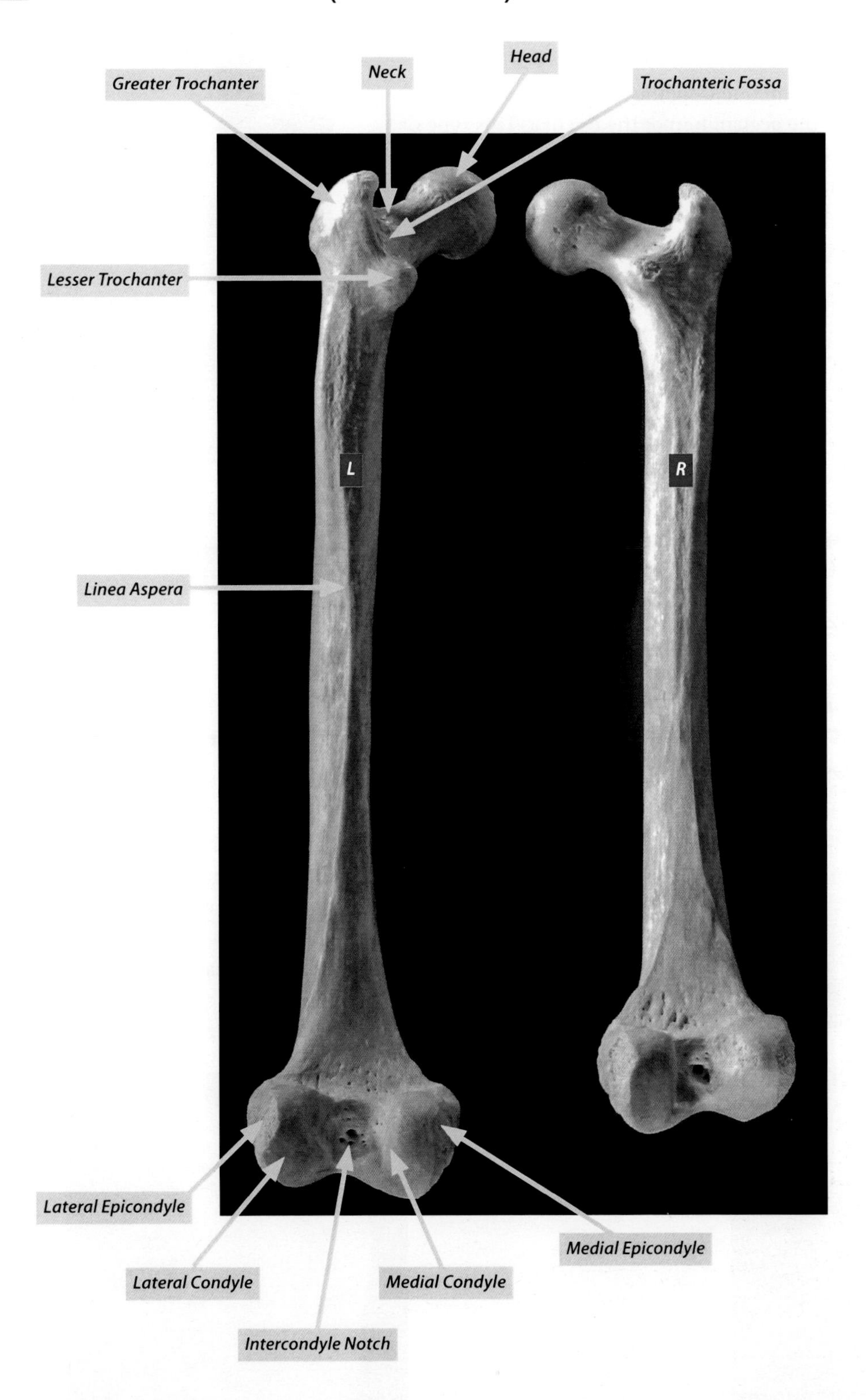

# Femur

## Head (Femur)

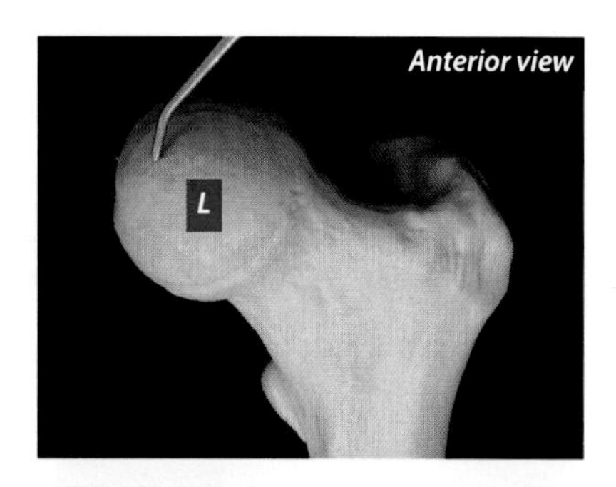


The head of the femur is of functional significance because it forms a ball and socket joint with the acetabulum of the os coxa. This type of joint allows for flexion/extension, abduction/adduction (and therefore circumduction), and rotation. The head of the femur is held in place by a ligament called the ligament of the head of the femur (ligamentum teres). You should be able to see the pit (fovea capitis) on real femurs where this ligament attached. Remember that the head points medially, which should help you distinguish left femurs from right ones.

## Neck (Femur)

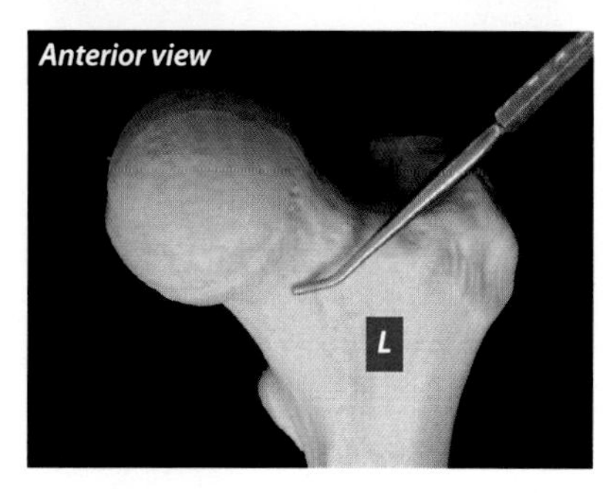


The neck of the femur is important functionally because it does not descend vertically; rather, it angles laterally to attach to the shaft of the bone. This is necessary because the articulation with the acetabulum is on the lateral aspect of the os coxa instead of on the inferior portion of the bones. This is the weakest part of the femur and is often the part that fractures when a person "breaks a hip."

## Greater Trochanter (Femur)

The greater trochanter is the insertion for:

1. gluteus medius and
2. gluteus minimus.

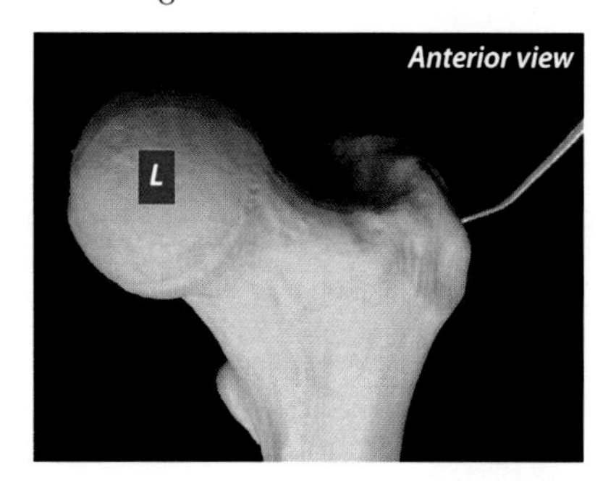


## Lesser Trochanter (Femur)

The lesser trochanter is the insertion for:

1. iliopsoas.

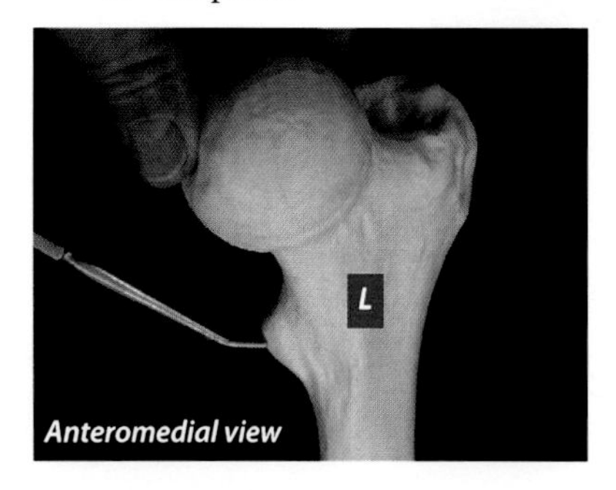


## Trochanteric Fossa (Femur)

The trochanteric fossa is also sometimes referred to as the intertrochanteric fossa. It is a deep depression located on the posterior and medial surface of the greater trochanter. This landmark serves as the insertion point for four of the six lateral (external) rotators of the hip (obturator externus, obturator internus, gemellus superior, and inferior gemellus). You will not be responsible for knowing these muscles.

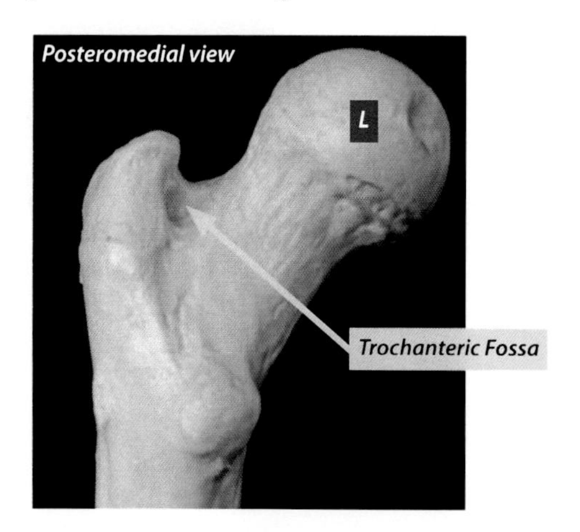

# Femur

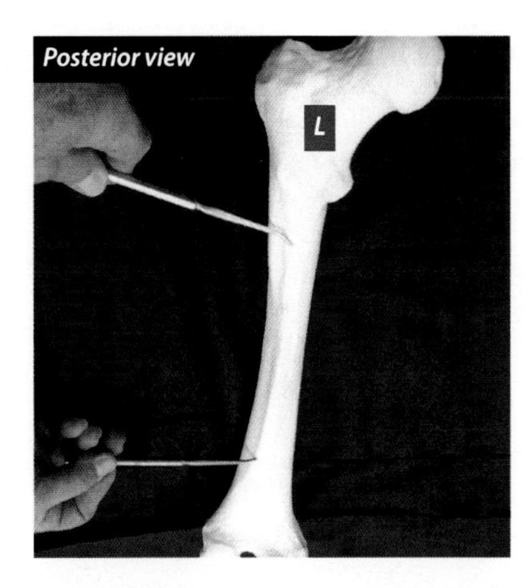


## Linea Aspera (Femur)

The linea aspera is on the posterior side of the femur. Translated, its name means "rough line." The linea aspera is the origin for:

1. a portion of biceps femoris,

and the insertion for:

1. adductor magnus,
2. adductor brevis, and
3. adductor longus.

The pectineus inserts proximal to the linea aspera along the pectineal line of the femur.

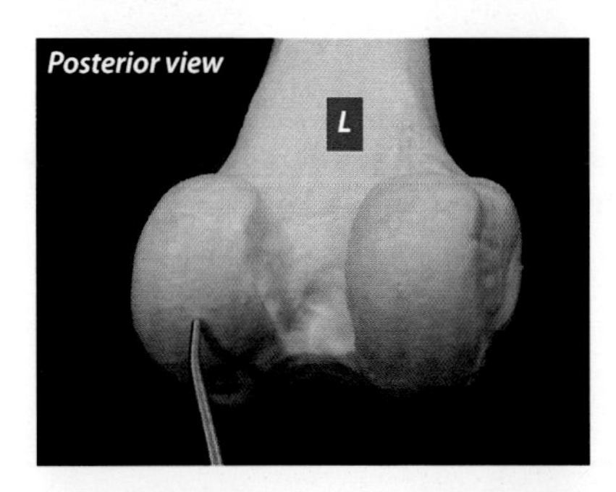


## Lateral Condyle (Femur)

The lateral condyle is the origin for:

1. the lateral head of gastrocnemius.

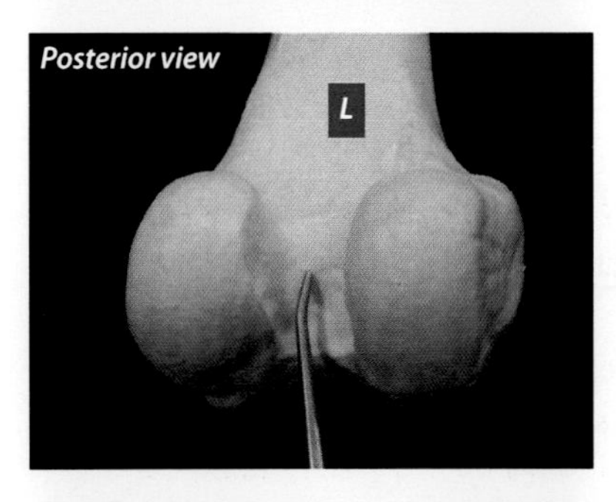


## Intercondylar Notch

The intercondylar notch of the femur is also known as the intercondylar fossa. Within this notch are facets for the attachment of the anterior and posterior cruciate ligaments. This notch is narrower in adult females than it is in adult males. This may contribute to the higher incidence of tears to one or both of these ligaments in adult females as compared with adult males.

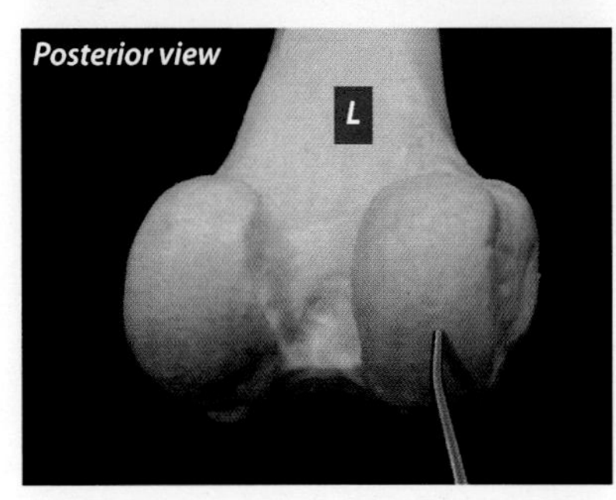


## Medial Condyle (Femur)

The medial condyle is the origin for:

1. the medial head of gastrocnemius,

and the insertion for:

1. adductor magnus.

# Femur

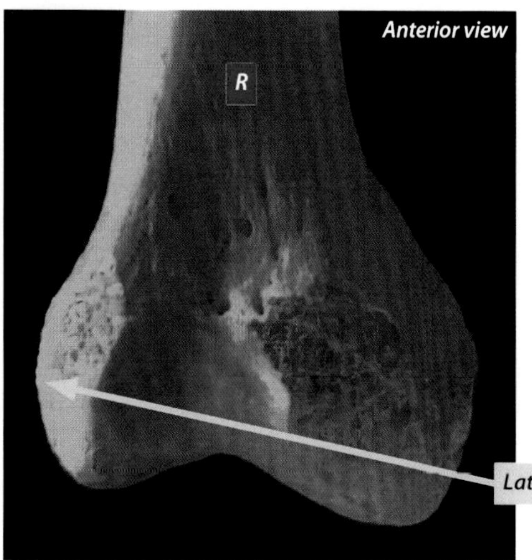


## Lateral Epicondyle (Femur)

The lateral epicondyle is an elevation where collateral ligaments attach. These ligaments help stabilize the knee. This surface does not articulate with any bone.

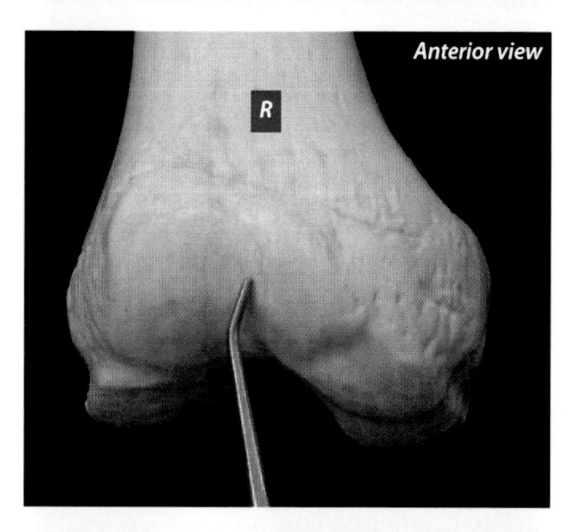


## Patellar Surface (Femur)

The patella articulates with this smooth surface between the condyles.

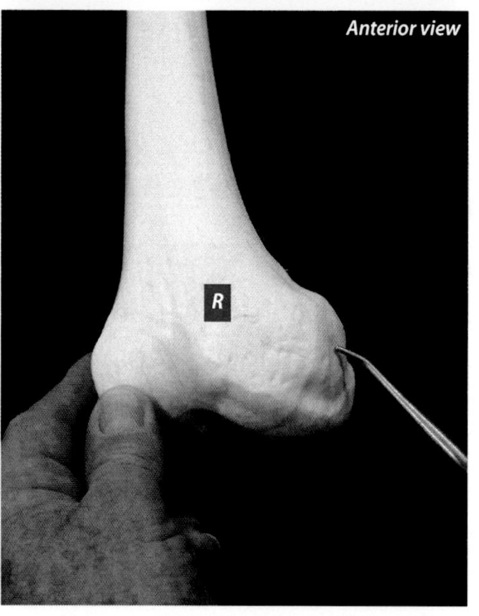


## Medial Epicondyle (Femur)

The medial epicondyle is an elevation where collateral ligaments attach. These ligaments help stabilize the knee. This surface does not articulate with any bone.

## Leg and Ankle (Overview)

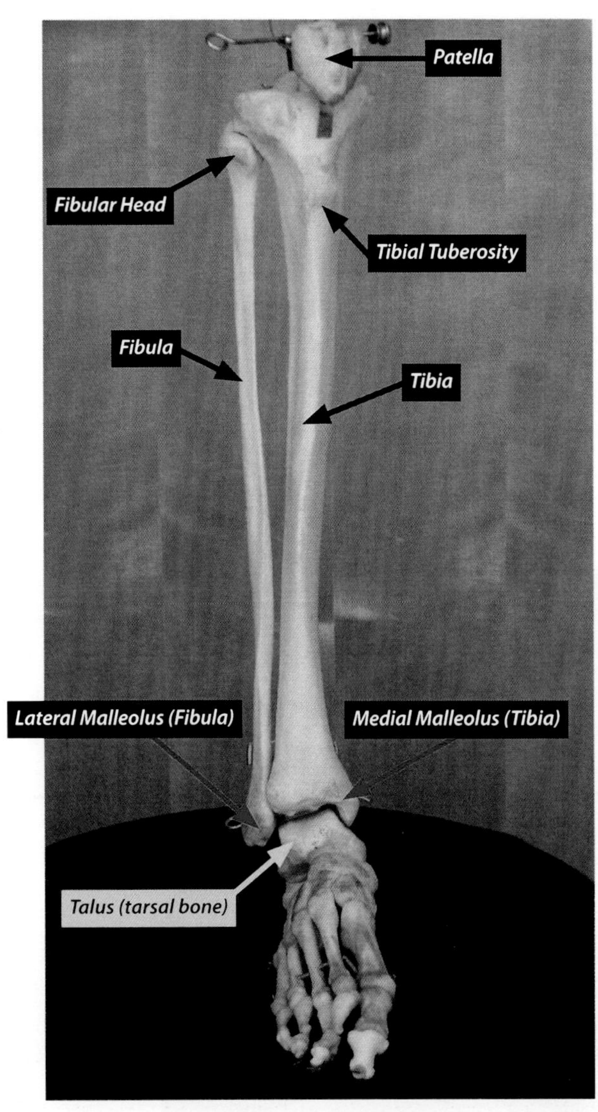


*Anterior view, right side*

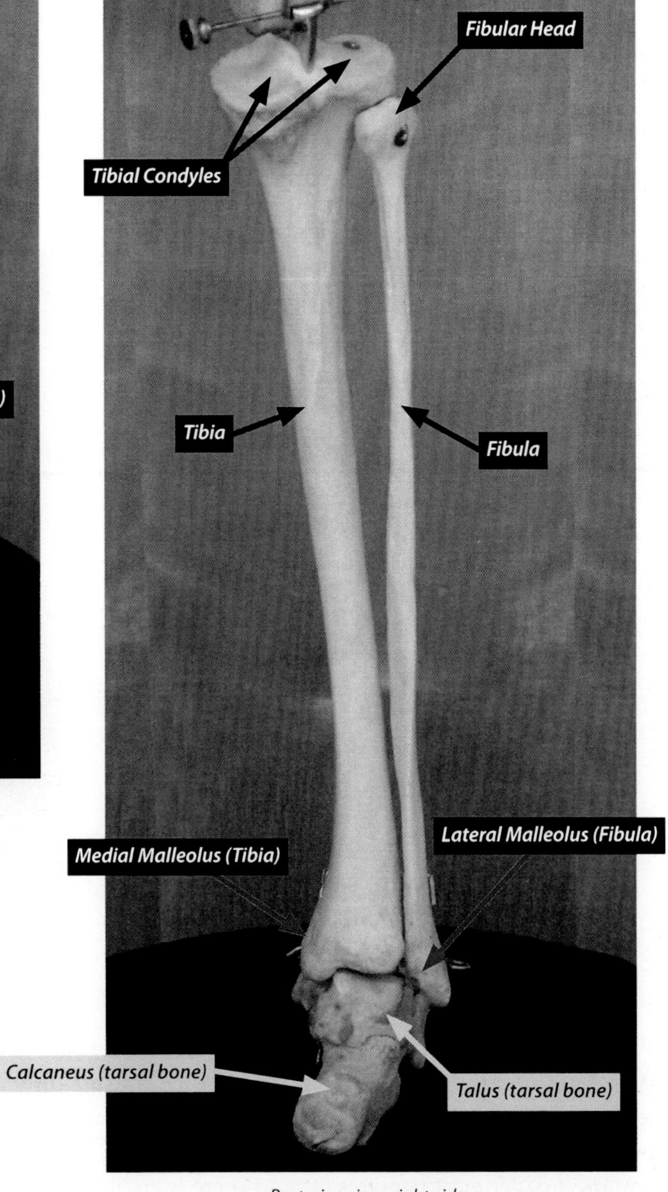


*Posterior view, right side*

## Patella

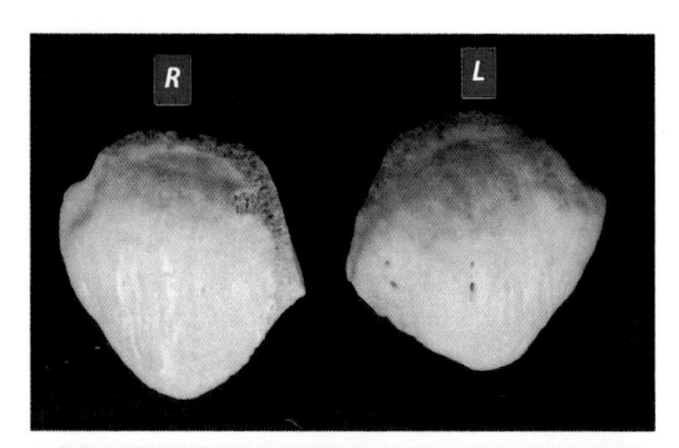


*Here each apex points to the bottom of the picture. In the anatomical position each apex points inferiorly.*

The patella is the largest sesamoid bone in the body and does not completely ossify until we are from three to five years old. It is functionally important because it is the anatomical insertion for the quadriceps muscle. It is attached to the tibia by three ligaments: the patellar ligament and the medial and lateral patellar retinacula. This association increases the leverage of the muscles in the thigh that pull across the anterior surface of the knee, thereby extending the leg. Deep to the patella is the patellar surface of the femur. Dr. J has a trick for telling the left from the right patella. Place the patella on a horizontal, flat surface. Direct the apex (in nursery school it was the pointy end) away from you and release the patella. It will tip toward the side it came from. How cool is that?

## Fibula

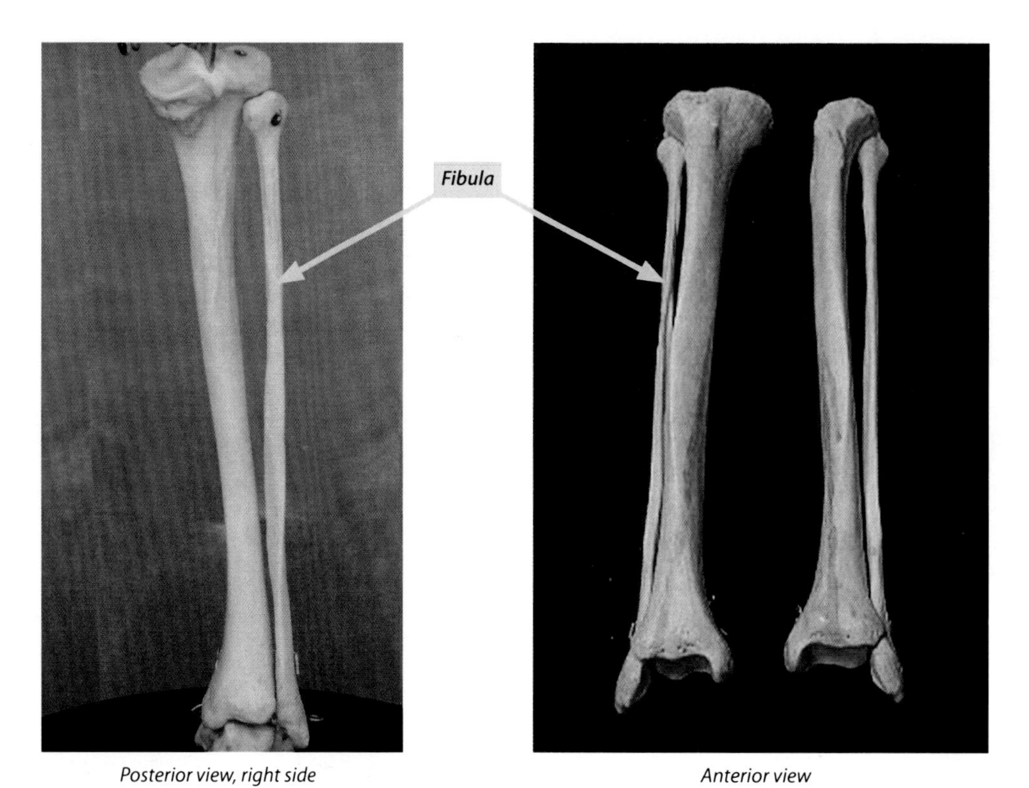


*Posterior view, right side* *Anterior view*

The fibula is the lateral bone of the leg. Although it bears only 10 percent (or less) of our weight, it is important as the origin of several muscles. There are only two landmarks that you will be required to know. Ask Dr. J for his tip on how to distinguish the left fibula from the right fibula. You may find this tip to be helpful during a practical exam. Dr. J wants you to spell this correctly. It is lateral, and the first letter of lateral is "l" and there is an "l" in fibula, but not in the name tibia! Wow!

# Fibula

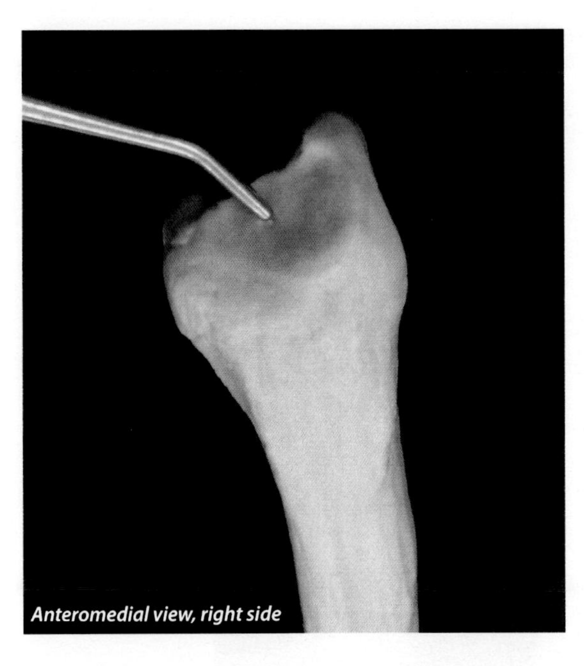
Anteromedial view, right side

## Head (Fibula)

The head of the fibula is a proximal feature and articulates posterolaterally with the tibia. It is the insertion for:

1. a portion of biceps femoris (the lateral side of the head of the fibula).

## Lateral Malleolus (Fibula)

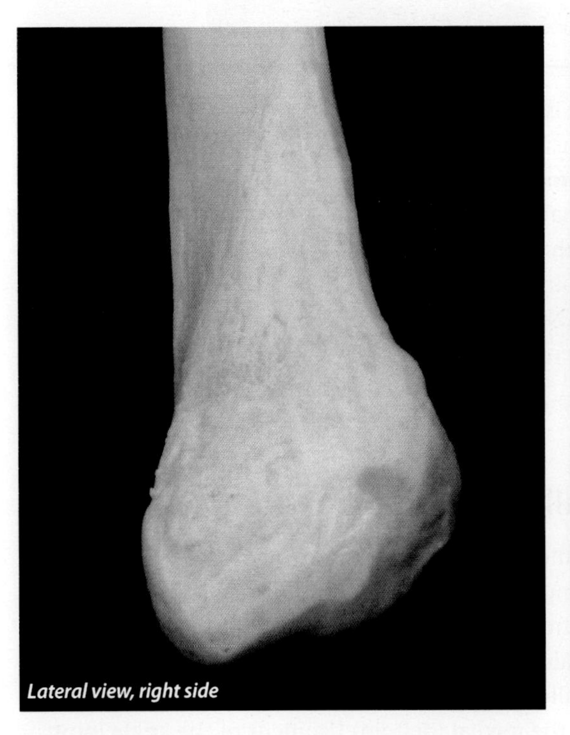
Lateral view, right side

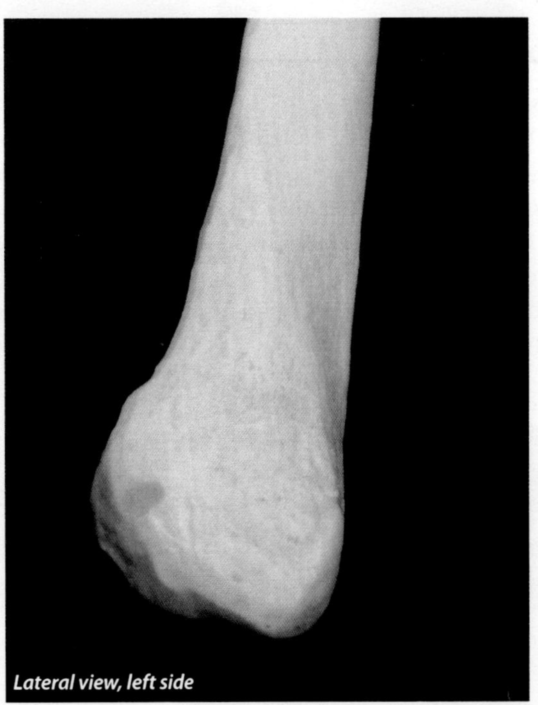
Lateral view, left side

The lateral malleolus of the fibula is a distal feature, which articulates with the lateral side of the talus to form the lateral bulge of the ankle. It also forms the lateral side of a mortise and tenon joint with the talus and the tibia.

# Tibia

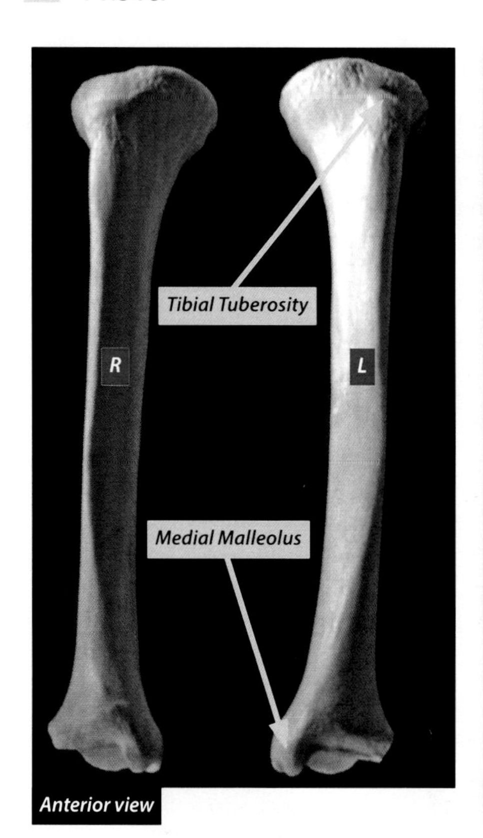


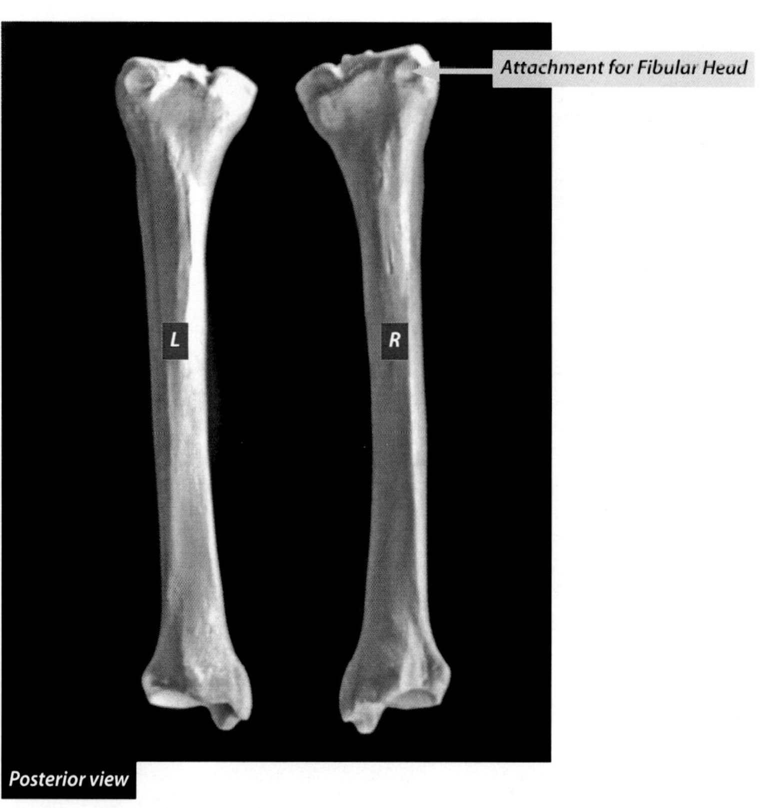


The tibia is named after the Greek word for a type of flute. It is not just the larger of the two bones in the leg; it also bears 90 percent or more of the weight that is transferred from the femur. It is medial to the fibula. The patellar ligament attaches to the tibia (tibial tuberosity), which is the functional insertion of the quadriceps muscles. The distal end of the tibia (medial malleolus) forms part of a mortise and tenon joint with the talus. Dr. J's tip for distinguishing left from right is to find the tibial tuberosity, a large bump on the anterior proximal tibia. Keep that close to you and point the distal end of the tibia away from you. The medial malleolus is medial—another Grant thing! (Note: the lateral malleolus is on the fibula, which is always the lateral bone.) That makes it simple to identify whether a tibia is a left or right.

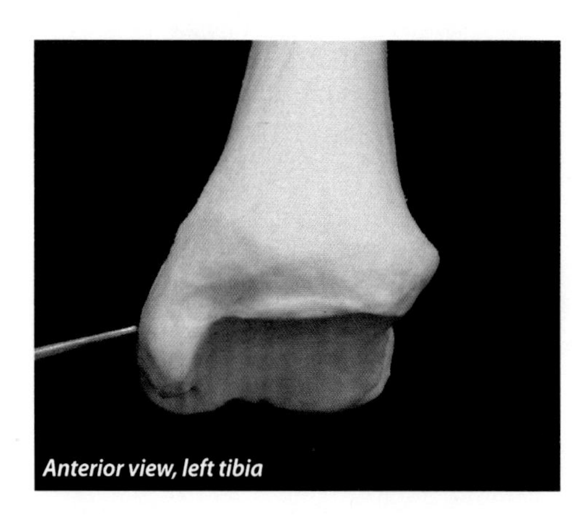


## Medial Malleolus (Tibia)

The medial malleolus of the tibia forms the medial side of a mortise and tenon joint. In doing so, it forms the medial bulge of the ankle. The lateral malleolus of the fibula forms the lateral side, and the talus (a tarsal bone) fits into the space they create. The medial malleolus of the tibia also provides a surface for the attachment of the medial (deltoid) ligament of the ankle joint.

# Tibia

## Lateral Condyle (Tibia)

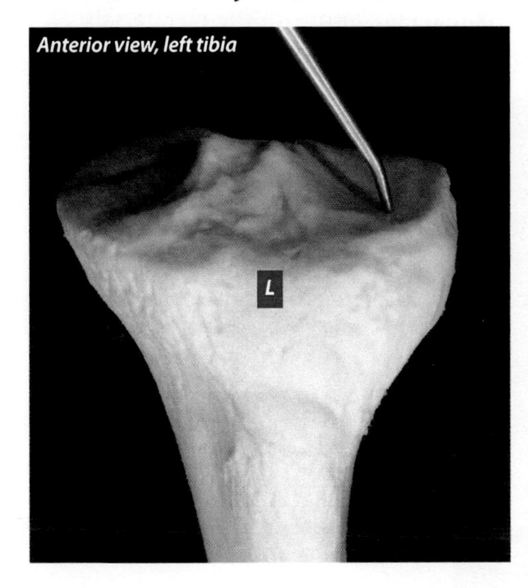


## Medial Condyle (Tibia)

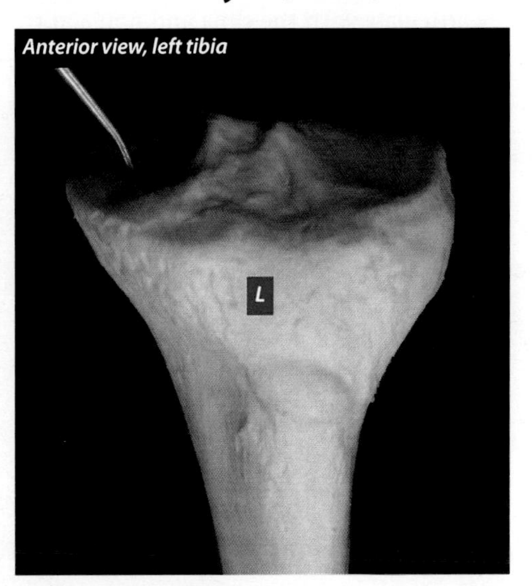


The lateral and medial condyles of the tibia (tibial condyles) are located proximally. They articulate with the lateral and medial condyles of the femur (femoral condyles) respectively, to form a hinge joint. The "C"-shaped menisci rest on the superior surfaces of the tibial condyles and together they help stabilize the knee joint, distribute weight evenly, and guide the condyles during flexion and extension. Additional stability is provided by the anterior and posterior cruciate ligaments and medial (tibial) and lateral (fibular) collateral ligaments of the knee.

The lateral condyle of the tibia is the insertion for:

1. a portion of biceps femoris.

The medial condyle of the tibia is the insertion for:

1. semimembranosus (posterior surface of medial condyle).

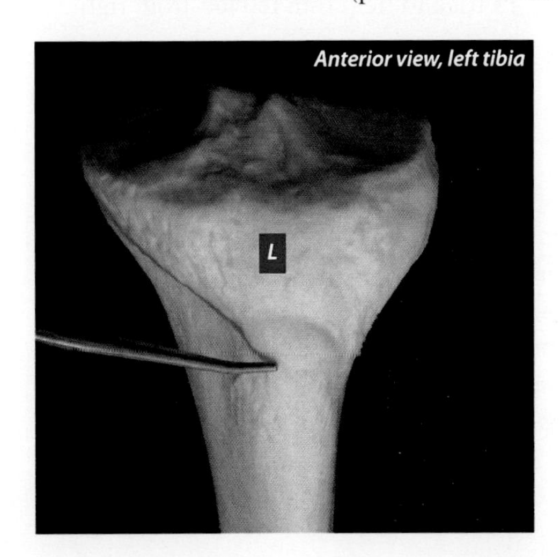


## Tibial Tuberosity (Tibia)

The tibial tuberosity is of functional significance because this is where the patellar ligament attaches; therefore, functionally, it is where the quadriceps inserts.

# Tarsals (Specific Bones)

There are seven tarsals, and they make up the posterior portion of the foot. The tarsals correspond to the carpals of the wrist. They articulate with the tibia and fibula at the proximal end, and with the metatarsals at the distal end. You will only have to know two tarsals by name—the talus and the calcaneus.

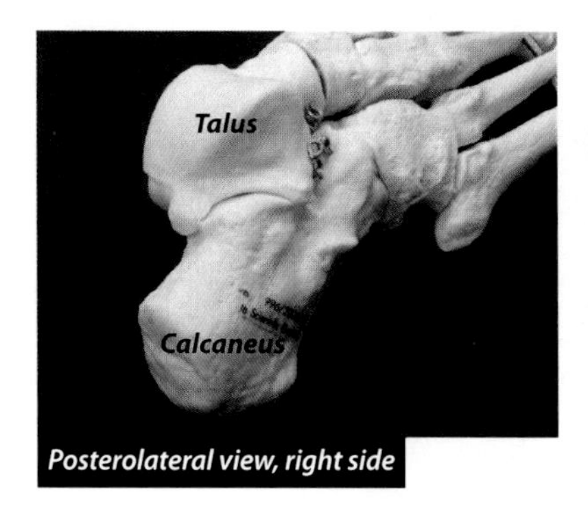

Posterolateral view, right side

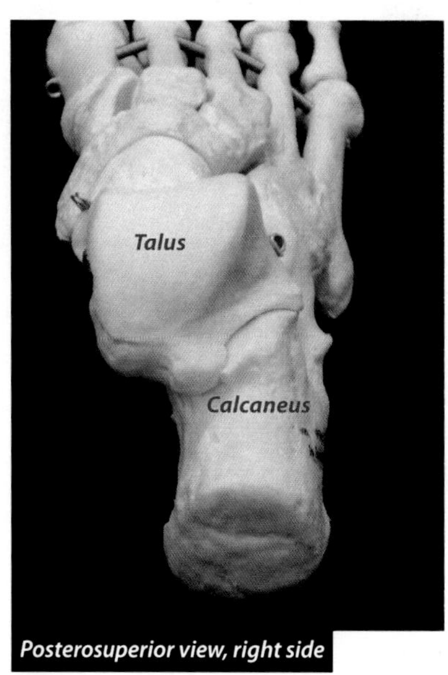

Posterosuperior view, right side

## Calcaneus (Tarsals)

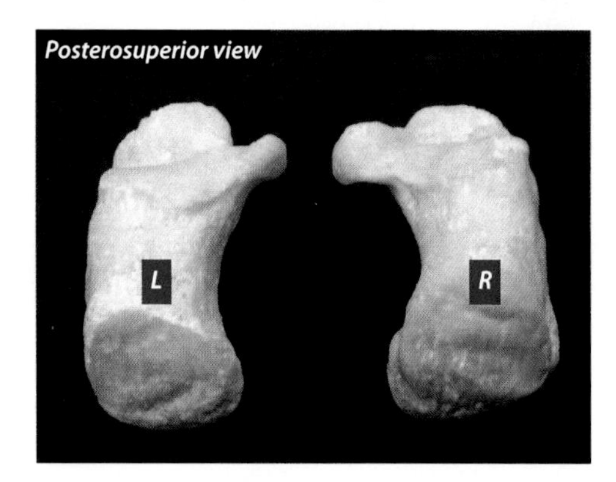

Posterosuperior view

The calcaneus is the insertion for:

1. gastrocnemius (via the calcaneal tendon).

Note that when the calcaneus is in the anatomical position, the lateral side is relatively flat while the medial side is not. Also, there is a projection on the superior surface of its medial side. This is similar to your big toe, which is also on the medial side of your foot. This is Dr. J's trick to help him tell left from right.

## Talus (Tarsals)

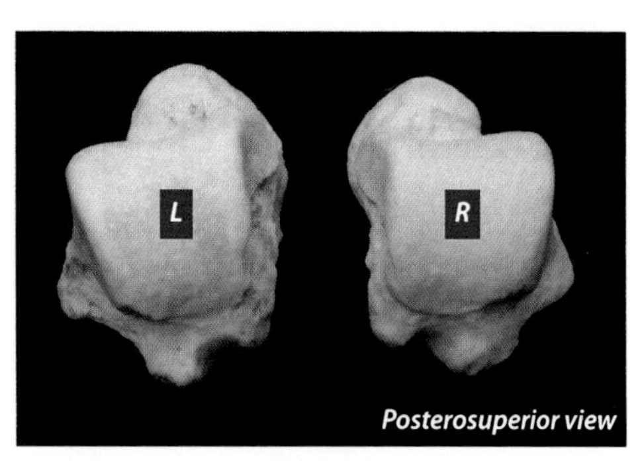

Posterosuperior view

Many of Dr. J's students think the talus looks a little like a turtle. If you put the part that looks like the head at the anterior end (in the above pictures, the end that looks like the head is at the top of the picture), the medial side is relatively flat while the lateral side is not. So, on the left-hand side of the above picture, the talus on the left is a left talus and the talus on the right is a right talus. This is Dr. J's trick to help him tell left from right.

## Metatarsals

The five metatarsals are distal to the tarsal bones and proximal to the phalanges of the foot. Therefore, they are sometimes called the bones of the intermediate portion of the foot. They correspond to the metacarpals of the hand. They are numbered from medial to lateral. The first metatarsal articulates with the base of the big toe and is functionally important because it helps support the weight of the body. The head of the first metatarsal is enlarged, and this is known as the ball of the foot.

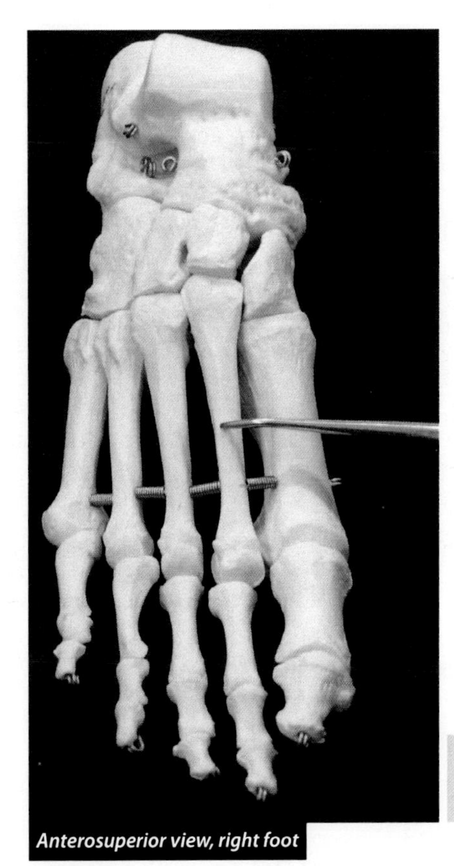

Anterosuperior view, right foot

*Probe is pointing to the metatarsal of second digit, right side.*

## Phalanges

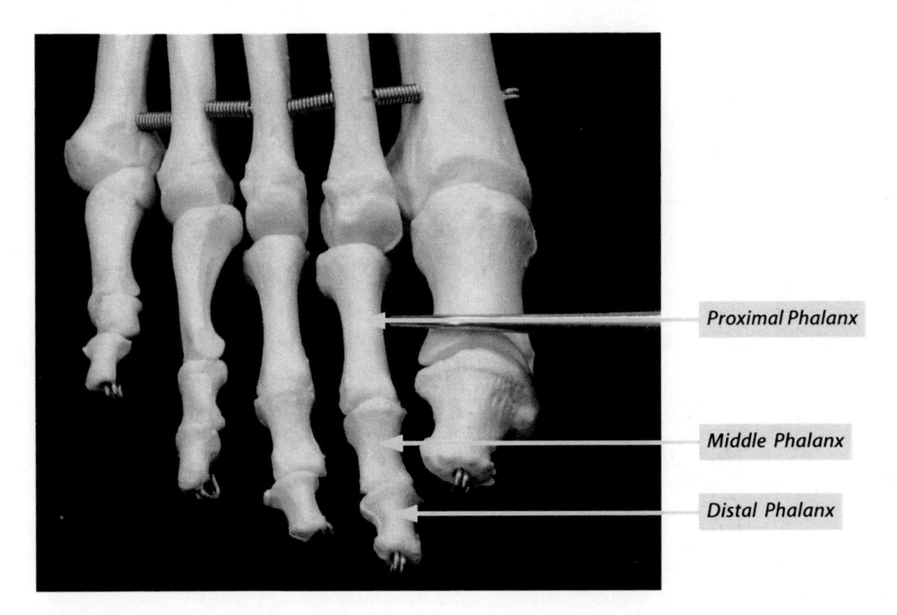


There are fourteen phalanges of the foot, and they are smaller and less agile than those of the hand. There are two phalanges for the big toe and three for each of the other four toes. You will not need to know these by specific names.

# MUSCLES

## Muscles of the Hip and Thigh

### Caudofemoralis (Cat)

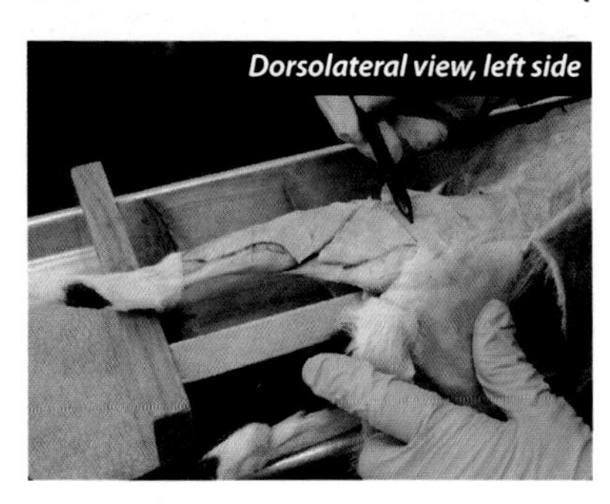
Dorsolateral view, left side

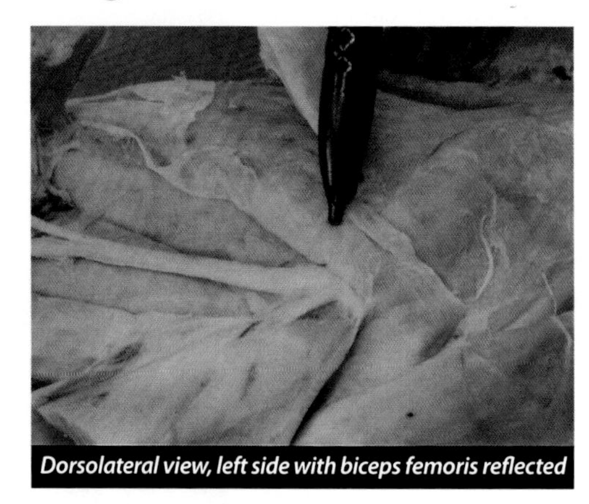
Dorsolateral view, left side with biceps femoris reflected

***Cat Information:***

**origin:** transverse processes of second and third caudal vertebrae
**insertion:** middle of lateral border of the patella tibia and the lateral aspect of patella
**nerve:** caudal gluteal
**action:** abduction of thigh, extension of leg

The caudofemoralis is not found in humans, although it is found in cats.

### Tensor Fasciae Latae

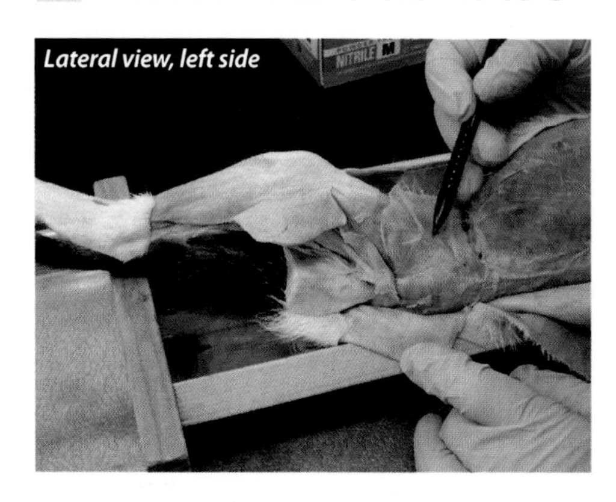

Lateral view, left side

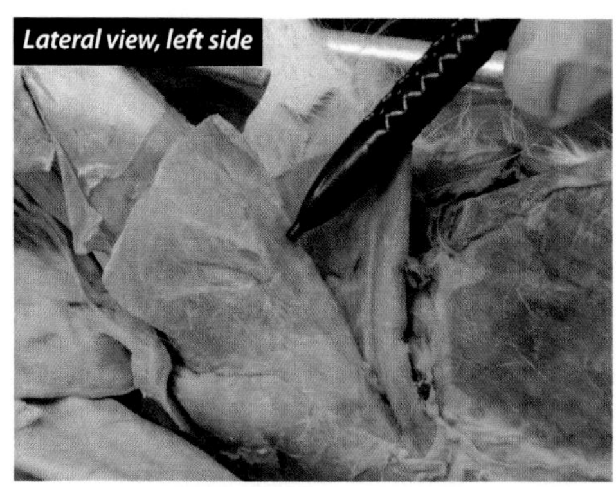
Lateral view, left side

#### Fascia Lata

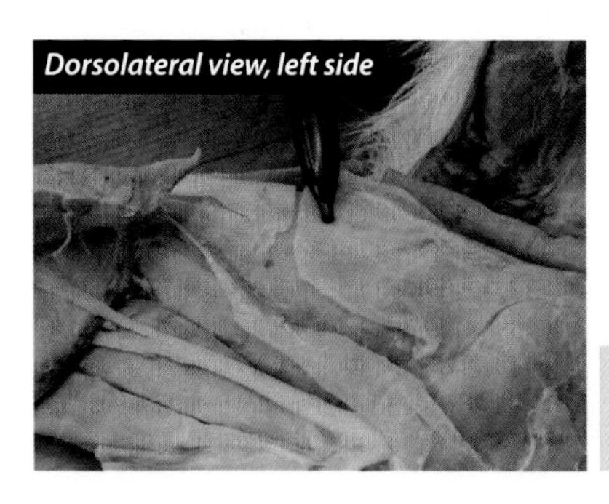
Dorsolateral view, left side

***Human Information:***

**origin:** anterior superior iliac spine (ASIS) and iliac crest
**insertion:** iliotibial tract (IT band)
**nerve:** superior gluteal
**action:** abducts, flexes, and medially rotates thigh

*The fascia lata forms part of the iliotibial band. This aponeurosis anchors the tensor fasciae latae muscle to its insertion.*

# Tensor Fasciae Latae

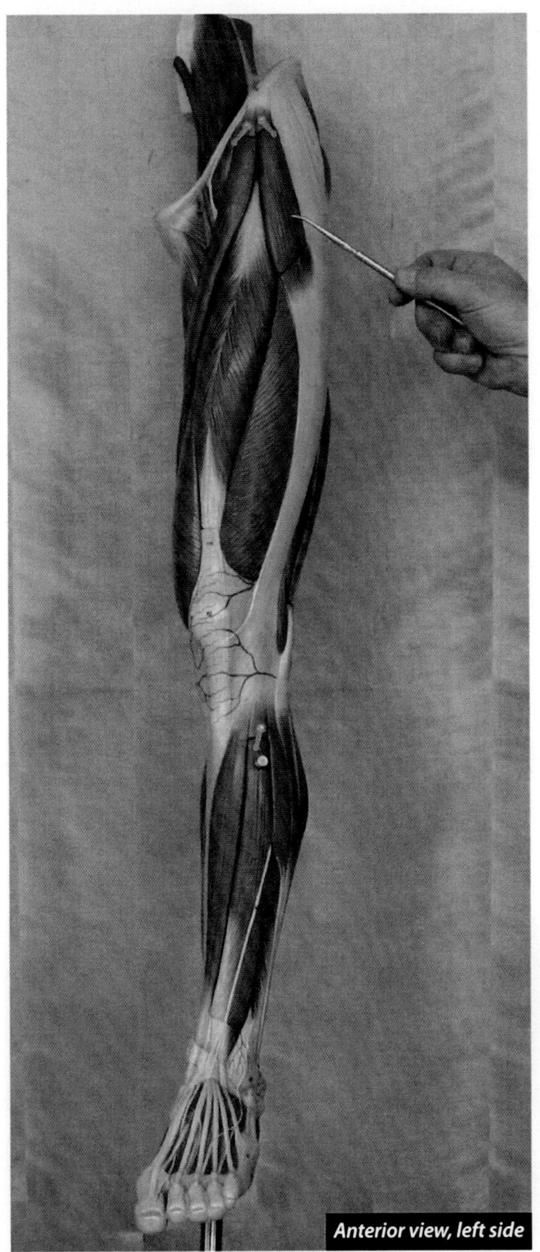
Anterior view, left side

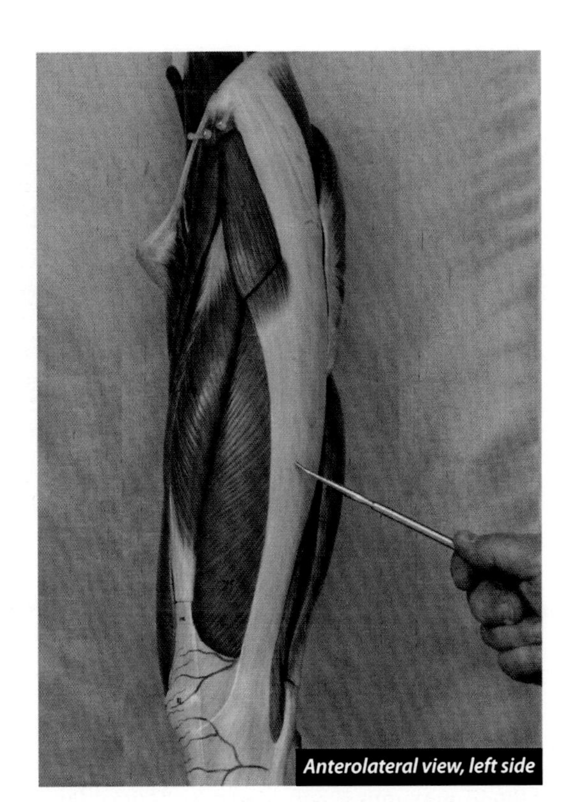
Anterolateral view, left side

The cat's tensor fasciae latae is shaped like an asymmetrical heart with its apex at the proximal end. It is superficial on the lateral side of the human thigh. It is a synergist to the gluteus medius, the gluteus minimus, and the proximal fibers of the gluteus maximus.

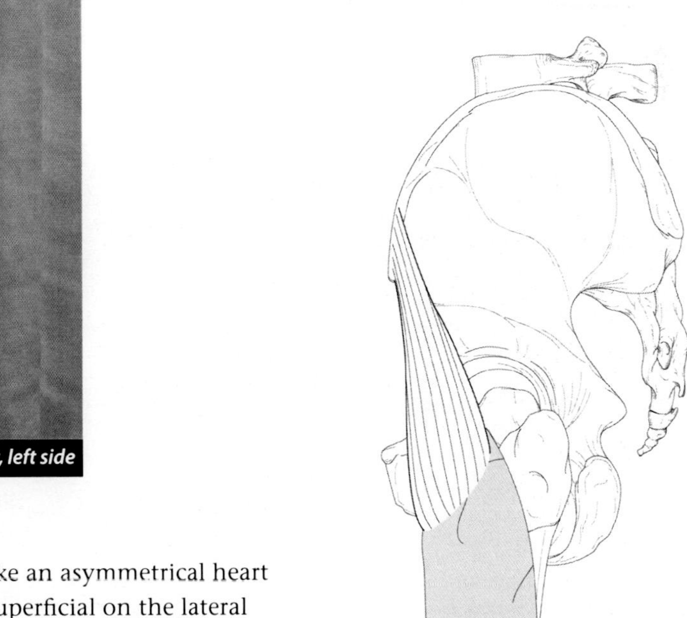
Lateral view, left side

# Gluteals

## Gluteus Maximus

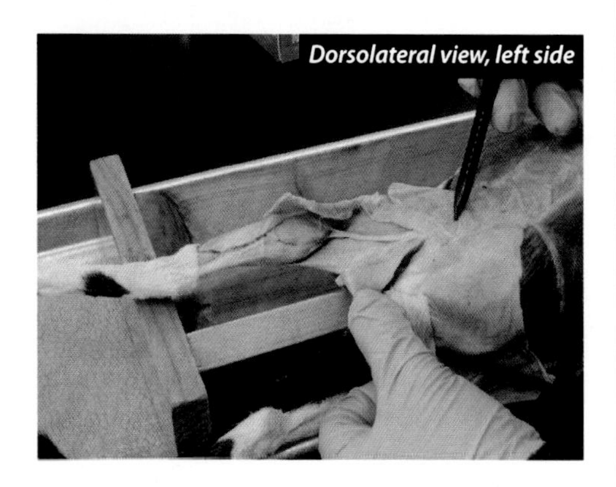
Dorsolateral view, left side

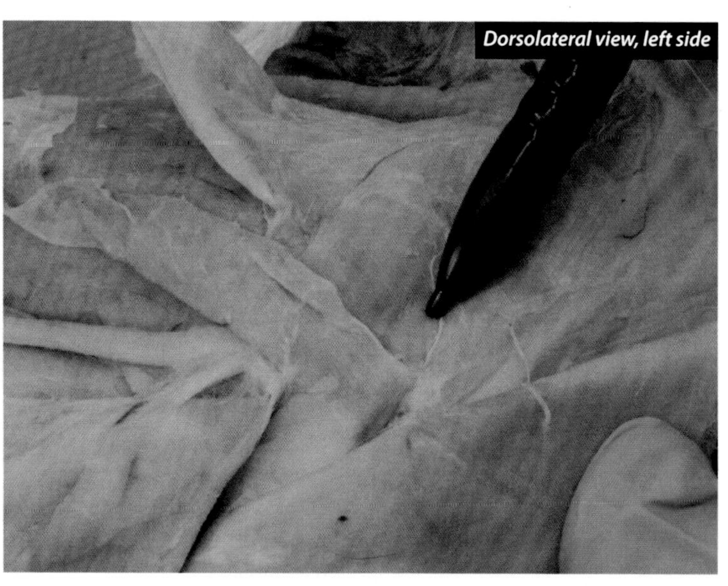
Dorsolateral view, left side

***Human Information:***

**origin:** posterior surface of ilium, sacrum, and coccyx
**insertion:** proximal posterior femur, iliotibial tract (IT band)
**nerve:** inferior gluteal
**action:** extends and laterally rotates thigh

Note that in the cat, the gluteus maximus is more of an abductor than an extender of the thigh. But in the human, it is more important as an extender of the thigh because of its different position. The gluteus maximus is considered by "Trivial Pursuit" to be the largest muscle in the human body. Dr. J thinks the quadriceps is the largest muscle in the human body if you consider it as one muscle.

## Gluteus Medius

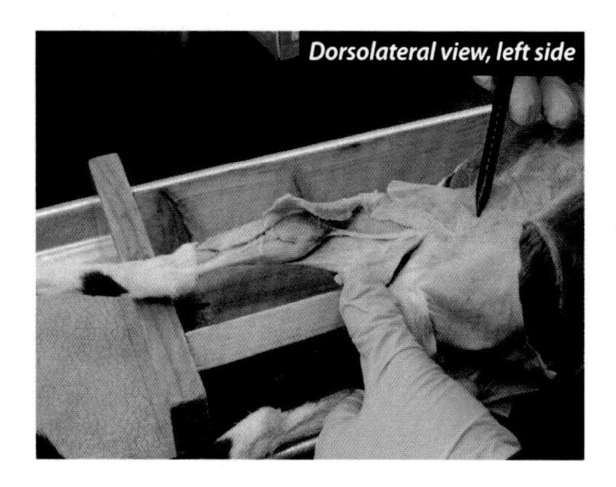
Dorsolateral view, left side

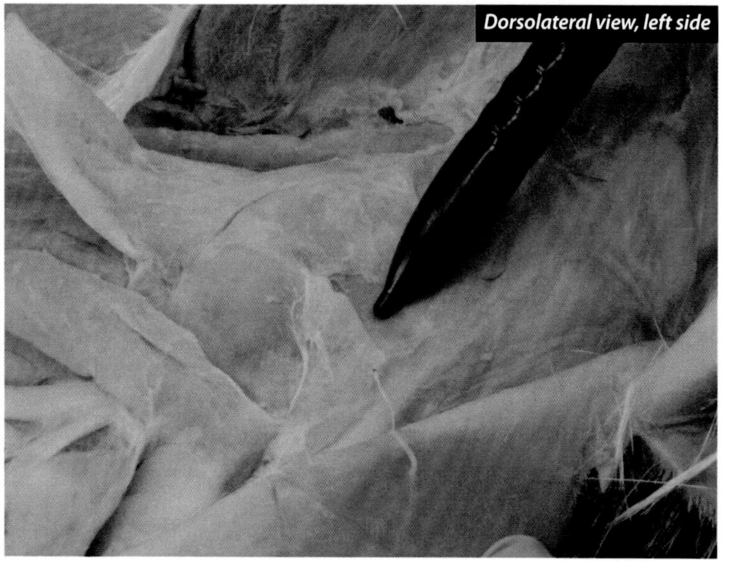
Dorsolateral view, left side

***Human Information:***

**origin:** superficial surface of ilium
**insertion:** greater trochanter of femur
**nerve:** superior gluteal
**action:** abducts and medially rotates thigh

In humans, the gluteus medius is a major pelvic stabilizer working opposite the quadratus lumborum when we walk.

## Gluteus Maximus

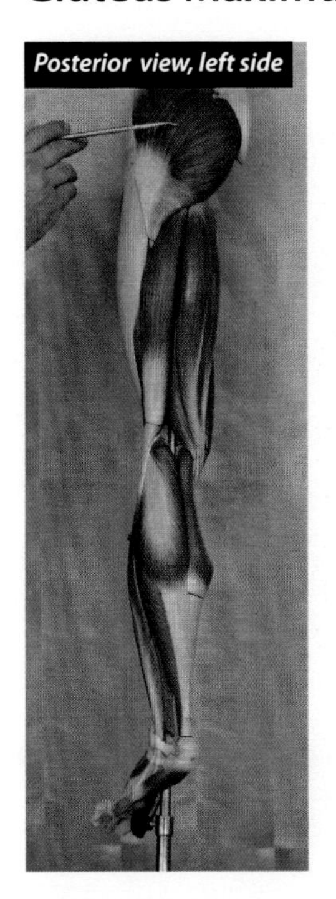


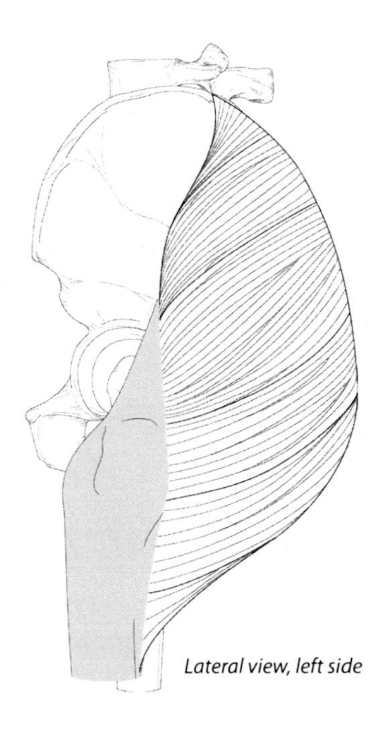


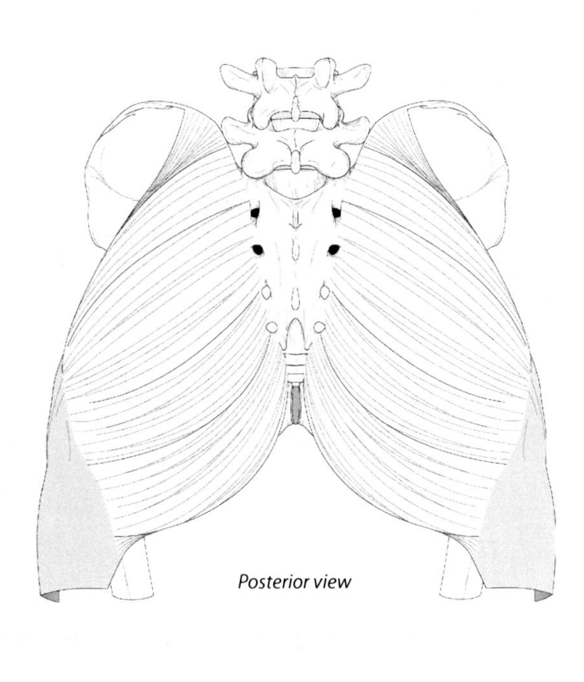


## Gluteus Medius

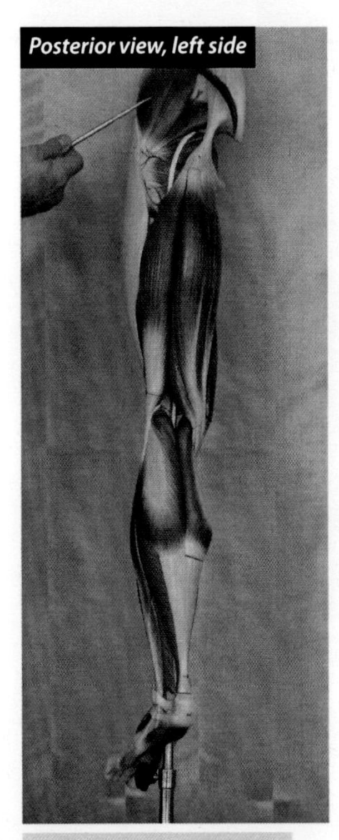


Gluteus maximus removed.

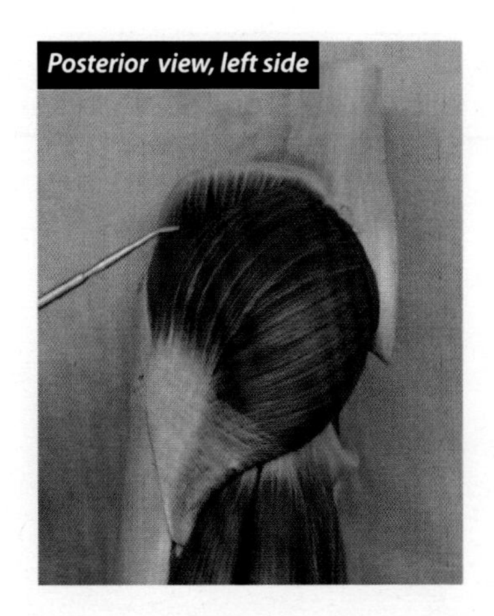


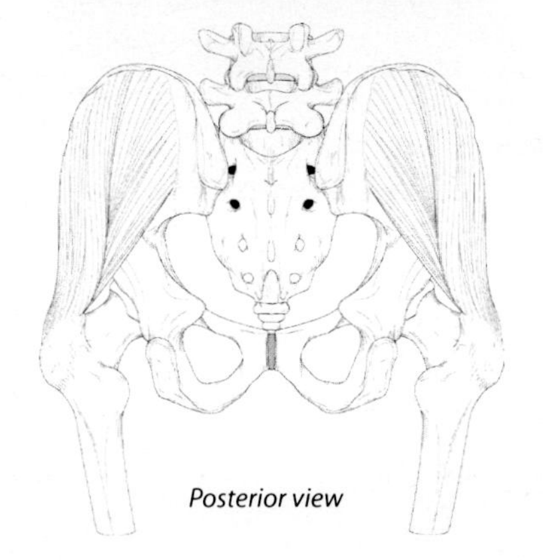


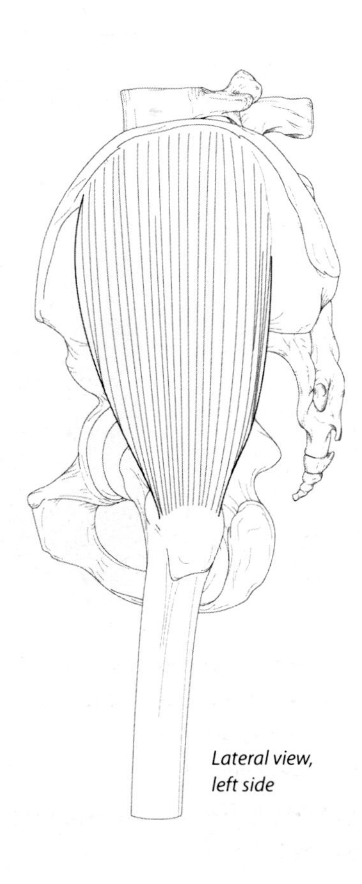

# Hamstrings

## Biceps Femoris

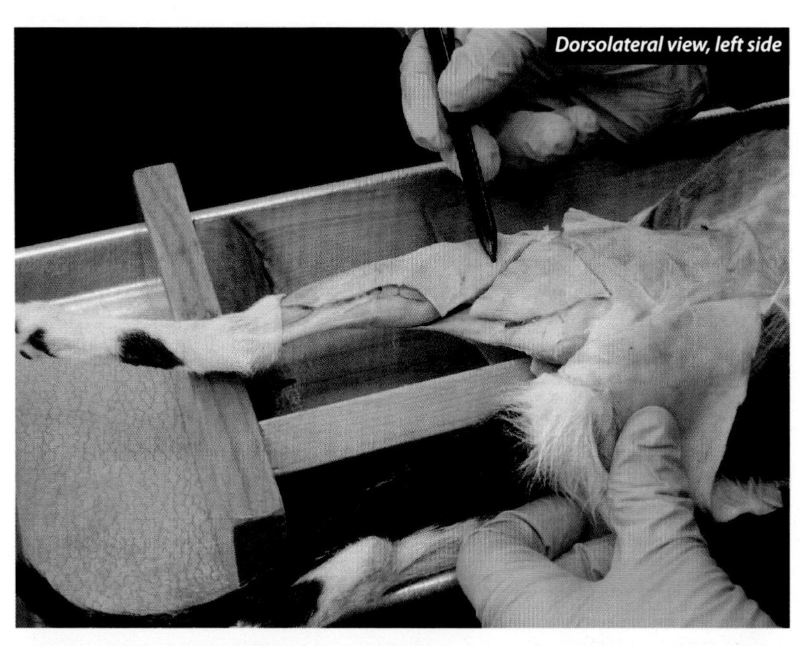


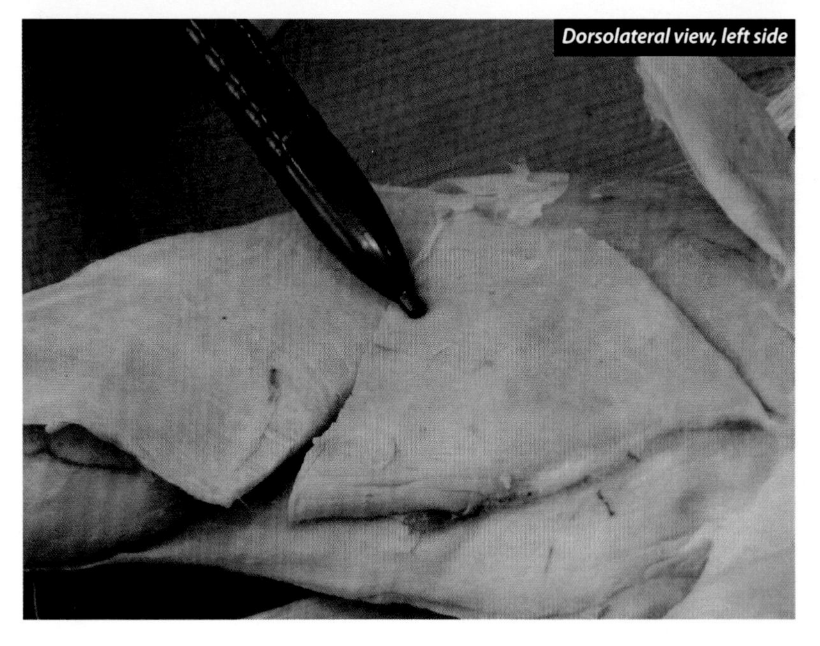


***Human Information:***

**origin:** ischial tuberosity; linea aspera
**insertion:** fibular head and lateral tibial condyle
**nerve:** sciatic
**action:** extends thigh; flex and laterally rotates leg.

## Biceps Femoris

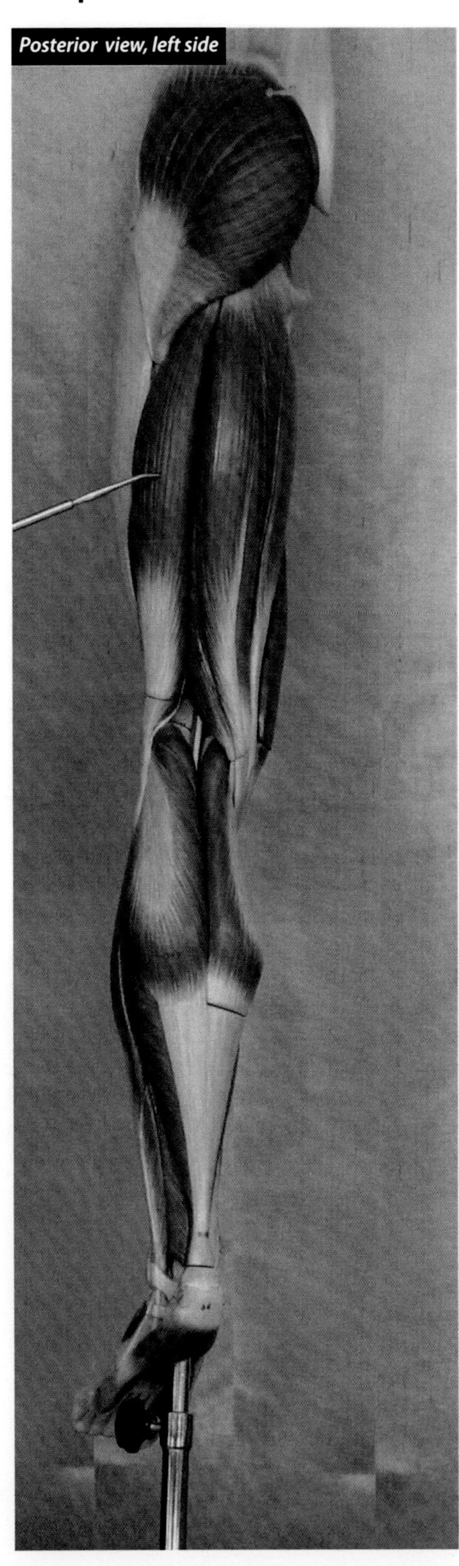
Posterior view, left side

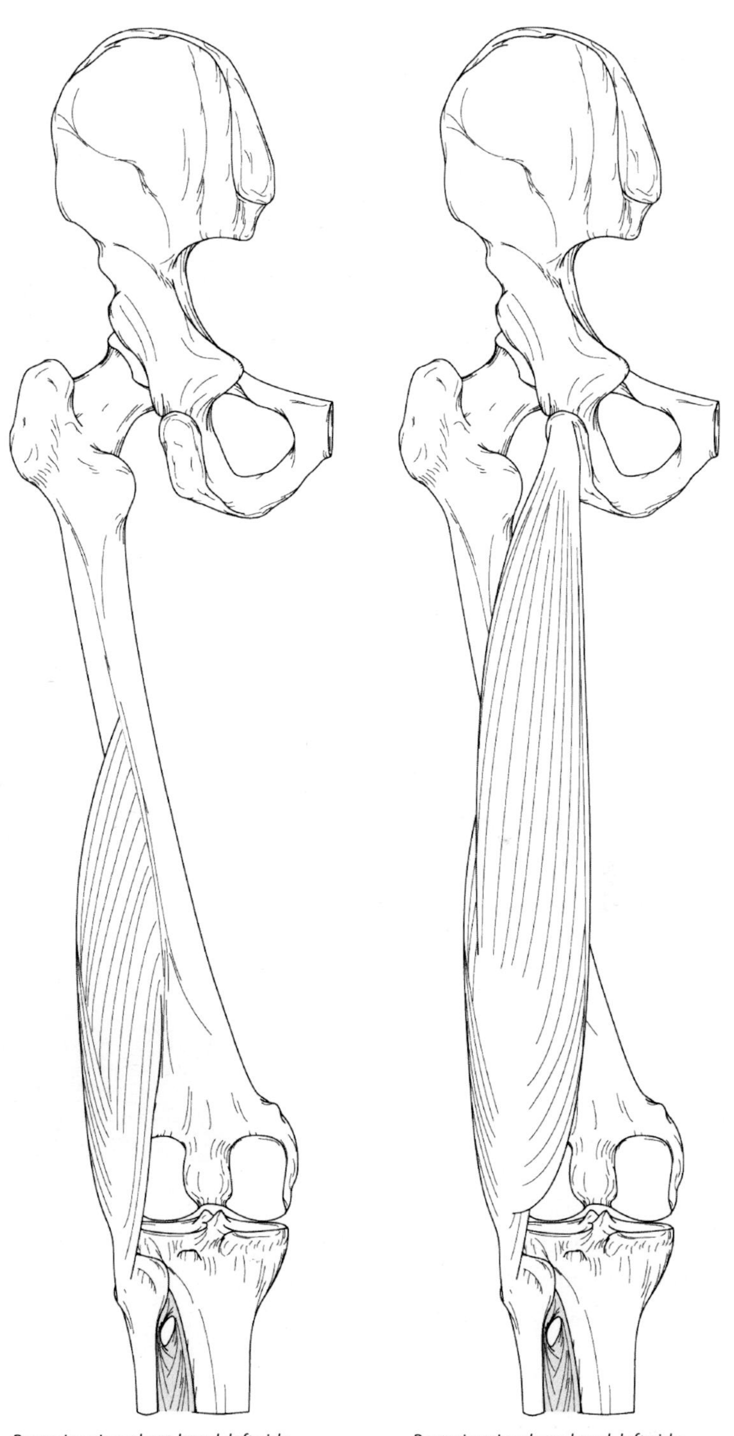
*Posterior view short head, left side*

*Posterior view long head, left side*

Biceps femoris is the largest of the hamstring muscles. These muscles are in the posterior compartment of the thigh.

## Semitendinosus

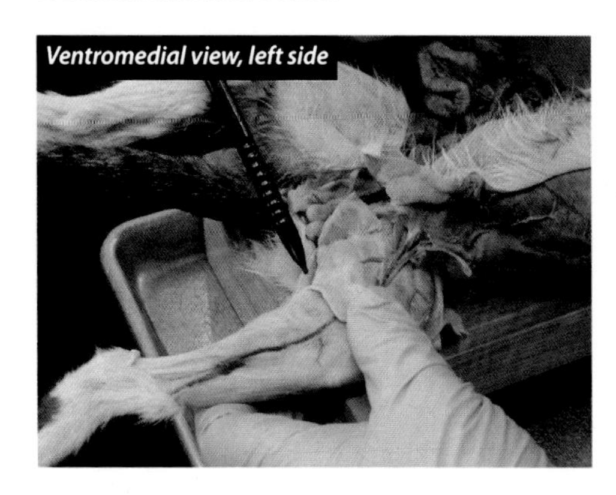


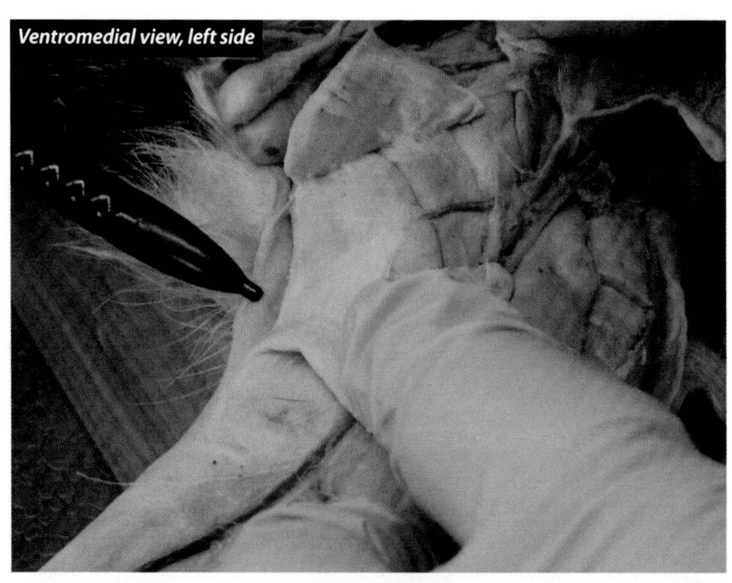


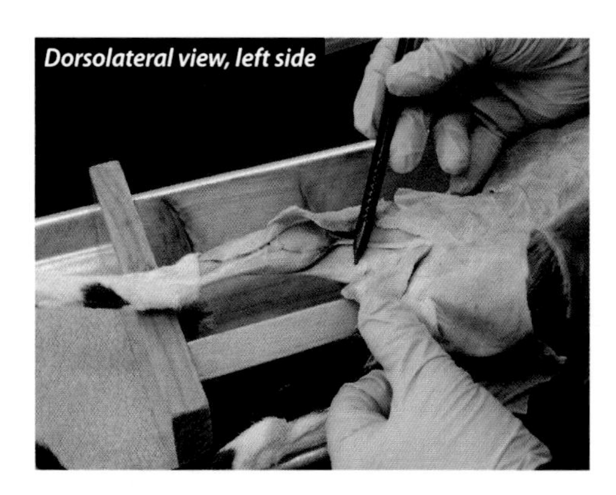


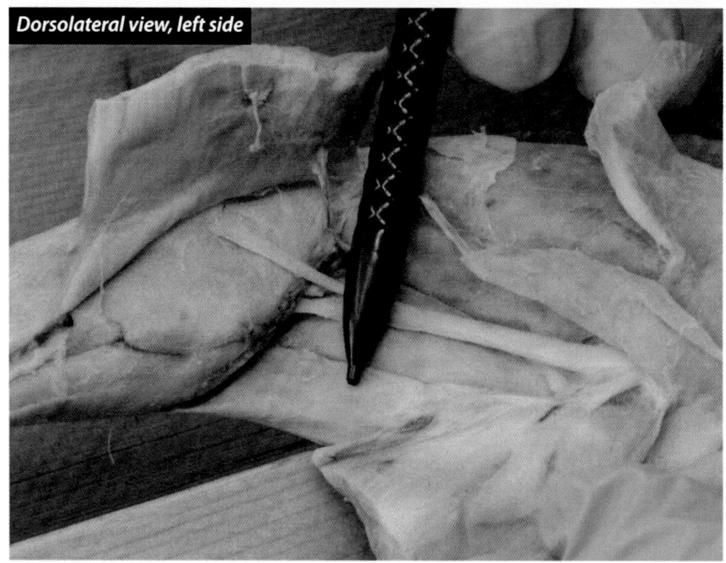


### *Human Information:*

**origin:** ischial tuberosity
**insertion:** proximal anteromedial tibia, (pes anserine)
**nerve:** sciatic
**action:** extends thigh; flexes and medially rotates leg

## Semitendinosus

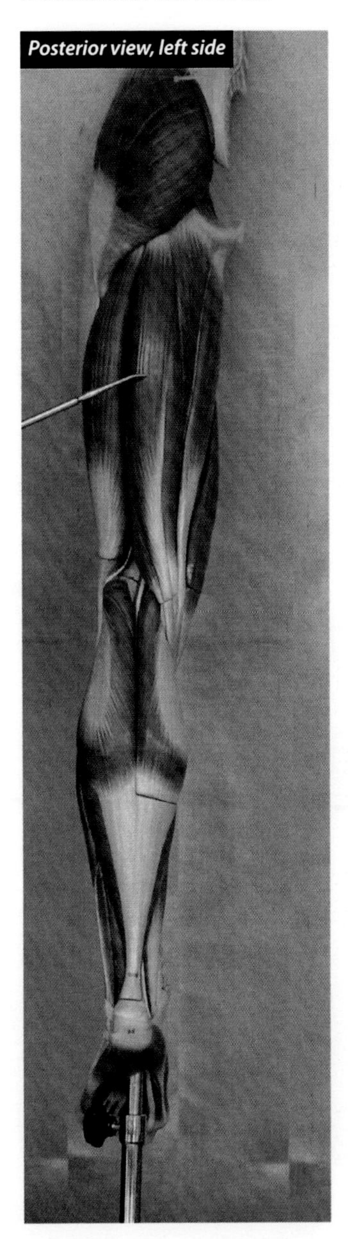


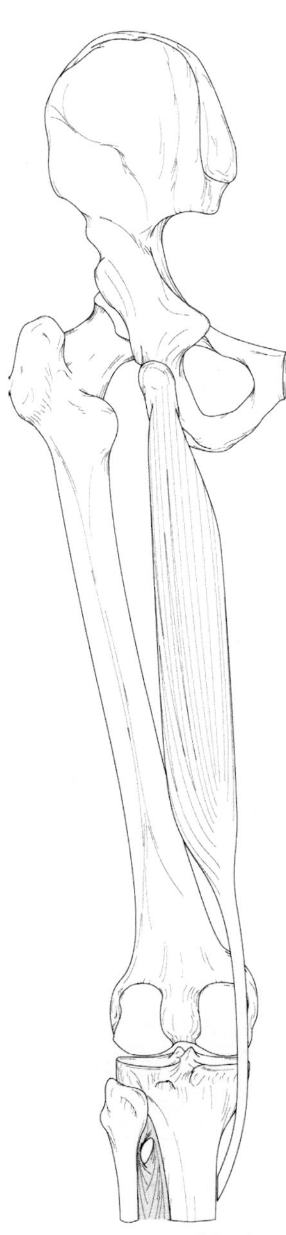

Posterior view, left side

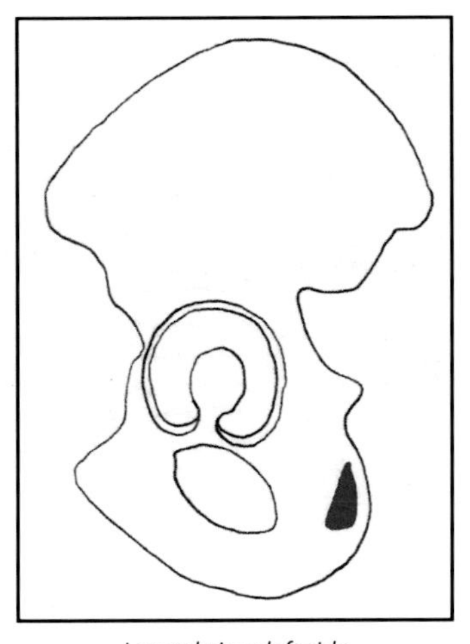
Lateral view, left side

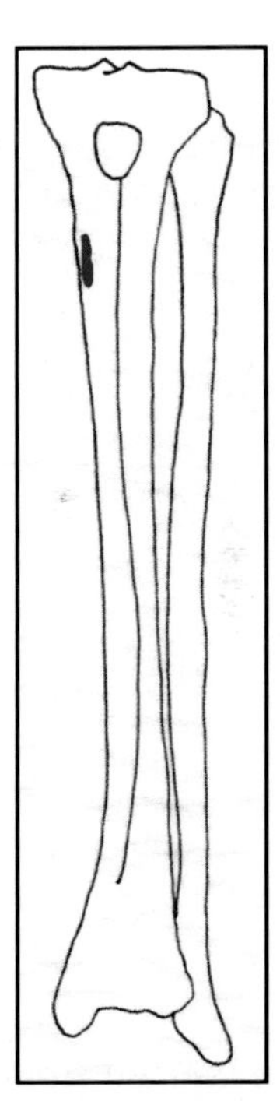
Anterior view, left side

*The above drawings of the origin and insertion might help you visualize this information (red is the origin, blue the insertion).*

The semitendinosus is one of the hamstring muscles. These muscles are in the posterior compartment of the thigh. It is a synergist to the semimembranosus, the biceps femoris, the gastrocnemius, the gracilis, and the sartorius. Dr. J's students sometimes refer to this muscle as the Three Dog Night muscle because of that group's famous song "Try a Little Semitendinosus." Wow. If you know the mnemonic for the thigh, this is "Some" from "Some Sailors Admire Attractive Parrots."

## Semimembranosus

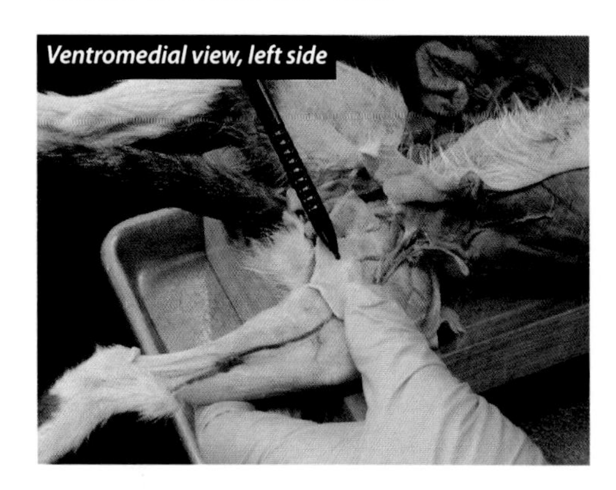


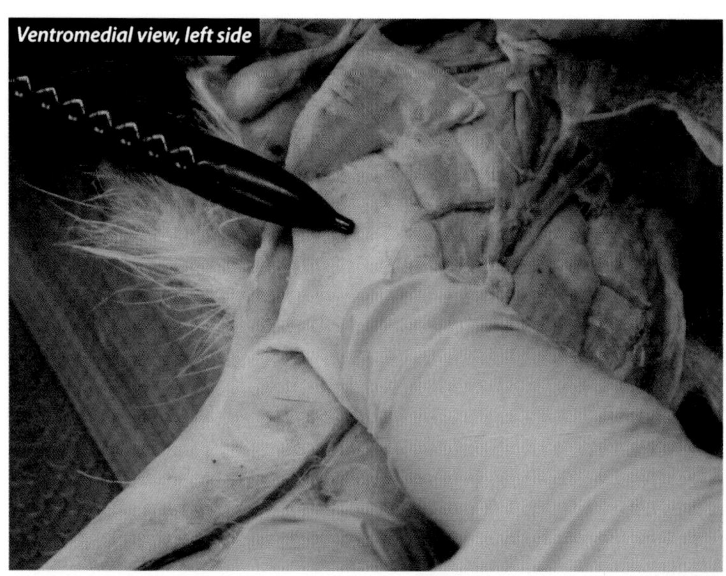


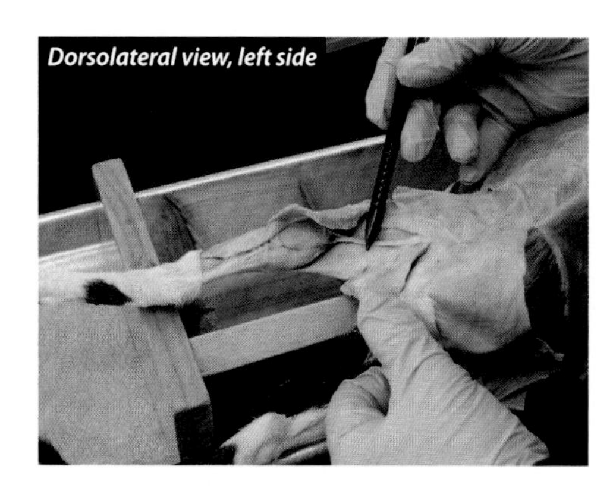


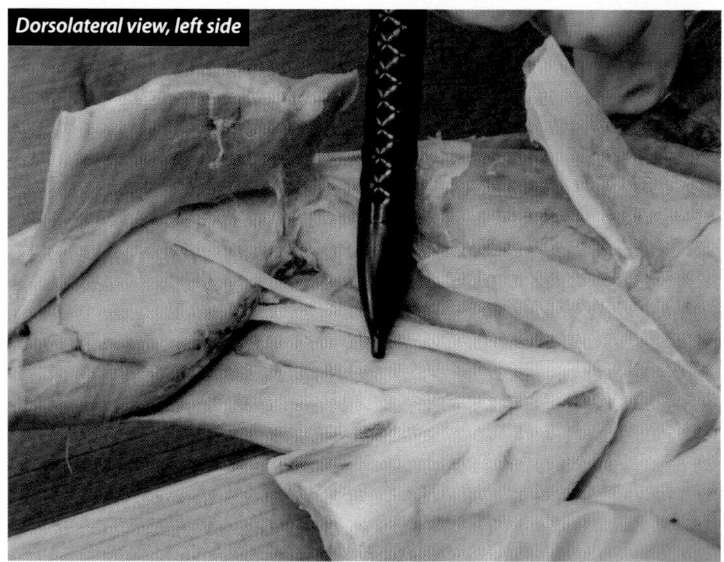


### *Human Information:*

**origin:** ischial tuberosity
**insertion:** medial tibial condyle (posterior surface)
**nerve:** sciatic
**action:** extends thigh; flexes and medially rotates leg

## Semimembranosus

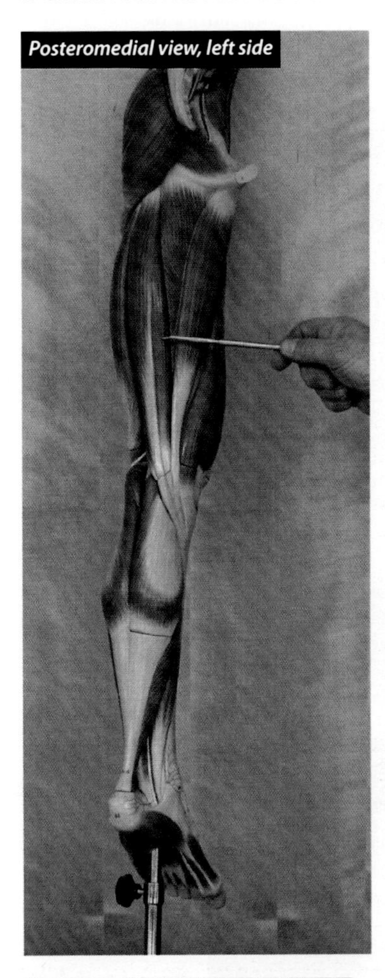
*Posteromedial view, left side*

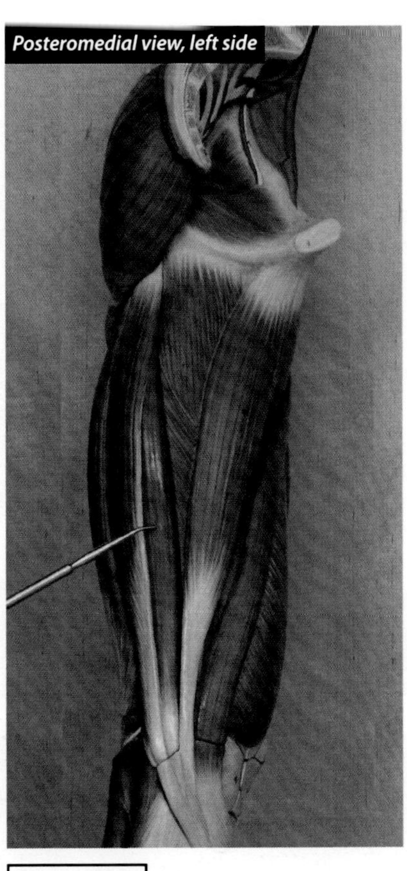
*Posteromedial view, left side*

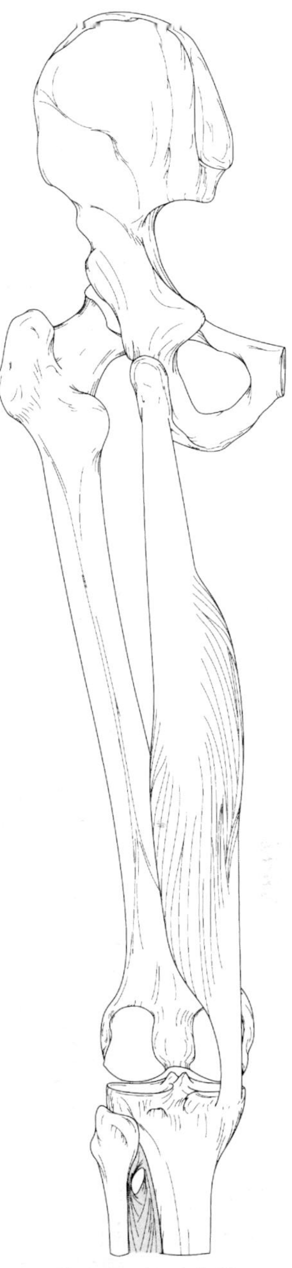
*Posterior view, left side*

*Lateral view, left side*

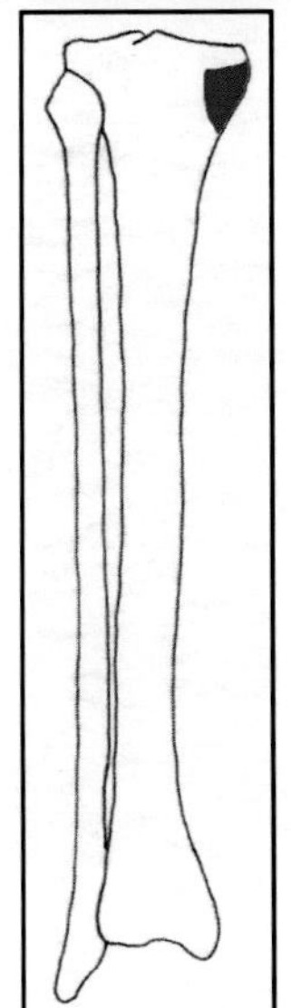
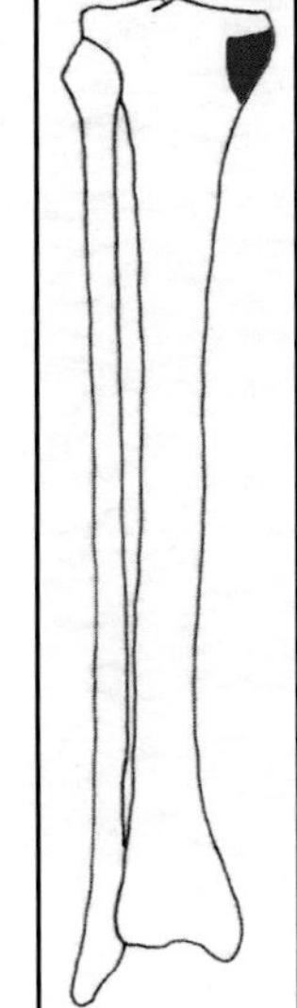
*Posterior view, left side*

*These drawings of the origin and insertion might help you visualize this information (red is the origin, blue the insertion).*

The semimembranosus is one of the hamstring muscles. These muscles are in the posterior compartment of the thigh. It is a synergist to the semitendinosus, the biceps femoris, the gastrocnemius, the gracilis, and the sartorius. If you know the mnemonic for the thigh, this is "Sailors" from "Some Sailors Admire Attractive Parrots."

# Adductors

## Gracilis

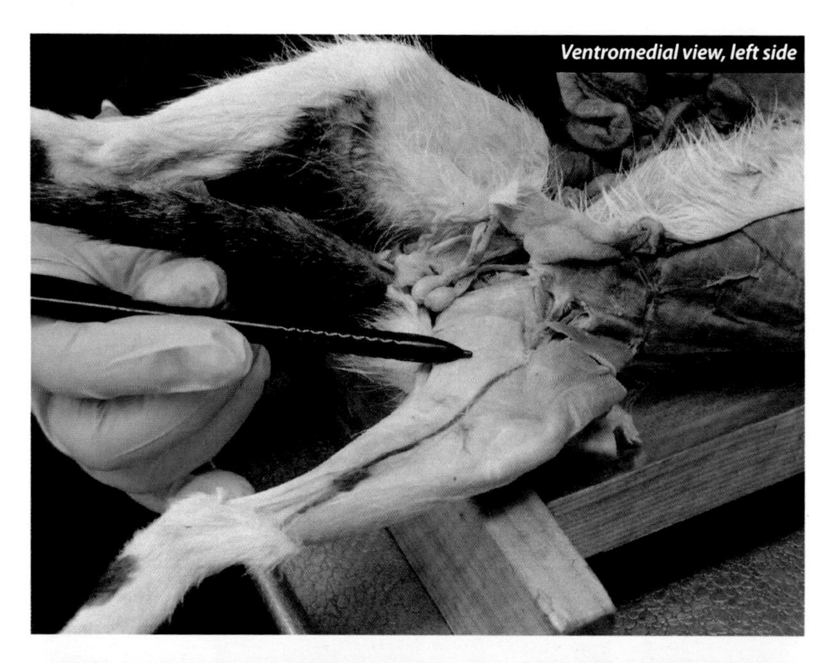


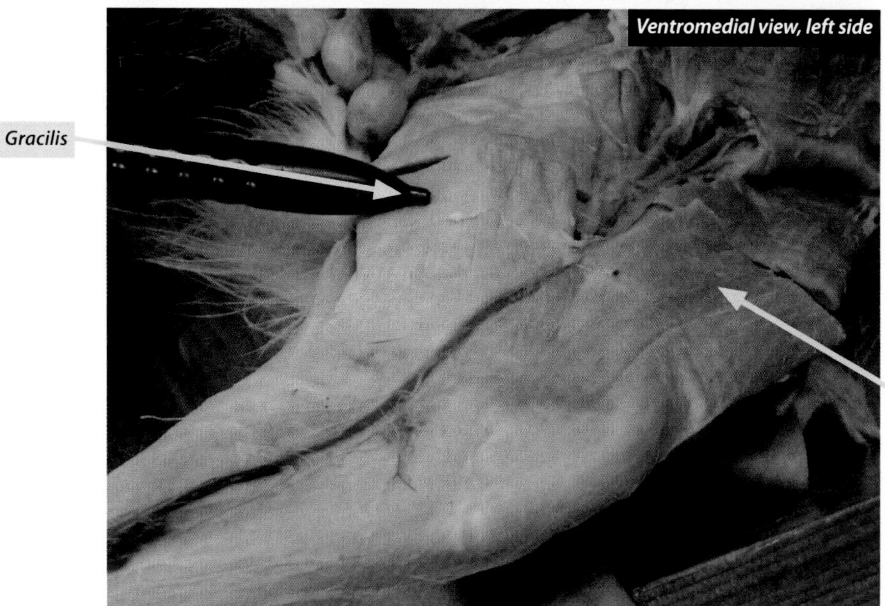


***Human Information:***

**origin:** pubis

**insertion:** proximal anteromedial tibia (pes anserine)

**nerve:** obturator

**action:** adducts, flexes, and medially rotates thigh; flexes leg

## Gracilis

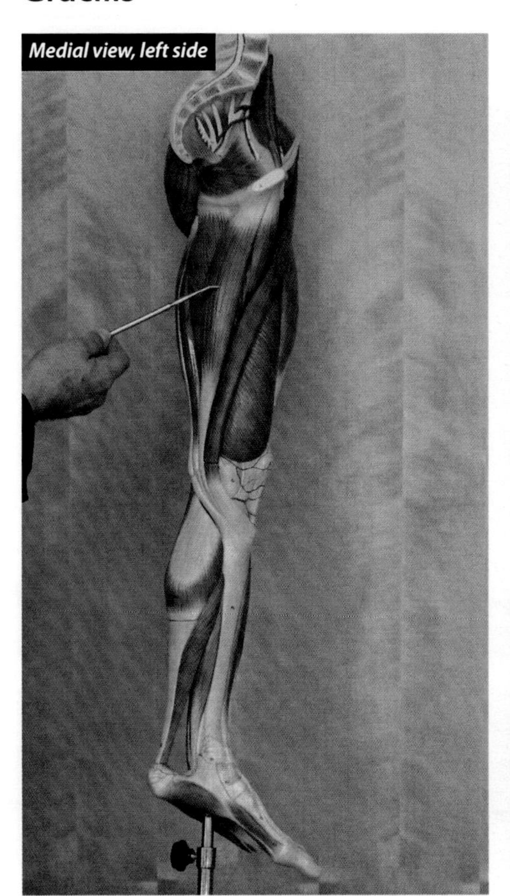


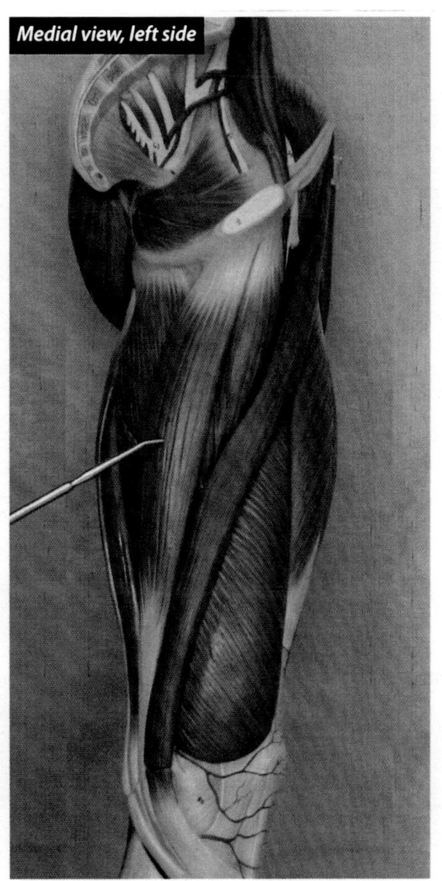


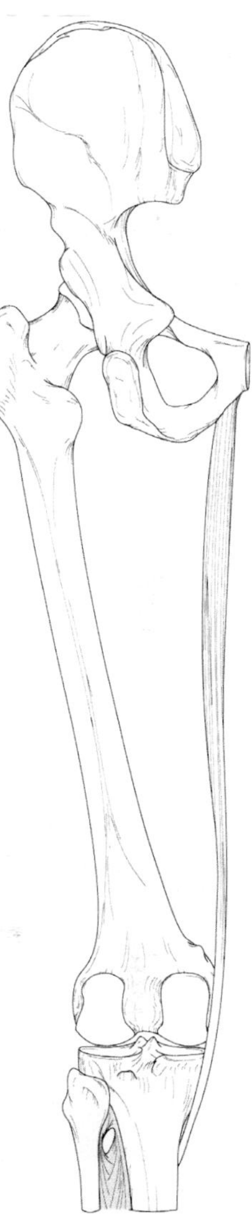
Posterior view, left side

The gracilis is one of the muscles in the pes anserine group. Literally meaning "goose foot," this term refers to a group of three muscles whose tendons share a common bursa and area of insertion on the proximal, anteromedial tibia: sartorius, gracilis, and semitendinosus. They are of functional importance because these muscles flex the knee and also influence tibial rotation. Together, they stabilize the knee medially.

# Adductors

## Adductor Femoris (Magnus et Brevis)

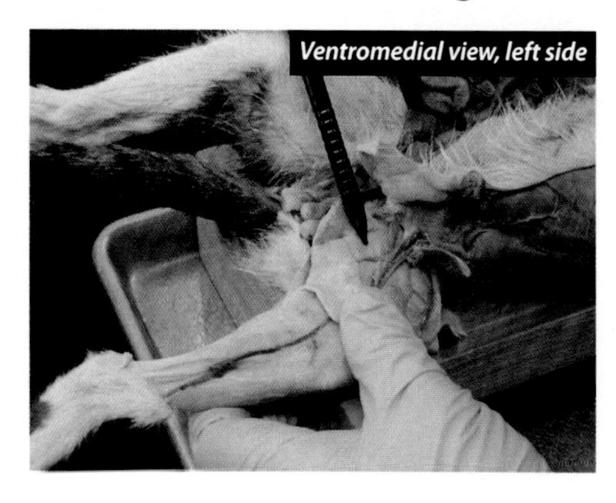


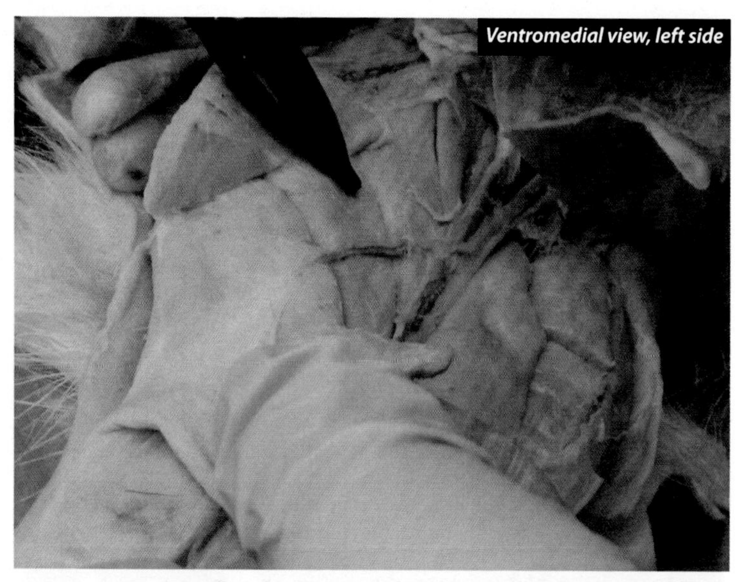


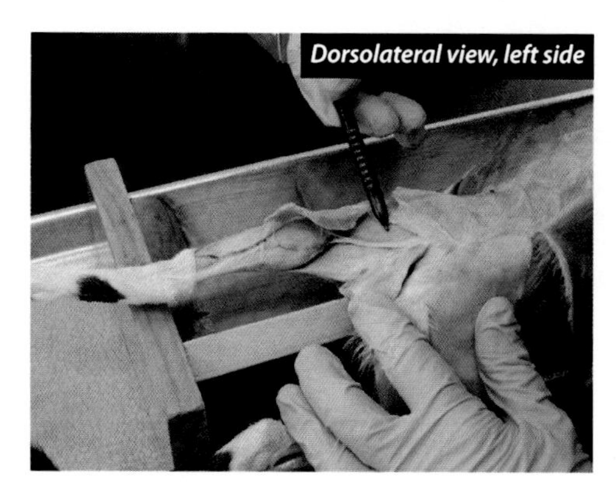


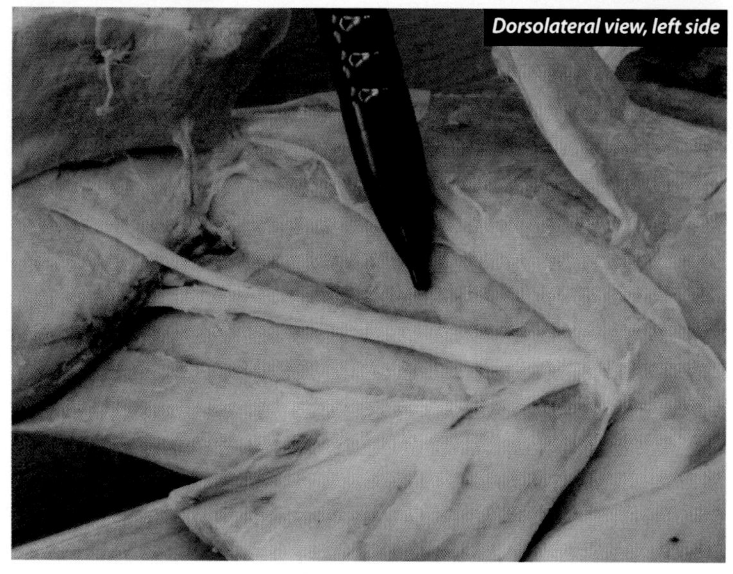


***Human Information:***

**origin:** pubis and ischial tuberosity
**insertion:** linea aspera
**nerve:** obturator
**action:** adducts and flexes thigh

## Adductor Femoris (Magnus et Brevis)

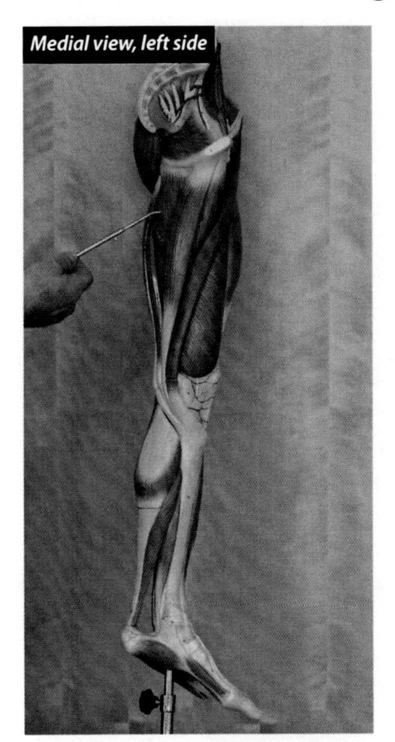


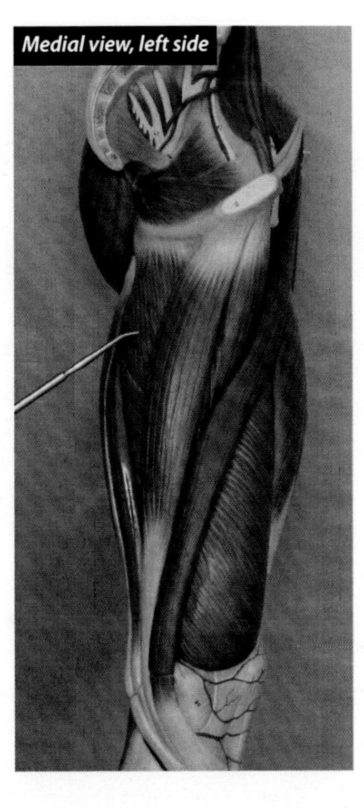


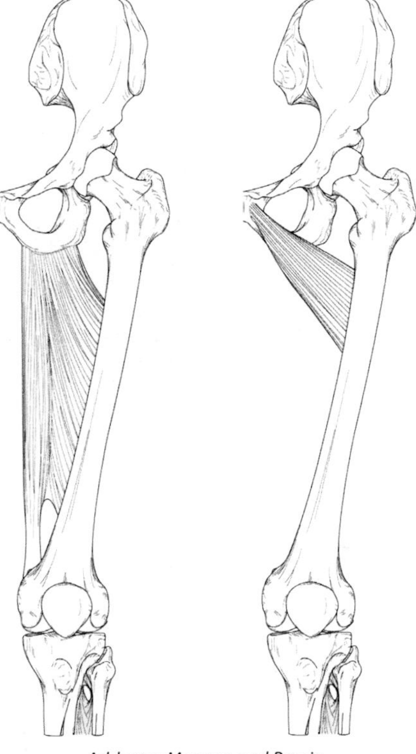

Adductor Magnus and Brevis
Anterior view, left side

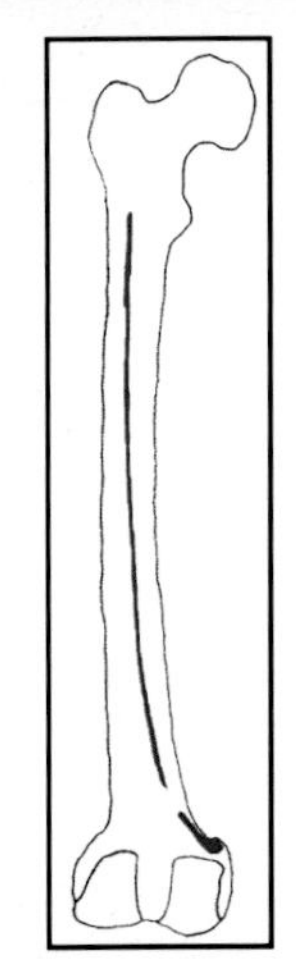

Posterior view, left side

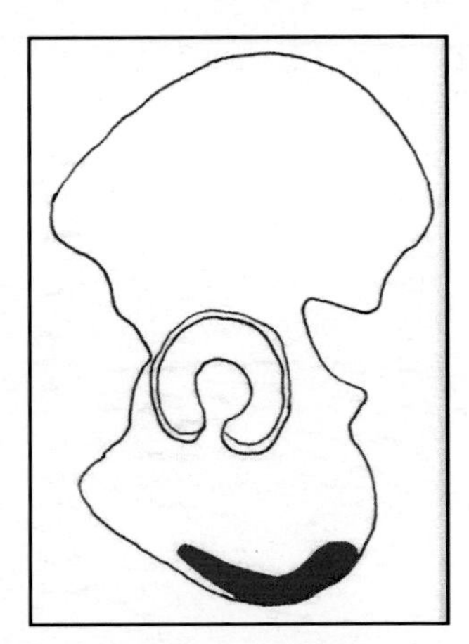

Lateral view, left side

*These drawings of the origin and insertion for adductor femoris magnus might help you visualize this information (red is the origin, blue the insertion).*

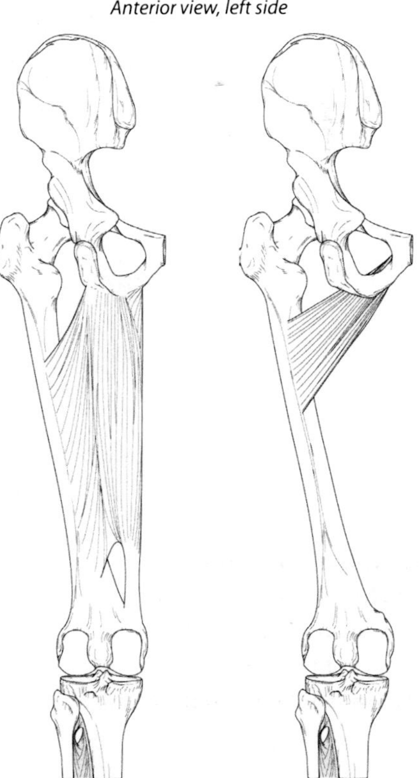

Adductor Magnus and Brevis
Posterior view, left side

The adductor femoris is in the medial compartment of the thigh. If you know the mnemonic for the thigh, this is "Admire" from "Some Sailors Admire Attractive Parrots." It is sometimes called the Tom Selleck muscle because of the association with his TV series *Magnus PI.* It is also occasionally associated with the cartoon *Brevis and Butthead.*

## Adductor Longus

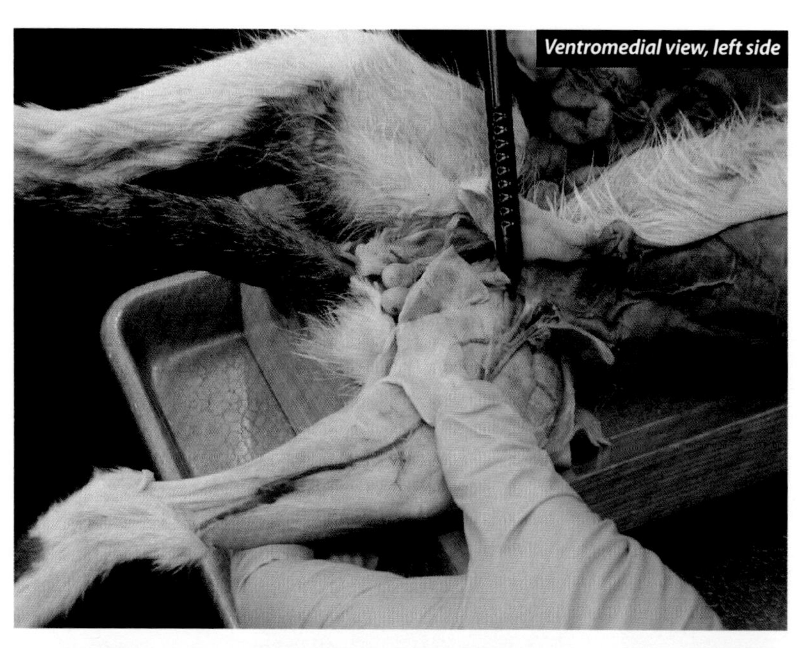


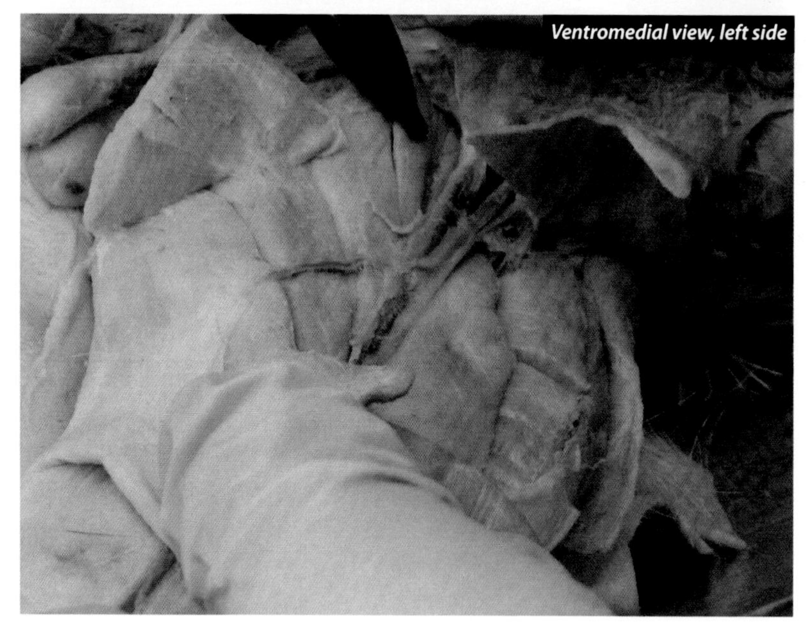


***Human Information:***

**origin:** pubis
**insertion:** linea aspera
**nerve:** obturator
**action:** adducts and flexes thigh

## Adductor Longus

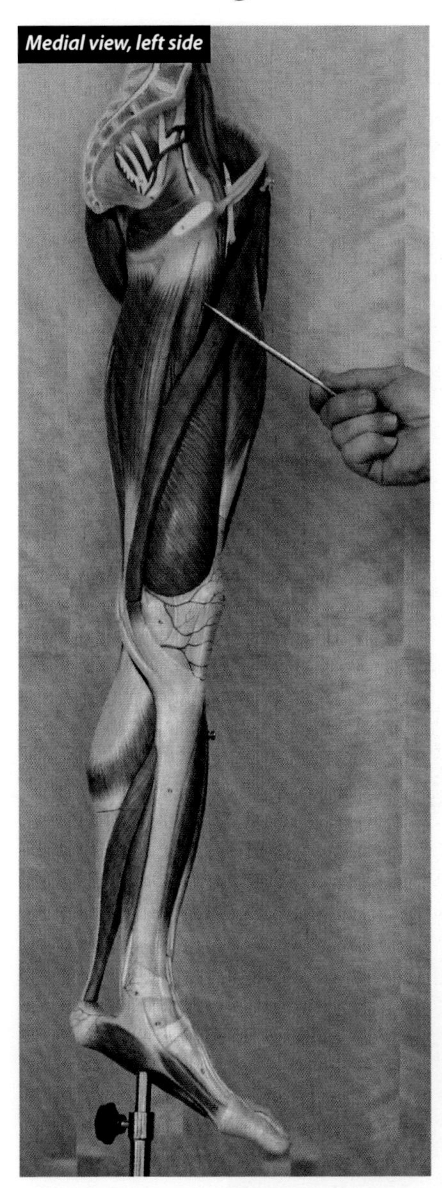


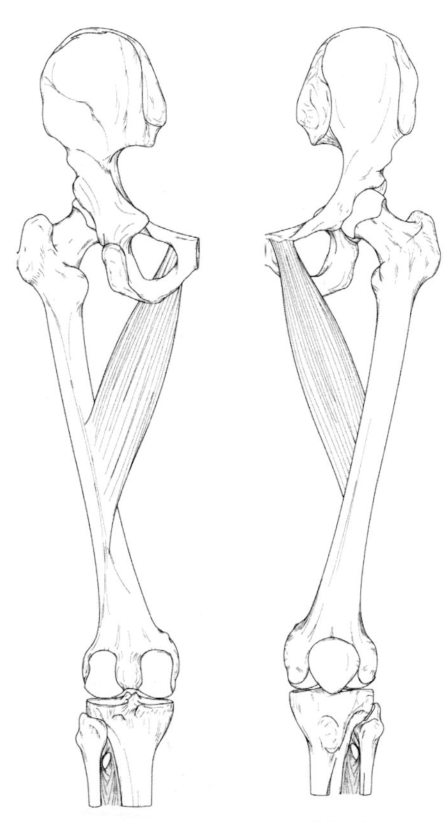

Posterior and anterior views, left side

The adductor longus is in the medial compartment of the thigh. If you know the mnemonic for the thigh, this is "Attractive" from "Some Sailors Admire Attractive Parrots."

## Pectineus

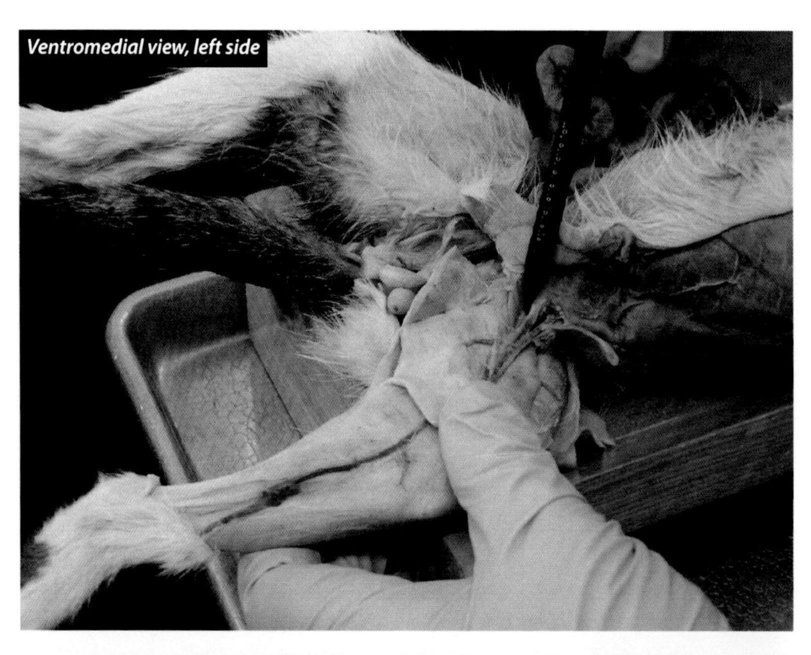


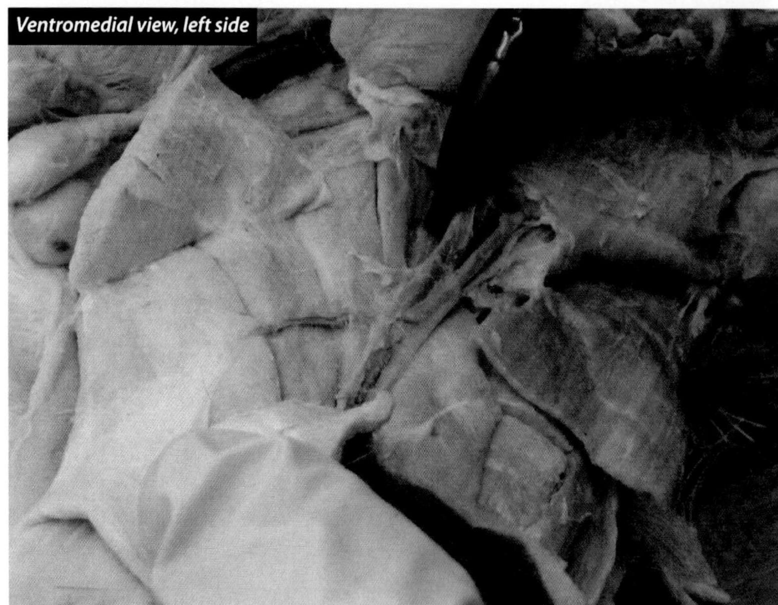


***Human Information:***

**origin:** pubis

**insertion:** line from lesser trochanter to linea aspera (pectineal line).

**nerve:** femoral

**action:** adducts and flexes thigh

## Pectineus

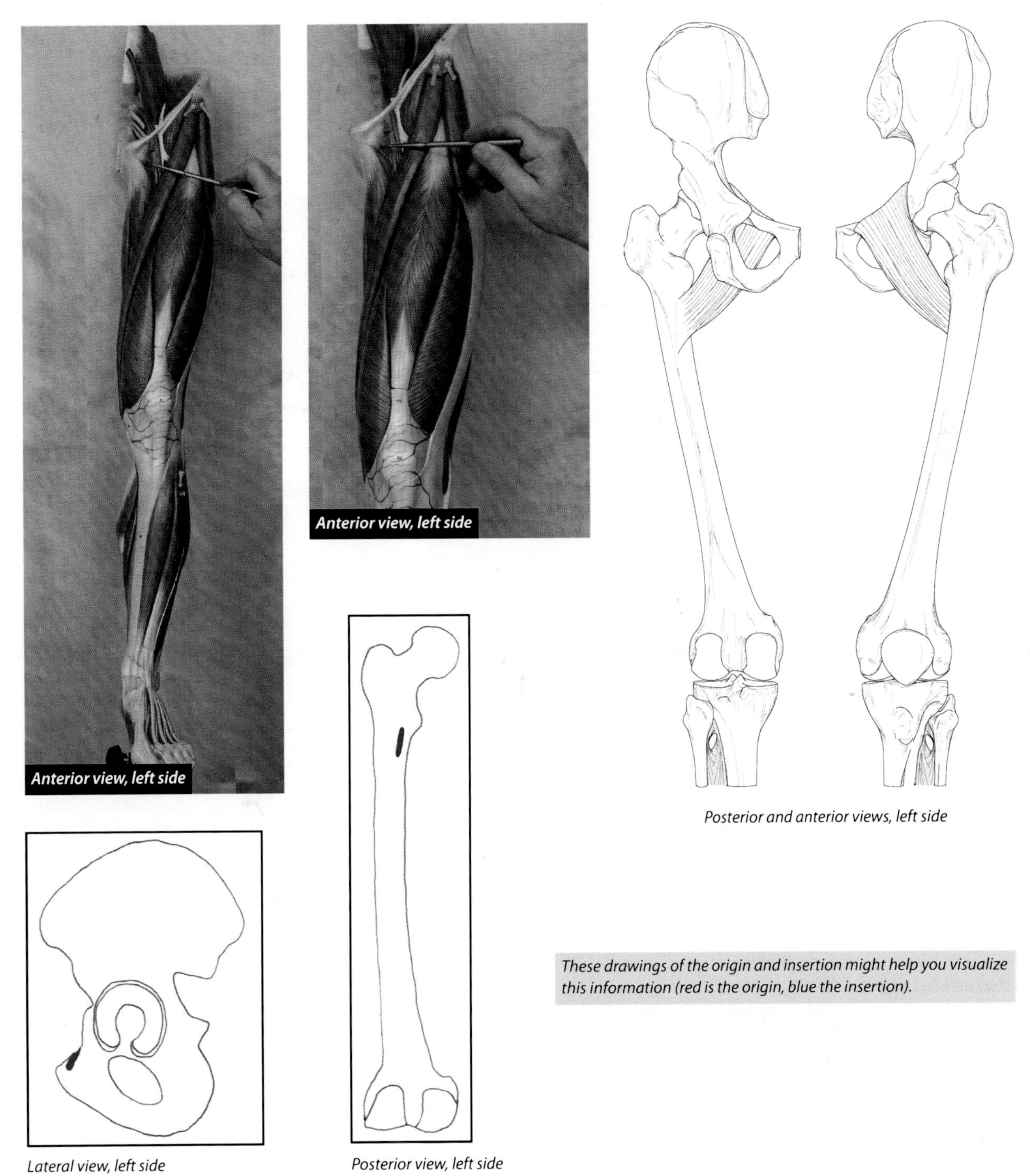

*Anterior view, left side*

*Anterior view, left side*

*Posterior and anterior views, left side*

*Lateral view, left side*

*Posterior view, left side*

*These drawings of the origin and insertion might help you visualize this information (red is the origin, blue the insertion).*

The pectineus is listed by Zanella (*Owner's Manual*) and Snell (medical student text we used in the past) as being in the anterior compartment of the thigh, while Drake et al. list it in the medial compartment of the thigh. Since we are using Drake as a text, we will go with medial compartment. If you know the mnemonic for the thigh, this is "Parrots" from "Some Sailors Admire Attractive Parrots."

## Iliopsoas (Psoas Major and Iliacus)

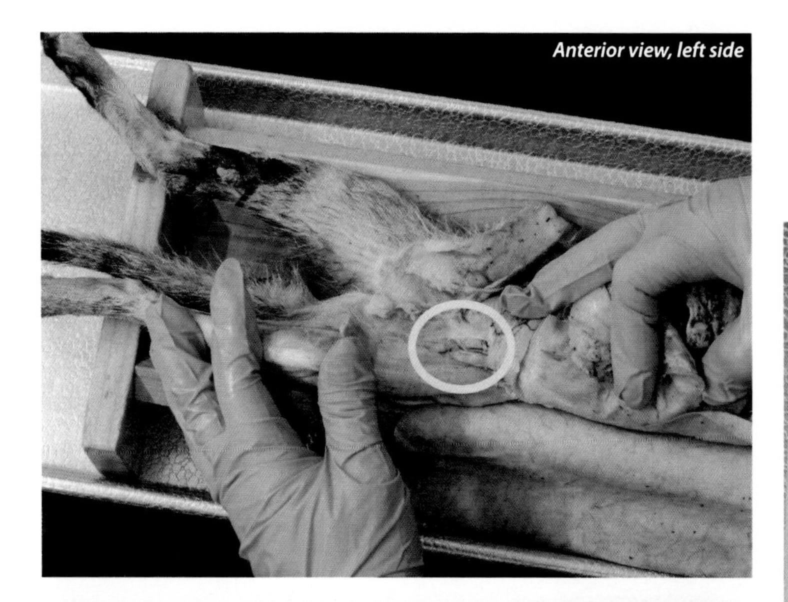

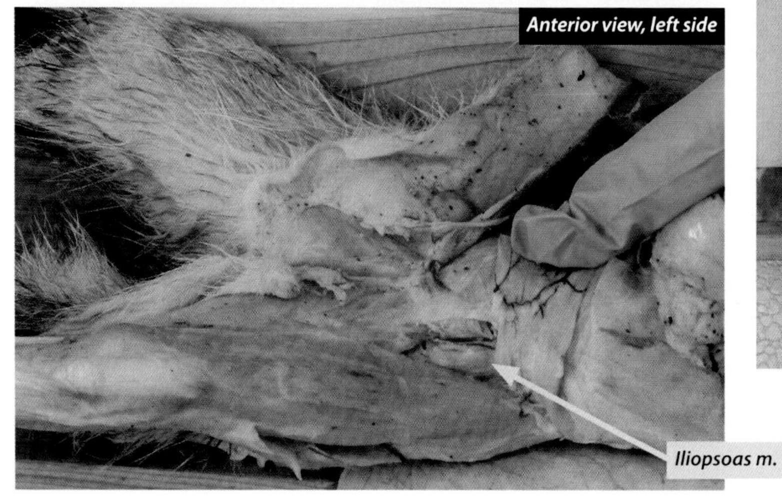


### *Human Information:*

**Psoas Major**
**origin:** lateral surfaces of T12 through L5
**insertion:** lesser trochanter of femur
**nerve:** anterior rami L1–L3
**action:** prime flexor of thigh, flexes and laterally bends lumbar region (trunk)

**Iliacus**
**origin:** iliac fossa
**insertion:** lesser trochanter of femur
**nerve:** femoral
**action:** prime flexor of thigh, flexes, and laterally bends lumbar region (trunk)

# Iliopsoas (Psoas Major and Iliacus)

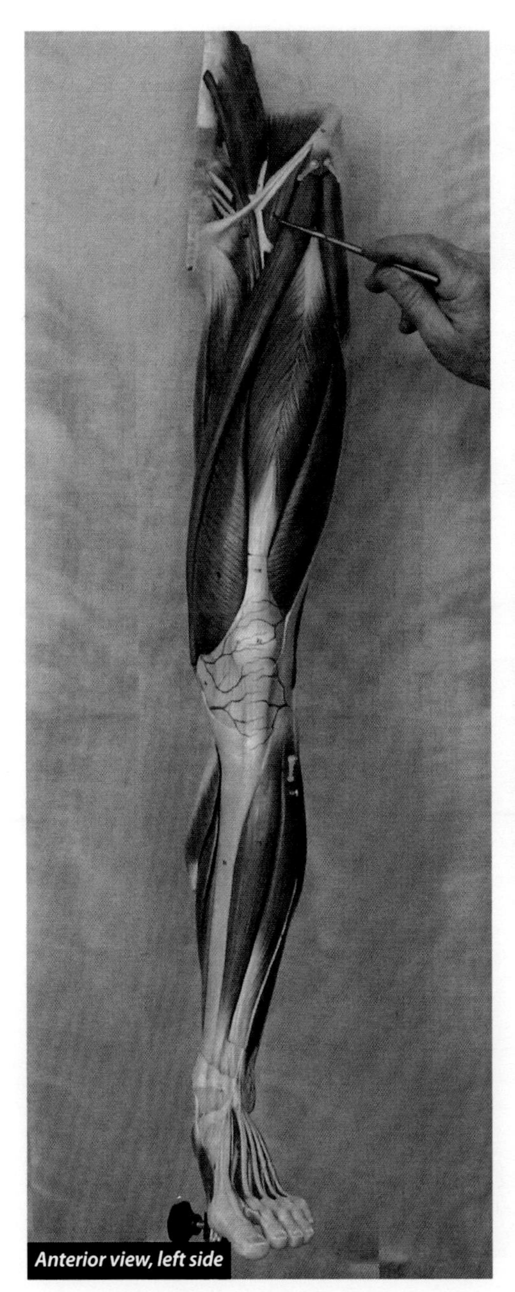

Anterior view, left side

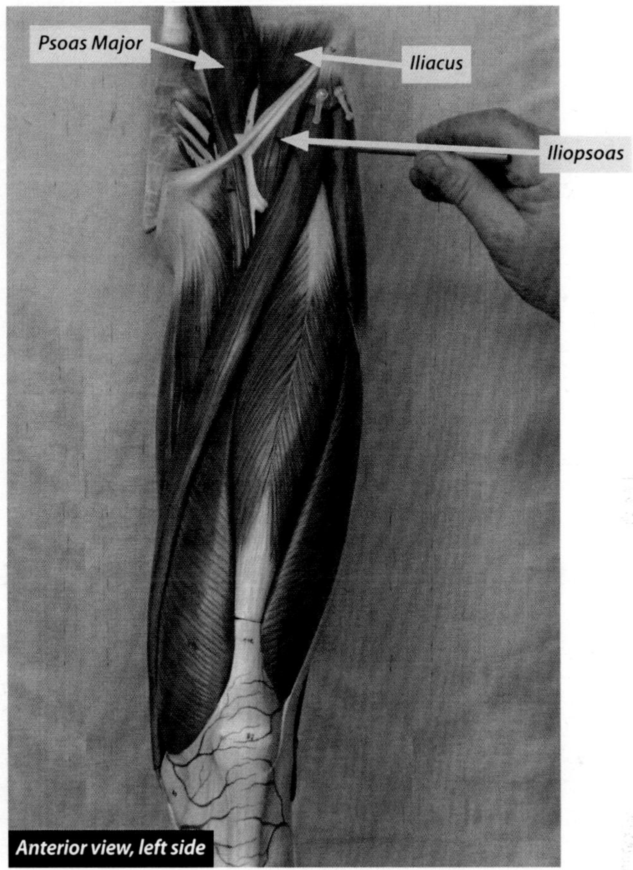


Anterior view, left side

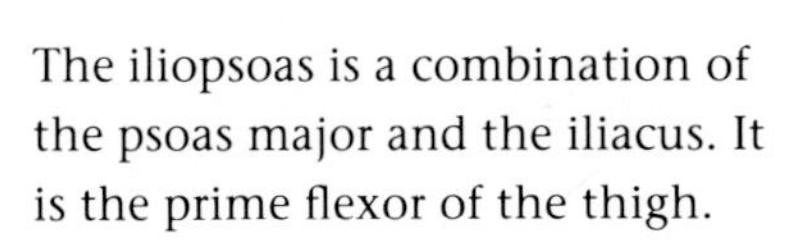

The iliopsoas is a combination of the psoas major and the iliacus. It is the prime flexor of the thigh.

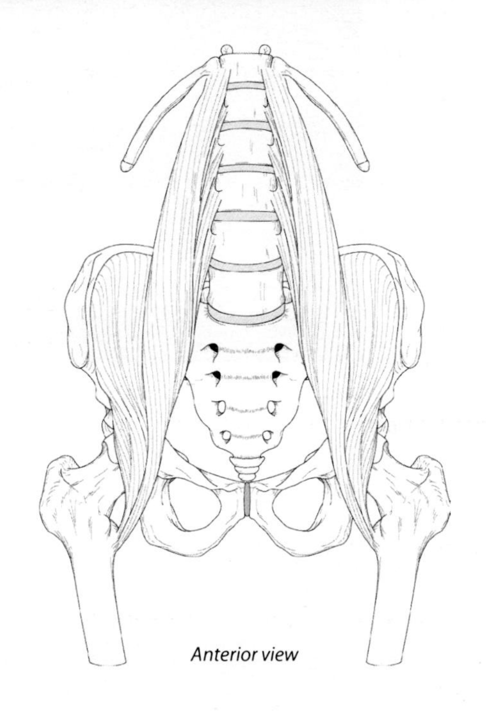

Anterior view

## Sartorius

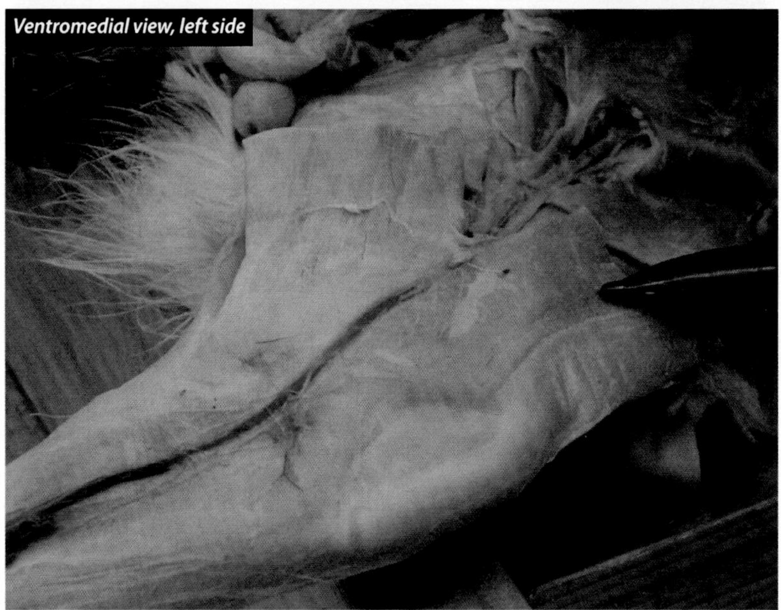


***Human Information:***

**origin:** anterior superior iliac spine (ASIS)

**insertion:** proximal anteromedial tibia (pes anserine)

**nerve:** femoral

**action:** flexes, abducts, and laterally rotates thigh; flexes leg

# Sartorius

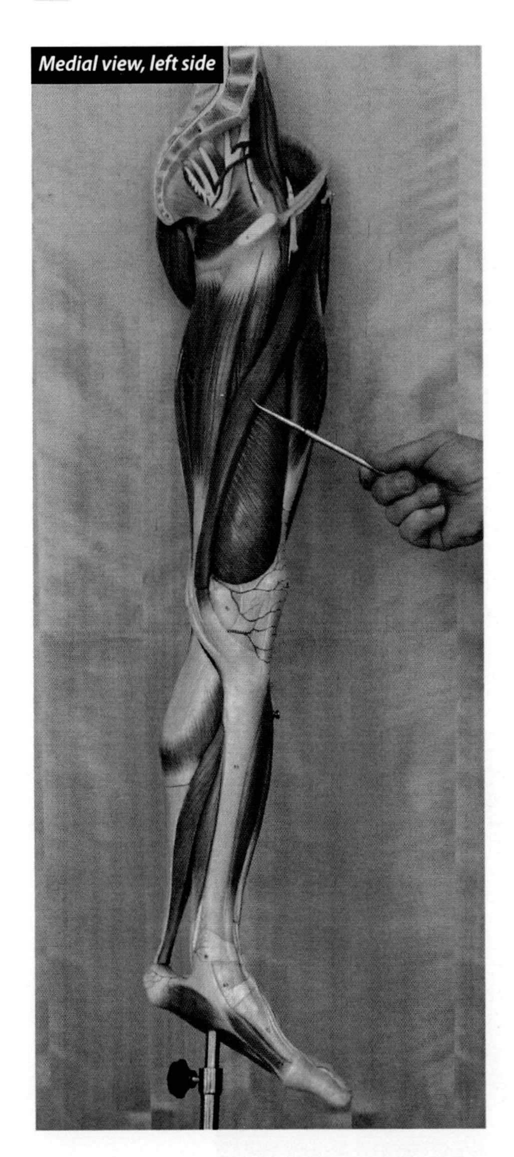


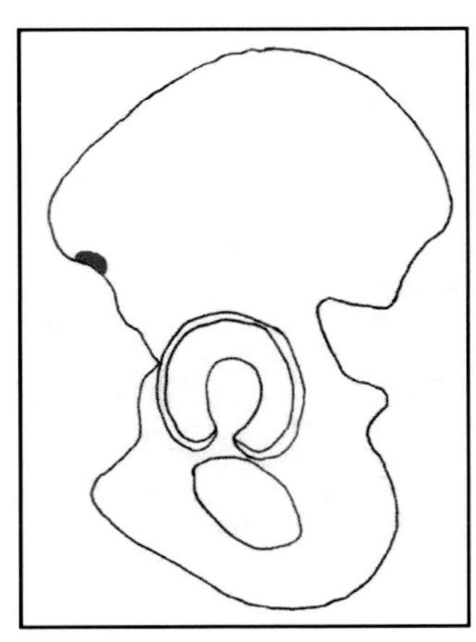

*Lateral view, left side*

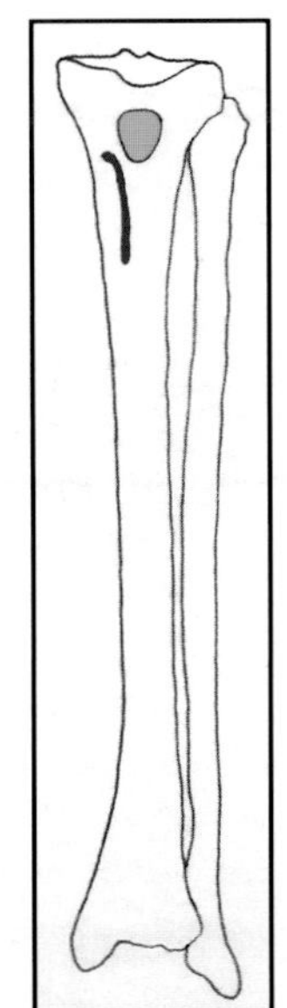

*Anterior view, left side*

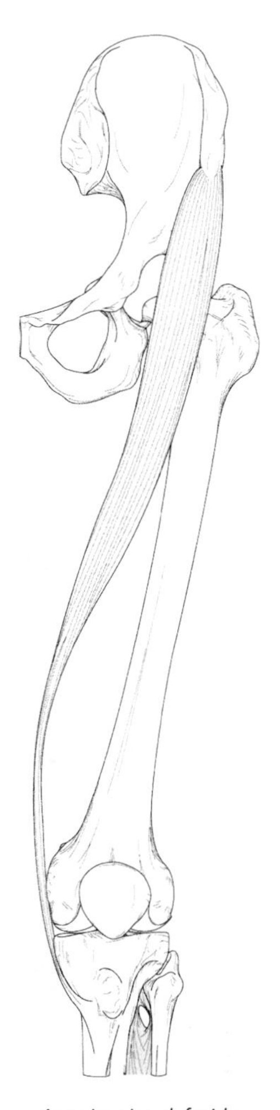

*Anterior view, left side*

*These drawings below of the origin and insertion might help you visualize this information (red is the origin, blue on the tibia is the insertion). The green on the tibia is where the patellar ligament attaches to the tibia. The functional insertion is the tibial tuberosity via the patellar ligament.*

In humans, sartorius is a long, thin, strap-like muscle sometimes classified as the longest muscle in the body. It runs obliquely from the lateral hip to the proximal tibia, just medial to the tibial tuberosity. It is known as the tailor's muscle because it allows one to place an ankle on the thigh to make a workbench, as tailors used to do when they were working. The word "sartorial" also relates to clothing and tailoring; this obscure detail may come in handy on a national exam someday. This muscle is best at performing its actions in combination: hip flexion, abduction, and external rotation, paired with knee flexion. Another analogy that can be used for the actions of this muscle is when you look at the sole of your shoe to see if you have stepped in something—gum, maybe. In the cat, it is a broad, thin muscle on the medial thigh. Sartorius also belongs to the pes anserine group. Literally meaning "goose foot," this term refers to a group of three muscles whose tendons share a common bursa and area of insertion on the proximal, anteromedial tibia: sartorius, gracilis, and semitendinosus. They flex the knee and also influence tibial rotation. Together, these muscles stabilize the knee medially.

# Quadriceps

## Rectus Femoris

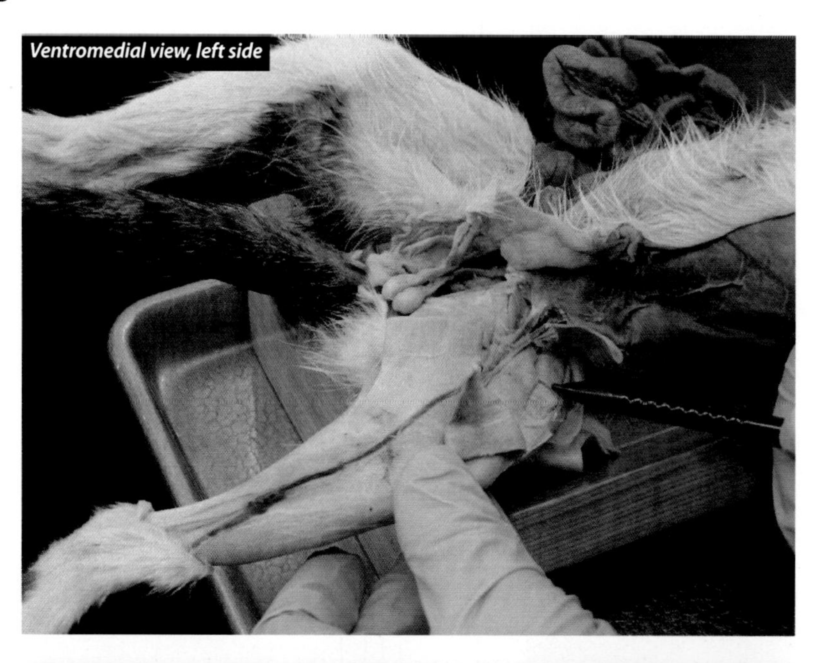
Ventromedial view, left side

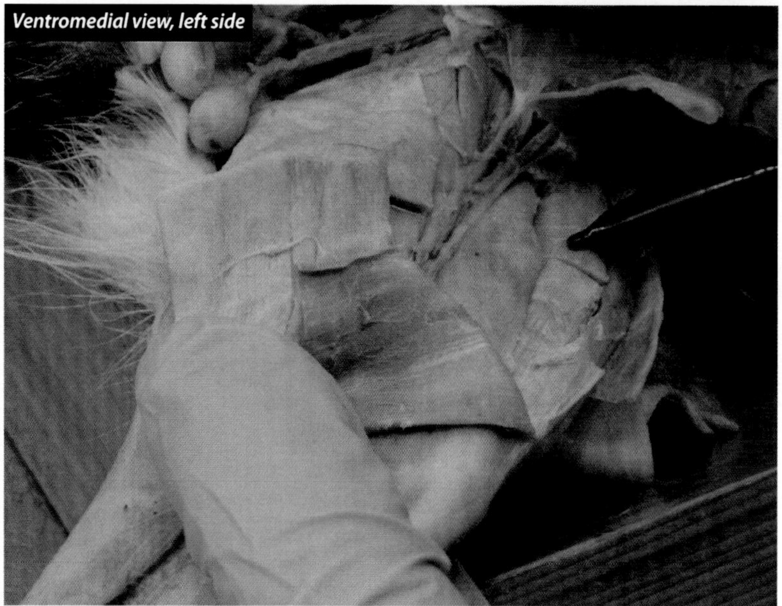
Ventromedial view, left side

***Human Information:***

**origin:** anterior inferior iliac spine (AIIS)
**insertion:** tibial tuberosity via the patellar ligament
**nerve:** femoral
**action:** flexes thigh; extends leg

## Rectus Femoris

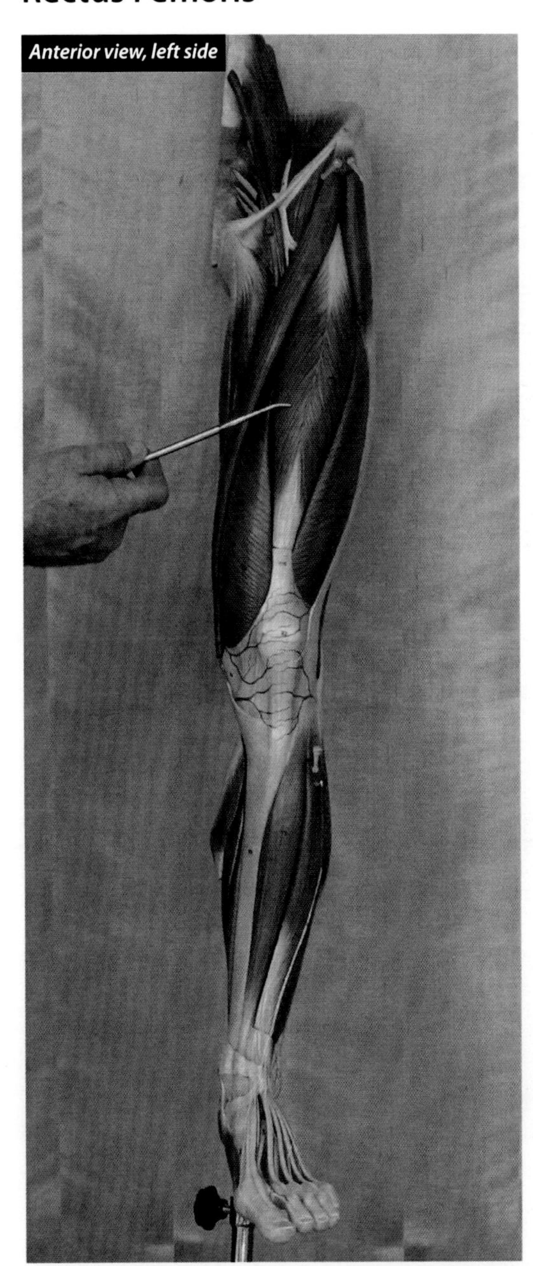
Anterior view, left side

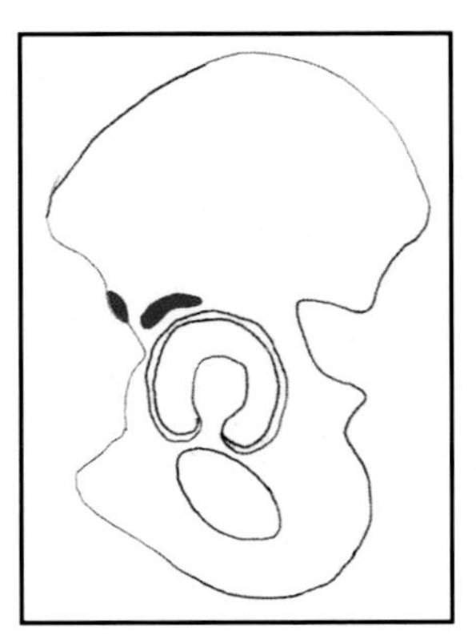
Lateral view, left side

Anterior view, left side

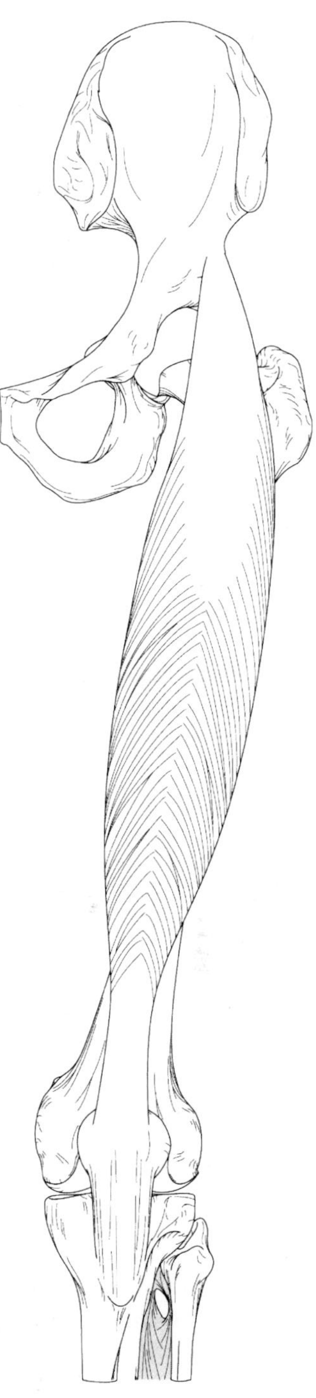
Anterior view, left side

*These drawings of the origin and insertion might help you visualize this information (red is the origin, blue on the tibia is the insertion). The green on the patella is where the patellar ligament attaches to the patella. The functional insertion is on the tibial tuberosity via the patellar ligament.*

The rectus femoris is one of the quadriceps muscles and is in the anterior compartment of the thigh. The quadriceps is functionally important as it is the only extensor of the leg. The hamstring muscles (posterior compartment of thigh) are antagonistic to it. In Dr. J's lab, it is often called the hotdog muscle because of the uncanny similarity to a cocktail wiener! In fact, Dr. J suspects this may be where cocktail wieners come from.

## Vastus Lateralis

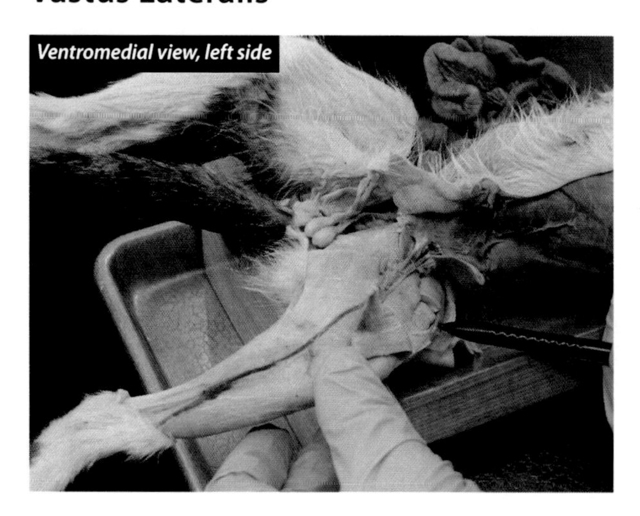


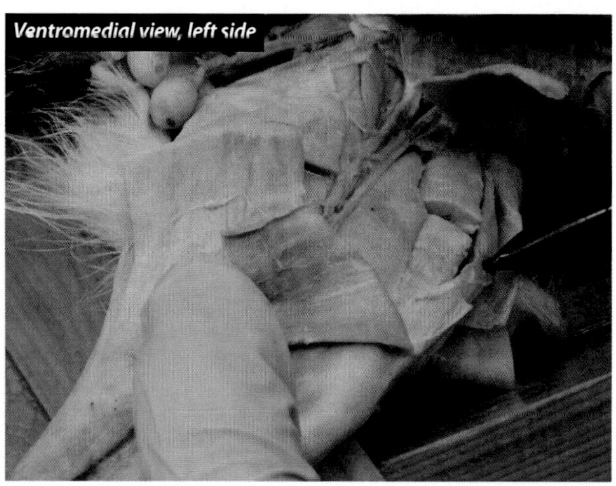


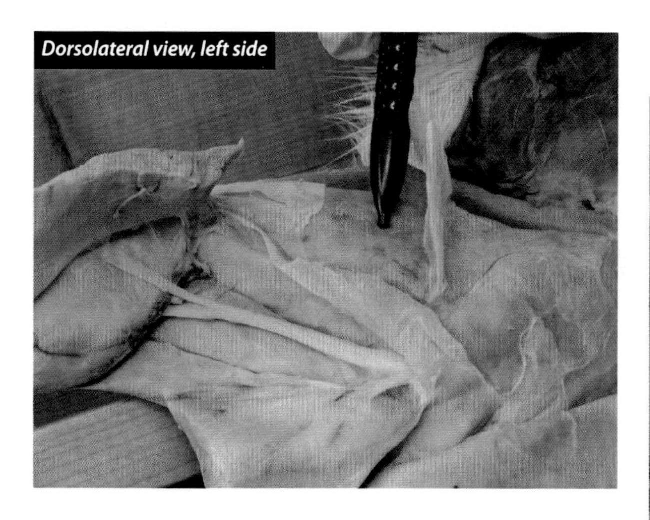


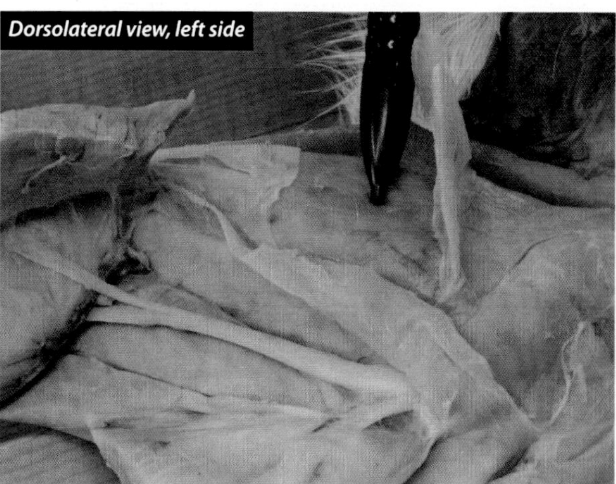


### *Human Information:*

**origin:** linea aspera of femur
**insertion:** tibial tuberosity via the patellar ligament
**nerve:** femoral
**action:** extends leg

## Vastus Lateralis

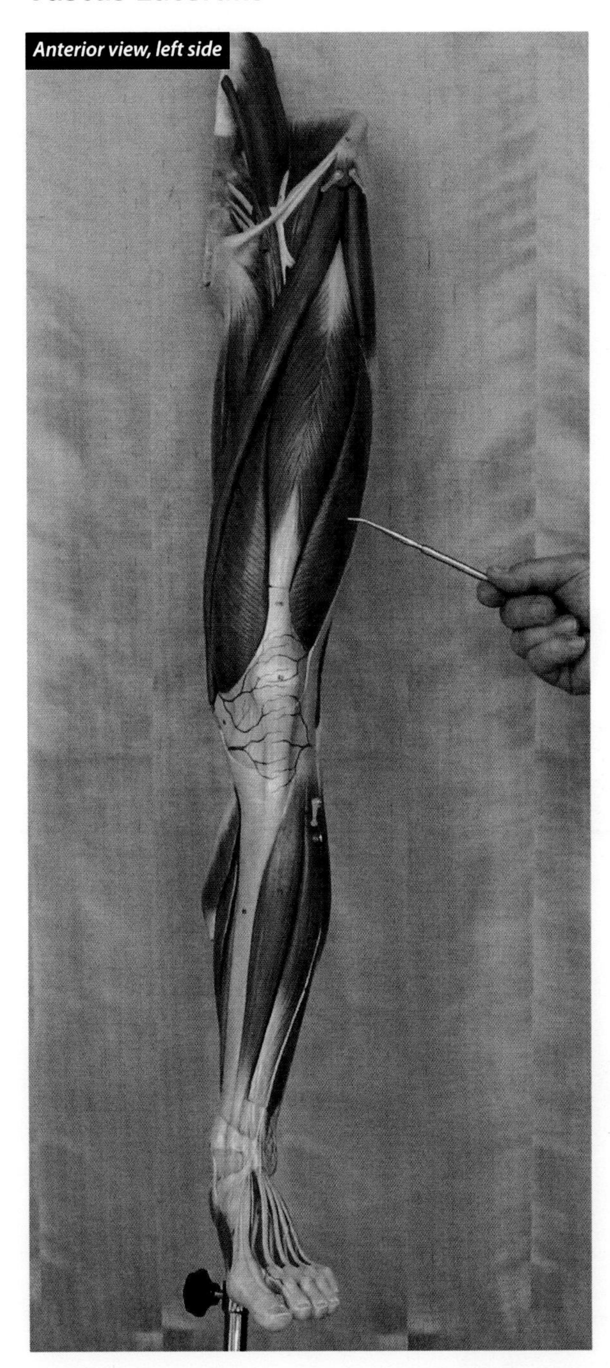


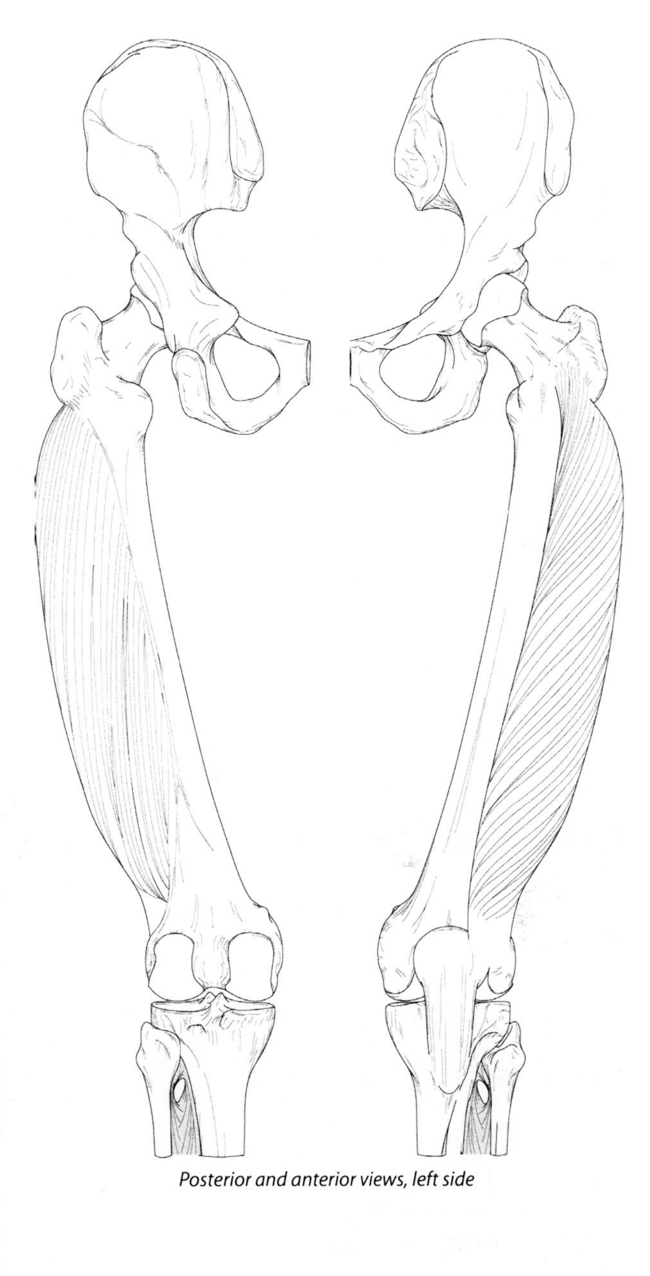

Posterior and anterior views, left side

Vastus lateralis is the lateral head of quadriceps femoris. That would make it a Grant thing. It is the largest of the four heads.

## Vastus Medialis

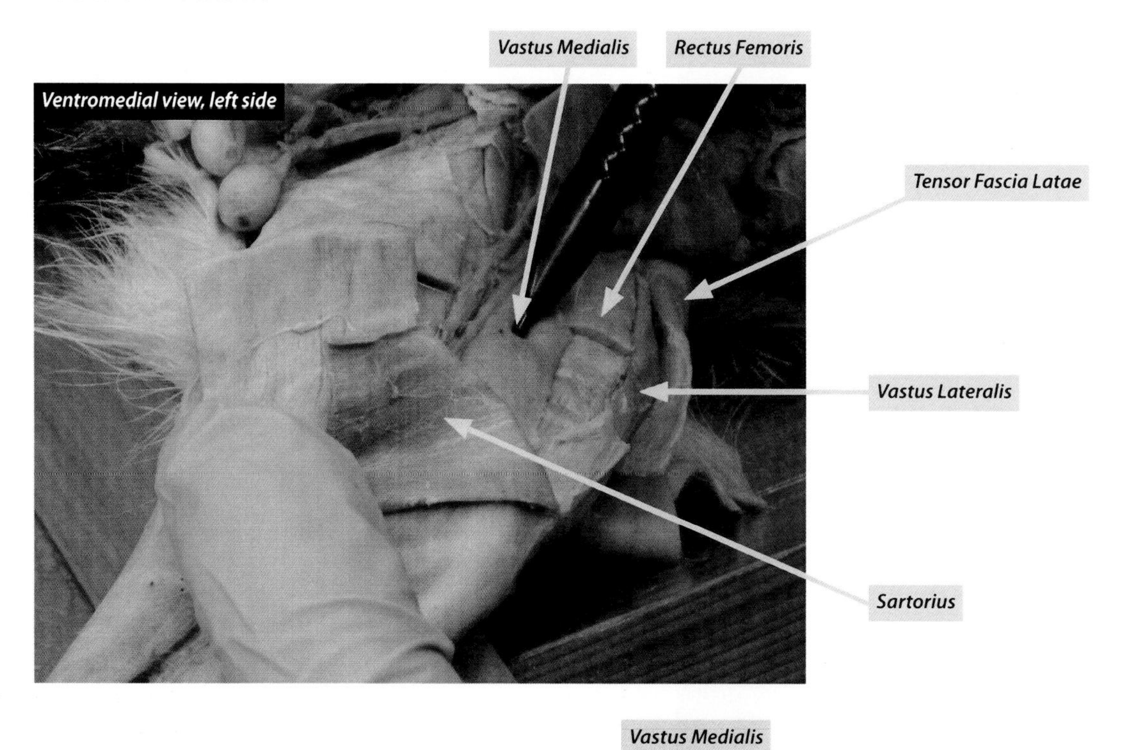


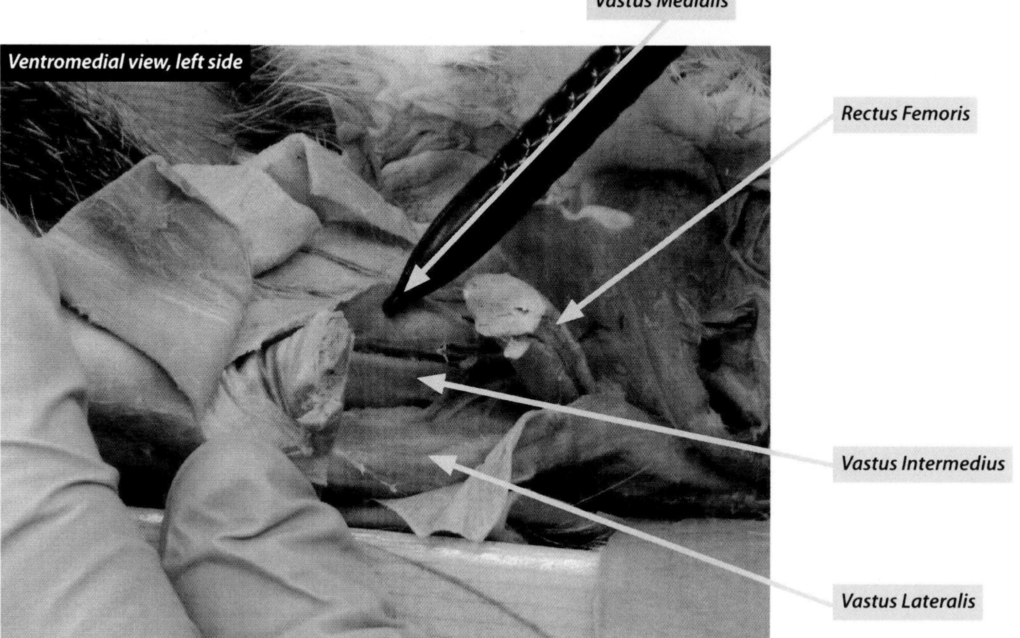


### *Human information:*

**origin:** linea aspera of femur
**insertion:** tibial tuberosity via the patellar ligament
**nerve:** femoral
**action:** extends leg

## Vastus Medialis

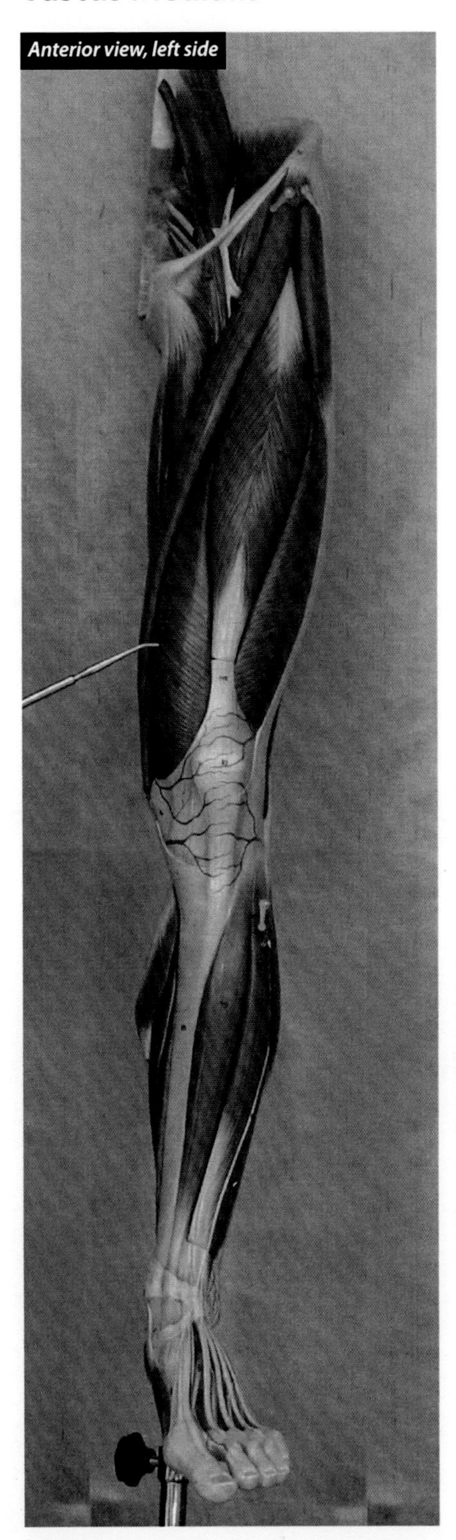


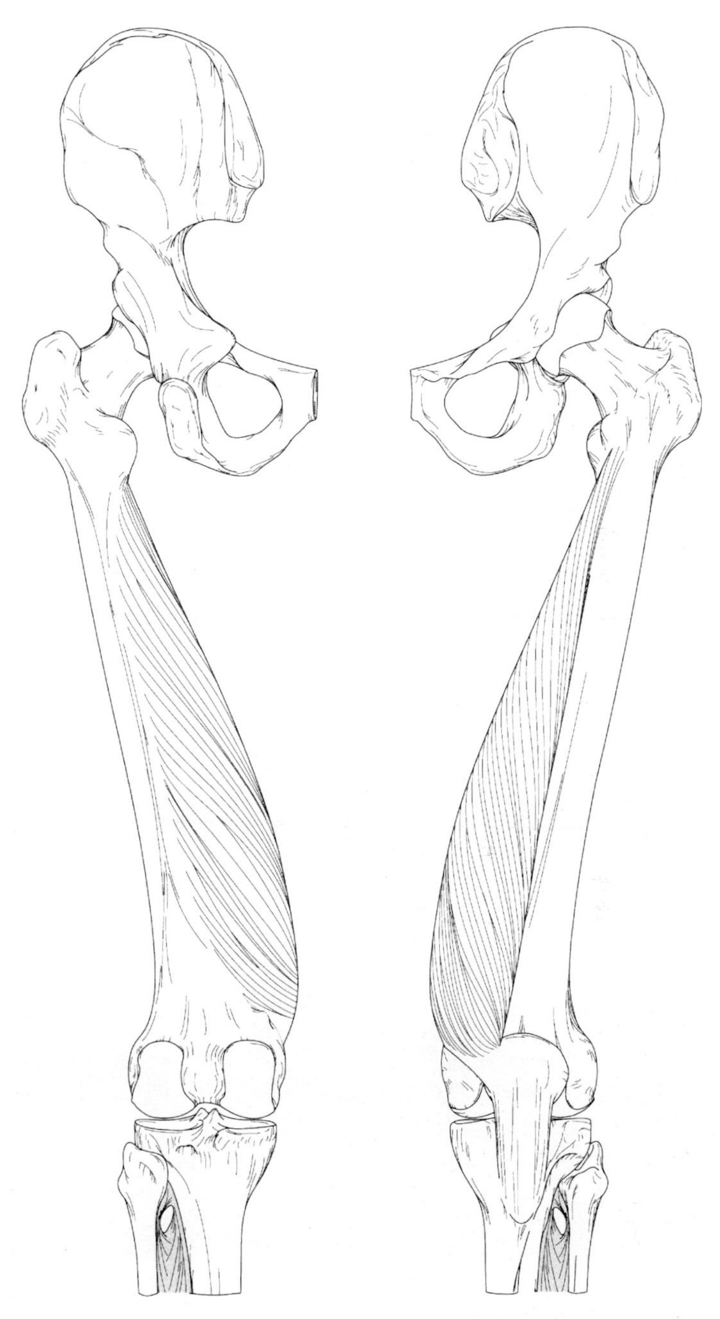

*Posterior and anterior views, left side*

Vastus medialis is the medial head of quadriceps femoris. That would make it a Grant thing.

## Vastus Intermedius

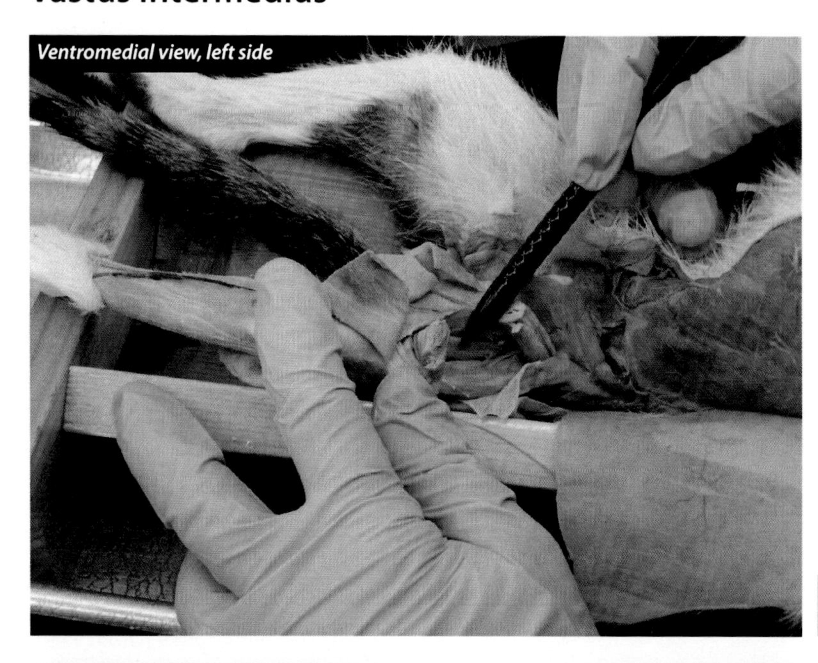
*Ventromedial view, left side*

*Note that rectus femoris is dissected and reflected to reveal vastus intermedius.*

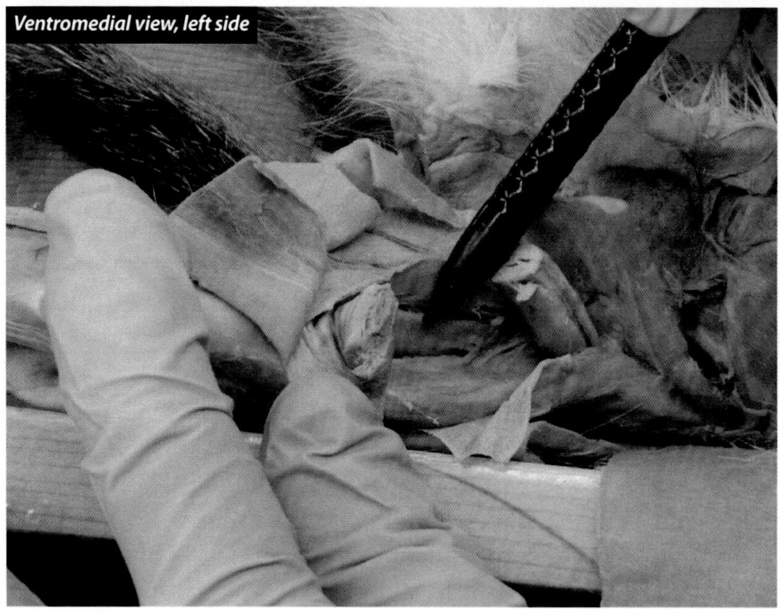
*Ventromedial view, left side*

### *Human Information:*

**origin:** anterior shaft of femur
**insertion:** tibial tuberosity via the patellar ligament
**nerve:** femoral
**action:** extension of leg

## Vastus Intermedius

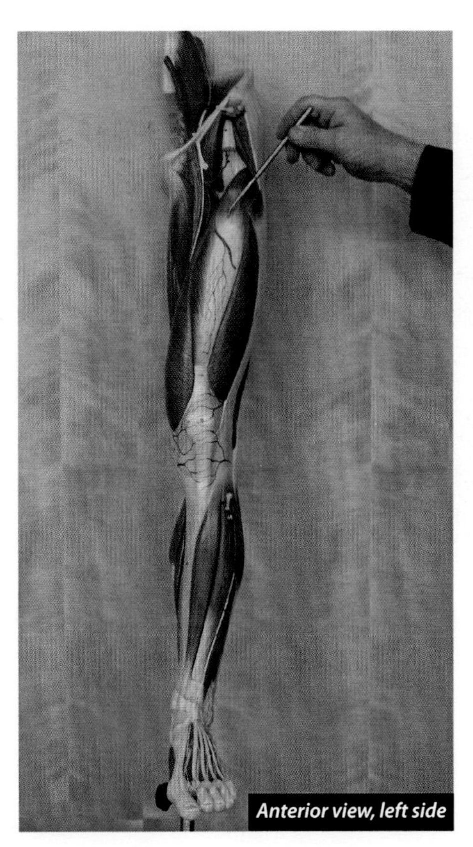

*Anterior view, left side*

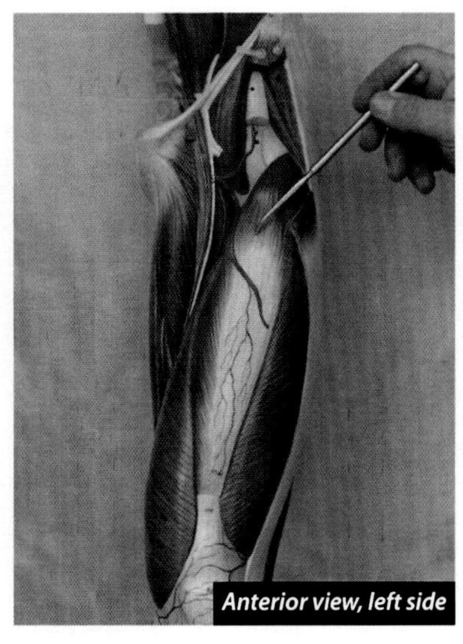

*Anterior view, left side*

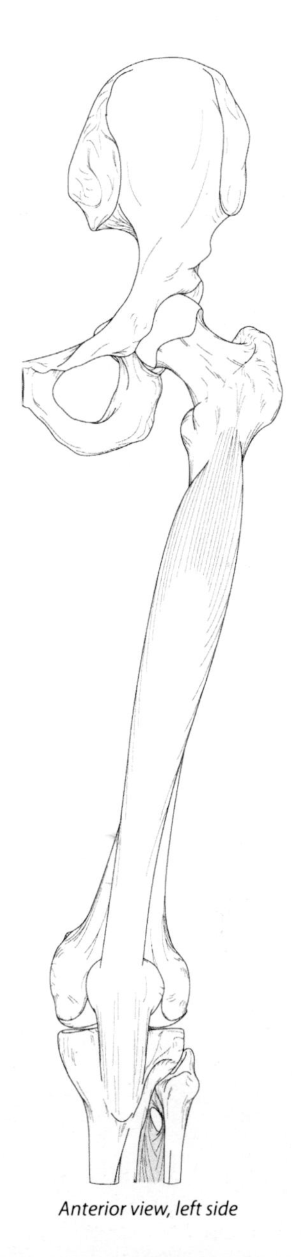

*Anterior view, left side*

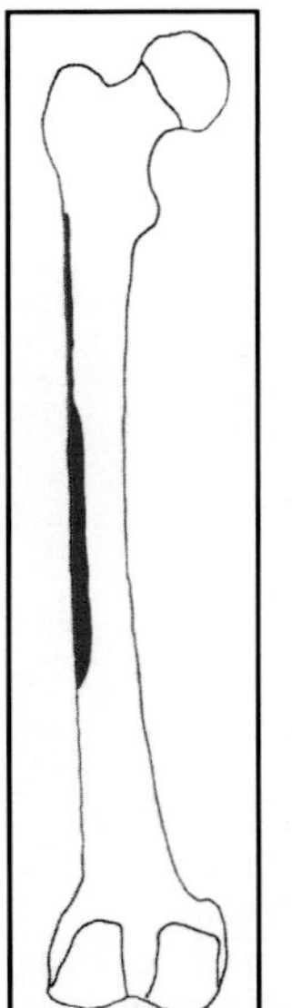

*Posterior view, left side*

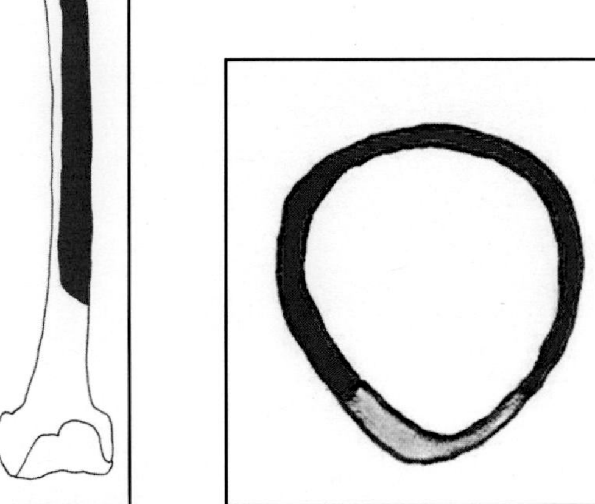

*Anterior view, left side*

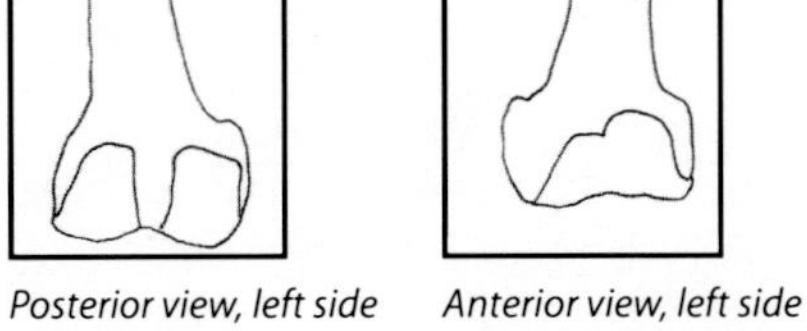

*Anterior view, left side*

*These drawings of the origin and insertion might help you visualize this information (red is the origin, blue on the tibia is the insertion). The green on the patella is where the patellar ligament attaches to the patella. The functional insertion is on the tibial tuberosity via the patellar ligament.*

The vastus intermedius is deep to the rectus femoris so it can only be seen when the rectus femoris is displaced. In the analogy of vastus muscles forming a "hot dog bun," it is the floor of the bun.

# Muscles of the Leg

## Gastrocnemius

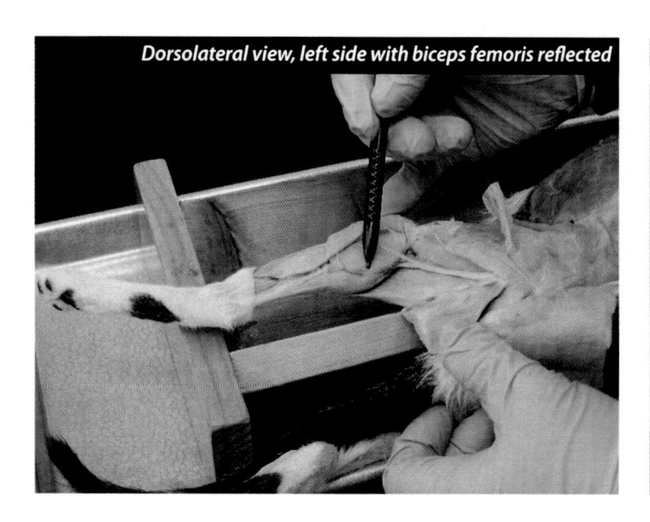
Dorsolateral view, left side with biceps femoris reflected

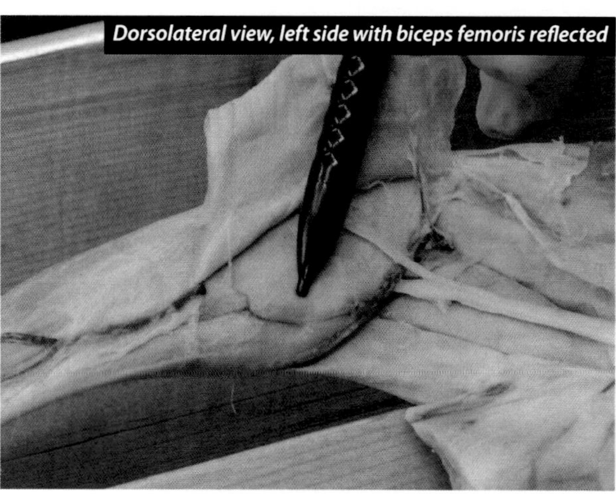
Dorsolateral view, left side with biceps femoris reflected

### Human Information:

**origin:** medial and lateral femoral condyles
**insertion:** calcaneus via calcaneal tendon
**nerve:** tibial
**action:** flexes leg; plantar flexes foot

Some students refer to this as the high-heel muscle.

## Calcaneal (Achilles) Tendon (Gastrocnemius)

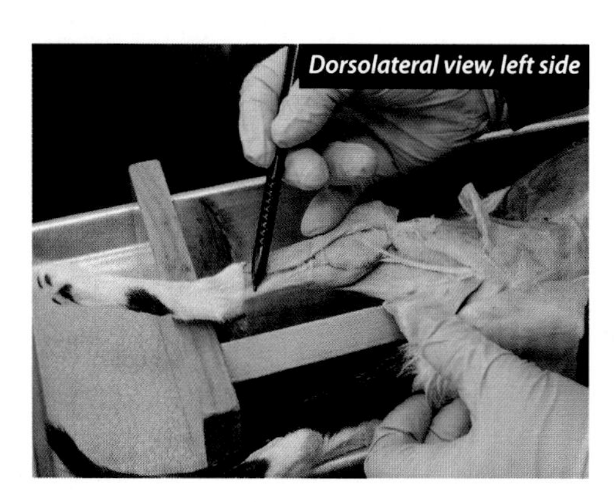
Dorsolateral view, left side

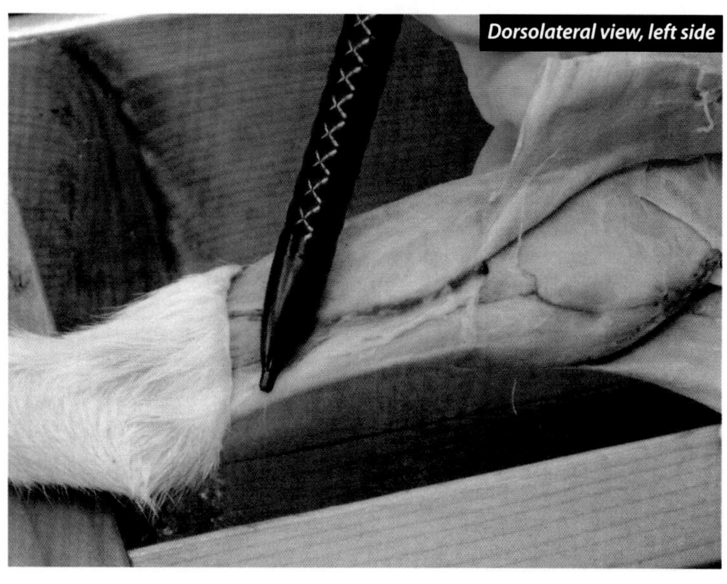
Dorsolateral view, left side

## Gastrocnemius

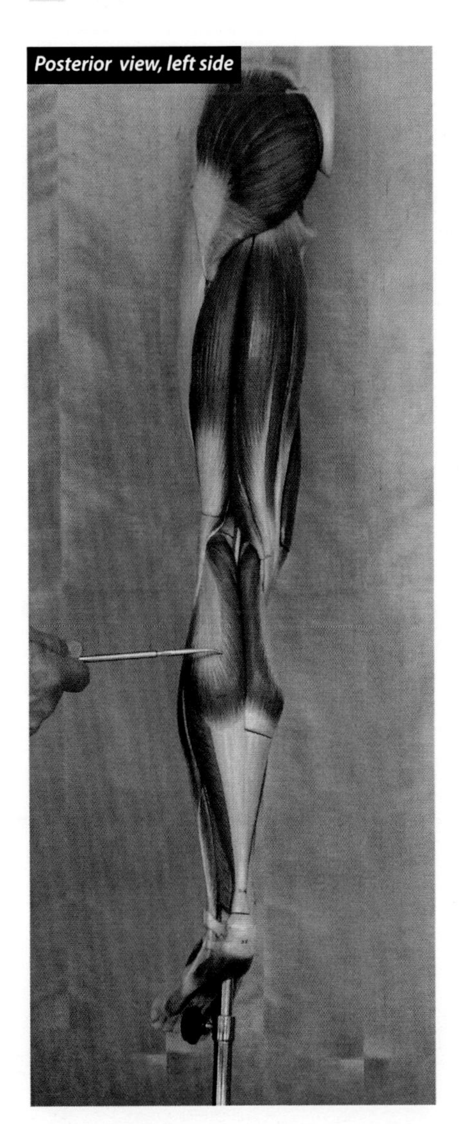


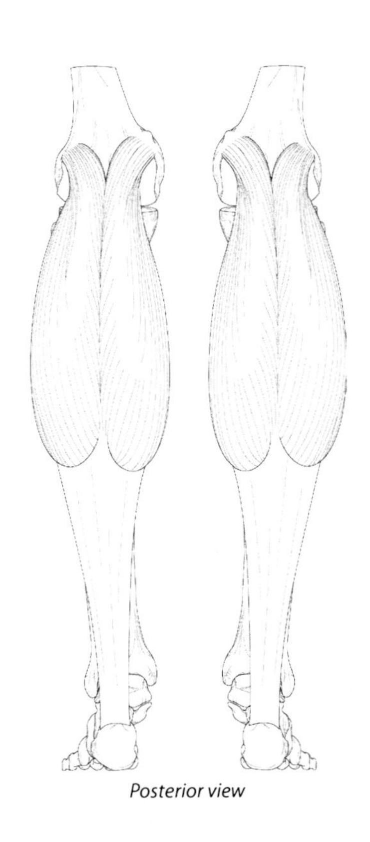

Posterior view

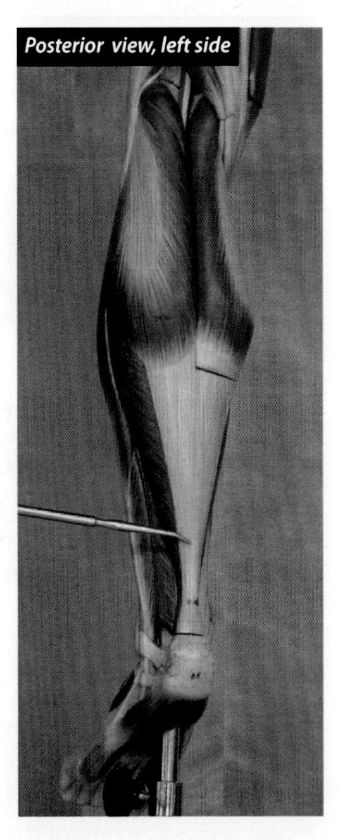


## Calcaneal (Achilles) Tendon (Gastrocnemius)

The calcaneal tendon (Achilles tendon or tendo calcaneus) connects the gastrocnemius to its insertion, the calcaneus. It is the strongest tendon in the human body. In your physiology class, you may make observations of a reflex involving this tendon. Rupture of this tendon requires surgery for recovery.

# NERVES

## Sciatic (Ischiadicus) Nerve

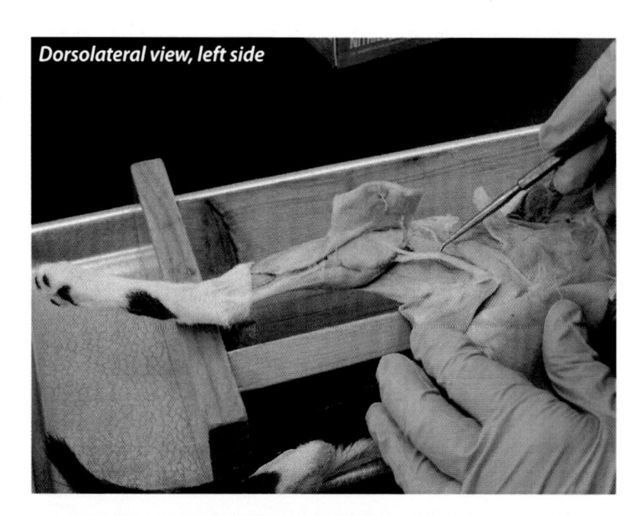


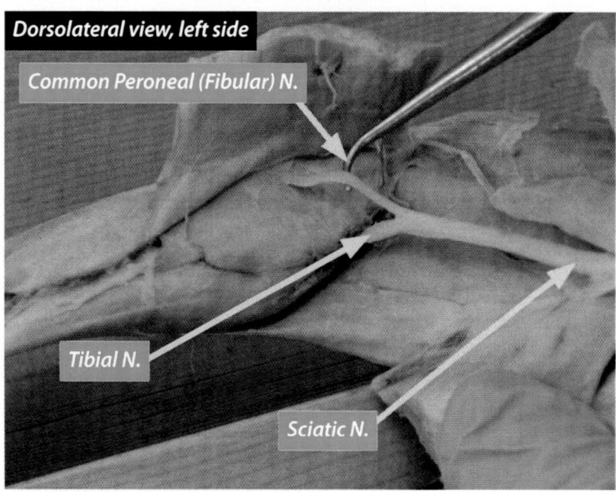


Dr. J's students often call the **sciatic nerve** the "biggest, hugest, nervus, in the thighus." It is bigger than the **radial nerve** in the arm; in fact, it is the longest and thickest nerve in the body! When dissecting, one must be careful not to cut this nerve while transecting the biceps femoris because the nerve is held to the deep surface of that muscle by fascia. The **sciatic nerve** is the largest branch of the **sacral plexus**. Approximately two thirds of the way towards the knee from the hip, the **sciatic nerve** bifurcates to give rise to the **tibial nerve** and the **common peroneal (fibular) nerve**. The **sciatic nerve** primarily serves the posterior compartment of the thigh. Specifically it serves:

1. posterior fibers of adductor magnus (**tibial** portion of the nerve)
2. biceps femoris
3. semimembranosus (**tibial** portion of the nerve)
4. semitendinosus (**tibial** portion of the nerve)

## Common Peroneal (Fibular) Nerve

The **common peroneal (fibular) nerve** is a branch of the **sciatic nerve** and runs laterally toward the knee. It will give rise to the **deep peroneal nerve** that serves the anterior compartment of the leg. It also gives rise to the **superficial peroneal nerve** that serves the lateral compartment of the leg. We will not study those two compartments of the leg for lack of time.

## Tibial Nerve

The **tibial nerve** is a branch of the **sciatic nerve**. It serves the gastrocnemius, as well as most muscles of the posterior compartment of the leg. It passes into the foot and bifurcates to form the **medial** and **lateral plantar nerves**, which are the main nerves of the sole of the foot. It also serves the skin of the sole of the foot and the skin of the posterior side of the leg.

## Femoral Nerve

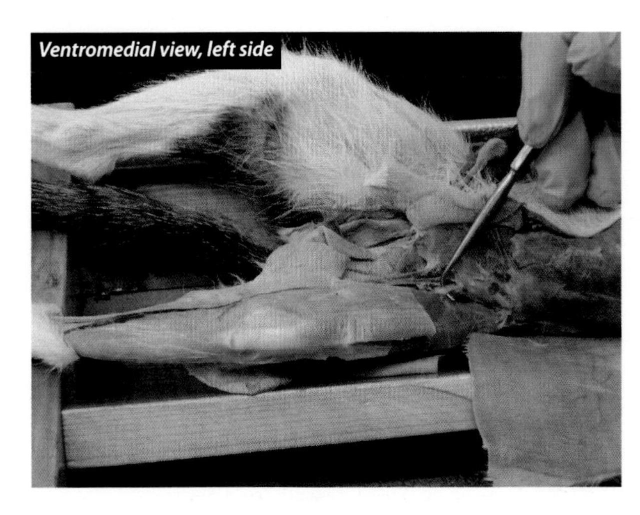


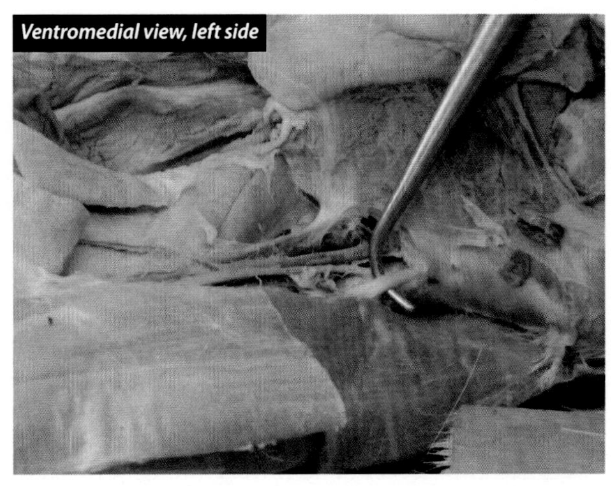


The **femoral nerve** is the largest terminal branch of the lumbar plexus. It can be found in the abdomen as it emerges from the psoas major. It runs deep to the inguinal ligament and enters the thigh where it primarily serves the muscles in the anterior compartment of the thigh. Specifically, it serves:

1. iliacus (that forms part of iliopsoas)
2. pectineus
3. rectus femoris
4. sartorius
5. vastus intermedius
6. vastus lateralis
7. vastus medialis

## Saphenous Nerve

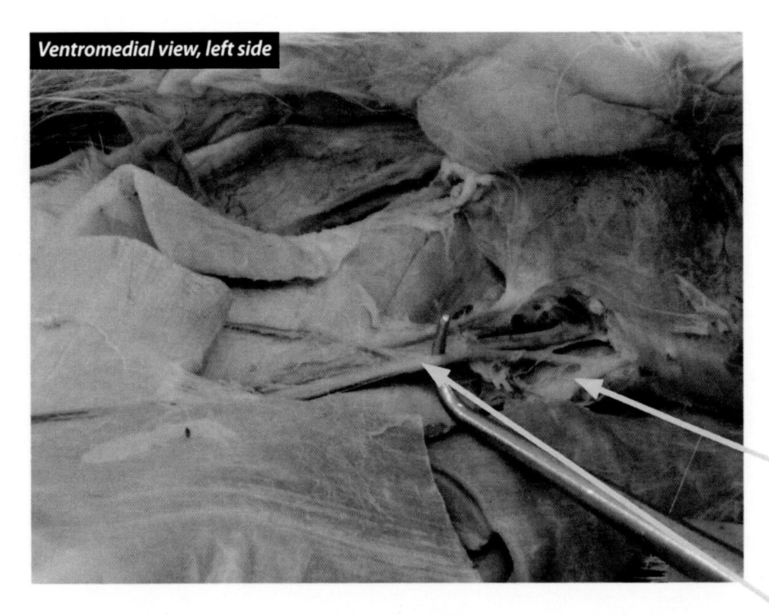


The **saphenous nerve** is a branch of the **femoral nerve**. Initially, it runs with the **femoral artery** and **vein** (Grant, Grant, Lincoln thing), but then after the **muscular artery**, it marks the end of the **femoral artery**. The **saphenous nerve** runs with the **saphenous artery** and **vein** (Grant, Grant, Grant thing). This nerve serves the skin of the anteromedial and posteromedial surfaces of the leg and the medial surface of the foot.

# VESSELS

## External Iliac Artery

The **external iliac artery** in the cat is the last lateral branch of the **aorta**. It passes deep to the inguinal ligament into the thigh. In humans, the **external iliac artery** is a branch of the **common iliac artery** in the pelvis on each side.

## Deep Femoral Artery

The **deep femoral artery** branches medially from the **external iliac artery**, marking the end of the **external iliac artery** and the beginning of the **femoral artery**.

## Femoral Artery and Vein

The **femoral artery** begins in the proximal thigh where the **deep femoral artery** branches from the **external iliac artery**. It ends when the **muscular artery** branches from it. This marks the beginning of the **saphenous artery**. A posterior branch becomes the **popliteal artery**. Since the femoral artery is superficial, it is a useful site from which to take a pulse. To find the femoral pulse, one would press slightly inferior to the midpoint of the inguinal ligament (between the ASIS and the pubic tubercle). Unfortunately, the lack of protection also makes this artery prone to injury. In humans, the **femoral vein** branches off the (**great**) **saphenous vein** in the proximal thigh. In cats, this **branch** is more distal, near the **muscular artery**.

## Lateral Femoral Circumflex Artery

The **lateral femoral circumflex artery** is the first significant lateral branch of the **femoral artery** and the only lateral branch we will study. This artery forms collateral circulation with the **medial femoral circumflex artery** for the proximal portion of the thigh. The **lateral femoral circumflex artery** also supplies blood to the vastus lateralis muscle.

## Muscular Artery

The **muscular artery** is the medial, terminal branch of the **femoral artery**, which also marks the beginning of the **saphenous artery**. The **muscular artery** passes deep to the gracilis muscle and serves the adductor femoris, the gracilis, and the semimembranosus.

## Saphenous Artery and Vein

The **saphenous artery** and **vein** begin when the **muscular artery** branches from the **femoral artery**. They run superficially into the leg with the **saphenous nerve**. This is a Grant, Grant, Grant thing. In humans, the **saphenous vein** and the **saphenous nerve** run together, but there is no comparable **saphenous artery** of any significance. The **saphenous vein** of humans is also known as the **great saphenous vein**. It is the longest vein in the human body, extending from the foot to the proximal thigh, and joining the **femoral vein** proximal to where it does in cats.

Medically, the **saphenous vein** is important as it can be used in cardiac bypass surgery to replace the **coronary arteries** that have become blocked. Although I have not been present when this is done, I have been told that they turn the vessel inside out so that the semilunar valves will not cause any problems with flow, and so that they actually make the walls stronger to contain the arterial pressure.

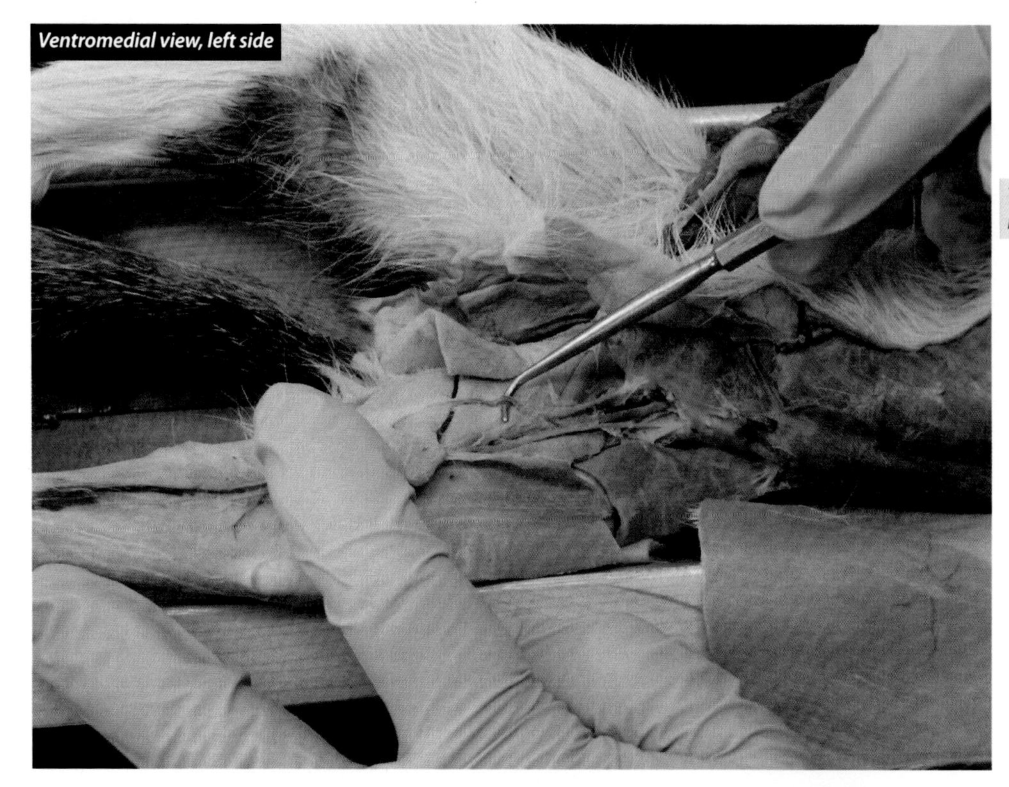


The probe is pointing to the **muscular artery**.

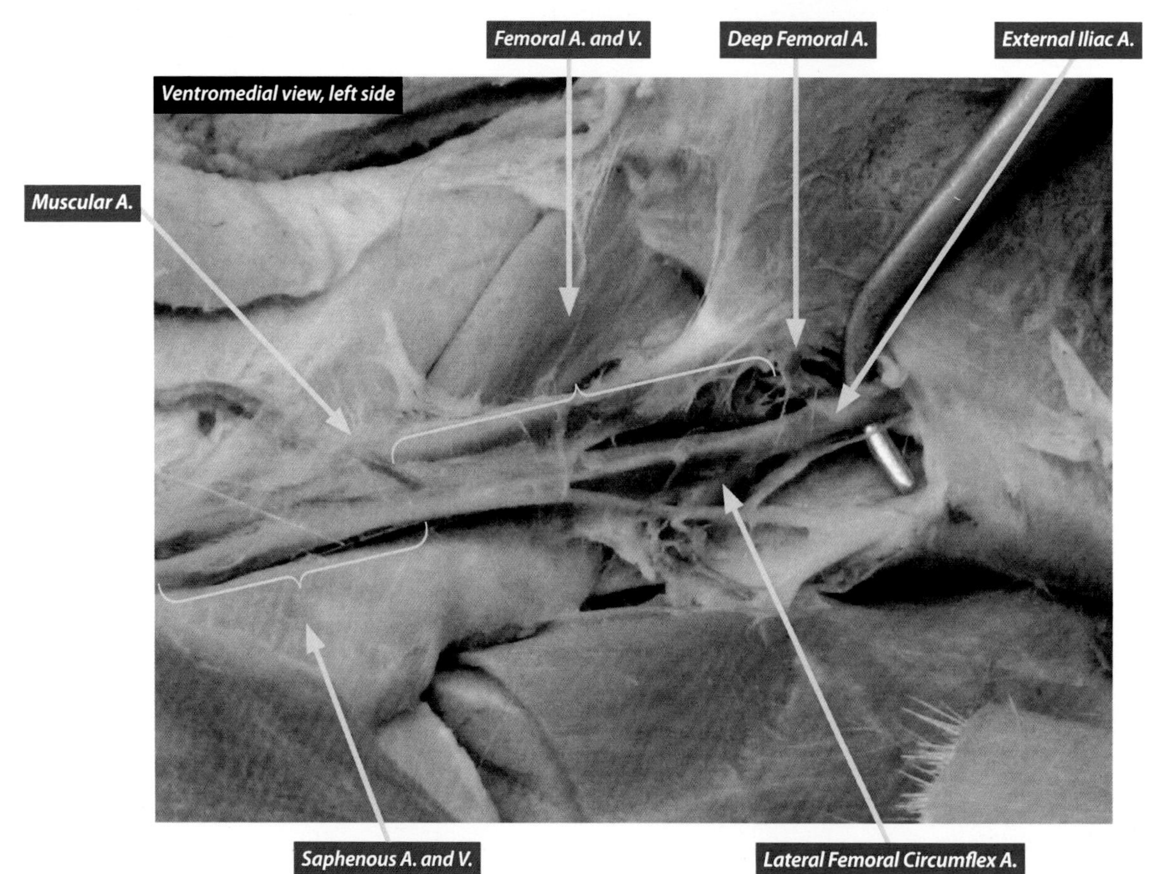

## Muscles of the Lower Appendage

<table>
<tr><th colspan="4">POSTERIOR = Mostly Thigh (Hip) Extensors & Leg (Knee) Flexors</th></tr>
<tr><th>MUSCLE</th><th>ORIGIN</th><th>INSERTION</th><th>ACTION</th></tr>
<tr><td>Gluteus maximus<br>Inferior gluteal n.</td><td>Posterior surface of ilium, sacrum, and coccyx</td><td>Proximal posterior femur and iliotibial tract (IT band)</td><td>Extends and laterally rotates thigh</td></tr>
<tr><td>Biceps femoris<br>Sciatic n.</td><td>Ischial tuberosity; linea aspera</td><td>Fibular head and lateral tibial condyle</td><td>Extends thigh; flexes and laterally rotates leg</td></tr>
<tr><td>Semitendinosus<br>Sciatic n.</td><td rowspan="2">Ischial tuberosity</td><td>Proximal anteromedial tibia (pes anserine)</td><td rowspan="2">Extends thigh; flexes and medially rotates leg</td></tr>
<tr><td>Semimembranosus<br>Sciatic n.</td><td>Medial tibial condyle (posterior surface)</td></tr>
</table>

<table>
<tr><th colspan="4">ANTERIOR = Mostly Thigh (Hip) Flexors & Leg (Knee) Extensors</th></tr>
<tr><th>MUSCLE</th><th>ORIGIN</th><th>INSERTION</th><th>ACTION</th></tr>
<tr><td>Iliopsoas<br>Iliacus—Femoral n.<br><br>Psoas major—Anterior rami of L1–L3</td><td>Iliac fossa and lateral surfaces of T12–L5</td><td>Lesser trochanter of femur</td><td>Prime flexor of thigh; flexes and laterally bends lumbar region (trunk)</td></tr>
<tr><td>Sartorius<br>Femoral n.</td><td>ASIS</td><td>Proximal anteromedial tibia (pes anserine)</td><td>Flexes, ABducts, and laterally rotates thigh; flexes leg</td></tr>
<tr><td>Rectus femoris<br>Femoral n.</td><td>AIIS</td><td rowspan="4">Tibial tuberosity via patellar ligament</td><td>Flexes thigh; extends leg</td></tr>
<tr><td>Vastus lateralis<br>Femoral n.</td><td rowspan="2">Linea aspera</td><td rowspan="3">Extends the leg</td></tr>
<tr><td>Vastus medialis<br>Femoral n.</td></tr>
<tr><td>Vastus intermedius<br>Femoral n.</td><td>Anterior shaft of femur</td></tr>
</table>

## Muscles of the Lower Appendage

| MEDIAL= Mostly Thigh (Hip) ADductors | | | |
|---|---|---|---|
| **MUSCLE** | **ORIGIN** | **INSERTION** | **ACTION** |
| **Gracilis**<br>**Obturator n.** | Pubis | Proximal anteromedial tibia (pes anserine) | ADducts, flexes, and medially rotates thigh; flexes leg |
| **Adductor longus**<br>**Obturator n.** | | Linea aspera | ADducts and flexes thigh |
| **Adductor Femoris (magnus et brevis)**<br>**Obturator n.** | Pubis and ischial tuberosity | | |
| **Pectineus**<br>**Femoral n.** | Pubis | Line from lesser trochanter to linea aspera (pectineal line) | |

| LATERAL = Mostly Thigh (Hip) ABductors | | | |
|---|---|---|---|
| **MUSCLE** | **ORIGIN** | **INSERTION** | **ACTION** |
| **Gluteus medius**<br>**Superior gluteal n.** | Superficial surface of ilium | Greater trochanter of femur | ABducts and medially rotates thigh |
| **Gluteus minimus**<br>**Superior gluteal n.** | | | |
| **Tensor fasciae latae**<br>**Superior gluteal n.** | ASIS and iliac crest | Iliotibial tract (IT band) | ABducts, flexes, and medially rotates thigh |

## Muscles of the Foot (Ankle)

| COMPARTMENT | ACTION | NERVE |
|---|---|---|
| **Anterior** | Dorsiflexes and inverts foot; extends toes | **Deep peroneal n.** |
| **Lateral** | Plantar flexes and everts foot | **Superficial peroneal n.** |
| **Posterior** | Flexes leg; plantar flexes and inverts foot; flexes toes | **Tibial n.** |
| **Note:** The **Gastrocnemius m. originates** on the medial and lateral femoral condyles and **inserts** on the calcaneus via calcaneal tendon | Flexes leg; plantar flexes foot | **Tibial n.** |

Source: Louis J. Zanella, *The Human Body—An Owners Manual—1995.* Modified by William C. Johnson (Dr. J) and Lisa Miller—1/8/2013